CANCER IMMUNOLOGY
AND IMMUNOTHERAPY

CANCER IMMUNOLOGY AND IMMUNOTHERAPY

Volume 1 of Delivery Strategies and Engineering Technologies in Cancer Immunotherapy

Edited by

MANSOOR M. AMIJI
*University Distinguished Professor, Professor of Pharmaceutical Science,
and Professor of Chemical Engineering, Northeastern University, Boston,
Massachusetts, United States*

LARA SCHEHERAZADE MILANE
*Bouvé College of Health Sciences Distinguished Educator,
Assistant Teaching Professor, Department of Pharmaceutical Sciences
Northeastern University, Boston, Massachusetts, United States*

ELSEVIER

ACADEMIC PRESS
An imprint of Elsevier

Academic Press is an imprint of Elsevier
125 London Wall, London EC2Y 5AS, United Kingdom
525 B Street, Suite 1650, San Diego, CA 92101, United States
50 Hampshire Street, 5th Floor, Cambridge, MA 02139, United States
The Boulevard, Langford Lane, Kidlington, Oxford OX5 1GB, United Kingdom

Notices
Knowledge and best practice in this field are constantly changing. As new research and experience broaden our understanding, changes in research methods, professional practices, or medical treatment may become necessary.

Practitioners and researchers must always rely on their own experience and knowledge in evaluating and using any information, methods, compounds, or experiments described herein. In using such information or methods they should be mindful of their own safety and the safety of others, including parties for whom they have a professional responsibility.

To the fullest extent of the law, neither the Publisher nor the authors, contributors, or editors, assume any liability for any injury and/or damage to persons or property as a matter of products liability, negligence or otherwise, or from any use or operation of any methods, products, instructions, or ideas contained in the material herein.

Library of Congress Cataloging-in-Publication Data
A catalog record for this book is available from the Library of Congress

British Library Cataloguing-in-Publication Data
A catalogue record for this book is available from the British Library

ISBN 978-0-12-823397-9

For information on all Academic Press publications
visit our website at https://www.elsevier.com/books-and-journals

Publisher: Andre Gerhard Wolff
Acquisitions Editor: Erin Hill-Parks
Editorial Project Manager: Susan Ikeda
Production Project Manager: Kiruthika Govindaraju
Cover Designer: Matthew Limbert

Typeset by STRAIVE, India

Working together
to grow libraries in
developing countries

www.elsevier.com • www.bookaid.org

Dedication

This book is dedicated to my teachers and mentors, especially Professor Kinam Park and Dr. Haesun Park, who remain an inspiration for their intellect, passion, generosity, and kindness. I stand on the shoulders of these giants!

Mansoor M. Amiji

I dedicate this work in loving memory to my twin sister, Samantha Tari Jabr—thank you for being my soulmate and for your unwavering love that keeps me going. I also dedicate this work to my daughter, Mirabella—you are the light of my life and I thank you for your eternal brilliance.

Lara Scheherazade Milane

Contents

Contributors

Gulzar Ahmad
Repertoire Immune Medicines, Cambridge, MA, United States

Nadia Tasnim Ahmed
Department of Pharmaceutics and Center for Pharmaceutical Engineering and Sciences, School of Pharmacy, Virginia Commonwealth University, Richmond, VA, United States

Amedeo Amedei
Department of Experimental and Clinical Medicine, University of Florence, Florence, Italy

Mansoor M. Amiji
School of Pharmacy, Bouve College of Health Sciences, Northeastern University, Boston, MA, United States

Kirtika H. Asrani
Oncology Research, Bluebird Bio, Cambridge, MA, United States

Federico Boem
Department of Literature and Philosophy, University of Florence, Florence, Italy

Kevin A. Cassady
Department of Pediatrics, Division of Pediatric Infectious Diseases, Nationwide Children's Hospital; The Research Institute at Nationwide Children's Hospital, Center for Childhood Cancer and Blood Diseases, Division of Pediatric Infectious Diseases, The Ohio State University College of Medicine, Pelotonia Institute for Immuno-Oncology, Columbus, OH, United States

Yu Chen
Department of Medical Oncology, Chongqing University Cancer Hospital, Chongqing, China

Patty A. Culp
Maverick Therapeutics, Brisbane, CA, United States

Jeremiah D. Degenhardt
Maverick Therapeutics, Brisbane, CA, United States

Danielle E. Dettling
Maverick Therapeutics, Brisbane, CA, United States

Mustafa B.A. Djamgoz
Imperial College London, Department of Life Sciences, South Kensington Campus, London, United Kingdom; Biotechnology Research Centre, Cyprus International University, North Cyprus, Mersin, Turkey

Aatman S. Doshi
Bioscience, AstraZeneca R&D Boston, Waltham; Department of Pharmaceutical Sciences, Bouvé College of Health Sciences, Northeastern University, Boston, MA, United States

Eyad Elkord
Natural and Medical Sciences Research Center, University of Nizwa, Nizwa, Oman; Biomedical Research Center, School of Science, Engineering and Environment, University of Salford, Manchester, United Kingdom

Laetitia Firmenich
Imperial College London, Department of Life Sciences, South Kensington Campus; University College London, Faculty of Medical Sciences, London, United Kingdom

Lauren M. Gauthier
Drug Safety Research and Evaluation, Takeda Pharmaceuticals, Cambridge, MA, United States

Hazem E. Ghoneim
Department of Microbial Infection and Immunity, College of Medicine; The James Comprehensive Cancer Center, Pelotonia Institute for Immuno-Oncology, The Ohio State University, Columbus, OH, United States

Taichiro Goto
Lung Cancer and Respiratory Disease Center, Yamanashi Central Hospital, Yamanashi, Japan

Ilse Hernandez-Aguirre
Medical Scientist Training Program, The Ohio State University Wexner Medical Center; The Research Institute at Nationwide Children's Hospital, Center for Childhood Cancer and Blood Diseases, The Ohio State University College of Medicine, Columbus, OH, United States

Hae Lin Jang
Center for Engineered Therapeutics, Department of Medicine, Brigham and Women's Hospital, Harvard Medical School, Boston, MA, United States

Yongsheng Li
Department of Medical Oncology, Chongqing University Cancer Hospital, Chongqing, China

Xiang Liu
Department of Pharmaceutics and Center for Pharmaceutical Engineering and Sciences, School of Pharmacy, Virginia Commonwealth University, Richmond, VA, United States

Gerardo G. Mackenzie
Department of Nutrition, University of California at Davis, Davis, CA, United States

Chad May
Maverick Therapeutics, Brisbane, CA, United States

Lara Scheherazade Milane
Department of Pharmaceutical Sciences, School of Pharmacy, Northeastern University, Boston, MA, United States

Edda Russo
Department of Experimental and Clinical Medicine, University of Florence, Florence, Italy

Abbey A. Saadey
Department of Microbial Infection and Immunity, College of Medicine, The Ohio State University, Columbus, OH, United States

Reem Saleh
Sir Peter MacCallum Department of Oncology, University of Melbourne, Parkville, VIC, Australia

Varun Sasidharan Nair
Department of Experimental Immunology, Helmholtz Centre for Infection Research, Braunschweig, Germany

Shiladitya Sengupta
Center for Engineered Therapeutics, Department of Medicine, Brigham and Women's Hospital, Harvard Medical School, Boston, MA, United States

Ting Su
Department of Pharmaceutics and Center for Pharmaceutical Engineering and Sciences, School of Pharmacy, Virginia Commonwealth University, Richmond, VA, United States

Salman M. Toor
College of Health and Life Sciences (CHLS), Hamad Bin Khalifa University (HBKU), Doha, Qatar

Amir Yousif
Department of Microbial Infection and Immunity, College of Medicine, The Ohio State University, Columbus, OH, United States

Shurong Zhou
Department of Pharmaceutics and Center for Pharmaceutical Engineering and Sciences, School of Pharmacy, Virginia Commonwealth University, Richmond, VA, United States

Guizhi Zhu
Department of Pharmaceutics and Center for Pharmaceutical Engineering and Sciences, School of Pharmacy, Virginia Commonwealth University, Richmond, VA, United States

Preface

We are currently in the *Age of Immunology*. Never before has there been such rampant discoveries and advancements in immunotherapies. For example, on March 11th 2020, the World Health Organization declared a COVID-19 pandemic. Less than one year after, hundreds of SARS-CoV-2 (causative virus of COVID-19) vaccines have been developed and four have already received emergency use approval around the globe. This rapid and effective vaccine development clearly depicts the current state and capabilities in the field of immunotherapy. Tremendous advancements are being made every day, not only in vaccine and drug development for the current COVID-19 pandemic but also for cancer. It is an exciting and critical time for the development of immunotherapies for treating cancer. Cancer immunotherapy lies at the intersection of many disciplines including oncology, immunology, pharmaceutical science, biomedical engineering, and clinical research. This series *Delivery Strategies and Engineering Technologies in Cancer Immunotherapy* includes three volumes; Volume One: Cancer Immunology and Immunotherapy, Volume Two: Systemic Drug Delivery Strategies, and Volume Three: Engineering Technologies and Clinical Translation.

Volume One of this series dives deep into cancer immunology and the opportunities for immunotherapy intervention. On multiple levels, cancer is a disease of immune dysfunction. First and foremost, cancer occurs due to the ineffectiveness of the immune system in recognizing and eliminating an abnormal cell. Secondly, the immune cells within the tumor microenvironment are transformed and contribute to chronic inflammation. Third, this chronic inflammation propagates tumor growth, survival, and multidrug resistance. The process of immune-driven aggression of a tumor is referred to as immunoediting. There is a cycle of immune priming where the signaling interplay of immune cells, cancer cells, and fibroblasts escalates the inflammatory response within a tumor until there is a state of chronic inflammation. This chronic inflammatory state (that is a product of immunoediting) then becomes a driver for a higher state of tumor aggression and resistance. Additional drivers include metabolic, genomic, and epigenetic reprogramming as well as tumor heterogeneity. Metabolic transformations include increased glycolysis in cancer cells and altered nutrient metabolism and mitochondrial alterations in all cells (immune cells, cancer cells, and supportive cells) within the tumor microenvironment. Drivers of tumor aggression also contribute to the response of a tumor to immunotherapy. Immunotherapy approaches include interferon therapy (such as Intron A), oncolytic viruses such as Imlygic that increase tumor immunogenicity, epigenetic modifiers such as DNA methyltransferase inhibitors, pattern-recognition receptor agonists including stimulator of interferon genes (STING) agonists, macrophage reprogramming, and antigen/antibody approaches. One of the earliest immunotherapy approaches include Chimeric Antigen Receptor (CAR) T-cell therapy that has now been expanded to CAR-natural killer cell and CAR-macrophage therapies. Another immunotherapy strategy discussed in this Volume is the use of T-cell engaging bispecific antibodies. These antibodies are designed to bind to both T-cells and cancer cells with the objective of attaining T-cell-induced cytotoxicity through perforin and granzyme release. Although immunotherapy is promising, the challenge of overcoming drug resistance is significant and toxicity profiles are also a concern. To overcome resistance, combination immunotherapy regimens are being explored. A challenge in developing and optimizing immunotherapies and in predicting immune-related adverse events is the limitations of current animal, cell, and computer models. Advancements in human immune modeling are critical to improving the value of preclinical immunotherapy studies. Volume One discusses the immune–tumor interface and key topics in immunotherapy, providing a solid foundation of knowledge upon which Volumes Two and Three build upon.

Acknowledgments

First and foremost, we are deeply grateful to all the individual chapter authors who have contributed immensely to the success of *Delivery Strategies and Engineering Technologies in Cancer Immunotherapy*. The three-volume series came about because of the time and effort that these authors put into their respective chapters. Each one of them is a leader in his or her field and we are forever in their debt. Special thanks to the amazing team at Elsevier Press starting with Erin Hill-Parks, Susan Ikeda, and Kavitha Balasundaram who took an idea and made it into reality.

We hope that the three-volume series *Delivery Strategies and Engineering Technologies in Cancer Immunotherapy* will ignite more questions and create greater research opportunities in the field of cancer immunotherapy and lead to the development of novel solutions that can improve patients' outcomes.

Mansoor M. Amiji
Lara Scheherazade Milane

CHAPTER ONE

The hallmarks of cancer and immunology

Lara Scheherazade Milane

Department of Pharmaceutical Sciences, School of Pharmacy, Northeastern University, Boston, MA, United States

Contents

1. Introduction

Cancer is a disease of molecular evolution and immune dysfunction. A cancer cell is most simply a cell that will adapt and change however necessary for it to survive. Cancer is a disease of immune dysfunction for three primary reasons. First, an adequate immune response would recognize and kill a cancer cell before the establishment of a tumor. It is the failure of an effective immune response that enables the initiation of a tumor. Second, as a tumor grows, a dysfunctional immune response contributes to the growth and propagation of a tumor. Indeed, tumors that have been immunoedited and immune-primed are more aggressive and drug resistant [1]. Third, the immune cells in the tumor microenvironment are transformed cells that behave differently from normal immune cells; cancer associated macrophages are the most characterized transformed immune cells in the tumor microenvironment [2]. As illustrated in Fig. 1, the hallmarks of cancer have been progressively identified over the years. In 2000, in a seminal review paper, Hanahan and Weinberg characterized the first six hallmarks of cancer: apoptotic resistance, angiogenesis, replicative immortality, metastasis and invasion, evading growth suppression, and

Cancer Immunology and Immunotherapy
https://doi.org/10.1016/B978-0-12-823397-9.00001-6

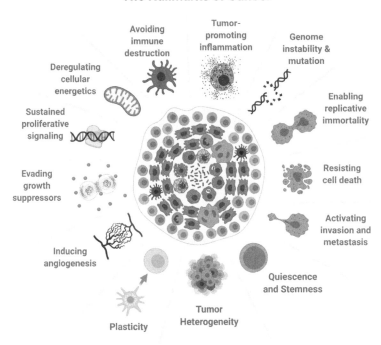

Fig. 1 The hallmarks of cancer create a disease of molecular evolution. Cancer is a disease of molecular evolution; a disease of rapid adaptation and survival. The 13 hallmarks of cancer work together to enable tumor progression and survival.

sustaining proliferative signaling [3]. In 2011, Hanahan and Weinberg composed another seminal review identifying four new hallmarks; tumor-promoting inflammation, genome instability and mutation, avoiding immune destruction, and deregulated energetics [4]. In 2017, I characterized three additional hallmarks of cancer: quiescence and stemness, plasticity, and tumor heterogeneity [5]. Beyond the two direct immune hallmarks (tumor-promoting inflammation and avoiding immune destruction), immunology is related to each hallmark of cancer and will be discussed in the following sections.

2. Activating invasion and metastasis

Metastasis is the movement of cancer cells from the site of the primary tumor to secondary sites in the body. The occurrence of metastasis is actually how cancer acquired its name credited to Hippocrates (460–370 BC) referring to the disease as carcinoma (Greek for Crab) and Celsus (28–50 BC), who later translated this into the Latin word for crab "cancer" [6]. The core concept of this hallmark is that metastasis is a symptom of cancer cells becoming deregulated. This hallmark is of extreme importance as metastatic cancer is responsible for most of the disease morbidity [7]. There are four main components of metastasis and each involve immune dysfunction or manipulation.

The first element of metastasis is loss of cellular polarity and migration [8]. Depolarization of a cell is the first cellular event leading to metastasis [8]. Atypical protein kinase C (aPKC) and partitioning defective protein (PAR) mediate the dissolution of tight junctions, E-cadherin is endocytosed and degraded, and adherens junctions are disrupted by p120 catenin [8]. Following loss of polarity, the cell undergoes enhanced proliferation and increased receptor recycling [8]. While the cell is undergoing these early changes, the tissue resident immune cells fail to recognize and eliminate the transforming cell for two reasons: the changes do not trigger a normal immune response and the tumor resident immune cells are already transformed. Following enhanced proliferation, cell migration begins as integrin recycling increases, Rab (GTPases) protein expression increases, and basement membrane invasion occurs [8].

The second element of metastasis is epithelial-to-mesenchymal transition and mesenchymal-to-epithelial transition (EMT to MET) [9]. For a cancer cell to migrate through systemic circulation, the cell needs to lose its epithelial characteristics and assume mesenchymal characteristics. Mesenchymal cells are derived from the mesoderm germ layer of an embryo [10]. Mesenchymal stem cells (MSCs) are multipotent and can differentiate into many tissues including connective tissue, blood vessels, lymphatic tissue, and skeletal tissue [10]. MSCs are important for maintaining immune cell homeostasis in the marrow and for immunoregulation [10, 11]. Direct cell-to-cell contact or signaling from MSC exosomes, growth factors, and cytokines can stimulate T-cells and macrophages or inhibit T-cells and B-cells [11]. Importantly, while MSCs can regulate the immune response, they are often referred to as being "immune privileged" as they have low immunogenicity [11]. *In vivo* studies and clinical trials of allogenic MSCs suggest that MSCs may not be completely immune privileged as humoral and cellular immune responses have been reported [11]. The process of EMT in metastasis is direct exploitation of the immune privilege by cancer cells. Becoming mesenchymal manipulates the immune escape of MSCs while the metastasizing cancer cell migrates through systemic or lymphatic circulation. Mesenchymal character also allows platelets to adhere to the migrating cancer cell in systemic circulation, further protecting the cell from natural killer (NK) cell recognition [9]. Once the metastatic cell reaches the metastatic site, the cancer cell undergoes MET to establish a secondary tumor. Epithelial character is necessary for proliferation and growth of the secondary tumor.

The third component of the metastatic process is exosome signaling. Exosome signaling is an important form of communication in the tumor microenvironment. Exosomes are variably sized extracellular vesicles that are released as a passive mass balance counter-process to endocytosis and as an active/deliberate signaling process. Exosome signaling from cancer cells, cancer-associated fibroblasts, and immune cells (dendritic cells and cancer-associated macrophages) within the microenvironment are drivers for growth and survival within the tumor microenvironment. Extracellular and intercellular communication in the forms of exosomes, tunneling nanotubes, cytokines, chemokines, and growth factors is important in unifying the cells as a collective organ-mimicking tumor. Exosome signaling from immune cells also contributes to immunoediting of tumors (discussed later). The process of exosome formation in the cells within the tumor microenvironment is detailed in Fig. 1. Endocytosis or membrane remodeling can trigger exosome formation and release, or the process can occur beginning at the level of the endosomal sorting complex required for transcript (ESCRT) machinery without endo-cytotic involvement [12]. Trafficking endosomes, lysosomes, and signaling vesicles are transient organelles. Referring to them as vesicles downplays the intracellular signaling and interactions of these organelles. For example, endosomes with ESCRTs are loaded with specific content from the cytosol, the endoplasmic reticulum, and the Golgi appa-ratus (Fig. 1, Step 7) [12].

Concerning metastasis, exosomes play three critical roles: (1) As discussed, they are an important form of communication between cells within the tumor microenvironment; (2) exosomes are loaded with matrix metalloproteinases (MMPs) capable of degrading the extracellular matrix and are released from the leading edge of the migrating cell; and (3) exosomes released from the tumor microenvironment enter systemic circulation and can prime a new site as a metastatic niche. The release of exosomes from the leading edge of the migrating cell has a bulldozer effect, clearing the way for the cell to move through the extracellular matrix and the basement membrane [12]. Exosomes from the tumor micro-environment have been shown to travel through systemic circulation and establish a premetastatic niche that is primed to attract migrating cancer cells to undergo extrava-sation at the site [9].

The fourth element of metastasis is organtropism or "soil and seed." In 1889, Stephen Paget, then an assistant surgeon in West London, established the "soil and seed" theory of metastasis [13]. His theory stated that not only did the metastasizing cancer cell (the seed) have to have the necessary qualities for metastasis to occur but the metastatic site (the soil) also had to have the necessary qualities to receive the seed [13]. Since Paget's soil and seed theory originated, it has morphed into the study of organtropism in metastasis. Organtropism is the preferred or dominance of secondary metastatic sites associated with certain primary tumors [14]. For example, 71% of breast cancer metastasize to both bone and lungs, 62% to the liver, and 22% to the brain [14, 15]. On the other hand, 11.2% of ovarian cancers metastasize to bone, 33.9% to the lung, 47.9% to the liver, 3% to the brain, and 83.6% to the peritoneum [14, 16]. We discussed migration of the "seed" earlier in the EMT and exosome processes. For the soil to be efficient for establishment of a metastatic site, the immune cells at that site must be controlled and deflected so the metas-tasizing cell can adhere and undergo MET without immune clearance [14]. This process begins with exosome priming of the premetastatic niche [14]. In cases where metastatic sites have not been "primed," exosomes and signaling factors from the cancer cell modulate immune cells [14]. For example, exosomes that prime liver metastatic sites stimulate Kupffer cells to release transforming growth factor-β (TGFβ) and pro-inflammatory S100A8, which promotes extracellular matrix deposition and activates hepatic stellate cells promoting metastasis [14]. In the scenario of bone metastasis, the invading cell expresses osteomimetic genes to avoid immune recognition and factors such as lysyl oxidase (LOX) to remodel osteoblasts [14]. EMT is critical to the ability of a cancer cell to invade a metastatic site and undergo genetic mimicry of the tissue to evade immune surveillance.

3. Resistance to cell death

When Hanahan and Weinberg characterized the first six hallmarks of cancer, they characterized this hallmark as "apoptotic resistance" [3]. I have renamed this hallmark "resistance to cell death." In the 20 + years since the hallmark was first characterized, much progress has been made in the biology of cell death and we now know that there are at least 12 different forms of cell death and cancer is resistant to all of these [17]. The core concept of this hallmark is that the primary goal of a cancer cell is to survive and evade cell death. A cancer cell must make sure it not only survives with its current cellular conditions (nutrients, oxygen supply, etc.) but also make sure it evades immune recognition and clearance. During tumor initiation, the events that occur are transformative—the cancer cells begin working together to transform local fibroblasts and immune cells. After transformation of the local supporting and immune cells, the events that occur ensure maintenance and propagation. This is where immunoediting drives cell-death resistance.

The first element of this hallmark is the cell-death threshold. The 12 different forms of cell death each have a threshold of activation—above which cell death will be executed. These forms of cell death do not work in isolation. For example, a cell undergoing autophagy may approach the threshold for autophagic cell death but then, through external stimuli, undergo inflammation causing breakdown of the plasma membrane and crossing the threshold for activation of necroptosis (inflammatory cell death). Immuno-regulation within the tumor microenvironment is critical to maintaining the cell-death threshold below activation for all cell-death pathways. During tumor initiation, cancer cells avoid immune clearance via multiple mechanisms including the expression of neoantigens, death receptor mutations, alterations in the local pH and extracellular matrix, and having an elevated threshold of activation for cell death through alterations in the expression of proteins in the cell death signaling cascade pathways [1, 18]. This elevated threshold allows a cancer cell to accommodate to local immune signaling and insults and adapt to confer survival.

The second element of this hallmark is multidrug resistance (MDR). MDR is expected after the initiation of chemotherapy (acquired MDR) but can also occur before exposure to any therapeutics (innate MDR). Michael Gottesman has made significant contributions to the study of MDR in cancer. In one of his review articles he detailed the mechanisms of MDR as decreasing drug influx, increasing drug efflux, activation of detoxification systems, activation of DNA repair, and blocked apoptosis [19]. Overexpression of drug efflux pumps such as P-glycoprotein (P-gp; multidrug-resistant protein 1, MDR1) is a significant mechanism of MDR [19]. In the early tumor, these occur in response to immune or environmental insults. The main mechanisms of initial immune escape include: (1) decreased expression of MHC I molecules that decreases priming and activation of dendritic cells and CD8$^+$ T-cells and (2) decreased expression of tumor-associated antigens [20]. As the tumor develops increased mutations of oncogenes confer resistance to immunity effectors and increase the production and response to growth factors [20]. In the MDR tumor, either immunosuppression or tumor-promoting inflammation occurs. This occurs through cytokine and metabolic signaling, recruitment of regulatory T-cells (T-regs) and myeloid-derived suppressor cells (MDSCs), or through ligation of inhibitory receptors on immune effectors [20]. In the transformed tumor, inflammation perpetuates the mechanisms of MDR. Inflammatory cancers are notably more aggressive and more drug resistant than noninflammatory cancers [21].

4. Evading growth suppression and sustaining proliferative signaling

The core concept of the evading growth suppression hallmark is that cancer cells ignore stop signs; they propel past cell-cycle checkpoints without recognition and grow over each other and surrounding cells without stopping. The core concept of the sustaining proliferative signaling hallmark is that cancer cells promote their own growth by controlling receptors, ligands, and checkpoints. As mentioned at the beginning of this chapter, cancer cells are merely cells that will evolve and adapt in any way necessary to confer survival. Considering this, it is easy to conceive why many proteins have pleiotropic effects in cancer. These pleiotropic molecules are responding to different states of a tumor. An example of a pleiotropic protein that regulates evasion of growth suppression is TGFβ, which is a well-known immune regulator. In the early tumor TGFβ inhibits growth, whereas in the late tumor TGFβ promotes invasion and metastasis [22]. TGFβ mediates the inflammatory response and endogenous processes such as embryonic development, as such TGFβ has a role in many pathologies [23]. In the early tumor, TGFβ suppresses tumor growth through the upregulation of cyclin kinase inhibitors [22]. The role of TGFβ signaling in the late tumor is illustrated in Fig. 2. In the late tumor, TGFβ inhibits CD8$^+$ T-cells and NK cells while promoting M2 macrophage polarization, transformation of fibroblasts into cancer-associated fibroblasts, recruitment of granulocytes and platelets [23]. Although I am referring to the early and late tumor, as

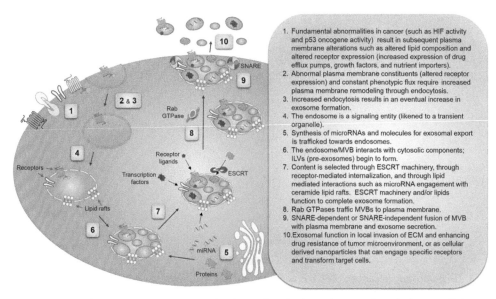

1. Fundamental abnormalities in cancer (such as HIF activity and p53 oncogene activity) result in subsequent plasma membrane alterations such as altered lipid composition and altered receptor expression (increased expression of drug efflux pumps, growth factors, and nutrient importers).
2. Abnormal plasma membrane constituents (altered receptor expression) and constant phenotypic flux require increased plasma membrane remodeling through endocytosis.
3. Increased endocytosis results in an eventual increase in exosome formation.
4. The endosome is a signaling entity (likened to a transient organelle).
5. Synthesis of microRNAs and molecules for exosomal export is trafficked towards endosomes.
6. The endosome/MVB interacts with cytosolic components; ILVs (pre-exosomes) begin to form.
7. Content is selected through ESCRT machinery, through receptor-mediated internalization, and through lipid mediated interactions such as microRNA engagement with ceramide lipid rafts. ESCRT machinery and/or lipids function to complete exosome formation.
8. Rab GTPases traffic MVBs to plasma membrane.
9. SNARE-dependent or SNARE-independent fusion of MVB with plasma membrane and exosome secretion.
10. Exosomal function in local invasion of ECM and enhancing drug resistance of tumor microenvironment, or as cellular derived nanoparticles that can engage specific receptors and transform target cells.

Fig. 2 Exosome formation. Exosomes can be formed and released passively to balance cellular and membrane expansion that occurs during endocytotic processes (Steps 1–4) or exosomes can be formed and released as a purely deliberate and active process beginning at the level of the endosomal sorting complex required for transcript (ESCRT) machinery and trafficking of content into the ESCRT (Step 5). During Step 6 intraluminal vesicles (ILVs) begin to form. During Step 7 the ESCRT and other receptors and channels actively load content into the trafficking endosome with exosomes and multivesicular bodies (MVB) form as Rab GTPases traffic MVB to the plasma membrane (Step 8). Soluble NSF attachment protein receptors (SNAREs) fuse the MVBs with the plasma membrane and exosomes are released for local and distal signaling as the exosomes transverse into systemic circulation (Step 10). *(Reprinted from Milane L, et al. Exosome mediated communication within the tumor microenvironment. J Control Release 2015;219:278–94 with permission from Elsevier.)*

is common practice, it is important to note that this is a reference to phenotype and not a time-dependent reference. As such, an "early tumor" can certainly be more aggressive and have the characteristics of a "late" tumor. Likewise, a tumor may persist in a slowly evolving state and never reach what we refer to as the "late" aggressive state. We could easily interchange early and late with drug-sensitive and MDR (Fig. 3).

Another element that contributes to the evasion of growth suppression hallmark is the retinoblastoma (RB) tumor suppressor gene. Inactivation of RB is associated with many tumors [24]. RB is mainly a repressive transcription factor that binds to E2F transcription factors or promoter genes and inhibits the factors or initiates remodeling of chromatin [24]. RB also serves as an adaptor protein to mark proteins for destruction [24]. RB also functions as an activating cofactor to other transcription factors including hypoxia-inducible factor 1α (HIF-1α) [24]. Loss of RB pushes a cancer cell past cell-cycle arrest induced by replicative and oncogenic signaling [24]. Loss of contact inhibition also contributes to the evasion of

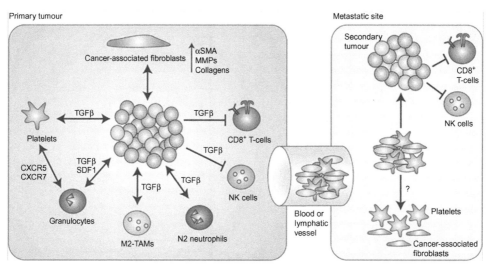

Fig. 3 TGFβ signaling in cancer. TGFβ has an important role in immune evasion and activation by cancer cells. Cells with bidirectional arrows are both activated by TGFβ signaling from cancer cells and reciprocate the signaling with activating, tumor-promoting activity. Additional cytokines and chemokines are listed along the arrows. CD8+ T-cells and natural killer (NK) cells are inhibited by tumor TGFβ signaling. Activation of cancer associated fibroblasts increases matrix deposition which promotes aggressive, MDR cancer. Please see text for additional abbreviations. *(Reprinted from Bellomo C, Caja L, Moustakas A. Transforming growth factor β as regulator of cancer stemness and metastasis. Br J Cancer 2016;115(7):761–9 under the Creative Commons Attribution-Non-Commercial-Share Alike 4.0 International License. A copy of the license can be accessed: http://creativecommons.org/licenses/by-nc-sa/4.0/.)*

growth suppression hallmark. To evade growth suppression, cells can migrate and metastasize. Loss of contact inhibition begins with loss of E-cadherin, gain of N-cadherin, and polarization of RAC1 (GTPase) to the leading edge of the moving cell [25].

An element that overlaps with both evading growth suppression and sustaining proliferative signaling is p53. p53 is sometimes referred to as the master tumor suppressor gene as p53 mutations are associated with many cancers [26]. MDM2 is a negative regulator of p53 [26]. In normal cells stress factors inactivate MDM2, which allows p53 to be activated; p53 is activated by cell stressors such as hypoxia, nucleotide depletion, DNA damage, telomere erosion, and microtubule inhibition [27]. p53 can be thought of as a master decision maker that attempts to correct the cell stressor that activated it and decide on the final cellular response to the correction. The final response may be cell-cycle arrest, differentiation, DNA repair, apoptosis, on senescence [27]. Inactivation of p53 occurs through mutations in the p53 gene (mainly point mutations), activation of MDM2, or disruption of downstream p53 responses [27]. Proliferative signaling would normally trigger p53 cellular maintenance and suppress growth. p53 mutations contribute to both hallmarks. Importantly, the majority of p53 mutations in cancer inactivate the tumor suppressor function of p53 and overactivate a beneficial function such as increased capacity for oxidative phosphorylation

[28]. p53 loss results in dramatic inflammation through the loss of p53 regulation of the transcription factor nuclear factor-κB (NF-κB) [28]. Depending on the mutation, mutant p53 can activate inflammation in various ways. For example, some p53 mutants inhibit directly interact with and promote NF-κB activity, while other mutants increase toll-like receptor 3 (TLR3) cytokine responses and inhibit secretion of interleukin-1 receptor antagonist (sIL-1Ra); TLR3 is a nucleotide-sensing molecular pattern recognition receptor while IL-1 is a pro-inflammatory cytokine [28]. Loss of p53 and mutant p53 increase tumor-promoting inflammation.

Oncogene and tumor suppressor activity converge to promote sustaining proliferative signaling at the level of cell-cycle checkpoint regulation. Cyclins and cyclin-dependent kinases (CDKs) drive the progression of the cell cycle, while mitogenic signaling, DNA damage, cell stressors, and grow inhibitory or stimulatory signals trigger RB, p53, and checkpoint inhibitor activation if necessary. Tumor-promoting inflammation favors inhibition of the negative regulators of the cell cycle, while overexpression of growth factor receptors and signaling drives propagation [28].

5. Genome instability and mutation and replicative immortality

The core concept of the genome instability and mutation hallmark is that genetic manipulation allows cancer cells to directly select for phenotypes that confer survival. This is evident from the previous discussion of p53 mutations. The core concept of the replicative immortality hallmark is that cancer cells are cells that can replicate forever if given a supportive microenvironment. These hallmarks are discussed together as perpetually mutation and adaptation is necessary to maintain replicative immortality.

A key element of genome instability is that transcriptional regulation, epigenetics, and microRNA interplay to confer survival. Aberrant or null expression of tumor suppressors and oncogenes can confer survival at different stages. When considering mutations, it is important to consider that the tissue context is significant. Tissue-specific tumorigenesis studies identify that different tissue-derived cancers have predominant gene mutations [29]. For example, according to data obtained from the Welcome Trust Sanger Institute's Catalogue of Somatic Mutations in Cancer, p53 mutations are associated with 48% of ovarian cancer cases, 24% of breast cancer cases, 44% of pancreatic ductal adenocarcinoma cases, 36% of non-small cell lung cancer cases, and 5% of kidney cancer cases, to name a few [29]. On the other hand, 43% of kidney cancer cases had mutation in the von Hippel-Lindau (VHL) [29]. Mutational drivers for tumor progression are tissue specific.

In addition to somatic mutations, epigenetic regulation and microRNA regulation contribute to phenotype control. The most common epigenetic changes in cancer occur as DNA promoter hypermethylation, abnormal chromatin and abnormal histone modifications, and genome-wide hypomethylation [30]. The DNA promoter hypermethylation suppress gene expression [30]. The abnormal chromatin and histone modifications lead to alterations in cell-cycle progression, DNA repair, and transcription [30]. The genome-wide hypomethylation results in genome instability [30]. Various cancers have been reported to have hypermethylation of chemokine promoters, decreasing their activity and altering the immune-cancer interface [31]. DNA promoter hypermethylation is associated with chronic inflammatory diseases and infections [31]. Translational repression by microRNA is exploited by many cancers as cancer cells tend to have a higher level of microRNA and even microRNAs that have not been detected in normal cells [31]. Micro-RNA translational control of expression is a very important mechanism for cancer cells as it allows the cell to repress protein expression even if gene expression is intact. This is critical to the rapid cellular evolution and adaption required by these cells in response to their continually changing microenvironments. Owing to constant vascular remodeling and destruction, a cancer cell may experience sudden loss of nutrients and oxygen. The ability to rapidly adapt and alter expression through microRNA is perhaps one of the most critical tools that enables cancer cell adaption. Some microRNAs have been classified as oncogenic microRNA [32]. For example, inflammation induces the expression of oncogenic micro-RNAs miR-21 and miR-155 [32]. miR-21 is associated with primary inflammatory diseases, increased cell proliferation, and inhibits apoptosis, while miR-155 appears to target p53 downstream effectors [32].

The two components of replicative immortality are telomerase and cell-cycle checkpoint escape (discussed earlier; sustaining proliferative signaling). It is well known that telomeres function as biological clocks at the end of human chromosomes. Telomeres consist of DNA (TTAGGG double stranded repeats with a 3′ overhang), a protective shelter in protein complex, and a telomerase complex with a telomerase reverse transcriptase enzyme and a telomerase RNA template component [33]. Telomerase is upregulated in many cancers enabling the cells to replenish telomeres and avoid cell death [33]. The upregulated telomerase contributes to the genomic instability and mutation in cancer cells as perpetual telomerase activity results in rearrangement, dicentric chromosomes, and loss of heterozygosity [33]. Telomerase upregulation is often regarded as a simple process that allows cancer cells to have perpetual growth but the result is dramatic genomic alterations that cannot be corrected with the loss or alteration of p53 and RB [33].

6. Angiogenesis

The core concept of the angiogenesis hallmark is that due to constant vascular deconstruction (from tumor growth) there is a constant demand for vascular reconstruction and neovascularization. A key element of this hallmark is that the drivers for angiogenesis are the diffusion limit of oxygen and the diffusion limit of glucose. The haphazard anatomy of a tumor occurs because of loss of contact inhibition. Cells grow over each other and crush new and existing vasculature. There is as much competition as there is cooperation within the tumor microenvironment. When cells are beyond the diffusion limit of oxygen the transcription factor, HIF-1α, is activated [34]. Upon activation, HIF-1α translocates from the cytoplasm to the nucleus where it binds with HIF-1β forming an active transcription factor that upregulates targets genes includes growth factor receptors, multidrug-resistant efflux pumps, cell proliferation signals, vascular endothelial growth factor (VEGF), and proteins that control glucose metabolism, to name a few [34]. Hypoxic transformation contributes to MDR and disease aggression. Cells that are beyond the diffusion limit of glucose may enter G0 or quiescence (discussed later), which makes these cells even more resistant and difficult to treat. Oxygen and glucose depletion drive acidosis and are selection pressures for cancer.

HIF-1α activation of VEGF is significant as VEGF activates angiogenesis. A second element of this hallmark is that the angiogenetic switch is always on. Perpetual vascular deconstruction from the solid tumor anatomy (loss of contact inhibition and evading growth suppression) drives this switch. Angiogenesis occurs in tumors by the same three mechanisms as in normal tissue, but tumors have three additional unique mechanisms of vessel formation [35]. The normal processes of vessel formation are by sprouting angiogenesis via a tip cell, vasculogenesis from a bone-marrow derived progenitor cell, and intussusception that results in vessel splitting [35]. The three cancer-specific mechanisms of vessel formation are vessel co-option where cancer cells reside inside of existing vessels, vascular mimicry where cancer cells line vessels and appear to repair or restore vessel gaps, or endothelial cell differentiation where cancer stem cells differentiate into endothelial cells [35]. The last two mechanisms attest to the dramatic plasticity in cancer (hallmark 11). There is a phenomenon referred to as the enhanced permeability and retention (EPR) effect. The enhanced permeability is the result of leaky, disorganized vasculature. The retention effect occurs because once therapeutics are localized to the tumor microenvironment, they are retained due to the lack of lymphatic drainage in the tumor. Nanomedicines are the ideal size to exploit the EPR effect. The leaky vasculature is also likely to contribute to the dynamic communication between cancer cells and immune cells in the tumor microenvironment. The tumor microenvironment is often compared to wound healing; the same components are involved, just less organized and with excessive stressors.

There is also a feedback loop of activation between HIF-1α and NF-κB [35]. This reveals the relationship between hypoxia, drug resistance, aggressiveness, and inflammation. NF-κB is a prominent transcription factor that mediates immunity and inflammation [35].

7. Avoiding immune destruction and tumor-promoting inflammation

The core concept of the avoiding immune destruction hallmark is that immune cells within the tumor microenvironment are transformed and help cancer survive.

The core concept of tumor-promoting inflammation is that inflammation is a direct cancer driver and selection pressure for cancer. Cancer is very much an immune disease. The tumor microenvironment consists of cancer cells, cancer-associated fibroblasts, cancer-associated macrophages, the extracellular matrix, vasculature, and lymphatic vessels. There are also some infiltrating or resident T-regs, MDSC, T-cells, and B-cells [9]. Avoiding immune destruction is most important for the early tumor. Once a tumor is established and progresses, the strive for survival and the hallmarks of cancer assist to aid the tumor in avoiding immune destruction. All of the hallmarks intersect and overlap but there is very strong overlap of this hallmark with the resisting cell-death hallmark. As was mentioned in Section 3, the early tumor avoids immune destruction by expressing neoantigens, mutating death receptors and elevating the threshold activation for cell death [1, 18]. Neoantigens are generated due to the high mutation rate in cancer [1]. These point mutations, insertions, deletions, or frame shift mutations can lead to the presentation of surface molecules that are immunogenic or nonimmunogenic [1]. As cancer is a disease of molecular evolution, due to survival of the fittest within the tumor microenvironment the cells with the most nonimmunogenic mutations are selected for clonal expansion [1]. The rapid molecular evolution that occurs enables cancer cells in the early tumor to avoid immune destruction.

A key element of both hallmarks is the transformation of macrophages within the tumor microenvironment [2, 36]. Fig. 4 illustrates macrophage polarization states. One primary theme of cancer is molecular evolution. A second theme is that pleiotropy enables plasticity and many cancer hallmarks. The pleiotropic states of macrophages are central to tumor development. M1 macrophages are also known as classical macrophages and are associated with type I inflammation and killing of cancer cells and pathogens [2, 36]. M1 macrophages are stimulated by Interferon gamma (IFN-γ), lipopolysaccharide (LPS), and tumor necrosis factor (TNF) and cytokine signaling includes IL-12, TNF, and IL-6 [2, 36]. There are three M2 states. M2a alternative macrophages are stimulated by IL-4 and IL-13 and release IL-10 and polyamine [2, 36]. M2a macrophages are associated with type II inflammation, Th2 responses, allergy, and parasite killing [2, 36]. M2b type II macrophages are associated with Th2 activation and immunoregulation [2, 36]. M2b macrophages are polarized by immune complexes and toll-like receptors as well as IL-1R ligands; cytokines released include IL-10, TNF, and IL-6 [2, 36]. M2c deactivated macrophages are activated by IL-10 and are associated with immunoregulation, tissue remodeling, and matrix deposition [2, 36]. M2c macrophages release IL-10, TGF-β, and matrix components [2, 36]. The polarization that occurs in the tumor microenvironment is from the M1 state into one of the M2 states [2, 36]. Each M2 state has a vital role in tumor progression. The M2a phenotype are critical to creating tumor-promoting inflammation, while the M2b macrophages help to maintain this inflammation [2, 36]. The M2c polarized cells are often overlooked in cancer but these cells are critical in depositing extracellular matrix components and remodeling the tumor [2, 36].

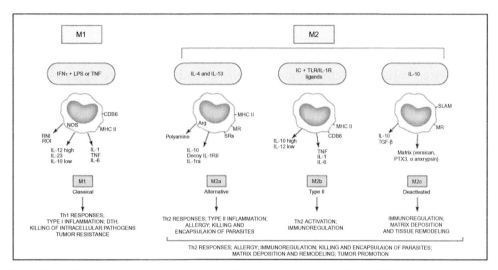

Fig. 4 Macrophage polarization states. Macrophages can be polarized in an M1, type 1 inflammation state or into one of three M2 states. M2a alternative macrophages are associated with type II inflammation where as M2b type II macrophages are associated with Th2 activation, M2c deactivated macrophages are associated with matrix deposition and tissue remodeling. The cytokines that induce each phenotype appear in yellow above each polarization state while the secreted factors surround the cell. *(Adapted and Reprinted from Martinez FO, Gordon S. The M1 and M2 paradigm of macrophage activation: time for reassessment. F1000Prime Rep 2014:6;13; Mantovani A, et al. The chemokine system in diverse forms of macrophage activation and polarization. Trends Immunol 2004;25(12):677–86 through Elsevier License and under the Creative Commons Attribution-Non-Commercial-Share Alike 4.0 International License. A copy of the license can be accessed: http://creativecommons.org/licenses/by-nc-sa/4.0/.)*

The M2c are the most dramatic polarization as M2c cells can function with cancer-associated fibroblasts to support the physical establishment of the tumor microenvironment and the extracellular matrix [2, 36].

Immunoediting is a second key element of both hallmarks. Immunoediting refers to an iterative process of immune signaling in the tumor microenvironment that culminates with tumor-promoting inflammation [37]. The process involves elimination, equilibrium, and escape but these stages can revert and are not linear as the tumor is continually evolving [37]. The elimination phase occurs (first) during early tumor formation and involves recognition and destruction of cancer cells by the immune system [37]. The equilibrium state first occurs when cancer cells that were not destroyed in the elimination phase undergo clonal expansion [37]. Although these cells have replicated, the immune system is able to contain the clonally expanded cells in a dormancy state [37]. As the cancer cells evolve and select for subclones capable of tipping the equilibrium balance, the escape phase begins [37]. The tumor has escaped the containment of the immune system [37]. The next step, which is often considered separately is immune priming [1]. During the escape phase, the tumor can have minimal immune involvement, which results in less aggressive disease [1]. After escape, the tumor can be associated with strong stromal inter-actions and have less immune interface, which also results in less aggressive disease [1]. Lastly, after escape, there can be a perpetual interplay between the immune system and the escaped tumor so the tumor is "primed" and has heighted aggression as it responds to myeloid suppressors, T-cells, and B-cells until inflammation transforms many of the cells in the tumor microenvironment and actually drives the aggressiveness of the tumor [1, 37]. Immunoediting and immune priming continue to occur as the tumor evolves. These events are critical to promoting tumor survival.

8. Deregulated energetics

The core concept of the deregulated energetics hallmarks is that aerobic glycolysis is protective and metabolically advantageous to cancer cells. In the angiogenesis section we discussed the role of hypoxia and HIF-1α. Target genes that are upregulated by HIF-1α include the proteins of the glycolytic pathway [38]. In the 1920s, Otto Warburg observed that cancer cells undergo glycolysis even in the presence of oxygen [38]. Normal cells only undergo glycolysis under conditions of oxygen depletion such as excessive exercise: anaerobic glycolysis. The process of aerobic glycolysis in cancer cells has since been termed the Warburg effect. Aerobic glycolysis is highly ineffective com-pared to oxidative phosphorylation, yet this form of energy acquisition can be beneficial to a cancer cell [39]. The benefits are faster energy supply (occurs in cytoplasm verses mitochondria), safer (less free radical production), high yield of metabolic precursors, altered pH that can enhance tumor protection, survival, and invasion [39]. The Warburg effect may also assist cancer cells as they progress through immunoediting and immune priming by making these cells less susceptible to the accumulation and effects of reactive oxygen species and by acidifying the pH, which leads to changes in the plasma membrane and extracellular matrix.

9. Plasticity, tumor heterogeneity, quiescence, and stemness

The core concept of the plasticity hallmark is that survival within the continually changing microenvironment of a tumor requires rapid, flexible, and sometimes pleiotro-pic cellular responses to microenvironmental selection pressures. The core concept of tumor heterogeneity is that due to cellular plasticity and solid tumor anatomy (constant vascular deconstruction and reconstruction), the cells in a solid tumor mass have different molecular signatures. The core concept of the quiescence and stemness hallmark is that cancer cells lose their molecular signatures and differentiation becoming "stem-like" and cancer cells can enter a state of cellular rest to protect them from death.

Cancer cells are malleable in their response to challenges but consistently respond to confer survival. All cells within the tumor microenvironment have a high level of plasticity (including immune cells). Macrophage repolarization within the tumor microenvironment is a form of plasticity, as is EMT to MET (and *vice versa*), and MDR. The molecular evolution of cancer requires that all cells in the tumor microenvironment have this trait. The cells that do not, will not adapt to the changing stressors. Tumor heterogeneity is the tissue level of cellular plasticity. Clonality and heterogeneity are not opposing concepts. As the local microenvironmental selection pressures are different for different cells within a tumor, cells need to response differently to achieve survival and both these cells may continue to be selected for and propagate. The high mutation rate also drives heterogeneity. Yet, the anatomy of a solid tumor is the strongest pressor for tumor heterogeneity to achieve survival. Immune cells are part of this anatomy and heterogeneity. Transitioning from the immunoediting state of equilibrium to escape and immune priming will result in greater heterogeneity as cells respond to local signaling.

Considering quiescence and stemness, these changes can be a response to immune challenge and a type of escape. Stemness is associated with survival and MDR and stemness serves multiple functions, which includes allowing vascular repair through differentiation into endothelial cells. Quiescence is also protective and a critical cellular decision that prevents cell death. Quiescence is the key to re-occurrence. These cells enter into a protected offshoot of G0 and are extremely difficult to treat. In the continual process of immunoediting and immune priming as well as challenging tumors with chemotherapeutics, a cancer cell may enter quiescence as a form of protection and emerge much later to initiate the process again. The tumor microenvironment is both cooperative and competitive and this drives both inflammatory-driven propagation and survival strategies such as stemness and quiescence.

10. Conclusion

The 13 hallmarks of cancer are the survival skills of the disease. These skills and abilities enable rapid and effective adaptation to changing cell stressors. The immune interface is important for each of these hallmarks as immune dysfunction is responsible for initiation and propagation of the disease. Understanding the significance of inflammation and the immune component of each hallmark is critical for advancing the study of the disease and designing new therapeutic strategies that overcome the strong immune dysfunction and motivation of cancer.

References

[1] Chen DS, Mellman I. Elements of cancer immunity and the cancer-immune set point. Nature 2017;541(7637):321–30.
[2] Martinez FO, Gordon S. The M1 and M2 paradigm of macrophage activation: time for reassessment. F1000Prime Rep 2014;6:13.
[3] Hanahan D, Weinberg RA. The hallmarks of cancer. Cell 2000;100(1):57–70.
[4] Hanahan D, Weinberg RA. Hallmarks of cancer: the next generation. Cell 2011;144(5):646–74.

[5] Milane L. Cancer. In: Amiji M, Milane L, editors. Nanomedicine for inflammatory diseases. Boca Raton, FL: CRC Press; 2017. p. 319–31.

[6] Society, A.C. Available from: https://www.cancer.org/cancer/breast-cancer/understanding-a-breast-cancer-diagnosis/types-of-breast-cancer/triple-negative.html; 2020.

[7] MacCarthy-Morrogh L, Martin P. The hallmarks of cancer are also the hallmarks of wound healing. Sci Signal 2020;13(648), eaay8690.

[8] Mosesson Y, Mills GB, Yarden Y. Derailed endocytosis: an emerging feature of cancer. Nat Rev Cancer 2008;8(11):835–50.

[9] Quail DF, Joyce JA. Microenvironmental regulation of tumor progression and metastasis. Nat Med 2013;19(11):1423–37.

[10] Pittenger MF, et al. Mesenchymal stem cell perspective: cell biology to clinical progress. npj Regen Med 2019;4(1):22.

[11] Ankrum JA, Ong JF, Karp JM. Mesenchymal stem cells: immune evasive, not immune privileged. Nat Biotechnol 2014;32(3):252–60.

[12] Milane L, et al. Exosome mediated communication within the tumor microenvironment. J Control Release 2015;219:278–94.

[13] Paget S. The distribution of secondary growths in cancer of the breast. Lancet 1889;133(3421):571–3.

[14] Gao Y, et al. Metastasis organotropism: redefining the congenial soil. Dev Cell 2019;49(3):375–91.

[15] Lee Y-TM. Patterns of metastasis and natural courses of breast carcinoma. Cancer Metastasis Rev 1985;4(2):153–72.

[16] Rose PG, et al. Metastatic patterns in histologic variants of ovarian cancer. An autopsy study. Cancer 1989;64(7):1508–13.

[17] Ke B, et al. Targeting programmed cell death using small-molecule compounds to improve potential cancer therapy. Med Res Rev 2016;36(6):983–1035.

[18] Labi V, Erlacher M. How cell death shapes cancer. Cell Death Dis 2015;6(3):e1675.

[19] Gottesman MM, Fojo T, Bates SE. Multidrug resistance in cancer: role of ATP-dependent transporters. Nat Rev Cancer 2002;2(1):48–58.

[20] Teng MWL, et al. From mice to humans: developments in cancer immunoediting. J Clin Invest 2015;125(9):3338–46.

[21] Huang A, Cao S, Tang L. The tumor microenvironment and inflammatory breast cancer. J Cancer 2017;8(10):1884–91.

[22] Prud'homme GJ. Pathobiology of transforming growth factor β in cancer, fibrosis and immunologic disease, and therapeutic considerations. Lab Investig 2007;87(11):1077–91.

[23] Bellomo C, Caja L, Moustakas A. Transforming growth factor β as regulator of cancer stemness and metastasis. Br J Cancer 2016;115(7):761–9.

[24] Burkhart DL, Sage J. Cellular mechanisms of tumour suppression by the retinoblastoma gene. Nat Rev Cancer 2008;8(9):671–82.

[25] Stramer B, Mayor R. Mechanisms and in vivo functions of contact inhibition of locomotion. Nat Rev Mol Cell Biol 2017;18(1):43–55.

[26] Vogelstein B, Lane D, Levine AJ. Surfing the p53 network. Nature 2000;408(6810):307–10.

[27] Vousden KH, Lu X. Live or let die: the cell's response to p53. Nat Rev Cancer 2002;2(8):594–604.

[28] Uehara I, Tanaka N. Role of p53 in the regulation of the inflammatory tumor microenvironment and tumor suppression. Cancer 2018;10(7):219.

[29] Schneider G, et al. Tissue-specific tumorigenesis: context matters. Nat Rev Cancer 2017;17 (4):239–53.

[30] Pfister SX, Ashworth A. Marked for death: targeting epigenetic changes in cancer. Nat Rev Drug Discov 2017;16(4):241–63.

[31] Yasmin R, et al. Epigenetic regulation of inflammatory cytokines and associated genes in human malignancies. Mediat Inflamm 2015;2015:201703.

[32] Schetter AJ, Heegaard NHH, Harris CC. Inflammation and cancer: interweaving microRNA, free radical, cytokine and p53 pathways. Carcinogenesis 2010;31(1):37–49.

[33] Maciejowski J, de Lange T. Telomeres in cancer: tumour suppression and genome instability. Nat Rev Mol Cell Biol 2017;18(3):175–86.

[34] Harris AL. Hypoxia—a key regulatory factor in tumour growth. Nat Rev Cancer 2002;2(1):38–47.

[35] Carmeliet P, Jain RK. Molecular mechanisms and clinical applications of angiogenesis. Nature 2011;473(7347):298–307.

[36] Mantovani A, et al. The chemokine system in diverse forms of macrophage activation and polarization. Trends Immunol 2004;25(12):677–86.

[37] O'Donnell JS, Teng MWL, Smyth MJ. Cancer immunoediting and resistance to T cell-based immunotherapy. Nat Rev Clin Oncol 2019;16(3):151–67.

[38] Semenza GL. Targeting HIF-1 for cancer therapy. Nat Rev Cancer 2003;3(10):721–32.

[39] Vander Heiden MG, Cantley LC, Thompson CB. Understanding the Warburg effect: the metabolic requirements of cell proliferation. Science 2009;324(5930):1029–33.

CHAPTER TWO

Innate and adaptive immunity in cancer

Aatman S. Doshi[a,b] and Kirtika H. Asrani[c]
[a]Bioscience, AstraZeneca R&D Boston, Waltham, MA, United States
[b]Department of Pharmaceutical Sciences, Bouvé College of Health Sciences, Northeastern University, Boston, MA, United States
[c]Oncology Research, Bluebird Bio, Cambridge, MA, United States

Contents

Cancer Immunology and Immunotherapy
https://doi.org/10.1016/B978-0-12-823397-9.00025-9

1. Introduction

One of human's greatest health challenges is cancer disease that kills millions of people around the world every year, triggering the urge to eradicate this disease as rapidly as possible [1]. Cancer disease is characterized by the uncontrolled proliferation of abnormal cells with the ability to spread to healthy tissues and organs [2]. In the late 19th century, it was believed that cancer is a clonal disease arising from a single cell of origin. Its progression is based on a sequential selection of more aggressive sublines from acquired genetic variability within the original single-cell clone [3]. However, Gloria Heppner and her colleagues at Roger Williams General Hospital reported the isolation of four distinct tumor subpopulations from a single spontaneously arising mouse mammary cancer describing tumor heterogeneity along with many other researchers at that time [4]. Tumor heterogeneity can be attributed to the varying growth rate, ability to metastasize, drug resistance, and failure in therapy among the many characteristics of them being of an aggressive phenotype. Genetic and epigenetic diversity in these heterogeneous tumors can accelerate their evolutionary fitness [5–9]. Yet for cancer to propagate further with its heterogeneity, it has to face the task of being recognized by immune cells as foreign [9]. There is a huge interplay between cancer and immune cells with a large effort being put in by the research community to develop effective immunotherapies for the treatment of cancer patients [10]. This has been described with the help of the cancer immunity cycle in Fig. 1.

Cancer immunotherapy is going through very exciting times and has reached an important inflection point. Interestingly, the ability of cancers to evade or escape the

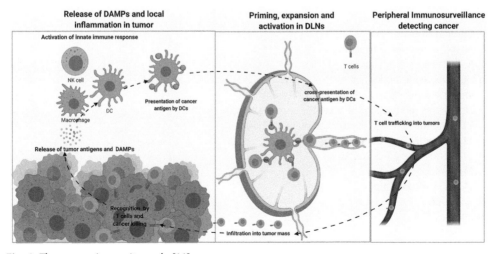

Fig. 1 The cancer-immunity cycle [11].

immune response is now recognized to be one of the most distinguished cancer hallmarks, which provides the platform for treatments within the context of immunotherapies.

Rudolph Virchow established the relationship between the immune system and cancer over a century ago [12]. The defensive mechanism of our body, *i.e.,* the immune system, relies on three main principles: detection of "non-self" antigens from either pathogens or malignant/tumor cells; encompasses effector functions to target and destroy the infected/malignant cells while protecting the host, and it develops immunological memory via the adaptive immune responses for subsequent defense mechanisms against these signals or antigens [13].

In this chapter, we will touch on the introduction and basic functions of innate and adaptive immunity in cancer and its protumor or antitumor functions. Also, we will discuss some of the mechanisms and current therapeutic approaches being explored preclinically or clinically.

The cancer-immunity cycle initiates with cancer cells releasing cancer neoantigens to oncogenesis or due to dying cells, which are recognized as danger-associated molecular patterns (DAMPs) by innate immune cells, leading to their activation followed by capturing of antigens by antigen-presenting cells (APCs) such as dendritic cells (DCs). One of the dendritic cell subtypes known as conventional DC (cDC) 1 subset travels to tumor-draining lymph nodes (TDLNs) and presents the antigen on their surface to T-cells. This further leads to priming, expansion, and activation of tumor antigen-specific effector T-cell responses. These effector T-cells hit the peripheral blood circulation for immunosurveillance to detect tumor cells. They then infiltrate into tumor mass, detect the cancer cells followed by their activation. Once they are activated, they perform the killing of cancer cells by releasing cytotoxic granules such as perforin and granzymes [11, 14].

2. Innate immunity

The innate immune system is the body's first line of defense against invading pathogens, which is nonspecific and consists of physical, chemical, and cellular defenses with the key aim of providing a fast response to avoid spreading of infection to other parts of the body [13]. Innate immunity involves various types of cells of the myeloid lineage, including DCs, monocytes, macrophages, polymorphonuclear leukocytes (PMNs), mast cells, natural killer (NK) cells, and natural killer T (NKT) cells [14, 15]. Innate immunity orchestrates an immune response through a specialized set of receptors, namely pattern recognition receptors (PRR), designed to distinguish modifications in the cells that pose a danger, thereby triggering the elimination by the immune system [16]. PRR can sense DAMPs or pathogen-associated molecular patterns (PAMPs), which is a sign of the presence of infection or damage in tissue. This leads to downstream maturation, activation, and production of soluble factors (chemokines, cytokines), leading to an increase in immune cells and proinflammation in the local area initiating an immune response against the infection [17]. Given the role of innate immunity in recognizing and clearing the transformed cells, it plays an important role in the immunobiology of tumors [18].

2.1 Natural killer cells

NK cells are effector lymphocytes of the innate immune compartment capable of controlling several types of infections and cancer [19]. They were historically described as large granular lymphocytes with natural cytotoxicity against cancerous cells. They further were defined as a separate lineage of lymphocytes that had the dual capacity of cytotoxicity and releasing effector cytokines [19–21]. The frequency of NK cells in human peripheral blood ranges between 5% and 15% [22, 23]. To perform their function of immunosurveillance throughout the vasculature, they are detected both in primary and secondary lymphoid organs such as the spleen, bone marrow, lymph nodes, and peripheral blood [24].

NK cells were highlighted for their natural cytotoxicity against tumor cells and metastasis without any priming or prior activation, claiming a key advantage over its counterpart, *i.e.*, adaptive immunity cells like T-cells or B-cells, which needs priming by APCs. They belong to a subset of innate lymphoid cells (ILCs) that express the transcription factor E4BP4/Nfil3 [25]. NK cells while on immunosurveillance are in contact with other cells constantly and their activation depends on the balance of different types of receptors on the NK cell surface. They express an array of receptors including inhibitory, activating, homing, and cytokine/chemokine receptors (Table 1). The NK cells are switched on when they come across activating receptors on the surface of cancer or infected cells [26]. Major histocompatibility complex class I (MHC I) is expressed on the majority of healthy cells that are marked as "self" and can be recognized by inhibitory receptors on NK cells keeping a check on their killing (Fig. 2). There are various MHC I specific inhibitory receptors including killer cell immunoglobulin-like receptors (KIRs)

Table 1 Activating, inhibitory, homing, and cytokine and chemokine receptors for NK cells [26–29].

Activating receptors	NKp46, CD16, NKp44, NKp30, NKp80, NKG2D, 2B4, NTB-A, CD84, Ly9, CRACC, CD94/NKG2C, Act., killer cell immunoglobulin receptor-S (KIR-S), NKR-P1C, DNAM-1, CD59
Inhibitory receptors	NKR-P1B, NKR-P1D, KLRG1, Siglec-7, LAIR-1, CEACAM-1, PILRα, TACTILE (CD96), TIGIT, NKG2A, Ly49, KIR-L, LILRB1, IL-1R8
Homing receptors	CCR2, CCR5, CCR7, CXCR3, CXCR6, L-Selectin, CXCR1, CXCR2, CX3CR1, ChemR23, CXCR4, S1P5, c-Kit
Cytokine and chemokine receptors	IL-1R, IL-2R, IL-12R, IL-15R, IL-18R, IL-21R, IFNAR, IFNGR1, IFNGR2, IL-17RA, IL-27RA, IL-10R

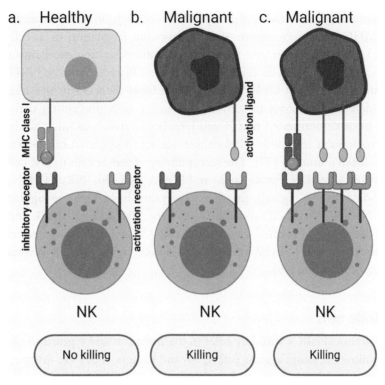

Fig. 2 NK cell-recognition mechanism of malignant cells. (A) Interaction of inhibitory receptors on NK cells with MHC I molecules without ligands for activating receptors ➔ no cytotoxicity is observed. (B) Interaction of activation receptors on NK cells with activating ligands in the absence of MHC I ➔ killing effect. (C) Despite an interaction of inhibitory receptors on NK cells with MHC I, due to the presence of excessive activating ligands on malignant cells, NK cells are induced to give a strong cytotoxic response ➔ killing effect [26, 30].

in humans, lectin-like Ly49 molecules in mice, and CD94/NKG2A heterodimers in both species that spare healthy cells expressing self-MHC class 1 [26, 31, 32]. In case of a viral infection or cancer, the cells are in "distress" leading to downregulation or lack of MHC class I expression ("missing self"), which can activate the NK cells [33]. Similar to cytotoxic T lymphocytes (CTLs), the mechanism of NK cell-mediated killing of target cells is through cytolytic granules secretion (granzymes, perforin), CD95L (FasL), and tumor necrosis factor (TNF)-related apoptosis-inducing ligand (TRAIL). Importantly, both human and mouse NK cells also express FcγRIIIA (CD16,) which enables NK cells to kill antibody-coated target cells through their Fc portion via antibody-dependent cell-mediated cytotoxicity (ADCC) [25].

NK cell reactivity toward tumor cells without prior sensitization has allowed us to explore them further for anticancer treatments. NK cells have been shown to reject tumors overexpressing natural-killer group 2D (NKG2D) ligands or costimulatory signals or lack of MHC class I expression, with downstream facilitation of T-cell antitumor immunity [34, 35]. However, the tumors have immune escape mechanisms to evade NKG2D-mediated detection such as secretion of these ligands. Secreted NKG2D ligands such as soluble MICA and ULBP proteins have been detected in the sera of cancer patients, which acts as decoys to subvert NK and T-cells preventing their activation [36]. It has been demonstrated that B-cell lymphoma arising in mice lacking perforin and β2 microglobulin and methylcholanthrene-induced sarcomas can be fought off by NK cell-mediated immunity [37]. The susceptibility of tumor cells to NK cell-mediated killing involves several receptors such as NKp46, NKp30, NKp44, DNAM1, and NKG2D. Some of the ligands for these receptors have been recognized such as B7-H6 for NKp30, MICA/B, and the ULBPs for NKG2D and PVR and Nectin-2 for DNAM-1 expressed on transformed tumor cells [26, 38]. Clinically a better prognosis has been associated with NK cell infiltration in human tumors such as cell lung, gastric, and colorectal carcinoma [39].

2.2 Dendritic cells

DCs are professional APCs with key roles in the initiation and regulation of innate and adaptive immune responses against pathogens and tumors. They are diverse, which are rare within the lymphoid organs and tumors but play a central role in antigen-specific immunity and tolerance [40, 41]. DCs in mouse develop from common DC precursors (CDPs) in the bone marrow, which comprise of two main subsets, namely more common $CD11b^+$ conventional DC (cDC) 2 subset and $CD8\alpha^+$ and/$CD103^+$ cDC1 subset [40, 42, 43]. $B220^+$ plasmacytoid DCs (pDCs) can be developed from CDP as well as lymphoid progenitors [40]. CC-chemokine receptor 2 (CCR2)-dependent recruitment of monocytes from the blood to the site of inflammation can lead to differentiation into monocyte-derived DCs (MoDCs) [44]. Mouse DC counterparts in humans, namely $CD141^+$ cDC1s, $CD123^+$ pDCs, and $CD1c^+$ cDC2s, can be found in the blood [44, 45].

DC functions are determined by the incorporation of various local signals sensed by intracellular and surface PRRs for PAMPs and DAMPs. In the context of the tumor microenvironment (TME), when DCs in their immature state sense DAMPs they take advantage of their strong phagocytic capabilities leading to their maturation. This mature state then loses its adhesion molecule expression with a cytoskeleton reorganization followed by migration to the nearest tumor-draining lymph nodes (TDLNs). Activated DC's are characterized by low endocytic and phagocytic receptors, increased expression of major histocompatibility complex II (MHC class II), costimulatory molecules, and CC-chemokine receptor type 7 (CCR7) and high ability to produce cytokines [40, 46]. The mature DCs become proficient in processing the captured antigens, which they present on their surface with MHC molecules (signal 1), provide costimulation (signal 2)

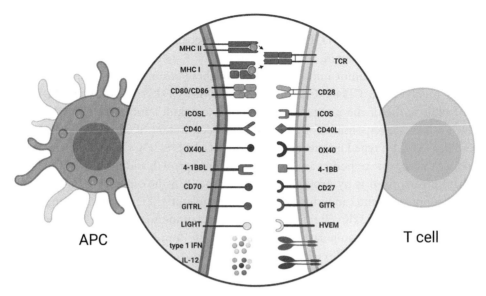

Fig. 3 Dendritic cell-T-cell interaction for activation of T-cells.

and soluble factors (signal 3) to orchestrate the T-cell responses generating antigen-specific antitumor immunity (Fig. 3) [40, 47, 48].

DCs in the TME can substantially influence the functions of T-cells because in addition to the lymphoid organs they can present tumor antigens to T-cells in the TME as well [46, 49]. cDC1s are the key cells that excel in presenting the exogenous antigens on MHC class I molecules to CD8$^+$ T-cells and also possess the ability to prime type 1 T-helper cell (T$_H$1 cell) responses [40, 42, 46, 50]. cDC1s (CD8α^+ and/CD103$^{+)}$ depend on interferon regulatory factor 8 (Irf8) and basic leucine zipper transcriptional factor ATF-like 3 (BATF3) [51, 52]. It has been shown by multiple groups that BATF3-dependent cDC1s are essential for effective T-cell trafficking, adoptive transfer T-cell therapy, and for driving an effective antitumor immunity [51, 53, 54]. It has been also shown that checkpoint inhibitor efficacy relies on the presence of CD103$^+$ DCs in the TME [55]. There are also reports that both cDC2s and MoDCs can also cross-present tumor antigens to CD8$^+$ T-cells directly [56]. In humans, BATF3-dependent DC is characterized by the CD141 (BDCA3), clec9a, and XCR1 [57]. The presence of CD141 DCs correlates with better clinical outcomes in patients with tumors establishing their critical role in antitumor immunity in humans [58–61]. Barry *et al.* have shown that intratumor NK cell expression of FMS-like tyrosine kinase 3 ligand (FLT3L) controls the levels of intratumoral CD103$^+$BDCA3$^+$ cDC1s. They went on to further validate this finding in patients with melanoma where FLT3L and intratumoral cDC1 levels correlated with NK cells. This correlation of NK cells and intratumoral cDC1s was predictive of responses to antiprogrammed cell death protein 1 (PD1) therapy [61].

CCR7 is upregulated on mature DCs after they are activated in the TME, which allows them to migrate to TDLNs [62, 63]. The costimulatory signals coming from mature DCs play a very significant role in priming of T-cells by providing them a signal 2 and generating antitumor immunity. Some of these costimulatory surface activation markers involved are CD40, CD80, CD83, CD86, CD137, OX40L, GITRL, and CD70 [40]. Some of the potent soluble factors (cytokines and chemokines) secreted by DCs which are responsible for activation and recruitment of effector T-cells include CXCL9, CXCL19, type I interferons (IFN), and IL-12 [48]. DCs have also been known to induce tolerogenic responses and can induce tumor growth and metastasis. One of the mechanisms involved is by producing indoleamine 2, 3-dioxygenase (IDO) which is immunosuppressive and can induce FOXP3$^+$ T regulatory (Treg) cells and can suppress the effector T-cell and NK cell responses [64]. A deep understanding of DC biology is of utmost importance to manipulate them for successful immunotherapy.

Dendritic cells activate T-cells in form of three signals. Signal 1 is given by the interaction between TCR and MHC I or MHC II complexed with tumor antigen. Signal 2 is given by costimulatory molecules interaction between DCs and T-cells. Signal 3 is given by the release of soluble factors from DCs such as IL-12 and type I IFN [40, 65].

2.3 Macrophages

Macrophages are a type of white blood cells of the mononuclear phagocyte immune system which serves as the first line of defense against infections and cancer, maintains tissue homeostasis, and protects our body through functions by engulfing harmful invading foreign substances [66, 67]. Similar to DCs, macrophages are initiators of innate immunity and can be mediators of adaptive immunity [68]. Macrophages can comprise up to 50% of the tumor mass and can traffic into the TME from bone-marrow-derived precursors or can differentiate from circulating monocytes or can proliferate from tissue-resident precursors [67–70]. Macrophages mainly develop into two main subtypes with contrasting functions in the TME, namely classically activated macrophages (M1) and alternatively activated macrophages based on the environmental cues in the TME. M1 macrophages can produce proinflammatory cytokines such (IL-12, TNF-α, CXCL10, IL-1β, IL-2, IL-6, IL-12, IL-23) in response to IFN gamma (IFN-γ), have cytotoxic activity against tumors and can also present antigens to T-cells [67, 68, 71]. Additionally, they can also trigger antibody-dependent cell phagocytosis (ADCP) of tumor cells directly [72]. Through their mechanism of killing microbes, they can also kill tumor cells with the release of nitric oxide (NO) and reactive oxygen species (ROS) [71]. M2 macrophages are described to have a role in tumor progression and metastasis by promoting angiogenesis, matrix remodeling, and suppression of adaptive immunity [73]. Thus, they have been a heavy topic of research to harness them for cancer immunotherapy. Macrophages with M2 phenotype can facilitate tissue remodeling by TH$_2$ responses, release antiinflammatory cytokines (IL–10, TGF-β) capable of inhibiting CD8$^+$ cytotoxic T-cell responses, produce extracellular components and promote immune tolerance [71, 73, 74]. The M2 phenotype is promoted by overactivation of STAT3 and by secretion of IL-4, IL-6, IL-10, prostaglandin E2, and TGF-β [69]. The beauty of macrophages is that they are plastic in their phenotype and are interchangeable into M1 or M2 phenotype by manipulating their environmental cues. Hence, many treatment options are being pursued actively to directly deplete M2 macrophages, reprogram or reeducate M2 macrophages to M1 phenotype, inhibiting active recruitment into tumors and block M2 phenotype [71].

2.4 Neutrophils

Neutrophils are also known as polymorphonuclear leukocytes (PMNs) and are the most abundant white blood cells with more than 70% of all leukocytes in composition. They have a very short half-life of 7–10 h in both humans and mice [75, 76]. They are the first line of defense for microbial infections and facilitators of wound healing [75]. PMN generation and differentiation are dependent on its master regulator, *i.e.,* granulocyte-colony stimulating factor (G-CSF) [77–79]. Neutrophils are recruited into tumors with help of various neutrophil-attracting CXC-chemokines [80]. CXCL8-CXCR2 signaling is one of the key signaling pathways involved in neutrophil recruitment [81]. Tumor-associated neutrophils (TANs) have been reported to have a poor prognosis in human tumors such as in ovarian, non-small-cell lung cancer (NSCLC), breast cancer, metastatic melanoma cancer, prostate cancer, colorectal cancer, hepatocellular carcinoma, and intrahepatic cholangiocarcinoma patients [82–91]. TANs are described to have both pro-tumor and antitumor functions [75]. Tumor-induced TAN polarization and activation are dependent on the specific tumor-derived factors such as TGF-β, G-CSF, and interferon-β (IFNβ) [75]. TGF-β or G-CSF acts as tumor-promoting whereas IFNβ can act as a negative regulator of pro-tumorigenic TANs [92–94]. TANs are believed to have several roles in cancer including tumor initiation and growth and metastasis [75]. In contrasting reports groups have reported depletion of neutrophils increases metastasis [95, 96]. Besides they can limit metastasis through the production of H_2O_2, through their expression of thrombospondin 1 (TSP1) and MET [75, 95, 97].

2.5 Natural killer T-cells

NKT cells are a unique subset of T-cells that recognize glycolipid antigens within the context of CD1d, a nonclassical MHC class I-like molecule, and are an important link between innate and adaptive immunity [98–100]. They are categorized into two subsets, *i.e.,* invariant (iNKT) or diverse (dNKT), of which iNKT are the most well-characterized being able to produce Th1-, Th2-, and Th17-type cytokines in large amounts post-stimulation. NKT cells can mount effective antitumor responses and have been manipulated for their potential use in immunotherapy. α-galactosylceramide (α-GalCer) is a potent activator of iNKT cells which can induce them to produce cytokines and eventually stimulate important immune cells such as NK cells, macrophages, DCs, B-cells, and T-cells. They can directly kill cancer cells [100, 101]. iNKT cells are being currently explored for their use in chimeric antigen receptor (CAR) modification because their presence in patients tumors correlate with better outcomes and also its ability to recognize antigen by CD1D restriction making potential for lower off-target toxicity [99].

3. Effector mechanisms and immunotherapies modulating innate immunity pathway

3.1 Targeting nucleic acid-mediated immunity

Toll-like receptors (TLRs) are critical PRRs required for initiation and induction of innate immune effector responses by detecting harmful pathogens. Until now there have been at least 11 members in this group. They can be expressed both intracellularly and on the cell surface [102–104]. Nucleic acid-sensing TLRs whose roles are found to be essential are TLR3, TLR7, TLR8, and TLR9 [103]. TLR3 is expressed in most innate immune cells [105, 106], TLR7 in pDCs and B-cells [107], TLR8 in macrophages, monocytes, and cDCs [107], and TLR9 in pDCs and B-cells [108]. TLR agonism has been explored for its ability to activate innate immunity to generate an antitumor response both preclinically and clinically. Sabbatini *et al.* have shown activation of NK cells, cytotoxic T-lymphocytes (CTLs), and NKT cells by agonism of TLR3 and melanoma differentiation-associated protein 5 (MDA5) with poly-ICLC (Hiltonol) [109]. Hammerich *et al.* have shown by combining FLT3L, radiotherapy, and TLR3 agonist with an *in situ* vaccination approach, there can be the recruitment of antigen-loaded and activated intratumoral, cross-presenting DCs, which are critical for a T-cell-mediated antitumor immunity. This approach induced CD8+ T-cell responses and abscopal cancer remission in patients with advanced indolent non-Hodgkin's lymphomas in an ongoing trial (NCT01976585) [109]. Imiquimod is a US Food and Drug Administration (FDA) approved TLR7 agonist for the treatment of genital warts, actinic keratosis, and superficial basal cell carcinomas [110]. Ma *et al.* have demonstrated imiquimod and TLR7/8 agonist gardiquimod can enhance the activation of splenic NKT cells, NK cells, and T-cells. Additionally, they increase the cytolytic activity of splenocytes against MCA-38 and B16 cell lines and also increase IL-12 secretion by macrophages and DCs thus providing a potent antitumor activity of tumor lysate-loaded DCs [111]. Another TLR7/8 agonist resiquimod that has been shown to provide antitumor efficacy *in vivo* with two pancreatic ductal adenocarcinoma, with increased CD8+ T-cell infiltration and activity, and decreased Treg frequency [112]. IMO-2125 (tilsotolimod) is a TLR9 agonist that is currently being tested in clinic in combination with ipilimumab or pembrolizumab or nivolumab (NCT02644967, NCT03865082), which is well tolerated in phase I/II trials and actively promotes antitumor responses in primary (injected) and secondary (uninjected) melanoma in combination with Ipilimumab [104, 113].

Stimulator of interferon genes (STING) is a double-stranded DNA (dsDNA) sensor that has been very critical in battling DNA viruses and bacterial pathogens. dsDNA can be transformed into cyclic dinucleotide (CDN) by cyclic GMP-AMP synthase (cGAS). These CDNs can bind to STING and can lead to downstream signaling events leading to the release of type I IFNs. This mechanism has been explored in anticancer immune responses since type I IFNs can have pleiotropic effects on multiple immune cell activation [104, 114]. Corrales *et al.* have demonstrated by direct injections of synthetic CDNs in the tumors that lead to potent antitumor regressions in multiple mouse syngeneic tumor models [115]. Since then many companies have started clinical trials using synthetic CDNs or non-CDNs both via intratumor or intravenous dosing as a monotherapy or in combination with checkpoint inhibitors [14, 104, 114].

Retinoic acid-inducible gene I (RIG-I)-like receptors is another pathway that is of interest in being manipulated toward generating antitumor immune responses. They are categorized into RIG-I, melanoma differentiation-associated protein 5 (MDA5), and laboratory of genetics and physiology 2 (LGP2), which are expressed ubiquitously in most cell types necessary for recognition of exogenous nuclear-like viral RNAs. They have an important role as an antiviral sensor inducing innate immune responses. The binding of the viral RNAs to either RIG-I or MDA-5 can depend on their length. Following the recognition of dsRNAs by RIG-I or MDA-5 it will activate downstream adapter mitochondrial antiviral signaling (MAVS) to activate interferon regulatory factor 3 (IRF3) and nuclear factor-κB (NFκB) to stimulate the transcription of type I IFNs and antimicrobial inflammatory cytokines [104]. RIG-I like receptor ligands have been explored for its ability to induce strong innate immune responses and directly inducing apoptosis in cancer cells [104, 116–118]. Heidegger *et al.* have demonstrated therapeutic targeting of RIG-I with 5′-triphosphorylated RNA in combination with CTLA4 and PD1 checkpoint blockade in several preclinical cancer models. Mechanistically they have shown increased cross-presentation of tumor-associated antigens (TAAs) by CD103$^+$ DCs, caspase-3-mediated apoptosis, and expansion of tumor antigen-specific CD8$^+$ T-cell responses leading to increased infiltration in tumors [119].

3.2 Targeting immunosuppressive TME

The immunosuppressive phenotype of neutrophils termed PMN myeloid-derived suppressor cells (MDSCs) can exert suppressive effects on T-cells and NK cells by secreting arginase (ARG) 1 enzyme, which can deplete L-arginine in the TME. L-Arginine is a semiessential amino acid required by optimal proliferation and function of T-cells and NK cells. One of the approaches to relieve this immunosuppression is by inhibiting ARG, thus increasing the

L-arginine concentrations locally and systemically, which will eventually boost the T-cell responses. Calithera Biosciences in collaboration with Incyte is currently evaluating a small molecule inhibitor against ARG is currently in phase 1 and 2 clinical trials [120].

Another immunosuppressive mechanism that is being explored is the tryptophan-kynurenine-aryl hydrocarbon receptor (Trp-Kyn-AhR) pathway [121]. IDO presence in the TME catalyzes tryptophan into kynurenine, thus depleting the amino acid locally. Tryptophan depletion and production of kynurenine can suppress T-cells directly, induce Treg cells and MDSCs [121–123]. Kynurenine can have direct immunosuppression on T-cells and NK cells [124]. IDO can be expressed by tumor cells as well as suppressive myeloid cells [123]. Based on the important role of IDO in immunosuppression and tumor development, IDO has been an attractive approach to be pursued by many companies clinically for its inhibition [122]. Multiple companies such as Incyte, Newlink Genetics, IO Biotech, Eli Lilly and Company, Kyowa HakkoKirin Pharma, Bristol-Myers Squibb, Shanghai Denovo Pharmatech, and Hengrui Therapeutics have tried to clinically test the hypothesis of IDO inhibition through small molecule inhibitors to relieve immunosuppression [14]. Despite a nice proof-of-concept shown *in vivo* with multiple *in vivo* studies preclinically, there has not been much success around this hypothesis in providing a benefit to clinical outcomes [122, 125].

Adenosine is generated in the hypoxic areas of TME. ATP is degraded into ADP and AMP by CD39. AMP is dephosphorylated to adenosine by CD73. Adenosine receptors (A2AR, A2BR) are expressed on multiple immune cell subsets including T-cells, NK cells, macrophages, and DCs [126–128]. Adenosine can engage with their receptors and activate the downstream signaling, which can have immunosuppressive effects such as induction of immunosuppressive cytokines by DCs, suppression of proliferation of $CD8^+$ T-cells and NK cells, inhibition of antigen presentation by DCs, and enhancing the immunosuppressor function of $CD4^+FOXP3^+$ T-cells [129–132]. Adenosine signaling in tumor-associated macrophages (TAMs), MDSCs, or tumor-associated dendritic cells can promote the expression of immunosuppressive cytokine IL-10, which can further negatively regulate the T- and NK cells [133]. Genetic ablation or blocking of CD39/CD73 activity with targeted antibodies have been shown to generate tumor regressions preclinically in several syngeneic models [134–138]. Small molecule inhibitors of A2AR signaling have also shown antitumor responses in multiple syngeneic mouse tumor models in combination with anti-PDL1 and anti-CTLA4 [128, 139, 140]. These multiple ways of attacking this pathway are currently under investigation in clinical trials by numerous organizations [14].

3.3 Oncolytic viruses

Oncolytic viruses (OVs) is a novel promising immunotherapeutic approach taking advantage of the natural propensity of malignant cells to be infected by OVs. OVs can directly kill tumor cells through immunogenic cell death (ICD) producing antitumor immunity and can additionally generate antiviral immunity because of its viral capacity. OVs can be DNA or RNA viruses and they can bind directly to nucleic-acid receptors such as TLRs, RLRs, cGAS-STING, etc., which eventually lead to activation of IRF3 and NFκB to stimulate the transcription of type I IFNs and antimicrobial inflammatory cytokines leading to potent antitumor immunity in addition to ICD of tumors. ICD of tumors leads to release of PAMPs and DAMPs, which can further activate the innate immunity leading to activated DCs, enhanced cross-presentation of tumor antigens to T-cell in TDLN, enhancement of MHC class I, and enhancement of PD1, PDL1, and CTLA4 through type I IFNs [104, 141, 142]. Ribas *et al.* in phase 1b trials tested OV therapy with talimogene laherparepvec (modified herpes simplex virus type 1) in combination with anti-PD1 antibody pembrolizumab. They showed responses in patients with melanoma with increased $CD8^+$ T-cells, elevated PDL1 expression, and IFNγ gene expression on several cell subsets in tumors [143]. With multiple clinical trials underway, many efforts have been put in the next generation with novel designs engineering oncolytic viruses that express immunomodulatory transgenes to express cytokines, chemokines, immune stimulators, tumor suppressors, and immune checkpoint inhibitors [144].

3.4 Cancer vaccination

Vaccines have been the greatest triumphs of medicines since its use by Edward Jenner and his contemporaries, to prevent infectious disease by educating the immune system to act rapidly and accurately in eradicating the pathogens [145, 146]. This application has been explored in cancer as a therapeutic by delivery of TAAs, oncogenic viral antigens, and neoantigens to APCs in the presence of adjuvants [146, 147]. These APCs presents antigen to $CD4^+$ T-cells and $CD8^+$ T-cells locally in TME but also eventually traffic to DLNs priming a new T-cell response. For a cancer vaccine to be effective, it is necessary to have an effective APC activation, trafficking, and T-cell priming, subsequently leading to T-cell proliferation and its infiltration in tumors leading to antigen-specific cancer cell killing [148]. Generally, cancer vaccines can be categorized into molecular vaccines (peptides, RNA, or DNA), cellular vaccines, and virus vector vaccines [146].

Recently, Bharadwaj *et al.* reported the phase II trial (NCT02129075) testing the hypothesis in high-risk melanoma patients with pretreatment of FLT3L and poly-ICLC in combination with anti-DEC-205-NY-ESO-1, a fusion antibody targeting CD205 linked to NY-ESO-1. There were increases in monocytes and cDCs (cDC1, cDC2) and pDC along with increased activation of DCs, T-cells, and NK cells [149]. Another recent study by Sahin *et al.* used an RNA vaccine in combination with a checkpoint inhibitor in melanoma. They reported the results of a phase I dose-escalation trial in patients with advanced melanoma (Lipo-MERIT trial) patients of which the eligible ones for efficacy irRECIST 1.1 had received prior checkpoint inhibitor therapy. They intravenously administered liposomal RNA (RNA-LPX) vaccine (FixVac), which targets four nonmutated TAAs (NY-ESO-1, melanoma-associated antigen A3 (MAGE-A3), transmembrane phosphatase with tensin homology (TPTE) and tyrosinase) that are prevalent in melanoma. This was further combined with blockade of the checkpoint inhibitor PD1. FixVac was capable of producing strong specific cytotoxic T-cell responses against melanoma, resulting in partial responses in a group of preselected patients with ICI-resistant diseases. Upon peripheral blood immunological analyses of T-cells reactivity *ex vivo*, it was demonstrated that these TAA-specific T-cells secreted IFNγ and TNF. Overall, they tested the clinical utility of nonmutated vaccines in combination with CPI in patients who had already received CPI therapy [150]. One interesting approach used in cellular vaccines is by using DC-based approaches. DCs can be loaded with peptide antigens or transfected with interested antigen genes and be administered to cancer patients generating a robust antitumor immunity. One of the FDA-approved cancer vaccine sipuleucel-T (Provenge) is using the approach of enriching DCs from metastatic castration-resistant prostate cancer patients, followed by *ex vivo* activation with GM-CSF fused to antigen PAP. These were given back to patients showing a significant median overall survival in comparison to placebo in pivotal phase III clinical trial, thus sowing the benefit of autologous DC vaccines [151].

3.5 Chimeric antigen receptor therapy with innate immune cells

CARs are engineered molecules combining the specificities of antibodies with TCR signaling downstream nodes. CAR-based T-cell immunotherapy has revolutionized the field of cancer immunotherapy and is one of the greatest innovations of the century. CAR-T cell therapy (discussed further in Section 4) despite its successes in blood malignancies is still a big challenge in solid cancers due to its immunosuppressive TME and inability of CAR-Ts to penetrate the solid tumors. Besides they hit a bottleneck with loss of antigens, low response rates in solid cancers, and toxicity issues (on-target/off-target and cytokine release syndrome) [152, 153]. Thus, there is room for novel approaches with alternate cell types for effective CAR-based therapies, where innate immune (NK cells, macrophages, NKT cells) cells come into play, which possess multiple mechanisms of killing cancer cells and the inherent ability to break the barrier of solid cancers [152, 153]. For example, NK cells when engineered with CAR can kill tumor cells with CAR specificity and through natural cytotoxicity receptors and ADCC mechanisms [154]. Another big advantage is the potential use of CAR-based innate immune cells in allogeneic applications in contrast to autologous products for CA-T cells [153]. NK cells' short lifespan (~2 weeks *in vivo*) and different cytokine profiles make them an attractive candidate for CAR engineering. Initial preclinical studies have shown modest results with CAR-NKs directed against CD19 and CD20 malignancies with significantly better results with the addition of costimulatory domain to the CAR construct

[155–157]. Since then a lot of advances have been made with respect to using different generations of CAR constructs and the addition of genes expressing cytokines has had profound effects in having an antitumor efficacy leading to initiation of testing in the clinic [154]. The first clinical trial in phase I/II study assessed the safety of CAR CD19-CD28-zeta-2A-iC9-IL-15-transduced human leukocyte antigen (HLA)-mismatched cord blood NK cells in patients with relapsed/refractory CD19$^+$ B-lymphoid malignancies (NCT03056339). Initial results have been promising showing 8 out of 11 patients achieving complete responses with no cytokine release syndrome, neurotoxicity, or graft-vs-host disease [158]. Since then, multiple number of clinical trials have been initiated including CAR NK targeting CD19 in B-cell malignancies (NCT02892695), CD33 in AML (NCT02944162), mucin 1 (MUC1) in colorectal cancer, gastric cancer, pancreatic cancer, breast cancer, non-small cell lung cancer (NSCLC) and glioma (NCT02839954) and mesothelin in ovarian cancer (NCT03692637), to name a few [154]. Macrophages form a big chunk of solid tumor and central effectors and regulators of innate immunity can secret pro-inflammatory cytokines, perform phagocytosis, and can promote adaptive immune responses. Considering these characteristics of macrophages, Klichinsky *et al.* genetically engineered these cell types to make CAR-macrophages (CAR-Ms) direct their phagocytic activity against tumors. They were able to show that CAR-Ms' gene expression was closely related to M1 macrophage antitumor phenotype in a SKOV3 *in vivo* xenograft model and additionally they were able to influence the bystander myeloid cells toward a M1 phenotype. Additionally, they were able to show enhanced DC activation and increased infiltration of CD8$^+$ T-cells enhancing the antitumor immunity. This is the first demonstration of CAR-M utility as an anticancer therapy [159].

4. Adaptive immunity

Innate and adaptive immunity are two types of responses to protect and maintain the host's normal state of homeostasis [160]. As discussed in the previous section, innate immune response control pathogens that induce interferons and other nonspecific secreted defenses and those that have specific signaling molecular patterns [161]. These defenses are coded in the genome and hence do not form memory [161]. When innate immune cells continue to detect changes at the molecular level due to microbial infections, they also launch and amplify the adaptive immune response recruiting T- and B-cells while increasing their effector response. [14]. Adaptive immunity develops after sustained exposure to an antigen when the innate immune response is insufficient to control the infection. After the first line of defense by the innate immune system, adaptive immunity develops over time and is characterized by adaptive immune cells getting the message and differentiating into mature effector T-cells or antibody-secreting B-cells. [13]

While innate immunity is nonspecific and immediate, adaptive immunity provides a long-lasting response to specifically attack a foreign antigen (nonself-proteins, pathogens, allergenic antigens) that meets the host [161]. The memory of long-lasting protection from reinfection on reexposure and the specificity to the pathogen are the most important features of adaptive immunity. The memory allows for the host's body to react quickly and produce a quantitatively better immune response against the previously encountered antigen [162, 163]. This pathogen-specific response is a result of immunoglobin genes at the lymphocyte level [163].

4.1 Interplay between innate and adaptive immunity

The interplay between adaptive and innate immunity is demonstrated by the immune memory on how it continues to protect the host against re-infections much better [163]. This connection between two immune systems is due to the involvement of APCs such as dendritic cells, macrophages, and complement protein. [13]. In the innate immune system, DCs and macrophages function to phagocytose pathogens, but in adaptive immunity, they act as APCs to activate a T-cell response. In innate immunity, the complement system eradicates cancer cells via a membrane attack complex or by complement-dependent cytotoxicity (CDC) [13]. However, in adaptive immunity, complement system activation can lead to downstream activation and differentiation of T-cells. Also, it can reduce the T-cell-activation thresholds and can activate B-cell responses of the immune system. Thus, it would be accurate to call these APCs interfaces of evolution between innate and adaptive immunity [164–167].

4.2 Tumor-associated antigens

The adaptive immune system comprises of different cell types that have a goal to eradicate cancer cells or to inhibit their proliferation [11]. This anticancer response initiates in these cells when these cells encounter a specific antigen present on the cancer cells called TAAs. This phenomenon is called antigenicity and it is the most crucial step in triggering an adaptive response. Adaptive immune cells utilize the effector function of antibodies, T-cells, B-cells, and APCs to identify TAAs [168, 169]. TAAs are antigens overexpressed in tumors but may or may not be expressed in normal tissue, and these can be due to occurring genetic mutations in the cells. Some of these genes might be aberrantly expressed genes of fetal origin such as CEA (CD66), AFP [170], or from immune-privileged tissue such as testis that is abnormally expressed in case of cancer. These include melanoma antigen gene (MAGE) family [171], preferentially expressed antigen of melanoma (PRAME) family [172], and New York esophageal squamous cell carcinoma 1 (NYESO) [173]. Certain malignant cells express differentiation antigens that are also expressed on healthy cells of the same lineage, such as CD19 on B-cells, MART-1,

Table 2 Summary of tumor-associated antigens (TAAs) that can trigger an adaptive immune response.

Category	Example antigens	Reference
Aberrantly expressed cellular proteins	NYESO, PRAME, MAGE	[175]
Oncofetal antigens	CEA, AFP	[170, 176]
Tissue-specific differentiation	CD19, CD22, MART-1, gp100	[177–179]
Antigens of oncogenic viruses	HPV, CMV, EBV	[180, 181]
Neoantigens	KRAS, NRAS, HRAS TP53	[182–184]

and gp100 are also considered tumor targets. Viral gene products such as HPV, EBV, and CMV are also classified as nonself-antigens. The last type is neoantigens that are hotspot mutations as a consequence of somatic mutations in the genomic regions in cancer cells and thus are highly specific to tumor cells [174] (Table 2). There are many ways in which the tumor tries to get rid of certain TAAs to escape the immune response. Genes that are overexpressed in cancer are not essential for tumor survival; hence, one mechanism of the tumor to evade the immune response is to exert high selective pressure on tumor cells that have lost or have very low expression of those genes. This is commonly seen in melanomas where cancer progression due to resistance and immunosuppression is supported by the disablement of melanoma recognition by immune cells [185].

Adaptive immune responses occur in an organized peripheral lymphoid tissue against these different types of TAAs and are facilitated by two pathways: cell-mediated that are carried out by activated T-cells and humoral carried out by activated B-cells and antibodies.

4.3 T Lymphocytes

4.3.1 CD8$^+$ cytotoxic T lymphocytes

CTLs are the most important mediators of tumor-killing presenting on MHC I and have been identified as the population that has a direct positive correlation with survival in several cancers including ovarian and colorectal [186, 187]. The priming of CD8$^+$ T-cells occurs by the successful delivery of CD4$^+$ and CD8$^+$ recognized antigens on the same DC. As mentioned previously, CD8$^+$ T-cells need priming and are activated from naïve to effector phenotype for tumor killing by signals from TCR, costimulatory molecules, and the third signal by cytokines IL-12 or IFN-α. CTLs only develop in the presence of cytokine signals. IL-2 is required for CD8$^+$ T-cell proliferation and expansion [188]. Persistent TCR signaling can lead the CTLs to a nonresponsive anergic state, where they stop expanding. Thus, in this state CD4$^+$ helper T-cells produce the necessary IL-2 for continued expansion and differentiation into memory cells [189]. CTL killing is dictated by TCR and required constant cell to cell contact-mediated by the release of granzymes and perforins or IFN-γ. CTLs also induce apoptosis of target cells via CD95L (FasL) and TRAIL activating death receptors CD95 (Fas), TRAILR1, or

TRAILR2 on the tumor cells [190]. TNF receptor CD27 is responsible for CTL priming, hence various studies have been done to use CD27 agonism to enhance the efficacy of cancer vaccines [188]. High levels of T-cell infiltrates are not only present in the invasive edge of the tumor but also in its hypoxic core [191, 192], with high levels of $CD8^+$ T-cell infiltration, which also is a favorable prognosis in many cancer types such as melanoma [193] and breast [194], lung [195], ovarian [196, 197], colorectal [198], renal [197], prostate [199], and gastric cancer [200, 201]. Van De Ven *et al.* have been shown to engage costimulatory receptors CD27/CD70 pathway and to block coinhibitory receptors aiming to retain CTL activation for cancer cell immunosensitive [202]. Studies have shown mammalian target of rapamycin (mTOR) pathway negatively regulates $CD8^+$ T-cell activation, and there is a positive link between this signaling and PD1 expression, thus targeting this pathway for stimulation of PD1 low CTLs has been studies [203]. Immune checkpoint blockade discussed in detail later is considered as a breakthrough in immunotherapy; however, these therapies can also trigger autoimmunity.

Prolonged overexpression of costimulatory receptors on $CD8^+$ T-cells could lead to anergy, dysfunction, and T-cell exhaustion resulting in an inability to combat cancer. Thus, it is essential to consider optimal stimulation parameters to case a cytotoxic response without causing T-cell exhaustion [204].

4.3.2 CD4$^+$ helper T-lymphocytes

CTLs alone cannot maintain a sustained antitumor response without the help of $CD4^+$ helper T-cells. $CD4^+$ helper T-cells regulate several aspects of adaptive immunity [205]. Several mouse studies have demonstrated how $CD4^+$ T-cells are necessary for $CD8^+$ T-cell's cytotoxic responses [206] and they generate a much more durable antitumor response [207]. $CD4^+$ T-cells attack tumor cells in multiple ways: direct elimination through cytolytic mechanisms or indirectly by TME modulation [208]. To prevent the deleterious effects of autoimmunity due to checkpoint blockade, $CD4^+$ helper T-cells can play a role in overcoming this negative regulation. As previously mentioned, $CD4^+$ T-cells promote priming CTLs for both effector and memory functions, thus enhancing T-cell-mediated killing. It also facilitates the recruitment of CTLs to the tumor site and tumor cell recognition. Several studies have shown the development of peptide vaccine to enhance antitumor $CD8^+$ cytotoxic responses for tumor regression [208]. Her2 peptide-pulsed type 1 DC vaccine administered intranodal for invasive breast cancer showed durable $CD4^+$ T-cell response in 80% of patients [209]. In several mouse model studies, combination therapy with CD27 antibody and PD1 inhibition has captured the effects and contribution of $CD4^+$ T-cells [210]. Memory T-cells are formed when the activation of $CD4^+$ T-cells takes place by MHC class II cells on APCs, thus priming them for subsequent exposures to that specific peptide: MHC class II complex. [11, 211]. This provides long-term immunity against the specific antigens and offers crucial information for cancer vaccine development [212].

5. Recognition of antigens

To combat TAAs, activation of naïve T-cells is initiated to produce effector T-cells on the encounter of the specific antigen in the form of a peptide:MHC complex on the surface of an activated APC. The APCs also express specific costimulatory signals that are required to pair up with MHC class complexes to activate naïve T-cells [13]. Dendritic cells are highly specialized and function by either ingestion or for processing and presenting the endogenous antigens to T-cells in the context of MHC I molecules. Macrophages also phagocytic cells are a type of APC that can be activated to process and present the exogenous peptide on MHC class II to activate naïve T-cells. This is called cell-mediated immunity. B-cells are lymphocytes that secrete antibodies and are also highly efficient at presenting antigens that bind to their surface immunoglobulins; thus can also serve as APCs in some instances [13, 161, 169]. Owing to this highly efficient mechanism, B-cells endogenously have high expression levels of MHC class II molecules and can help to activate and differentiating antigen-specific $CD4^+$ helper T-cells along with naïve T-cells to gain a $CD8^+$ cytotoxic T-cell effector phenotype [161]. Thus, dendritic cells, macrophages, and B-cells can be termed as professional APCs.

6. Costimulatory signals

The processed TAAs facilitated by costimulatory signals are presented on MHC class I (endogenous peptides) and MHC class II (exogenous peptides) molecules on the APCs to the antigen-specific T-cell receptors on $CD8^+$ T-cells and $CD4^+$ T-cells, respectively [11]. Activation of $CD8^+$ T-cells (cytotoxic T-cells) is initiated by engagement of antigen-specific T-cell receptors with MHC class I: tumor antigen complexes. Activated $CD8^+$ T-cells produce IL-2, driving its proliferation and differentiation [213, 214]. Antigen recognition in the absence of co-stimulation inactivates naive T-cells, inducing a state known as anergy. The most important change in anergic T-cells is their inability to produce IL-2 that facilitates T-cell proliferation and differentiation from naïve to effector [161]. Some $CD8^+$ T-cells that recognize antigen on cells that express a weak co-stimulatory signal need a synergistic effect of $CD4^+$ T-cells bound to the same APC. This action is driven by effector $CD4^+$ T-cells that recognize the antigen on the APC, trigger increased costimulatory signals on the APC and activate the $CD8^+$ T-cells synthesizing IL-2. This interaction is a result of CD40 ligand on $CD4^+$ T-cells to CD40 receptors on the APC that induces B7 expression on DCs enabling APCs to costimulate $CD8^+$ T-cells directly [215].

7. Adaptive immune activation

T-cell differentiation is dictated by the cytokines present in the tumor microenvironment (TME) at the time of CD4$^+$ T-cell activation. Activated CD4$^+$ T-cells can also activate naïve B-cells [216, 217]. This process is known as thymus-dependent activation of B-cells and encompasses two types of signals between CD4$^+$ T helper cells and B-cells: peptide:MHC class II with tumor antigen and a costimulatory signal between CD40 ligand and CD40 as mentioned previously [215, 218]. In absence of these signals, B-cells will not get activated or proliferate. The other type of B-cell activation is thymus independent activation and it takes place through antigens with highly repetitive structures [218]. This results in B-cells expressing high-affinity antibodies that differentiate into antibody-secreting plasma cells that secrete antibodies and create memory B-cells mediating humoral immunity against pathogens [219]. This initiates tumor cell lysis via ADCC or CDC by binding to Fc receptors on NK cells [220]. Naive CD4$^+$ T-cells can differentiate upon activation into either TH1 or TH2 cells, which differ in the cytokines they produce and thus in their function. The consequences of inducing TH1 *vs* TH2 cells are profound: the selective production of TH1 cells leads to cell-mediated immunity, whereas predominant production of TH2 cells provides humoral immunity [161, 168].

In the context of adaptive immune responses, multiple regulatory mechanisms act as immune checks for cancer progression or regression. Because dysfunctional T-cell responses can be detrimental to the host [221–223], cell-intrinsic and cell-extrinsic mechanisms exist to suppress their function [224]. As mentioned earlier, the costimulatory signal is one such mechanism that mediates proper signal is delivered between the APC and T-cell, *i.e.,* peptide:MHC bound complexes with T-cells receptors and costimulatory signals [13, 223, 225]. Costimulatory signals include CD28, ICOS, and CD80 (B7.1)/CD86 (B7.2) [226, 227]. In absence of these signals, T-cell activation and proliferation are in checkpoint and regulated [222]. Hence, for a regulated adaptive response for cancer therapy, it is essential to have a regulated costimulatory signal to avoid T-cell energy that could lead to immune tolerance and escape of TAAs [228, 229].

8. Adaptive immune regulation

Regulatory cells (Tregs) are a specialized subset of CD4$^+$ T-cells that suppress T-cell responses by either enhancing or inhibiting their suppressive activities based on environmental stimuli. Tregs can utilize multiple mechanisms to maintain immune homeostasis and to suppress conventional T-cell responses. These include cell-contact-dependent mechanisms mediated by surface receptors, such as CTLA4, ICOS, CD103, GITR, LAG-3, and Nrp1, which can modulate the functions of T-cells or other immune cells, such as APCs, to suppress T-cell responses [224, 230]. Additionally, Tregs also suppress T-cell responses by secreting antiinflammatory cytokines and disrupting metabolic responses such that conventional T-cell proliferation and activation are impaired. CTLA4 and PDL1 are two critical regulatory molecules known to stimulate immune suppression by binding to T-cell activating signals and are facilitated by Tregs [224, 225, 231, 232]. CTLA4 expressed on T-cells binding to CD80/CD86 proteins on APCs, blocks its binding to CD28, its natural receptor for CD4$^+$ T-cells activation, resulting in T-cell inhibition. Thus, the expression of CTLA4 on several cancers/tumors is a mechanism of cancer evasion giving rise to CTLA4 blockers as an innovation in cancer immunotherapy [226, 227]. Expression of PD1 (programmed cell death protein 1) on T-cells and it is binding with its ligand PDL1 on APCs mediates T-cell inactivation, another common mechanism of immunosuppression [226]. PD1 is also expressed on other immune cell types such as B-cells, NK cells, monocytes, DCs, and Tregs [129, 226]. Literature has shown the immunosuppressive properties of Tregs are responsible for immune escape in solid tumors and hematological cancers. Tregs prevent or delay inflammation that is hypothesized as the driving force of malignant transformation [224] and thus promote tumor development [224, 230]. Treg presence influences and informs the response to treatment. IL-2 is an approved therapy for metastatic melanoma and renal cell carcinoma (RCC); however, it also increases the expansion of Tregs including ICOS$^+$ Tregs with an activated phenotype [224, 233]. ICOS$^+$ Tregs are associated with poorer prognosis [233–235]. Thus, treatment strategies used in such cases employ antibodies or recombinant proteins such as anti-CTLA4 antibody, anti-CD25 antibody, recombinant IL-2, and anti-IL-2 antibody [224].

9. Humoral mediated immunity

B-cells or Bursal-derived lymphocytes are key players of humoral immunity and they facilitate this by antibody production. B-cell receptors or BCR is made of membrane-anchored immunoglobin and coreceptor molecules [236]. On activation by an antigen that binds to the surface immunoglobins, naïve B-cells are activated to proliferate and differentiate into plasma cells that are responsible for antibody secretion.

The role of B-cells in antitumor immunity has recently been studied extensively. B-cells are not only involved in the secretion of antibodies and cytokines, but they also modulate cell-mediated and innate immune signals alongside recognizing antigens and regulating its processing and presentation [237]. As described earlier, TAAs can trigger a humoral response as a result of mutations (neoantigens), aberrant posttranslation modification, overexpression of genes, expression of specific differentiation marker (CD19 and CD20 in B-related lymphoma), or expression of a marker normally found in restricted tissues (*e.g.,* cancer-testis antigens) [238]. Antibodies bind to Fc receptors (FcRs) on macrophages, neutrophils, and natural killer cells (NK) activating the CDC or ADCC, thereby inducing death [239]. Antibodies also increase the afferent arm of adaptive immunity by activating and increasing the efficacy of phagocytosis and antigen processing and presentation, thus activating CD8$^+$ T-cells [170].

In addition to mediating tumor growth by acting as APCs, antibody and cytokine secretion alongside T-cell modulation, B-cells have a direct ability to kill tumor cells by production of IL-17A or by CpG-activated B-cells activating TRAIL/Apo-2L-related pathway [236].

B-cells are divided into different subsets in terms of their location and by the way, they are activated in a T-cell-dependent or T-independent way. Like Tregs, Bregs are regulatory cells of B-cell origin that are a poorly characterized population and are functionally responsible to inhibit T-cell-mediated immunity [236]. Bregs produce suppressive cytokines such as IL-10, IL-35, and TGF-β and/or high level of expression of negative immune checkpoint molecules such as PDL1. Several different mechanisms used by Bregs are used to inhibit other immune cells, but their functional characterization is limited in the context of cancer [240].

The different B-cell subpopulations can behave differently based on their locations, justifying the role of B-cells in either pro- or antitumorigenic immunity. The tumor microenvironment (TME) plays a vital for B-cells as well. As we understand, TME is a highly immunosuppressive environment due to sustained cell-cell contact, cytokine production, and metabolic factors. Tregs also play a crucial role in B-cell regulation. Tregs preferentially suppress antigen-pulsed B-cells that act as APCs and inhibit its proliferation by activating granzyme-dependent cell death [241]. MDSCs are another cell type that secretes IL-7 that decrease antibody production in B-cells [242]. Besides Tregs and MDSCs, different types of cells in the TME including but not limited to T-cells, DCs, stromal cells, myeloid cells, and NK cells target B-cells to reduce its function and efficacy of secreting specific cytokines [243]. Tumor cells themselves can also secrete cytokines and metabolites differentiating B-cells into Bregs, thus promoting cancer progression and metastasis through transcriptional changes in the B-cell population [236]. Sometimes these effects are antigen-dependent as all these changes of B-cells happen in the TME and not to peripheral B-cells [244].

Immune checkpoint signaling also has an inhibitory effect on the B-cell population resulting in tumor-induced immunosuppression. Immune checkpoint molecules such as PDL1/PD1, galectin-9/TIM3, IDO1, lymphocyte-activation gene 3 (LAG-3), and cytotoxic T-lymphocyte-associated protein 4 (CTLA4) are responsible for the inhibition of activation of effector CD8$^+$ T-cell population. These immune checkpoint molecules also suppress antigen-specific proliferation of B-cells [245]. The hypoxic environment of TME also affects B-cell proliferation and antibody production by deletion of Glut1 [246]. Thus, TME can play a crucial role in B-cell suppression leading to tumor progression.

B-cell presence in tumors has been a predictor of increased patient survival, but also have roles in both antitumor and protumor immunity [247]. B-cell signatures are enriched in responding *vs* nonresponding patients, and its presence is shown to be positively associated with chemotherapy response in various cancers. Evidence suggests localization of B-cells in tertiary lymphoid structures (TLS) are hallmarks of positive prognosis and response to immunotherapy. There is potential for these to be developed into biomarkers and therapeutics [248–250].

10. Immune suppression and cancer progression

As discussed previously, the protumor role of B-cells can take place through many different mechanisms; immunosuppression led by IL-10 and TGF-β or via direct stimulation of tumor cells by IL-35 secreted by B-cell in human pancreatic neoplasms and KRAS-driven pancreatic neoplasms in mice [251–253].

As mentioned previously, immune suppression in the tumors can take place systemically and in the TME where tumor cells are trained to avoid recognition by immune cells. Several immunosuppressive molecules and enzymes such as transforming growth factor β (TGF-β), soluble Fas ligand, IL-10, IL-35, and IDO are produced in the tumor cells to escape the immune response [236, 254]. Checkpoint immune molecules, hypoxic environments, and glucose depletion result in lactic acid production and accumulation leading to low pH in TEMs that suppresses effector T-cells by impairing lytic activity with reduced IL-2 and IFN-G secretion [254, 255]. Apart from dysfunctional cytotoxic T-cell population, cancer-associated fibroblasts (CAFs), myeloid-derived suppressor cells (MDSCs), and M2 subtypes of TAMs in the tumor microenvironment are reported to restrict infiltration of cytotoxic T-lymphocytes (CTLs) and immune cells that contribute to immunosuppression in the tumor [236, 256]. High numbers of these cells can be detected in non-small-cell lung cancer and ovarian cancer [256–258].

IL-10 is responsible for the downregulation of immunogenic responses by decreasing IL-12 production in macrophages and DCs, thus decreasing IFN-γ production by NK cells and T-cells. It also inhibits MHC class II expression on APCs and the expression of costimulatory molecules [259, 260]. Inflammatory markers in TEM such as prostaglandin E2 (PGE2) can affect effector T-cells depending on IL-6, chemokine (C-X-C motif) ligand 1 (CXCL1), and granulocyte-colony stimulating factor (G-C SF) [255]. The other ways tumors can disguise the immune response are by upregulation of antigen escape, DNA mutation or splicing-mediated antigen loss, diminished antigen expression, or lineage switch [254].

The immune-suppressive effects can also be systemic. The status of immune function is largely responsible for an antitumor response [254]. An increase in the number of Tregs or Bregs is one way where the immune cells cause immunosuppression. An increase in regulatory T-cells has been observed in the peripheral blood of patients with head and neck cancer [256]. An increase in MDSCs and granulocytes also suppress tumor-specific T-cells in mice. Tumor growth impairs T-cell expansion, persistence, and cytotoxicity with different mechanisms. This has always resulted in poor cancer prognosis [254]. Transcriptomic analysis has revealed increased expression of key regulators of late/ memory T-cell differentiation markers resulting in immunosuppression. In addition to impaired cytotoxicity and expansion, T-cell exhaustion is another mechanism of immune resistance of tumors as observed in melanoma patients [261]. T-cell exhaustion is a dysfunctional state of T-cells resulting in decreased effector population and increased expression of inhibitory receptors including PD1 and CTLA4 due to chronic pro-tumorigenic antigen stimulation [174, 262, 263]. Other negative regulatory coreceptors are TIM3, LAG3, VISTA, CD244, CD160, and BTLA-4 [174]. One major readout of T-cell exhaustion is overexpression of PD1, thus checkpoint blockade of either PD1 (T-cells) or PDL1 (tumor cells) successfully reactivates T-cell function. Several studies in various cancer types (discussed in detail later) have shown successful reversal of T-cell exhaustion with checkpoint blockade; however, this is not a long-lasting response and might be due to the interplay of multiple inhibitory receptors in T-cells, warranting a need for combination therapies [264].

11. Treatments modulating the adaptive immune system

11.1 Immune checkpoint blocking with antibodies

As discussed earlier, immune checkpoint signals are essential to regulate the magnitude of T-cell response. Cytotoxic T-lymphocyte-associated protein 4 (CTLA4), PD1, LAG3, TIGIT, and TIM3 are a few of the many checkpoint signaling molecules that are extensively studied for their role in T-cell signaling pathways and how to leverage them as potential immunotherapies for cancer [265]. Several early preclinical evidence suggested that immune checkpoint activation was a mechanism of immune escape to evade antigen-specific CTL response [266]. Current immune checkpoint blockade therapies with antibodies aim at the receptor-ligand interaction, thus modulating the surface expression and intracellular expression of these checkpoint molecules, resulting in activating the antitumor immunity. PD1 signaling triggers with the engagement with PDL1 and PDL2, that are present on the APCs or tumor cells. On binding with the ligand, PD1 inhibits both the antigen and costimulatory signal for T-cell activation [267]. Monoclonal antibodies (mAbs) for targeting immune checkpoints, *i.e.,* anti-CTLA4 and anti-PD1 or anti-PDL1 have shown efficacy in several cancers including advanced metastatic melanoma, NSCLC, and RCC [268]. In 2014, FDA approved pembrolizumab followed by nivolumab, as the first PD1 inhibitors as a breakthrough in the field of cancer immunotherapy to treat several malignancies from lymphoma to head and neck squamous cell carcinoma (HNSCC) [269]. Following the first two path-breaking checkpoint inhibitor antibodies, the development of other similar molecules gained resulting in the development of atezolizumab in 206, avelumab in 2017, and durvalumab in 2017. Currently, FDA has approved 10 cancer types for PD1/PDL1 inhibition including hematological malignancies, melanoma, NSCLC, HNSCC, Hodgkin lymphoma, urothelial carcinoma, tissue agnostic (MSI-H), gastric cancer, colorectal cancer, RCC, and hepatocellular carcinoma (HCC) [266].

CTLA4 engages with CD80/86 with high affinity and inhibits the costimulation by ligand competition. [270]. CTLA4 blockade is suggested to significantly reduce T-cell activation threshold, by affecting the immune priming phase and reduced Treg-mediated suppression. *In vitro* studies have demonstrated that the antitumor activity of CTLA4 blockade is due to macrophages expressing Fc gamma receptor. Ipilimumab was approved by the FDA in 2011 for late-stage metastatic melanoma [271]. There is evidence of tumor rejection in all mice that were challenged with murine melanoma tumor lines after the administration of CTLA4 inhibitors with GM-CSF secreting cell vaccines. The synergy demonstrated by the combination was hypothesized to be due to vaccine-activated improved cross-priming of T-cells by host APCs and CTLA4 inhibited enhanced T-cell effects [272]. In several clinical trials, ipilimumab was administered with gp100 multipeptide vaccine that resulted in responses longer than 2 years in some patients [271]. In a recent report, an increase in the heterogeneity of peripheral T-cell population has been reported following CTLA4 blockade in melanoma patients. An ipilimumab study with melanoma and prostate cancer suggests reading baseline T-cell profile to learn more about the turnover of T-cell repertoire and overall TCR diversity that has a positive correlation with the treatment. Overall survival was associated with maintaining a high frequency of clones at baseline [269].

Thus, blocking these checkpoint immune molecules blocks the tumor from T-cell attack making them a state-of-art immune therapy. It has been observed that immune checkpoints expression is not restricted to T-cells alone [273]. These molecules are also found on the surface of B-cells, implying a direct correlation of checkpoint blockade with B-lymphocyte activity [236].

Humoral immune response was also detected in metastatic cancer patients treated with immune checkpoint blockade immunotherapy (anti-CTLA4, anti-PD1). The response was characterized by somatic hypermutation, IgG class switch, and clonal expansion of plasmablasts [273, 274].

Combination immunotherapies are being used to combat and halt resistance development. Combination therapies with TIM3, LAG3, and VISTA are currently being studied to overcome resistance due to PD1 and other checkpoint inhibitors and create synergy are obtained in preclinical models [275, 276] (Table 3). However, despite advances and initially observed success, published results have addressed the outcome of PD1 and CTLA4 inhibition with relapse due to resistance.

11.2 Adoptive cell transfer

A significant amount of effort has gone into the evaluation of adoptive T-cell transfer to treat cancer. Strategies include but are not limited to using the patient's T-cells isolated from tumor-infiltrating lymphocytes (TILs) [284] and *ex vivo* expansion of peripheral

Table 3 Overview of checkpoint receptor with their ligands highlighting their location and role in T-cell inhibition and/or exhaustion.

Checkpoint receptor	Checkpoint receptor-ligand	Cellular location	T-cell role	Reference
CTLA4	CD80, CD86	T cell (CD4$^+$, CD8$^+$) Treg	Inhibition	[277]
PD1	PDL1, PDL2	T cell (CD4$^+$, CD8$^+$), B-cell, NK cell	Inhibition	[277, 278]
LAG3	LSECtin, Galectin-3, MHC II	T cell (CD4$^+$, CD8$^+$), NK cell, Tregs, DCs	Inhibition Exhaustion	[279, 280]
TIM3	CEACAM1, Galectin-9, HHMCB1	T cell (CD4$^+$, CD8$^+$), DC, monocytes	Inhibition	[281, 282]
VISTA	VISTA, VSIG-3	T cell (CD4$^+$, CD8$^+$), neutrophils, macrophages	Inhibition	[283]

T-lymphocytes [285]. This approach does not generate a plethora of antigen-specific T-cells alone, but enriches a population of activated T-cells with lowered triggering thresholds [286]. In the early years, Eberlein *et al.* injected *in vitro* cultured immune cells supplemented in IL-2 to successfully treat FBL-3 lymphoma in mice. He also performed subsequent studies of IL-2 administration to enhance the response from these IL-2 dependent cells *in vivo* [287, 288]. Groundbreaking improvement in the efficacy of adoptive cell transfer (ACT) came with the incorporation of an immunodepleting regimen before the cell transfer, facilitating clonal repopulation in patients. Several preclinical studies provided increasing evidence of prior lymphodepletion to remove Tregs as well as normal endogenous lymphocytes competing with ACT products for better efficacy [289].

Several clinical trials have reported successful expansion of functional T-cells rapidly for ACT in hematological malignancies [290]. ACT has also shown complete and long-lasting durable clinical responses in patients with late-stage, chemo-resistant leukemias [291, 292]. As an anticancer therapy, ACT has demonstrated robust T-cell expansion *in vivo*, potent antitumor activity, and persistence of memory T-cell subset [286].

However, where lack of efficacy has been observed in clinical trials, the data has suggested poor *in vivo* persistence of the infused T-cells. The lower persistence can be controlled by several factors including but not limited to *ex vivo* culture conditions, lack of long-term transgene expression, poor effector functionality, T-cell exhaustion, and development of antifused humoral or cellular immune responses [286]. A set of adverse reactions have been recently reported in the clinic associated with ACT and attributed to the initiation of the proinflammatory immune state of these T-cells [284, 292–294]. Autologous TIL reinfusion has caused substantial hindrance due to these factors in other types of tumors in the clinic apart from melanomas [295].

Moreover, the logistics of infused cell product, quality, quantity, and nature of *ex vivo* manipulation are important to address. An opportunity for adoptive T-cell therapy will be strategies to combine with other antitumor therapies. Therapeutic vaccination, checkpoint inhibition, agonistic antibodies, small molecule inhibitors of tumors, and targeting of tumor stroma and neovasculature may augment current adoptive transfer technology [296].

11.3 Engineered T-cells: CAR and TCR therapies

As discussed, TILs endogenously isolated from patients are typically ineffective after initial success to prevent tumor growth. Multiple factors including immunosuppressive TME, tolerance to self, and T-cell exhaustion are responsible for this immune escape and resistance to therapy. Moreover, the ACT needs to be balanced with the issue of long-term persisting memory T-cells and tumor dormancy. All the data suggests that we have peaked the existing field of adoptive cell transfer therapy. The new age of synthetic immunology of CAR and TCR therapies may increase the potency for such approaches that target cancer [297–299].

CARs classically contain scFvs targeting extracellular antigens of cell-surface proteins expressed by cancer cells, thus enabling major histocompatibility complex (MHC)-independent T-cell activation; however, MHC-dependent, T-cell receptor (TCR)-mimic CARs that enable the recognition of intracellular TAAs have also been studied [300].

Genetically engineered T-lymphocytes that may possess TCRs overexpressing tumor-specific antigens are also widely studied. These TCRs recognize tumor antigens in the complexes of HLAs [301, 302]. Currently, numerous tumor antigens have been investigated both clinically and preclinically for therapeutic effectiveness against cancer in the CAR and TCR space. CD19 in B-cell malignancies was the first tumor antigen targeted with this approach and became the first therapeutic to be approved by the FDA for use in the United States [303].

There was CR observed with this therapeutic modality ranging from 70% to 90% in patients with resistant and released acute lymphoblastic leukemia, and 57% in patients with end-stage advanced CLL [304]. Today, we have about 8 years of follow up data of the earliest recipients of CD19 CAR T-cells in the clinic; 60% of patients with B-cell lymphomas are resistant to treatment with CD19 CAR T-cells [305]. These are a result of various mechanisms that tumor adapts to escape the immune reaction through innate and adaptive resistance [306]. This resistance could be driven by T-cell mechanisms by, tumor mechanisms or dictated by the TME. The tumor also elicits escape, downregulation, or loss of tumor antigen and HLA via many different mechanisms to escape CAR and TCR therapies [254]. The tumor microenvironment is a very complex immunosuppressive area that is being extensively studied and poses a great challenge for CAR T and TCR therapies specifically in solid tumors. Thus, a greater understanding of improving the efficacy and potency of these therapies is needed.

Alongside understanding CAR T-cell-centric and tumor-centric mediators of CAR T-cell cytotoxicity, it has also become essential to have an immunomechanical understanding of CAR T and TCR therapies [234, 298]. It has been studied that CAR T-cells enforce multiple mechanical forces through the immunological synapse (IS), a dynamically organized macromolecular membrane assembly formed between an activated T-cell and a target cancer cell to have a successful cytotoxic immune function [307]. Recent studies have also shown how CARs are largely different compared to the conventional T-cell receptors and become more different from genetic engineering in terms of mechanosensing and mechanotransductive mechanisms [298]. TCR and CAR therapies not only have to overcome biological issues such as physical barriers by tumors for infiltration but also antigen tumor heterogeneity. If the CARs are efficacious, the immunosuppressive factors in the TME impair the CAR T functions [300].

Using autologous T-cells obtained from the intended patient is the FDA-approved CAR T therapy. Thus, although successful, this possesses several hindrances in the number of patients who can receive this treatment. Chemotherapy that is the primary first-line treatment decreases the percentage of naïve T-cells in peripheral blood that is necessary for the clinical efficacy of CAR T-cells. The quality and characteristics of the apheresis product are also critical for the success of CAR T-cell production and therapy [300].

CAR T-cells from allogeneic donor T-cells present an alternative to this approach to overcome manufacturing issues. HSCT [308], NK cells [309], gamma delta T-cells [310] are all being considered for potential CAR therapies. Thus, various engineering strategies are yet to be explored to fully capitalize on the potential of engineered T-cell therapies.

Data has shown, despite high induction remission rates, 50% of patients with B-cell acute lymphoblastic leukemia (B-ALL) treated with CAR T-cells for CD19 and CD22 eventually relapse [311]. It is widely accepted that improving the efficacy of these therapies requires combination therapies of this modality with synergistic combinations. In multiple myeloma, reports from both preclinical and clinical experiences suggest that combining a gamma-secretase inhibitor with a B-cell maturation antigen (BCMA) to enhance the activity of CAR T-cells seems promising (NCT03502577) [312]. Safety and associated toxicities with CAR T therapies include graft vs. host disease (GvHD), macrophage activation syndrome (MAS), and cytokine release syndrome (CRS) which also pose to be a major hindrance to this therapeutic class [305, 313].

Thus, combination therapies of CAR T and TCR therapies with mAbs, bispecific antibodies, vaccines, and immune-checkpoint inhibitors, small molecules are essential for overall cancer progression [314, 315]. Genomic CRISPR screens have been performed to profile potential small-molecule candidates and combine them with existing CAR T therapies to enhance CAR T-cell toxicity [306, 316].

11.4 Antibodies

11.4.1 Monoclonal antibodies

Targeting tumor antigens with monoclonal antibodies is a known treatment option for both solid and hematological malignancies. For hematological malignancies, antibodies are raised against the cluster of differentiation (CD) markers on T- and B-cells, whereas for solid tumors HER2, EGFR, RANKL, VEGF are the targets of choice for these therapeutics [317, 318]. As discussed previously, antibody targets tumors through direct killing through apoptosis by blocking the tumor receptors, immune-mediated killing via ADCC and CDC and by modulation along with recruitment and activation of cellular mediated immunity by activation of T-cells by inhibition of checkpoint molecules [319]. CD20 antibody along with two radioimmunoconjugates has shown significant antitumor activity and have improved survival in an aggressive state of B-cell non-Hodgkin's lymphoma. CD33, overexpressed in acute myeloid leukemia has an approved antibody for its use in refractory acute myeloid leukemia. A Her2/neu antibody has been widely used alone and in combination with chemotherapy in breast cancer. Several other monoclonal antibodies are in development in either preclinical or clinical space [320]. Several antibody therapeutics are used in combination therapies with other modalities to promote adaptive tumor antigen-specific immunity.

11.4.2 Bispecific antibody

Monoclonal antibodies are highly specific, potent, and long-lasting modalities for cancer treatment [320]. However, there are several limitations to this class of therapeutics as well. Since these agents are single agents, they possess a challenge to tackle cancer that adapts several immunosuppressive mechanisms to combat single antibody-based therapy. Thus, the majority of patients, after an initial successful outcome has shown relapse of cancer due to drug resistance caused by gene mutation, escape, or alteration of signaling pathway activation. Bispecific antibodies have thus shown success in overcoming the limitations of a single antibody therapeutic [321]. In bispecific antibodies, dual-specificity is achieved in a structure that is much smaller than a traditional antibody [322]. Blinatumomab, a T-lymphocyte engaging bispecific antibody, binds to CD3[+] cytotoxic cells and CD19[+] malignant cells and is an approved therapy for relapsed/refractory B actuate lymphoblastic leukemia [321]. Catumaxomab targets EpCAM and CD3[+] T-cells and is clinically approved for EpCAM positive tumors with malignant ascites [323]. With time, the number of bispecific antibodies in the clinic is increasing, validating the therapeutic efficacy of this class. Bispecific antibodies have a short serum half-life than traditional antibodies because they do not engage with FcR that undergo the FcR recycling mechanism for persistence. Safety and toxicity studies for this class are still under study, but the most common adverse effects are lymphopenia and leukopenia. Although lymphodepleting preconditioning is not required for bispecific antibodies therapy, toxicities after blinatumomab treatment are common, but manageable [321, 324, 325].

A promising approach to deliver therapeutic bispecific antibodies is through nonprotein delivery via mRNA-coded or DNA-coded formats in lipid nanoparticle systems [326]. Stadler and colleagues have shown the use of fragment-based bispecific antibodies for $CD3^+$ T-cells and CDLN6 through protein engineering using lipid-based formulation. They achieved expression and targeting in the liver after IV administration and thus effectively producing purified protein like in any other mRNA-based rare disease delivery [327–329]. These mRNAs can also be engineered to create stable mRNA variants for better translation and/or protein variants imparting prolonged efficacy [330, 331]. Because manufacturing of pharmaceutical-grade mRNA and DNA is fast, the investigators argue that this approach could accelerate the clinical development of novel bispecific antibodies [326, 332].

11.5 Cytokine therapy

Several cytokines are known to limit tumor growth and proliferation through their pro-apoptotic activity. As we have discussed, they are also known to engage with the other immune cells to stimulate a cytotoxic reaction. Only two cytokines to date have been modestly successful in the clinic IL-2 and IFN-α and have received approval from FDA to treat several malignant diseases. IL-2 was approved for the treatment of advanced RCC and metastatic melanoma, whereas IFN-α was approved for the treatment of hairy cell leukemia, follicular non-Hodgkin lymphoma, melanoma, and AIDS-related Kaposi's sarcoma [243].

ACT and CAR T therapies need cytokines for *ex vivo* expansions and *in vivo* persistence of T-cells—hence they have been extensively used in this field and have received FDA approval [333]. Recently, several engineering strategies are using the incorporation of these cytokines genes in the CAR lentiviral structure [334]. As we have discussed the role of IL-2 in the expansion of T-cells, various second generations IL-2-based compounds have been engineered to increase the safety of this molecule. Low response rate and high toxicity have been associated with both these therapies warranting them to be used as a combination with immune checkpoint therapies [335, 336]. Engineered IL-2 cytokines are in clinical trials with checkpoint inhibitors (NCT02983045, NCT03282344, and NCT03435640) and its use concomitantly nivolumab plus ipilimumab (NCT02983045) [337, 338].

Other potential cytokines that are in preclinical and clinical development are IL-15, IL-10 IL-21, and GM-CSF. Cytokines can be absolute partners in terms of combination therapy with other modalities such as gene therapy, cell therapy, and monoclonal antibody-based therapies for synergistic actions that will be very efficacious for their anti-tumor activity [243, 335].

12. Conclusion

The role of the immune system in the treatment of cancer has been appreciated for over a century; however, the diversity in cancer and the specific immune responses each trigger has challenged the field of immuno-oncology to evolve continuously. Despite breakthrough therapies have been approved by the FDA in this space, there is a lot of potential for several other conceptual immunotherapies that are actively being studied. Ongoing characterization of tumor and TME in cancer patients have shown evidence for specific immunologic phenotypes differentiated based on the presence or absence of T-cell-based inflammation. Synthetic engineering has allowed for further enhancement of such modalities to combat several biologically identified limitations such as efficacy, persistence, and safety to combat immune inhibitory mechanisms of cancer. Several genomic- and proteomic-based studies are informing detection of predictive biomarkers for response to immunotherapies or guiding the invention of new immunotherapeutic modalities. These approaches would also assist in addressing non–T-cell infiltrated tumors where additional innate immune activation would be necessary to promote an antitumor response. In conclusion, immunotherapy is continuously evolving, but as cancer evasion mechanisms continue to outsmart the scientific interventions, combination immuno-therapies have shown encouraging clinical outcomes in numerous early-phase clinical trial datasets.

References

[1] Torre LA, et al. Global cancer incidence and mortality rates and trends—an update. Cancer Epidemiol Biomark Prev 2016;25(1):16–27.
[2] Hanahan D, Weinberg RA. Hallmarks of cancer: the next generation. Cell 2011;144(5):646–74.
[3] Nowell PC. Tumors as clonal proliferation. Virchows Arch B Cell Pathol 1978;29(1–2):145–50.
[4] Heppner GH. Tumor heterogeneity. Cancer Res 1984;44(6):2259–65.
[5] McGranahan N, Swanton C. Clonal heterogeneity and tumor evolution: past, present, and the future. Cell 2017;168(4):613–28.
[6] Tabassum DP, Polyak K. Tumorigenesis: it takes a village. Nat Rev Cancer 2015;15(8):473–83.
[7] Gerlinger M, et al. Cancer: evolution within a lifetime. Annu Rev Genet 2014;48:215–36.
[8] McGranahan N, Swanton C. Biological and therapeutic impact of intratumor heterogeneity in cancer evolution. Cancer Cell 2015;27(1):15–26.
[9] Chen DS, Mellman I. Elements of cancer immunity and the cancer-immune set point. Nature 2017;541(7637):321–30.
[10] Gajewski TF, Schreiber H, Fu YX. Innate and adaptive immune cells in the tumor microenviron-ment. Nat Immunol 2013;14(10):1014–22.
[11] Chen DS, Mellman I. Oncology meets immunology: the cancer-immunity cycle. Immunity 2013;39 (1):1–10.
[12] Kotoula V, Fountzilas G. Overview of advances in cancer immunotherapy. Ann Transl Med 2016;4 (14):260.
[13] Pandya PH, et al. The immune system in cancer pathogenesis: potential therapeutic approaches. J Immunol Res 2016;2016:4273943.
[14] Demaria O, et al. Harnessing innate immunity in cancer therapy. Nature 2019;574(7776):45–56.
[15] Medzhitov R, Janeway Jr CA. Innate immune induction of the adaptive immune response. Cold Spring Harb Symp Quant Biol 1999;64:429–35.
[16] Takeuchi O, Akira S. Pattern recognition receptors and inflammation. Cell 2010;140(6):805–20.
[17] Garg AD, Agostinis P. Cell death and immunity in cancer: from danger signals to mimicry of pathogen defense responses. Immunol Rev 2017;280(1):126–48.
[18] Dar TB, Henson RM, Shiao SL. Targeting innate immunity to enhance the efficacy of radiation ther-apy. Front Immunol 2018;9:3077.
[19] Trinchieri G. Biology of natural killer cells. Adv Immunol 1989;47:187–376.
[20] Rosenberg EB, et al. Lymphocyte cytotoxicity reactions to leukemia-associated antigens in identical twins. Int J Cancer 1972;9(3):648–58.

[21] Herberman RB, et al. Natural cytotoxic reactivity of mouse lymphoid cells against syngeneic and allogeneic tumors. II. Characterization of effector cells. Int J Cancer 1975;16(2):230–9.

[22] Angelo LS, et al. Practical NK cell phenotyping and variability in healthy adults. Immunol Res 2015;62(3):341–56.

[23] Mah AY, Cooper MA. Metabolic regulation of natural killer cell IFN-gamma production. Crit Rev Immunol 2016;36(2):131–47.

[24] Ivanova D, et al. NK cells in mucosal defense against infection. Biomed Res Int 2014;2014:413982.

[25] Vivier E, et al. Targeting natural killer cells and natural killer T cells in cancer. Nat Rev Immunol 2012;12(4):239–52.

[26] Vivier E, et al. Innate or adaptive immunity? The example of natural killer cells. Science 2011;331 (6013):44–9.

[27] Chen Y, et al. Research progress on NK cell receptors and their signaling pathways. Mediat Inflamm 2020;2020:6437057.

[28] Sivori S, et al. Human NK cells: surface receptors, inhibitory checkpoints, and translational applications. Cell Mol Immunol 2019;16(5):430–41.

[29] Seillet C, Belz GT, Huntington ND. Development, homeostasis, and heterogeneity of NK cells and ILC1. Curr Top Microbiol Immunol 2016;395:37–61.

[30] Langers I, et al. Natural killer cells: role in local tumor growth and metastasis. Biol Theory 2012;6:73–82.

[31] Moretta A, et al. Receptors for HLA class-I molecules in human natural killer cells. Annu Rev Immunol 1996;14:619–48.

[32] Karlhofer F, Ribaudo R, Yokoyama W. MHC class I alloantigen specificity of Ly-49 + IL-2-activated natural killer cells. Nature 1992;358(6381):66–70.

[33] Karre K, et al. Selective rejection of H-2-deficient lymphoma variants suggests alternative immune defence strategy. Nature 1986;319(6055):675–8.

[34] Diefenbach A, et al. Rae1 and H60 ligands of the NKG2D receptor stimulate tumour immunity. Nature 2001;413(6852):165–71.

[35] Smyth MJ, et al. NKG2D function protects the host from tumor initiation. J Exp Med 2005;202 (5):583–8.

[36] Hastings KT. Innate and adaptive immune responses to cancer. In: Fundamentals of cancer prevention. Springer; 2019.

[37] Terme M, et al. Natural killer cell-directed therapies: moving from unexpected results to successful strategies. Nat Immunol 2008;9(5):486–94.

[38] Bottino C, et al. Cellular ligands of activating NK receptors. Trends Immunol 2005;26(4):221–6.

[39] Coca S, et al. The prognostic significance of intratumoral natural killer cells in patients with colorectal carcinoma. Cancer 1997;79(12):2320–8.

[40] Wculek SK, et al. Dendritic cells in cancer immunology and immunotherapy. Nat Rev Immunol 2020;20(1):7–24.

[41] Steinman RM. Decisions about dendritic cells: past, present, and future. Annu Rev Immunol 2012;30:1–22.

[42] Merad M, et al. The dendritic cell lineage: ontogeny and function of dendritic cells and their subsets in the steady state and the inflamed setting. Annu Rev Immunol 2013;31:563–604.

[43] Mildner A, Jung S. Development and function of dendritic cell subsets. Immunity 2014;40(5):642–56.

[44] Schlitzer A, McGovern N, Ginhoux F. Dendritic cells and monocyte-derived cells: two complementary and integrated functional systems. Semin Cell Dev Biol 2015;41:9–22.

[45] Villani AC, et al. Single-cell RNA-seq reveals new types of human blood dendritic cells, monocytes, and progenitors. Science 2017;356(6335):aah4573.

[46] Veglia F, Gabrilovich DI. Dendritic cells in cancer: the role revisited. Curr Opin Immunol 2017;45:43–51.

[47] Tai Y, et al. Molecular mechanisms of T cells activation by dendritic cells in autoimmune diseases. Front Pharmacol 2018;9:642.

[48] Curtsinger JM, Mescher MF. Inflammatory cytokines as a third signal for T cell activation. Curr Opin Immunol 2010;22(3):333–40.

[49] Liu Y, Cao X. Intratumoral dendritic cells in the anti-tumor immune response. Cell Mol Immunol 2015;12(4):387–90.

[50] Chiang MC, et al. Differential uptake and cross-presentation of soluble and necrotic cell antigen by human DC subsets. Eur J Immunol 2016;46(2):329–39.

[51] Hildner K, et al. Batf3 deficiency reveals a critical role for CD8alpha+ dendritic cells in cytotoxic T cell immunity. Science 2008;322(5904):1097–100.

[52] Hacker C, et al. Transcriptional profiling identifies Id2 function in dendritic cell development. Nat Immunol 2003;4(4):380–6.

[53] Spranger S, et al. Tumor-residing Batf3 dendritic cells are required for effector T cell trafficking and adoptive T cell therapy. Cancer Cell 2017;31(5):711–23 [e4].

[54] Martinez-Lopez M, et al. Batf3-dependent CD103+ dendritic cells are major producers of IL-12 that drive local Th1 immunity against *Leishmania major* infection in mice. Eur J Immunol 2015;45 (1):119–29.

[55] Sanchez-Paulete AR, et al. Cancer immunotherapy with immunomodulatory anti-CD137 and anti-PD-1 monoclonal antibodies requires BATF3-dependent dendritic cells. Cancer Discov 2016;6 (1):71–9.

[56] Binnewies M, et al. Unleashing Type-2 dendritic cells to drive protective antitumor CD4(+) T cell immunity. Cell 2019;177(3). 556–71.e16.

[57] Poulin LF, et al. Characterization of human DNGR-1+ BDCA3+ leukocytes as putative equivalents of mouse CD8alpha+ dendritic cells. J Exp Med 2010;207(6):1261–71.

[58] Sluijter BJ, et al. Arming the melanoma sentinel lymph node through local administration of CpG-B and GM-CSF: recruitment and activation of BDCA3/CD141(+) dendritic cells and enhanced cross-presentation. Cancer Immunol Res 2015;3(5):495–505.

[59] Broz ML, Krummel MF. The emerging understanding of myeloid cells as partners and targets in tumor rejection. Cancer Immunol Res 2015;3(4):313–9.

[60] Broz ML, et al. Dissecting the tumor myeloid compartment reveals rare activating antigen-presenting cells critical for T cell immunity. Cancer Cell 2014;26(5):638–52.

[61] Barry KC, et al. A natural killer-dendritic cell axis defines checkpoint therapy-responsive tumor microenvironments. Nat Med 2018;24(8):1178–91.

[62] MartIn-Fontecha A, et al. Regulation of dendritic cell migration to the draining lymph node: impact on T lymphocyte traffic and priming. J Exp Med 2003;198(4):615–21.

[63] Hirao M, et al. CC chemokine receptor-7 on dendritic cells is induced after interaction with apoptotic tumor cells: critical role in migration from the tumor site to draining lymph nodes. Cancer Res 2000;60(8):2209–17.

[64] Munn DH, Mellor AL. IDO in the tumor microenvironment: inflammation, counter-regulation, and tolerance. Trends Immunol 2016;37(3):193–207.

[65] Bakdash G, et al. The nature of activatory and tolerogenic dendritic cell-derived signal II. Front Immunol 2013;4:53.

[66] Haniffa M, Bigley V, Collin M. Human mononuclear phagocyte system reunited. Semin Cell Dev Biol 2015;41:59–69.

[67] De Palma M, Lewis CE. Macrophage regulation of tumor responses to anticancer therapies. Cancer Cell 2013;23(3):277–86.

[68] Zhou J, et al. Tumor-associated macrophages: recent insights and therapies. Front Oncol 2020;10:188.

[69] Knowles HJ, Harris AL. Macrophages and the hypoxic tumour microenvironment. Front Biosci 2007;12:4298–314.

[70] Pathria P, Louis TL, Varner JA. Targeting tumor-associated macrophages in cancer. Trends Immunol 2019;40(4):310–27.

[71] Genard G, Lucas S, Michiels C. Reprogramming of tumor-associated macrophages with anticancer therapies: radiotherapy versus chemo- and immunotherapies. Front Immunol 2017;8:828.

[72] Overdijk MB, et al. Antibody-mediated phagocytosis contributes to the anti-tumor activity of the therapeutic antibody daratumumab in lymphoma and multiple myeloma. MAbs 2015;7(2): 311–21.

[73] Sica A, Mantovani A. Macrophage plasticity and polarization: in vivo veritas. J Clin Invest 2012;122 (3):787–95.

[74] Qian BZ, Pollard JW. Macrophage diversity enhances tumor progression and metastasis. Cell 2010;141(1):39–51.

[75] Coffelt SB, Wellenstein MD, de Visser KE. Neutrophils in cancer: neutral no more. Nat Rev Cancer 2016;16(7):431–46.

[76] Shaul ME, Fridlender ZG. Tumour-associated neutrophils in patients with cancer. Nat Rev Clin Oncol 2019;16(10):601–20.

[77] Lieschke GJ, et al. Mice lacking granulocyte colony-stimulating factor have chronic neutropenia, granulocyte and macrophage progenitor cell deficiency, and impaired neutrophil mobilization. Blood 1994;84(6):1737–46.

[78] Liu F, et al. Impaired production and increased apoptosis of neutrophils in granulocyte colony-stimulating factor receptor-deficient mice. Immunity 1996;5(5):491–501.

[79] Richards MK, et al. Pivotal role of granulocyte colony-stimulating factor in the development of progenitors in the common myeloid pathway. Blood 2003;102(10):3562–8.

[80] Mantovani A, et al. Neutrophils in the activation and regulation of innate and adaptive immunity. Nat Rev Immunol 2011;11(8):519–31.

[81] Jablonska J, et al. CXCR2-mediated tumor-associated neutrophil recruitment is regulated by IFN-beta. Int J Cancer 2014;134(6):1346–58.

[82] Kuang DM, et al. Peritumoral neutrophils link inflammatory response to disease progression by fostering angiogenesis in hepatocellular carcinoma. J Hepatol 2011;54(5):948–55.

[83] Wislez M, et al. Hepatocyte growth factor production by neutrophils infiltrating bronchioloalveolar subtype pulmonary adenocarcinoma: role in tumor progression and death. Cancer Res 2003;63 (6):1405–12.

[84] Charles KA, et al. The tumor-promoting actions of TNF-alpha involve TNFR1 and IL-17 in ovarian cancer in mice and humans. J Clin Invest 2009;119(10):3011–23.

[85] Peng B, et al. Prognostic significance of the neutrophil to lymphocyte ratio in patients with non-small cell lung cancer: a systemic review and meta-analysis. Int J Clin Exp Med 2015;8(3):3098–106.

[86] Krenn-Pilko S, et al. The elevated preoperative platelet-to-lymphocyte ratio predicts poor prognosis in breast cancer patients. Br J Cancer 2014;110(10):2524–30.

[87] Schmidt H, et al. Elevated neutrophil and monocyte counts in peripheral blood are associated with poor survival in patients with metastatic melanoma: a prognostic model. Br J Cancer 2005;93 (3):273–8.

[88] Gu X, et al. Prognostic significance of neutrophil-to-lymphocyte ratio in prostate cancer: evidence from 16,266 patients. Sci Rep 2016;6:22089.

[89] Grenader T, et al. Derived neutrophil lymphocyte ratio is predictive of survival from intermittent therapy in advanced colorectal cancer: a post hoc analysis of the MRC COIN study. Br J Cancer 2016;114(6):612–5.

[90] Terashima T, et al. Blood neutrophil to lymphocyte ratio as a predictor in patients with advanced hepatocellular carcinoma treated with hepatic arterial infusion chemotherapy. Hepatol Res 2015;45(9):949–59.

[91] Lin G, et al. Elevated neutrophil-to-lymphocyte ratio is an independent poor prognostic factor in patients with intrahepatic cholangiocarcinoma. Oncotarget 2016;7(32):50963–71.

[92] Casbon AJ, et al. Invasive breast cancer reprograms early myeloid differentiation in the bone marrow to generate immunosuppressive neutrophils. Proc Natl Acad Sci USA 2015;112(6):E566–75.

[93] Waight JD, et al. Tumor-derived G-CSF facilitates neoplastic growth through a granulocytic myeloid-derived suppressor cell-dependent mechanism. PLoS One 2011;6(11), e27690.

[94] Fridlender ZG, et al. Polarization of tumor-associated neutrophil phenotype by TGF-beta: "N1" versus "N2" TAN. Cancer Cell 2009;16(3):183–94.

[95] Tachiyama G, et al. Endogenous endotoxemia in patients with liver cirrhosis—a quantitative analysis of endotoxin in portal and peripheral blood. Jpn J Surg 1988;18(4):403–8.

[96] Lopez-Lago MA, et al. Neutrophil chemokines secreted by tumor cells mount a lung antimetastatic response during renal cell carcinoma progression. Oncogene 2013;32(14):1752–60.

[97] Catena R, et al. Bone marrow-derived Gr1 + cells can generate a metastasis-resistant microenvironment via induced secretion of thrombospondin-1. Cancer Discov 2013;3(5):578–89.

[98] Godfrey DI, Kronenberg M. Going both ways: immune regulation via CD1d-dependent NKT cells. J Clin Invest 2004;114(10):1379–88.

[99] Patel S, et al. Beyond CAR T cells: other cell-based immunotherapeutic strategies against cancer. Front Oncol 2019;9:196.

[100] Webb TJ, et al. Editorial: NKT cells in cancer immunotherapy. Front Immunol 2020;11:1314.

[101] Lam PY, Nissen MD, Mattarollo SR. Invariant natural killer T cells in immune regulation of blood cancers: harnessing their potential in immunotherapies. Front Immunol 2017;8:1355.

[102] Broz P, Monack DM. Newly described pattern recognition receptors team up against intracellular pathogens. Nat Rev Immunol 2013;13(8):551–65.

[103] Aderem A, Ulevitch RJ. Toll-like receptors in the induction of the innate immune response. Nature 2000;406(6797):782–7.

[104] Chen M, et al. Targeting nuclear acid-mediated immunity in cancer immune checkpoint inhibitor therapies. Signal Transduct Target Ther 2020;5(1):270.

[105] Alexopoulou L, et al. Recognition of double-stranded RNA and activation of NF-kappaB by Toll-like receptor 3. Nature 2001;413(6857):732–8.

[106] Matsumoto M, et al. Establishment of a monoclonal antibody against human Toll-like receptor 3 that blocks double-stranded RNA-mediated signaling. Biochem Biophys Res Commun 2002;293 (5):1364–9.

[107] Tatematsu M, et al. Toll-like receptor 3 recognizes incomplete stem structures in single-stranded viral RNA. Nat Commun 2013;4:1833.

[108] Ohto U, et al. Structural basis of CpG and inhibitory DNA recognition by Toll-like receptor 9. Nature 2015;520(7549):702–5.

[109] Sabbatini P, et al. Phase I trial of overlapping long peptides from a tumor self-antigen and poly-ICLC shows rapid induction of integrated immune response in ovarian cancer patients. Clin Cancer Res 2012;18(23):6497–508.

[110] Hanna E, Abadi R, Abbas O. Imiquimod in dermatology: an overview. Int J Dermatol 2016;55 (8):831–44.

[111] Ma F, et al. The TLR7 agonists imiquimod and gardiquimod improve DC-based immunotherapy for melanoma in mice. Cell Mol Immunol 2010;7(5):381–8.

[112] Michaelis KA, et al. The TLR7/8 agonist R848 remodels tumor and host responses to promote survival in pancreatic cancer. Nat Commun 2019;10(1):4682.

[113] Hamid O, Ismail R, Puzanov I. Intratumoral immunotherapy—update 2019. Oncologist 2020;25(3): e423–38.

[114] Le Naour J, et al. Trial watch: STING agonists in cancer therapy. Oncoimmunology 2020;9 (1):1777624.

[115] Corrales L, et al. Direct activation of STING in the tumor microenvironment leads to potent and systemic tumor regression and immunity. Cell Rep 2015;11(7):1018–30.

[116] Elion DL, et al. Therapeutically active RIG-I agonist induces immunogenic tumor cell killing in breast cancers. Cancer Res 2018;78(21):6183–95.

[117] Poeck H, et al. 5′-Triphosphate-siRNA: turning gene silencing and Rig-I activation against melanoma. Nat Med 2008;14(11):1256–63.

[118] Ruzicka M, et al. RIG-I-based immunotherapy enhances survival in preclinical AML models and sensitizes AML cells to checkpoint blockade. Leukemia 2020;34(4):1017–26.

[119] Heidegger S, et al. RIG-I activation is critical for responsiveness to checkpoint blockade. Sci Immunol 2019;4(39):aau8943.

[120] Steggerda SM, et al. Inhibition of arginase by CB-1158 blocks myeloid cell-mediated immune suppression in the tumor microenvironment. J Immunother Cancer 2017;5(1):101.

[121] Labadie BW, Bao R, Luke JJ. Reimagining IDO pathway inhibition in cancer immunotherapy via downstream focus on the tryptophan-kynurenine-aryl hydrocarbon axis. Clin Cancer Res 2019;25 (5):1462–71.

[122] Liu M, et al. Targeting the IDO1 pathway in cancer: from bench to bedside. J Hematol Oncol 2018;11(1):100.

[123] Holmgaard RB, et al. Tumor-expressed IDO recruits and activates MDSCs in a Treg-dependent manner. Cell Rep 2015;13(2):412–24.

[124] Terness P, et al. Inhibition of allogeneic T cell proliferation by indoleamine 2,3-dioxygenase-expressing dendritic cells: mediation of suppression by tryptophan metabolites. J Exp Med 2002;196(4):447–57.

[125] Long GV, et al. Epacadostat plus pembrolizumab versus placebo plus pembrolizumab in patients with unresectable or metastatic melanoma (ECHO-301/KEYNOTE-252): a phase 3, randomised, double-blind study. Lancet Oncol 2019;20(8):1083–97.

[126] Young A, et al. A2AR adenosine signaling suppresses natural killer cell maturation in the tumor microenvironment. Cancer Res 2018;78(4):1003–16.

[127] Fredholm BB. Adenosine receptors as drug targets. Exp Cell Res 2010;316(8):1284–8.

[128] Borodovsky A, et al. Small molecule AZD4635 inhibitor of A2AR signaling rescues immune cell function including CD103(+) dendritic cells enhancing anti-tumor immunity. J Immunother Cancer 2020;8(2):e000417.

[129] Borsellino G, et al. Expression of ectonucleotidase CD39 by Foxp3 + Treg cells: hydrolysis of extra-cellular ATP and immune suppression. Blood 2007;110(4):1225–32.

[130] Novitskiy SV, et al. Adenosine receptors in regulation of dendritic cell differentiation and function. Blood 2008;112(5):1822–31.

[131] Panther E, et al. Expression and function of adenosine receptors in human dendritic cells. FASEB J 2001;15(11):1963–70.

[132] Panther E, et al. Adenosine affects expression of membrane molecules, cytokine and chemokine release, and the T-cell stimulatory capacity of human dendritic cells. Blood 2003;101(10):3985–90.

[133] Cekic C, et al. Myeloid expression of adenosine A2A receptor suppresses T and NK cell responses in the solid tumor microenvironment. Cancer Res 2014;74(24):7250–9.

[134] Wang L, et al. CD73 has distinct roles in nonhematopoietic and hematopoietic cells to promote tumor growth in mice. J Clin Invest 2011;121(6):2371–82.

[135] Perrot I, et al. Blocking antibodies targeting the CD39/CD73 immunosuppressive pathway unleash immune responses in combination cancer therapies. Cell Rep 2019;27(8):2411–25 [e9].

[136] Allard B, et al. Targeting CD73 enhances the antitumor activity of anti-PD-1 and anti-CTLA-4 mAbs. Clin Cancer Res 2013;19(20):5626–35.

[137] Stagg J, et al. CD73-deficient mice have increased antitumor immunity and are resistant to experimental metastasis. Cancer Res 2011;71(8):2892–900.

[138] Hay CM, et al. Targeting CD73 in the tumor microenvironment with MEDI9447. Oncoimmunology 2016;5(8), e1208875.

[139] Willingham SB, et al. A2AR antagonism with CPI-444 induces antitumor responses and augments efficacy to anti-PD-(L)1 and anti-CTLA-4 in preclinical models. Cancer Immunol Res 2018;6(10):1136–49.

[140] Leone RD, et al. Inhibition of the adenosine A2a receptor modulates expression of T cell coinhibitory receptors and improves effector function for enhanced checkpoint blockade and ACT in murine cancer models. Cancer Immunol Immunother 2018;67(8):1271–84.

[141] Lemos de Matos A, Franco LS, McFadden G. Oncolytic viruses and the immune system: the dynamic duo. Mol Ther Methods Clin Dev 2020;17:349–58.

[142] Bommareddy PK, Shettigar M, Kaufman HL. Integrating oncolytic viruses in combination cancer immunotherapy. Nat Rev Immunol 2018;18(8):498–513.

[143] Ribas A, et al. Oncolytic virotherapy promotes Intratumoral T cell infiltration and improves anti-PD-1 immunotherapy. Cell 2018;174(4):1031–2.

[144] Zhang Q, Liu F. Correction: advances and potential pitfalls of oncolytic viruses expressing immuno-modulatory transgene therapy for malignant gliomas. Cell Death Dis 2020;11(11):1007.

[145] Roden RBS, Stern PL. Opportunities and challenges for human papillomavirus vaccination in cancer. Nat Rev Cancer 2018;18(4):240–54.

[146] Hollingsworth RE, Jansen K. Turning the corner on therapeutic cancer vaccines. npj Vaccines 2019;4:7.

[147] Peng M, et al. Neoantigen vaccine: an emerging tumor immunotherapy. Mol Cancer 2019;18(1):128.

[148] Sayour EJ, Mitchell DA. Manipulation of innate and adaptive immunity through cancer vaccines. J Immunol Res 2017;2017:3145742.

[149] Bhardwaj N, Friedlander PA, Pavlick AC, et al. Flt3 ligand augments immune responses to anti-DEC-205-NY-ESO-1 vaccine through expansion of dendritic cell subsets. Nat Cancer 2020;1:1204–17.

[150] Sahin U, et al. An RNA vaccine drives immunity in checkpoint-inhibitor-treated melanoma. Nature 2020;585(7823):107–12.

[151] Kantoff PW, et al. Sipuleucel-T immunotherapy for castration-resistant prostate cancer. N Engl J Med 2010;363(5):411–22.

[152] Lin C, Zhang J. Chimeric antigen receptor engineered innate immune cells in cancer immunotherapy. Sci China Life Sci 2019;62(5):633–9.

[153] Cortes-Selva D, et al. Innate and innate-like cells: the future of chimeric antigen receptor (CAR) cell therapy. Trends Pharmacol Sci 2020;42(1):45–59.

[154] Rafei H, Daher M, Rezvani K. Chimeric antigen receptor (CAR) natural killer (NK)-cell therapy: leveraging the power of innate immunity. Br J Haematol 2020;193(2):216–30.

[155] Imai C, Iwamoto S, Campana D. Genetic modification of primary natural killer cells overcomes inhibitory signals and induces specific killing of leukemic cells. Blood 2005;106(1):376–83.

[156] Watzl C, Long EO. Signal transduction during activation and inhibition of natural killer cells. Curr Protoc Immunol 2010;Chapter 11 [Unit 11 9B].

[157] Chu Y, et al. Targeting CD20+ aggressive B-cell non-Hodgkin lymphoma by anti-CD20 CAR mRNA-modified expanded natural killer cells in vitro and in NSG mice. Cancer Immunol Res 2015;3(4):333–44.

[158] Liu E, et al. Use of CAR-transduced natural killer cells in CD19-positive lymphoid tumors. N Engl J Med 2020;382(6):545–53.

[159] Klichinsky M, et al. Human chimeric antigen receptor macrophages for cancer immunotherapy. Nat Biotechnol 2020;38(8):947–53.

[160] Chaplin DD. Overview of the immune response. J Allergy Clin Immunol 2010;125(2 Suppl 2):S3–23.

[161] Weaver C, Murphy K. Janeway's immunobiology. 9th ed. Garland Science, Taylor & Francis Group; 2017.

[162] Janeway Jr CA. How the immune system recognizes invaders. Sci Am 1993;269(3):72–9.

[163] Mulder WJM, et al. Therapeutic targeting of trained immunity. Nat Rev Drug Discov 2019;18(7):553–66.

[164] Dunkelberger JR, Song WC. Complement and its role in innate and adaptive immune responses. Cell Res 2010;20(1):34–50.

[165] Kwan WH, van der Touw W, Heeger PS. Complement regulation of T cell immunity. Immunol Res 2012;54(1–3):247–53.

[166] Deseke M, Prinz I. Ligand recognition by the gammadelta TCR and discrimination between homeostasis and stress conditions. Cell Mol Immunol 2020;17(9):914–24.

[167] Waldhauer I, Steinle A. NK cells and cancer immunosurveillance. Oncogene 2008;27(45):5932–43.

[168] Marshall JS, et al. An introduction to immunology and immunopathology. Allergy Asthma Clin Immunol 2018;14(Suppl 2):49.

[169] Warrington R, et al. An introduction to immunology and immunopathology. Allergy Asthma Clin Immunol 2011;7(Suppl 1):S1.

[170] Hastings K. Innate and adaptive immune responses to cancer. In: Fundamentals of cancer prevention. Berlin, Heidelberg: Springer; 2008. p. 79–108.

[171] Ilyas S, Yang JC. Landscape of tumor antigens in T cell immunotherapy. J Immunol 2015;195(11):5117–22.

[172] Xu Y, et al. The role of the cancer testis antigen PRAME in tumorigenesis and immunotherapy in human cancer. Cell Prolif 2020;53(3), e12770.

[173] Raza A, et al. Unleashing the immune response to NY-ESO-1 cancer testis antigen as a potential target for cancer immunotherapy. J Transl Med 2020;18(1):140.

[174] Bastien JP, et al. Cellular therapy approaches harnessing the power of the immune system for personalized cancer treatment. Semin Immunol 2019;42:101306.

[175] Tessari A, et al. Expression of NY-ESO-1, MAGE-A3, PRAME and WT1 in different subgroups of breast cancer: an indication to immunotherapy? Breast 2018;42:68–73.

[176] Sarandakou A, Protonotariou E, Rizos D. Tumor markers in biological fluids associated with pregnancy. Crit Rev Clin Lab Sci 2007;44(2):151–78.

[177] Abramson JS. Anti-CD19 CAR T-cell therapy for B-cell non-Hodgkin lymphoma. Transfus Med Rev 2020;34(1):29–33.

[178] Pehlivan KC, Duncan BB, Lee DW. CAR-T cell therapy for acute lymphoblastic leukemia: transforming the treatment of relapsed and refractory disease. Curr Hematol Malig Rep 2018;13 (5):396–406.

[179] Mohanty R, et al. CAR T cell therapy: a new era for cancer treatment (review). Oncol Rep 2019;42 (6):2183–95.

[180] Polz-Gruszka D, et al. EBV, HSV, CMV and HPV in laryngeal and oropharyngeal carcinoma in polish patients. Anticancer Res 2015;35(3):1657–61.

[181] Vranic S, et al. The role of Epstein-Barr virus in cervical cancer: a brief update. Front Oncol 2018;8:113.

[182] Cicenas J, et al. KRAS, TP53, CDKN2A, SMAD4, BRCA1, and BRCA2 mutations in pancreatic cancer. Cancers (Basel) 2017;9(5):42.

[183] Dong ZY, et al. Potential predictive value of TP53 and KRAS mutation status for response to PD-1 blockade immunotherapy in lung adenocarcinoma. Clin Cancer Res 2017;23(12):3012–24.

[184] Cavalieri S, et al. Efficacy and safety of single-agent pan-human epidermal growth factor receptor (HER) inhibitor dacomitinib in locally advanced unresectable or metastatic skin squamous cell cancer. Eur J Cancer 2018;97:7–15.

[185] Passarelli A, et al. Immune system and melanoma biology: a balance between immunosurveillance and immune escape. Oncotarget 2017;8(62):106132–42.

[186] Sato E, et al. Intraepithelial CD8+ tumor-infiltrating lymphocytes and a high CD8+/regulatory T cell ratio are associated with favorable prognosis in ovarian cancer. Proc Natl Acad Sci USA 2005;102(51):18538–43.

[187] Kikuchi T, et al. A subset of patients with MSS/MSI-low-colorectal cancer showed increased CD8(+) TILs together with up-regulated IFN-gamma. Oncol Lett 2019;18(6):5977–85.

[188] Farhood B, Najafi M, Mortezaee K. CD8(+) cytotoxic T lymphocytes in cancer immunotherapy: a review. J Cell Physiol 2019;234(6):8509–21.

[189] Topalian SL, et al. Mechanism-driven biomarkers to guide immune checkpoint blockade in cancer therapy. Nat Rev Cancer 2016;16(5):275–87.

[190] Hassin D, et al. Cytotoxic T lymphocyte perforin and Fas ligand working in concert even when Fas ligand lytic action is still not detectable. Immunology 2011;133(2):190–6.

[191] Halama N, et al. Natural killer cells are scarce in colorectal carcinoma tissue despite high levels of chemokines and cytokines. Clin Cancer Res 2011;17(4):678–89.

[192] Mlecnik B, et al. The tumor microenvironment and immunoscore are critical determinants of dissemination to distant metastasis. Sci Transl Med 2016;8(327). 327ra26.

[193] Clemente CG, et al. Prognostic value of tumor infiltrating lymphocytes in the vertical growth phase of primary cutaneous melanoma. Cancer 1996;77(7):1303–10.

[194] Oldford SA, et al. Tumor cell expression of HLA-DM associates with a Th1 profile and predicts improved survival in breast carcinoma patients. Int Immunol 2006;18(11):1591–602.

[195] Dieu-Nosjean MC, et al. Long-term survival for patients with non-small-cell lung cancer with intratumoral lymphoid structures. J Clin Oncol 2008;26(27):4410–7.

[196] Kusuda T, et al. Relative expression levels of Th1 and Th2 cytokine mRNA are independent prognostic factors in patients with ovarian cancer. Oncol Rep 2005;13(6):1153–8.

[197] Andersen R, et al. Tumor infiltrating lymphocyte therapy for ovarian cancer and renal cell carcinoma. Hum Vaccin Immunother 2015;11(12):2790–5.

[198] Tosolini M, et al. Clinical impact of different classes of infiltrating T cytotoxic and helper cells (Th1, th2, treg, th17) in patients with colorectal cancer. Cancer Res 2011;71(4):1263–71.

[199] Vesalainen S, et al. Histological grade, perineural infiltration, tumour-infiltrating lymphocytes and apoptosis as determinants of long-term prognosis in prostatic adenocarcinoma. Eur J Cancer 1994;30A(12):1797–803.

[200] Ubukata H, et al. Evaluations of interferon-gamma/interleukin-4 ratio and neutrophil/lymphocyte ratio as prognostic indicators in gastric cancer patients. J Surg Oncol 2010;102(7):742–7.

[201] Ju X, et al. Predictive relevance of PD-L1 expression with pre-existing TILs in gastric cancer. Oncotarget 2017;8(59):99372–81.

[202] van de Ven K, Borst J. Targeting the T-cell co-stimulatory CD27/CD70 pathway in cancer immunotherapy: rationale and potential. Immunotherapy 2015;7(6):655–67.

[203] Yang J, et al. MiR-15a/16 deficiency enhances anti-tumor immunity of glioma-infiltrating CD8+ T cells through targeting mTOR. Int J Cancer 2017;141(10):2082–92.

[204] Chihara N, et al. Induction and transcriptional regulation of the co-inhibitory gene module in T cells. Nature 2018;558(7710):454–9.

[205] Knutson KL, Disis ML. Augmenting T helper cell immunity in cancer. Curr Drug Targets Immune Endocr Metabol Disord 2005;5(4):365–71.

[206] Kumai T, et al. Optimization of peptide vaccines to induce robust antitumor CD4 T-cell responses. Cancer Immunol Res 2017;5(1):72–83.

[207] Spitzer MH, et al. Systemic immunity is required for effective cancer immunotherapy. Cell 2017;168 (3). 487–502.e15.

[208] Melssen M, Slingluff Jr CL. Vaccines targeting helper T cells for cancer immunotherapy. Curr Opin Immunol 2017;47:85–92.

[209] Koski GK, et al. A novel dendritic cell-based immunization approach for the induction of durable Th1-polarized anti-HER-2/neu responses in women with early breast cancer. J Immunother 2012;35(1):54–65.

[210] Ahrends T, et al. CD27 agonism plus PD-1 blockade recapitulates CD4+ T-cell help in therapeutic anticancer vaccination. Cancer Res 2016;76(10):2921–31.

[211] Harris TJ, Drake CG. Primer on tumor immunology and cancer immunotherapy. J Immunother Cancer 2013;1:12.

[212] Laidlaw BJ, Craft JE, Kaech SM. The multifaceted role of CD4(+) T cells in CD8(+) T cell memory. Nat Rev Immunol 2016;16(2):102–11.

[213] Minami Y, et al. The IL-2 receptor complex: its structure, function, and target genes. Annu Rev Immunol 1993;11:245–68.

[214] Ross SH, Cantrell DA. Signaling and function of Interleukin-2 in T lymphocytes. Annu Rev Immunol 2018;36:411–33.

[215] Raval RR, et al. Tumor immunology and cancer immunotherapy: summary of the 2013 SITC primer. J Immunother Cancer 2014;2:14.

[216] Maddaly R, et al. Receptors and signaling mechanisms for B-lymphocyte activation, proliferation and differentiation—insights from both in vivo and in vitro approaches. FEBS Lett 2010;584 (24):4883–94.

[217] Kwun J, et al. Crosstalk between T and B cells in the germinal center after transplantation. Transplantation 2017;101(4):704–12.

[218] Janeway Jr C, Travers P, Walport M, et al. B-cell activation by armed helper T cells. In: Immunobiology. 5th ed. New York, NY: Garland Science; 2001.

[219] De Silva NS, Klein U. Dynamics of B cells in germinal centres. Nat Rev Immunol 2015;15 (3):137–48.

[220] Sun JC, Lanier LL. Natural killer cells remember: an evolutionary bridge between innate and adaptive immunity? Eur J Immunol 2009;39(8):2059–64.

[221] Yi JS, Cox MA, Zajac AJ. T-cell exhaustion: characteristics, causes and conversion. Immunology 2010;129(4):474–81.

[222] Smith-Garvin JE, Koretzky GA, Jordan MS. T cell activation. Annu Rev Immunol 2009;27:591–619.

[223] Zeng H, Chi H. mTOR and lymphocyte metabolism. Curr Opin Immunol 2013;25(3):347–55.

[224] Chapman NM, Chi H. mTOR signaling, Tregs and immune modulation. Immunotherapy 2014;6 (12):1295–311.

[225] Podojil JR, Miller SD. Molecular mechanisms of T-cell receptor and costimulatory molecule ligation/blockade in autoimmune disease therapy. Immunol Rev 2009;229(1):337–55.

[226] Floess S, et al. Epigenetic control of the foxp3 locus in regulatory T cells. PLoS Biol 2007;5(2), e38.

[227] Pandiyan P, et al. CD4+CD25+Foxp3+ regulatory T cells induce cytokine deprivation-mediated apoptosis of effector CD4+ T cells. Nat Immunol 2007;8(12):1353–62.

[228] Bopp T, et al. Cyclic adenosine monophosphate is a key component of regulatory T cell-mediated suppression. J Exp Med 2007;204(6):1303–10.

[229] Schwartz RH. T cell anergy. Annu Rev Immunol 2003;21:305–34.

[230] Baron U, et al. DNA demethylation in the human FOXP3 locus discriminates regulatory T cells from activated FOXP3(+) conventional T cells. Eur J Immunol 2007;37(9):2378–89.

[231] Sharpe AH. Mechanisms of costimulation. Immunol Rev 2009;229(1):5–11.

[232] Schwartz JC, et al. Structural mechanisms of costimulation. Nat Immunol 2002;3(5):427–34.

[233] Sim GC, et al. IL-2 therapy promotes suppressive ICOS+ Treg expansion in melanoma patients. J Clin Invest 2014;124(1):99–110.

[234] Li DY, Xiong XZ. ICOS(+) Tregs: a functional subset of Tregs in immune diseases. Front Immunol 2020;11:2104.

[235] Amatore F, et al. ICOS is widely expressed in cutaneous T-cell lymphoma, and its targeting promotes potent killing of malignant cells. Blood Adv 2020;4(20):5203–14.

[236] Largeot A, et al. The B-side of cancer immunity: the underrated tune. Cell 2019;8(5):449.

[237] Andreu P, et al. FcRgamma activation regulates inflammation-associated squamous carcinogenesis. Cancer Cell 2010;17(2):121–34.

[238] Da Gama Duarte J, Peyper JM, Blackburn JM. B cells and antibody production in melanoma. Mamm Genome 2018;29(11−12):790–805.

[239] Chen X, et al. FcgammaR-binding is an important functional attribute for immune checkpoint antibodies in cancer immunotherapy. Front Immunol 2019;10:292.

[240] Boldison J, et al. Dendritic cells license regulatory B cells to produce IL-10 and mediate suppression of antigen-specific CD8 T cells. Cell Mol Immunol 2020;17(8):843–55.

[241] Lindner S, et al. Interleukin 21-induced granzyme B-expressing B cells infiltrate tumors and regulate T cells. Cancer Res 2013;73(8):2468–79.

[242] Wang Y, et al. Myeloid-derived suppressor cells impair B cell responses in lung cancer through IL-7 and STAT5. J Immunol 2018;201(1):278–95.

[243] Berraondo P, et al. Cytokines in clinical cancer immunotherapy. Br J Cancer 2019;120(1):6–15.

[244] Han S, et al. Glioma cell-derived placental growth factor induces regulatory B cells. Int J Biochem Cell Biol 2014;57:63–8.

[245] Haro MA, et al. PD-1 suppresses development of humoral responses that protect against Tn-bearing tumors. Cancer Immunol Res 2016;4(12):1027–37.

[246] Caro-Maldonado A, et al. Metabolic reprogramming is required for antibody production that is suppressed in anergic but exaggerated in chronically BAFF-exposed B cells. J Immunol 2014;192(8):3626–36.

[247] Bruno TC, et al. Antigen-presenting intratumoral B cells affect CD4(+) TIL phenotypes in non-small cell lung cancer patients. Cancer Immunol Res 2017;5(10):898–907.

[248] Cabrita R, et al. Tertiary lymphoid structures improve immunotherapy and survival in melanoma. Nature 2020;577(7791):561–5.

[249] Petitprez F, et al. B cells are associated with survival and immunotherapy response in sarcoma. Nature 2020;577(7791):556–60.

[250] Helmink BA, et al. B cells and tertiary lymphoid structures promote immunotherapy response. Nature 2020;577(7791):549–55.

[251] Schioppa T, et al. B regulatory cells and the tumor-promoting actions of TNF-alpha during squamous carcinogenesis. Proc Natl Acad Sci USA 2011;108(26):10662–7.

[252] Olkhanud PB, et al. Tumor-evoked regulatory B cells promote breast cancer metastasis by converting resting CD4(+) T cells to T-regulatory cells. Cancer Res 2011;71(10):3505–15.

[253] Pylayeva-Gupta Y, et al. IL35-producing B cells promote the development of pancreatic neoplasia. Cancer Discov 2016;6(3):247–55.

[254] Cheng J, et al. Understanding the mechanisms of resistance to CAR T-cell therapy in malignancies. Front Oncol 2019;9:1237.

[255] Gajewski TF, Louahed J, Brichard VG. Gene signature in melanoma associated with clinical activity: a potential clue to unlock cancer immunotherapy. Cancer J 2010;16(4):399–403.

[256] Finn OJ. Cancer immunology. N Engl J Med 2008;358(25):2704–15.

[257] Wu SG, Shih JY. Management of acquired resistance to EGFR TKI-targeted therapy in advanced non-small cell lung cancer. Mol Cancer 2018;17(1):38.

[258] Inoue S, et al. Inhibitory effects of B cells on antitumor immunity. Cancer Res 2006;66(15):7741–7.

[259] Liu J, et al. Interleukin-12: an update on its immunological activities, signaling and regulation of gene expression. Curr Immunol Rev 2005;1(2):119–37.

[260] Kershaw MH, et al. A phase I study on adoptive immunotherapy using gene-modified T cells for ovarian cancer. Clin Cancer Res 2006;12(20 Pt 1):6106–15.

[261] Baitsch L, et al. Exhaustion of tumor-specific CD8(+) T cells in metastases from melanoma patients. J Clin Invest 2011;121(6):2350–60.

[262] Wherry EJ, Kurachi M. Molecular and cellular insights into T cell exhaustion. Nat Rev Immunol 2015;15(8):486–99.

[263] van Deursen JM. The role of senescent cells in ageing. Nature 2014;509(7501):439–46.

[264] Fenwick C, et al. T-cell exhaustion in HIV infection. Immunol Rev 2019;292(1):149–63.

[265] Qin S, et al. Novel immune checkpoint targets: moving beyond PD-1 and CTLA-4. Mol Cancer 2019;18(1):155.

[266] Gong J, et al. Development of PD-1 and PD-L1 inhibitors as a form of cancer immunotherapy: a comprehensive review of registration trials and future considerations. J Immunother Cancer 2018;6(1):8.

[267] Arasanz H, et al. PD1 signal transduction pathways in T cells. Oncotarget 2017;8(31):51936–45.

[268] Pardoll DM. The blockade of immune checkpoints in cancer immunotherapy. Nat Rev Cancer 2012;12(4):252–64.

[269] Buchbinder EI, Desai A. CTLA-4 and PD-1 pathways: similarities, differences, and implications of their inhibition. Am J Clin Oncol 2016;39(1):98–106.

[270] Schildberg FA, et al. Coinhibitory pathways in the B7-CD28 ligand-receptor family. Immunity 2016;44(5):955–72.

[271] Savoia P, Astrua C, Fava P. Ipilimumab (anti-Ctla-4 Mab) in the treatment of metastatic melanoma: effectiveness and toxicity management. Hum Vaccin Immunother 2016;12(5):1092–101.

[272] van Elsas A, Hurwitz AA, Allison JP. Combination immunotherapy of B16 melanoma using anti-cytotoxic T lymphocyte-associated antigen 4 (CTLA-4) and granulocyte/macrophage colony-stimulating factor (GM-CSF)-producing vaccines induces rejection of subcutaneous and metastatic tumors accompanied by autoimmune depigmentation. J Exp Med 1999;190(3):355–66.

[273] DeFalco J, et al. Non-progressing cancer patients have persistent B cell responses expressing shared antibody paratopes that target public tumor antigens. Clin Immunol 2018;187:37–45.

[274] Yuen GJ, Demissie E, Pillai S. B lymphocytes and cancer: a love-hate relationship. Trends Cancer 2016;2(12):747–57.

[275] Haibe Y, et al. Resisting resistance to immune checkpoint therapy: a systematic review. Int J Mol Sci 2020;21(17):6176.

[276] Barrueto L, et al. Resistance to checkpoint inhibition in cancer immunotherapy. Transl Oncol 2020;13(3):100738.

[277] Miller PL, Carson TL. Mechanisms and microbial influences on CTLA-4 and PD-1-based immunotherapy in the treatment of cancer: a narrative review. Gut Pathog 2020;12:43.

[278] Sun L, et al. Clinical efficacy and safety of anti-PD-1/PD-L1 inhibitors for the treatment of advanced or metastatic cancer: a systematic review and meta-analysis. Sci Rep 2020;10(1):2083.

[279] Andrews LP, et al. LAG3 (CD223) as a cancer immunotherapy target. Immunol Rev 2017;276(1):80–96.

[280] Ruffo E, et al. Lymphocyte-activation gene 3 (LAG3): the next immune checkpoint receptor. Semin Immunol 2019;42:101305.

[281] Joller N, Kuchroo VK. Tim-3, Lag-3, and TIGIT. Curr Top Microbiol Immunol 2017;410:127–56.

[282] He Y, et al. TIM-3, a promising target for cancer immunotherapy. Onco Targets Ther 2018;11:7005–9.

[283] ElTanbouly MA, et al. VISTA: a novel immunotherapy target for normalizing innate and adaptive immunity. Semin Immunol 2019;42:101308.

[284] Rosenberg SA, et al. Adoptive cell transfer: a clinical path to effective cancer immunotherapy. Nat Rev Cancer 2008;8(4):299–308.

[285] Rapoport AP, et al. Restoration of immunity in lymphopenic individuals with cancer by vaccination and adoptive T-cell transfer. Nat Med 2005;11(11):1230–7.

[286] Kalos M, June CH. Adoptive T cell transfer for cancer immunotherapy in the era of synthetic biology. Immunity 2013;39(1):49–60.

[287] Eberlein TJ, Rosenstein M, Rosenberg SA. Regression of a disseminated syngeneic solid tumor by systemic transfer of lymphoid cells expanded in interleukin 2. J Exp Med 1982;156(2):385–97.

[288] Donohue JH, et al. The systemic administration of purified interleukin 2 enhances the ability of sensitized murine lymphocytes to cure a disseminated syngeneic lymphoma. J Immunol 1984;132 (4):2123–8.

[289] Gattinoni L, et al. Removal of homeostatic cytokine sinks by lymphodepletion enhances the efficacy of adoptively transferred tumor-specific CD8+ T cells. J Exp Med 2005;202(7):907–12.

[290] Laport GG, et al. Adoptive transfer of costimulated T cells induces lymphocytosis in patients with relapsed/refractory non-Hodgkin lymphoma following CD34+-selected hematopoietic cell transplantation. Blood 2003;102(6):2004–13.

[291] Maude SL, et al. Chimeric antigen receptor T cells for sustained remissions in leukemia. N Engl J Med 2014;371(16):1507–17.

[292] Kalos M, et al. T cells with chimeric antigen receptors have potent antitumor effects and can establish memory in patients with advanced leukemia. Sci Transl Med 2011;3(95). 95ra73.

[293] Kochenderfer JN, et al. B-cell depletion and remissions of malignancy along with cytokine-associated toxicity in a clinical trial of anti-CD19 chimeric-antigen-receptor-transduced T cells. Blood 2012;119 (12):2709–20.

[294] Brentjens RJ, et al. CD19-targeted T cells rapidly induce molecular remissions in adults with chemotherapy-refractory acute lymphoblastic leukemia. Sci Transl Med 2013;5(177). 177ra38.

[295] Farkona S, Diamandis EP, Blasutig IM. Cancer immunotherapy: the beginning of the end of cancer? BMC Med 2016;14:73.

[296] Berger C, et al. Adoptive transfer of effector CD8+ T cells derived from central memory cells establishes persistent T cell memory in primates. J Clin Invest 2008;118(1):294–305.

[297] June CH, et al. CAR T cell immunotherapy for human cancer. Science 2018;359(6382):1361–5.

[298] Li D, et al. Genetically engineered T cells for cancer immunotherapy. Signal Transduct Target Ther 2019;4(1):35.

[299] Rossi JF. Targeted therapies in adult B-cell malignancies. Biomed Res Int 2015;2015:217593.

[300] Rafiq S, Hackett CS, Brentjens RJ. Engineering strategies to overcome the current roadblocks in CAR T cell therapy. Nat Rev Clin Oncol 2020;17(3):147–67.

[301] Gonzalez H, Hagerling C, Werb Z. Roles of the immune system in cancer: from tumor initiation to metastatic progression. Genes Dev 2018;32(19–20):1267–84.

[302] Zhao L, Cao YJ. Engineered T cell therapy for cancer in the clinic. Front Immunol 2019; 10:2250.

[303] Prasad V. Tisagenlecleucel—the first approved CAR-T-cell therapy: implications for payers and policy makers. 15. Nature Reviews Clinical Oncology; 2018. p. 11–2.

[304] Chen KH, et al. A compound chimeric antigen receptor strategy for targeting multiple myeloma. Leukemia 2018;32(2):402–12.

[305] Zhang X, et al. Efficacy and safety of anti-CD19 CAR T-cell therapy in 110 patients with B-cell acute lymphoblastic leukemia with high-risk features. Blood Adv 2020;4(10):2325–38.

[306] Dufva O, et al. Integrated drug profiling and CRISPR screening identify essential pathways for CAR T-cell cytotoxicity. Blood 2020;135(9):597–609.

[307] Liu D, et al. The role of immunological synapse in predicting the efficacy of chimeric antigen receptor (CAR) immunotherapy. Cell Commun Signal 2020;18(1):134.

[308] Dimitriou F, et al. Long-term disease control after allogeneic hematopoietic stem cell transplantation in primary cutaneous T-cell lymphoma; results from a single institution analysis. Front Med (Lausanne) 2020;7:290.

[309] Lupo KB, Matosevic S. Natural killer cells as allogeneic effectors in adoptive cancer immunotherapy. Cancers (Basel) 2019;11(6):769.

[310] Kabelitz D, et al. Cancer immunotherapy with gammadelta T cells: many paths ahead of us. Cell Mol Immunol 2020;17(9):925–39.

[311] Wang Y, et al. A retrospective comparison of CD19 single and CD19/CD22 bispecific targeted chimeric antigen receptor T cell therapy in patients with relapsed/refractory acute lymphoblastic leukemia. Blood Cancer J 2020;10(10):105.

[312] Pont MJ, et al. γ-Secretase inhibition increases efficacy of BCMA-specific chimeric antigen receptor T cells in multiple myeloma. Blood 2019;134(19):1585–97.

[313] Yu S, et al. Next generation chimeric antigen receptor T cells: safety strategies to overcome toxicity. Mol Cancer 2019;18(1):125.

[314] Xu J, et al. Combination therapy: a feasibility strategy for CAR-T cell therapy in the treatment of solid tumors. Oncol Lett 2018;16(2):2063–70.

[315] Grosser R, et al. Combination immunotherapy with CAR T cells and checkpoint blockade for the treatment of solid tumors. Cancer Cell 2019;36(5):471–82.

[316] Huang R, et al. Recent advances in CAR-T cell engineering. J Hematol Oncol 2020;13(1):86.

[317] Trenevska I, Li D, Banham AH. Therapeutic antibodies against intracellular tumor antigens. Front Immunol 2017;8:1001.

[318] Spurrell EL, Lockley M. Adaptive immunity in cancer immunology and therapeutics. Ecancermedicalscience 2014;8:441.

[319] Verma S, et al. Trastuzumab emtansine for HER2-positive advanced breast cancer. N Engl J Med 2012;367(19):1783–91.

[320] Weiner LM, Dhodapkar MV, Ferrone S. Monoclonal antibodies for cancer immunotherapy. Lancet 2009;373(9668):1033–40.

[321] Chen S, et al. Bispecific antibodies in cancer immunotherapy. Hum Vaccin Immunother 2016;12 (10):2491–500.

[322] Smits NC, Sentman CL. Bispecific T-cell engagers (BiTEs) as treatment of B-cell lymphoma. J Clin Oncol 2016;34(10):1131–3.

[323] Jager M, et al. Immunomonitoring results of a phase II/III study of malignant ascites patients treated with the trifunctional antibody catumaxomab (anti-EpCAM x anti-CD3). Cancer Res 2012;72(1):24–32.

[324] Sedykh SE, et al. Bispecific antibodies: design, therapy, perspectives. Drug Des Devel Ther 2018;12:195–208.

[325] Lejeune M, et al. Bispecific, T-cell-recruiting antibodies in B-cell malignancies. Front Immunol 2020;11:762.

[326] Labrijn AF, et al. Bispecific antibodies: a mechanistic review of the pipeline. Nat Rev Drug Discov 2019;18(8):585–608.

[327] Roseman DS, et al. G6PC mRNA therapy positively regulates fasting blood glucose and decreases liver abnormalities in a mouse model of glycogen storage disease 1a. Mol Ther 2018;26(3):814–21.

[328] Asrani KH, et al. Arginase I mRNA therapy—a novel approach to rescue arginase 1 enzyme deficiency. RNA Biol 2018;15(7):914–22.

[329] Connolly B, et al. SERPINA1 mRNA as a treatment for alpha-1 antitrypsin deficiency. J Nucleic Acids 2018;2018:8247935.

[330] Asrani KH, et al. Optimization of mRNA untranslated regions for improved expression of therapeutic mRNA. RNA Biol 2018;15(6):756–62.

[331] Farelli JD, et al. Leveraging rational protein engineering to improve mRNA therapeutics. Nucleic Acid Ther 2018;28(2):74–85.

[332] Shim H. Bispecific antibodies and antibody-drug conjugates for cancer therapy: technological considerations. Biomolecules 2020;10(3):360.

[333] Stock S, Schmitt M, Sellner L. Optimizing manufacturing protocols of chimeric antigen receptor T cells for improved anticancer immunotherapy. Int J Mol Sci 2019;20(24):6223.

[334] Spangler JB, et al. Insights into cytokine-receptor interactions from cytokine engineering. Annu Rev Immunol 2015;33:139–67.

[335] Waldmann TA. Cytokines in cancer immunotherapy. Cold Spring Harb Perspect Biol 2018;10(12): a028472.

[336] Sharma P, Allison JP. The future of immune checkpoint therapy. Science 2015;348(6230):56–61.

[337] Sun Z, et al. A next-generation tumor-targeting IL-2 preferentially promotes tumor-infiltrating CD8 (+) T-cell response and effective tumor control. Nat Commun 2019;10(1):3874.

[338] Scheller J, et al. Immunoreceptor engineering and synthetic cytokine signaling for therapeutics. Trends Immunol 2019;40(3):258–72.

CHAPTER THREE

Inflammation and cancer

Gerardo G. Mackenzie

Department of Nutrition, University of California at Davis, Davis, CA, United States

Contents

1. Introduction

Inflammation is a physiological process, which includes the activation, recruitment, and function of cell of innate and adaptive immunity. Inflammation has a critical role in host defense against pathogens, as well as in tissue repair, regeneration, and remodeling, and for the regulation of tissue homeostasis, which requires inflammatory signals [1]. Interestingly, the relationship between inflammation and carcinogenesis is not new. Galen (129–199 AD), an ancient Greek physician, had initially described the resemblance between cancer and inflammation, and the possibility that cancers might develop from inflammatory lesions [2]. Indeed, Galen used the word cancer to describe certain inflammatory breast tumors in which superficial veins appeared swollen and radiated somewhat like the claws of a crab (*karkinos*) [2]. However, it was Virchow in 1863 [3], who formally proposed a connection between chronic inflammation and cancer, following his observation of a "lymphoreticular infiltrate" in tumors and deduced them as a reflection of the origin of cancer at sites of inflammation. One hundred years later, it was Dvorak who described that inflammation and cancer share many basic mechanisms (angiogenesis) and tissue-infiltrating cell types (lymphocytes, macrophages, and mast cells) and that tumors act like "wounds that do not heal" [4].

Cancer Immunology and Immunotherapy
https://doi.org/10.1016/B978-0-12-823397-9.00003-X

During the past decades, there has been significant progress understanding the molecular/cellular pathways involved in inflammation as it relates to cancer and the role of innate and adaptive immune cells in facilitating or preventing tumor initiation, progression, and metastasis [5–7]. Moreover, the "tumor cell intrinsic" vision, which was a standard view several decades ago, has shifted to a more inclusive concept in which the cancer cell is placed within a network of stromal cells that are comprised of fibroblasts and vascular cells and inflammatory immune cells that all together form the tumor microenvironment. Indeed, the generation of an inflammatory stimulus has a great effect on the composition of the tumor microenvironment and particularly on the plasticity of, both, the tumor and stromal cells.

It is now clear that the immune system can have anti- and protumorigenic roles at all stages of the tumorigenesis [8–11]. The antitumorigenic role of the immune system is endogenous, *i.e.,* being exerted normally in response to transformed and tumor cells. In addition, a recent prominent development in the field of cancer treatment is the successful implementation of various cancer immunotherapies, which use various strategies to activate the immune system toward the recognition, confining, and killing of cancer cells [12–14]. In contrast, the protumorigenic inflammation promotes cancer growth by blocking antitumor immunity, influencing the tumor microenvironment toward a more tumor-permissive state and by exerting direct tumor-promoting signals and functions onto epithelial and cancer cells.

An extensive body of literature has implicated inflammation and tissue repair immune responses to enhanced tumor incidence, growth, and progression. This evidence includes large clinical studies on "non-specific" inhibition of inflammation with nonsteroidal antiinflammatory drugs (NSAIDs) such as aspirin, reducing incidence and mortality in many cancers [15, 16] or specific inhibition of cytokines such as interleukin 1β (IL-1β), which significantly reduces the risk of lung cancer development [17]. Also, organ- and site-specific chronic inflammation predisposes to cancer development at the same site [18]. Certainly, the immune system plays a critical role during tumor initiation, promotion, and progression, often referred to as "cancer-promoting inflammation." Although tumors might not be characterized by a prominent T-cell infiltration or their functional activation, these tumors might still upregulate inflammatory mediators and recruit other immune cells, *i.e.,* macrophages, monocytes, or neutrophils, which can exert tumor-promoting properties.

This chapter focuses primarily on the tumor-promoting role of immune and inflammatory responses in cancer. It compares tumor-promoting inflammation and "physiological" functions of inflammation in tissue regeneration and host defense. It discusses the potential roles and timing of inflammatory action in cancer, describes possible sources of inflammation-initiating stimuli, and outlines underlying mechanisms of how inflammation can promote cancer at different stages of tumor progression.

2. Similarities and differences between inflammatory responses in "Physiological" conditions and "Inflammation in Cancer"

Inflammatory responses are important mechanisms for the regeneration of normal tissue. The low-level "physiological" inflammatory response in the context of tissue homeostasis as well as the inflammation in altered tissues [19] functions to sense and react against potential insults such as stresses, tissue damage, infection, metabolic alterations, and other changes to homeostasis (Fig. 1). By responding to these insults, inflammation is able to restore homeostasis and to prevent loss of tissue function [1].

Local tissue-associated macrophages represent major tissue guardians. These macrophages are activated by tissue alterations and their functions are to clear dying apoptotic cells, generate chemotactic molecules for the recruitment of additional cell types, and modulate immune responses and barrier functions. These tissue inflammatory responses

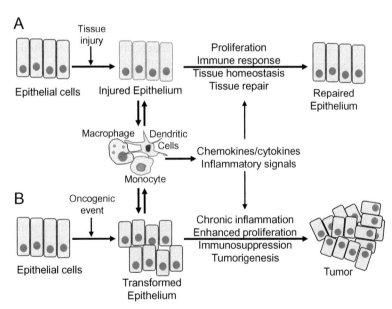

Fig. 1 Functional similarities and differences between inflammation during tissue injury and inflammation in cancer. (A) Inflammation during tissue injury: Signals to epithelial cells caused by injury or infection lead to activation of myeloid cells, which will produce inflammatory cytokines to activate innate and adaptive immunity to get rid of the insult, to activate epithelial cell proliferation and finally, it will repair the tissue injury. (B) Inflammation in cancer: In the event that a disturbance of epithelial cell homeostasis is initially caused by an oncogenic event, the immunity is unable to eliminate this early insult and the chronic inflammation can enhance proliferation driven primarily by cytokines and other inflammatory signals, and this aids tumor growth. Figure was adapted from [122].

are initiated and sustained by, at least, these key mechanisms. First, when the degree of the initiating insult itself is not strong, the local tissue macrophages and dendritic cells (DCs) can increase their numbers. As a response to robust perturbations of tissue homeostasis, immune cells from bone marrow (monocytes, neutrophils, and monocyte-derived cells) and secondary lymphoid tissues (lymphoid cells) will be recruited. Then, recruited or locally amplified inflammatory cells can further undergo local activation, differentiation, and polarization, instructed by the signals from the microenvironment [1, 20].

Besides the critical role played by inflammatory responses in resolving tissue insults, inflammation can also promote tumorigenesis. In the context of cancer, it is possible that the uncontrolled proliferation of epithelial cells may also induce a response that lead to the increase in the numbers of macrophages and fibroblasts, connected to the epithelial tissues as basic "tissue blocks." This is due to the existence of a cell-to-cell signaling communication through multiple cytokines, chemokines, and growth factors reciprocally produced by a certain cell and targeting another [21]. This mechanism is likely to be common between tissue amplification for organism growth, tissue repair after wounding or infection, and tumor growth. Indeed, simply put, epithelial and cancer cell growth trigger a process of acquiring more macrophages and fibroblasts in the tissue proximity. When the initial levels of stress, hypoxia, and other tumor-specific insults are low or moderate, the majority of the tumor-associated macrophages are recruited from local proliferation and migration of tissue macrophages [22, 23].

There are major differences between "physiological" inflammatory responses and "inflammation in cancer." In contrast to infections and wound healing that would clear after immune cell recruitment and epithelial cell proliferation, growing tumors have many signals that provide a positive-feeding loop of inflammation-induced signaling and inflammatory cell recruitment. These signals include oncogene-derived stress, cell-death signals, and microbial signals that can contribute by themselves or together to the persistent inflammatory signals (Fig. 1). Of note, during tumor growth and progression, the emergence of monocyte-derived cells, such as monocytes themselves and neutrophils, which lack their local precursors residing in the tissue, is triggered through the recruitment of hematopoietic-derived (spleen, blood, and bone marrow) cellular precursors.

3. Sources of inflammation during tumorigenesis

Multiple sources enhance inflammation in tumors. These include carcinogenic microbes, environmental pollutants (particles and smoke), and low-grade inflammation associated with obesity, among others (Fig. 2). Although the stimuli of inflammation in cancer might be different, it appears that the induction of inflammation is strongly linked with the appearance of factors absolutely needed for the oncogenic process-alterations in oncogenes and tumor suppressors, infections for microbial-induced cancers, or

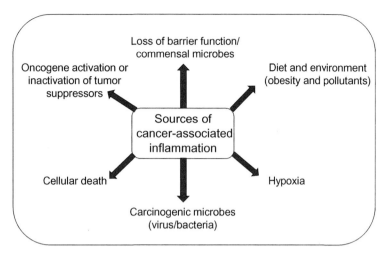

Fig. 2 Cell intrinsic, host dependent or environmental sources that may contribute to tumor-associated inflammation.

deterioration of barrier function because of transformation-induced loss of tissue organization. Because inflammation is generated in response to the loss of tissue homeostasis, the induction of inflammation is, in many circumstance/s, "pre-encoded" in genetic and transcriptional programs essential for oncogenic transformation.

One of the hallmarks of cancer is the loss of cell-intrinsic tumor suppressor functions. One of the most commonly mutated tumor suppressors is Tp53, encoding for p53 protein. p53 protein has multifaceted functions to regulate cellular homeostasis, and one of them is its transcriptional antagonism with nuclear factor-κB (NF-κB) [24, 25], a key regulator of inflammation. Given that NF-κB activating signals are always present within the tumor microenvironment, and even in a normal tissue, loss of functional p53 increases NF-κB-regulated inflammatory gene expression. Indeed, NF-kB is aberrantly expressed in many inflammation-dependent cancers. In colorectal cancer, this inflammatory signature contributes to tumor progression and metastasis [25–27]. Loss of tumor suppressors can also inhibit proper DNA repair and accelerate DNA damage, which can trigger DNA-damage-induced inflammatory pathways [28].

Besides the loss of tumor suppressor function, the activation of oncogenes is critical for tumor development. Of note, oncogene activation is mechanistically linked to the increased cytokines and chemokines production, followed by the recruitment of myeloid cells, which are responsible for either directly tumor promoting or immunosuppressive effects. For instance, enhanced Kras activation regulates the expression of CXCL3, a main chemokine for myeloid cell recruitment [29]. Furthermore, Kras activation enhances the release of cytokines and chemokines related to "senescence-associated secretory phenotype" [30], including IL-1α, IL-1β, CCL2, and CXCL1. In addition,

Kras cooperates with c-Myc in inducing CCL9, IL-23, and other inflammatory entities in pancreatic cancer [31]. Indeed, the mechanisms in which the activation of an oncogene leads to the production of an excessive amount of inflammatory cytokines and chemokines might be a unifying mechanism for how inflammation is triggered in many cancers.

Presence of certain chemokines, cytokines, and myeloid cell subsets correlates with poor prognosis in colon cancer [32, 33], and potentially other tumor types as well. Interestingly, in animal models, the inhibition of inflammatory responses in cancers that are "noninflammatory" reduces tumor growth and progression. Moreover, these "noninflammatory" cancers support tumor growth by recruiting immune cells and increasing expression of inflammatory mediators to their benefit, in a term commonly referred to as "tumor-elicited inflammation" [18, 34].

Given the worldwide obesity pandemic, chronic low-grade inflammation associated with obesity currently represents a major driver of protumorigenic inflammation. Indeed, low-grade chronic inflammation induced by obesity, type II diabetes, and excess body mass is mainly of systemic nature and, as a result, can promote or increase risk of many different cancers, including colon, pancreatic, breast, and liver, among others [35, 36]. Although the exact molecular mechanisms underlying the detrimental effects of obesity as a risk factor for developing cancer remain unknown, probable mediators include insulin, estrogens, and inflammatory molecules such as adipokines. For example, in pancreatic cancer, lipocalin 2, an adipokine, which is secreted by adipose tissue in obese subjects, has been linked to promoting the development of obesity-associated pancreatic cancer and stimulating a receptor-mediated proinflammatory response in the tumor microenvironment [37]. In addition, diet-induced obesity can act as a proinflammatory stimulus to enhance active Kras signaling. In this context, cyclooxygenase 2 appears to be critical in the inflammatory loop that connects inflammation, increased fibrosis, active Kras signaling, and pancreatic tumor progression in preclinical models of pancreatic cancer [38]. Another factor that has been suggested as a potential link between obesity and pancreatic cancer may involve the Receptor for AGE (RAGE). This factor, present in pancreatic cancer, maintains chronic inflammation pathways, activates Kras, and is involved in the Ras-induced inflammation feed-forward loop that is seen due to obesity-associated pancreatic cancer [37].

In the case of carcinogenic microbes, select pathogens such as human papilloma virus (HPV), *Helicobacter pylori*, hepatitis C virus (HCV), and hepatitis B virus (HBV) are known to drive tumorigenesis. In this scenario, the recognition of persistent pathogens will of course promote distinct inflammatory responses, via molecular pattern-recognition receptors, which recognize specific molecular patterns and activate innate inflammatory responses. For example, sensors such as Toll-like receptor 2 (TLR2) and TLR4, stimulator of interferon genes (STING), cyclic GMP-AMP synthase (cGAS), as well as multiple sensors associated with inflammasomes, sense oncogenic bacteria and viruses [39]. Besides this classical receptor pathway, an alternative emerging paradigm is that many cancers might be promoted by commensal microbiota, either by translocation and adherence of microbes to cancer cells or by the distant release of inflammation-activating microbial metabolites. These microbes and microbial products can even travel with tumors to the site of metastasis and serve as a source of inflammation in metastasis [40]. Pancreatic cancer is often associated with chronic pancreatitis, which might be sometimes associated with infection but even in mouse models is driven by microbial-induced Th17 responses [41]. Nonalcoholic steatohepatitis (NASH) and fibrosis, which underlie hepatocellular carcinoma (HCC) development are actively promoted by intestinal bacteria and their products [42, 43]. Moreover, the microbiota in the lungs might play a role in the induction of inflammation and tumorigenesis [44, 45]. Similarly, commensal microbiota in the intestine can also influence colon tumor growth. When a deterioration of the intestinal barrier occurs, the hyperproliferating cells fail to properly differentiate and form protective tight and adherens junctions and well-developed mucus layer, isolating immune compartment from bacteria [34]. Treatment with broad-spectrum antibiotics has been shown to reduce inflammation and tumor growth, even in animal models where potential pathogens are absent. Moreover, diet-induced changes in the microbiome promote tumor progression in the presence of activating Kras mutations [46]. In human colon cancer, several species of bacteria have been suggested to be preferentially associated with adenomas and carcinomas, including *Bacteroides fragilis* [47], *Fusobacterium nucleatum* [48], and some subspecies of *Escherichia coli* [49]. Given these species, it is conceivable to imagine that the tumor-promoting actions of bacteria and other microbes are, in part, mediated by the inflammation they generate.

In addition, multiple environmental factors can influence and promote cancer risk by inducing chronic inflammation. In this scenario, inflammation usually precedes and/or go hand-in-hand with the development of the tumor. For example, tobacco smoke and the inhalation of fine particulate matter would cause lung and airway inflammation and promote lung cancer [50, 51]. Moreover, smoking is also a well-known risk factor for chronic pancreatitis and pancreatic cancer. In smoking-induced progression of experimental chronic pancreatitis, there is a crosstalk between immune cells and pancreatic stellate cells, mediated, in part, through IL-22 signaling [52]. Other proposed mechanisms of smoking-related effects involve pathways such as IL-6 and histone deacetylases in immune and cancer cell interactions [53]. Therefore, these findings further stress the importance of immune-related signals and the complex multicellular interactions that take place in linking inflammation and cancer progression.

Even though the capacity of tumor cells to avoid cell death is a hallmark of cancer [54], it is essential to discriminate "who dies, when, and how." Remarkably, cell death is not only important in the setting of therapy-induced inflammation, but cell death of untransformed cells in close proximity to tumor niche might also be important for tumor growth [55]. This may be explained through clearance of the adjacent space and the induction of proliferation of tumor clones; other mechanisms clearly involve immuno-logical recognition of cell death and induction of protumorigenic immune responses. For instance, necroptotic pancreatic cancer cells induce CXCL1, which leads to the recruit-ment of myeloid cells, inhibition of antitumor T-cells responses and increases in tumor growth [56]. The signal that triggers cell death, which may be due to therapy, infection, hypoxia, or metabolic-induced stress in growing tumors, appears to be important for the consequences of cell death and inflammation. For example, the type of cell death appears to be critical, with necrosis and necroptosis being more potent inflammatory inducers, resulting in the release of damage-associated molecular patterns (DAMPs), whereas apoptosis and autophagy being less inflammatory.

Given that approximately 20% of all cancer cases are preceded by chronic inflamma-tion, or autoimmunity at the same tissue or organ site [18, 57], an important aspect is to understand how inflammation in cancer is initially induced and maintained. Understand-ing when, where, and the strength/duration of the inflammatory stimulus is critical to decipher the roles and the mechanisms of action of inflammation and cancer. In many of these cases, the inflammation that enhances cancer risk is generated earlier than the formation of the tumor. Some examples include: (1) inflammatory bowel disease (IBD), which increases the risk of colon cancer; (2) chronic hepatitis, which increases liver cancer risk; and (3) Helicobacter-induced gastritis, which increases the risk for stom-ach cancer [58]. Besides its role in early events, systemic inflammation can have an effect even during the late stages of tumor development. For example, inflammation induced by tobacco-smoke or obesity can activate neutrophils and their function of extracellular DNA trap formation that assists in breast cancer metastasis into the lungs [59]. In the fol-lowing sections, I discuss the role of inflammation in tumor initiation, progression, and metastasis, and some of the associated mechanisms.

4. Role of inflammation in tumor initiation

Two key interdependent events are essential for successful tumor initiation. The first one is the accumulation of mutations and/or epigenetic alterations in genes involved in oncogenic (activation) and tumor suppression (inactivation) pathways. Besides envi-ronmental factors (carcinogens and/or variable radiation), which can lead to intrinsic errors in DNA repair and replication, inflammatory responses harbor multiple mecha-nisms that can also lead to accumulation of mutation and various epigenetic changes in cells. For instance, macrophages and neutrophils are strong producers of reactive oxy-gen and nitrogen species (RONS), which induce mutations. Moreover, inflammatory cytokines such as IL-22 can induce expression of DNA damage response (DDR) genes to counteract possible genotoxic insult caused by inflammation [60]. Regarding epige-netic mechanisms, cytokines secreted by inflammatory cells (*i.e.,* IL-6, TNF-α, and

IL-1β) may activate epigenetic machinery, including components of DNA and histone modifications (Dnmt1, Dnmt3, and disruptor of telomeric silencing 1 [DOTL1]), micro-RNA (miRNA)- and long noncoding RNA (lncRNA)-modulating expression levels of oncogenes, and tumor suppressors [61]. Thus, the increase in inflammation can lead to increased mutagenesis, predisposing to accumulation of mutations in normal tissue. For example, chronic inflammation in the intestinal milieu has been shown to lead to mutations in Tp53 and other cancer-related genes in intestinal epithelial cells [62–66].

Given that stem cells are the proposed "cells of origin" for cancer initiation, inflammatory processes can trigger de-differentiation of postmitotic epithelia into tumor-initiating stem-like cells [67]. Another way that inflammation might contribute to tumor initiation is by inducing tissue damage, weakening barrier function, and bringing stem cells closer to environmental carcinogens or genotoxic compounds. Finally, in microbial-rich cancers, chronic inflammation can shape the qualitative characteristics of epithelial-adhesive microbiota, enriching the content of species harboring genotoxic gene products, such as colibactin in some strains of E. coli [49, 68] capable of inflicting mutations in host cells.

The second interdependent mechanism by which inflammation can meaningfully contribute to tumor initiation is by stimulating the growth of a newly transformed and/or malignant clone into an established tumor. For example, in transformed clones, pro-survival pathways, such as NF-κB, STAT3, and other types of signaling can be activated via cytokine receptor signaling [69–71], thereby increasing proliferation and survival probability of malignant cells. In particular, it is possible that these early inflammatory signaling pathways are critical for a successful tumor initiation, given that these initial malignant cells are unable to establish a full-scale tumor microenvironment, capable of feeding them sufficient tumor-supporting growth factors. Moreover, these inflammation-driven cell survival signals are also critical in the setting of cancer immunosurveillance [72, 73]. For example, signals activating STAT3 can guard epithelial cells from CD8 T cytotoxic cell attack [74, 75]. In addition, IFNγ signaling can upregulate expression of T-cell exhaustion, which can lead to the induction of molecule programmed death ligand 1 (PD-L1) on transformed epithelium, which is recognized by T-cells.

Similarly, inflammatory signals might decrease the expression of "stress ligands" on cancer cells, which is a critical mechanism for proper recognition, cell repair, and turnover in tissues, thus providing an opportunity for the development of the malignant clones [55, 76, 77]. This is particularly evident in skin and liver cancer, where inflammation-induced tissue injury and cell death appear to be required for tumor outgrowth, since the death of normal cells is essential for compensatory proliferation of neighboring transformed clones. In addition, the expression of inflammatory mediators as well as some of the immunological mechanisms of Crohn's disease and ulcerative colitis might be similar, both known to increase the risk for colon cancer [78]. Of note, while the simple presence of inflammatory cytokines is needed for tumor initiation, cytokines are also important in inflammation-induced tissue damage and repair.

Finally, it is important to stress the importance played by the tumor microenvironment in tumor initiation and expansion, since malignant clones with the exact same genetic alterations may have different propensities for outgrowth and proliferation, depending on the tumor microenvironment these clones are placed into [79–81]. While inflamed tissue might be favorable to stimulating tumor growth, normal, unaffected tissue might inhibit it. Likewise, the presence of enhanced inflammation, in part, through the ability of cytokines to promote survival and proliferation, can induce the outgrowth of dormant clones, both in the case of primary tumors and in metastasis.

5. Role of inflammation in tumor progression

Significant research has shed light on the role of inflammation in tumor progression. Understanding how inflammation drives tumor growth is critical since inhibition of inflammation can block tumor growth, may provide valuable information for early detection of cancers, and may inform us on how metastatic tumor proliferate once established. The progress in this area has been driven by multiple animal models of inflammation and cancer, which have aided in identifying most of the mechanisms linking inflammation to cancer. As in the case of tumor initiation, enhanced inflammation can serve as direct fuel for growing tumors. Moreover, inflammatory factors are required to shape cell plasticity within the tumor microenvironment, which further modulates tumor growth by several mechanisms.

A major signaling pathway linking inflammation in tumor promotion is the NF-κB pathway. For example, in a model of colitis-associated cancer, the inactivation of NF-κB activation in myeloid cells, led to a reduction in tumor growth [82]. Interestingly, in immune cells, NF-κB is a master regulator of cytokines, which promote survival and proliferation on epithelial in cancer cells. Moreover, NF-κB also regulates chemokine expression, which is essential for cell recruitment and reshaping of the tumor microenvironment. Indeed, NF-κB-dependent cytokines convene distinctive roles in tumor growth as their pharmacological inhibition reduce tumor growth, acting primarily via the modulation of additional oncogenic pathways, such as STAT3, extracellular signal-regulated kinase (ERK), and receptor Tyr kinases (RTK) [70, 83–86]. Moreover, inflammatory cytokines, such as IL-6, IL-17, and IL-11, trigger signals that lead to increases in cancer cell proliferation, especially in *in vivo* conditions, such as hypoxia and lack of nutrients.

Additional functions of the inflammatory cells and their mediators during tumor growth include the stimulation of angiogenesis, antagonizing potential antitumor immunity as well as the recruitment of fibroblasts and other stromal cells, which exert tumor-supporting functions. Finally, inflammatory signals can affect metabolic and mechanical properties of stromal cells in the tumor microenvironment by regulating formation and consistency of extracellular matrix and availability of growth factors, regulation of consumption, and the availability of key metabolites, including those involved in amino acid and redox metabolism [18]. Overall, these effects highlight the strong and diverse role of inflammation in tumor promotion.

6. Role of inflammation in tumor metastasis

Besides its role in tumor initiation and promotion, inflammation plays a major role in tumor metastasis. Indeed, understanding inflammatory mechanisms that mediate the process of metastasis is critical, since approximately 90% of cancer-related deaths are due to metastasis. Indeed, many clinical studies supporting the role of inflammation in cancer mortality underscore the critical role of inflammation in regulation of metastasis [16], rather than earlier stages of tumorigenesis, since metastasis is the main cause of cancer-related deaths. For example, long-term use of aspirin use can reduce overall colon cancer mortality and distant metastasis. Moreover, in experimental models, inhibiting inflammatory responses leads to poor colonization and eradicates micrometastases [87].

The process of metastasis initiates with the acquisition of an epithelial to mesenchymal transition (EMT) phenotype followed by the invasion of tumor cells into the neighboring tissues. Although the EMT cell transformation may be often temporary and partial [88], this process allows tumor cells to break through the basal membrane, reach the lymphatic system or blood vessels for further dissemination into neighboring tissue. Inflammation can influence cancer invasion, EMT, as well as cell migration at multiple levels. For example, select cytokines, such as TNF and IL-1β can directly modulate the expression of EMT-inducing transcription factors Twist and Slug [89, 90]. Furthermore, the recruitment of TGFβ-driven fibroblasts is dependent on IL-11, and this recruitment enhances tumor invasion and immune escape in colon cancer [91]. Likewise, in breast cancer, IL-11 within the tumor microenvironment drives clonal selection of the most invasive and malignant cancer cells [92]. In addition, an increase in myeloid cells in close proximity to the tumor cells leads to the synthesis of matrix metalloproteinases (MMPs), which remodel the extracellular matrix and enable cell migration [93, 94]. Besides producing MMPs, these myeloid cells, denoted as myeloid-derived suppressor cells, contribute to the suppression of antitumor responses [95, 96]. Of note, a patients' better prognosis has been shown to correlate with an increased accumulation of cytotoxic T and natural killer cells at the invasive margin of the primary tumor, along with decreased presence of myeloid cells [32, 97]. However, it remains unclear if myeloid-induced immunosuppression is essential at the invasion stage or at preventing metastatic outgrowth. It is possible that macrophages may play many roles in immunosuppression, and damage of extracellular matrix as the invasion continues. Moreover, increased production of inflammatory entities causing enhanced recruitment of myeloid cell into the invading tumors might support tumor outgrowth via a variety of mechanisms that are evolutionarily similar to those observed in organogenesis.

Cancer stem cells are critical for tumor metastasis and resistance to therapy. In particular, in contrast to "bulk" cancerous cells, cancer stem cells are more effective in their capacity to serve as metastatic seeds [98]. It is noteworthy that cancer stem cells are also transcriptionally and functionally similar to mesenchymal cells than "bulk" tumor cells. This highlights their location in the niche rich in mesenchymal growth and differentiation factors [99]. However, it is noteworthy that the amount and proportion of cancer stem cells in tumors are variable. Interestingly, inflammatory signals through NF-κB and STAT3 can drive cancer stemness, increase the percentage of cancer stem cells among the tumor cell population and thus raise the invasive potential [67, 100]. Similar effects can originate from mesenchymal-cancer interactions [101, 102], and inflammation can also regulate mesenchymal and stromal cells.

Given that metastasis usually occurs through the lymphatic system or blood stream, the processes of intravasation and extravasation are essential. These processes, mediated by adhesion molecules and integrins, permit cell-cell interaction, adhesion, and movement. Inflammatory cytokines act as strong inducers of integrins, selectins, and adhesion molecules including intercellular adhesion molecule 1 (ICAM-1) and vascular cell adhesion molecule 1 (VCAM-1). Interestingly, the expression of tissue-specific adhesion molecules may also determine the organ-specific tropism of metastasis and therefore, it is conceivable that specific inflammatory stimulus can potentially have an effect to which distant organ, a particular tumor seed it will mainly spread to. Furthermore, inflammatory cells, such as neutrophils and monocytes, can also help in the process of adhesion and extravasation [103], which might assist in the adhesion and translocation of metastatic seeds though the blood vessel wall, and establishing and maintaining the metastatic niche [104–106]. The interactions between the metastatic tumor cells themselves, or monocyte-cancer cells, neutrophil-cancer cells also aid to defend the metastatic seeds from the immunosurveillance when the metastasis is being establish, which is a vulnerable moment away from established immunosuppressive microenvironment of the primary tumor.

Finally, it is important to note that select inflammatory stimuli, ranging from those induced by obesity or tumor-specific inflammatory signals that enhance these cell-cell interactions, increase the rate of metastasis. For example, IL-17-dependent activation of neutrophils drives breast cancer metastasis [107], whereas lipopolysaccharide (LPS) enhances lung colonization [108]. Furthermore, inflammation induced by obesity, tobacco smoke, or microbial compounds can also increase lung metastatic burden by activating neutrophils and increasing neutrophil-cancer cell interactions [59]. Of note, the process of metastatic growth from smaller nodules to larger ones might also be promoted by inflammatory cytokines and growth factors produced by immune cells, similar to what occurs in primary tumors [109]. However, it is still unclear if there are differences among the different tumor types, if the need of inflammatory signals and growth factors is similar for the growth of the primary tumor and/or for the metastatic growth at different secondary organ sites.

7. Inflammation elicited due to cancer therapy

Inflammation that develops following therapy might play a critical role in the determination of therapy efficacy or resistance to therapy. For this reason, understanding the exact signals that induce inflammation during tumor treatment will undoubtedly assist to fill in a broader picture on tumor evolution and potentially treatment strategies. Although it may not be present originally in intact tumors, therapy-associated inflammation may develop in response to various anticancer therapies, including chemo- and radiotherapy, or immune infiltration caused by immunotherapies. Indeed, the activation of the immune system in the tumor microenvironment upon treatment is the cornerstone for current immunotherapies [13]. This process certainly can be beneficial for stimulation of antitumor immune responses and assist in developing combination with standard chemotherapies.

In certain cases, the release of DAMPs from dying tumor cells can induce the making of IL-1α and other immunostimulatory cytokines. The release of DAMPs together with the increase in tumor neo-antigens might stimulate de novo antitumor T-cell responses [110] or possibly lead to immunosuppression [111]. However, the overall outcome may likely be tumor-specific and vary across tumor types or even among individual tumors [112]. Moreover, this might also be influenced by the cytotoxic regimens or radiotherapies being administered, which will affect activation and function of the cells of anticancer immunity.

Furthermore, a significant number of tumors present deficiencies in mechanisms driving apoptosis and, hence, cell death by necrosis might be more immunostimulatory [113, 114]. For instance, necroptotic cell death of tumor cells may induce local antitumor immunity [115]. Importantly, in many instances, chemo- or radiotherapies can release dead-cell material from the tumors that may have immunosuppressive effects [111]. These therapies might stimulate an inflammatory response similar to an injury to normal tissue, followed by wound healing and tissue repair. In a cancer scenario, the generation of dying tumor cells would stimulate the production of cytokines and growth factors (tumor necrosis factor (TNF), epidermal growth factor, IL-6, and others) by the cells of the tumor microenvironment (such as myeloid cells, macrophages, and fibroblasts) and additional recruitment of these cells. In most cases, these growth factors tend to limit the efficiency of the therapy being used, since they generally serve as cell-extrinsic anticell-death signals. For instance, secretion of TNF by macrophages or fibroblasts is a key factor of therapy resistance in cancer [116].

Cytokines such as IL-22, IL-11, and IL-6 can also play a role modulating the phenotype of cancer stem cells. By turning down the metabolic and proliferative machinery, cancer stem cells are less sensitive to chemotherapy and radiotherapy. Furthermore, resistance of pancreatic cancers to gemcitabine is observed in tumors with enhanced myeloid cells recruitment, since myeloid cells in the microenvironment can secrete pyrimidine nucleotides that confer resistance to gemcitabine treatment [117], highlighting a metabolic role of inflammatory cells in chemotherapy resistance. Moreover, cytokines, such as IL-17, have been shown to provide colon cancer cells resistance to anticolon cancer therapy with 5-fluorouracil (5-FU) [118] being inflammatory signaling targeting remaining tumor cells, a critical driver of chemotherapy resistance [119, 120]. Finally, chemotherapies also induce tissue damage in normal tissues, leading to inflammation. For example, at the level of the intestine, damage of the normal epithelium can translate to an increase in inflammatory microbial products, and activation of systemic inflammation, which can subsequently induce tumorigenesis [121], and/or accelerate metastases [59].

8. Conclusions

The "physiological" inflammatory mechanisms that mediate immunity against infection and promote tissue homeostasis can also be protumorigenic, and tumors use these mechanisms toward their benefit. The induction of inflammation in the tumor as well as in the tumor microenvironment follows distinct timing. It may occur prior to tumor initiation, during tumor initiation, or it may happen only during tumor progression or metastasis (later stages of tumorigenesis). Thus, the contribution of protumorigenic inflammation may occur very early or emerge at late stages of metastasis or therapy resistance, and this may depend on types of tumors and/or the stimuli generating the inflammation. Regarding this last point, multiple distinct stimuli are known to enhance inflammation in tumors, with many of them being potentially targetable. For example, carcinogenic microbes, environmental pollutants (particles and smoke), and low-grade inflammation associated with obesity, are key inflammatory stimuli and represent targets for cancer prevention, through the reduction of tumor-initiating inflammation or by limiting or blocking the original stimulus. Indeed, successful cancer prevention that targets protumorigenic inflammation can be attained through dietary interventions, select antibiotics usage, vaccinations, and better environmental protection.

Even though we have made progress in the understanding of the molecular and cellular mechanisms for inflammation and immune cells in tumor progression, additional research is still needed to decipher the roles of inflammation in tumor initiation and metastasis, in which faithful *in vivo* models have only recently been available. Moreover, although many animal models correctly represent the tumor microenvironment observed in human patients [11, 12], it is important to consider additional challenges to fully translate animal models findings into human cancers. For example, human cancers present an additional degree of complexity, including a vast genetic variability, different dietary habits, history of environmental exposures, and exposure to commensal and pathogenic microbe and virus composition. In this context, the development and implementation of "omics" techniques in the last decade has increased our understanding of transcriptional and other signaling programs governing cellular plasticity within the tumor microenvironment. This will further aid the investigations on the intricate crosstalk and functional diversification of multiple cellular types that surround and interact with the tumor cells. Using these newer technologies will allow the creation of new fate mapping and the identification of targets common for many types of cancer or, alternatively, rapidly amenable for precision therapies based on robust molecular and genetic analyses of the primary and metastatic tumors of individual patients.

In summary, inflammation plays a distinct role during tumor initiation, promotion, and progression, which is often referred as "cancer-promoting inflammation." We have made great progress in understanding several basic principles and mechanisms of how inflammation promotes cancer. However, the new discoveries in the field using an array of modern techniques will likely add additional clarity to several additional mechanistic insights governing molecular and cellular mechanisms of tumor-promoting inflammation. As the ways we treat cancer continue to increase in complexity, *i.e.,* immunotherapies, new combination therapies as well as vaccines, it will be imperative to uncover the role of immune and inflammatory pathways in therapy resistance, which will likely aid in our understanding of a patient's response or resistance to therapy.

Acknowledgments

This work was supported by funds from the University of California, Davis and NIFA-USDA (CA-D-NUT-2397-H) to G.G.M.

Conflict of interest disclosure

The author declares no conflict of interest.

References

[1] Medzhitov R. Origin and physiological roles of inflammation. Nature 2008;454(7203):428–35.

[2] Reedy J. Galen on cancer and related diseases. Clio Med 1975;10(3):227–38.

[3] Balkwill F, Mantovani A. Inflammation and cancer: back to Virchow? Lancet 2001;357 (9255):539–45.

[4] Dvorak HF. Tumors: wounds that do not heal. Similarities between tumor stroma generation and wound healing. N Engl J Med 1986;315(26):1650–9.

[5] Colotta F, et al. Cancer-related inflammation, the seventh hallmark of cancer: links to genetic instability. Carcinogenesis 2009;30(7):1073–81.

[6] Coussens LM, Werb Z. Inflammation and cancer. Nature 2002;420(6917):860–7.

[7] Schetter AJ, Heegaard NH, Harris CC. Inflammation and cancer: interweaving microRNA, free radical, cytokine and p53 pathways. Carcinogenesis 2010;31(1):37–49.

[8] Koebel CM, et al. Adaptive immunity maintains occult cancer in an equilibrium state. Nature 2007;450(7171):903–7.

[9] McGranahan N, Swanton C. Cancer evolution constrained by the immune microenvironment. Cell 2017;170(5):825–7.

[10] Rosenthal R, et al. Neoantigen-directed immune escape in lung cancer evolution. Nature 2019;567 (7749):479–85.

[11] Zilionis R, et al. Single-cell transcriptomics of human and mouse lung cancers reveals conserved myeloid populations across individuals and species. Immunity 2019;50(5):1317–1334.e10.

[12] Binnewies M, et al. Understanding the tumor immune microenvironment (TIME) for effective therapy. Nat Med 2018;24(5):541–50.

[13] Sharma P, Allison JP. The future of immune checkpoint therapy. Science 2015;348(6230):56–61.

[14] Topalian SL, Drake CG, Pardoll DM. Immune checkpoint blockade: a common denominator approach to cancer therapy. Cancer Cell 2015;27(4):450–61.

[15] Rothwell PM, et al. Effect of daily aspirin on long-term risk of death due to cancer: analysis of individual patient data from randomised trials. Lancet 2011;377(9759):31–41.

[16] Rothwell PM, et al. Effect of daily aspirin on risk of cancer metastasis: a study of incident cancers during randomised controlled trials. Lancet 2012;379(9826):1591–601.

[17] Ridker PM, et al. Effect of interleukin-1β inhibition with canakinumab on incident lung cancer in patients with atherosclerosis: exploratory results from a randomised, double-blind, placebo-controlled trial. Lancet 2017;390(10105):1833–42.

[18] Grivennikov SI, Greten FR, Karin M. Immunity, inflammation, and cancer. Cell 2010;140 (6):883–99.

[19] Aran D, et al. Widespread parainflammation in human cancer. Genome Biol 2016;17(1):145.

[20] Okabe Y, Medzhitov R. Tissue biology perspective on macrophages. Nat Immunol 2016;17(1):9–17.

[21] Zhou X, et al. Circuit design features of a stable two-cell system. Cell 2018;172(4):744–757.e17.

[22] Loyher PL, et al. Macrophages of distinct origins contribute to tumor development in the lung. J Exp Med 2018;215(10):2536–53.

[23] Zhu Y, et al. Tissue-resident macrophages in pancreatic ductal adenocarcinoma originate from embryonic hematopoiesis and promote tumor progression. Immunity 2017;47(2):323–338.e6.

[24] Komarova EA, et al. p53 is a suppressor of inflammatory response in mice. FASEB J 2005;19 (8):1030–2.

[25] Schwitalla S, et al. Loss of p53 in enterocytes generates an inflammatory microenvironment enabling invasion and lymph node metastasis of carcinogen-induced colorectal tumors. Cancer Cell 2013;23 (1):93–106.

[26] Elyada E, et al. CKIα ablation highlights a critical role for p53 in invasiveness control. Nature 2011;470(7334):409–13.

[27] Pribluda A, et al. A senescence-inflammatory switch from cancer-inhibitory to cancer-promoting mechanism. Cancer Cell 2013;24(2):242–56.

[28] Andriani GA, et al. Whole chromosome instability induces senescence and promotes SASP. Sci Rep 2016;6:35218.

[29] Liao W, et al. KRAS-IRF2 axis drives immune suppression and immune therapy resistance in colorectal cancer. Cancer Cell 2019;35(4):559–572.e7.

[30] Davalos AR, et al. Senescent cells as a source of inflammatory factors for tumor progression. Cancer Metastasis Rev 2010;29(2):273–83.

[31] Kortlever RM, et al. Myc cooperates with Ras by programming inflammation and immune suppression. Cell 2017;171(6):1301–1315.e14.

[32] Mlecnik B, et al. Integrative analyses of colorectal cancer show immunoscore is a stronger predictor of patient survival than microsatellite instability. Immunity 2016;44(3):698–711.

[33] Tosolini M, et al. Clinical impact of different classes of infiltrating T cytotoxic and helper cells (Th1, Th2, Treg, Th17) in patients with colorectal cancer. Cancer Res 2011;71(4):1263–71.

[34] Grivennikov SI, et al. Adenoma-linked barrier defects and microbial products drive IL-23/IL-17-mediated tumour growth. Nature 2012;491(7423):254–8.

[35] Quail DF, Dannenberg AJ. The obese adipose tissue microenvironment in cancer development and progression. Nat Rev Endocrinol 2019;15(3):139–54.

[36] Quail DF, et al. Obesity alters the lung myeloid cell landscape to enhance breast cancer metastasis through IL5 and GM-CSF. Nat Cell Biol 2017;19(8):974–87.

[37] Gomez-Chou SB, et al. Lipocalin-2 promotes pancreatic ductal adenocarcinoma by regulating inflammation in the tumor microenvironment. Cancer Res 2017;77(10):2647–60.

[38] Philip B, et al. A high-fat diet activates oncogenic Kras and COX2 to induce development of pancreatic ductal adenocarcinoma in mice. Gastroenterology 2013;145(6):1449–58.

[39] Woo SR, Corrales L, Gajewski TF. Innate immune recognition of cancer. Annu Rev Immunol 2015;33:445–74.

[40] Bullman S, et al. Analysis of Fusobacterium persistence and antibiotic response in colorectal cancer. Science 2017;358(6369):1443–8.

[41] McAllister F, et al. Oncogenic Kras activates a hematopoietic-to-epithelial IL-17 signaling axis in preinvasive pancreatic neoplasia. Cancer Cell 2014;25(5):621–37.

[42] Dapito DH, et al. Promotion of hepatocellular carcinoma by the intestinal microbiota and TLR4. Cancer Cell 2012;21(4):504–16.

[43] Shalapour S, et al. Inflammation-induced IgA+ cells dismantle anti-liver cancer immunity. Nature 2017;551(7680):340–5.

[44] Greathouse KL, et al. Interaction between the microbiome and TP53 in human lung cancer. Genome Biol 2018;19(1):123.

[45] Jin C, et al. Commensal microbiota promote lung cancer development via γδ T cells. Cell 2019;176 (5):998–1013.e16.

[46] Schulz MD, et al. High-fat-diet-mediated dysbiosis promotes intestinal carcinogenesis independently of obesity. Nature 2014;514(7523):508–12.

[47] Wu S, et al. A human colonic commensal promotes colon tumorigenesis via activation of T helper type 17 T cell responses. Nat Med 2009;15(9):1016–22.

[48] Kostic AD, et al. *Fusobacterium nucleatum* potentiates intestinal tumorigenesis and modulates the tumor-immune microenvironment. Cell Host Microbe 2013;14(2):207–15.

[49] Arthur JC, et al. Intestinal inflammation targets cancer-inducing activity of the microbiota. Science 2012;338(6103):120–3.

[50] Kadariya Y, et al. Inflammation-related IL1β/IL1R signaling promotes the development of asbestos-induced malignant mesothelioma. Cancer Prev Res (Phila) 2016;9(5):406–14.

[51] Takahashi H, et al. Tobacco smoke promotes lung tumorigenesis by triggering IKKbeta- and JNK1-dependent inflammation. Cancer Cell 2010;17(1):89–97.

[52] Xue J, et al. Aryl hydrocarbon receptor ligands in cigarette smoke induce production of interleukin-22 to promote pancreatic fibrosis in models of chronic pancreatitis. Gastroenterology 2016;151 (6):1206–17.

[53] Edderkaoui M, et al. HDAC3 mediates smoking-induced pancreatic cancer. Oncotarget 2016;7 (7):7747–60.

[54] Hanahan D, Weinberg RA. Hallmarks of cancer: the next generation. Cell 2011;144(5):646–74.

[55] Kuraishy A, Karin M, Grivennikov SI. Tumor promotion via injury- and death-induced inflammation. Immunity 2011;35(4):467–77.

[56] Seifert L, et al. The necrosome promotes pancreatic oncogenesis via CXCL1 and Mincle-induced immune suppression. Nature 2016;532(7598):245–9.

[57] Mantovani A, et al. Cancer-related inflammation. Nature 2008;454(7203):436–44.

[58] Trinchieri G. Cancer and inflammation: an old intuition with rapidly evolving new concepts. Annu Rev Immunol 2012;30:677–706.

[59] Albrengues J, et al. Neutrophil extracellular traps produced during inflammation awaken dormant cancer cells in mice. Science 2018;361(6409), eaao4227.

[60] Gronke K, et al. Interleukin-22 protects intestinal stem cells against genotoxic stress. Nature 2019;566 (7743):249–53.

[61] Grivennikov SI. Inflammation and colorectal cancer: colitis-associated neoplasia. Semin Immunopathol 2013;35(2):229–44.

[62] Canli Ö, et al. Myeloid cell-derived reactive oxygen species induce epithelial mutagenesis. Cancer Cell 2017;32(6):869–883.e5.

[63] Chang WC, et al. Loss of p53 enhances the induction of colitis-associated neoplasia by dextran sulfate sodium. Carcinogenesis 2007;28(11):2375–81.

[64] Hussain SP, Hofseth LJ, Harris CC. Radical causes of cancer. Nat Rev Cancer 2003;3(4):276–85.

[65] Robles AI, et al. Whole-exome sequencing analyses of inflammatory bowel disease-associated colorectal cancers. Gastroenterology 2016;150(4):931–43.

[66] Meira LB, et al. DNA damage induced by chronic inflammation contributes to colon carcinogenesis in mice. J Clin Invest 2008;118(7):2516–25.

[67] Schwitalla S, et al. Intestinal tumorigenesis initiated by dedifferentiation and acquisition of stem-cell-like properties. Cell 2013;152(1–2):25–38.

[68] Wilson MR, et al. The human gut bacterial genotoxin colibactin alkylates DNA. Science 2019;363 (6428):eaar7785.

[69] Dmitrieva-Posocco O, et al. Cell-type-specific responses to interleukin-1 control microbial invasion and tumor-elicited inflammation in colorectal cancer. Immunity 2019;50(1). 166–80.e7.

[70] Grivennikov S, et al. IL-6 and Stat3 are required for survival of intestinal epithelial cells and development of colitis-associated cancer. Cancer Cell 2009;15(2):103–13.

[71] Karin M, Greten FR. NF-kappaB: linking inflammation and immunity to cancer development and progression. Nat Rev Immunol 2005;5(10):749–59.

[72] Dunn GP, Old LJ, Schreiber RD. The three Es of cancer immunoediting. Annu Rev Immunol 2004;22:329–60.

[73] Schreiber RD, Old LJ, Smyth MJ. Cancer immunoediting: integrating immunity's roles in cancer suppression and promotion. Science 2011;331(6024):1565–70.

[74] Yu H, Pardoll D, Jove R. STATs in cancer inflammation and immunity: a leading role for STAT3. Nat Rev Cancer 2009;9(11):798–809.

[75] Ziegler PK, et al. Mitophagy in intestinal epithelial cells triggers adaptive immunity during tumorigenesis. Cell 2018;174(1):88–101.e16.

[76] Iannello A, Raulet DH. Immunosurveillance of senescent cancer cells by natural killer cells. Oncoimmunology 2014;3(1), e27616.

[77] Lam AR, et al. RAE1 ligands for the NKG2D receptor are regulated by STING-dependent DNA sensor pathways in lymphoma. Cancer Res 2014;74(8):2193–203.

[78] Ullman TA, Itzkowitz SH. Intestinal inflammation and cancer. Gastroenterology 2011;140 (6):1807–16.

[79] DeGregori J. Connecting cancer to its causes requires incorporation of effects on tissue microenvironments. Cancer Res 2017;77(22):6065–8.

[80] Henry CJ, et al. Aging-associated inflammation promotes selection for adaptive oncogenic events in B cell progenitors. J Clin Invest 2015;125(12):4666–80.

[81] Rozhok A, DeGregori J. A generalized theory of age-dependent carcinogenesis. eLife 2019;8.

[82] Greten FR, et al. IKKbeta links inflammation and tumorigenesis in a mouse model of colitis-associated cancer. Cell 2004;118(3):285–96.

[83] Becker C, et al. TGF-beta suppresses tumor progression in colon cancer by inhibition of IL-6 trans-signaling. Immunity 2004;21(4):491–501.

[84] Huber S, et al. IL-22BP is regulated by the inflammasome and modulates tumorigenesis in the intestine. Nature 2012;491(7423):259–63.

[85] Popivanova BK, et al. Blocking TNF-alpha in mice reduces colorectal carcinogenesis associated with chronic colitis. J Clin Invest 2008;118(2):560–70.

[86] Putoczki TL, et al. Interleukin-11 is the dominant IL-6 family cytokine during gastrointestinal tumorigenesis and can be targeted therapeutically. Cancer Cell 2013;24(2):257–71.

[87] Panigrahy D, et al. Preoperative stimulation of resolution and inflammation blockade eradicates micrometastases. J Clin Invest 2019;129(7):2964–79.

[88] Varga J, Greten FR. Cell plasticity in epithelial homeostasis and tumorigenesis. Nat Cell Biol 2017;19 (10):1133–41.

[89] Francart ME, et al. Epithelial-mesenchymal plasticity and circulating tumor cells: travel companions to metastases. Dev Dyn 2018;247(3):432–50.

[90] Suarez-Carmona M, et al. EMT and inflammation: inseparable actors of cancer progression. Mol Oncol 2017;11(7):805–23.

[91] Calon A, et al. Dependency of colorectal cancer on a TGF-β-driven program in stromal cells for metastasis initiation. Cancer Cell 2012;22(5):571–84.

[92] Marusyk A, et al. Non-cell-autonomous driving of tumour growth supports sub-clonal heterogeneity. Nature 2014;514(7520):54–8.

[93] Akkari L, et al. Distinct functions of macrophage-derived and cancer cell-derived cathepsin Z combine to promote tumor malignancy via interactions with the extracellular matrix. Genes Dev 2014;28(19):2134–50.

[94] Sevenich L, et al. Analysis of tumour- and stroma-supplied proteolytic networks reveals a brain-metastasis-promoting role for cathepsin S. Nat Cell Biol 2014;16(9):876–88.

[95] Veglia F, Perego M, Gabrilovich D. Myeloid-derived suppressor cells coming of age. Nat Immunol 2018;19(2):108–19.

[96] Yang L, et al. Abrogation of TGF beta signaling in mammary carcinomas recruits Gr-1+CD11b+ myeloid cells that promote metastasis. Cancer Cell 2008;13(1):23–35.

[97] Bindea G, et al. Spatiotemporal dynamics of intratumoral immune cells reveal the immune landscape in human cancer. Immunity 2013;39(4):782–95.

[98] de Sousa e Melo F, et al. A distinct role for Lgr5(+) stem cells in primary and metastatic colon cancer. Nature 2017;543(7647):676–80.

[99] Dominguez C, David JM, Palena C. Epithelial-mesenchymal transition and inflammation at the site of the primary tumor. Semin Cancer Biol 2017;47:177–84.

[100] Kryczek I, et al. IL-22(+)CD4(+) T cells promote colorectal cancer stemness via STAT3 transcription factor activation and induction of the methyltransferase DOT1L. Immunity 2014;40(5):772–84.

[101] Del Pozo Martin Y, et al. Mesenchymal cancer cell-stroma crosstalk promotes niche activation, epithelial reversion, and metastatic colonization. Cell Rep 2015;13(11):2456–69.

[102] Malanchi I, et al. Interactions between cancer stem cells and their niche govern metastatic colonization. Nature 2011;481(7379):85–9.

[103] Kersten K, et al. Mammary tumor-derived CCL2 enhances pro-metastatic systemic inflammation through upregulation of IL1β in tumor-associated macrophages. Oncoimmunology 2017;6(8), e1334744.

[104] Aceto N, et al. Circulating tumor cell clusters are oligoclonal precursors of breast cancer metastasis. Cell 2014;158(5):1110–22.

[105] Szczerba BM, et al. Neutrophils escort circulating tumour cells to enable cell cycle progression. Nature 2019;566(7745):553–7.

[106] Wolf MJ, et al. Endothelial CCR2 signaling induced by colon carcinoma cells enables extravasation via the JAK2-Stat5 and p38MAPK pathway. Cancer Cell 2012;22(1):91–105.

[107] Coffelt SB, et al. IL-17-producing γδ T cells and neutrophils conspire to promote breast cancer metastasis. Nature 2015;522(7556):345–8.

[108] Luo JL, et al. Inhibition of NF-kappaB in cancer cells converts inflammation-induced tumor growth mediated by TNFalpha to TRAIL-mediated tumor regression. Cancer Cell 2004;6(3):297–305.

[109] Krall JA, et al. The systemic response to surgery triggers the outgrowth of distant immune-controlled tumors in mouse models of dormancy. Sci Transl Med 2018;10(436):eaan3464.

[110] Ghiringhelli F, et al. Activation of the NLRP3 inflammasome in dendritic cells induces IL-1beta-dependent adaptive immunity against tumors. Nat Med 2009;15(10):1170–8.

[111] Hou J, Greten TF, Xia Q. Immunosuppressive cell death in cancer. Nat Rev Immunol 2017;17 (6):401.

[112] Ciampricotti M, et al. Chemotherapy response of spontaneous mammary tumors is independent of the adaptive immune system. Nat Med 2012;18(3):344–6 [author reply 346].

[113] Galluzzi L, et al. Molecular mechanisms of cell death: recommendations of the Nomenclature Committee on Cell Death 2018. Cell Death Differ 2018;25(3):486–541.

[114] Weinlich R, et al. Necroptosis in development, inflammation and disease. Nat Rev Mol Cell Biol 2017;18(2):127–36.

[115] Snyder AG, et al. Intratumoral activation of the necroptotic pathway components RIPK1 and RIPK3 potentiates antitumor immunity. Sci Immunol 2019;4(36):eaaw2004.

[116] Srivatsa S, et al. EGFR in tumor-associated myeloid cells promotes development of colorectal cancer in mice and associates with outcomes of patients. Gastroenterology 2017;153(1):178–190.e10.

[117] Halbrook CJ, et al. Macrophage-released pyrimidines inhibit gemcitabine therapy in pancreatic cancer. Cell Metab 2019;29(6):1390–1399.e6.

[118] Wang K, et al. Interleukin-17 receptor a signaling in transformed enterocytes promotes early colorectal tumorigenesis. Immunity 2014;41(6):1052–63.

[119] Jinushi M, et al. Tumor-associated macrophages regulate tumorigenicity and anticancer drug responses of cancer stem/initiating cells. Proc Natl Acad Sci USA 2011;108(30):12425–30.

[120] Malesci A, et al. Tumor-associated macrophages and response to 5-fluorouracil adjuvant therapy in stage III colorectal cancer. Oncoimmunology 2017;6(12), e1342918.

[121] Meisel M, et al. Microbial signals drive pre-leukaemic myeloproliferation in a Tet2-deficient host. Nature 2018;557(7706):580–4.

[122] Greten FR, Grivennikov SI. Inflammation and Cancer: Triggers, Mechanisms and Consequences. Immunity 2019;51(1):27–41.

CHAPTER FOUR

Novel immunotherapeutic approaches to cancer: Voltage-gated sodium channel expression in immune cells and tumors

Mustafa B.A. Djamgoz[a,b] and Laetitia Firmenich[a,c]
[a]Imperial College London, Department of Life Sciences, South Kensington Campus, London, United Kingdom
[b]Biotechnology Research Centre, Cyprus International University, North Cyprus, Mersin, Turkey
[c]University College London, Faculty of Medical Sciences, London, United Kingdom

Contents

Abbreviations

Ab	antibody
ADC	antibody–drug conjugate
AutoAb	autoantibody

Cancer Immunology and Immunotherapy
https://doi.org/10.1016/B978-0-12-823397-9.00004-1

CAR	chimeric antigen receptor
CRAC	calcium release-activated channel
CSC	cancer stem cell
CTL	cytotoxic T-lymphocyte
DP	double positive
ECM	extracellular matrix
EMT	epithelial-mesenchymal transition
ENaC	epithelial sodium channel
ERK	extracellular signal-regulated kinase
HIF-1α	hypoxia-inducible factor 1-alpha
HLA	human leukocyte antigen
IgG	immunoglobulin
iPSC	induced pluripotent stem cell
K_{Ca}	calcium-activated potassium channel
Kv	voltage-gated potassium channel
LES	Lambert-Eaton syndrome
mAb	monoclonal antibody
MHC	major histocompatibility complex
miR	microRNA (subtypes exist)
Nav	voltage-gated sodium channel (subtypes exist)
NHE1	sodium-hydrogen exchanger 1
NKC	natural killer cell
nNav1.5	neonatal voltage-gated sodium channel (splice variant)
PBMC	peripheral blood mononuclear cell
PD-1	programmed cell-death protein 1
PND	paraneoplastic disorder
SCLC	small-cell lung cancer
TAM	tumor-associated macrophage
TCR	T-cell receptor
TIL	tumor infiltrating lymphocyte
TME	tumor microenvironment
TTX	tetrodotoxin
VGSC	voltage-gated sodium channel
VGSCα	voltage-gated sodium channel alpha subunit

1. Introduction

Cancer is a major health issue and will continue to be a problem for the foreseeable future, current incidence rates being expected to increase by some 70% in the next two decades [1]. Unfortunately, several problems remain in the clinical management of cancers including limited functional diagnosis, undesirable side effects of available therapies, and frequent onset of drug resistance. In fact, some current treatments (such as chemotherapy) can in the long term even make cancer worse (*e.g.,* Refs. [2–4]). One of the most exciting recent developments in cancer treatment is immunotherapy. Although this field has its roots in the 19th century, it has only relatively recently become a clinical reality by the identification of specific mechanisms of immune regulation of cancer and development of appropriate drugs, including immune checkpoint inhibitors (*e.g.,* Refs. [5, 6]). However, even this modality suffers from various shortcomings such as off-target effects, severe adverse reactions, development of autoimmunity, highly personalized nature of the treatments, and, ultimately, cost. Consequently, there is major interest to develop novel immunotherapies by discovering new targets and management modalities as well as making existing therapies more efficient. In these respects, ionic mechanisms operating in cells of the immune system and/or cancer cells, alone or in combination, offer significant potential [7, 8]. Indeed, functional expression of ion channels such as Kv1.3, KCa3.1, and Stim/Orai has already received significant attention in immune cell functioning [9]. Here, we evaluate the case for voltage-gated sodium channels (VGSCs) as regards (i) their functional role in immune and cancer cells and (ii) potential use in immunotherapy. First, we give an overview of the relevant essentials of the cancer process.

1.1 Primary *vs* secondary tumorigenesis

Most commonly, cancer starts as a primary tumor that results from uncontrolled proliferation of cells in a given part of the body. More than 90% of cancers are carcinomas arising in epithelial tissues. Classically, it is assumed that upon repeated division, which may result in accumulated mutations, cancerous cells lose their genetical stability and homeostatic balance, resulting in uncontrolled hyperactive behavior and invasion of their surroundings. Upon encountering a blood vessel and circulating around the body, this may ultimately lead to metastasis (distal or proximal formation of secondary tumors). Although extremely complex, metastasis can be considered in a reductionist setting to comprise a series of basic cellular behaviors such as motility, secretion, adhesion, etc. (*e.g.,* Ref. [10]). However, metastasis may also occur without the formation of a primary tumor, leading to the clinical distinction of "carcinoma of unknown primary" (*e.g.,* Ref. [11]). In fact, the genes controlling primary and secondary tumorigenesis can be different, even independent [12]. Furthermore, these can be specific to the organ(s) being metastasized [13]. These characteristics could have important implications for the immunobiology/therapy of cancer since the molecular architecture and, hence, strategy for exact targeting of tumors may depend on disease stage (as well as treatment history).

1.2 Epigenetics

Cancer is mainly an epigenetic disease, *i.e.,* it is the result of changes in the "epigenome" resulting in aberrant expression of otherwise normal genes. Importantly, the latter include "protein-making" genes that, however, constitute only some 2% of the genome. In other words, as also highlighted earlier, it is loss of genetic "homeostasis" that causes cells to lose control of their "normality." This coupled with the deregulation of the appropriate apoptotic machinery eventually leads to full-blown cancer. Gene mutations that are heritable play a role in only a minority (10%–15%) of cancer cases. Crucially, protein-gene expression is controlled or regulated by "noncoding" genes or sequences that constitute the remaining *ca.* 98% of our genetic material. Since the latter themselves are prone to mutations, however, cancer can ultimately be deemed a genetic disease [14]. In terms of therapy overall, therefore, both epigenetic and homeostatic mechanisms represent viable targets. VGSC expression in cancer is primarily an epigenetic phenomenon.

1.3 Stemness

It is well established that at least a subpopulation of cancer cells has "stemness" and this is responsible for the general aggressiveness, including drug resistance, of tumors [15]. Interestingly, cancer stem cells (CSCs) are regulated by the neuronal component of the tumor microenvironment (TME) [16, 17]. The stemness of cancer also manifests itself in the form of "oncofetal" gene expression whereby several genes in cancer cells are expressed in their embryonic forms [18]. These include some well-known and clinically used genes/proteins like the carcinoembryonic antigen. Among oncofetal genes also is the VGSC subtype Nav1.5. This is expressed in breast, colon cancers, melanoma, and astrocytoma as its developmentally regulated 5' "neonatal" splice variant [19–23]. Indeed, this splice variant was first demonstrated in a cancer (neuroblastoma) cell line [24]. As an embryonic gene expressed in the adult body, neonatal Nav1.5 (nNav1.5) is potentially cancer specific [25]. Thus, it could offer significant potential in immune therapy of the cancers expressing it, including by antibody-based approaches.

1.4 Tumor microenvironment

Tumors are heterogenous in their cellular make-up (Fig. 1). Alongside the cancer cells are innate immune cells (*e.g.,* macrophages, natural killer cells, and myeloid-derived suppressor cells), adaptive immune cells (*e.g.,* T- and B-lymphocytes), endothelial cells, fibroblasts, and neurons, some interconnected by gap junctions and tumor nano/microtubes [5,16,17,26]. The whole cellular complex is embedded in an extracellular matrix (ECM), altogether forming the TME. The width of the extracellular space within the TME does not appear to have been quantified but it can be assumed to be similar to brain, *i.e., ca.* 40 nm [27]. Because of this narrowness, ions and metabolites released from active cells can reach significant concentrations that, in turn, exert significant regulatory influence upon the cancer cells. Indeed, all cells within the TME contribute to the cancer process. The ECM regulates epithelial-mesenchymal transition (EMT), an early event in invasiveness and metastasis [28]. In addition, tumors are extrinsically innervated, and this input can play a significant role in the progression of cancer [17].

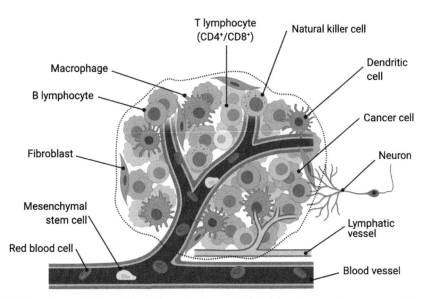

Fig. 1 Cellular makeup of the tumor microenvironment. The tumor microenvironment contains a variety of cells and vessels involved collectively in tumorigenesis. Cancer cells interact with infiltrated cancer-associated fibroblasts, adipocytes, pericytes, CSCs, neurons, and blood vessels. Associated lymphatic vessels can house a variety of innate and adaptive immune cells. These include myeloid cells (tumor-associated macrophages, dendritic cells, myeloid-derived suppressor cells) and lymphoid cells (CD4[+] T-lymphocytes, CD8[+] T-lymphocytes, B-lymphocytes, NK cells, dendritic cells, and plasma cells). Intrinsic neurons and extrinsic nerve input may be associated with tumor progression by secreting neurotransmitters in the TME that can stimulate growth and metastasis. Exosomes are concentrated within the extracellular matrix of the TME.

1.5 Exosomes

Exosomes, small (∼100 nm) diameter lipid bilayer vesicles secreted by tumor cells, transport and transmit large amounts of genetic information and thus mediate extensive intercellular communication. Exosomes travel to various sites around the body through the blood circulation and can be differentiated via the surface proteins they express. Within the TME, exosomes are involved in the maintenance of tissues homeostasis, tumor lesions, and the formation of metastatic niches via manipulation of immune evasion [29]. They also play a functional role in "long distance" communication between tumor cells and immunocytes, promoting invasiveness and immunosuppression [30]. The genetic material transported by exosomes include microRNAs some of which may impact on VGSCs. Examples include regulation of Nav1.1 and Nav1.2 protein expression and trafficking by miR-9 and miR-132 [31,32]. Nav1.7 expression is regulated by miR-146 [33]. Interestingly, miR-146 is also elevated in prostate cancer, where Nav1.7 is the predominant VGSC [34,35]. Thus, exosomes could potentially be exploited as a vehicle for the regulation of VGSC expression and functioning in various cancer types. More broadly, exosome microRNA signatures could also be used as cancer biomarkers for noninvasive diagnosis and targeted therapeutics [36].

1.6 Hypoxia

Hypoxia is an inherent property of growing tumors. Once a tumor reaches a diameter of 0.5–1 mm, diffusion of oxygen becomes limited and the core of the tumor becomes hypoxic (pO_2 = 1–10 mmHg $cf.$ 100–150 mmHg in healthy lungs). In extreme cases, the tumor core can even become anoxic and necrotic. Hypoxia is a major trigger for the expression of a range of genes (>70) via the transcription factor HIF-1α. Among these are genes that represent the hallmarks of cancer, including angiogenesis that enables renewed supply of oxygen and nutrients to the growing tumor, as well as the disposal of undesirable metabolic by-products. Histone deacetylases and other epigenetic mechanisms are also altered in hypoxia [37]. Under hypoxic conditions, cancer cells become generally more aggressive and invasive [38]. Hypoxia can also have a significant effect on VGSC functioning, especially by increasing its open time and promoting a "persistent current" component [21,39]. The hypoxic region is surrounded by an area of acidified pH, which initiates extracellular proteolysis and facilitates invasion [40].

1.7 Voltage-gated sodium channels

VGSCs are made up of an alpha subunit (VGSCα) and one or two auxiliary beta subunit(s). VGSCαs can form a functional channel by themselves, but their expression and activity are modulated by the beta-subunit(s). VGSCαs are a part of the voltage-gated ion channel superfamily and include nine functional members: Nav1.1-Nav1.9. A given VGSCα comprises four homologous domains (DI–DIV) each with six transmembrane segments (S1–S6). Of the latter, the positively charged S4 segments mainly give the channel its voltage sensitivity while the S1–S2 regions are responsible for creating the pore through which Na^+ flows into cells. In physiological conditions, channel opening lasts only for about a millisecond, inactivation being controlled by the DIII–DIV region. Under hypoxia, however, the channel remains open for 100–1000 s of milliseconds thus giving rise to a "persistent current" (I_{NaP}) resulting in substantial influx of Na^+ into the cell [21,39,41]. Importantly, VGSCαs are subject to a hierarchy of regulatory mechanisms, ranging from transcriptional to posttranslational [42]. In particular, splicing of exon 6 is developmentally regulated giving rise to a 3′ form, which is expressed in the adult and a 5′ form which is "neonatal." This has been studied mostly in Nav1.5 and affects the DI:S3/S4 loop region of the channel, the two splice variants differing by six amino acids in an extracellular part of the protein (Fig. 2A). Among these, importantly, the aspartate at position 211 switches to a lysine resulting in charge reversal from negative to positive (Fig. 2B).

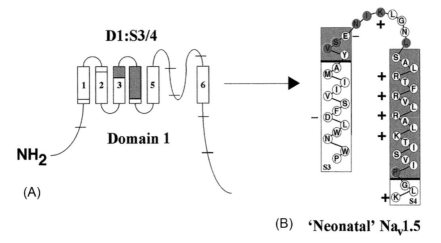

Fig. 2 Spliced domain 1 region of the voltage-gated sodium channel (Nav1.5). (A) The shaded area (S3–S4) denotes the spliced region. (B) A more detailed view of the S3–S4 spliced region. The amino acids denoted in *red* are those that change as a result of the splicing. The charges of the amino acids are also indicated.

2. VGSC expression and function in cells of the immune system

VGSCs have been shown to be expressed in several types of cell in the immune system [9]. Here, we give an overview of the expression and the channels' contributions to various aspects of immune function.

2.1 T-lymphocytes

Expression of functional VGSCs in lymphocytes isolated from healthy human blood was reported originally by Cahalan *et al.* [43]. Later, using the Jurkat cell model of lymphocytes, Fraser *et al.* [44] showed that a subpopulation (*ca.* 10%) of these cells indeed expressed a functional VGSC, thought to be Nav1.5. A similar observation was made in other model cells and extended to peripheral blood mononuclear cells (PBMCs) by Huang *et al.* [45]. Taken together, as well as Nav1.5, several other VGSCα (Nav1.3, 1.6 and 1.7) mRNAs were found to be present, similar in the cell lines and PBMCs. Consistent with this "mixed" expression pattern, tetrodotoxin (TTX) blocked the VGSC current in MOLT-4 cells completely at 2 μM and demonstrated an IC_{50} of *ca.* 900 nM in Jurkat cells [42,45].

2.1.1 T-cell selection

"Lymphocyte selection" is the complex process of differentiation and genetic rearrangement of multipotent thymocytes ultimately to produce different classes of lymphocytes. Relevant here is the generation of CD4$^+$ and CD8$^+$ T-lymphocytes. CD4$^+$ T-cells are "helpers" determining how other parts of the immune system may respond (*e.g.*, via cytokine signaling) to perceived threats, ultimately leading to the killing of cells identified as "foreign." These cells are also regulatory; Treg cells facilitate normalization of function following an immune response and can appear immunosuppressive [46]. CD8$^+$ cells (often called "cytotoxic T-lymphocytes" or CTLs) play a critical role in tumor surveillance and in the production/release of cytotoxic granules for direct killing of virus-infected cells and cancer cells [47]. Indeed, high-level occurrence of CD8$^+$ tumor infiltrating lymphocytes (TILs) in tumors generally correlates positively (even "completely") with response to primary systemic therapy [48]. CD4$^+$ T-cells promote the generation of CTL memory [49]. Thus, CD4$^+$ and CD8$^+$ work closely together in the TME to combat cancerous cells. The CD4$^+$:CD8$^+$ ratio in the blood and in tumor tissues is an important indicator of immune activity and, consequently, cancer prognosis [50,51]. A low CD4$^+$:CD8$^+$ ratio in the TME is associated with poor outcome in many cancer types [52]. Accordingly, understanding the mechanism(s) controlling the CD4$^+$-CD8$^+$ selection process is centrally important to cancer immunology and could ultimately facilitate generation of novel immunotherapies. Positive or negative selection will occur by weak or strong "TCR-MHC-self-antigen" binding, respectively [53]. Functional VGSC expression plays a significant, stage-dependent role in the process of T-lymphocyte selection by facilitating the positive selection of CD4$^+$ T cells from CD4$^+$/CD8$^+$ double-positive (DP) lymphocytes. The intermediary signal is intracellular Ca^{2+} that can be controlled, in part, by VGSC activity (Fig. 3). The VGSC in the DP lymphocytes is Nav1.5 (gene: *SCN5A*) with the regulatory subunit is β4 (gene: *SCN4B*) [54]. A key downstream player is extracellular signal-regulated kinase (ERK) that can mediate different antiproliferative events, including apoptosis, depending on the cell type and stimulus pattern [55].

Positive selection. Weak TCR-antigen binding, known to lead to positive selection of CD4$^+$ T-lymphocytes, generates an initial maintained rise in intracellular Ca^{2+}. This occurs probably by release from intracellular stores and/or influx through calcium release-activated channels (CRACs) in plasma membrane. This result is brief, low-level activation of ERK signaling [56]. In turn, the expression of *SCN5A* and *SCN4B* is upregulated. VGSC activity further facilitates and maintains the intracellular Ca^{2+} level, possibly by depolarizing the membrane potential and leading to the opening of voltage-gated Ca^{2+} channels and/or reversed Na$^+$-Ca^{2+} exchange in mitochondria [57]. This area needs further investigation. In addition, the β4 subunit co-expression could serve to promote the associated Nav1.5 activity [58]. Nevertheless, it is the generation of a *maintained* intracellular Ca^{2+} level that is essential for the survival of the developing CD4$^+$ cells. Accordingly, blocking VGSC activity with TTX or silencing *SCN5A* prevented the positive selection of CD4$^+$ (but not CD8$^+$) T-cells [54].

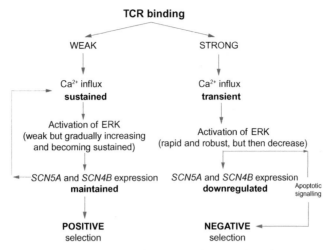

Fig. 3 T-lymphocyte selection and its control by VGSC. Selection of CD4[+] T-lymphocyte is initiated by the binding of the T-cell receptor (TCR) with the dendritic cell MHC-self-antigen complex. Weak TCR-MHC binding leads to positive CD4[+] selection while strong binding leads to negative CD4[+] selection. Weak binding causes a sustained rise in the intracellular Ca^{2+} level of the naïve lymphocyte. This activates ERK (and dependent pathways) initially weakly but then ERK signaling gradually increases and becomes sustained. Sustained ERK signaling maintains the expression of *SCN5A* and *SCN4B*, two components of the VGSC essential for the positive selection of CD4[+] T-lymphocytes. It is possible that the maintenance of the signaling involves a positive feedback effect *(dotted line)*. In contrast, strong TCR-MHC binding in naïve lymphocytes causes a transient rise in the intracellular level of Ca^{2+}, which then rapidly diminishes. Although this initially leads to a rapid and robust activation of ERK, signaling then weakens briskly. This leads to apoptotic cell death, *i.e.*, CD4[+] lymphocytes are eliminated ("negatively selected"). Diminished ERK signaling also downregulates expression of *SCN5A* and *SCN4B*, which are no longer involved in the selection process.

Negative selection. This is initiated by strong TCR–antigen binding, leading only to a transient increase in intracellular Ca^{2+}. This level and pattern of Ca^{2+} result in a brisk and robust activation of the ERK pathway that then diminishes rapidly. The result is apoptotic elimination of the cells. Concurrently, the expression of *SCN5A* and *SCN4B* is downregulated and these are not involved in the negative selection [54].

In conclusion, VGSCs are a target for modulating the ratio of CD4[+]:CD8[+] lymphocytes ultimately entering cancerous tissues. Other immunotherapeutic targets could focus on the pluripotent capabilities of CD4[+] TCLs to differentiate into various active subtypes in response context-dependent signals [59].

2.1.2 Lymphocyte proliferation and trafficking

Circulating mature lymphocytes can exit the vasculature and penetrate into tissues as a part of the adaptive immune response. Such tissues include tumors, and TILs are an integral part of the TME (Fig. 1). In breast cancer, for example, TILs are comprised primarily of $CD8^+$ and $CD4^+$ T-cells, and a smaller proportion of B-cells and natural killer (NK) cells [60]. In order to be able to actively infiltrate the target tissues lymphocytes need basic invasion "machinery," including a dynamic cytoskeleton that is metabolically controlled in line with local environmental cues [61]. Indeed, in Jurkat and MOLT-4 cells, complete blockage of the VGSC activity with TTX suppressed the cells' invasiveness through Matrigel strongly by *ca.* 90% [44,45]. Na^+ influx through VGSCs can regulate cell volume in Jurkat cells and, in turn, cell volume control is integral to the invasion process (*e.g.,* Refs. [62,63]). Importantly, in contrast, TTX did not affect the cells' proliferation or transverse migration (not involving Matrigel) [45]. These results clearly suggest that the role of the VGSC in promoting invasiveness is mainly due to increased proteolytic activity rather than motility per se. This is similar to the role of the VGSC in cancer cell invasiveness [64,65]. Thus, the channel activity appears to control secretion of a proteolytic enzyme and/or its activation by acidification of the pericellular space, as in the case of metastasizing tumor cells [64,65]. In conclusion, while VGSC-dependent activity of TILs can initially produce anticancer effects, chronic inflammation can increase cancer risk. The latter would involve deep penetration and increased activity of immune cells within the TME, leading to overproduction of cytokines, DNA damage, and loss of tissue homeostasis [66,67].

2.2 NK cell cytotoxicity

Natural killer cells (NKCs), a subtype of lymphocyte, are a part of the innate immune system and, as such, respond quickly and aggressively to a wide variety of pathological conditions, including cancer [68]. NKCs kill by secreting "granzymes," toxic granular molecules that activate apoptotic pathways in target cells [69]. In addition, NKCs also interact with macrophages, dendritic cells, endothelial cells, and T-lymphocytes to shape the innate and adaptive immune systems [68,70]. NKCs are regulated by two types of surface receptors: (1) activating receptors (that recognize antigens associated with oncogenic transformation) and (2) inhibitory receptors (that prevent the killing of healthy self-cells) [71]. Cancer is known to evade the innate killing mechanisms of NKCs by misleadingly expressing inhibitory surface receptors (such as MHC-I or HLA). Activated NKCs (such as chimeric antigen receptor (CAR)-modified NK-92 cells) are induced genetic variants of NKCs that lack inhibitory surface receptors and that can therefore be used to target tumors [50,51]. NKCs can also destroy tumor cells by recognizing specific antibodies bound to cancer antigens. Mandler *et al.* [72] was first to report expression of functional VGSCs in a human NK-enriched cell preparation (CD16) (also, Ref. [73]). In the earlier study, flow cytometry and a voltage-sensitive dye (oxonol) were used to characterize the effects of VGSC modulators on NKC cytotoxicity against human myeloid leukemia (K562) tumor cells in co-culture. Activating the VGSC (by pretreatment with veratridine) led to the membrane depolarization and a subsequent reduction in NKC cytotoxic potency (reduced granule release). This effect was blocked by TTX [72]. These findings suggested that VGSC plays the role of a "brake" on NKC activity in the TME. Unfortunately, the effects of VGSCs in this study were conducted without quantifying the expression of activating and inhibiting NKC surface receptors. Additionally, the VGSC-associated mechanism(s) of NKC inhibition of the tumor-cell killing was not studied. Furthermore, the consequence(s) *in vivo* of the targeted tumor cells themselves expressing a functional VGSC could not be considered. Clearly, much more work is needed to understand the molecular nature and functional role of VGSC expression in the cytotoxic activity of NKCs [74].

2.3 Macrophage activity

Macrophages, when recruited to tumors, can accelerate cancer progression. Using zebrafish and mouse models of melanoma, Roh-Johnson *et al.* [75] showed that recruited macrophages transferred their cytoplasm onto melanoma cells. Contributed chemicals include cytokines, chemokines, and growth factors, as well as triggered release of inhibitory immune checkpoint proteins in T-cells. Thus, tumor-associated macrophages (TAMs) create an immunosuppressive TME that facilitates metastasis [76]. Functional VGSC (mainly Nav1.5 and Nav1.6) expression occurs in macrophages [77–81]. Intracellular Nav1.6, in association with the F-actin cytoskeleton was shown to promotes macrophage motility [79]. The underlying mechanism involved release of Na^+ from vesicular intracellular stores, uptake by mitochondria and extrusion of Ca^{2+} from mitochondria, leading to formation of invadopodia [77]. Carrithers *et al.* [80] demonstrated expression of Nav1.5 within phagosomes of activated human macrophages. Expression was restricted to late endosomes and was not detected in early endosomes or on the macrophage plasma membrane. Mechanistically, the channels would provide a route for Na^+ efflux, which would counterbalance proton influx, and thereby maintain electroneutrality during acidification, which is one of the final stages of phagocytosis [77]. Accordingly, blocking VGSC activity would suppress the macrophage-dependent component of cancer progression.

3. Potential of VGSC (nNav1.5) expression in immunotherapy

Ion channels are the second largest membrane protein drug target class after G-protein coupled receptors, representing *ca.* 18% of the total "small molecule drug" targets developed and listed in the ChEMBL database [82]. In fact, dozens of ion channels, including VGSCs, are currently under investigation as targets in drug, including antibody, development programs [17,83]. Neonatal Nav1.5 (nNav1.5) is a cancer-specific VGSC exclusively expressed on the surface of several metastatic cancer types [23]. As such, nNav1.5 can be described as a cancer-specific antigen providing a tumor-specific binding site for immunotherapeutic targets. Furthermore, the functional role of VGSCs in immunocyte development, regulation, and functioning, as described in Sections 2.1–2.3, indicates that VGSCs could also be used as genetic targets for immunotherapeutic drug design. Accordingly, ion channels, and more notably cancer-specific ion channels, such as nNav1.5, could offer significant promise for antibody-based drug design.

3.1 Antibody drugs

Antibodies (Abs), also known as immunoglobulins (IgGs), are large (*ca.* 150 kDa) plasma proteins that are an active part of the normal adaptive immune system and can exist in various forms (*e.g.,* monoclonal or polyclonal). Abs circulate freely in blood but their entrance to tumors can be challenging. The unique amino acid structure of the spliced region of nNav1.5 led to the production of polyclonal antibody, NESOpAb [19, 20]. This blocked nNav1.5 with two orders of magnitude more effectively compared with aNav1.5 expressed mainly in adult cardiac tissue [19]. Furthermore, NESOpAb also suppressed the invasiveness of human breast cancer, MDA-MB-231, cells as effectively as the highly specific VGSC blocker, TTX [84]. Furthermore, NESOpAb was instrumental in demonstrating the cancer-specificity of nNav1.5 protein expression in adult human tissues [25]. In an early application, Gómez-Varela *et al.* [85] used monoclonal Abs (mAbs) to block the activity of the human voltage-gated potassium channel, Eag1, and showed that this would lead to inhibited tumor cell growth *in vitro* and *in vivo*. In a comparable way, a humanized mAb specific for nNav1.5 could serve as an antimetastatic antibody drug. Another interesting possibility would be use of bi-specific mAbs (bi-mAbs) [86,87]. Such a bi-mAb could be one targeting nNav1.5 and an allosterically associated membrane protein, such as NHE1 [88].

3.2 Antibody-drug conjugates

Antibody-drug conjugates (ADCs) are a highly developed class of antibody-based drugs that incorporate elements of a mAb, stable linkers, and a cytotoxin [89,90]. The mAb binds to a target antigen on a cancer cell, while the stable linker enables the controls of precise cytotoxic agent delivery into a target cell. A wide range of cytotoxins exits, including auristatins, maytansinoids, calicheamicins, and amatoxins. Importantly, the ADC-specific anticancer cytotoxic agents can be a thousand times more potent than current marketed immunotherapies [91]. Once bound to an antigen, the ADC can be integrated into a tumor cell via "receptor-mediated endocytosis" where it is then gradually degraded by lysosomal degradation [92]. This allows the release of the anticancer cytotoxic agent inside the target cell, where it can disrupt normal cellular function and activate apoptosis mechanisms. Since 2009, FDA has approved 10 ADCs and more than

80 others are in active clinical studies. Nevertheless, limitations remain in ADC applications, including toxicity and associated resistance mechanisms. Consequently, novel cancer-specific mAbs are needed. In this regard, nNav1.5 again represents a viable target. First, the cancer-specific epitope is in an extracellular part of the protein so it can readily be accessed through the circulation [19, 20]. Second, the channel protein has a turnover time of 32–36 h, so ADC internalization would readily be possible [93]. Importantly, also, nNav1.5-based ADCs could offer the major advantage of recognizing and killing micrometastases, one of the most difficult problems in clinical management of cancer (*e.g.,* Ref. [94]).

3.3 Chimeric antigen receptor therapy

CAR therapy utilizes the properties of a patient's own living immunocytes to proficiently eradicate autoimmune diseases and cancers [95]. In CAR therapy, most commonly the patient's cytotoxic (CD8$^+$) T-cells or, more recently, NKCs are harvested and genetically modified to recognize and destroy tumors [96]. It is also possible to engineer both cell types to attack cancer in congruence [97]. A primary requisite is the availability of a cancer-specific antigen. Yamaci *et al.* [25] showed among a range of adult tissues that nNav1.5 protein expression is restricted to carcinomas and could satisfy this prerequisite. Accordingly, a monoclonal antibody specific for nNav1.5 will (1) need to be produced and (2) engineered so as to be expressed in extracted CTLs and/or NKCs. Then, these can be enriched and infused back into the patient to seek out and destroy the nNav1.5-expressing tumors. Importantly, Nav1.5 expression is an early event in the acquisition of invasiveness, as demonstrated for colorectal cancer [98]. Thus, CAR-T involving nNav1.5 could be a very efficient way of eradicating early-stage tumors, *i.e.,* before systemic disease occurs.

Interestingly, dual-CAR-T therapies are also being considered. Thus, in one application, Shan *et al.* [99] obtained extraordinary effects by using bispecific anti-CD-19 and anti-CD-20 CAR-T cells against CD-19 positive B-cell cancers (non-Hodgkin lymphoma and chronic lymphocytic leukemia). This approach can readily be extended to nNav1.5 as one arm of the bispecific CAR-T treatment.

Finally, new technologies are being coupled to CAR-T cell therapy. First, "mechanogenetics" can be used for remote activation of CAR-T cells via ultrasonic microbubble amplification [100]. Here, CAR-T cells are genetically engineered to incorporate Piezo-1 mechanosensitive Ca^{2+} channels and a signal transduction module that enables the precise activation of CAR-T cells in tumors but not in healthy tissues [100]. Interestingly, Nav1.5 is inherently mechanosensitive and may be exploited similarly [101]. Second, optogenetics (combination of optics and genetics) has been used to generate light-activatable ion channels expressed in T-lymphocytes and this has been extended to CAR-T [102]. Also, CRISPR could enable engineering of T-cells from healthy donors, potentially creating more accessible medicines [103,104]. Finally, the quantity of available NKCs can be enhanced by reprogramming iPSCs [105].

3.4 Vaccines

Immunotherapeutic cancer vaccines can take several different forms, from DNA to whole tissue. The ability of the immune system to recognize ion channels as "foreign" antigens is apparent from the occurrence of certain paraneoplastic disorders (PNDs) in which the body produces an (auto)antibody (autoAb) to a "neuronal" antigen (*e.g.,* Refs. [106,107]). Such an antigen can be a cancer-associated ion channel. The best-known example of the latter is the Lambert-Eaton syndrome (LES) [108,109]. LES is associated most commonly with small-cell lung cancer (SCLC), an aggressive cancer that can metastasize before it can be detected [110]. In this syndrome, the immune system reacts to the tumor by producing antibodies that impede skeletal neuromuscular transmission [111]. Interestingly, SCLC patients with LES survive longer than patients without the syndrome [112]. Importantly, antisera or IgG obtained from cancer patients with LES can block VGSC currents recorded from human SCLC cells [113,114]. These results, taken together, have two important implications. First, a VGSC is expressed in the tumor and elicits an immune response in the form of autoantibodies to the channel. Presumably, these autoAbs similarly suppress VGSC activity also in the cancer cells and thus suppress tumor progression and improve prognosis. Second, the antigenicity of the tumor VGSC implies that it is different from the normal VGSCs present in the adult body. One possibility is that this is a neonatal splice variant, in line with the dedifferentiated nature and stemness of cancer. Indeed, a recent study showed that autoAb's to nNav1.5 do occur in sera of breast cancer patients and their level drops following treatment [115]. In SCLC, the predominant VGSC is presumed to be TTX-sensitive since it is blocked by 5 µM TTX [113,116]. This would exclude it being Nav1.5. It would be interesting to determine the predominant VGSC subtype(s) expressed in SCLC cells and to characterize the autoAb.

In terms of possible VGSC-based vaccines, the most direct application would be a "peptide vaccine" incorporating the unique sequence of the spliced region of nNav1.5: "**VSENIK**LGN**L**SALRC" [19, 20]. Here, the amino acids in bold are those that are different from adult Nav1.5; the underlined *lysine* (K) is a "focal" residue at position 211, which switches from an *aspartate* (D) in the adult, *i.e.,* a double-charge change (Fig. 2B) [22]. Lys_{211} is essential for the binding of NESOpAb and could, therefore, represent an antigenic "hotspot." Blast analysis suggests that this sequence is unique in the human genome. The other most common VGSC in cancers is Nav1.7 [23]. For this, spliced region in the neonatal form would include the following sequence: YLTEFV**N**LGN. Here, the conserved aspartate at position 211 switches to an *asparagine* (N), *i.e.,* single-charge change. Also, Val_{206} switches to Leu_{206} (neutral) but this is likely to be buried in the membrane and hence may not contribute to the possible antigenicity (Fig. 2B). It is possible that the predominant VGSC in SCLC is Nav1.7, as in non-SCLC [117]. If so, it would follow that even one amino acid change may give rise to an autoimmune response like LES.

Exploiting, again, the unique genomic property of nNav1.5, the "peptide" approach can readily be extended to RNA- or DNA-based vaccines and may involve nanoparticles or exosomes (e.g., Refs. [118,119]). Such vaccines can be personalized by predetermining the possible VGSC subtype expression in the original tumor, if present, or as a precautionary measure if the individual is somehow predisposed to a particular cancer, e.g., through genetic makeup and/or lifestyle. Furthermore, possible presence of polymorphisms in the VGSC gene may personalize the vaccination [120].

3.5 Systemic treatments and combinations

Recent work emphasized improving clinical immunotherapies by combination with small-molecule inhibitors [121]. We can envisage two distinct ways (first order and second order) in which combination therapies can be developed in association with VGSC inhibition (Fig. 4).

3.5.1 First order

In this mode, a VGSC blocker can be combined with an immunotherapeutic agent. In such an approach, the VGSC blocker itself would have both direct and indirect effects on tumorigenesis.

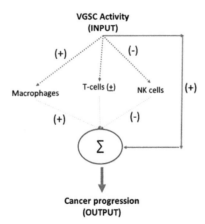

Fig. 4 Cellular effects of VGSC in tumor progression. Pathways relating to "indirect" effects mediated by immune cells are indicated by dotted lines. Solid pathway indicates the direct effect of the VGSC in promoting tumor progression. (+), potentiation; (−), inhibition. T-cells are indicated as (±) due to the possible stage-dependence of their involvement. "$\sum$" represents the conceptual summation of all the VGSC-dependent effects.

Direct. VGSC blockers have already been shown to be antimetastatic *in vivo* (see Djamgoz *et al.* [23] for a review of the evidence). The original study was done on the Dunning rat model of prostate cancer where local, intratumor injection of the highly specific VGSC blocker, TTX, suppressed metastasis to lungs and prolonged survival by *ca.* 20% [122]. This experiment was designed to ensure (1) that the VGSC blocker used (TTX) was the most effective, (2) that the animal would not suffer any toxic effect, and (3) that access to the primary tumor, from where metastasis would start, was guaranteed. Further *in vivo* studies on various breast cancer models, involving systemic application of different pharmacological blockers of VGSCs supported this finding [123–125].

Indirect. Mechanisms that dictate the balance between cancer-promoting *vs* cancer-inhibitory inflammation within the TME are not clear [126,127]. VGSC blockage would be expected to have a wave of "dynamic" effects on the immune cells. In the first instance, the resulting effects on tumorigenesis could be as follows (Fig. 4):

(1) *T-cells.* Although the signaling mechanisms that instruct the establishment of T-cell-inflamed tumors are not well understood, the $CD4^+$:$CD8^+$ ratio will be decreased as a result of VGSC blockage. This will result in an increase in the $CD8^+$ CTL population that will drive early immunosurveillance antitumor responses. However, this is likely to be short-lived since, ultimately, depleted $CD4^+$ T-cells would reduce immunological memory and compromise the tumor rechallenge by CTLs [128]. Irrespective of the subtype of T-cells present, however, the TIL population would be reduced and, ultimately, chronic inflammation would be suppressed.

(2) *NKCs.* Their cytotoxic activity will be promoted by VGSC blockage, thus suppressing tumorigenesis.

(3) *Macrophages.* The infiltration of tumors by TAMs will be inhibited also resulting suppression of tumorigenesis.

Overall, although the net effect of VGSC blockage on tumorigenesis through 1–3 is likely to be highly dynamic and not clear at present, it is likely to be broadly anticancer. Indeed, an insight for this can be obtained from *in vivo* experimental observations. Treating tumor-bearing animals systemically with VGSC (mainly I_{NaP}) blocker drugs such as phenytoin, ranolazine, or the mexiletine-analog RS100642 produced significant anti-tumor/metastatic effects [123–125,129]. This would imply either (1) that the net effect of the *indirect* (immunological) mechanisms was antitumor (as indicated previously) and/or (2) that the *direct* effect dominated. Although some of these tests were performed on immunodeficient mice, the anticancer effect was also seen in animal models, including a syngeneic one, with intact immune systems [123,129].

Combination. Since the net effect of VGSC inhibition *in vivo* clearly is to suppress tumorigenesis, VGSC blockers may be combined with immunotherapy. In the first instance, this could be considered for PD-1/PD-L1 blockade (to dampen TME immunosuppression and overcome drug resistance) and/or CAR-T/NK cell therapy. In an interesting recent study, it was shown in hand and neck cancer patients responding to PD-1 blockade that pembrolizumab caused increased Kv1.3 channel activity (and Ca^{2+} fluxes) leading to increased chemotaxis and cytotoxicity of $CD8^+$ TILs [130]. A synergistic immunotherapy-ion channel combination has also been seen for a vaccine. Thus, Geng *et al.* [131] showed that amiloride-enhanced DNA vaccine entry into APCs (but not somatic cells) *in vitro* and *in vivo* and thus potentiate both innate and adaptive immune responses. Amiloride is an inhibitor of the epithelial sodium channel, ENaC. It will be interesting to determine if VGSC blockers would produce the same effect especially under hypoxic conditions when a persistent current develops, and sodium influx is enhanced. A further possibility would be to use VGSC blockers to enhance CAR-NK cell therapy since both arms of the treatment would suppress tumorigenesis. In addition, VGSC blockers would suppress the macrophage-dependent component of the pro-cancer immune response and could thus represent a viable combination therapy. We should note, however, that any combinatorial vaccine strategy involving immature dendritic cells may need special attention since these express a functional VGSC (Nav1.7) [132].

A possible novel immunotherapy-VGSC combination would arise from work associating checkpoint inhibition with EMT that occurs at the start of the metastatic cascade [133]. Working on breast cancer, Dongre *et al.* [134] related the efficiency of anti-CTLA4 immunotherapy to EMT status whereby "epithelial" tumors were found to be more susceptible than "mesenchymal" tumors (Fig. 5) Even more strikingly, in tumors arising from a mixture of both cell types, a minority population (10%) of mesenchymal cells could cross-protect the vast majority (90%) of their epithelial neighbors. More recently, adenosinergic signaling was shown to mediate this effect [135]. EMT has also been associated with activation of other checkpoint molecules, including PD-L1 [136]. Importantly, EMT is controlled in part by VGSC activity [137,138]. Taken together, the available evidence would suggest that checkpoint inhibition combined with VGSC blockage could be an effective way of eradicating early-stage solid tumors, including by using repurposed drugs [39,139].

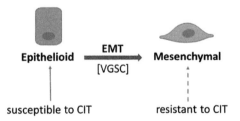

Fig. 5 Epithelial-mesenchymal transition (EMT) and checkpoint immunotherapy (CIT). Therapy is more effective on tumors in epithelioid state. Since voltage-gated sodium channel (VGSC) activity can drive EMT, combination of VGSC blockers with CIT can be an effective strategy for eradicating solid tumors at an early stage.

3.5.2 Second order

In this situation, a VGSC-associated, downstream ionic mechanisms (as outlined in Section 2.3) may synergize with immunotherapy. In this regard, a promising co-target is the sodium-hydrogen exchanger (NHE1), which is well known to control the pH level of solid carcinomas [140]. In turn, NHE1 activity is potentiated by VGSC co-expression [64,88,141]. A "proof of concept" of such a second-order approach was demonstrated in a preclinical mouse study by Pilon-Thomas *et al.* [142]. Thus, it was found (1) that bicarbonate ("alkaline") water could suppress growth of some melanoma cells and (2) that it could potentiate the effectiveness of PD1-based immunotherapy against an induced pancreatic tumor. Accordingly, VGSC blockers could also potentiate the effectiveness of immunotherapy through alkalization. Overall, tumor acidity was deemed a central regulator of cancer immunity that orchestrates both local and systemic immunosuppression and that may offer a broad panel of therapeutic targets [143,144]. The importance of body alkalization was demonstrated recently in another setting whereby sodium bicarbonate was found to reprogram the metabolic profile of T-cells in patients with acute myeloid leukemia and prevent relapse after hematopoietic stem-cell transplant [145].

4. Conclusions and future perspectives

In conclusion, functional VGSC expression plays a significant dual role (1) in promoting metastasis and (2) in modulating the functioning of immune cells. The evidence for (1) is quite substantial [23]. On the other hand, compared with other ion channels and transporters, the role of VGSCs in immune cells is complex and much less well understood. Consequently, much more work is needed to evaluate their potential in immunotherapy both alone and in combination with other targeted drugs.

Importantly, nevertheless, the neonatal nature of Nav1.5 expression in several carcinomas offers distinct immunotherapeutic advantages. Such direct approaches would include CAR-T and CAR-NK cell therapies and vaccines. Much more work is required, however, to elucidate how the functional roles of the VGSCs in cancer and immune cells "merge" mechanistically *in situ* and how such "mergers" could be exploited clinically. There are promising examples involving related ionic mechanisms such as ENaC, Kv1.3, K_{Ca} and pH regulation (HCO_3^- transport) to boost the effectiveness of immunotherapy [146,147]. It will be interesting to extend such experiments to VGSC blockers.

As the field of immunobiology-therapy of cancer advances rapidly, many questions, as follows, remain as regards the role of the functional VGSCs in both cancer and immune cells. (1) What are the specific variants and the molecular characteristics of VGSCs expressed in the different immunocytes? (2) What are the functional roles of these VGSCs in the immunocytes? (3) What are the dynamics (in space and time) and hierarchy of VGSC-dependent behavior of different immunocytes? (4) Can functional VGSC expression in various subpopulations of immune cells be used to manipulate their quantity and quality so as to enhance ultimately the effectiveness of their use in immunotherapy? (5) How could VGSC blockers be combined most effectively with immune checkpoint inhibitors? (6) Interestingly, the sodium level of body fluids can impact upon cancer progression, but the reported effects are anomalous and difficult to interpret at present (*e.g.*, Refs. [148–151]). It would be worthwhile investigating this area further and gain insights both mechanistically and as regards dietary considerations. (7) What about the potential of other ion channels and transporters as immunotherapy targets or co-targets, even in possible combination with VGSC expression?

Elucidation of such questions will help fulfill the potential of VGSC, especially nNav1.5, expression in immunological therapies of cancer. This effort could be facilitated by modern methods of artificial intelligence [152].

Acknowledgments

Our research—*neuroscience solutions to cancer*—overall has been supported over many years by the Pro Cancer Research Fund (PCRF). We thank Drs Annarosa Arcangeli, Will Brackenbury, Cristina Lo Celso, Dina Pospori, and Cesare Sala for reading and commenting on the manuscript.

Authors' declaration

M.B.A.D. conceived and planned the paper. M.B.A.D. and L.F. did the literature searches and shared the writing.

Conflict of interest

M.B.A.D. is involved in a small biotech company aiming to exploit the clinical potential of voltage-gated sodium channel expression in cancer.

References

[1] Bray F, Ferlay J, Soerjomataram I, Siegel RL, Torre LA, Jemal A. Global cancer statistics 2018: GLOBOCAN estimates of incidence and mortality worldwide for 36 cancers in 185 countries. CA Cancer J Clin 2018;68(6):394–424. https://doi.org/10.3322/caac.21492.

[2] Gatenby RA, Brown JS. Integrating evolutionary dynamics into cancer therapy. Nat Rev Clin Oncol 2020;17(11):675–86. https://doi.org/10.1038/s41571-020-0411-1.

[3] Karagiannis GS, Condeelis JS, Oktay MH. Chemotherapy-induced metastasis: molecular mechanisms, clinical manifestations, therapeutic interventions. Cancer Res 2019;79(18):4567–76. https://doi.org/10.1158/0008-5472.CAN-19-1147.

[4] Menu-Branthomme A, Rubino C, Shamsaldin A, Hawkins MM, Grimaud E, Dondon MG, Hardiman C, Vassal G, Campbell S, Panis X, Daly-Schveitzer N, Lagrange JL, Zucker JM, Chavaudra J, Hartman O, de Vathaire F. Radiation dose, chemotherapy and risk of soft tissue sarcoma after solid tumours during childhood. Int J Cancer 2004;110(1):87–93. https://doi.org/10.1002/ijc.20002.

[5] Marshall HT, Djamgoz MBA. Immuno-oncology: emerging targets and combination therapies. Front Oncol 2018;8:315. https://doi.org/10.3389/fonc.2018.00315.

[6] Varadé J, Magadán S, González-Fernández Á. Human immunology and immunotherapy: main achievements and challenges. Cell Mol Immunol 2020;2:1–24. https://doi.org/10.1038/s41423-020-00530-6.

[7] Feske S, Concepcion AR, Coetzee WA. Eye on ion channels in immune cells. Sci Signal 2019;12 (572). https://doi.org/10.1126/scisignal.aaw8014, eaaw8014.

[8] Firmenich L, Djamgoz MBA. Ion channels and transporters in immunity, inflammation and antitumor immunity. Bioelectricity 2020;4:418–23. https://doi.org/10.1089/bioe.2020.0045 [in press].

[9] Feske S, Wulff H, Skolnik EY. Ion channels in innate and adaptive immunity. Annu Rev Immunol 2015;33:291–353. https://doi.org/10.1146/annurev-immunol-032414-112212.

[10] Payne SL, Levin M, Oudin MJ. Bioelectric control of metastasis in solid tumors. Bioelectricity 2019;1 (3):114–30. https://doi.org/10.1089/bioe.2019.0013.

[11] Moran S, Martinez-Cardús A, Boussios S, Esteller M. Precision medicine based on epigenomics: the paradigm of carcinoma of unknown primary. Nat Rev Clin Oncol 2017;14(11):682–94. https://doi.org/10.1038/nrclinonc.2017.97.

[12] Ohgaki H, Kleihues P. Genetic pathways to primary and secondary glioblastoma. Am J Pathol 2007;170(5):1445–53. https://doi.org/10.2353/ajpath.2007.070011.

[13] Riggi N, Aguet M, Stamenkovic I. Cancer metastasis: a reappraisal of its underlying mechanisms and their relevance to treatment. Annu Rev Pathol 2018;13:117–40. https://doi.org/10.1146/annurev-pathol-020117-044127.

[14] Pfister SX, Ashworth A. Marked for death: targeting epigenetic changes in cancer. Nat Rev Drug Discov 2017;16(4):241–63. https://doi.org/10.1038/nrd.2016.256.

[15] Prager BC, Xie Q, Bao S, Rich JN. Cancer stem cells: the architects of the tumor ecosystem. Cell Stem Cell 2019;24(1):41–53. https://doi.org/10.1016/j.stem.2018.12.009.

[16] Boilly B, Faulkner S, Jobling P, Hondermarck H. Nerve dependence: from regeneration to cancer. Cancer Cell 2017;31(3):342–54. https://doi.org/10.1016/j.ccell.2017.02.005.

[17] Hutchings C, Phillips JA, Djamgoz MBA. Nerve input to tumours: pathophysiological consequences of a dynamic relationship. Biochim Biophys Acta Rev Cancer 2020;1874(2):188411. https://doi.org/10.1016/j.bbcan.2020.188411.

[18] Zaidi SK, Frietze SE, Gordon JA, Heath JL, Messier T, Hong D, Boyd JR, Kang M, Imbalzano AN, Lian JB, Stein JL, Stein GS. Bivalent epigenetic control of oncofetal gene expression in cancer. Mol Cell Biol 2017;37(23). https://doi.org/10.1128/MCB.00352-17, e00352-17.

[19] Chioni AM, Fraser SP, Pani F, Foran P, Wilkin GP, Diss JK, Djamgoz MB. A novel polyclonal antibody specific for the Na(v)1.5 voltage-gated Na(+) channel 'neonatal' splice form. J Neurosci Methods 2005;147(2):88–98. https://doi.org/10.1016/j.jneumeth.2005.03.010.

[20] Fraser SP, Diss JK, Chioni AM, Mycielska ME, Pan H, Yamaci RF, Pani F, Siwy Z, Krasowska M, Grzywna Z, Brackenbury WJ, Theodorou D, Koyutürk M, Kaya H, Battaloglu E, De Bella MT, Slade MJ, Tolhurst R, Palmieri C, Jiang J, Latchman DS, Coombes RC, Djamgoz MB. Voltage-gated sodium channel expression and potentiation of human breast cancer metastasis. Clin Cancer Res 2005;11(15):5381–9. https://doi.org/10.1158/1078-0432.CCR-05-0327.

[21] Guzel RM, Ogmen K, Ilieva KM, Fraser SP, Djamgoz MBA. Colorectal cancer invasiveness in vitro: predominant contribution of neonatal Nav1.5 under normoxia and hypoxia. J Cell Physiol 2019;234 (5):6582–93. https://doi.org/10.1002/jcp.27399.

[22] Onkal R, Djamgoz MB. Molecular pharmacology of voltage-gated sodium channel expression in metastatic disease: clinical potential of neonatal Nav1.5 in breast cancer. Eur J Pharmacol 2009;625 (1–3):206–19. https://doi.org/10.1016/j.ejphar.2009.08.040.

[23] Djamgoz MBA, Fraser SP, Brackenbury WJ. In vivo evidence for voltage-gated sodium channel expression in carcinomas and potentiation of metastasis. Cancers (Basel) 2019;11(11):1675. https://doi.org/10.3390/cancers11111675.

[24] Ou SW, Kameyama A, Hao LY, Horiuchi M, Minobe E, Wang WY, Makita N, Kameyama M. Tetrodotoxin-resistant Na+ channels in human neuroblastoma cells are encoded by new variants of Nav1.5/SCN5A. Eur J Neurosci 2005;22(4):793–801. https://doi.org/10.1111/j.1460-9568.2005.04280.x.

[25] Yamaci RF, Fraser SP, Battaloglu E, Kaya H, Erguler K, Foster CS, Djamgoz MBA. Neonatal Nav1.5 protein expression in normal adult human tissues and breast cancer. Pathol Res Pract 2017;213 (8):900–7. https://doi.org/10.1016/j.prp.2017.06.003.

[26] Lou E, Fujisawa S, Morozov A, Barlas A, Romin Y, Dogan Y, Gholami S, Moreira AL, Manova-Todorova K, Moore MA. Tunneling nanotubes provide a unique conduit for intercellular transfer of cellular contents in human malignant pleural mesothelioma. PLoS One 2012;7(3). https://doi.org/10.1371/journal.pone.0033093, e33093.

[27] Nicholson C, Hrabětová S. Brain extracellular space: the final frontier of neuroscience. Biophys J 2017;113(10):2133–42. https://doi.org/10.1016/j.bpj.2017.06.052.

[28] Henke E, Nandigama R, Ergün S. Extracellular matrix in the tumor microenvironment and its impact on cancer therapy. Front Mol Biosci 2020;6:160. https://doi.org/10.3389/fmolb.2019.00160.

[29] Osaki M, Okada F. Exosomes and their role in cancer progression. Yonago Acta Med 2019;62 (2):182–90. https://doi.org/10.33160/yam.2019.06.002.

[30] Greening DW, Gopal SK, Xu R, Simpson RJ, Chen W. Exosomes and their roles in immune regulation and cancer. Semin Cell Dev Biol 2015;40:72–81. https://doi.org/10.1016/j.semcdb.2015.02.009.

[31] Hu XL, Wang XX, Zhu YM, Xuan LN, Peng LW, Liu YQ, Yang H, Yang C, Jiao L, Hang PZ, Sun LH. MicroRNA-132 regulates total protein of Nav1.1 and Nav1.2 in the hippocampus and cortex of rat with chronic cerebral hypoperfusion. Behav Brain Res 2019;366:118–25. https://doi.org/10.1016/j.bbr.2019.03.026.

[32] Sun LH, Yan ML, Hu XL, Peng LW, Che H, Bao YN, Guo F, Liu T, Chen X, Zhang R, Ban T, Wang N, Liu HL, Hou X, Ai J. MicroRNA-9 induces defective trafficking of Nav1.1 and Nav1.2 by targeting Navβ2 protein coding region in rat with chronic brain hypoperfusion. Mol Neurodegener 2015;10:36. https://doi.org/10.1186/s13024-015-0032-9.

[33] Wang Y, Jiang W, Xia B, Zhang M, Wang Y. MicroRNA-146a attenuates the development of morphine analgesic tolerance in a rat model. Neurol Res 2020;42(5):415–21. https://doi.org/10.1080/01616412.2020.1735818.

[34] Porzycki P, Ciszkowicz E, Semik M, Tyrka M. Combination of three miRNA (miR-141, miR-21, and miR-375) as potential diagnostic tool for prostate cancer recognition. Int Urol Nephrol 2018;50 (9):1619–26. https://doi.org/10.1007/s11255-018-1938-2.

[35] Diss JK, Stewart D, Pani F, Foster CS, Walker MM, Patel A, et al. A potential novel marker for human prostate cancer: voltage-gated sodium channel expression in vivo. Prostate Cancer Prostatic Dis 2005;8(3):266–73. https://doi.org/10.1038/sj.pcan.4500796.

[36] Chen M, Xu R, Rai A, Suwakulsiri W, Izumikawa K, Ishikawa H, Greening DW, Takahashi N, Simpson RJ. Distinct shed microvesicle and exosome microRNA signatures reveal diagnostic markers for colorectal cancer. PLoS One 2019;14(1). https://doi.org/10.1371/journal.pone.0210003, e0210003.

[37] Choudhry H, Harris AL. Advances in hypoxia-inducible factor biology. Cell Metab 2018;27 (2):281–98. https://doi.org/10.1016/j.cmet.2017.10.005.

[38] Chang Q, Jurisica I, Do T, Hedley DW. Hypoxia predicts aggressive growth and spontaneous metastasis formation from orthotopically grown primary xenografts of human pancreatic cancer. Cancer Res 2011;71(8):3110–20. https://doi.org/10.1158/0008-5472.CAN-10-4049.

[39] Djamgoz MB, Onkal R. Persistent current blockers of voltage-gated sodium channels: a clinical opportunity for controlling metastatic disease. Recent Pat Anticancer Drug Discov 2013;8 (1):66–84. https://doi.org/10.2174/15748928130107.

[40] Pedersen SF, Novak I, Alves F, Schwab A, Pardo LA. Alternating pH landscapes shape epithelial cancer initiation and progression: focus on pancreatic cancer. Bioessays 2017;39(6). https://doi.org/10.1002/bies.201600253.

[41] Leslie TK, James AD, Zaccagna F, Grist JT, Deen S, Kennerley A, Riemer F, Kaggie JD, Gallagher FA, Gilbert FJ, Brackenbury WJ. Sodium homeostasis in the tumour microenvironment. Biochim Biophys Acta Rev Cancer 2019;1872(2):188304. https://doi.org/10.1016/j.bbcan.2019.07.001.

[42] Fraser SP, Ozerlat-Gunduz I, Brackenbury WJ, Fitzgerald EM, Campbell TM, Coombes RC, Djamgoz MB. Regulation of voltage-gated sodium channel expression in cancer: hormones, growth factors and auto-regulation. Philos Trans R Soc Lond Ser B: Biol Sci 2014;369(1638). https://doi.org/10.1098/rstb.2013.0105, 20130105.

[43] Cahalan MD, Chandy KG, DeCoursey TE, Gupta S. A voltage-gated potassium channel in human T lymphocytes. J Physiol 1985;358:197–237. https://doi.org/10.1113/jphysiol.1985.sp015548.

[44] Fraser SP, Diss JK, Lloyd LJ, Pani F, Chioni AM, George AJ, Djamgoz MB. T-lymphocyte invasiveness: control by voltage-gated Na+ channel activity. FEBS Lett 2004;569(1–3):191–4. https://doi.org/10.1016/j.febslet.2004.05.063.

[45] Huang W, Lu C, Wu Y, Ouyang S, Chen Y. Identification and functional characterization of voltage-gated sodium channels in lymphocytes. Biochem Biophys Res Commun 2015;458(2):294–9. https://doi.org/10.1016/j.bbrc.2015.01.103.

[46] Romano M, Fanelli G, Albany CJ, Giganti G, Lombardi G. Past, present, and future of regulatory t cell therapy in transplantation and autoimmunity. Front Immunol 2019;10:43. https://doi.org/10.3389/fimmu.2019.00043.

[47] Farhood B, Najafi M, Mortezaee K. CD8$^+$ cytotoxic T lymphocytes in cancer immunotherapy: a review. J Cell Physiol 2019;234(6):8509–21. https://doi.org/10.1002/jcp.27782.

[48] Seo AN, Lee HJ, Kim EJ, Kim HJ, Jang MH, Lee HE, Kim YJ, Kim JH, Park SY. Tumour-infiltrating CD8 + lymphocytes as an independent predictive factor for pathological complete response to primary systemic therapy in breast cancer. Br J Cancer 2013;109(10):2705–13. https://doi.org/10.1038/bjc.2013.634.

[49] Ahrends T, Busselaar J, Severson TM, Bąbała N, de Vries E, Bovens A, Wessels L, van Leeuwen F, Borst J. CD4$^+$ T cell help creates memory CD8$^+$ T cells with innate and help-independent recall capacities. Nat Commun 2019;10(1):5531. https://doi.org/10.1038/s41467-019-13438-1.

[50] Zhang C, Ding H, Huang H, Palashati H, Miao Y, Xiong H, Lu Z. TCR repertoire intratumor heterogeneity of CD4$^+$ and CD8$^+$ T cells in centers and margins of localized lung adenocarcinomas. Int J Cancer 2019;144(4):818–27. https://doi.org/10.1002/ijc.31760.

[51] Zhang J, Zheng H, Diao Y. Natural killer cells and current applications of chimeric antigen receptor-modified NK-92 cells in tumor immunotherapy. Int J Mol Sci 2019;20(2):317. https://doi.org/10.3390/ijms20020317.

[52] Das D, Sarkar B, Mukhopadhyay S, Banerjee C, Biswas Mondal S. An altered ratio of CD4 + And CD8 + T lymphocytes in cervical cancer tissues and peripheral blood—a prognostic clue? Asian Pac J Cancer Prev 2018;19(2):471–8. https://doi.org/10.22034/APJCP.2018.19.2.471.

[53] Gorentla BK, Zhong XP. T cell receptor signal transduction in T lymphocytes. J Clin Cell Immunol 2012;2012(Suppl. 12):5. https://doi.org/10.4172/2155-9899.S12-005.

[54] Lo WL, Donermeyer DL, Allen PM. A voltage-gated sodium channel is essential for the positive selection of CD4(+) T cells. Nat Immunol 2012;13(9):880–7. https://doi.org/10.1038/ni.2379.

[55] Cagnol S, Chambard JC. ERK and cell death: mechanisms of ERK-induced cell death—apoptosis, autophagy and senescence. FEBS J 2010;277(1):2–21. https://doi.org/10.1111/j.1742-4658.2009.07366.x.

[56] McNeil LK, Starr TK, Hogquist KA. A requirement for sustained ERK signaling during thymocyte positive selection in vivo. Proc Natl Acad Sci U S A 2005;102(38):13574–9. https://doi.org/10.1073/pnas.0505110102.

[57] Opuni K, Reeves JP. Feedback inhibition of sodium/calcium exchange by mitochondrial calcium accumulation. J Biol Chem 2000;275(28):21549–54. https://doi.org/10.1074/jbc.M003158200.

[58] Bon E, Driffort V, Gradek F, Martinez-Caceres C, Anchelin M, Pelegrin P, Cayuela ML, Marionneau-Lambot S, Oullier T, Guibon R, Fromont G, Gutierrez-Pajares JL, Domingo I, Piver E, Moreau A, Burlaud-Gaillard J, Frank PG, Chevalier S, Besson P, Roger S. SCN4B acts as a metastasis-suppressor gene preventing hyperactivation of cell migration in breast cancer. Nat Commun 2016;7:13648. https://doi.org/10.1038/ncomms13648.

[59] Tay RE, Richardson EK, Toh HC. Revisiting the role of CD4$^+$ T cells in cancer immunotherapy-new insights into old paradigms. Cancer Gene Ther 2020. https://doi.org/10.1038/s41417-020-0183-x.

[60] Pruneri G, Vingiani A, Denkert C. Tumor infiltrating lymphocytes in early breast cancer. Breast 2018;37:207–14. https://doi.org/10.1016/j.breast.2017.03.010.

[61] Vuononvirta J, Marelli-Berg FM, Poobalasingam T. Metabolic regulation of T lymphocyte motility and migration. Mol Asp Med 2020;16:100888. https://doi.org/10.1016/j.mam.2020.100888.

[62] Bortner CD, Cidlowski JA. Uncoupling cell shrinkage from apoptosis reveals that Na$^+$ influx is required for volume loss during programmed cell death. J Biol Chem 2003;278(40):39176–84. https://doi.org/10.1074/jbc.M303516200.

[63] Soroceanu L, Manning Jr TJ, Sontheimer H. Modulation of glioma cell migration and invasion using Cl(−) and K(+) ion channel blockers. J Neurosci 1999;19(14):5942–54. https://doi.org/10.1523/JNEUROSCI.19-14-05942.1999.

[64] Brisson L, Gillet L, Calaghan S, Besson P, Le Guennec JY, Roger S, Gore J. Na(V)1.5 enhances breast cancer cell invasiveness by increasing NHE1-dependent H(+) efflux in caveolae. Oncogene 2011;30 (17):2070–6. https://doi.org/10.1038/onc.2010.574.

[65] Busco G, Cardone RA, Greco MR, Bellizzi A, Colella M, Antelmi E, Mancini MT, Dell'Aquila ME, Casavola V, Paradiso A, Reshkin SJ. NHE1 promotes invadopodial ECM proteolysis through acidification of the peri-invadopodial space. FASEB J 2010;24(10):3903–15. https://doi.org/10.1096/fj.09-149518.

[66] Grivennikov SI, Greten FR, Karin M. Immunity, inflammation, and cancer. Cell 2010;140 (6):883–99. https://doi.org/10.1016/j.cell.2010.01.025.

[67] Karan D. Inflammasomes: emerging central players in cancer immunology and immunotherapy. Front Immunol 2018;9:3028. https://doi.org/10.3389/fimmu.2018.03028.

[68] Rosenberg J, Huang J. CD8+ T cells and NK cells: parallel and complementary soldiers of immunotherapy. Curr Opin Chem Eng 2018;19:9–20. https://doi.org/10.1016/j.coche.2017.11.006.

[69] Shimasaki N, Jain A, Campana D. NK cells for cancer immunotherapy. Nat Rev Drug Discov 2020;19(3):200–18. https://doi.org/10.1038/s41573-019-0052-1.

[70] Brilot F, Strowig T, Munz C. NK cells interactions with dendritic cells shape innate and adaptive immunity. Front Biosci 2008;13:6443–54. https://doi.org/10.2741/3165.

[71] Sivori S, Vacca P, Del Zotto G, Munari E, Mingari MC, Moretta L. Human NK cells: surface receptors, inhibitory checkpoints, and translational applications. Cell Mol Immunol 2019;16(5):430–41. https://doi.org/10.1038/s41423-019-0206-4.

[72] Mandler RN, Seamer LC, Whitlinger D, Lennon M, Rosenberg E, Bankhurst AD. Human natural killer cells express Na+ channels. A pharmacologic flow cytometric study. J Immunol 1990;144 (6):2365–70.

[73] DeCoursey TE, Chandy KG, Gupta S, Cahalan MD. Voltage-dependent ion channels in T-lymphocytes. J Neuroimmunol 1985;10(1):71–95. https://doi.org/10.1016/0165-5728(85) 90035-9.

[74] Schlichter LC, MacCoubrey IC. Interactive effects of Na and K in killing by human natural killer cells. Exp Cell Res 1989;184(1):99–108. https://doi.org/10.1016/0014-4827(89)90368-6.

[75] Roh-Johnson M, Shah AN, Stonick JA, Poudel KR, Kargl J, Yang GH, et al. Macrophage-dependent cytoplasmic transfer during melanoma invasion in vivo. Dev Cell 2017;43(5):549–562.e6. https://doi.org/10.1016/j.devcel.2017.11.003.

[76] Lin Y, Xu J, Lan H. Tumor-associated macrophages in tumor metastasis: biological roles and clinical therapeutic applications. J Hematol Oncol 2019;12(1):76. https://doi.org/10.1186/s13045-019-0760-3.

[77] Black JA, Waxman SG. Noncanonical roles of voltage-gated sodium channels. Neuron 2013;80 (2):280–91. https://doi.org/10.1016/j.neuron.2013.09.012.

[78] Carrithers LM, Hulseberg P, Sandor M, Carrithers MD. The human macrophage sodium channel NaV1.5 regulates mycobacteria processing through organelle polarization and localized calcium oscillations. FEMS Immunol Med Microbiol 2011;63(3):319–27. https://doi.org/10.1111/j.1574-695X.2011.00853.x.

[79] Carrithers MD, Chatterjee G, Carrithers LM, Offoha R, Iheagwara U, Rahner C, Graham M, Waxman SG. Regulation of podosome formation in macrophages by a splice variant of the sodium channel SCN8A. J Biol Chem 2009;284(12):8114–26. https://doi.org/10.1074/jbc.M801892200.

[80] Carrithers MD, Dib-Hajj S, Carrithers LM, Tokmoulina G, Pypaert M, Jonas EA, Waxman SG. Expression of the voltage-gated sodium channel NaV1.5 in the macrophage late endosome regulates endosomal acidification. J Immunol 2007;178(12):7822–32. https://doi.org/10.4049/jimmunol.178.12.7822.

[81] Schmidtmayer J, Jacobsen C, Miksch G, Sievers J. Blood monocytes and spleen macrophages differentiate into microglia-like cells on monolayers of astrocytes: membrane currents. Glia 1994;12 (4):259–67. https://doi.org/10.1002/glia.440120403.

[82] Santos R, Ursu O, Gaulton A, Bento AP, Donadi RS, Bologa CG, Karlsson A, Al-Lazikani B, Hersey A, Oprea TI, Overington JP. A comprehensive map of molecular drug targets. Nat Rev Drug Discov 2017;16(1):19–34. https://doi.org/10.1038/nrd.2016.230.

[83] Bajaj S, Ong ST, Chandy KG. Contributions of natural products to ion channel pharmacology. Nat Prod Rep 2020;37(5):703–16. https://doi.org/10.1039/c9np00056a.

[84] Brackenbury WJ, Chioni AM, Diss JK, Djamgoz MB. The neonatal splice variant of Nav1.5 potentiates in vitro invasive behaviour of MDA-MB-231 human breast cancer cells. Breast Cancer Res Treat 2007;101(2):149–60. https://doi.org/10.1007/s10549-006-9281-1.

[85] Gómez-Varela D, Zwick-Wallasch E, Knötgen H, Sánchez A, Hettmann T, Ossipov D, Weseloh R, Contreras-Jurado C, Rothe M, Stühmer W, Pardo LA. Monoclonal antibody blockade of the human Eag1 potassium channel function exerts antitumor activity. Cancer Res 2007;67(15):7343–9. https://doi.org/10.1158/0008-5472.CAN-07-0107.

[86] Duranti C, Arcangeli A. Ion channel targeting with antibodies and antibody fragments for cancer diagnosis. Antibodies (Basel) 2019;8(2):33. https://doi.org/10.3390/antib8020033.

[87] Florian P, Flechsenhar KR, Bartnik E, Ding-Pfennigdorff D, Herrmann M, Bryce PJ, et al. Translational drug discovery and development with the use of tissue-relevant biomarkers: towards more physiological relevance and better prediction of clinical efficacy. Exp Dermatol 2020;29(1):4–14. https://doi.org/10.1111/exd.13942.

[88] Brisson L, Driffort V, Benoist L, Poet M, Counillon L, Antelmi E, Rubino R, Besson P, Labbal F, Chevalier S, Reshkin SJ, Gore J, Roger S. NaV1.5 Na$^+$ channels allosterically regulate the NHE-1 exchanger and promote the activity of breast cancer cell invadopodia. J Cell Sci 2013;126(Pt 21):4835–42. https://doi.org/10.1242/jcs.123901.

[89] Joubert N, Beck A, Dumontet C, Denevault-Sabourin C. Antibody-drug conjugates: the last decade. Pharmaceuticals (Basel) 2020;13(9):245. https://doi.org/10.3390/ph13090245.

[90] Peters C, Brown S. Antibody-drug conjugates as novel anti-cancer chemotherapeutics. Biosci Rep 2015;35(4). https://doi.org/10.1042/BSR20150089, e00225.

[91] Collins DM, Bossenmaier B, Kollmorgen G, Niederfellner G. Acquired resistance to antibody-drug conjugates. Cancers (Basel) 2019;11(3):394. https://doi.org/10.3390/cancers11030394.

[92] Kalim M, Chen J, Wang S, Lin C, Ullah S, Liang K, Ding Q, Chen S, Zhan J. Intracellular trafficking of new anticancer therapeutics: antibody-drug conjugates. Drug Des Devel Ther 2017;11:2265–76. https://doi.org/10.2147/DDDT.S135571.

[93] Maltsev VA, Kyle JW, Mishra S, Undrovinas A. Molecular identity of the late sodium current in adult dog cardiomyocytes identified by Nav1.5 antisense inhibition. Am J Physiol Heart Circ Physiol 2008;295(2):H667–76. https://doi.org/10.1152/ajpheart.00111.2008.

[94] Akhtar M, Haider A, Rashid S, Al-Nabet ADMH. Paget's "seed and soil" theory of cancer metastasis: an idea whose time has come. Adv Anat Pathol 2019;26(1):69–74. https://doi.org/10.1097/PAP.0000000000000219.

[95] Namdari H, Rezaei F, Teymoori-Rad M, Mortezagholi S, Sadeghi A, Akbari A. CAR T cells: living HIV drugs. Rev Med Virol 2020;26. https://doi.org/10.1002/rmv.2139, e2139.

[96] Liu D. CAR-T "the living drugs", immune checkpoint inhibitors, and precision medicine: a new era of cancer therapy. J Hematol Oncol 2019;12(1):113. https://doi.org/10.1186/s13045-019-0819-1.

[97] Rosenberg SA, Restifo NP. Adoptive cell transfer as personalized immunotherapy for human cancer. Science 2015;348(6230):62–8. https://doi.org/10.1126/science.aaa4967.

[98] House CD, Vaske CJ, Schwartz AM, Obias V, Frank B, Luu T, Sarvazyan N, Irby R, Strausberg RL, Hales TG, Stuart JM, Lee NH. Voltage-gated Na+ channel SCN5A is a key regulator of a gene transcriptional network that controls colon cancer invasion. Cancer Res 2010;70(17):6957–67. https://doi.org/10.1158/0008-5472.CAN-10-1169.

[99] Shah NN, Johnson BD, Schneider D, Zhu F, Szabo A, Keever-Taylor CA, et al. Bispecific anti-CD20, anti-CD19 CAR T cells for relapsed B cell malignancies: a phase 1 dose escalation and expansion trial. Nat Med 2020;26(10):1569–75. https://doi.org/10.1038/s41591-020-1081-3.

[100] Pan Y, Yoon S, Sun J, Huang Z, Lee C, Allen M, Wu Y, Chang YJ, Sadelain M, Shung KK, Chien S, Wang Y. Mechanogenetics for the remote and noninvasive control of cancer immunotherapy. Proc Natl Acad Sci U S A 2018;115(5):992–7. https://doi.org/10.1073/pnas.1714900115.

[101] Beyder A, Rae JL, Bernard C, Strege PR, Sachs F, Farrugia G. Mechanosensitivity of Nav1.5, a voltage-sensitive sodium channel. J Physiol 2010;588(Pt 24):4969–85. https://doi.org/10.1113/jphysiol.2010.199034.

[102] Yang X, Ma G, Zheng S, Qin X, Li X, Du L, Wang Y, Zhou Y, Li M. Optical control of CRAC channels using photoswitchable azopyrazoles. J Am Chem Soc 2020;142(20):9460–70. https://doi.org/10.1021/jacs.0c02949.

[103] Huang D, Miller M, Ashok B, Jain S, Peppas NA. CRISPR/Cas systems to overcome challenges in developing the next generation of T cells for cancer therapy. Adv Drug Deliv Rev 2020. https://doi.org/10.1016/j.addr.2020.07.015. pii: S0169-409X(20)30099-5.

[104] Miri SM, Tafsiri E, Cho WCS, Ghaemi A. Correction to: CRISPR-Cas, a robust gene-editing technology in the era of modern cancer immunotherapy. Cancer Cell Int 2020;20:521. https://doi.org/10.1186/s12935-020-01609-w.

[105] Zhu H, Blum RH, Bernareggi D, Ask EH, Wu Z, Hoel HJ, Meng Z, Wu C, Guan KL, Malmberg KJ, Kaufman DS. Metabolic reprograming via deletion of CISH in human iPSC-derived NK cells promotes in vivo persistence and enhances anti-tumor activity. Cell Stem Cell 2020;27(2):224–237.e6. https://doi.org/10.1016/j.stem.2020.05.008.

[106] Blaes F, Tschernatsch M. Paraneoplastic neurological disorders. Expert Rev Neurother 2010;10(10):1559–68. https://doi.org/10.1586/ern.10.134.

[107] Williams JP, Carlson NG, Greenlee JE. Antibodies in autoimmune human neurological disease: pathogenesis and immunopathology. Semin Neurol 2018;38(3):267–77. https://doi.org/10.1055/s-0038-1660501.

[108] Schoser B, Eymard B, Datt J, Mantegazza R. Lambert-Eaton myasthenic syndrome (LEMS): a rare autoimmune presynaptic disorder often associated with cancer. J Neurol 2017;264(9):1854–63. https://doi.org/10.1007/s00415-017-8541-9.

[109] Titulaer MJ, Lang B, Verschuuren JJ. Lambert-Eaton myasthenic syndrome: from clinical characteristics to therapeutic strategies. Lancet Neurol 2011;10(12):1098–107. https://doi.org/10.1016/S1474-4422(11)70245-9.

[110] Li C, Wang X, Sun L, Deng H, Han Y, Zheng W. Anti-SOX1 antibody-positive paraneoplastic neurological syndrome presenting with Lambert-Eaton myasthenic syndrome and small cell lung cancer: a case report. Thorac Cancer 2020;11(2):465–9. https://doi.org/10.1111/1759-7714.13290.

[111] Huang K, Luo YB, Yang H. Autoimmune channelopathies at neuromuscular junction. Front Neurol 2019;10:516. https://doi.org/10.3389/fneur.2019.00516.

[112] Maddison P, Newsom-Davis J, Mills KR, Souhami RL. Favourable prognosis in Lambert-Eaton myasthenic syndrome and small-cell lung carcinoma. Lancet 1999;353(9147):117–8. https://doi.org/10.1016/S0140-6736(05)76153-5.

[113] Blandino JK, Viglione MP, Bradley WA, Oie HK, Kim YI. Voltage-dependent sodium channels in human small-cell lung cancer cells: role in action potentials and inhibition by Lambert-Eaton syndrome IgG. J Membr Biol 1995;143(2):153–63. https://doi.org/10.1007/BF00234661.

[114] Viglione MP, Blandino JK, Kim SJ, Kim YI. Effects of Lambert-Eaton syndrome serum and IgG on calcium and sodium currents in small-cell lung cancer cells. Ann N Y Acad Sci 1993;681:418–21. https://doi.org/10.1111/j.1749-6632.1993.tb22925.x.

[115] Rajaratinam H, Rasudin NS, Al Astani TAD, Mokhtar NF, Yahya MM, Zain WZW, Asma-Abdullah N, Fuad WEM. Breast cancer therapy affects the expression of antineonatal Nav1.5 antibodies in the serum of patients with breast cancer. Oncol Lett 2021;21(2):108. https://doi.org/10.3892/ol.2020.12369.

[116] Onganer PU, Djamgoz MB. Small-cell lung cancer (human): potentiation of endocytic membrane activity by voltage-gated Na(+) channel expression in vitro. J Membr Biol 2005;204(2):67–75. https://doi.org/10.1007/s00232-005-0747-6.

[117] Campbell TM, Main MJ, Fitzgerald EM. Functional expression of the voltage-gated Na^+-channel Nav1.7 is necessary for EGF-mediated invasion in human non-small cell lung cancer cells. J Cell Sci 2013;126(Pt 21):4939–49. https://doi.org/10.1242/jcs.130013.

[118] Jahanafrooz Z, Baradaran B, Mosafer J, Hashemzaei M, Rezaei T, Mokhtarzadeh A, Hamblin MR. Comparison of DNA and mRNA vaccines against cancer. Drug Discov Today 2020;25(3):552–60. https://doi.org/10.1016/j.drudis.2019.12.003.

[119] Naseri M, Bozorgmehr M, Zöller M, Ranaei Pirmardan E, Madjd Z. Tumor-derived exosomes: the next generation of promising cell-free vaccines in cancer immunotherapy. Onco Targets Ther 2020;9(1):1779991. https://doi.org/10.1080/2162402X.2020.1779991.

[120] Benhaim L, Gerger A, Bohanes P, Paez D, Wakatsuki T, Yang D, Labonte MJ, Ning Y, El-Khoueiry R, Loupakis F, Zhang W, Laurent-Puig P, Lenz HJ. Gender-specific profiling in SCN1A polymorphisms and time-to-recurrence in patients with stage II/III colorectal cancer treated with adjuvant 5-fluoruracil chemotherapy. Pharmacogenomics J 2014;14(2):135–41. https://doi.org/10.1038/tpj.2013.21.

[121] Sinha D, Smith C, Khanna R. Joining forces: improving clinical response to cellular immunotherapies with small-molecule inhibitors. Trends Mol Med 2020. https://doi.org/10.1016/j.molmed.2020.09.005. pii: S1471-4914(20)30222-7.

[122] Yildirim S, Altun S, Gumushan H, Patel A, Djamgoz MBA. Voltage-gated sodium channel activity promotes prostate cancer metastasis in vivo. Cancer Lett 2012;323(1):58–61. https://doi.org/10.1016/j.canlet.2012.03.036.

[123] Batcioglu K, Uyumlu AB, Satilmis B, Yildirim B, Yucel N, Demirtas H, Onkal R, Guzel RM, Djamgoz MB. Oxidative stress in the in vivo DMBA rat model of breast cancer: suppression by a voltage-gated sodium channel inhibitor (RS100642). Basic Clin Pharmacol Toxicol 2012;111(2):137–41. https://doi.org/10.1111/j.1742-7843.2012.00880.x.

[124] Driffort V, Gillet L, Bon E, Marionneau-Lambot S, Oullier T, Joulin V, Collin C, Pagès JC, Jourdan ML, Chevalier S, Bougnoux P, Le Guennec JY, Besson P, Roger S. Ranolazine inhibits NaV1.5-mediated breast cancer cell invasiveness and lung colonization. Mol Cancer 2014;13:264. https://doi.org/10.1186/1476-4598-13-264.

[125] Nelson M, Yang M, Millican-Slater R, Brackenbury WJ. Nav1.5 regulates breast tumor growth and metastatic dissemination in vivo. Oncotarget 2015;6(32):32914–29. https://doi.org/10.18632/oncotarget.5441.

[126] Bonavita E, Bromley CP, Jonsson G, Pelly VS, Sahoo S, Walwyn-Brown K, Mensurado S, Moeini A, Flanagan E, Bell CR, Chiang SC, Chikkanna-Gowda CP, Rogers N, Silva-Santos B, Jaillon S, Mantovani A, Reis E, Sousa C, Guerra N, Davis DM, Zelenay S. Antagonistic inflammatory phenotypes dictate tumor fate and response to immune checkpoint blockade. Immunity 2020. https://doi.org/10.1016/j.immuni.2020.10.020. pii: S1074-7613(20)30461-1.

[127] Shalapour S, Karin M. Pas de deux: control of anti-tumor immunity by cancer-associated inflammation. Immunity 2019;51:15–26.

[128] Jing W, Gershan JA, Johnson BD. Depletion of CD4 T cells enhances immunotherapy for neuroblastoma after syngeneic HSCT but compromises development of antitumor immune memory. Blood 2009;113(18):4449–57. https://doi.org/10.1182/blood-2008-11-190827.

[129] Bugan I, Kucuk S, Karagoz Z, Fraser SP, Kaya H, Dodson A, et al. Anti-metastatic effect of ranolazine in an in vivo rat model of prostate cancer, and expression of voltage-gated sodium channel protein in human prostate. Prostate Cancer Prostatic Dis 2019;22(4):569–79. https://doi.org/10.1038/s41391-019-0128-3.

[130] Newton HS, Gawali VS, Chimote AA, Lehn MA, Palackdharry SM, Hinrichs BH, Jandarov R, Hildeman D, Janssen EM, Wise-Draper TM, Conforti L. PD1 blockade enhances K + channel activity, Ca^{2+} signaling, and migratory ability in cytotoxic T lymphocytes of patients with head and neck cancer. J Immunother Cancer 2020;8(2). https://doi.org/10.1136/jitc-2020-000844, e000844.

[131] Geng S, Zhong Y, Wang S, Liu H, Zou Q, Xie X, Li C, Yu Q, He Z, Wang B. Amiloride enhances antigen specific CTL by faciliting HBV DNA vaccine entry into cells. PLoS One 2012;7(3). https://doi.org/10.1371/journal.pone.0033015, e33015.

[132] Zsiros E, Kis-Toth K, Hajdu P, Gaspar R, Bielanska J, Felipe A, Rajnavolgyi E, Panyi G. Developmental switch of the expression of ion channels in human dendritic cells. J Immunol 2009;183(7):4483–92. https://doi.org/10.4049/jimmunol.0803003.

[133] Derynck R, Weinberg RA. EMT and cancer: more than meets the eye. Dev Cell 2019;49(3):313–6. https://doi.org/10.1016/j.devcel.2019.04.026.

[134] Dongre A, Rashidian M, Reinhardt F, Bagnato A, Keckesova Z, Ploegh HL, Weinberg RA. Epithelial-to-mesenchymal transition contributes to immunosuppression in breast carcinomas. Cancer Res 2017;77(15):3982–9. https://doi.org/10.1136/jitc-2020-SITC2020.0232.

[135] Dongre A, Weinberg R, Rashidian M, Eaton E, Reinhardt F, Thiru P, Zagorulya M, Nepal S, Banaz T, Martner A, Spranger S. The epithelial-to-mesenchymal transition (EMT) contributes to

immunosuppression in breast carcinomas and regulates their response to immune checkpoint block-ade. J Immunother Cancer 2020;8. https://doi.org/10.1136/jitc-2020-SITC2020.0232.

[136] Jiang Y, Zhan H. Communication between EMT and PD-L1 signaling: new insights into tumor immune evasion. Cancer Lett 2020;468:72–81. https://doi.org/10.1016/j.canlet.2019.10.013.

[137] Gradek F, Lopez-Charcas O, Chadet S, Poisson L, Ouldamer L, Goupille C, Jourdan ML, Chevalier S, Moussata D, Besson P, Roger S. Sodium channel Nav1.5 controls epithelial-to-mesenchymal transition and invasiveness in breast cancer cells through its regulation by the salt-inducible kinase-1. Sci Rep 2019;9(1):18652. https://doi.org/10.1038/s41598-019-55197-5.

[138] Luo Q, Wu T, Wu W, Chen G, Luo X, Jiang L, Tao H, Rong M, Kang S, Deng M. The functional role of voltage-gated sodium channel nav1.5 in metastatic breast cancer. Front Pharmacol 2020;11:1111. https://doi.org/10.3389/fphar.2020.01111.

[139] Ramesh V, Brabletz T, Ceppi P. Targeting EMT in cancer with repurposed metabolic inhibitors. Trends Cancer 2020;6(11):942–50. https://doi.org/10.1016/j.trecan.2020.06.005.

[140] Stock C, Pedersen SF. Roles of pH and the Na^+/H^+ exchanger NHE1 in cancer: from cell biology and animal models to an emerging translational perspective? Semin Cancer Biol 2017;43:5–16. https://doi.org/10.1016/j.semcancer.2016.12.001.

[141] Mao W, Zhang J, Körner H, Jiang Y, Ying S. The emerging role of voltage-gated sodium channels in tumor biology. Front Oncol 2019;9:124. https://doi.org/10.3389/fonc.2019.00124.

[142] Pilon-Thomas S, Kodumudi KN, El-Kenawi AE, Russell S, Weber AM, Luddy K, et al. Neutralization of tumor acidity improves antitumor responses to immunotherapy. Cancer Res 2016;76 (6):1381–90. https://doi.org/10.1158/0008-5472.CAN-15-1743.

[143] Huber V, Camisaschi C, Berzi A, Ferro S, Lugini L, Triulzi T, Tuccitto A, Tagliabue E, Castelli C, Rivoltini L. Cancer acidity: an ultimate frontier of tumor immune escape and a novel target of immunomodulation. Semin Cancer Biol 2017;43:74–89. https://doi.org/10.1016/j. semcancer.2017.03.001.

[144] Jentzsch V, Davis JAA, Djamgoz MBA. Pancreatic cancer (PDAC): introduction of evidence-based complementary measures into integrative clinical management. Cancers (Basel) 2020;12(11):3096. https://doi.org/10.3390/cancers12113096.

[145] Uhl FM, Chen S, O'Sullivan D, Edwards-Hicks J, Richter G, Haring E, Andrieux G, Halbach S, Apostolova P, Büscher J, Duquesne S, Melchinger W, Sauer B, Shoumariyeh K, Schmitt-Graeff A, Kreutz M, Lübbert M, Duyster J, Brummer T, Boerries M, Madl T, Blazar BR, Groß O, Pearce EL, Zeiser R. Metabolic reprogramming of donor T cells enhances graft-versus-leukemia effects in mice and humans. Sci Transl Med 2020;12(567). https://doi.org/10.1126/scitranslmed.abb8969, eabb8969.

[146] Eil R, Vodnala SK, Clever D, Klebanoff CA, Sukumar M, Pan JH, Palmer DC, Gros A, Yamamoto TN, Patel SJ, Guittard GC, Yu Z, Carbonaro V, Okkenhaug K, Schrump DS, Linehan WM, Roychoudhuri R, Restifo NP. Ionic immune suppression within the tumour microenvironment limits T cell effector function. Nature 2016;537(7621):539–43. https://doi.org/10.1038/nature19364.

[147] Panyi G, Beeton C, Felipe A. Ion channels and anti-cancer immunity. Philos Trans R Soc Lond Ser B: Biol Sci 2014;369(1638):20130106. https://doi.org/10.1098/rstb.2013.0106.

[148] Amara S, Tiriveedhi V. Inflammatory role of high salt level in tumor microenvironment (review). Int J Oncol 2017;50(5):1477–81. https://doi.org/10.3892/ijo.2017.3936.

[149] Sharif K, Amital H, Shoenfeld Y. The role of dietary sodium in autoimmune diseases: the salty truth. Autoimmun Rev 2018;17(11):1069–73. https://doi.org/10.1016/j.autrev.2018.05.007.

[150] Willebrand R, Hamad I, Van Zeebroeck L, Kiss M, Bruderek K, Geuzens A, Swinnen D, Côrte-Real BF, Markó L, Lebegge E, Laoui D, Kemna J, Kammertoens T, Brandau S, Van Ginderachter JA, Kleinewietfeld M. High salt inhibits tumor growth by enhancing anti-tumor immunity. Front Immunol 2019;10:1141. https://doi.org/10.3389/fimmu.2019.01141.

[151] Willebrand R, Kleinewietfeld M. The role of salt for immune cell function and disease. Immunology 2018;154(3):346–53. https://doi.org/10.1111/imm.12915.

[152] Kuenzi BM, Park J, Fong SH, Sanchez KS, Lee J, Kreisberg JF, Ma J, Ideker T. Predicting drug response and synergy using a deep learning model of human cancer cells. Cancer Cell 2020. https://doi.org/10.1016/j.ccell.2020.09.014. pii: S1535-6108(20)30488-8.

CHAPTER FIVE

Immunoediting and cancer priming

Taichiro Goto

Lung Cancer and Respiratory Disease Center, Yamanashi Central Hospital, Yamanashi, Japan

Contents

Abbreviations

APCs	antigen-presenting cells
CCL	CC-chemokine ligand
CTLA-4	cytotoxic T-lymphocyte associated protein 4
DCs	dendritic cells
HLA	human leukocyte antigen
ICIs	immune checkpoint inhibitors
Ig	immunoglobulin
LOH	loss of heterozygosity
MHC I	major histocompatibility complex class I
NSCLCs	nonsmall-cell lung cancers
PD-L1	programmed cell death ligand 1
TCGA	The Cancer Genome Atlas
TCR	T-cell antigen receptor
TGFβ	transforming growth factor- β
TMB	tumor mutation burden

Cancer Immunology and Immunotherapy
https://doi.org/10.1016/B978-0-12-823397-9.00005-3

TRACERx   TRAcking Cancer Evolution through Therapy (Rx)
VEGF   vascular endothelial growth factor

1. Introduction

In 1891, William Coley noticed several clinical cases of spontaneous regression and disappearance of tumors in patients following infection [1, 2]. He conceived the idea of injecting bacteria into his cancer patients, which showed promising results and became known as Coley's toxin. In 1909, Paul Ehrlich postulated the concept of cancer immunosurveillance wherein subclinical tumor cells are continuously and spontaneously eradicated by the activated immune system [3]. This concept is now clearly established—primarily through demonstrations of increased cancer incidence in immunodeficient mice and humans [4]. However, cancer develops despite this immunosurveillance. The paradigm of cancer immunoediting reconciles the mechanisms of tumor surveillance with the presence of clinically evident cancers. Tumors derived from mutagenesis in immunodeficient mice are more immunogenic compared to those from immunocompetent mice [4, 5]. This phenomenon indicates that tumors develop differently in the presence of immune system. Specifically, the immune system applies a selective pressure to tumor development and growth, and tumors are forced to edit the surrounding immune system, which is known as cancer immunoediting [6, 7]. Cancer immunoediting involves three sequential phases within the context of the immune system: the elimination phase, whereby immunogenic tumors are eradicated; the equilibrium phase, whereby tumor cells and immune cells coexist; and, finally, the escape phase, wherein tumor escape from immune control. Clinically detectable cancers often represent this escape phase.

2. Cancer-immunity cycle

2.1 Cancer-immunity cycle outlined

The concept of a "cancer-immunity cycle," demonstrating the dynamic antitumor immune responses of cancer in seven steps, has been proposed (Fig. 1) [8]. Cancer antigens are released by tumor cells during apoptosis or cell death (Step 1). The antigens released are captured by dendritic cells (DCs), which then mature and simultaneously migrate to lymph nodes (Step 2). At the lymph nodes, the DCs present the captured cancer antigen to the major histocompatibility complex class (MHC) I molecule, resulting in the priming of T-cells (Step 3). The activated T-cells begin the migration to the tumor (Step 4), infiltrate the tumor tissue (Step 5), recognize cancer cells (Step 6), and injure them (Step 7). Steps 3 and 6 serve as immune checkpoints at which the activation of

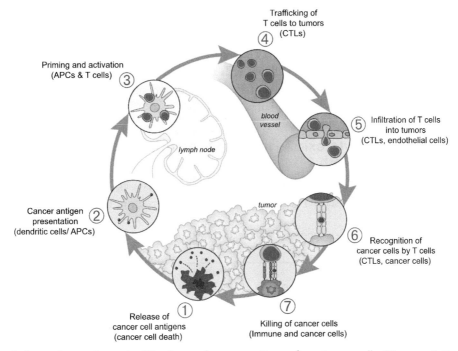

Fig. 1 Cancer-immunity cycle: (1) release of cancer antigens from tumor cells; (2) presentation of cancer antigens on the major histocompatibility complex class (MHC) by antigen-presenting cells; (3) recognition of cancer antigens on the MHC by the T-cell receptor, resulting in T-cell activation; (4) trafficking of activated T-cells; (5) infiltration into the tumor; (6) recognition of cancer antigens on the MHC within the tumor; (7) attack on tumor cells, resulting in tumor cell injury/death. *(Adopted from Chen DS, Mellman I. Oncology meets immunology: the cancer-immunity cycle. Immunity 2013;39 (1):1–10).*

T-cells is controlled. From the cancer cells injured by T-cells in Step 7, cancer antigens are further released, and the cancer-immunity cycle returns to Step 1 and continues. In many cancer patients, one or more of these steps can be interrupted, resulting in ineffective immune responses to cancer.

2.2 CD8-positive cytotoxic T-cells

A key step of antitumor immune responses is the activation of CD8-positive cytotoxic T-cells (CTLs). Cytotoxic T-cells, activated by the recognition of cancer-specific antigens presented by antigen-presenting cells (APCs), begin attacking cancer cells [9]. It has been shown that if the immune checkpoint molecules (co-suppression molecules) expressed on immunocompetent cells, such as CTLs are suppressed or the immune co-stimulatory molecules are stimulated, the CTLs are activated, resulting in the reinforcement of antitumor immune responses [10].

The action of co-stimulatory and co-suppression molecules involved in T-cell reactions and their action on effector T-cells and Tregs is explained in Table 1. In addition, the relationship between co-stimulatory and co-suppression molecules to their ligands is classified by the priming (naïve T-cells are activated by antigen presentation by APCs and proliferate; Fig. 2) and effector phases (the activated and proliferating cancer-specific T-cells gather in the cancer cells and attack; Fig. 3).

2.3 CD28 family: Cytotoxic T-lymphocyte associated protein 4 and programmed cell death 1

Cytotoxic T-lymphocyte-associated protein 4 (CTLA-4) is a co-suppression molecule not expressed on steady-state T-cells, but induced in a dynamic manner following the

Table 1 Co-stimulatory and co-suppression molecules involved in cancer immunity and their activities.

Co-Stimulatory Molecule	Ligand	Expression on Teff	Effects on Teff Function	Expression on Treg	Effects on Treg Function
CTLA-4	CD80, CD86	induced by stimulation	↓	constitutively expressed	↑
PD-1	PD-L1	induced by stimulation	↓	constitutively expressed	↑
ICOS	ICOSL	induced by stimulation	↑	constitutively expressed	↑
TIGIT	CD155	induced by stimulation	↓	constitutively expressed	↑
CD27	CD70	constitutively expressed	↑	constitutively expressed	↑
4-1BB	4-1BBL	induced by stimulation	↑	-	-
OX40	OX40L	induced by stimulation	↑	constitutively expressed	↓
GITR	GITRL	induced by stimulation	↑	constitutively expressed	↓
CD226	CD155	constitutively expressed	↑	-	-
BTLA	HVEM	constitutively expressed	↓	-	-
TIM-3	galectin-9	induced by stimulation	↓	constitutively expressed	↑
LAG-3	MHC-II	induced by stimulation	↓	induced by stimulation	↑

Red arrows indicate the signals reinforcing the immune responses to cancer, while *blue arrows* indicate the signals causing attenuation of such responses. *Teff*, effector T-cell; *Treg*, regulatory T-cell.

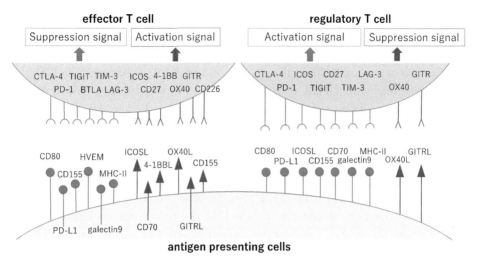

Fig. 2 Major co-stimulatory and co-suppression molecules involved in the priming phase of cancer immunity. The cancer antigen presented by antigen-presenting cells in the lymph nodes is recognized via the T-cell antigen receptor (TCR) by cancer-specific T-cells (signal 1, main signal). Major co-stimulatory and co-suppression molecules producing auxiliary stimuli (signal 2, subsignal) during this step and their ligands are shown here. The signals reinforcing the immune responses to cancer are shown in *red*, while those the attenuation of these responses are shown in *blue*. *(Adopted from my previous work Kunimasa K, Goto T. Immunosurveillance and immunoediting of lung cancer: current perspectives and challenges. Int J Mol Sci 2020;21(2)).*

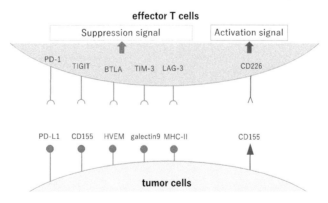

Fig. 3 Major co-stimulatory and co-suppression molecules involved in the effector phase of cancer immunity. This figure shows major co-stimulatory and co-suppression molecules (involved in the attack against the cancer cells by activated and proliferated effector T-cells after antigen presentation and recognition) and their ligands. The signals reinforcing the immune responses to cancer are shown in *red*, while the signals causing attenuation of such responses are shown in *blue*. *(Adopted from my previous work Kunimasa K, Goto T. Immunosurveillance and immunoediting of lung cancer: current perspectives and challenges. Int J Mol Sci 2020;21(2)).*

activation of effector CD4+ or CD8+ T-cells [11]. CD80 and CD86 expressed primarily on APCs serve as a ligand for CTLA-4 and CD28 (a representative co-stimulatory molecule) [12] (Fig. 4). CD28 activates the effector T-cells after recognition of a specific antigen and its activation induces the expression of CTLA-4 that binds to CD80 in a manner competing with CD28 with a 10–100-fold higher binding affinity [12, 13]. Co-stimulation by CD28 is thus attenuated, resulting in the suppression of immune responses. The CTLA-4 on Treg causes the attenuation of APC function indirectly by emitting a reverse signal to CD80 and CD86 or effector T-cell function, directly [14]. The development of systemic autoimmune diseases and death 2 months thereafter of CTLA-4 knock-out mice also supports its immunosuppressive activity [15].

Programmed cell death 1 (PD-1) is a cosuppression molecule expressed dynamically following the activation of effector T-cells [16]. Its representative ligand is PD-L1, expressed primarily on APCs and tumor cells (Fig. 4). The expression of PD-L1 on the tumor is induced by IFN-γ released from activated effector T-cells following antigen recognition via T-cell antigen receptor (TCR) [17]. In the tumor tissue, primarily CD8-positive cytotoxic lymphocytes having undergone activation show high levels of PD-1 expression. If PD-L1 binds to these lymphocytes, tyrosine dephosphorylases (SHP-1, SHP-2, etc.) gather into the intracellular part of the immune checkpoint molecules

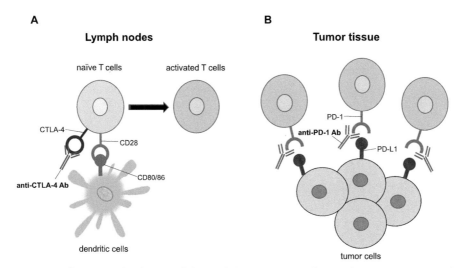

Fig. 4 Targets of immune checkpoint inhibitors. (A) Sensitization phase. When anti-CTLA-4 antibody binds to CTLA-4, the inhibitory effect on T-cells is suppressed. Consequently, effector T-cells are induced. (B) Effector phase. PD-1 blockade reverses immune evasion mediated by the interaction of PD-1$^+$ immune cells and PD-L1$^+$ tumor cells. *CTLA-4*, cytotoxic T-lymphocyte-associated protein 4; *PD-1*, programmed cell death-1; *PD-L1*, programmed cell death-ligand 1. *(Adopted from my previous work Kunimasa K, Goto T. Immunosurveillance and immunoediting of lung cancer: current perspectives and challenges. Int J Mol Sci 2020;21(2)).*

and inhibit tyrosine phosphorylation of ZAP70, resulting in T-cell dysfunction and the induction of apoptosis, thereby suppressing excessive antitumor immune responses [18]. Cancer antigen-specific CTLs invading tumor cells are selectively suppressed by PD-1, which is constitutively expressed on Treg, suggesting its role in Treg differentiation and function [19]. Supporting this, studies have shown that PD-1 knock-out mice develop diverse autoimmune diseases (e.g., nephritis, arthritis, and myocarditis) spontaneously [20, 21].

2.4 TNF receptor superfamily: OX-40 and glucocorticoid-induced TNFR-related (GITR) gene

OX-40 is a costimulatory molecule; its ligand is OX-40L, which is induced primarily following activation of T-cells and APCs [22]. The stimulatory signals of OX-40 promote activation, survival, and proliferation of effector cells and the production of cytokines [23]. OX-40 is constitutively expressed on Treg and involved in the formation and function of Treg. In many animal models, the anti-OX-40 agonist antibody has been reported to manifest potent antitumor activity by inducing the activation of effector T-cells and attenuation of Treg function [23].

GITR is expressed on effector T-cells at low levels, serving as a costimulatory molecule whose expression increases following effector T-cell activation [22]. Its ligand, GITRL, is expressed constitutively on APCs and endothelial cells [22]. GITR signals promote the activation and proliferation of effector T-cells and suppression of apoptosis. GITR is constitutively expressed on Treg, and GITR signals cause the attenuation of Treg's suppressive activity [24]. In many animal models, the anti-GITR agonist antibody is known to manifest antitumor activity by causing the activation of effector T-cells and attenuation of Treg's suppressive activity [24].

3. Tumors escape immunosurveillance through immunoediting

As revealed by the presence of occult and clinically imperceptible tumors, tumor cells that successfully evaded immunosurveillance can enter tumor dormancy (the equilibrium phase) [25] (Fig. 5). In the equilibrium phase, the adaptive immune system prevents tumor invasion and outgrowth and sculpts tumor immunogenicity. However, as a consequence of constant immune selection imposed on genetically unstable tumor cells held in equilibrium, tumor cells can acquire the ability to circumvent immune recognition, thereby avoiding destruction and eventually developing into clinically detectable tumors (the escape phase) [26, 27]. The mechanisms controlling tumor escape from immunosurveillance are diverse and include downregulation or loss of expression of MHC I molecules, which are essential for CD8+ cytotoxic T-cell recognition [28, 29], and increased expression of cytotoxic T-cell inhibitory ligands, such as programmed cell death ligand 1 (PD-L1), which suppresses cytotoxic T-cell attack [30] (Fig. 3).

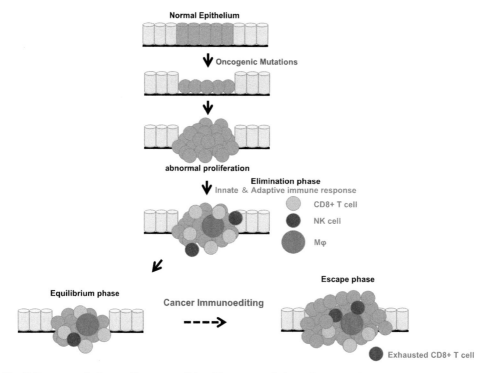

Fig. 5 Tumor evolution and immunoediting. The accumulation of oncogenic mutations initiates tumor evolution. The immune surveillance system, including cytotoxic CD8 positive T-cells, natural killer cells, and macrophages, recognize tumor-associated neoantigens presented by MHC molecules on antigen-presenting cells. The balance between signals from the tumor microenvironment and the immune system shifts during tumor elimination, equilibrium, and escape. *(Adopted from my previous work Kunimasa K, Goto T. Immunosurveillance and immunoediting of lung cancer: current perspectives and challenges. Int J Mol Sci 2020;21(2)).*

Cytotoxic T-cell activation involves multiple steps generated through the immune synapse between tumor and immune cells. The first signal is transduced through the binding of a T-cell receptor to its cognate peptide located on surface-expressed MHC molecules. This interaction assigns specificity to the consequence of downstream T-cell subsets. Many negative regulatory checkpoints are present to inhibit overactive immune responses and set a limit on T-cell activation. In order to escape detection by the immune system, tumor cells upregulate the surface expression of inhibitory molecules, including PD-L1, cytotoxic T-lymphocyte associated protein 4 (CTLA-4), LAG-3, TIM-3, and 4-1BB [31–35]. As a result of the constant exposure to antigens in the tumor microenvironment, intratumoral T-cells display a broad spectrum of suppressed and dysfunctional states, called T-cell exhaustion [36]. Thus, tumor cells evade immunosurveillance by editing their microenvironment, including immune cells and their neighboring normal cells. The escape phase has been an area of intense investigation in the field of tumor immunology over the past decades. Many studies have demonstrated that tumors in the escape phase evade the immune system through direct and/or indirect mechanisms to aid in their growth and metastases [26]. Focused work is now being undertaken to devise strategies that can target these mechanisms of escape since they represent a means to treat tumors using novel cancer immunotherapies [37, 38]. This review provides multiple examples of tumor immunoediting mainly in lung cancer.

4. Human leukocyte antigen (HLA) loss and immune escape in lung cancer evolution

The MHC, a set of genes that code for cell surface proteins, helps the immune system recognize foreign substances. MHC proteins are inherent in all higher vertebrates. The human MHC complex is synonymous with the human HLA complex. Silencing, downregulation, or loss of HLA alleles inhibits peptide antigen presentation and facilitates tumor cells to escape from immunosurveillance [28, 29, 39] (Fig. 6).

A recent report elucidated loss of heterozygosity (LOH) at the HLA alleles; in particular, loss of HLA-C*08:02 was observed in a resistant lesion treated with tumor-infiltrating lymphocytes composed of cytotoxic T-cell clones targeting the KRAS G12D mutation [40]. Since the presence of the HLA-C*08:02 allele is requisite for the presentation of the neoantigen KRAS G12D and tumor recognition by

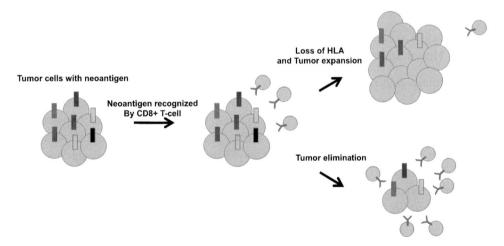

Fig. 6 Tumor escape through human leukocyte antigen (HLA) loss. HLA loss allows tumor cells to escape the immune system. During tumor evolution, the accumulation of tumor neoantigens induces local immune infiltration. Tumor cells with HLA loss can be positively selected by avoiding CD8 T-cell recognition. *(Adopted from my previous work Kunimasa K, Goto T. Immunosurveillance and immunoediting of lung cancer: current perspectives and challenges. Int J Mol Sci 2020;21(2)).*

T–lymphocytes, its loss was supposed to directly cause immune evasion. In hepatocellular carcinoma, tumors with HLA LOH exhibited significant association with an increased recurrence, indicating that immune escape caused by HLA LOH may accelerate tumor progression [41]. Downregulation of HLA in vulvar intraepithelial neoplasia was also associated with the development of recurrences [42]. Clinically, patients with high evasion capacity tend to have a higher recurrence rate compared to those with low evasion capacity [41, 42]. However, the magnitude and significance of the loss of the HLA haplotype has not been systematically evaluated in human cancers, because the polymorphic feature of the HLA locus precludes alignment of sequence reads to the human reference genome and estimation of copy number. To overcome this challenge, McGranahan *et al.* recently developed a new computational tool using next-generation sequencing data to estimate the allele-specific copy number of the HLA locus; loss of heterozygosity in the human leukocyte antigen (LOH HLA) was found to occur in approximately 40% of nonsmall-cell lung cancers (NSCLCs) and was associated with high PD-L1 expression [43]. The subclonal frequency of HLA LOH, which occurs in a subset of tumor cells and is located on the branches of evolutionary phylogenetic trees, suggests that HLA LOH is usually a late event in tumor phylogeny and that immune microenvironment in the tumor tissue may function as a key selective pressure in shaping branched tumor evolution. In a cohort of primary NSCLC tumors with matched brain metastasis, HLA LOH was detected in 47% of the cases, which occurred subclonally and preferentially at the metastatic brain sites. Thus, HLA LOH is a common feature of NSCLC and facilitates immune escape from immunosurveillance and significant immune editing, leading to subclonal genome evolution.

5. Heterogenous immunoediting in lung cancer

Multidimensional datasets, such as The Cancer Genome Atlas (TCGA), International Cancer Genome Consortium (ICGC), and Gene Expression Omnibus (GEO), enable the study of relationships between different biological processes, *e.g.*, genome-wide DNA sequencing, DNA methylation, gene expression, and copy number change, and the leveraging of multiple data types to draw inferences about biological systems [44–46]. Recent TCGA datasets have associated the genomic profile of tumors with tumor immunity, which implicates the neoantigen burden in promoting T-cell responses [47] and identifies somatic mutations in relation to immune infiltrates [48]. The TRACERx [TRAcking Cancer Evolution through therapy (Rx)] lung study is a multimillion pound research project taking place since April 2014 [49], which will transform our understanding of NSCLC and take a practical step toward an era of precision medicine. The study will reveal mechanisms of cancer evolution by analyzing the intratumor heterogeneity in lung tumors from approximately 850 patients and tracking its evolutionary trajectory from diagnosis through to relapse. The project enrolled patients to obtain samples of surgically resected NSCLC tumors in stages IA through IIIA for high-depth, multiregion whole-exome sequencing. TRACERx recently demonstrated that the immunological profiles can vary dramatically among different regions of the same early-stage tumor, similar to genomic heterogeneity, which can affect the prognosis of patients [50]. They also assessed immune cell infiltration in 258 samples from 88 treatment-naïve, early-stage, lung cancers by histological analyses and RNA sequencing. Intriguingly, heterogenous immune cell infiltration was associated with genomic heterogeneity; intratumor variations in tumor mutation burden (TMB), may confound some putative biomarkers for precision medicine. To examine the influence of tumor-infiltrating immune cells on tumor evolution, neoantigen burden was compared and analyzed. An analysis of nonneoantigenic, nonsynonymous mutations confirmed a significant reduction in expressed neoantigens—this depletion was limited to tumors with intact HLA alleles and, again, to tumors with a high level of immune cell infiltration, suggesting that these alterations are mediated by immune cells. These data suggest that the immune profile can vary markedly within each tumor and that there is a heterogeneity of immunoediting within the tumor, which affects tumor evolution.

The heterogeneity of immunoediting reflects the mutational burden of lung cancer [51], and the TCR repertoire represents the breadth and strength of the T-cell immune response [52]. The enriched TCR repertoire in both tumors and the surrounding normal tissue are composed of thousands of different TCR repertoires. Many of them are present at low frequency and a fraction contains bystander T-cell populations [53]. In contrast, the expanded intratumor TCR repertoire is differentially expressed in the tumor compared to the adjacent normal tissue. Intriguingly, the number of expanded intratumor TCR repertoires significantly correlates with the number of nonsynonymous mutations, and the heterogeneity of intratumor TCR repertoire correlates with spatial mutational heterogeneity. Moreover, the number of ubiquitous and regional TCRs correlates with the number of ubiquitous and regional nonsynonymous mutations, respectively. Thus, heterogeneity of the mutation landscape can affect immune landscape heterogeneity.

The immune microenvironment also affects tumor evolution at different metastatic organs [54, 55]. Examining the influence of the tumor microenvironment on the metastatic potential of tumor cells has been hindered by the vast diversity of infiltrating immune cells [56] as well as the extensive interactions between tumor cells and neighboring normal cells [57]. Extensive analysis of a patient with metastatic colorectal cancer during an 11-year-old spatiotemporal follow-up revealed that cancer clonal evolution patterns during metastatic progression depend on the immune context at metastatic sites [58]. Moreover, neoantigen depletion was observed in metastatic sites with high levels of tumor-infiltrating T-cells [58]. The immunoedited tumor clones were eliminated, whereas the progressive clones were immune privileged in spite of the presence of tumor-infiltrating immune cells [58, 59]. Tumor immunoediting influences both intra- and intertumor heterogeneity and sculpts tumor clonal evolution.

6. Neoantigen derived from mutation

6.1 Tumor antigens

Tumor antigens can be differentiated into five categories: (1) viral antigens; (2) differentiation antigens; (3) cancer–germline antigens; (4) overexpressed antigens; and (5) neoantigens (Fig. 7) [60–63]. As antigens (1)–(4) can be expressed in normal tissues as well as cancer tissues, these antigens are more likely to promote immunological tolerance and are less likely to bring about effective antitumor immune responses [64].

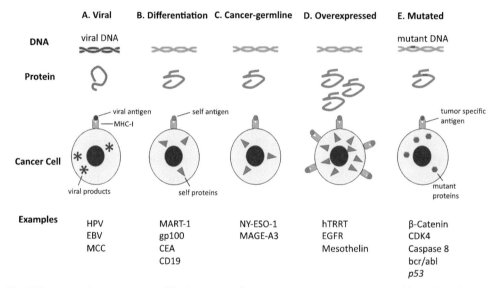

Fig. 7 Tumor antigens recognized by immune cells. Tumor antigens are categorized based on the pattern of gene expression. The production of antigenic peptides by cancer cells is illustrated herein. (A) Viral antigens are only expressed in virus-infected cells. (B) Differentiation antigens are encoded by genes with tissue-specific expression. (C) Cancer-germline genes are expressed in tumors or germ cells because of whole-genome demethylation. (D) Some genes are overexpressed in tumors owing to increased transcription or gene amplification. The resulting peptides are upregulated on these tumors, but also show a low level of expression in some noncancerous tissues. (E) However, mutated genes may yield a mutant peptide (neoantigen), which is recognized as nonself by immune cells. *CEA*, carcinoembryonic Ag; *EGFR*, epidermal growth factor receptor; *EBV*, Epstein-Barr virus; *HPV*, human papillomavirus; *hTERT*, human telomerase reverse transcriptase; *MAGE-A3*, melanoma-associated antigen 3; *MART-1*, melanoma antigen recognized by T-cells-1; *MCC*, Merkel cell carcinoma. *(Adapted from our previous article Goto T. Radiation as an in situ auto-vaccination: current perspectives and challenges. Vaccines (Basel) 2019;7(3). Adopted from my previous work Kunimasa K, Goto T. Immunosurveillance and immunoediting of lung cancer: current perspectives and challenges. Int J Mol Sci 2020;21(2)).*

However, neoantigens are supposed to be expressed exclusively in cancer cells due to genomic mutations altering the amino acid sequence. This type of antigen is tumor-specific and can cause an immune response sufficient to kill tumor cells when activated [65–68].

Thus, tumor-specific antigens originating from cancer cell-specific gene mutation exist as nonself, because they are absent in normal cells and present in cancer cells alone and are called "neoantigens." The peptide fragment (neoantigen) containing amino acid mutants processed from the mutant protein of cancer cells binds to the MHC molecule and is presented on the cell surface. It can efficiently induce highly reactive specific T-cells without undergoing immunotolerance. Gene mutations of cancer cells can be divided into driver mutations (directly associated with cellular oncogenesis) and passenger mutations (not associated with cellular oncogenesis), and both can yield neoantigens [69]. Neoantigens originating from driver mutations of *BRAF, KRAS,* and *p53* have a high potential of being distributed extensively among cancer patients [70]. A vaccine with wide applicability can be developed using these antigens as the target [71]. If the driver mutation is set as the target, immune escape through the loss of antigen from cancer cells is less likely to occur, and higher clinical efficacy is expected. However, it is incorrect to conclude that the entire peptide sequence, including the part of the driver mutation, is presented by APCs and recognized by T-cells. In fact, driver mutations containing peptide sequences less likely to be presented as the antigens are found more frequently in cancer cells [72]. In contrast, neoantigens originating from passenger mutation occur at a much higher frequency in cancer cells. However, interindividual variations in passenger mutations among patients make their detection difficult using conventional technology. Recently, the development of next-generation sequencers enables easier detection through whole-exome analysis [73, 74]. In addition, gene fusions are also identified as a source of immunogenic neoantigens, which can mediate anticancer immune responses [75, 76]. Their computational prediction from DNA or RNA sequencing data necessitates specialized bioinformatics expertise to assemble a computational workflow including the prediction of translated peptide and peptide-HLA binding affinity [73, 76]. Thus, personalized cancer immunotherapy may be developed by identifying neoantigen from the gene mutations (mostly passenger mutations), which vary from one case to another and setting a target of treatment at the identified neoantigen.

6.2 Antitumor immune responses by neoantigen-specific T-cells

In recent years, the clinical efficacy of immune checkpoint inhibitors has been demonstrated, motivating the clinical use of these inhibitors in patients with various cancers [77, 78]. However, since the response rate to these inhibitors is low, exploration of efficacy-predictive biomarkers identifying patients expected to respond to these inhibitors has been conducted worldwide, and close attention has been paid to the tumor mutational burden as one possible predictor [79, 80]. The responses to immune checkpoint inhibitors correlate positively with the total number of gene mutations, and therapies using these inhibitors have been reported to be particularly effective against cancers involving several gene mutations due to extrinsic factors (ultraviolet ray, smoking, etc.) such as malignant melanomas and squamous cell carcinomas of the lungs [81, 82]. Furthermore, as an intrinsic factor, it has been reported that patients with cancers involving the accumulation of gene mutations due to deficient mismatch repairs (dMMR) respond more markedly to the anti-PD-1 antibody [83]. This antibody has been used extensively in the clinical practice against many types of solid cancers, which often shows microsatellite instability (MSI), a marker of dMMR [84]. It has been estimated that an increase in the number of gene mutations in cancer cells is associated with an increase in the number of neoantigens formed from such mutations, resulting in an increase in neoantigen-specific T-cells, which are activated by immune checkpoint inhibitors and manifest antitumor activity [83, 85].

Recently, there has been an increase in the number of reports directly suggesting the presence of neoantigen-specific T-cells among cancer patients and the clinical significance of the presence of such cells [86]. Zacharakis et al. infused tumor-infiltrating lymphocytes, containing four types of neoantigen-specific T-cell clones, into patients with breast cancer and concomitantly administered immune checkpoint inhibitors to these patients and reported that the metastatic foci subsided and the cancer was eradicated completely [87]. Moreover, several studies have also shown that when the antigenic specificity of infused lymphocytes was investigated in cancer patients having survived years following T-cell infusion therapy, the neoantigen-recognizing T-cell clones were identified with high frequency [88]. Thus, neoantigen-specific T-cells are believed to play a central role in antitumor immune responses.

In addition, Anagnostou et al. demonstrated that among the patients with NSCLC who responded to immune checkpoint inhibitors, the disappearance of a total of 41 neoantigens (7–18 antigens per case) was noted in the four cases where the disease recurred [52]. The specific T-cells against the disappearing neoantigens were detected during the effective period, but decreased during disease progression, suggesting that tumor reduction in response to immune checkpoint inhibitors is mediated by immune responses to neoantigens and that the disappearance of neoantigens serves as one possible mechanism for the development of resistance to therapy [52, 89].

The immunosurveillance and immunoediting mechanisms of cancer exist, but the likelihood of the manifestation of these mechanisms can vary depending on the cancer development process or tumor microenvironments of different types of cancer [90, 91]. Immunotherapy using immune checkpoint inhibitors can trigger therapy-induced immunoediting (immune reconstruction) in some cancers, possibly leading to the restoration of cancer elimination/equilibrium, while stimulating therapy-induced tumor escape in others [52]. It is, therefore, necessary to understand the mechanisms involved in immunoediting by cancer or therapy-induced immunoediting and devise a strategy capable of preventing the disappearance of cancer antigens and avoiding therapy-induced tumor escape.

7. Treg and tumor immunity

The immune system distinguishes self from nonself-molecules. Immunotolerance to self-molecules, *i.e.*, a lack of an immune response to molecules, allows the normal self-tissue to escape immune system attacks. Lymphocytes possessing receptors binding strongly to self-antigens undergo apoptosis following negative selection in the thymus, resulting in the elimination before maturation [92]. However, stimulation of peripheral blood mononuclear cells by the tumor-self antigen can induce self-antigen specific CD8 + T-cells. Thus, elimination does not occur for all self-reactive T-cells. Based on these findings, it was anticipated that self-reactive T-cells are regulated by some mechanisms and later that Treg is involved in this mechanism [93, 94].

Regulatory T-cells (Treg) are a group of CD4 T-cells that suppress various immune responses. They express high levels of IL-2 receptor α chain (CD25), CTLA-4, and transcription factor FoxP3 [94]. CD25 + CD4 + T-cells account for approximately 10% of all CD4 + T-cells found in the peripheral blood of normal individuals [95]. Tregs are mostly formed in the thymus as a subset of T-cells specialized in immunosuppression. Part of Treg undergoes differentiation from naïve T-cells in peripheral lymph tissue (intestinal lymph tissue among others) under certain conditions. It enables self-immunotolerance through the suppression of self-reactive immune responses and thus, plays an important role in the suppression of autoimmune disease [96]. Thymus-derived Foxp3 + CD25 + CD4 + Treg is indispensable for the introduction and maintenance of peripheral immune self-tolerance. For example, if peripheral T-cells of the same strain of normal mice are transferred into T-cell-deficient nude mice after removal of CD25 + CD4 + T-cells (removal of Treg from normal animals), autoimmune diseases such as thyroiditis and gastritis develop spontaneously [97]. The onset of these diseases can be prevented by the replenishment of the Treg. Furthermore, if the aforementioned mice are then inoculated with tumor cells originating from the same strain of mouse, potent antitumor immune responses are induced [98]. If the proliferation of allografted antigen-specific Treg is attempted in the same experimental system, immunotolerance for the organ graft can be induced [99].

In case of malignant tumors, Treg suppresses the antitumor immune responses, thus contributing to the proliferation of tumor cells. The tumor tissues often have a large number of Treg that are activated after stimulation with the antigen, and the presence of Treg in the tumor serves as a biomarker of a poor prognosis [100].

Tregs manifest immunosuppressive functions via multiple mechanisms, with the most important mechanism being the inhibition of T-cell activation through the suppression of APCs [101, 102]. Tregs constitutively express CTLA-4, which binds to the

CD80/CD86 of APCs, causing their downregulation and the suppression of their T-cell activating capability [103, 104]. In addition, as Treg expresses high levels of CD25 (IL-2 receptor α chain), it has a high affinity for IL-2. Owing to the transcription factor FoxP3 suppressing the expression of the IL-2 gene, Treg cannot produce IL-2 on its own. Thus, it survives by consuming the IL-2 produced by other cells [105], resulting in the depletion of IL-2 in local environments and making it impossible for effector T-cells to become sufficiently activated. Tregs also suppress immune responses by producing immunosuppressive cytokines such as transforming growth factor-β (TGF-β) and IL-10 [100, 106].

Therefore, if immunosuppression by Treg is controlled, anticancer immunotherapy may become more effective. With this expectation, various attempts at removing Treg by targeting IL-2 and CD25 have been made but have been clinically ineffective. To the best of our knowledge, no agent capable of removing Treg in a reliable manner has been developed to date [107].

In humans, FoxP3 can also be induced by stimulating naïve T-cells using antigens; hence, it is not completely Treg-specific. The appropriate definition of a Treg is essential in the development of new drugs. Depending on the combination of CD45RA (a marker of naïve T-cell) and FoxP3, FoxP3 positive T-cells can be divided into three fractions: (1) CD45RA+FoxP3^{low}CD4+ T-cell (naïve type Treg); (2) CD45RA-FoxP3^{high}CD4 + T-cell (effector type Treg); and (3) CD45RA-FoxP3^{low}CD4+ T-cell (non-Treg without immunosuppressive activity) (Fig. 8) [101, 108]. Of these fractions, the effector type Treg

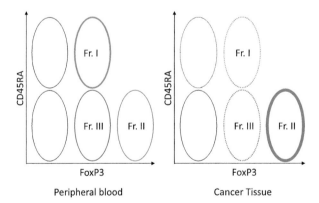

Fig. 8 Foxp3 + T-cell subset in the blood and tumor tissue. FoxP3^{low}CD45RA+ naïve Treg (Fr. I) can be viewed as cells that have just been supplied from the thymus. If these naïve Treg are activated by the TCR signal, CD45RA becomes negative and FoxP3 is upregulated, yielding FoxP3^{high}CD45RA- effector Treg (Fr. II). Effector Treg (Fr. II) expresses high levels of CCR4. Regarding FoxP3+ cells, the CD45RA negative FoxP3^{low} CD4+ T-cell (Fr. III) which has no immunosuppressive activity is not Treg, but a helper T-cell producing IFN-γ or IL-17 when activated. *(Adopted from my previous work Kunimasa K, Goto T. Immunosurveillance and immunoediting of lung cancer: current perspectives and challenges. Int J Mol Sci 2020;21(2)).*

has the most potent immunosuppressive activity. Most of the Treg invading the tumor tissue is the effector-type Treg, while naïve-type Treg is abundantly seen in peripheral blood [109]. At present, attempts are being made to develop new means of treatment by which the peripheral naïve-type Treg is preserved (by setting a target at the CCR4 selectively expressed on the effector-type Treg) and antitumor immune responses are activated by setting a target only at the effector-type Treg in the tumor tissue [109, 110].

8. Tumor-associated macrophages (TAMs) and tumor immunity

Macrophages are cells, found in various tissues of the body, possessing diverse functions such as ontogeny, homeostasis of organisms, tissue repair, and immune responses to pathogen infection. They are conventionally viewed as differentiating from monocytes among leukocytes. In recent years, however, it has been shown in multiple tissues that the macrophages indigenous in the tissue during the fetal period undergo proliferation in that tissue [111, 112]. More importantly, the macrophages that invade the tumor stroma are called TAM and stimulate the development, progression, and metastasis of tumors (Fig. 9) [113–115]. In many tumors, TAM is recruited into microenvironments by CCL2, CSF-1, IL-10, TGF-β, etc. secreted from cancer cells. CSF-1 plays a particularly important role in TAM as it contributes not only to TAM differentiation, but also invasion into tumor tissue [116]. Owing to these activities, the proliferation of Lewis lung carcinoma cells implanted subcutaneously is suppressed in mice functionally lacking CSF-1 [117].

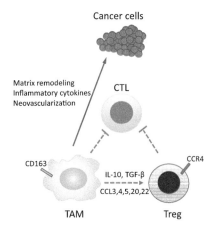

Fig. 9 Roles of TAM in the tumor environment. TAM has the nature of M2-macrophage, directly suppressing the CTL, but also producing mediators involved in neovascularization, tumor infiltration/metastasis, and Treg migration/maintenance. The immunosuppressive activity is represented as a *blue line* and the tumor-stimulating activity as a *red line*. (Adopted from my previous work Kunimasa K, Goto T. Immunosurveillance and immunoediting of lung cancer: current perspectives and challenges. Int J Mol Sci 2020;21(2)).

In a mouse model of breast cancer, defects of CSF-1 caused delays in the development of metastatic tumors [118].

Two representative mechanisms are known for the immunosuppressive activity of TAM. The first involves regulation through direct interactions with immunocompetent cells [119]. Here, TAM suppresses the activation of the immune mechanism by the direct transmission of negative signals to the immunocompetent cells [119]. It expresses immune checkpoint molecules, such as PD-L1 and CD80/CD86, which are recognized by PD-1 and CTLA-4, respectively, present on the CTL, resulting in inactivation of these CTLs [120]. Furthermore, TAM expresses HLA-G and HLA-E, which are suppressive MHC-1 molecules [121]. HLA-G is recognized by LIT-2 (leukocyte immunoglobulin-like receptor 2) expressed on CD4+ T-cells, resulting in suppressed activation of these cells [120]. Similarly, HLA-E is recognized by NKG2 on natural killer cells, resulting in suppressed migration of these cells and suppressed secretion of IFN-γ [120]. The second mechanism involves the recruitment of Treg. TAM regulates immune responses by recruiting Treg into the tumor microenvironment [121]. TAMs are known to recruit natural Treg into tumor tissues by producing chemokines (CC-chemokine ligands: CCL3, CCL4, CCL5, CCL20, CCL22) [122]. Furthermore, by producing IL-10 and TGF-β, TAM activates the transcription factor Foxp3 of CD4+ T-cells, possibly contributing to the induction of inducible Treg [121, 123].

Other than these mechanisms, immune responses are regulated by chemokines, cytokines, etc. secreted from TAM and cancer cells. For example, chronic pathogen infections stimulate the continuous production of inflammatory cytokines (IFN-γ, TNF, IL-6, etc.) by macrophages, resulting in chronic inflammation [124]. Such a state of chronic inflammation may stimulate tumor progression. Furthermore, TAM induces cancer cell migration, infiltration, intravascular invasion, and neovascularization needed for tumor growth, and this can lead to tumor metastasis [122]. For example, in breast cancer, CSF-1 secreted from cancer cells and vascular endothelial growth factor (VEGF) secreted from TAM mutually stimulate the secretion of these factors, resulting in the acceleration of tumor infiltration and intravascular invasion [125]. TAM is also involved in remodeling of the tumor microenvironment through the expression of proteases such as matrix metalloproteinase (MMP), cathepsin S, urokinase type plasminogen activation factor, and matrix remodeling enzymes, such as lysyl oxidase and SPARC [122]. These proteases cut the extracellular matrix and create free space around the tumor, thus stimulating the secretion of growth factors, such as heparin-binding epidermal growth factor (HB-EGF) and resulting in tumor infiltration and metastasis [122].

Macrophages also play an important role in tumor neovascularization. TAMs secrete neovascularization factors, such as VEGF, TNF, IL-1β, IL-8, PDGF, and FGF and are thus, involved in the stimulation of neovascularization around the tumor [126]. In the tumor microenvironment, a low oxygen supply (hypoxia) can arise from malnutrition, extracellular pH reduction, and insufficient blood flow. For cancer cells to survive, hypoxia needs to be avoided by inducing neovascularization and increasing blood flow to the tumor environment—the hypoxia inducible factor (HIF) plays an important role in this process. Excessive activation of HIF by TAM is often seen, resulting in the stimulation of VEGF expression and enhancement of neovascularization [127]. The molecular mechanism for this process has been increasingly elucidated. Recent studies have demonstrated that the lactic acid formed by cancer cell's energy metabolism stabilizes TAM, resulting in the induction of VEGF expression [128].

The involvement of macrophages in tumor metastasis has also been observed in various studies. For example, in the foci of lung cancer metastasis, a small blood clot collecting cancer cells is first formed. Then, CCL2 is secreted from these cancer cells and recruits CCR2+Ly6c+ inflammatory monocytes, resulting in differentiation into Ly6c- metastasis associated macrophage (MAM) [129, 130]. The monocytes and MAM recruited into the foci of metastasized tissue stimulates extravascular migration of cancer cells through VEGF expression. Moreover, cancer cell VECAM1 (CD106) is linked to MAM's integrin α4 (VECAM1 counter-receptor), and MAM is involved in increases in metastatic cancer cell viability and tumor growth [129, 130]. Likewise, defects in MAM inhibit tumor growth [125]. As illustrated in Fig. 9, TAM enables the tumor to escape immunity through diverse mechanisms.

9. Conclusions

The cancer immunosurveillance and immunoediting mechanism hypotheses began with the proposal of the attractive concept "recognition of the immune system by cancer" and has stimulated discussion over the effective, but incomplete control of cancer by the immune system. Thus, the importance of neoantigens based on individual cancer mutations has begun attracting close attention. The extreme success of ICIs in advanced tumors, especially in lung cancer and melanoma, has elucidated how our immune system can be leveraged to eradicate cancers. ICIs have broken new ground in the cancer therapy field. However, there are still many patients suffering from advanced cancers with no hopeful treatment options. Moreover, a new challenge is that tumors can acquire resistance to ICIs. To overcome these problems, we will need to more deeply understand the fundamental mechanisms of cancer escape from the immune system and develop new treatment strategies. The fight against cancer to utilize the immune system has just begun and is still budding. We anticipate that the next generation of cancer immunotherapies will be one step closer to the eradication of cancer.

Funding
This research received no external funding.

Conflicts of interest

The authors declare no conflict of interest.

References

[1] Nauts HC, Swift WE, Coley BL. The treatment of malignant tumors by bacterial toxins as developed by the late William B. Coley, M.D., reviewed in the light of modern research. Cancer Res 1946;6:205–16.

[2] McCarthy EF. The toxins of William B. Coley and the treatment of bone and soft-tissue sarcomas. Iowa Orthop J 2006;26:154–8.

[3] Krause RM. Paul Ehrlich and O.T. Avery: pathfinders in the search for immunity. Vaccine 1999;17 (Suppl. 3):S64–7.

[4] Shankaran V, Ikeda H, Bruce AT, White JM, Swanson PE, Old LJ, Schreiber RD. IFNgamma and lymphocytes prevent primary tumour development and shape tumour immunogenicity. Nature 2001;410(6832):1107–11.

[5] Svane IM, Engel AM, Thomsen AR, Werdelin O. The susceptibility to cytotoxic T lymphocyte mediated lysis of chemically induced sarcomas from immunodeficient and normal mice. Scand J Immunol 1997;45(1):28–35.

[6] Dunn GP, Old LJ, Schreiber RD. The three Es of cancer immunoediting. Annu Rev Immunol 2004;22:329–60.

[7] Schreiber RD, Old LJ, Smyth MJ. Cancer immunoediting: integrating immunity's roles in cancer suppression and promotion. Science 2011;331(6024):1565–70.

[8] Chen DS, Mellman I. Oncology meets immunology: the cancer-immunity cycle. Immunity 2013;39 (1):1–10.

[9] Higuchi R, Goto T, Hirotsu Y, Nakagomi T, Yokoyama Y, Otake S, Amemiya K, Oyama T, Omata M. PD-L1 expression and tumor-infiltrating lymphocytes in thymic epithelial neoplasms. J Clin Med 2019;8(11).

[10] Lesokhin AM, Callahan MK, Postow MA, Wolchok JD. On being less tolerant: enhanced cancer immunosurveillance enabled by targeting checkpoints and agonists of T cell activation. Sci Transl Med 2015;7(280):280sr1.

[11] Pentcheva-Hoang T, Corse E, Allison JP. Negative regulators of T-cell activation: potential targets for therapeutic intervention in cancer, autoimmune disease, and persistent infections. Immunol Rev 2009;229(1):67–87.

[12] Sharma P, Allison JP. The future of immune checkpoint therapy. Science 2015;348(6230):56–61.

[13] Collins AV, Brodie DW, Gilbert RJ, Iaboni A, Manso-Sancho R, Walse B, Stuart DI, van der Merwe PA, Davis SJ. The interaction properties of costimulatory molecules revisited. Immunity 2002;17 (2):201–10.

[14] Wing K, Onishi Y, Prieto-Martin P, Yamaguchi T, Miyara M, Fehervari Z, Nomura T, Sakaguchi S. CTLA-4 control over Foxp3+ regulatory T cell function. Science 2008;322(5899):271–5.

[15] Callahan MK, Wolchok JD, Allison JP. Anti-CTLA-4 antibody therapy: immune monitoring during clinical development of a novel immunotherapy. Semin Oncol 2010;37(5):473–84.

[16] Francisco LM, Sage PT, Sharpe AH. The PD-1 pathway in tolerance and autoimmunity. Immunol Rev 2010;236:219–42.

[17] Blank C, Kuball J, Voelkl S, Wiendl H, Becker B, Walter B, Majdic O, Gajewski TF, Theobald M, Andreesen R, Mackensen A. Blockade of PD-L1 (B7-H1) augments human tumor-specific T cell responses in vitro. Int J Cancer 2006;119(2):317–27.

[18] Okazaki T, Chikuma S, Iwai Y, Fagarasan S, Honjo T. A rheostat for immune responses: the unique properties of PD-1 and their advantages for clinical application. Nat Immunol 2013;14(12):1212–8.

[19] Inozume T, Hanada K, Wang QJ, Ahmadzadeh M, Wunderlich JR, Rosenberg SA, Yang JC. Selection of CD8+PD-1+ lymphocytes in fresh human melanomas enriches for tumor-reactive T cells. J Immunother 2010;33(9):956–64.

[20] Nishimura H, Nose M, Hiai H, Minato N, Honjo T. Development of lupus-like autoimmune diseases by disruption of the PD-1 gene encoding an ITIM motif-carrying immunoreceptor. Immunity 1999;11(2):141–51.

[21] Nishimura H, Okazaki T, Tanaka Y, Nakatani K, Hara M, Matsumori A, Sasayama S, Mizoguchi A, Hiai H, Minato N, Honjo T. Autoimmune dilated cardiomyopathy in PD-1 receptor-deficient mice. Science 2001;291(5502):319–22.

[22] Croft M, Duan W, Choi H, Eun SY, Madireddi S, Mehta A. TNF superfamily in inflammatory disease: translating basic insights. Trends Immunol 2012;33(3):144–52.

[23] Melero I, Hirschhorn-Cymerman D, Morales-Kastresana A, Sanmamed MF, Wolchok JD. Agonist antibodies to TNFR molecules that costimulate T and NK cells. Clin Cancer Res 2013;19 (5):1044–53.

[24] Knee DA, Hewes B, Brogdon JL. Rationale for anti-GITR cancer immunotherapy. Eur J Cancer 2016;67:1–10.

[25] Loeb LA, Loeb KR, Anderson JP. Multiple mutations and cancer. Proc Natl Acad Sci U S A 2003;100 (3):776–81.

[26] Beatty GL, Gladney WL. Immune escape mechanisms as a guide for cancer immunotherapy. Clin Cancer Res 2015;21(4):687–92.

[27] Vinay DS, Ryan EP, Pawelec G, Talib WH, Stagg J, Elkord E, Lichtor T, Decker WK, Whelan RL, Kumara H, Signori E, Honoki K, Georgakilas AG, Amin A, Helferich WG, Boosani CS, Guha G, Ciriolo MR, Chen S, Mohammed SI, Azmi AS, Keith WN, Bilsland A, Bhakta D, Halicka D, Fujii H, Aquilano K, Ashraf SS, Nowsheen S, Yang X, Choi BK, Kwon BS. Immune evasion in cancer: mechanistic basis and therapeutic strategies. Semin Cancer Biol 2015;35(Suppl.):S185–98.

[28] Campoli M, Ferrone S. HLA antigen changes in malignant cells: epigenetic mechanisms and biologic significance. Oncogene 2008;27(45):5869–85.

[29] Hiraki A, Fujii N, Murakami T, Kiura K, Aoe K, Yamane H, Masuda K, Maeda T, Sugi K, Darzynkiewicz Z, Tanimoto M, Harada M. High frequency of allele-specific down-regulation of HLA class I expression in lung cancer cell lines. Anticancer Res 2004;24(3a):1525–8.

[30] Hicklin DJ, Marincola FM, Ferrone S. HLA class I antigen downregulation in human cancers: T-cell immunotherapy revives an old story. Mol Med Today 1999;5(4):178–86.

[31] Leach DR, Krummel MF, Allison JP. Enhancement of antitumor immunity by CTLA-4 blockade. Science 1996;271(5256):1734–6.

[32] Hurwitz AA, Yu TF, Leach DR, Allison JP. CTLA-4 blockade synergizes with tumor-derived granulocyte-macrophage colony-stimulating factor for treatment of an experimental mammary carcinoma. Proc Natl Acad Sci U S A 1998;95(17):10067–71.

[33] Ishida Y, Agata Y, Shibahara K, Honjo T. Induced expression of PD-1, a novel member of the immunoglobulin gene superfamily, upon programmed cell death. EMBO J 1992;11(11):3887–95.

[34] Iwai Y, Hamanishi J, Chamoto K, Honjo T. Cancer immunotherapies targeting the PD-1 signaling pathway. J Biomed Sci 2017;24(1):26.

[35] Crespo J, Sun H, Welling TH, Tian Z, Zou W. T cell anergy, exhaustion, senescence, and stemness in the tumor microenvironment. Curr Opin Immunol 2013;25(2):214–21.

[36] Wherry EJ, Kurachi M. Molecular and cellular insights into T cell exhaustion. Nat Rev Immunol 2015;15(8):486–99.

[37] Mellman I, Coukos G, Dranoff G. Cancer immunotherapy comes of age. Nature 2011;480 (7378):480–9.

[38] Sharma P, Wagner K, Wolchok JD, Allison JP. Novel cancer immunotherapy agents with survival benefit: recent successes and next steps. Nat Rev Cancer 2011;11(11):805–12.

[39] Mehta AM, Jordanova ES, Kenter GG, Ferrone S, Fleuren GJ. Association of antigen processing machinery and HLA class I defects with clinicopathological outcome in cervical carcinoma. Cancer Immunol Immunother 2008;57(2):197–206.

[40] Tran E, Robbins PF, Lu YC, Prickett TD, Gartner JJ, Jia L, Pasetto A, Zheng Z, Ray S, Groh EM, Kriley IR, Rosenberg SA. T-cell transfer therapy targeting mutant KRAS in cancer. N Engl J Med 2016;375(23):2255–62.

[41] Klippel ZK, Chou J, Towlerton AM, Voong LN, Robbins P, Bensinger WI, Warren EH. Immune escape from NY-ESO-1-specific T-cell therapy via loss of heterozygosity in the MHC. Gene Ther 2014;21(3):337–42.

[42] van Esch EM, Tummers B, Baartmans V, Osse EM, Ter Haar N, Trietsch MD, Hellebrekers BW, Holleboom CA, Nagel HT, Tan LT, Fleuren GJ, van Poelgeest MI, van der Burg SH, Jordanova ES. Alterations in classical and nonclassical HLA expression in recurrent and progressive HPV-induced usual vulvar intraepithelial neoplasia and implications for immunotherapy. Int J Cancer 2014;135(4):830–42.

[43] McGranahan N, Swanton C. Cancer evolution constrained by the immune microenvironment. Cell 2017;170(5):825–7.

[44] Vogelstein B, Papadopoulos N, Velculescu VE, Zhou S, Diaz Jr LA, Kinzler KW. Cancer genome landscapes. Science 2013;339(6127):1546–58.

[45] Cancer Genome Atlas Research N. Comprehensive molecular profiling of lung adenocarcinoma. Nature 2014;511(7511):543–50.

[46] Cancer Genome Atlas Research N. Comprehensive genomic characterization of squamous cell lung cancers. Nature 2012;489(7417):519–25.

[47] Brown SD, Warren RL, Gibb EA, Martin SD, Spinelli JJ, Nelson BH, Holt RA. Neo-antigens predicted by tumor genome meta-analysis correlate with increased patient survival. Genome Res 2014;24(5):743–50.

[48] Rutledge WC, Kong J, Gao J, Gutman DA, Cooper LA, Appin C, Park Y, Scarpace L, Mikkelsen T, Cohen ML, Aldape KD, McLendon RE, Lehman NL, Miller CR, Schniederjan MJ, Brennan CW, Saltz JH, Moreno CS, Brat DJ. Tumor-infiltrating lymphocytes in glioblastoma are associated with specific genomic alterations and related to transcriptional class. Clin Cancer Res 2013;19 (18):4951–60.

[49] Jamal-Hanjani M, Wilson GA, McGranahan N, Birkbak NJ, Watkins TBK, Veeriah S, Shafi S, Johnson DH, Mitter R, Rosenthal R, Salm M, Horswell S, Escudero M, Matthews N, Rowan A, Chambers T, Moore DA, Turajlic S, Xu H, Lee SM, Forster MD, Ahmad T, Hiley CT, Abbosh C, Falzon M, Borg E, Marafioti T, Lawrence D, Hayward M, Kolvekar S, Panagiotopoulos N, Janes SM, Thakrar R, Ahmed A, Blackhall F, Summers Y, Shah R, Joseph L, Quinn AM, Crosbie PA, Naidu B, Middleton G, Langman G, Trotter S, Nicolson M, Remmen H, Kerr K, Chetty M, Gomersall L, Fennell DA, Nakas A, Rathinam S, Anand G, Khan S, Russell P, Ezhil V, Ismail B, Irvin-Sellers M, Prakash V, Lester JF, Kornaszewska M, Attanoos R, Adams H, Davies H, Dentro S, Taniere P, O'Sullivan B, Lowe HL, Hartley JA, Iles N, Bell H, Ngai Y, Shaw JA, Herrero J, Szallasi Z, Schwarz RF, Stewart A, Quezada SA, Le Quesne J, Van Loo P, Dive C, Hackshaw A, Swanton C, Consortium TR. Tracking the evolution of non-small-cell lung cancer. N Engl J Med 2017;376(22):2109–21.

[50] Rosenthal R, Cadieux EL, Salgado R, Bakir MA, Moore DA, Hiley CT, Lund T, Tanic M, Reading JL, Joshi K, Henry JY, Ghorani E, Wilson GA, Birkbak NJ, Jamal-Hanjani M, Veeriah S, Szallasi Z, Loi S, Hellmann MD, Feber A, Chain B, Herrero J, Quezada SA, Demeulemeester J, Van Loo P, Beck S, McGranahan N, Swanton C, Consortium TR. Neoantigen-directed immune escape in lung cancer evolution. Nature 2019;567(7749):479–85.

[51] Joshi K, Robert de Massy M, Ismail M, Reading JL, Uddin I, Woolston A, Hatipoglu E, Oakes T, Rosenthal R, Peacock T, Ronel T, Noursadeghi M, Turati V, Furness AJS, Georgiou A, Wong YNS, Ben Aissa A, Werner Sunderland M, Jamal-Hanjani M, Veeriah S, Birkbak NJ, Wilson GA, Hiley CT, Ghorani E, Guerra-Assuncao JA, Herrero J, Enver T, Hadrup SR, Hackshaw A, Peggs KS, McGranahan N, Swanton C, Consortium TR, Quezada SA, Chain B. Spatial heterogeneity of the T cell receptor repertoire reflects the mutational landscape in lung cancer. Nat Med 2019;25(10): 1549–59.

[52] Anagnostou V, Smith KN, Forde PM, Niknafs N, Bhattacharya R, White J, Zhang T, Adleff V, Phallen J, Wali N, Hruban C, Guthrie VB, Rodgers K, Naidoo J, Kang H, Sharfman W, Georgiades C, Verde F, Illei P, Li QK, Gabrielson E, Brock MV, Zahnow CA, Baylin SB, Scharpf RB, Brahmer JR, Karchin R, Pardoll DM, Velculescu VE. Evolution of neoantigen landscape during immune checkpoint blockade in non-small cell lung cancer. Cancer Discov 2017;7(3):264–76.

[53] Simoni Y, Becht E, Fehlings M, Loh CY, Koo SL, Teng KWW, Yeong JPS, Nahar R, Zhang T, Kared H, Duan K, Ang N, Poidinger M, Lee YY, Larbi A, Khng AJ, Tan E, Fu C, Mathew R, Teo M, Lim WT, Toh CK, Ong BH, Koh T, Hillmer AM, Takano A, Lim TKH, Tan EH, Zhai W, Tan DSW, Tan IB, Newell EW. Bystander CD8(+) T cells are abundant and phenotypically distinct in human tumour infiltrates. Nature 2018;557(7706):575–9.

[54] Otake S, Goto T. Stereotactic radiotherapy for oligometastasis. Cancers (Basel) 2019;11(2).

[55] Yachida S, Iacobuzio-Donahue CA. Evolution and dynamics of pancreatic cancer progression. Oncogene 2013;32(45):5253–60.

[56] Bindea G, Mlecnik B, Tosolini M, Kirilovsky A, Waldner M, Obenauf AC, Angell H, Fredriksen T, Lafontaine L, Berger A, Bruneval P, Fridman WH, Becker C, Pages F, Speicher MR, Trajanoski Z, Galon J. Spatiotemporal dynamics of intratumoral immune cells reveal the immune landscape in human cancer. Immunity 2013;39(4):782–95.

[57] Labiano S, Palazon A, Melero I. Immune response regulation in the tumor microenvironment by hypoxia. Semin Oncol 2015;42(3):378–86.

[58] Angelova M, Mlecnik B, Vasaturo A, Bindea G, Fredriksen T, Lafontaine L, Buttard B, Morgand E, Bruni D, Jouret-Mourin A, Hubert C, Kartheuser A, Humblet Y, Ceccarelli M, Syed N, Marincola FM, Bedognetti D, Van den Eynde M, Galon J. Evolution of metastases in space and time under immune selection. Cell 2018;175(3):751–65. e16.

[59] van der Heijden M, Miedema DM, Waclaw B, Veenstra VL, Lecca MC, Nijman LE, van Dijk E, van Neerven SM, Lodestijn SC, Lenos KJ, de Groot NE, Prasetyanti PR, Arricibita Varea A, Winton DJ, Medema JP, Morrissey E, Ylstra B, Nowak MA, Bijlsma MF, Vermeulen L. Spatiotemporal regulation of clonogenicity in colorectal cancer xenografts. Proc Natl Acad Sci U S A 2019;116(13):6140–5.

[60] Hutchison S, Pritchard AL. Identifying neoantigens for use in immunotherapy. Mamm Genome 2018;29(11−12):714–30.

[61] Ilyas S, Yang JC. Landscape of tumor antigens in T cell immunotherapy. J Immunol 2015;195(11):5117–22.

[62] Vigneron N. Human tumor antigens and cancer immunotherapy. Biomed Res Int 2015;2015:948501.

[63] Goto T. Radiation as an in situ auto-vaccination: current perspectives and challenges. Vaccines (Basel) 2019;7(3).

[64] Cloosen S, Arnold J, Thio M, Bos GM, Kyewski B, Germeraad WT. Expression of tumor-associated differentiation antigens, MUC1 glycoforms and CEA, in human thymic epithelial cells: implications for self-tolerance and tumor therapy. Cancer Res 2007;67(8):3919–26.

[65] Aleksic M, Liddy N, Molloy PE, Pumphrey N, Vuidepot A, Chang KM, Jakobsen BK. Different affinity windows for virus and cancer-specific T-cell receptors: implications for therapeutic strategies. Eur J Immunol 2012;42(12):3174–9.

[66] Lennerz V, Fatho M, Gentilini C, Frye RA, Lifke A, Ferel D, Wolfel C, Huber C, Wolfel T. The response of autologous T cells to a human melanoma is dominated by mutated neoantigens. Proc Natl Acad Sci U S A 2005;102(44):16013–8.

[67] Segal NH, Parsons DW, Peggs KS, Velculescu V, Kinzler KW, Vogelstein B, Allison JP. Epitope landscape in breast and colorectal cancer. Cancer Res 2008;68(3):889–92.

[68] Tan MP, Gerry AB, Brewer JE, Melchiori L, Bridgeman JS, Bennett AD, Pumphrey NJ, Jakobsen BK, Price DA, Ladell K, Sewell AK. T cell receptor binding affinity governs the functional profile of cancer-specific CD8+ T cells. Clin Exp Immunol 2015;180(2):255–70.

[69] Nakagomi T, Goto T, Hirotsu Y, Shikata D, Yokoyama Y, Higuchi R, Otake S, Amemiya K, Oyama T, Mochizuki H, Omata M. Genomic characteristics of invasive mucinous adenocarcinomas of the lung and potential therapeutic targets of B7-H3. Cancers (Basel) 2018;10(12).

[70] Lu YC, Robbins PF. Targeting neoantigens for cancer immunotherapy. Int Immunol 2016;28(7):365–70.

[71] Goto T, Hirotsu Y, Mochizuki H, Nakagomi T, Shikata D, Yokoyama Y, Oyama T, Amemiya K, Okimoto K, Omata M. Mutational analysis of multiple lung cancers: discrimination between primary and metastatic lung cancers by genomic profile. Oncotarget 2017;8(19):31133–43.

[72] Marty R, Kaabinejadian S, Rossell D, Slifker MJ, van de Haar J, Engin HB, de Prisco N, Ideker T, Hildebrand WH, Font-Burgada J, Carter H. MHC-I genotype restricts the oncogenic mutational landscape. Cell 2017;171(6):1272–83. e15.

[73] Jiang T, Shi T, Zhang H, Hu J, Song Y, Wei J, Ren S, Zhou C. Tumor neoantigens: from basic research to clinical applications. J Hematol Oncol 2019;12(1):93.

[74] Yi M, Qin S, Zhao W, Yu S, Chu Q, Wu K. The role of neoantigen in immune checkpoint blockade therapy. Exp Hematol Oncol 2018;7:28.

[75] Fotakis G, Rieder D, Haider M, Trajanoski Z, Finotello F. NeoFuse: predicting fusion neoantigens from RNA sequencing data. Bioinformatics 2020;36(7):2260–1.

[76] Yang W, Lee KW, Srivastava RM, Kuo F, Krishna C, Chowell D, Makarov V, Hoen D, Dalin MG, Wexler L, Ghossein R, Katabi N, Nadeem Z, Cohen MA, Tian SK, Robine N, Arora K, Geiger H, Agius P, Bouvier N, Huberman K, Vanness K, Havel JJ, Sims JS, Samstein RM, Mandal R, Tepe J, Ganly I, Ho AL, Riaz N, Wong RJ, Shukla N, Chan TA, Morris LGT. Immunogenic neoantigens derived from gene fusions stimulate T cell responses. Nat Med 2019;25(5):767–75.

[77] Motzer RJ, Escudier B, McDermott DF, George S, Hammers HJ, Srinivas S, Tykodi SS, Sosman JA, Procopio G, Plimack ER, Castellano D, Choueiri TK, Gurney H, Donskov F, Bono P, Wagstaff J, Gauler TC, Ueda T, Tomita Y, Schutz FA, Kollmannsberger C, Larkin J, Ravaud A, Simon JS, Xu LA, Waxman IM, Sharma P, CheckMate I. Nivolumab versus Everolimus in advanced renal-cell carcinoma. N Engl J Med 2015;373(19):1803–13.

[78] Reck M, Rodriguez-Abreu D, Robinson AG, Hui R, Csoszi T, Fulop A, Gottfried M, Peled N, Tafreshi A, Cuffe S, O'Brien M, Rao S, Hotta K, Leiby MA, Lubiniecki GM, Shentu Y, Rangwala R, Brahmer JR, Investigators K. Pembrolizumab versus chemotherapy for PD-L1-positive non-small-cell lung cancer. N Engl J Med 2016;375(19):1823–33.

[79] Galuppini F, Dal Pozzo CA, Deckert J, Loupakis F, Fassan M, Baffa R. Tumor mutation burden: from comprehensive mutational screening to the clinic. Cancer Cell Int 2019;19:209.

[80] Klempner SJ, Fabrizio D, Bane S, Reinhart M, Peoples T, Ali SM, Sokol ES, Frampton G, Schrock AB, Anhorn R, Reddy P. Tumor mutational burden as a predictive biomarker for response to immune checkpoint inhibitors: a review of current evidence. Oncologist 2020;25(1):e147–59.

[81] Snyder A, Makarov V, Merghoub T, Yuan J, Zaretsky JM, Desrichard A, Walsh LA, Postow MA, Wong P, Ho TS, Hollmann TJ, Bruggeman C, Kannan K, Li Y, Elipenahli C, Liu C, Harbison CT, Wang L, Ribas A, Wolchok JD, Chan TA. Genetic basis for clinical response to CTLA-4 blockade in melanoma. N Engl J Med 2014;371(23):2189–99.

[82] Yarchoan M, Hopkins A, Jaffee EM. Tumor mutational burden and response rate to PD-1 inhibition. N Engl J Med 2017;377(25):2500–1.

[83] Le DT, Durham JN, Smith KN, Wang H, Bartlett BR, Aulakh LK, Lu S, Kemberling H, Wilt C, Luber BS, Wong F, Azad NS, Rucki AA, Laheru D, Donehower R, Zaheer A, Fisher GA, Crocenzi TS, Lee JJ, Greten TF, Duffy AG, Ciombor KK, Eyring AD, Lam BH, Joe A, Kang SP, Holdhoff M, Danilova L, Cope L, Meyer C, Zhou S, Goldberg RM, Armstrong DK, Bever KM, Fader AN, Taube J, Housseau F, Spetzler D, Xiao N, Pardoll DM, Papadopoulos N, Kinzler KW, Eshleman JR, Vogelstein B, Anders RA, Diaz Jr LA. Mismatch repair deficiency predicts response of solid tumors to PD-1 blockade. Science 2017;357(6349):409–13.

[84] Mandal R, Samstein RM, Lee KW, Havel JJ, Wang H, Krishna C, Sabio EY, Makarov V, Kuo F, Blecua P, Ramaswamy AT, Durham JN, Bartlett B, Ma X, Srivastava R, Middha S, Zehir A, Hechtman JF, Morris LG, Weinhold N, Riaz N, Le DT, Diaz Jr LA, Chan TA. Genetic diversity of tumors with mismatch repair deficiency influences anti-PD-1 immunotherapy response. Science 2019;364(6439):485–91.

[85] Rizvi NA, Hellmann MD, Snyder A, Kvistborg P, Makarov V, Havel JJ, Lee W, Yuan J, Wong P, Ho TS, Miller ML, Rekhtman N, Moreira AL, Ibrahim F, Bruggeman C, Gasmi B, Zappasodi R, Maeda Y, Sander C, Garon EB, Merghoub T, Wolchok JD, Schumacher TN, Chan TA. Cancer immunology. Mutational landscape determines sensitivity to PD-1 blockade in non-small cell lung cancer. Science 2015;348(6230):124–8.

[86] Yamamoto TN, Kishton RJ, Restifo NP. Developing neoantigen-targeted T cell-based treatments for solid tumors. Nat Med 2019;25(10):1488–99.

[87] Zacharakis N, Chinnasamy H, Black M, Xu H, Lu YC, Zheng Z, Pasetto A, Langhan M, Shelton T, Prickett T, Gartner J, Jia L, Trebska-McGowan K, Somerville RP, Robbins PF, Rosenberg SA, Goff SL, Feldman SA. Immune recognition of somatic mutations leading to complete durable regression in metastatic breast cancer. Nat Med 2018;24(6):724–30.

[88] Tran E, Turcotte S, Gros A, Robbins PF, Lu YC, Dudley ME, Wunderlich JR, Somerville RP, Hogan K, Hinrichs CS, Parkhurst MR, Yang JC, Rosenberg SA. Cancer immunotherapy based on mutation-specific CD4+ T cells in a patient with epithelial cancer. Science 2014;344 (6184):641–5.

[89] Nejo T, Matsushita H, Karasaki T, Nomura M, Saito K, Tanaka S, Takayanagi S, Hana T, Takahashi S, Kitagawa Y, Koike T, Kobayashi Y, Nagae G, Yamamoto S, Ueda H, Tatsuno K, Narita Y, Nagane M, Ueki K, Nishikawa R, Aburatani H, Mukasa A, Saito N, Kakimi K. Reduced neoantigen

expression revealed by longitudinal multiomics as a possible immune evasion mechanism in glioma. Cancer Immunol Res 2019;7(7):1148–61.

[90] Aragon-Sanabria V, Kim GB, Dong C. From cancer immunoediting to new strategies in cancer immunotherapy: the roles of immune cells and mechanics in oncology. Adv Exp Med Biol 2018;1092:113–38.

[91] O'Donnell JS, Teng MWL, Smyth MJ. Cancer immunoediting and resistance to T cell-based immunotherapy. Nat Rev Clin Oncol 2019;16(3):151–67.

[92] Kisielow P. How does the immune system learn to distinguish between good and evil? The first definitive studies of T cell central tolerance and positive selection. Immunogenetics 2019;71(8–9):513–8.

[93] Sakaguchi S, Miyara M, Costantino CM, Hafler DA. FOXP3+ regulatory T cells in the human immune system. Nat Rev Immunol 2010;10(7):490–500.

[94] Sakaguchi S, Yamaguchi T, Nomura T, Ono M. Regulatory T cells and immune tolerance. Cell 2008;133(5):775–87.

[95] Liu Z, Kim JH, Falo Jr LD, You Z. Tumor regulatory T cells potently abrogate antitumor immunity. J Immunol 2009;182(10):6160–7.

[96] Horwitz DA, Fahmy TM, Piccirillo CA, La Cava A. Rebalancing immune homeostasis to treat autoimmune diseases. Trends Immunol 2019;40(10):888–908.

[97] Sakaguchi S, Sakaguchi N, Asano M, Itoh M, Toda M. Immunologic self-tolerance maintained by activated T cells expressing IL-2 receptor alpha-chains (CD25). Breakdown of a single mechanism of self-tolerance causes various autoimmune diseases. J Immunol 1995;155(3):1151–64.

[98] Shimizu J, Yamazaki S, Sakaguchi S. Induction of tumor immunity by removing CD25+CD4+ T cells: a common basis between tumor immunity and autoimmunity. J Immunol 1999;163 (10):5211–8.

[99] Verma ND, Robinson CM, Carter N, Wilcox P, Tran GT, Wang C, Sharland A, Nomura M, Plain KM, Bishop GA, Hodgkinson SJ, Hall BM. Alloactivation of naive CD4(+)CD8(−)CD25(+)T regulatory cells: expression of CD8alpha identifies potent suppressor cells that can promote transplant tolerance induction. Front Immunol 2019;10:2397.

[100] Nishikawa H, Sakaguchi S. Regulatory T cells in tumor immunity. Int J Cancer 2010;127(4):759–67.

[101] Miyara M, Yoshioka Y, Kitoh A, Shima T, Wing K, Niwa A, Parizot C, Taflin C, Heike T, Valeyre D, Mathian A, Nakahata T, Yamaguchi T, Nomura T, Ono M, Amoura Z, Gorochov G, Sakaguchi S. Functional delineation and differentiation dynamics of human CD4+ T cells expressing the FoxP3 transcription factor. Immunity 2009;30(6):899–911.

[102] Sakaguchi S, Wing K, Onishi Y, Prieto-Martin P, Yamaguchi T. Regulatory T cells: how do they suppress immune responses? Int Immunol 2009;21(10):1105–11.

[103] Qureshi OS, Zheng Y, Nakamura K, Attridge K, Manzotti C, Schmidt EM, Baker J, Jeffery LE, Kaur S, Briggs Z, Hou TZ, Futter CE, Anderson G, Walker LS, Sansom DM. Trans-endocytosis of CD80 and CD86: a molecular basis for the cell-extrinsic function of CTLA-4. Science 2011;332 (6029):600–3.

[104] Yamaguchi T, Kishi A, Osaki M, Morikawa H, Prieto-Martin P, Wing K, Saito T, Sakaguchi S. Construction of self-recognizing regulatory T cells from conventional T cells by controlling CTLA-4 and IL-2 expression. Proc Natl Acad Sci U S A 2013;110(23):E2116–25.

[105] Setoguchi R, Hori S, Takahashi T, Sakaguchi S. Homeostatic maintenance of natural Foxp3(+) CD25 (+) CD4(+) regulatory T cells by interleukin (IL)-2 and induction of autoimmune disease by IL-2 neutralization. J Exp Med 2005;201(5):723–35.

[106] Joosse ME, Nederlof I, Walker LSK, Samsom JN. Tipping the balance: inhibitory checkpoints in intestinal homeostasis. Mucosal Immunol 2019;12(1):21–35.

[107] Jacobs JF, Nierkens S, Figdor CG, de Vries IJ, Adema GJ. Regulatory T cells in melanoma: the final hurdle towards effective immunotherapy? Lancet Oncol 2012;13(1):e32–42.

[108] Nishikawa H, Sakaguchi S. Regulatory T cells in cancer immunotherapy. Curr Opin Immunol 2014;27:1–7.

[109] Sugiyama D, Nishikawa H, Maeda Y, Nishioka M, Tanemura A, Katayama I, Ezoe S, Kanakura Y, Sato E, Fukumori Y, Karbach J, Jager E, Sakaguchi S. Anti-CCR4 mAb selectively depletes effector-type FoxP3+CD4+ regulatory T cells, evoking antitumor immune responses in humans. Proc Natl Acad Sci U S A 2013;110(44):17945–50.

[110] Onishi Y, Fehervari Z, Yamaguchi T, Sakaguchi S. Foxp3+ natural regulatory T cells preferentially form aggregates on dendritic cells in vitro and actively inhibit their maturation. Proc Natl Acad Sci U S A 2008;105(29):10113–8.

[111] Bain CC, Hawley CA, Garner H, Scott CL, Schridde A, Steers NJ, Mack M, Joshi A, Guilliams M, Mowat AM, Geissmann F, Jenkins SJ. Long-lived self-renewing bone marrow-derived macrophages displace embryo-derived cells to inhabit adult serous cavities. Nat Commun 2016;7, ncomms11852.

[112] Guerriero JL. Macrophages: the road less traveled, changing anticancer therapy. Trends Mol Med 2018;24(5):472–89.

[113] Wynn TA, Chawla A, Pollard JW. Macrophage biology in development, homeostasis and disease. Nature 2013;496(7446):445–55.

[114] Qian BZ, Pollard JW. Macrophage diversity enhances tumor progression and metastasis. Cell 2010;141(1):39–51.

[115] Sharma SK, Chintala NK, Vadrevu SK, Patel J, Karbowniczek M, Markiewski MM. Pulmonary alveolar macrophages contribute to the premetastatic niche by suppressing antitumor T cell responses in the lungs. J Immunol 2015;194(11):5529–38.

[116] Quail DF, Joyce JA. Molecular pathways: deciphering mechanisms of resistance to macrophage-targeted therapies. Clin Cancer Res 2017;23(4):876–84.

[117] Nowicki A, Szenajch J, Ostrowska G, Wojtowicz A, Wojtowicz K, Kruszewski AA, Maruszynski M, Aukerman SL, Wiktor-Jedrzejczak W. Impaired tumor growth in colony-stimulating factor 1 (CSF-1)-deficient, macrophage-deficient op/op mouse: evidence for a role of CSF-1-dependent macrophages in formation of tumor stroma. Int J Cancer 1996;65(1):112–9.

[118] Lin EY, Nguyen AV, Russell RG, Pollard JW. Colony-stimulating factor 1 promotes progression of mammary tumors to malignancy. J Exp Med 2001;193(6):727–40.

[119] Wang J, Li D, Cang H, Guo B. Crosstalk between cancer and immune cells: role of tumor-associated macrophages in the tumor microenvironment. Cancer Med 2019;8(10):4709–21.

[120] Quaranta V, Schmid MC. Macrophage-mediated subversion of anti-tumour immunity. Cells 2019;8 (7), 747.

[121] Noy R, Pollard JW. Tumor-associated macrophages: from mechanisms to therapy. Immunity 2014;41(1):49–61.

[122] Wang J, Yang L, Yu L, Wang YY, Chen R, Qian J, Hong ZP, Su XS. Surgery-induced monocytic myeloid-derived suppressor cells expand regulatory T cells in lung cancer. Oncotarget 2017;8 (10):17050–8.

[123] Biswas SK, Allavena P, Mantovani A. Tumor-associated macrophages: functional diversity, clinical significance, and open questions. Semin Immunopathol 2013;35(5):585–600.

[124] Capece D, Fischietti M, Verzella D, Gaggiano A, Cicciarelli G, Tessitore A, Zazzeroni F, Alesse E. The inflammatory microenvironment in hepatocellular carcinoma: a pivotal role for tumor-associated macrophages. Biomed Res Int 2013;2013:187204.

[125] Qian BZ, Zhang H, Li J, He T, Yeo EJ, Soong DY, Carragher NO, Munro A, Chang A, Bresnick AR, Lang RA, Pollard JW. FLT1 signaling in metastasis-associated macrophages activates an inflammatory signature that promotes breast cancer metastasis. J Exp Med 2015;212(9):1433–48.

[126] Rahma OE, Hodi FS. The intersection between tumor angiogenesis and immune suppression. Clin Cancer Res 2019;25(18):5449–57.

[127] Ugel S, De Sanctis F, Mandruzzato S, Bronte V. Tumor-induced myeloid deviation: when myeloid-derived suppressor cells meet tumor-associated macrophages. J Clin Invest 2015;125(9):3365–76.

[128] Colegio OR, Chu NQ, Szabo AL, Chu T, Rhebergen AM, Jairam V, Cyrus N, Brokowski CE, Eisenbarth SC, Phillips GM, Cline GW, Phillips AJ, Medzhitov R. Functional polarization of tumour-associated macrophages by tumour-derived lactic acid. Nature 2014;513(7519):559–63.

[129] Kitamura T, Qian BZ, Soong D, Cassetta L, Noy R, Sugano G, Kato Y, Li J, Pollard JW. CCL2-induced chemokine cascade promotes breast cancer metastasis by enhancing retention of metastasis-associated macrophages. J Exp Med 2015;212(7):1043–59.

[130] Qian BZ, Li J, Zhang H, Kitamura T, Zhang J, Campion LR, Kaiser EA, Snyder LA, Pollard JW. CCL2 recruits inflammatory monocytes to facilitate breast-tumour metastasis. Nature 2011;475 (7355):222–5.

CHAPTER SIX

Metabolic reprogramming and immunity in cancer

Yu Chen and Yongsheng Li

Department of Medical Oncology, Chongqing University Cancer Hospital, Chongqing, China

Contents

Abbreviations

3-PG	3-phosphoglycerate
5-FU	5-fluorouracil
6-MP	6-mercaptopurine
ACC	acetyl-CoA carboxylase
acetyl-CoA	acetyl-coenzyme A
ACLY	ATP citrate lyase
ACSS2	acetyl-CoA synthetase 2
ACT	adoptive cell transfer
ADCC	antibody-dependent cell-mediated cytotoxicity
AhR	aromatic hydrocarbon receptor
ALL	acute lymphoblastic leukemia
AMPK	AMP-activated protein kinase
APC	antigen-presenting cells
Arg	arginine
ARG1	arginase 1

Cancer Immunology and Immunotherapy
https://doi.org/10.1016/B978-0-12-823397-9.00006-5

ASNS	asparagine synthase
Atg7	autophagy-related protein 7
ATP	adenosine triphosphate
BA	bile acid
BCG	bacille Calmette-Guérin
BMSC	bone marrow-derived mesenchymal stem cell
CAR-T	chimeric antigen receptor T-cell immunotherapy
CAT2	cationic amino acid transporter
CCR9	C-C motif chemokine receptor 9
CDC	complement-dependent cytotoxicity
CgA	chromogranin A
CIC	citrate carrier
CLL	chronic lymphocytic leukemia
CNS	central nervous system
COX-2	cyclooxygenase-2
CP	cisplatin-pemetrexed
CR	cisplatin-raltitrexed
CRC	colorectal cancer
CTLA-4	anticytotoxic T-lymphocyte antigen 4
CXCR3	chemokine (C-X-C motif) receptor 3
DAMP	damage-associated molecular pattern
DC	dendritic cell
DCA	dichloroacetic acid
Drp1	dynamin-related protein-1
EMT	epithelial-mesenchymal transformation
ER	endoplasmic reticulum
ETC	electron transport chain
F6P	fructose 6-phosphate
FA	fatty acid
FADH$_2$	flavin adenine dinucleotide
FAO	fatty acid oxidation
FAS	fatty acid synthesis
FASN	fatty acid synthase
FATP2	fatty acid transporter 2
FBP1	fructose-1,6-bisphosphatase
FMT	fecal microbiota transplantation
G6P	glucose 6-phosphate
GA	glutaminase
GF	germfree
Gln	glutamine
GLS	glutaminase
Glu	glutamate
GLUT1	glucose transporter 1
GPX	glutathione peroxidase
GSH	glutathione
GSSG	glutathione disulfide
GTPase	guanosine triphosphatase
HBP	hexosamine biosynthesis pathway
HC	hydroxylated cholesterol

HCC	hepatocellular carcinoma
HFD	high-fat diet
HGPRTase	hypoxanthine guanine phosphoribosyl transferase
HIF-1	hypoxia-inducible factor-1
HK2	hexokinase 2
HMGCR	3-hydroxy-3-methylglutaryl-CoA reductase
IBD	inflammatory bowel disease
IDH	isocitrate dehydrogenase
IDO	indoleamine 2,3-dioxygenase
IFN-γ	interferon γ
iNOS	endothelial nitric oxide synthase
irAEs	immune-related adverse events
IS	immune synapse
Kyn	kynurenine
LCA	lithocholic acid
LDH	lactate dehydrogenase
LPL	lipoprotein lipase
LPS	lipopolysaccharide
LXR	liver X receptors
Ly6c	lymphocyte antigen 6 complex locus C1
Ly6g	lymphocyte antigen 6 complex locus G6D
MDSC	myeloid-derived suppressor cell
Mfn	mitofusin
MHC	major histocompatibility complex
MP	mevalonate pathway
mtDNA	mitochondrial deoxyribonucleic acid
mtROS	mitochondrial reactive oxygen species
MTX	methotrexate
MUFA	monounsaturated fatty acid
NAD	nicotinamide adenine dinucleotide
NADPH	nicotinamide adenine dinucleotide phosphate
NK cell	natural killer cell
NKT	natural killer T-cells
NO	nitric oxide
NSCL	nonsmall-cell lung cancer
Opa1	optic atrophy 1
OXPHOS	oxidative phosphorylation
PD-1	programmed death-1
PDH	pyruvate dehydrogenase
PDK1	pyruvate dehydrogenase kinase 1
PD-L1	programmed death ligand-1
PFK	phosphofructokinase
PFKFB3	6-phosphofructo-2-Kinase/Fructose-2,6-Biphosphatase 3
PGE2	prostaglandin E2
PHGDH	phosphoglycerate dehydrogenase
PKM1	pyruvate kinase M1
PKM2	pyruvate kinase M2
PPAR	peroxisomal proliferator-activated receptor
PPP	pentose phosphate pathway

PUFA	polyunsaturated fatty acid
RIPK3	receptor interaction protein kinase 3
SCD	desaturase
SCFA	short-chain fatty acid
SDH	succinic acid dehydrogenase
Ser	serine
SOD	superoxide dismutase
SPM	specialized proresolving mediator
SREBP	sterol regulating element-binding protein
SSP	serine biosynthesis pathway
TADC	tumor-associated dendritic cell
TAG	triacylglycerol
TAM	tumor associate macrophage
TAN	tumor associate neutrophil
TCA	tricarboxylic acid
TCR-T	receptor-engineered T-cell therapy
T_{eff} cell	effector T-cell
Tfam	mitochondrial transcription factor
TGF	transforming growth factor
TGF-β	transforming growth factor β
TIL	tumor-infiltrating lymphocytes
TLR	toll-like receptor
TME	tumor microenvironment
T_{mem} cell	memory T-cell
T_{reg} cell	regulatory T-cell
Trp	tryptophan
UCP2	uncoupling protein 2
UDP-Glc	uridine 5′-diphosphate-glucose
VEGF	vascular endothelial growth factor
VLBCL	intravascular large B-cell lymphoma
Vps34	vacuolar protein sorting 34
α-KG	α-ketoglutarate

1. Introduction

The human immune system can recognize and remove pathogens and tumor cells, which functions as an important defense against cancer. However, tumor cells have developed various mechanisms to evade immune surveillance and undergo unlimited proliferation, known as the immune escape of tumor cells. The immune system kills tumor cells by activating the "tumor-immune cycle," and tumor immune escape involves the destruction of this cycle, which forms the theoretical basis of tumor immunotherapy. Tumor immunotherapy does not directly kill cancer cells but restores the normal antitumor immune response of the body by restarting and maintaining the "tumor–immune cycle." In recent years, this method is commonly used for the treatment of advanced malignant tumors. Although the latest advances in immunotherapy have been encouraging, with significant benefits for numerous cancer patients, some patients remain unresponsive to immunotherapy owing to the diverse immune escape mechanisms of cancer cells.

Metabolism is known to regulate the phenotypic and biological functions of cells, and metabolic reprogramming is one of the top 10 characteristics of malignant tumors. Therefore, elucidation of the regulatory mechanisms of metabolic reprogramming in cancer and immune cells in the tumor microenvironment (TME) and the molecular mechanisms by which metabolism affects cell differentiation and function are crucial to the exploitation of the potential targets in accurate cancer immunotherapy.

The most prominent feature of cancer metabolism is high glycolysis; however, the metabolism of cancer cells varies across the four stages of cancer progression (tumorigenesis, epithelial-mesenchymal transformation (EMT), metastasis, and ectopic colonization) to support their need for proliferation, movement, and tissue attachment (as you can see in Fig. 1). The oncogenic transformation of cancer cells involves influencing the metabolic reprogramming and the functions of immune cells through various mechanisms such as nutrient competition, cytokine and metabolite secretion, and TME condition changes. In addition, nutrient (glucose, lipids, and amino acids) metabolism (Fig. 2), mitochondrial metabolism (Fig. 3), and microbiotas (Table 1) also play important roles in the regulation of tumor progression. Glycolysis provides energy by burning glucose as well as carbon-containing intermediates and reduced nicotinamide adenine dinucleotide phosphate (NADPH), which promote the anabolism of lipids, amino acids, nucleic acids, and other macromolecules. The oxidation of fatty acids also provides energy, and most lipids (such as fatty acids, phospholipids, and cholesterol) are the major components of cell membranes and participate in various signal transduction pathways. Amino acids not only are the raw materials for protein synthesis but also participate in the synthesis of other biological macromolecules; their metabolites also have important effects on the regulation of cell functions. Mitochondria are the energy factories of cells and are involved in the regulation of cell death, cell proliferation, and cell migration.

The role of microorganisms in the immune system and cancer cells has also been a recent research hotspot. Symbiotic microorganisms change the host immune system directly or indirectly, and their metabolites regulate the effect of immunotherapy. For example, fecal microbiota transplantation (FMT) has been shown to be a safe and effective treatment for various diseases [25].

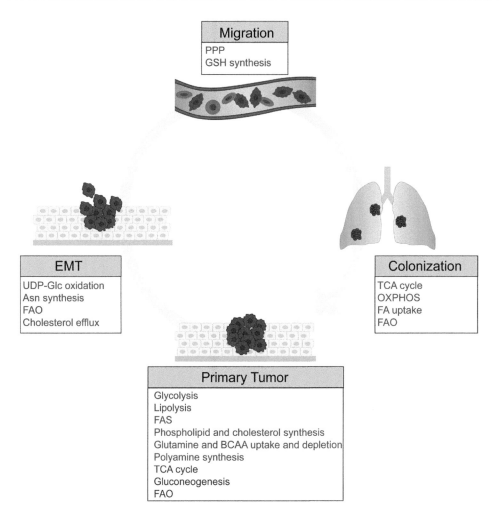

Fig. 1 Metabolic reprogramming of cancer cells. This schematic depicts the metabolic reprogramming of cancer cells during cancer progression, the upregulated pathways are in *red* and the downregulated ones are in *blue*. *Abbreviations*: *FAS*, fatty acid synthesis; *TCA cycle*, tricarboxylic acid cycle; *FAO*, fatty acid oxidation; *EMT*, epithelia-mesenchymal transformation; *UDP-Glc*, uridine 5′-diphosphate-glucose; *PPP*, pentose phosphate pathway, *FA*, fatty acid.

Therefore, achieving the metabolic modification of cells and rationally utilizing the metabolites of symbiotic microorganisms would be important strategies to suppress the immune escape of cancer cells and improve the anticancer ability of the immune system. In this section, we discuss the nature of cancer immunotherapy and progress; the metabolic reprogramming of cancer cells and immune cells; the regulatory function of glucose, lipids, amino acids, and mitochondrial metabolism in tumor immunotherapy; and

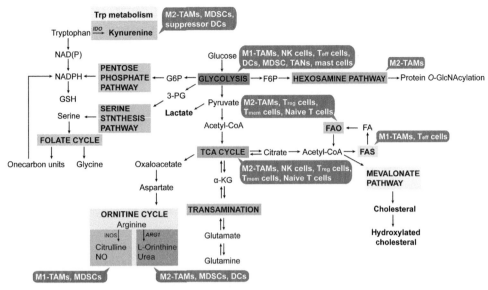

Fig. 2 Metabolic reprogramming of tumor associate immunity cells. This diagram depicts the cellular metabolic pathways in tumor associate immunity cells. *Abbreviations*: *M1-TAMs*, M1 type tumor associate macrophages; *M2-TAMs*, M2 type tumor associate macrophages; *MDSCs*, myeloid-derived suppressor cells; *DCs*, dendritic cells; *NK cells*, natural killer cells; *TANs*, tumor associate neutrophils; T_{eff} *cells*, effector T cells; T_{reg} *cells*, regulatory T cells; T_{mem} *cells*, memory T cells; *3-PG*, 3-phosphoglycerate; *F6P*, fructose 6-phosphate; *G6P*, glucose 6-phosphate; *α-KG*, α-ketoglutarate; *NO*, nitric oxide; *FA*, fatty acid; *GSH*, glutathione; *NAD*, nicotinamide adenine dinucleotide; *NADPH*, nicotinamide adenine dinucleotide phosphate; *IDO*, indoleamine 2,3-dioxygenase; *iNOS*, endothelial nitric oxide synthase; *ARG1*, arginase 1.

the influence of symbiotic microbes on the development of the host immune system and cancer.

2. Cancer immunity and immunotherapy

2.1 Immunotherapy sheds a new light in cancer treatment

Before the 21st century, the main treatment strategies for cancer were surgery, radiotherapy, chemotherapy, and targeted therapy. Radiotherapy uses radiation to physically destruct cancer cells, such as α, β, γ, and X rays; electron lines; and proton beams. Chemotherapy kills cancer cells with poisonous drugs such as carcinogens, antimetabolites, cisplatin, and paclitaxel. However, radiotherapy and chemotherapy can kill all cells, thus they are associated with major adverse effects. Conversely, targeted therapy inhibits or blocks the known oncogenic sites to control the growth and metastasis of tumor cells. Common drugs such as rituximab, trastuzumab, and imatinib are used for targeted therapy,

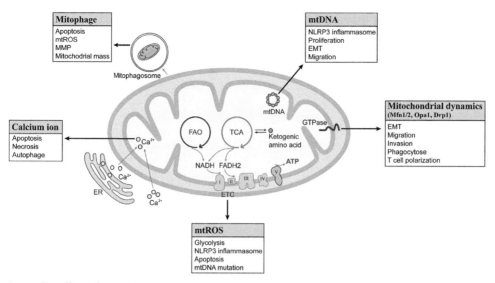

Fig. 3 The effect of mitochondria on tumor cells and immunity cells. This diagram depicts the functions of mitochondria in cellular activities. *Abbreviations*: *TCA*, tricarboxylic acid cycle; *FAO*, fatty acid oxidation; *NADPH*, nicotinamide adenine dinucleotide phosphate; *FADH$_2$*, flavin adenine dinucleotide; *I*, complex 1; *II*, complex 2; *III*, complex 3; *VI*, complex 4; *V*, ATP synthase; *ETC*, electron transport chain; *ATP*, adenosine triphosphate; *mtROS*, mitochondrial reactive oxygen species; *mtDNA*, mitochondrial deoxyribonucleic acid; *ER*, endoplasmic reticulum; *GTPase*, guanosine triphosphatase; *Mfn1/2*, mitofusin 1/2; *Opa1*, optic atrophy 1; *Drp1*, dynamin-related protein-1; *EMT*, epithelia-mesenchymal transformation.

but tumor cells often develop resistance to these drugs, and hence their therapeutic effect is poor.

Tumor immunotherapy is the most promising therapeutic approach; it eliminates tumors by activating the immune system and suppressing tumor immune escape. In 1863, Virchew first proposed "lymphoreticular infiltration" of cancer; since then, ipilimumab [anticytotoxic T-lymphocyte antigen 4 (CTLA-4)] was approved for the treatment of advanced melanoma in 2011, chimeric antigen receptor T-cell immunotherapy (CAR-T) was approved for marketing in 2017, and immunocheckpoint inhibitors [represented by CTLA-4 and programmed death-1/programmed death ligand-1 (PD-1/PD-L1) pathways] and cell therapies [such as CAR-T, T-cell receptor-engineered T-cell therapy (TCR-T), and CAR-natural killer T-cells (NKT)] were successfully used in the treatment of many tumors [*e.g.*, melanoma, nonsmall-cell lung cancer (NSCL), kidney cancer, prostate cancer, diffuse large B-cell lymphoma, and acute lymphoblastic leukemia (ALL)] in recent years [26–31]; after a decade of research, immunotherapy is now being used in tumor treatment. In December 2013, the *Science* journal considered cancer immunotherapy as the top 10 annual scientific research breakthroughs. *Nature* and *Science* journals successively published special issues on tumor immunotherapy

Table 1 The effect of microbiota on cancers and immunity cells.

Cancer types	Immune checkpoints	Immune cells	Microbiota metabolites	Microbiota types	References
Colorectal/colon cancer (H)		Treg cells, DCs, CD8$^+$ IFN-γ^+ T-	Secondary BAs (DCA, LCA), SCFAs (acetate, butyrate, propionate), indole-3-aldehyde	*Enterococcus faecalis, Lactobacillus, Methanobrevibacter smithii, Methanosphera stadmanae, A. muciniphila, Bifidobacterium* spp., *Prevotella* spp., *Blautia hydrogenotrophica, Clostridium* spp., *Streptococcus* spp., *Coprococcus comes, Coprococcus eutactus, Anaerostipes* spp., *Coprococcus catus, Eubacterium rectale, Eubacterium hallii, Faecalibacterium prausnitzii, Phascolarctobacterium succinatutens, Dialister* spp., *Veillonella* spp., *Megasphaera elsdenii, Salmonella* spp., *Roseburia inulinivorans, Ruminococcus obeum, B. fragilis,* psk$^+$ *E. coli, Fusobacterium nucleatum, Bacteroides* spp. *Enterobacteriaceae*	[1–9]
Liver cancer (H)		NKT cells	LPS, DCA	*C. scindens, Escherichia coli*	[10–13]
Pancreatic cancer (H)				*Fusobacterium nucleatum, Saccharopolyspora rectivirgula, Pseudomonas aeruginosa*	[7, 14, 15]
Breast cancer(H)				*Fusobacterium nucleatum, Bacillus, Escherichia coli, Staphylococcus epidermidis, Lactobacillus, Atopobium, Hydrogenophaga, Gluconacetobacter*	[7, 16, 17]
Melanoma (H)	PD-1, CTLA-4			*B. longum, C. aerofaciens, E. faecium, Ruminococcaceae, Bacteroidales, Faecalibacterium, B. fragilis, Bacterooides taiotaomicron, G. vaginalis*	[7, 18–21]

Continued

Table 1 The effect of microbiota on cancers and immunity cells—cont'd

Cancer types	Immune checkpoints	Immune cells	Microbiota metabolites	Microbiota types	References
Lung cancer (H)	PD-1			*A. muciniphila, Enterococcus hirae*	[18]
Renal carcinoma (H)	PD-1, CTLA-4			*A. muciniphila*	[18]
Urothelial carcinoma (H)	PD-1				[18]
Fibrosarcoma (M)	PD-1, CTLA-4	Th1, Tc1, T_{mem} cells, CCR9$^+$ CXCR3$^+$ CD4$^+$ T-cells		*A. muciniphila, Enterococcus hirae*	[18]
Melanoma (M)		CD45$^+$ CD11c$^-$ CD11b$^+$ Ly6c$^-$ Ly6g$^+$ TANs	Polysaccharide β-glucan		[22]
MC38 adenocarcinoma (M)		CD8$^+$ IFN-γ$^+$ T-cells		*Bacteroides* spp.	[23]
Intestinal and epithelial tumors (M)	PD-1, CTLA-4	CD4$^+$ IFN-γ$^+$ T cells, CD8$^+$ IFN-γ$^+$ T-cells	Inosine	*B. pseudolongum, L. johnsonii, Olsenella*	[24]

H, human; *M*, mice; T_{reg} *cells*, regulatory T-cells; T_{mem} *cells*, memory T-cells; *Th1*, type 1T helper cell; *Tc1*, type 1 CD8$^+$ T-cell; *DCs*, dendritic cells; *NKT cells*, natural killer T-cells; *TANs*, tumor associate neutrophils; *PD-1*, programmed cell death–1; *CTLA-4*, cytotoxic T-lymphocyte-associated protein 4; *LPS*, lipopolysaccharides; *BAs*, bile acids; *DCA*, deoxycholic acid; *LCA*, lithocholic acid; *SCFA*, short-chain fatty acid; *IFN-γ*, interferon-gamma; *Ly6c*, lymphocyte antigen 6 complex locus C1; *Ly6g*, lymphocyte antigen 6 complex locus G6D; *CCR9*, C-C motif chemokine receptor 9; *CXCR3*, chemokine (C–X–C motif) receptor 3; *A. muciniphila, Akkermansia muciniphila*; *B. longum, Bifidobacterium longum*; *C. aerofaciens, Collinsella aerofaciens*; *E. facium, Enterococcus facium*; *B. pseudolongum, Bifidobacterium pseudolongum*; *L. johnsonii, Lactobacillus johnsonii*; *B. fragilis, Bacteroides fragilis*; *C. scindens, Clostridium scindens*; *B. fragilis, Bacteroides fragilis*; *E. coli, Escherichia coli*; *G. vaginalis, Gardnerella vaginalis*.

in 2013 and 2015, marking the arrival of tumor immunotherapy 2.0 era; subsequently, immunotherapy became a new hotspot in tumor therapy. Studies using the immune system to fight cancer have been mainly focused on the following aspects: cancer vaccines and anticancer immune cells, anticancer antibodies, and various forms of immunotherapy and biotherapy [32].

At present, tumor immunotherapy is still at the preliminary stage of practice, and its extensive use has led to numerous immune-related adverse events (irAEs); thus, its side effects in many aspects need to be further reduced. For example, Fc monoclonal antibodies cause antibody-dependent cell-mediated cytotoxicity (ADCC) and complement-dependent cytotoxicity (CDC) [33]; ipilimumab (anti-CTLA-4) promoted colitis and hypophysitis [34, 35]; anti-PD-1 therapy improved the incidence of thyroiditis, pneumonia, and diabetes [34, 36, 37]; and CAR-T therapy could cause an immune storm, and the insufficient persistence of CAR-T cells may cause a higher recurrence rate of cancer [38]. Therefore, further studies are warranted to explore the anticancer effects of immunotherapy.

2.2 Theoretical basis of cancer immunotherapy: Immunoediting

Tumor immunotherapy was first introduced in 2002, when the American tumor biologist Schreiber proposed the hypothesis of "tumor immunoediting," indicating that the immune system has a dual role of both removing tumor cells and promoting tumor growth [39]. Whether mutated tumor cells form a tumor depends on the outcome of the interaction between tumor cells and the immune system. During this dynamic interaction process, the immune system removes some tumor cells, but remodels the biological characteristics of others, *i.e.*, reduces the antigenicity of some tumor cells.

According to the theory of tumor immunoediting, carcinogenesis was divided into three stages: immune clearance, immune balance, and immune escape [40–42]. In the first stage of elimination, the immune system recognizes and removes emerging "nonexisting" components, including viruses and tumor cells, with strong antigenicity. Nonspecific innate immune mechanisms (*e.g.*, macrophages and natural killer (NK) cells) and specific acquired immune mechanisms (such as $CD4^+$ and $CD8^+$ T-cells) are involved in the tumor cell clearance process. When the tumor cells are completely eliminated, the immune editing process stops. However, the immune system cannot always eliminate all the tumor cells effectively, especially those with low immunogenicity. This leads to the second state, "immune balance" [40, 43]. In this stage, the antigen expression of tumor cells is reduced and cannot be easily recognized by the immune system, although the tumor cells cannot overgrow under the immune surveillance. Specific acquired immunity (*e.g.*, $CD4^+$ and $CD8^+$ T-cells), and the cytokines interleukin (IL)-12 and interferon (IFN)-γ were suggested to be the main factors responsible for maintaining this equilibrium, and innate immunity is generally believed to be not involved in this process [40]. However, this balance is relatively dynamic: under the clearance pressure of the immune system, tumor cells with low immunogenicity and high degree of malignancy are selected and accumulate gradually. Similar to Darwin's theory of natural selection, this process is called immune remodeling. Surviving tumor cells could repeatedly stimulate the immune system and eventually induce immune tolerance and enter the state of "immune escape" [40]. In this stage, tumor cells show a series of malignant phenotypes to evade the recognition of T-cells, such as downregulation of major histocompatibility complex (MHC) molecules and tumor peptides and upregulation of factors such as PD-L1 and CD47. In addition, tumor cells undergo metabolic reprogramming, including glycolysis and pentose phosphate pathway (PPP) upregulation, glutamine (Gln) and fat decomposition, lipid synthesis, and mitochondrial biogenesis. During the rapid growth of tumors, the tumor cells contribute to a TME by releasing some immunosuppressive molecules such as transforming growth factor (TGF)-β, indoleamine-2,3-dioxygenase (IDO), IL-10, galectin 1, and vascular endothelial growth factor (VEGF). In this TME, protumor cells [$CD8^+$ CTLA 4^+ T cells, regulatory T-cells (Tregs), myeloid-derived suppressor cells (MDSCs), M2-tumor-associated macrophages (TAMs), and tumor-associated neutrophils (TANs)] were induced, whereas antitumor cells [effector T-cells (T_{eff} cells), NK cells, NKT cells, M1-TAMs, and antigen-presenting cells (APCs)] were inhibited, leading to the systemic immunosuppression and avoidance of the tumor-killing immune responses [44]. By this stage, the antitumor immune system becomes dysfunctional, and the tumor grows rapidly and begins to metastasize.

The three stages of immune editing are not fixed in the sequence of occurrence but are closely related to the malignancy of the primary tumor and the immune state of the body. When the immune function declines sharply, owing to factors such as the stress caused by emergency events, or long-term use of immunosuppressive drugs after organ transplantation, the tumor may pass the "clearance" stage, or even directly enter the "escape" stage. In contrast, immune editing can also cause a reverse action. In clinical practice, even patients with intermediate or advanced cancer were found to have been clinically cured after the appropriate clinical intervention. The essence of cancer immunotherapy is to promote the reversal of the immune editing process by interfering with the immune escape of the tumor. Therefore, tumor immunoediting is the theoretical basis of cancer immunotherapy.

2.3 T-cell immunity

Cancer immunotherapy should not only suppress the immune escape of cancer cells but also enhance the anticancer ability of the immune system. The concept of "tumor-immune cycle" proposed by Chen and Mellman in 2013 has revealed the mechanism of how T-cells kill tumor cells, providing another theoretical basis for cancer immunotherapy [45]. The tumor-immune cycle, also called "T cell immunity," mainly involves seven steps. (1) Tumor antigens are released during tumorigenesis, which promotes the recognition of tumor cells and initiates cytotoxicity. (2) Tumor antigens such as MHCI and MHCII are captured and processed by dendrite cells (DCs) or APCs, and then presented to T-cells. (3) Tumor-specific antigen response is initiated and activated by effector T-cells. (4) The effector T-cells migrate to the tumor. (5) They infiltrate into the TME and (6) specifically recognize and bind tumor cells through the interaction between the TCR and tumor antigen. (7) Finally, the effector T-cells kill target cancer cells. The dead tumor cells then release the relevant antigens, restarting the cycle. Tumor immune escape becomes possible owing to some abnormalities in some steps of the "tumor-immune cycle," such as low tumor immunogenicity, inability of the DCs or APCs to recognize tumor antigens, inability of T-cells to accurately locate tumors, and inhibition of effector T-cells in the TME. Hence, the primary target of cancer immunotherapy is to start or restart a continuous tumor-immune cycle, so that the immune system can adequately kill all tumor cells without producing unlimited autoimmune inflammatory response [45]. The activation of T-cells is the key to cancer immunotherapy.

The level and quality of T-cell activation depended on the balance between the activation and inhibitory signals. The inhibitory signals that impede T-cells and prevent a strong immune response that can destroy healthy cells in the body are called immune checkpoints, such as CTLA-4, PD-1, PD-L1, LAG3, B7-H3, B7-H4, TIM3, and TIGIT [46]. As the mediators for immune escape in tumors, immune checkpoints have become important targets for cancer immunotherapy. For example, in melanoma, ipilimumab (the anti-CTLA-4 monoclonal antibody) promotes the recognition of CD80/86 and CD28 by blocking the binding of CD80/86 to CTLA-4 and reverses the functional inhibition of T-cells [47, 48]. The anti-PD-1/PD-L1 monoclonal antibody can relieve the binding of PD-1 in T-cells to PD-L1 in tumor cells and activate the recognition of tumor cells by T-cells. Nivolumab (the anti-PD-1 monoclonal antibody) combined with ipilimumab can suppress the melanoma more significantly [49]. In breast cancer patients, anti-PD-1/PD-L1 monotherapy could achieve an objective response rate of 12%–21% [50]. However, monoclonal antibodies of immune checkpoints also have certain adverse effects. For example, the anti-CTLA-4 monoclonal antibodies can cause autoimmune inflammation [51], and *in vitro* use of them increases the number of Tregs [52]; however, its effects have been rarely investigated in clinical practice.

Immunotherapy is not sufficient to remove the "brake" effect of the immune checkpoints on T-cells [53]. Elimination of other immunosuppressive molecules in the TME can remarkably enhance the killing effect of T-cells, such as IDO in tumor cells and infiltrating bone marrow cells, and arginase[1] in M2-TAMs and MDSCs. In addition, some TNF receptor superfamily members on the surface of T-cell, such as GITR, CD27, 4-1BB, and OX40 are involved in the activation, inhibition, proliferation, and apoptosis of T-cells. Therefore, immunosuppressive molecules and TNF receptor superfamily members can be used as targets for cancer immunotherapy to enhance the functional activity of T-cells.

In recent years, cell immunotherapy targeting T-cell modification has become a hotspot in cancer immunotherapy. It is also known as adoptive cell transfer (ACT) that mainly includes tumor-infiltrating lymphocytes (TIL), TCR and CAR. The CAR-T cell therapy aims to improve the identification of tumor antigens by T-cells through loading receptors and stimulating molecules on the T-cells from patients *in vitro* and then reintroduced into the patients after their amplification *in vitro*. This treatment showed an early clinical trial success in patients with B-cell-ALL and B-cell lymphoma [38]. However, the persistence of CAR-T cells is poor, and the loss or inhibition of tumor antigens always leads to drug resistance of cancer cells. In addition, in a considerable number of patients, the remission effect is temporary, and hence CAR-T cell therapy still faces great challenges and requires improvements.

Thus, effector T-cells are the main tumor-killing cells in the immune system; they play an important role in anticancer therapy. T-cell immunity provides a strong theoretical basis and target for cancer immunotherapy.

3. Metabolic reprogramming and immunometabolism

Metabolomics is an emerging field in immune and cancer research [54, 55]. Any life activity of a cell depends on material and energy metabolism, which mainly includes glucose, lipid, amino acid, and nucleotide metabolism, and energy metabolism occurs with material metabolism.

3.1 Metabolic reprogramming of cancer cells

Unlike normal cells, cancer cells show remarkable changes in their metabolism with heterogenic characteristics—known as tumor metabolic reprogramming—which is one of the characteristics of malignant tumor, in order to adapt to the hypoxic microenvironment with low pH and nutrition, and realize rapid growth, unlimited proliferation, death resistance, immune escape, EMT, migration, and ectopic colonization. Berardinis *et al.* [56] indicated that, from precancerous lesions to locally invasive tumors and then to metastatic cancers, the metabolic phenotypes of tumor cells evolve with tumor progression, as well, carcinogenesis and cancer development also depend on the metabolic reprogramming of tumor cells.

During tumorigenesis, the rapid proliferation of tumor cells requires excessive amount of energy. Warburg suggested that, even in the presence of sufficient oxygen, glucose metabolism in cancer cells primarily occurs through the glycolytic pathway rather than the trichloroacetic acid (TCA) cycle; this is called the Warburg effect [57, 58]. Although the amount of ATP produced by glycolysis is lower than that by TCA (2 ATP molecules are produced per glucose molecule in glycolysis, whereas 36 ATP molecules are produced in TCA), the energy supply efficiency of glycolysis is greater, thereby better meeting the need for rapid proliferation and growth of cancer cells [59]. Furthermore, glucose metabolism of tumor cells is characterized by TCA retardation and gluconeogenesis enhancement. In addition, fatty acid metabolism plays an extremely important role in cancer: fatty acids act as not only the structural component of the cell membrane but also the secondary messenger involved in cell-signal transduction and fuel source for energy production [60]. Therefore, even in the presence of exogenous lipid sources, the fatty acid synthesis and triacylglycerol (TAG) decomposition pathways are upregulated to enhance fatty acid content and promote cell membrane synthesis [61, 62]. In cancer cells, fatty acid oxidation (FAO) is decreased, and phospholipid and cholesterol synthesis pathways are increased. Cancer cells can also obtain energy from high levels of Gln to support their proliferation [63]. The decomposition of Gln can reduce $NADP^+$ to NADPH in tumor cells, which provides electron donors for the reduction steps during lipid synthesis and nucleotide metabolism, and maintains the reduced state of glutathione (GSH) [64]. Therefore, the remarkably enhancement of Gln uptake and catabolism pathway plays an important role in the rapid and massive proliferation of tumor cells. In addition, cancer cells show enhanced uptake and oxidative decomposition of branched-chain amino acids and synthesis of arginine and polyamines.

During the metastasis stage, EMT is essential for tumor cells invading the blood or lymph from the primary site. The most obvious metabolic manifestation is increased nutrient consumption of sugars and amino acids. The 5′-diphosphate (UDP)-glucose-6-dehydrogenase-dependent glucose consumption can increase Snail, a transcription factor that promotes mesenchymal properties and increases the migration and metastasis of tumor cells in mice [65]. In addition, the amino acid synthesis ability of cancer cells is also enhanced during metastasis. For example, the upregulation of asparagine synthase (ASNS) can accelerate EMT and promote the invasion and metastasis of breast cancer cells [66]. In addition, the increase of cholesterol efflux can promote membrane fluidity and EMT of tumor cells [67]. After cancer cells escape from the primary site, NADPH, the production of PPP, is enhanced [68], leading to the increase of GSH synthesis [69], which can avoid the oxidative stress produced by matrix detachment. When metastatic cancer cells colonize at the new distal organs, their TCA and lipid oxidation levels are enhanced. For example, the upregulation of α-ketoglutarate (α-KG) stimulates collagen cross-linking by the α-KG-dependent enzyme prolyl-4-hydroxylase [70]. Fatty acid oxidation promotes the lymph node colonization of tumor cells [71]. Similarly, the uptake of fatty acids mediated by CD36 can promote the metastasis of oral cancer cells to the lymph nodes [72].

The metabolic activities of tumor cells may differ across diverse tumors or the diverse stages or parts of the same tumor. For example, significant differences exist in the metabolism of melanoma cells with different metastatic potentials [73]. The following factors influence the metabolic activity of tumor cells: (1) Characteristics of the primary lesion: tissue structure, resident cell type, epigenetic pattern, and transcriptional network;

(2) Intrinsic phenotype of cancer cells: oncogenotype, interference signal and gene expression, and dysplasia status; (3) Tumor microenvironment: altered tissue structure, changes of the interactions between cell and cell-matrix, and metabolic abnormalities; and (4) Metabolism of patients: genetic factors, other metabolic diseases such as obesity and diabetes, and dietary structure [56]. These abnormal molecules and factors can be used as targets for cancer diagnosis, detection, and treatment.

The metabolism of cancer cells is reprogrammed with cancer progression to promote their growth and development. Therefore, specific targeting of cancer cell metabolism and blocking of their reprogramming process can effectively inhibit the proliferation and metastasis of cancer cells. In other words, tumor metastasis can be prevented even in the early stage of cancer by altering the factors that affect the metabolism of cancer cells.

3.2 Cancer antimetabolites

3.2.1 Traditional antimetabolites

Considering the vigorous metabolism of cancer cells, antimetabolite drugs that can affect nucleic acid biosynthesis, have played a crucial role in the clinical suppression of cancer. Antimetabolites can prevent cell division and proliferation by specifically interfering with the substitution of nucleic acids required for DNA synthesis, such as folic acid, purine, and pyrimidine. Clinical antimetabolites account for about 40% of chemotherapy drugs, which are commonly used in conditions such as leukemia and villous epithelioma and have certain effects on some solid tumors. Traditional antimetabolites are mainly divided into the following categories: interfering with DNA and RNA synthesis, such as purine analogues (6-mercaptopurine, 6-tioguanine, 8-azaguanine, nelarabine, fludarabine, cladribine, clofarabine), pyrimidine analogues (capecitabine, 5-fluorouracil), deoxycytidine analogues (cytarabine, ancitabine, gemcitabine, troxacitabine, decitabine), amino acid analogues (azaserine), and folic acid analogues (methotrexate, methotrexate); fatty acid synthase inhibitors (TVB-2640); melatonin analogues or melatonin receptor agonists (Ramelteon, Tasimelteon); polyamine inhibitors (DFMO); and nonsteroidal drugs (Aspirin).

Among them, the functions of the most widely used antimetabolic drugs are as follows. Methotrexate (MTX) prevents one-carbon-unit transfer and interferes with the synthetic pathways of purine and pyrimidine nucleosides by the competitive inhibition of dihydrofolate reductase and folinic acid, thereby causing DNA synthesis disorders. It shows distinct curative effect in children with acute leukemia and can also be used for the treatment of chorionic carcinoma, malignant hydatidiform mole, ovarian cancer, head and neck cancer, and gastrointestinal cancer. The structure of 6-mercaptopurine (6-MP) is similar to that of adenine and hypoxanthine. 6-MP prevents the synthesis of purine nucleosides by competitively combining hypoxanthine guanine phosphoribosyl transferases (HGPRTases) and interferes with the function of RNA and DNA by inserting into them. Clinically, 6-MP is mainly used for treating children with ALL and acute or chronic nonlymphocytic leukemia, and it is often used in combination with MTX. 5-Fluorouracil (5-FU) can be transformed into fluorouracil deoxynucleotide (FdUMP) *in vivo*, which suppresses thymidylic acid synthase and prevents DNA thymidylic acid synthesis; the phosphorylated products of FdUTP, such as FdUMP, can insert into DNA and RNA. It was successfully used for the treatment of gastrointestinal cancer (esophageal cancer, gastric cancer, colon cancer, rectal cancer, pancreatic cancer, and liver cancer), breast cancer and ovarian cancer, cervical cancer, nasopharyngeal cancer, bladder cancer, and prostate cancer. Cytarabine (ArA-C) is converted to cytarabine triphosphate (ArA-CTP) in cells, which can inhibit DNA polymerase and become incorporated into DNA to inhibit the initiation and extension of the DNA chain. It is a common drug used for acute leukemia and can also be used for the prevention and treatment of central nervous leukemia. Gemcitabine can enter cells and be transformed into dFdCDP and dFdCTP. The former one disturbs the synthesis and repair of DNA by inhibiting ribonucleotide reductase. The latter one induces cell apoptosis by competitively constructing DNA with dNTP. It is often used in pancreatic cancer and NSCLC, and can also be used in ALL and chronic lymphocytic leukemia (CLL).

As early as in 1988, scientists observed the anticancer potential of aspirin in case-control studies [74]. According to incomplete statistics, more than 100 observational studies revealed the anticancer effect of aspirin [75, 76]. Follow-up studies of people at high risk of colorectal cancer with inherited mutations revealed that continuous aspirin use for more than 2 years reduced the risk of colorectal cancer by 50%, and the protective effect of aspirin persisted for up to 20 years [77].

Although antimetabolites are effective in the clinical treatment of many cancer types, they kill cancer cells as well as somatic cells, and some malignant cancer cells are insensitive to them. Therefore, new antimetabolites and cancer treatment strategies need to be developed.

3.2.2 Antimetabolites in trials

Given the importance of antimetabolites in cancer therapy, numerous clinical trials are ongoing with antimetabolites. In platinum-resistant high-grade serous ovarian cancer, the combination of berzosertib (a selective ATR inhibitor) and gemcitabine showed acceptable toxicity and superior efficacy to gemcitabine alone in phase II trial [78]. A study on advanced NSCLC suggested that the combination of nivolumab (anti-PD-1 Ab) and platinum-based chemotherapy, especially carboplatin, paclitaxel, and bevacizumab, could be promising for reducing the risk of death and improving the long-term survival rates [79]. In addition, triple-drug combination chemoradiotherapy exerted significant antitumor effect with sufficient safety in a phase II study of unresectable advanced esophageal cancer [80]. Moreover, cisplatin-pemetrexed (CP) and cisplatin-raltitrexed (CR) were shown to have comparable effect in the treatment of metastatic pleural mesothelioma, but CP showed a modest increased risk of grade III–IV adverse effects [81]. In a phase II trial of patients with intravascular large B-cell lymphoma (VLBCL) without apparent central nervous system (CNS) involvement at diagnosis, R–CHOP (rituximab, cyclophosphamide, doxorubicin, vincristine, and prednisolone) combined with rituximab and high-dose methotrexate plus intrathecal chemotherapy was found to be a safe and effective treatment [82]. A Swedish population-based study revealed that the use intensity of statin was positively correlated with the survival of patients with multiple myeloma of both sexes [83].

Various clinical trials are ongoing. For example, a phase III study exploring the effect of the addition of capecitabine to carboplatine-based chemotherapy in early "Triple Negative" breast cancer (NCT04335669); a phase II study of pembrolizumab plus pemetrexed for elderly patients with nonsquamous NSCLC with PD-L1 tumor proportion score of less than 50% (NCT04396457); a phase II trial of pegylated arginine deiminase (ADI-PEG 20) in combination with gemcitabine and docetaxel for the treatment of soft tissue sarcoma osteosarcoma, Ewing's sarcoma, and small cell lung cancer (NCT03449901); a phase II trial of neoadjuvant agen1884 plus agen2034 in combination with cisplatin-gemcitabine for muscle-invasive bladder cancer before radical cystectomy (NCT04430036); a phase II study of TGFβ type II receptor inhibitor ly2157299 with neoadjuvant chemoradiation in patients with locally advanced rectal adenocarcinoma (NCT02688712); randomized open-label trial of dose dense, fixed dose capecitabine compared to standard dose capecitabine in metastatic breast cancer and advanced/metastatic gastrointestinal cancers (NCT02595320); and adjuvant chemotherapy in elderly patients with colon cancer stage III, geriatric assessment and prognostic gene signatures (NCT02978612).

3.3 Effect of tumors on immune cell metabolism in the microenvironment

One of the main reasons for the failure of antitumor immunotherapy is the complexity and diversity of the TME, which is mainly composed of tumor cells, immune cells, endothelial cells, fibroblasts, extracellular matrix, as well as various cytokines and chemokines, and the TME is characterized by low oxygen, low pH, and poor nutrition. Tumor cells alter and maintain the microenvironment by consuming nutrients and oxygen as well as *via* autocrine and paracrine actions, which affect the metabolism, differentiation, and function of tumor-infiltrating immune cells (*e.g.*, TAMs, TANs, NK cells, DCs, mast cells, MDSCs, T-cells, and B-cells) [84, 85], in order to promote their immune escape and drug tolerance.

The first type of cells infiltrating in the TME are the innate immune cells. Dynamic changes in the TME can regulate the metabolism and function of TAMs [86]. In the TME, TNF-α and IL-1 can stimulate the activation of M1-TAMs and lead to increased glycolysis gene expression, glucose intake, and lactic acid secretion [87]. In addition, damage-associated molecular patterns (DAMPs) can activate the toll-like receptor (TLR) and the PI3K-Akt pathway of inflammatory M1-TAMs, leading to the upregulation of the proinflammatory capacity and glycolysis level in the cells [88]. However, hypoxia, lactic acid, IL-4, IL10, TGF-β, and CSF1 can promote the polarization of immunosuppressive M2-TAMs [89–92] with enhanced oxidative phosphorylation (OXPHOS) and FAO and reduced glycolysis [93, 94], as well as increase the expression of procancer factors such as Arg1 and VEGF in M2-TAMs [90]. TANs also contain two biological phenotypes: antitumor N1 and protumor N2; the latter produces prooncogenic factors such as ARG1, ROS, MMPs, IL-6, and IL-1 [95, 96]. In the TME, hypoxia promotes the upregulation of aerobic glycolysis in TANs [97], and tumor-derived TGF-β promotes the conversion of N1 to N2 [98]. NK cells prefer OXPHOS in the resting state and short-term activation state, whereas IL-15 and transient hypoxia can lead to enhanced glycolysis of NK cells and activate their tumor-killing function in the early stage of tumor development [99, 100]. However, TGF-β can stimulate the expression of fructose-1,6-bisphosphatase (FBP1) in NK cells, which inhibits glycolysis, resulting in their dysfunction and inactivation [101]. In addition, low arginine [102], high lactic acid and low pH [103, 104], and high adenosine [105] in the TME inhibit the cytotoxicity and proliferation of NK cells. In the resting state, DCs prefer OXPHOS, whereas the hypoxic TME drives the transcription of mTOR and glycolytic-related genes, thereby promoting glycolysis and maturation of DCs [106]. However, lactic acid in the TME can impede the glycolytic pathway in tumor-associated DCs (TADCs) and inhibit their antitumor function [107], and high levels of adenosine can also promote OXPHOS and inhibit glycolysis by inducing AMP-activated protein kinase (AMPK) [108]. In the TME, mast cells can be activated by extracellular vesicles derived from cancer cells, which promote tumor angiogenesis and immunosuppression [109]. In active mast cells, glycolysis is upregulated by activating the PI3K pathway [110]. Numerous cytokines released in the TME, such as VEGF, IL-6, IL-10, GM-CSF, and TGF-β, can induce MDSCs to infiltrate into tumor and promote its aggregation, metabolic reprogramming, and polarization [111]. Hypoxia promotes the differentiation of MDSCs into immunosuppressive TAMs, TANs and TADCs, and enhances the immunosuppressive function by mediating the expression of ROS, iNOS, ARG1, and PD-L1 [112, 113].

The TME also affects the metabolic reprogramming of acquired immune cells. High levels of lactic acid in the TME can inhibit the glycolytic pathway by blocking the transport of lactic acid in CD8$^+$ T-cells and inhibiting the PI3K-Akt-mTOR pathway, thereby downregulating their proliferation, cytokine production, and cytotoxicity [114–117]. In addition, nutritional deprivation and antiinflammatory factors (IL-4, IL-10, and TGF-β) can induce OXPHOS and Gln-dependent mitochondrial metabolism and thus inhibit the antitumor response of T-cells [118, 119]. Primary CD4$^+$ T-cells prefer glycolysis and fatty acid synthase (FAS), and preferentially differentiate to Th17 under the stimulation of TGF-β and IL-6 [120]. Hypoxic TME promotes Th17 differentiation and inhibits Tregs generation by enhancing the glycolytic pathway [121]. Tumor cells express high levels of B-cell activating factor, which induces glycolysis and antibody production in B-cells *via* the glucose transporter 1 (GLUT1)-dependent pathway [122]. Moreover, B-cells, stimulated by tumor cells, upregulate glycolysis, Gln decomposition, and mitochondrial biosynthesis by activating the PI3K-AKT pathway [123].

Since the effect of tumor cells and TME on the metabolism of tumor-infiltrating immune cells is the key mechanism to promote tumor escape, targeting the regulation of tumor cells and TME on immune cells or enhancing the ability of immune cells to evade these influencing factors may be an effective strategy for antitumor therapy.

3.4 Metabolic reprogramming of TAMs and MDSCs
3.4.1 TAMs

TAMs are macrophages infiltrated in the TME; they have an important influence on inflammation and tumor development. Depending on the influence on tumor growth, invasion, metastasis, angiogenesis, chemotherapy resistance, and immunosuppression, TAMs are traditionally divided into anticancer M1 type and cancer-promoting M2 type [86, 124]. Although some studies show that TAMs are composed of multiple cell subsets, some of which have the characteristics of both M1 and M2, M1/M2 is still the simplest method for classifying TAM functions [125–127]. Mantovani *et al.* [128] suggested that TAMs tended to transform from activated M1 type to M2 type and had strong immunosuppression, but both of M1 and M2 types coexist in the TME. M1-TAMs with inflammatory characteristics play an important role in antitumor immunity [129]. The main cellular signaling pathways affecting TAM differentiation include JNK, PI3K/Akt, notch, JAK/stat, irf5/irf4, and c/EBP; these pathways regulate M1-like (*inos*, *Il-6*, *Tnf-α*) and M2-like (*Arg1*, *Il-10*, *Ym1*, *Fizz1*, *Mgl1*) genes [130]. When TAMs differentiate, their metabolism also changes correspondingly.

The glucose metabolism of TAMs was significantly different in the two cell subtypes, with M1-TAMs tend to enhance glycolysis and M2-TAMs tend to increase OXPHOS [93]. After the activation of M1-TAMs, the PI3K-AKT pathway is upregulated, which further promotes the expression of glucose transporters (*e.g.*, GLUT1) and key glycolytic enzymes (*e.g.*, hexokinase and phosphofructose kinase-1), and glucose uptake and aerobic glycolysis is increased [87, 131, 132]. Moreover, the activation of the PI3K-AKT pathway can promote the antiangiogenesis ability of M1-TAMs [133]. The TCA cycle showed two interruptions in M1-TAMs: the first interruption involves isocitrate dehydrogenase (IDH), which results in an increase of citrate and itaconic acid. Phosphorylation and activation of citrate lyase can promote the conversion of citrate from the TCA cycle to acetyl-coenzyme A (acetyl-CoA) for the synthesis of fatty acids (FAs), prostaglandins, and nitric oxide (NO), among which NO is an important proinflammatory mediator in the antitumor response. The antimicrobial properties of itaconic acid can promote the removal of pathogens from M1-TAMs. The second interruption occurs with succinic acid dehydrogenase (SDH), which leads to elevated succinate levels and thus increased stability of hypoxia-inducible factor-1 (HIF-1), which promotes inflammatory progression and the production of key enzymes involved in glucose metabolism (GLUT1, HK2, and PGK1) [134] as well as IL-1 [135] and ROS/RNI [136]. In addition, increased succinate levels complement the TCA cycle, which further increases citrate levels and supplies the urea cycle, contributing to NO production. The PPP is highly activated in M1-TAMs [101], and M1-TAMs rely on PPP and maltase to produce NO, ROS, and NADPH, which not only promotes the killing of pathogens and tumor cells, but also protects themselves from the oxidative stress injury [137]. After the activation of M2-TAMs, the levels of STAT6, c-Myc, and AMPK are upregulated to promote M2 OXPHOS [93, 138], and the levels of ARG1, VEGF-β, Tie-2, and IL-10 are increased to promote the occurrence of TAM-related tumors [90, 139, 140]. The transcription factor c-Myc promotes the protumor function by increasing CCL18, TGF, VEGF, and MMPs [141]. Pyruvate kinase M2 (PKM2) differently affects cell metabolism in M1- and M2-TAMs *via* different forms. In M1-TAMs, PKM2 binds to HIF-1 in its inactive dimer form to promote glycolysis. In contrast, in M2-TAMs, the active tetramer form of PKM2 can inhibit glycolysis and enhance OXPHOS [142].

There were also significant differences in lipid metabolism between the two kinds of TAMs: FAS is increased in M1-TAMs, whereas M2-TAMs prefer FAO [143]. In the TME, IFN-γ produced by Th1 cells activates M1-TAMs and inhibits FA intake and oxidation, and M1 macrophages tend to store excess FA as TAG and cholesterol esters in lipid droplets [144]. However, the TCA cycle is disrupted in M1-TAMs, and thus citrate is converted to free fatty acids, which in turn activates the IKK and JNK1 signaling molecules of M1-macrophage, promoting the production of IL-1 and TNF and triggering an inflammatory response [145]. Increased FA uptake and TAG level in M2-TAMs play important roles in promoting FAO and M2 activation *via* the peroxisomal proliferator-activated receptors (PPARs) and liver X receptors (LXRs) pathways [143, 146, 147]. PPARs mediate M2 macrophage polarization and promote FAO as well as tumor progression and metastasis [148]. Hydroxylated cholesterol activates LXRs in macrophages, leading to ARG1 upregulation and promoting the M2 immunosuppressive phenotype [149].

Amino acid metabolism also shows significantly different between M1-TAMs and M2-TAMs. Arginine (Arg), as the substrate of iNOS to produce NO, is a key nutrient for proinflammatory M1 macrophages to promote pathogen clearance [150]. In contrast, L-Arg is decomposed by ARG1 into urea and ornithine in the immunosuppressive M2 macrophages [151] which play a role in inhibiting protein synthesis and anticancer activity of T-cells [152, 153]. Gln metabolism is another important pathway for the differentiation and function of TAMs. Key enzymes in Gln metabolism, such as AKG, GPT2, GLUL, and GA-TM, are enhanced in M2 macrophages. In addition, Gln synthase inhibitors can promote M1-like phenotypes to M2-polarized macrophages [154, 155]. Gln is converted to glutamate by Gln synthase, which can be further converted to α-KG. The α-KG can reduce the expression of the M1 polarization marker genes by inhibiting the mTORC1 signaling pathway [156].

3.4.2 MDSCs

MDSCs are a group of immature myeloid cells that resemble granulocyte-mononuclear progenitor cells that have not yet differentiated into macrophages, DCs, or granulocytes. MDSCs consist of two major groups of cells called granulocytic or polymorphonuclear cells (PMN-MDSCs) which are phenotypically and morphologically similar to neutrophils, and mononuclear cells (M-MDSCs) which are similar to monocytes in phenotype and morphology [157]. The main feature of MDSCs is immunosuppression, and the immunosuppressors include ARG1, iNOS, TGF-β, IL-10, COX2, IDO [157, 158]. In addition to immunosuppressive mechanisms, MDSCs promote tumor progression by influencing the remodeling of the TME and producing VEGF, bFGF, Bv8, and MMP9 [159–161]. Two type signals increase the accumulation of MDSCs, one promotes the expansion of MDSCs—STAT3, IRF8, C/EBP, Notch, adenosine receptor A2b signal, and NLRP3—and the other promotes the pathological activation of MDSCs—HMGB1, NF-κB, STAT1, STAT6, prostaglandin E_2 (PGE$_2$), and COX2 [162]. In addition, endoplasmic reticulum (ER) stress has been reported to increase MDSC accumulation and immunosuppressive activity, and its marker molecules are XBP1s and CHOP [163–165].

Recent studies have shown that the metabolic reprogramming of MDSCs plays an important role in the regulation of their immunosuppressive function. In the TME, glycolysis, TCA cycle, FA uptake, FAO, Gln decomposition, and arginine metabolism of MDSCs are all increased, and their immunosuppressive activity is enhanced [166, 167]. High level of glycolysis in MDSCs can lead to the production of a large amount of phosphoenolpyruvate, which acts as an antioxidant roles to weaken ROS-mediated oxidative damage [168]. Moreover, lactic acid, a glycolytic metabolite, can promote tumor infiltration of MDSCs [169]. L-ARG is an essential amino acid for T-lymphocyte activity. However, MDSCs in the TME increase the expression of cationic amino acid transporter 2 (CAT2) and compete with T-cells to absorb L-ARG [170]. Under the stimulation of Th2 cytokines such as IL-4, IL-10, and IL-13, MDSCs show a high level of ARG1. The overexpression of iNOS in MDSCs induced by Th1 cytokines such as TNF-α, IL-1, and IFN-γ catalyzes the production of NO and L-citrulline from L-Arg [171, 172]. In addition, under the stimulation of the inflammatory factor IFN-γ, MDSCs overexpress IDO, which metabolizes L-tryptophan to L-kynurenine (L-Kyn), thereby inducing Treg amplification and inhibiting T-cell proliferation in the TME [173]. In addition, in the case of Gln deficiency, iNOS activity in MDSCs was limited, but ARG1 activity was not affected [174].

In tumor-infiltrating MDSCs (T-MDSCs), the uptake of FAs and the expression of key FAO enzymes (including carnitine palmitoyl transferase 1, acyl coenzyme A dehydrogenase, peroxisome proliferator-activated receptor γ-coenzyme activator 1-β, and 3-hydroxyacyl-CoA dehydrogenase) are increased [167]. In addition, G-CSF and GM-CSF in the TME can enhance the levels of FA translocators CD36 and FAO and thus promote FA uptake and oxidation in T-MDSCs by inducing the activation of STAT3 and STAT5 signaling pathways [175]. In addition, MDSCs can use cyclooxygenase-2 (COX-2) to convert arachidonic acid to PGE$_2$, and then promotes the chemotaxis of MDSCs to the TME by inducing CXCL12, and upregulates the inhibitory function of MDSCs by increasing ARG1, iNOS, and IDO [176–178]. In addition, fatty acid transporter 2 (FATP2) was overexpressed in PMN-MDSCs, which upregulated arachidonic acid uptake and PGE2 synthesis and enhanced the inhibitory effect of PMN-MDSCs [179].

3.5 Metabolic reprogramming of tumor-associated T-cells

CD4$^+_{conv}$ and CD8$^+$ T$_{eff}$ cells are the main effector cells of antitumor immunity in the TME. CD4$^+$ T-cells can be differentiated into different subtypes to promote or inhibit the occurrence of tumors. CD4$^+$ Th1 cells produce IFN-γ, which promotes the non-specific clearance of tumor cells, and Th1 cell infiltration in the TME is closely related to clinical prognosis. In contrast, Th2 cells and Tregs secrete IL-4, IL-10, and TGF-β, which suppress inflammation and promote tumor immune escape. Th17 cells may promote or inhibit tumor progression in a tumor type- and tumor stage-dependent manner; Tfh promotes humoral immunity at germinal centers [180–182]. After activation by APCs, CD8$^+$ T-cells can migrate to tumor tissues and kill target cells through perforin (destroying cell membrane), granzyme (entering target cells and degrading DNA), and FasL. In addition, CD8$^+$ T-cells promote antitumor immune response by secreting cytokines such as IFN-γ and TNF-α [183].

T-cell metabolism is closely related to differentiation and functional activation. Naïve T-cells rely on OXPHOS, FAO, and low Gln decomposition levels for energy [184–186]. FAO and OXPHOS are needed for the production and survival of memory T-cells [186]. The metabolic characteristics of Tregs depend on activated AMPK and high levels of OXPHOS and FAO to support their survival and differentiation, and mitochondrial energy metabolism also plays an important role in the function and survival of Tregs [187–189].

In order to rapidly proliferate and release cytotoxic molecules, activated effector T-cells enhance aerobic glycolysis, PPP, FAS, and Gln decomposition, whereas FAO decreases [184-186, 190]. Recent studies have shown that TCA and OXPHOS are increased within 24 h after T-cell activation, which is also a key aspect of CD4$^+_{conv}$ and CD8$^+$ T$_{eff}$ cell activation [191, 192]. The activation of CD8$^+$ T, Th1, Th2, and

Th17 cells can stimulate intracellular PI3K-AKT, mTOR, c-Myc, and HIF1 signaling, thereby promoting the transcriptional upregulation of key glycolytic genes such as pyruvate kinase M1 (*PKM1*), hexokinase 2 (*HK2*), and *GLUT1* [193–196]. Moreover, when the glycolytic pathway is blocked in effector T-cells, their ability to secrete IFN-γ will be significantly reduced even when they are provided with sufficient costimulators and growth factors [197]. The PPP metabolizes glucose-6-phosphate to NADPH and ribose-5-phosphate, which is significantly upregulated after the activation of CD4$^+$ T cells [194]; NADPH is necessary for the synthesis of FAs and plasma membrane in activated CD8$^+$ T-cells [198]. In addition, the hexosamine biosynthesis pathway (HBP) is dependent on glucose and Gln metabolism, and its main substrate UDP-GlcNAc plays a crucial role in the expansion and functional activation of CD4$^+$ and CD8$^+$ T-cells [199].

The lipid metabolism of effector T-cells is also changed. The levels of sterol regulating element-binding proteins 1 and 2 (SREBP1 and SREBP2, respectively) are increased in effector T-cells, which promote the accumulation of FAs and cholesterol, and this is beneficial for cell membrane synthesis and promotes the accumulation of T-cell receptors on the cell membrane to enhance T-cell proliferation and tumor destruction ability [198, 200]. Moreover, amino acid transporters, including SLC7A5, SLC38A1, SLC38A2, and SLC1A5, are increased during the T-cell activation process [201–204]. In particular, the activation of mouse CD4$^+_{conv}$ T-cells significantly upregulates the expression of the Gln transporters SLC1A5 and SLC38A1 [203].

3.6 Tumor immunometabolism

3.6.1 Glucose metabolism and tumor immunity

Glucose is the most important nutrient for cell energy; its metabolism level in tumor cells is closely related to the progression of cancer. Glycolysis is the main glucose metabolism pathway in tumor cells. The TCA cycle in the mitochondria and PPP in the cytoplasm are also critical for tumor cells. In addition, recent studies have shown that the HBP can facilitate the O-GLCNAC modification of proteins by providing UDP-GLcNAc to promote carcinogenesis [205]; the activation of the serine biosynthesis pathway (SSP) can also increase the carcinogenic effect of tumor cells [206]. During glycolysis, the intermediate metabolite pyruvate can be converted to the final product lactic acid or can directly enter the TCA cycle to participate in OXPHOS. Glucose 6-phosphate and fructose 6-phosphate can enter the PPP to produce NADPH and promote GSH synthesis. Moreover, fructose 6-phosphate can participate in the HBP and synthesize UDP-GLcNAc, and 3-phosphoglycerate (3-PG) can be used as the initial substrate of SSP to upregulate serine level.

The significantly enhanced glycolysis level of tumor cells can provide not only ATP for tumor cells, but also more carbon intermediates for nucleotide, amino acid, and lipid biosynthesis to promote the anabolism of cancer cells [54, 207, 208]. Enhanced aerobic glycolysis results in increased lactic acid secretion in the TME, which can cause the death of immune cells such as CTLs, DCs, and APCs [117, 209, 210] and promote the activation of M2-TAMs, MDSCs, Tregs, and other immunosuppressive cells [90, 138, 211]. In addition, lactic acid can promote M1-TAMs to express M2-like genes *via* histone lactylation [212]. Lactic acid also plays an important role in energy metabolism [213] and provides essential fuel for tumor cells [214]. Lactic acid has been shown to promote tumor cell proliferation, metastasis, and immune escape [214, 215] and hence serves as a biomarker for tumor metastasis and overall survival [216, 217]. Low pH can change the interstitial interface of tumors and enhance the aggressiveness of cancer cells [218]. Besides, lactic acid can stimulate neutrophil mobilization in the TME, leading to inflammatory reaction [219].

The key enzymes involved in the glycolytic pathway are associated with the malignancy of cancer cells. Glut1 mediates the transmembrane transport of glucose, which is activated by oncogenes c-Myc, KRAS, and HIF-1α, but inhibited by the tumor suppressor gene p53 [54]. Activated AKT promotes the maintenance of Glut1 localization on cell surface and promotes gastric cancer progression [220, 221]. Inhibition of Glut1 has been reported to reduce tumor cell proliferation and invasion, and hence, it serves as a potential prognostic indicator for patients with breast cancer, prostate cancer, and colorectal cancer [222–225]. HK2 is the key rate-limiting enzyme in the glycolytic pathway; it can promote the growth and metastasis of pancreatic cancer cells by regulating the production of lactic acid [226]. In PTEN- and p53-deficiency-driven prostate cancer cells, the AKT-mTORC1-4EBP1 signaling pathway is activated, which in turn upregulates HK2, resulting in increased glycolysis levels [227]. LncRNA UCA1 can also promote HK2 activation and increase glycolysis in bladder cancer cells by activating mTOR-STAT3 [228]. In addition, HK2 can be used as a potential drug target for tumor therapy: ketoconazole and posaconazole can selectively target malignant glioma cells expressing HK2 [229]. Antisense oligonucleotides of HK2 can inhibit its activity and promote the death of HK1-HK^{2+} multiple myeloma [230]. Phosphofructokinase (PFK) catalysis is another rate-limiting step of glycolysis [231, 232]. Cancer cells have high fructose 2,6-bisphosphate level that stimulates the increase of 6-phosphofructo-2-kinase/fructose-2,6-biphosphatase 3 (PFKFB3), and PFKFB3 inhibitors show promising anticancer effects [233]. Pyruvate kinase (PK) converts phosphoenolpyruvate to pyruvate and is another major rate-limiting enzyme in the glycolytic pathway. The expression of PKM2 has been shown to be higher in various cancers such as colon, kidney, lung, and breast cancer [234]. Moreover, PKM2 can directly regulate the Warburg effect of tumor cells and promote glucose uptake and lactic acid generation, and thus stimulate cancer progression [234, 235]. Lactate dehydrogenase (LDH) is a rate-limiting enzyme in the last step of glycolysis. It consists of two subunits, LDHA and LDHB. LDHA catalyzes the conversion of pyruvate to L-lactic acid in the last step of anaerobic glycolysis, and LDHB is required for the reversible conversion of lactate and pyruvate. The small molecule inhibitor FX11 selectively inhibits LDHA and thus the glycolysis level and tumor progression; it has been shown to slow down tumor progression in human lymphoma, gastric cancer, neuroblastoma, and pancreatic cancer [236–239].

The mitochondrial TCA cycle is coupled with OXPHOS, which is the ATP production center. Moreover, various carb-containing intermediates involved in the TCA cycle can be converted into amino acids and FAs to promote the anabolism of cancer cells. Pyruvate dehydrogenase (PDH) is a key rate-limiting enzyme in the TCA cycle and catalyzes the conversion of glycolytic pyruvate to acetyl-CoA, which can not only be used as raw material for the TCA cycle and FAS activity, but also participate in the modification of protein acetylation, which plays an important regulatory role in tumor cells [240–242]. In hypoxic TME, activated AKT2 accumulates in the mitochondria and phosphorylates Thr346 of pyruvate dehydrogenase kinase 1 (PDK1), thereby inactivating the pyruvate dehydrogenase complex and inhibiting the entry of pyruvate into the TCA cycle [243]. Furthermore, the high level of HIF-1α directly upregulates LDH and PDK, thereby promoting glycolysis, inhibiting PDH activation, and impeding TCA cycle and OXPHOS [244, 245]. In contrast, glucose metabolites such as succinic acid, fumaric acid, pyruvate, lactic acid, and oxalylacetic acid enhance the activation and stabilization of HIF-1α by inhibiting PHDs, thereby inducing "HIF-1α-glycolysis" vicious circle in tumor cells [246, 247].

In cancer cells, high levels of ROS are produced due to rapid metabolism, which renders them more sensitive to the toxicity of oxidative stress [248]. Tumor cells can evade the oxidative killing effect of ROS *via* NADPH generated by PPP. Thus, inhibiting the PPP activity of cancer cells by using dichloroacetic acid (DCA) can reduce the glycolysis and proliferation of cancer cells [249]. Conversely, a high level of PPP can promote the metastasis and progression of cancers such as liver, breast, and lung cancer [250]. The tumor suppressor genes p53 and PTEN can inhibit PPP by binding G6PD [251, 252], whereas oncogenes PI3K, mTORC1, and K-ras^{G12D} can induce the upregulation of PPP [253, 254].

HBP can convert F6P into UDP-GlcNAc *via* various enzymes, and then produce O-GlcNAc under the catalysis of OGT, which produce protein O-GlcNAcylation. OGT promotes glycolysis in cancer cells by activating the PI3K/Akt/mTOR pathway and c-Myc, ChREBP, NF-kB, and HIF-1α signals; it is a potential target for cancer treatment owing to its correlation with the prognosis of breast cancer [255, 256]. Moreover, the main substrate produced by the HBP, UDP-GlcNAc, is important for effector CD4^{+} and CD8^{+} T-cell expansion and function [199]. In cancer cells, about 10% of 3-PG produced by glycolysis is utilized for the *de novo* synthesis of serine. Phosphoglycerate dehydrogenase (PHGDH), the first branch enzyme in SSP, catalyzes the conversion of 3-PG to 3-hydroxypyruvate [257]. PHGDH not only activates the AKT pathway, but also upregulates the expression of some protumor proteins by promoting the assembly of the eIF4F complex on their 5′-mRNA, thereby promoting the progression of pancreatic cancer [258]. Moreover, PHGDH can reduce α-KG to D-2-hydroxyglutaric acid—a tumor metabolite—which can cause cancer transformation and affect the methylation modification of proteins and DNA [259, 260]. In addition, PHGDH binds to the cancer factor FOXM1 to prevent its degradation by proteases and to induce the proliferation and invasion of glioma cells [261]. Furthermore, PHGDH is highly expressed in colorectal cancer, breast cancer, and melanoma and can promote tumor progression and metastasis [262–264].

3.6.2 Lipid metabolism and tumor immunity

Although studies on lipid metabolism in cancer cells are less than those on glucose metabolism, the increased FA level in cancer cells has been shown to meet the lipid demands for their cell membranes and signaling molecule synthesis, which significantly promotes cancer progression; in addition, FA synthesis enhancement in cancer cells has been considered to be an important metabolic reprogramming tag [216, 265, 266]. The level of intracellular FAs is closely associated with the survival of cancer cells, and their sources mainly include uptake, hydrolysis, synthesis, and endogenous transformation. Free FAs can be introduced into cells through the FA translocase CD36, and lipoprotein lipase (LPL) can hydrolyze triglycerides to produce FAs. LPL and CD36 proteins are both widely expressed in breast cancer, liposarcoma, and prostate tumor specimens [267]. Citrate, produced *via* the TCA cycle in the mitochondria, is exported across the inner mitochondrial membrane into the cytosol by the transport protein citrate carrier (CIC); it is then divided into acetyl CoA and oxaloacetic acid by ATP citrate lyase (ACLY). Acetyl CoA is involved in de novo FA synthesis pathway *via* acetyl-CoA carboxylase (ACC).

Several human cancers have high levels of CIC, the inhibition of which produces an antitumor effect without being toxic to normal adult tissues. Therefore, CIC is a potential therapeutic target for cancer and other human diseases [268]. The blocking of CIC in breast cancer was found to effectively limit the survival of tumor cells [269]. The upregulated ACLY can promote the growth of tumor cells, whereas downregulated ACLY can inhibit the growth of tumor cells [270, 271]. In addition, ACLY is increased in various cancers, including colorectal cancer, breast cancer, glioblastoma, and ovarian cancer [81, 272–274]. Acetyl-CoA synthetase 2 (ACSS2) converts acetate into acetyl-CoA and promotes FA synthesis in cancer cells. Silencing of ACSS2 can reduce the growth of xenograft tumors. Hypoxia can enhance the expression of ACSS2, which is related to the poor prognosis of breast cancer patients [275]. ACC has two subtypes. ACC1, which is located in the cytoplasm, catalyzes the conversion of acetyl-CoA to malonyl CoA and promotes the synthesis of FAs. ACC2 is located in the mitochondrial membrane and participates in the regulation of FAO [276]. ACC1 can promote the proliferation of liver cancer cells, and its expression is enhanced in liver cancer [277]. The activation of ACC1 has been shown to be associated with the metastasis and recurrence of breast and lung cancer [278]. Inhibition of ACC2 by Snail upregulates FAO to promote tumor progression [279]. In cetuximab-resistant cancer cells, the total ACC level is increased, which shifts cancer metabolism from glycolysis to fat generation [280]. Fatty acid synthase (FASN) catalyzes the continuous condensation reaction of malonyl CoA and acetyl CoA to produce saturated FAs. The high expression and activity of FASN and glycolysis are the main markers of tumor progression and metastasis as well as a common drug target for most malignant tumors [281–283]. The procancer effect of FASN does not depend on FA synthesis, but on the regulation of metabolic reprogramming of cancer cells and the signals of carcinogenesis and cancer development [284]. In primary tumors, the overexpression of monoacylglycerol lipase can drive oncogenic signaling in the FA network, promoting the migration, invasion, survival, and growth of tumors *in vivo* [285]. In addition, in mammals, desaturase (SCDs) is a key enzyme that mediates lipid desaturation and catalyzes the synthesis of monounsaturated FAs (MUFAs). SCD1 is upregulated in various cancers, and its inhibition can cause cancer cell death by reducing MUFAs, and can enhance the anticancer therapeutic effect of gefitinib in lung cancer [286]. The lack of MUFAs in tumor cells can cause the mTORC1-mediated endoplasmic reticulum stress, which further induces cell death [287]. However, polyunsaturated FAs (PUFAs) and the downstream metabolites [such as specialized proresolving mediators (SPMs)] have been shown to inhibit inflammation, angiogenesis, and cancer through various mechanisms [288], and PUFAs or their synthetic precursors (linoleic acid and linolenic acid) need to be supplemented *via* dietary intake.

Acetyl-CoA can also be used as a substrate for the mevalonate pathway (MP). Metabolites of MP, including sterol isoprenoids (such as cholesterol) and nonsterol isoprenoids (such as heme-A, ubiquinone, and dolichol), are involved in cell survival, proliferation, and metabolism. In addition, the metabolites farnesyl pyrophosphate and geranylgeranyl pyrophosphate, produced in the branched pathway, can serve as prenyl donors for a posttranslational modification at the C-terminus of various cellular proteins, which is defined as protein prenylation. 3-hydroxy-3-methylglutaryl-CoA reductase (HMGCR) is the key rate-limiting enzyme in the mevalonate pathway as well as an important regulatory target of cholesterol synthesis. By activating Hedgehog/Gli1 signaling pathway, HMGCR promotes the expression of Gli1 target gene to promote the growth and metastasis of cancer cells [289]. Cholesterol can also directly activate the Hedgehog oncopathway and induce mTORC1 signaling activation [290, 291]. In melanoma, enhanced expression of the cholesterol synthesis genes is associated with reduced survival rate [292]. AMPK inhibits cholesterol synthesis by phosphorylating and inactivating HMG-CoA reductase [293]. SREBP2 is a transcription factor of HMGCR, and mutated p53 can upregulate the MP and increase sterol levels in a SREBP2-dependent manner, promoting the progression of human breast cancer [294]. In addition, HMGCR is the target of statins, which have been shown to reduce the recurrence and mortality of several cancers, including breast, prostate, pancreatic, colorectal, and lung cancer [294–299]. However, the effects of statins on cancer treatment are still controversial, and they are not effective in ovarian, endometrial, and bladder cancer [297, 300]. In addition, Goldstein et al. [301] suggested that statins can cause the continuous increase of T-cells, because of which other immune responses against tumors might be weakened over time, and thus could lead to increased cancer risk. The inactivation of the tumor suppressor gene pRb can induce the abnormal expression of phenyl diphosphate synthase, isoprene transferase, and its upstream SREBPs in an E2F-dependent manner, leading to isopropylation and the activation of N-ras, thereby promoting the occurrence of C-cell adenocarcinoma [301].

Cholesterol has been shown to inhibit the cancer-fighting ability of T-cells. High cholesterol levels can promote the expression of immune checkpoints such as PD-1, 2B4, Tim-3, and Lag-3 by activating endoplasmic reticulum stress in T-cells and inducing their dysfunction [302]. Moreover, cholesterol can inhibit the polarization of $CD8^+$ T-cells toward Tc9 with high tumor-killing ability and promote its conversion to Tc1 with lower antitumor activity [303]. In addition, cholesterol can be reduced to hydroxylated cholesterol (HC) by cholesterol hydroxylase. The 24-HC, 25-HC, and 27-HC activate LXR and then upregulate the expression of ABC transporters (ABCA1 and ABCG1) and thus promote cholesterol efflux [304]; the increase of cholesterol efflux in cancer cells can accelerate cell-membrane fluidity and promote their metastatic ability. Among them, 27-HC is the ligand of LXR as well as estrogen receptor, which can increase estrogen receptor-dependent growth and LXR-dependent metastasis in mouse breast cancer model; 27-HC synthetase CYP27A1 is highly expressed in breast cancer cells and TAMs [305].

Furthermore, both cholesterol and 27-HC can promote thyroid carcinogenicity [306]. However, in gastric cancer, 27-HC can inhibit the proliferation and metastasis of cancer cells [307]. In contrast, 25-HC decreases the sensitivity of human gastric cancer cells to 5-FU and promotes GC cell invasion by upregulating TLR2/NF-κB-mediated MMP expression [308]. Furthermore, 25-HC can promote the metastasis and invasion of lung adenocarcinoma cells [309], and plays an important regulatory role in innate and acquired immunity, which promote the antivirus and antiinflammatory reactions of the body. In addition, 7α,25-dihydroxycholesterol functions as an immune cell guidance cue by engaging the G-protein-coupled receptor EBI2; it is required for increasing adaptive immune responses [310].

3.6.3 Amino acid metabolism and tumor immunity

Amino acid, the substrate for protein synthesis, in particular Gln, Arg, and tryptophan (Trp), have been shown to play an important role in the development of cancer. In cancer, Gln is considered to be the most important substrate except glucose, and it participates in a series of pathways, including energy production (*e.g.*, conversion to α–KG into TCA), macromolecular synthesis (*e.g.*, GSH), and signaling activation (*e.g.*, mTOR activation and protein synthesis) in cancer cells by providing nitrogen and carbon [309–311]. Moreover, the increase of Gln catabolism is one of the characteristics of the metabolic reprogramming in cancer cells. Glutaminase (GLS) is required for the first step in the catalytic conversion of Gln to glutamate (Glu) [314]. GLS is increased in many cancers, including breast, liver, colorectal, brain, cervical, and lung cancer as well as melanoma [313]. The oncogenic transcription factor c–Myc has been shown to directly upregulate the Gln metabolic enzymes glutaminase (GA) [315] and its function is directly regulated by HIF-1 and HIF-2 [316, 317]. Under hypoxia, IDH1-mediated reductive metabolism of Gln extensively increases adipogenesis in cancer cells [318]. The inhibition of GA can reduce the growth of IDH1-mutated glioma cells and primary acute myeloid leukemia cells [319, 320]. In contrast, the GLS isozyme GLS2 showed anticancer effect [321] and was increased by the tumor suppressor p53 [322, 323]. In addition, Gln has an important effect on the differentiation of effector T-cells, and limiting Gln during CD8$^+$ T-cells activation can change their differentiation to memory T cells (Tmem) [324].

Arg is involved in many important cellular metabolic pathways, including the urea cycle, and the biosynthesis of NO, nucleotides, proline, and glutamate [325]. Many malignant tumor cells die rapidly and selectively in Arg-deficient culture media [326]. For example, malignant melanoma cells and hepatocellular carcinoma cells cannot synthesize Arg by themselves as they lack arginine succinic acid synthase; thus, the targeted inhibition of Arg uptake in these tumors preferably induces an anticancer effect [327]. In addition, iNOS is highly expressed in M1 macrophages and can be used as a substrate to synthesize NO; it plays an anticancer role [150]. High levels of ARG1 in cancer cells [328], tumor-associated fibroblasts [329], MDSCs [330], M2 macrophages [331], and DCs [332] can reduce L-Arg in the TME and convert it to ornithine and urea, which are necessary for TCR activation in T-cells [333]. In addition, Arg deficiency leads to decreased protein synthesis and anticancer activity in T-cells [152,334]. Exogenous Arg supplementation during T-cell activation can promote the generation of central memory-like cells and enhance the antitumor activity of effector T-cells [153].

Trp is an important nutrient required for promoting T-cell proliferation and activation. Cancer cells, TAMs, MDSCs, and suppressor DCs, and tumor-associated fibroblasts can transform Trp to Kyn by IDO, and both the decrease in Trp and accumulation of Kyn synergistically inhibit the activation and proliferation of T-cells [335–337]. Thus, the upregulation of IDO is associated with T-cell aggregation, proliferation, impaired function, and poor prognosis in patients with cancer, including gastric cancer, colorectal cancer, NSCLC, and melanoma [338, 339]. IFN-γ produced by Th1 cells also stimulates IDO-1 production in APCs [340]. Kyn is a ligand of the aromatic hydrocarbon receptor (AhR), and targeted inhibition of the IDO1-Kyn-AhR pathway can significantly improve the efficacy of cancer immunotherapy [341, 342]. In addition, Kyn can inhibit the maturation of DCs, increase the proliferation of immunosuppressive cells (Tregs and MDSCs), and inhibit the proliferation of T-cells and NK cells [340]. Trp is an essential amino acid, and its depletion in the TME leads to T-cell apoptosis. Moreover, Trp is one of the amino acids required for the biosynthesis of NAD(P), which is essential for metabolic processes (glycolysis, gluconeogenesis, TCA cycle, and OXPHOS) as well as for NADPH production; NAD(P) contributes to $CD4^+$ T-cell differentiation and SIRT3-mediated mitochondrial genesis [343–345].

In addition to being derived from the de novo synthesis of glucose metabolism, serine (Ser) can be obtained through extracellular uptake. The conversion of Ser to glycine under the catalysis of SHMTs is regulated by c-Myc transcription [346]. Ser and Gly together provide necessary precursors for the synthesis of important proteins, nucleic acids, and lipids in cancer cells and influence the antioxidant capacity of cells to support tumor homeostasis [347, 348]. Besides, the proliferation and function of T_{eff} cells were both dependent on the sufficient serine metabolism *in vitro* and *in vivo* [349].

3.6.4 Mitochondrial and tumor immunity

Mitochondria, derived from endosymbiotic bacteria through natural selection, retained 16 kB of the genome that encodes tRNAs, rRNAs, and proteins necessary for respiration. Mitochondrial function is of remarkable significance for tumor and immune cell activity, because many cell metabolism signals appear in the mitochondria or are regulated by mitochondrial activity, including material energy metabolism (such as TCA, OXPHOS, FAO, and amino acid metabolism) [350–352], cell death (such as ROS-mediated oxidative stress, Ca^{2+} homeostasis disorder-mediated apoptosis) [353–355], cell proliferation (*e.g.*, mtDNA transcription-induced cell proliferation) [356, 357], and cell migration (*e.g.*, mitochondrial dynamics and mtDNA mutation-induced EMT and metastatic tumor cell progression) [358, 359].

Mitochondrial ROS (mtROS) is an important accessory substance of complex I and III in the electron transfer chain (ETC); they are related to oxidative damage [360–362]; however, some of them are also important for T-cell and B-cell activation [361, 363–366]. The Cu-Zn superoxide dismutase (SOD) can dismutate mtROS into hydrogen peroxide, which inhibits M1 macrophages and promotes the M2 phenotype [367]. mtROS also can activate NLRP3 inflammasomes and NF-κB signaling, and promote the production of IL-1 and IL-18 [362, 368–372]. GSH can reduce ROS to relieve cell damage, and is reversibly reduced to glutathione disulfide (GSSG) after oxidation. Under the catalysis of glutathione peroxidase (GPX), NADPH drived from the PPP is captured and transformed into GSH [373]. In addition to GSH, other antioxidant molecules also can clear ROS in cells, such as CO-Q, SIRT3 (deacetylation of acetyl-CoA dehydrogenase, manganese superoxide dismutase, and catalase) [374], and forkhead box O1. Uncoupling protein 2 (UCP2) helps the transportation of H^+ ions from the intermembrane space to the mitochondrial matrix, thereby reducing the membrane potential to inhibit excess ROS production [375]. Furthermore, blocking UCP2 results in the reduction of M2 macrophage activation induced by IL-4 [376]. In addition, rapamycin inhibits mtROS production by inhibiting complex I formation [377, 378], or by inducing autophagy and activating the P62 and NRF2 pathways [379]. The oxygen-free radicals produced by ETC complex II can produce the Fenton reaction with Fe^{3+}, which causes cell ferroptosis. Hydroxyl free radicals, the product of this reaction, are unstable and can destroy lipids, proteins, and DNA in cells [380].

Mitochondrial Ca^{2+} regulates metabolism, cell death, and cell signaling, and its uptake is mainly regulated through a calcium-selective ion channel complex in the mitochondrial membrane, the mitochondrial calcium unidirectional transporter [381]. Store-operated Ca^{2+} entry (SOCE) is a common Ca^{2+} influx pathway, and aspirin can inhibit tumor cell proliferation by interfering with mitochondrial Ca^{2+} uptake and disrupting SOCE homeostasis [382]. Ca^{2+} fluxes between the ER and mitochondria affect several cancer hallmarks, including apoptosis resistance, migration, and invasion [383]. Moreover, mitochondrial Ca^{2+} is involved in the regulation of various forms of cell death, including apoptosis, necrosis and autophagy [383].

Mitophagy, a form of autophagy, is regulated by PTEN-induced putative kinase 1 and Parkin, an E3-ubiquitin ligase [384]; it specifically eliminates damaged mitochondria to maintain their quality. T-cells utilize mitophagy to reduce increased ROS and apoptosis and maintain appropriate organelle homeostasis [385]. In autophagy-related protein 7 (Atg7)-deficient T-cells, the mitochondrial content, ROS level, and the expression of proapoptotic proteins such as Bak, cytochrome C, and AIF is increased [386]. In addition, vacuolar protein sorting 34 (Vps34), a member of the PI3K family, is the third type of lipid kinase. In Vps34-deficient $CD4^+$ and $CD8^+$ T-cells, ROS level, mitochondrial quality, and mitophagy of damaged mitochondria were upregulated [387]. Impaired mitophagy promotes the upregulation of mtROS in Tregs, thereby inducing cell death [388]. Moreover, after the autophagy agent Rapa is used to enhance mitochondrial autophagy, mitochondrial membrane potential will be increased, and the immunosuppressive function of Tregs will also be enhanced [389].

During mitochondrial dysfunction, mtDNA is released from the mitochondria into the cytosol, which binds and activates NLRP3 inflammasomes [368, 390]. Huang et al. [391] indicated that mtDNA inhibits yap-mediated endothelial cell proliferation by activating cGAS signal, thereby promoting inflammatory injury. Mitochondrial STAT3 was shown to regulate mtDNA transcription in zebrafish embryonic and juvenile stem cells, which is the basis for determining the proliferation rate of tissue stem cells [356]. Transferring mitochondria into bone marrow-derived mesenchymal stem cells (BMSCs) *in vitro* can promote their proliferation, migration, and osteogenic differentiation [392]. mtDNA has been shown to be enhanced during the activation of T-cells [393]. After the mitochondrial transcription factor (Tfam) was specifically knocked out in T-cells, the mtDNA and mitochondrial respiratory chain were missing, and their proliferation was less than that of wild-type [394]. Mitochondria are complex organelles that affect the initiation, growth, survival, and metastasis of cancer cells and actively promote the occurrence of cancer *via* many other aspects except energy production [395]. The changes in ROS levels lead to mutations in mtDNA, which disrupts the normal function of mitochondrial proteins [396]. Germ cells or somatic cells with mtDNA mutation have an increased risk of becoming cancerous [397], whereas cancer cells with mtDNA deletion or Tfam knocking down have reduced tumor formation in nude mice [398]. However, in eosinophilic thyroid tumors, mtDNA mutation produces the invasive phenotype in cancer cells, which leads to the bioenergy crisis [399]. Mambo *et al.* [400] found that mtDNA content was reduced in 80% of breast tumors relative to their corresponding normal. However, the mtDNA level was increased in papillary thyroid carcinomas, compared to the corresponding normal DNA obtained from the same individuals. These findings indicate that changes of mtDNA content during carcinogenesis may be regulated in a tumor-specific manner. The reduce of mtDNA content in human mammary epithelial cells induces EMT-like reprogramming to fibroblastic morphology, loss of cell polarity, contact inhibition and acquired migratory, and invasive phenotype; in addition, the reduction of mtDNA can promote the production of breast cancer stem cells [401]. The EMT induced by TGF-β can upregulate mtDNA copy number in NSCLC cells [402]. The destruction of mtDNA polymerase specifically in neutrophils also decreased the velocity of neutrophil interstitial migration [403].

The four major proteins involved in mitochondrial dynamics are members of dynamin-like guanosine triphosphatases (GTPases), including mitofusin 1 and 2 (Mfn1/2), optic atrophy 1 (Opa1), and dynamin-related protein-1 (Drp1) [404]. Perturbing the asymmetric distribution of mitochondria within migrating cells by interfering with mitochondrial fusion (opa-1) or fission (drp-1) proteins significantly reduces the number of cells with anterior localization of mitochondria and the velocity and directional persistence of the fastest moving cells [405]. Mitochondrial uncoupling agents or adenosine triphosphate synthesis inhibitors can be used to reduce EMT and the migration and invasion ability of breast cancer cells [406]. In addition, the phagocytosis of macrophages to apoptotic cells requires the division and increase of Drp1 [407]. Mitochondrial fission (opa-1) and fusion (drp-1) proteins are, respectively, associated with the polarization of activated T_{eff} cells and T_{mem} cells [408]. During the activation of antigen-specific T-cells and NK cells, their mitochondria move toward the immune synapse (IS) and promote the ATP production in IS [409, 410]. Receptor interaction protein kinase 3 (RIPK3) promotes NFAT translocation and Drp1 dephosphorylation in NKT cells by activating mitochondrial phosphoglycerate mutase 5 and enhances their immune response [411].

3.7 Microbiota and tumor immunity

The human gut microbes are the second human genome, and their gene expression and metabolism are linked to human health and disease. With the development of 16S rRNA or DNA/sequencing/metagenomic approaches in recent years, many studies have revealed some potential immune functions of the intestinal microbiome [412, 413]. Increasing evidence suggests that some intestinal microbiotas can promote carcinogenesis and help in the metastasis of cancer cells, leading to chemotherapy resistance and affecting the efficacy of immunotherapy [414–416]. For example, inflammatory bowel disease (IBD) is not sufficient to induce colorectal cancer (CRC) in the absence of intestinal microflora or microbial products [417]. Except for colorectal cancer, intestinal microbial metabolites also act on the development of gastrointestinal tumors (including esophageal, gastric, colorectal, liver, and pancreatic cancer) through the gut–liver axis [14, 418–422]. Moreover, the metabolites and metabolic changes influence the progression of breast cancer, stomach cancer, leukemia, and melanoma cancer [423–425].

The early implantation of microorganisms in the human body shows an important influence on the immune system [426]. The colonization of microorganisms in infancy affects the development of the immune system of mammals, and this influence lasts throughout the whole adulthood. In germfree (GF) mice, NKT cells tend to be higher in the lung and colon, which increases their susceptibility to colitis [427, 428]. Moreover, a few B-cell immunoglobulin A were found in the small intestine of GF mice [429], but excessive serum immunoglobulin E were found in the serum of GF mice that were prone to allergic reactions [430]. In addition, babies exposure to specific symbiotic microbial antigens during the first 2 weeks of life, but not thereafter, induces a Helios-negative (presumably induced) subset of Tregs from conventional $CD4^+$ T-cells, enabling them to recognize these antigens in adulthood and thus avoiding anaphylactic reactions [431]. Moreover, intestinal microbiotas can stimulate TAMs polarization to M2 type and the metastasis of cancer cells by increasing cathepsin K secretion in cancer cells [432].

Intestinal microorganisms influence the efficacy of checkpoint inhibitors for tumor immunotherapy. Zitvogeld *et al.* [18] found that, during PD-1 inhibitor treatment, patients who were administered antibiotics showed rapid recurrence, and their survival time was significantly shortened, compared to those with an intact intestinal flora. They also revealed that the absence of *Akkermansia muciniphila* remarkably reduced the efficacy of PD-1 inhibitors [18]. Intestinal melanoma patients containing numerous Clostridiales/Ruminococcaceae are more likely to respond to PD-1 antibody treatment, but those with more Bacteroidales are unlikely to respond to treatment, and those with higher *Faecalibacterium* species (belonging to the order Clostridiales) show significantly prolonged progression-free period [19]. Gajewski *et al.* [20] also confirmed that metastatic melanoma patients who had a rich abundance of *Bifidobacterium longum*, *Collinsella aerofaciens*, and *Enterococcus faecium* responded to anti-PD-1. In addition, oral bifidobacteria facilitate considerable anticancer effect with PD-L1 antibodies, and in their combination, bifidobacteria can also boost the antineoplastic effect of the PD-L1 antibody [433]. Furthermore, *Bifidobacterium pseudolongum*, *Lactobacillus johnsonii*, and *Olsenella* species significantly promote the efficacy of immune checkpoint inhibitors in the cancer mouse models [24]. In a recent study, 11 kinds of bacteria were screened that can enhance the immune response against infection and cancer. When the 11 strains were transplanted into tumor-burdened mice as a combination (11-mix), the effect of PD-1 antibodies and other immune checkpoint inhibitors was significantly enhanced, and 11-mix induced the increase of $CD8^+$ $IFN-\gamma^+$ T-cells to promote the recognition and removal of cancer cells [23]. In addition, another immune checkpoint inhibitor, CTLA-4 antibody, can promote the growth of *Bacterooides fragilis*, which has anticancer properties in melanoma patients, and both *Bacterooides thetaiotaomicron* and *B. fragilis* can activate T-cells and improve the therapeutic effect of the CTLA-4 antibody [21].

The toxins and metabolites of intestinal microbiota also play important roles in the development of cancer. Lipopolysaccharide (LPS) [10] and its metabolite deoxycholic acid (DCA) [11] are transmitted to the liver through the portal vein and promoted the development of chronic liver diseases and hepatocellular carcinoma (HCC). Ma *et al.* [12] found that *Clostridium scindens* controls cholic acid metabolism in the intestine of mice and downregulates CXCL16 expression in the liver through hepatointestinal circulation, thereby inhibiting the accumulation of NKT cells and promoting the growth of liver tumors. About 5%–10% of intestinal unreabsorbed bile acids (BAs) can serve as a substrate for microbial metabolism to produce secondary BAs, including DCA and lithocholic acid (LCA), which may further promote carcinogenesis [1, 2]. Short-chain fatty acids (SCFAs), a fermentation product derived from intestinal microorganisms, can regulate the number and function of Tregs in the colon and prevent colitis in mice in a Ffar2 (GPR43)-dependent manner [3]. Moreover, butyrate and propionate produced by commensal bacteria can enhance the generation of Tregs and induce the differentiation of colon Tregs [4, 5]. About 4%–6% of tryptophan is metabolized by the intestinal flora into indole, indican, tryptamine, skatole, and indoleic acid derivatives [434], which can target the host AhR and regulate the mucosal immune system [6, 435]. Moreover, bacterial tryptophan metabolites can cross the blood-brain barrier, and act on microglia and astrocytes in the CNS and limit inflammation and neurodegeneration through the AhR [436]. The fungal polysaccharide β-dextran and the bacille Calmette-Guérin (BCG) vaccine can enhance the phagocytic ability and antigen presentation of TANs and the generation of granulocyte cell-producing IFN-γ from bone marrow cells and neutrophil-mononuclear progenitor cells, thereby playing an antitumor role; the process is also called training innate immunity [22]. In addition, BCG vaccine induces hematopoietic stem cells to promote trained innate immunity [437], and it was used in bladder cancer treatment [438]. In contrast, *Mycobacterium tuberculosis* inhibits bone marrow formation, destroys trained innate immunity, and increases *Mycobacterium tuberculosis* susceptibility [437].

In addition, a study found that *Bacteroides fragilis* and PSK$^+$ *Escherichia coli* combined to form biofilms, which damaged the colonic mucosa and caused colon cancer [439]. The lack of TET2 leads to intestinal barrier defects in mice, allowing the bacteria to spread to the blood and peripheral organs, leading to the release of inflammatory molecules in the blood and promoting the proliferation of TET2-deficient hematopoietic stem cells and the occurrence of leukemia [440]. Moreover, the microbiota in the ileum and colon can directly affect the functions of the pancreas and liver by maintaining CART$^+$ intrinsic neurons in the intestinal tract, thereby affecting the occurrence and development of pancreatic cancer and liver cancer [441].

Microorganisms also exist in tumor tissues, including the TME, cancer cells, and immune cells; they influence tumor progression and treatment. Straussman *et al.* [7] found that, in melanoma, pancreatic cancer, lung cancer, ovarian cancer, glioblastoma, bone cancer, and breast cancer, microbes were extensively found in tumor tissues, and LPS and 16S ribosomal RNA were found in cancer cells and immune cells. The microfloral composition differs among various types of tumors. For example, the microbiota in bone cancer patients is more inclined to hydroxyproline degradation, and that in lung cancer patients is prone to show the enrichment of tobacco metabolism pathways. In addition, they found that the microbiome of melanoma patients that responded to immunotherapy was significantly different from that of nonresponders, with more clostridium bacteria in the former and more *Gardnerella vaginalis* in the latter. In addition, the bacteria in a tumor may influence the resistance of the tumor to the chemotherapy drug gemcitabine [442]. Therefore, not only gut microbes, but symbiotic microbes in all tissues of the human body should form the second genomic map, which needs to be more investigated.

Many factors affect the host microbiome, and the genetic factors of the host account for only 2%. In fact, 98% of the differences in the human intestinal microbiome are determined by diet and lifestyle [443]. Although the use of antibiotics may decrease the host microbial abundance, drugs such as proton pump inhibitors [444], metformin [445], statins [445], and laxatives [446] may have significant effects on intestinal microorganisms. A ketogenic diet with high fat and low carbohydrate content promotes the accumulation of probiotics such as *A. muciniphila* and *Parabacteroides*, which increases the levels of gamma-aminobutyric acid in the brain through the enteric-cerebral axis [447]. A high-fiber diet promotes the growth of 15 intestinal bacterial strains that produce SCFAs, butyric acid, and acetic acid, which creates a mildly acidic intestinal environment that reduces the number of harmful bacteria [448]. Tungstate (a soluble tungstate salt) in oral drinking water selectively prevented the proliferation of Enterobacteriaceae members in the gut without affecting the beneficial gut bacteria and normalized the intestinal microflora of the mouse colitis model, thereby reducing intestinal inflammation [449].

Zhernakova *et al.* [450] found that chromogranin A (CgA), a polypeptide secreted by nerve cells, endocrine cells, and immune cells, was associated with 61 intestinal microbiotas; the higher the diversity of the intestinal microbiota, the lower is the CgA concentration.

FMT involves the transplantation of gut microbiota from a healthy donor to patients through the upper and lower digestive tract in order to restore the gut microbial diversity; it has played positive roles in digestive diseases and cancer immunotherapy treatment [451]. Li *et al.* [452] found that the microbiotas of FMT donors could efficiently coexist with the original microbiotas of the recipient, leading to the significant increase in microbial abundance. According to the 2013 guidelines [453], FMT has been approved as a clinical method for treating recurrent *Clostridium difficile* infection with about 90% clinical efficacy [454]. Oral supplementation with *A. muciniphila* after FMT with nonresponder feces restored the efficacy of PD-1 blockade in an DCs-IL-12-dependent manner by increasing the recruitment of CCR9$^+$ CXCR3$^+$ CD4$^+$ T-lymphocytes into mouse tumor beds [18]. CRC mice transplanted with intestinal microorganisms from wild mice showed higher resistance and improved inflammation [455]. FMT also improved high-fat-diet (HFD)-induced steatohepatitis in mice and elevated the abundance of the beneficial bacteria including Christensenellaceae and *Lactobacillus* [456]. When FMT was applied to type 1 diabetes, the number of *Desulfovibrio piger* significantly increased and the levels of CD4$^+$ CXCR3$^+$ T-cells, CCL22, and CCL5 were downregulated to promote the functional stability of cells in the patients [457]. Transplantation of feces from patients responding to melanoma into mice was found to reduce tumor progression [19]; a clinical study is currently testing the effect of FMT on the intestinal tract of patients with melanoma not responding to PD-1 antibody [458]. In addition, the microbiota from pancreatic ductal adenocarcinoma mice accelerated tumor progression in GF mice [459].

4. Conclusion and perspectives

Tumor immunotherapy aims to suppress the immune escape of cancer cells and promote the tumor-killing ability of the immune system. Therefore, targeting the metabolic pathways of cancer and immune cells and regulating "cell fate" by changing "cellular substances" can provide a new strategy for tumor immunotherapy.

The metabolic reprogramming of cancer cells facilitates tumor progression; thus, targeting the metabolic pathways that promote the proliferation, EMT, and metastasis of cancer cells can effectively hinder tumor progression. During the tumorigenesis stage, inhibition of glycolysis can reduce the energy supply of cancer cells and inhibit their proliferation. Targeting the key rate-limiting enzymes in the glycolytic pathway, including Glut1 [87], HK2 [230], PFKFB3 [233], PKM2 [234], and LDH [236], can significantly inhibit glycolysis in cancer cells. In addition to oncogenes (such as c-Myc and Akt) that promote the upregulation of glycolysis [196, 220, 221, 460], HIF-1α—a characteristic molecule of the TME—increases the glycolysis pathway in cells [132, 461] and enhances cancer-related inflammation during initiation [461]. mTORC1 activates the expression of several glycolytic enzymes mediated by c-Myc and HIF1-1α [116, 462, 463]. The regulation of AMPK on glycolysis is controversial. Some studies have shown that AMPK inhibitors can decrease the aerobic glycolysis of cancer cells and inhibit the deterioration of cancer cells [464], whereas others have shown that AMPK is an inhibitor of the Warburg effect of cancer cells and can hinder the progression of cancer *in vivo* [465]. Hussien and Brooks found that two types of breast cancer cells can utilize lactic acid in different ways [466]. In the McF-7 cell line, MCT1 (the output of lactic acid) was abundant; LDHA, which converts pyruvate into lactic acid, was increased; and the lactic acid secreted into the TME played an immunosuppressive role [467]. In MDA-MB-231 cells, MCT4 (lactate uptake) was overexpressed, and LDHB, which converts lactic acid back to pyruvate, was increased, thereby enhancing the TCA cycle of cancer cells and providing raw materials for the synthesis of macromolecular substances such as lipids, amino acids, and nucleic acids of cancer cells [467]. Therefore, targeted lactic acid uptake and efflux can also block cancer cells. Targeting the lipid and amino acid metabolism of cancer cells can also inhibit tumor proliferation and progression. The inhibition of FAS pathway enzymes in cancer cells, including CIC [269], ACLY [271], ACC [468], and FASN [282], can play a role in cancer inhibition. ARG1 [328] and IDO [469], respectively, consume Arg and Trp in the TME and have toxic effects on T-cells. Targeted inhibition of these two enzymes can weaken the immune suppression of cancer cells and enhance the function of T-cells. In addition, targeting ASNS [66] and cholesterol effluents (ABCA1) [67] can inhibit the metastasis of cancer cells.

The TME heterogeneity caused by metabolic reprogramming of cancer cells plays an important role in the regulation of the metabolism, differentiation, and function of immune cells. The lack of nutrients inhibits antitumor immune cells such as $CD8^+$ T-cells, M1 macrophages, and N1 neutrophils, and promotes the differentiation and activation of protumor immune cells, including MDSCs, M2 macrophages, and Tregs [470, 471]. The strong aerobic glycolysis and proliferation of cancer cells facilitates the maintenance of hypoxic TME, which induces the cancer-promoting effect of immune cells [472]. The elevation of HIF-1α in TAMs, TADCs, MDSCs, and Tregs can promote tumor growth through PD-L1 expression, lactate release, and adenosine-adenosine receptor interaction, thereby contributing to immunosuppression and angiogenesis [113, 473–476]. Hypoxia induces the expression of the immune checkpoint v-domain Ig suppressor of T-cell activation [477, 478] and can enhance the immune checkpoint CD47 "don't eat me signal" in breast cancer cells, which mediates the escape of cancer cells from phagocytosis [479]. In TME, other factors and molecules such as noncoding RNA as well as complement- and coagulation-related factors, also regulate the survival, differentiation, and function of immune cells [480–482]. Therefore, targeting the metabolic reprogramming pathway of cancer cells and blocking its influence on immune cells can improve cancer immunotherapy. However, determining the mechanism that specifically changes the metabolism of cancer cells and reduce the inhibition on the metabolism of antitumor immune cells are the bottleneck of metabolism-targeting anticancer therapy.

In addition, the use of FMT to change the composition and metabolism of symbiotic microbiotas is also an emerging strategy to improve the efficacy of cancer immunotherapy. However, most of the reported studies of intestinal microbiome are limited to cell models and animal models, and some human clinical trials only exist in phase I or II clinical trials, and the number of patients involved in clinical trials is too small. However, as scientists continue to make new breakthroughs in the gut microbiome, the combination of microbiome regulation and immunotherapy such as CAR-T cell therapy and multiple immune checkpoint inhibitors are expected to benefit more patients in the future. The microbiome and their metabolites showed important role in regulating immunity system and cancer progression, and provide the biomarkers that can be tested for the risk of cancer and its progression; nevertheless, the types and functions of cancer-associated microbiotas and their metabolites are not clearly known, and the rational use of symbiotic microbiotas is still remarkably challenging.

References

[1] Ridlon JM, Kang DJ, Hylemon PB. Bile salt biotransformations by human intestinal bacteria. J Lipid Res 2006;47(2):241–59.
[2] Zeng H, Umar S, Rust B, Lazarova D, Bordonaro M. Secondary bile acids and short chain fatty acids in the colon: a focus on colonic microbiome, cell proliferation, inflammation, and cancer. Int J Mol Sci 2019;20(5):1214.
[3] Smith PM, Howitt MR, Panikov N, Michaud M, Gallini CA, Bohlooly-Y M, et al. The microbial metabolites, short-chain fatty acids, regulate colonic Treg cell homeostasis. Science 2013;341 (6145):569–73.
[4] Furusawa Y, Obata Y, Fukuda S, Endo TA, Nakato G, Takahashi D, et al. Commensal microbe-derived butyrate induces the differentiation of colonic regulatory T cells. Nature 2013;504 (7480):446–50.
[5] Arpaia N, Campbell C, Fan X, Dikiy S, van der Veeken J, DeRoos P, et al. Metabolites produced by commensal bacteria promote peripheral regulatory T-cell generation. Nature 2013;504(7480):451–5.
[6] Zelante T, Iannitti RG, Cunha C, De Luca A, Giovannini G, Pieraccini G, et al. Tryptophan catabolites from microbiota engage aryl hydrocarbon receptor and balance mucosal reactivity via interleukin-22. Immunity 2013;39(2):372–85.
[7] Nejman D, Livyatan I, Fuks G, Gavert N, Zwang Y, Geller LT, et al. The human tumor microbiome is composed of tumor type-specific intracellular bacteria. Science 2020;368(6494):973–80.
[8] Singh N, Thangaraju M, Prasad PD, Martin PM, Lambert NA, Boettger T, et al. Blockade of dendritic cell development by bacterial fermentation products butyrate and propionate through a transporter (Slc5a8)-dependent inhibition of histone deacetylases. J Biol Chem 2010;285(36):27601–8.
[9] Louis P, Hold GL, Flint HJ. The gut microbiota, bacterial metabolites and colorectal cancer. Nat Rev Microbiol 2014;12(10):661–72.
[10] Schroeder BO, Backhed F. Signals from the gut microbiota to distant organs in physiology and disease. Nat Med 2016;22(10):1079–89.
[11] Malhi H, Camilleri M. Modulating bile acid pathways and TGR5 receptors for treating liver and GI diseases. Curr Opin Pharmacol 2017;37:80–6.

[12] Ma C, Han M, Heinrich B, Fu Q, Zhang Q, Sandhu M, et al. Gut microbiome-mediated bile acid metabolism regulates liver cancer via NKT cells. Science 2018;360(6391):eaan5931.

[13] Chng KR, Chan SH, Ng A, Li C, Jusakul A, Bertrand D, et al. Tissue microbiome profiling identifies an enrichment of specific enteric bacteria in *Opisthorchis viverrini* associated cholangiocarcinoma. EBioMedicine 2016;8:195–202.

[14] Riquelme E, Zhang Y, Zhang L, Montiel M, Zoltan M, Dong W, et al. Tumor microbiome diversity and composition influence pancreatic cancer outcomes. Cell 2019;178(4):795–806. e12.

[15] Pandey KR, Naik SR, Vakil BV. Probiotics, prebiotics and synbiotics—a review. J Food Sci Technol 2015;52(12):7577–87.

[16] Hieken TJ, Chen J, Hoskin TL, Walther-Antonio M, Johnson S, Ramaker S, et al. The microbiome of aseptically collected human breast tissue in benign and malignant disease. Sci Rep 2016;6:30751.

[17] Urbaniak C, Gloor GB, Brackstone M, Scott L, Tangney M, Reid G. The microbiota of breast tissue and its association with breast cancer. Appl Environ Microbiol 2016;82(16):5039–48.

[18] Routy B, Le Chatelier E, Derosa L, Duong C, Alou MT, Daillere R, et al. Gut microbiome influences efficacy of PD-1-based immunotherapy against epithelial tumors. Science 2018;359(6371):91–7.

[19] Gopalakrishnan V, Spencer CN, Nezi L, Reuben A, Andrews MC, Karpinets TV, et al. Gut microbiome modulates response to anti-PD-1 immunotherapy in melanoma patients. Science 2018;359 (6371):97–103.

[20] Matson V, Fessler J, Bao R, Chongsuwat T, Zha Y, Alegre ML, et al. The commensal microbiome is associated with anti-PD-1 efficacy in metastatic melanoma patients. Science 2018;359(6371):104–8.

[21] Vetizou M, Pitt JM, Daillere R, Lepage P, Waldschmitt N, Flament C, et al. Anticancer immunotherapy by CTLA-4 blockade relies on the gut microbiota. Science 2015;350(6264):1079–84.

[22] Kalafati L, Kourtzelis I, Schulte-Schrepping J, Li X, Hatzioannou A, Grinenko T, et al. Innate immune training of granulopoiesis promotes anti-tumor activity. Cell 2020;183(3):771–85. e12.

[23] Tanoue T, Morita S, Plichta DR, Skelly AN, Suda W, Sugiura Y, et al. A defined commensal consortium elicits CD8 T cells and anti-cancer immunity. Nature 2019;565(7741):600–5.

[24] Mager LF, Burkhard R, Pett N, Cooke N, Brown K, Ramay H, et al. Microbiome-derived inosine modulates response to checkpoint inhibitor immunotherapy. Science 2020;369(6510):1481–9.

[25] Borody TJ, Khoruts A. Fecal microbiota transplantation and emerging applications. Nat Rev Gastroenterol Hepatol 2011;9(2):88–96.

[26] Khalil DN, Smith EL, Brentjens RJ, Wolchok JD. The future of cancer treatment: immuno-modulation, CARs and combination immunotherapy. Nat Rev Clin Oncol 2016;13(5):273–90.

[27] Sermer D, Brentjens R. CAR T-cell therapy: full speed ahead. Hematol Oncol 2019;37(Suppl. 1):95–100.

[28] Heczey A, Courtney AN, Montalbano A, Robinson S, Liu K, Li M, et al. Anti-GD2 CAR-NKT cells in patients with relapsed or refractory neuroblastoma: an interim analysis. Nat Med 2020;26 (11):1686–90.

[29] Geoerger B, Kang HJ, Yalon-Oren M, Marshall LV, Vezina C, Pappo A, et al. Pembrolizumab in paediatric patients with advanced melanoma or a PD-L1-positive, advanced, relapsed, or refractory solid tumour or lymphoma (KEYNOTE-051): interim analysis of an open-label, single-arm, phase 1–2 trial. Lancet Oncol 2020;21(1):121–33.

[30] Zhang J, Wang L. The emerging world of TCR-T cell trials against cancer: a systematic review. Technol Cancer Res Treat 2019;18, 1533033819831068.

[31] Keibel A, Singh V, Sharma MC. Inflammation, microenvironment, and the immune system in cancer progression. Curr Pharm Des 2009;15(17):1949–55.

[32] Greten FR, Grivennikov SI. Inflammation and cancer: triggers, mechanisms, and consequences. Immunity 2019;51(1):27–41.

[33] Liu L. Antibody glycosylation and its impact on the pharmacokinetics and pharmacodynamics of monoclonal antibodies and Fc-fusion proteins. J Pharm Sci 2015;104(6):1866–84.

[34] Boutros C, Tarhini A, Routier E, Lambotte O, Ladurie FL, Carbonnel F, et al. Safety profiles of anti-CTLA-4 and anti-PD-1 antibodies alone and in combination. Nat Rev Clin Oncol 2016;13 (8):473–86.

[35] Dillard T, Yedinak CG, Alumkal J, Fleseriu M. Anti-CTLA-4 antibody therapy associated autoimmune hypophysitis: serious immune related adverse events across a spectrum of cancer subtypes. Pituitary 2010;13(1):29–38.

[36] Osorio JC, Ni A, Chaft JE, Pollina R, Kasler MK, Stephens D, et al. Antibody-mediated thyroid dysfunction during T-cell checkpoint blockade in patients with non-small-cell lung cancer. Ann Oncol 2017;28(3):583–9.

[37] Gauci ML, Laly P, Vidal-Trecan T, Baroudjian B, Gottlieb J, Madjlessi-Ezra N, et al. Autoimmune diabetes induced by PD-1 inhibitor-retrospective analysis and pathogenesis: a case report and literature review. Cancer Immunol Immunother 2017;66(11):1399–410.

[38] Shah NN, Fry TJ. Mechanisms of resistance to CAR T cell therapy. Nat Rev Clin Oncol 2019;16 (6):372–85.

[39] Dunn GP, Bruce AT, Ikeda H, Old LJ, Schreiber RD. Cancer immunoediting: from immunosurveillance to tumor escape. Nat Immunol 2002;3(11):991–8.

[40] Vesely MD, Kershaw MH, Schreiber RD, Smyth MJ. Natural innate and adaptive immunity to cancer. Annu Rev Immunol 2011;29:235–71.

[41] Dunn GP, Old LJ, Schreiber RD. The immunobiology of cancer immunosurveillance and immunoediting. Immunity 2004;21(2):137–48.

[42] Dunn GP, Old LJ, Schreiber RD. The three Es of cancer immunoediting. Annu Rev Immunol 2004;22:329–60.

[43] Galon J, Angell HK, Bedognetti D, Marincola FM. The continuum of cancer immunosurveillance: prognostic, predictive, and mechanistic signatures. Immunity 2013;39(1):11–26.

[44] Inman BA, Frigola X, Dong H, Kwon ED. Costimulation, coinhibition and cancer. Curr Cancer Drug Targets 2007;7(1):15–30.

[45] Chen DS, Mellman I. Oncology meets immunology the cancer-immunity cycle. Immunity 2013;39 (1):1–10.

[46] Pardoll DM. The blockade of immune checkpoints in cancer immunotherapy. Nat Rev Cancer 2012;12(4):252–64.

[47] Hodi FS, O'Day SJ, McDermott DF, Weber RW, Sosman JA, Haanen JB, et al. Improved survival with ipilimumab in patients with metastatic melanoma. N Engl J Med 2010;363(8):711–23.

[48] Robert C, Thomas L, Bondarenko I, O'Day S, Weber J, Garbe C, et al. Ipilimumab plus dacarbazine for previously untreated metastatic melanoma. N Engl J Med 2011;364(26):2517–26.

[49] Larkin J, Chiarion-Sileni V, Gonzalez R, Grob JJ, Cowey CL, Lao CD, et al. Combined nivolumab and ipilimumab or monotherapy in untreated melanoma. N Engl J Med 2015;373(1):23–34.

[50] Adams S, Gatti-Mays ME, Kalinsky K, Korde LA, Sharon E, Amiri-Kordestani L, et al. Current landscape of immunotherapy in breast cancer: a review. JAMA Oncol 2019;5(8):1205–14.

[51] Blansfield JA, Beck KE, Tran K, Yang JC, Hughes MS, Kammula US, et al. Cytotoxic T-lymphocyte-associated antigen-4 blockage can induce autoimmune hypophysitis in patients with metastatic melanoma and renal cancer. J Immunother 2005;28(6):593–8.

[52] Barnes MJ, Griseri T, Johnson AM, Young W, Powrie F, Izcue A. CTLA-4 promotes Foxp3 induction and regulatory T cell accumulation in the intestinal lamina propria. Mucosal Immunol 2013;6 (2):324–34.

[53] Dougan M, Dranoff G, Dougan SK. Cancer immunotherapy: beyond checkpoint blockade. Annu Rev Cancer Biol 2019;3:55–75.

[54] Boroughs LK, DeBerardinis RJ. Metabolic pathways promoting cancer cell survival and growth. Nat Cell Biol 2015;17(4):351–9.

[55] Biswas SK. Metabolic reprogramming of immune cells in cancer progression. Immunity 2015;43 (3):435–49.

[56] Faubert B, Solmonson A, DeBerardinis RJ. Metabolic reprogramming and cancer progression. Science 2020;368(6487):eaaw5473.

[57] Warburg O. The metabolism of carcinoma cells. J Cancer Res 1925;9(1):148–63.

[58] Warburg O. On the origin of cancer cells. Science 1956;123(3191):309–14.

[59] Shestov AA, Liu X, Ser Z, Cluntun AA, Hung YP, Huang L, et al. Quantitative determinants of aerobic glycolysis identify flux through the enzyme GAPDH as a limiting step. Elife 2014;3:e03342.

[60] Koundouros N, Poulogiannis G. Reprogramming of fatty acid metabolism in cancer. Br J Cancer 2020;122(1):4–22.

[61] Rohrig F, Schulze A. The multifaceted roles of fatty acid synthesis in cancer. Nat Rev Cancer 2016;16 (11):732–49.

[62] Medes G, Thomas A, Weinhouse S. Metabolism of neoplastic tissue. IV. A study of lipid synthesis in neoplastic tissue slices in vitro. Cancer Res 1953;13(1):27–9.

[63] De Vitto H, Perez-Valencia J, Radosevich JA. Glutamine at focus: versatile roles in cancer. Tumour Biol 2016;37(2):1541–58.

[64] DeBerardinis RJ, Mancuso A, Daikhin E, Nissim I, Yudkoff M, Wehrli S, et al. Beyond aerobic glycolysis: transformed cells can engage in glutamine metabolism that exceeds the requirement for protein and nucleotide synthesis. Proc Natl Acad Sci U S A 2007;104(49):19345–50.

[65] Wang X, Liu R, Zhu W, Chu H, Yu H, Wei P, et al. UDP-glucose accelerates SNAI1 mRNA decay and impairs lung cancer metastasis. Nature 2019;571(7763):127–31.

[66] Knott SRV, Wagenblast E, Khan S, Kim SY, Soto M, Wagner M, et al. Asparagine bioavailability governs metastasis in a model of breast cancer. Nature 2018;554(7692):378–81.

[67] Zhao W, Prijic S, Urban BC, Tisza MJ, Zuo Y, Li L, et al. Candidate antimetastasis drugs suppress the metastatic capacity of breast cancer cells by reducing membrane fluidity. Cancer Res 2016;76 (7):2037–49.

[68] Meitzler JL, Konate MM, Doroshow JH. Hydrogen peroxide-producing NADPH oxidases and the promotion of migratory phenotypes in cancer. Arch Biochem Biophys 2019;675:108076.

[69] Bansal A, Simon MC. Glutathione metabolism in cancer progression and treatment resistance. J Cell Biol 2018;217(7):2291–8.

[70] Elia I, Rossi M, Stegen S, Broekaert D, Doglioni G, van Gorsel M, et al. Breast cancer cells rely on environmental pyruvate to shape the metastatic niche. Nature 2019;568(7750):117–21.

[71] Lee CK, Jeong SH, Jang C, Bae H, Kim YH, Park I, et al. Tumor metastasis to lymph nodes requires YAP-dependent metabolic adaptation. Science 2019;363(6427):644–9.

[72] Pascual G, Avgustinova A, Mejetta S, Martin M, Castellanos A, Attolini CS, et al. Targeting metastasis-initiating cells through the fatty acid receptor CD36. Nature 2017;541(7635):41–5.

[73] Tasdogan A, Faubert B, Ramesh V, Ubellacker JM, Shen B, Solmonson A, et al. Metabolic heterogeneity confers differences in melanoma metastatic potential. Nature 2020;577(7788):115–20.

[74] Kune GA, Kune S, Watson LF. Colorectal cancer risk, chronic illnesses, operations, and medications: case control results from the Melbourne Colorectal Cancer Study. Cancer Res 1988;48 (15):4399–404.

[75] Cuzick J, Otto F, Baron JA, Brown PH, Burn J, Greenwald P, et al. Aspirin and non-steroidal anti-inflammatory drugs for cancer prevention: an international consensus statement. Lancet Oncol 2009;10(5):501–7.

[76] Cuzick J, Thorat MA, Bosetti C, Brown PH, Burn J, Cook NR, et al. Estimates of benefits and harms of prophylactic use of aspirin in the general population. Ann Oncol 2015;26(1):47–57.

[77] Burn J, Sheth H, Elliott F, Reed L, Macrae F, Mecklin JP, et al. Cancer prevention with aspirin in hereditary colorectal cancer (Lynch syndrome), 10-year follow-up and registry-based 20-year data in the CAPP2 study: a double-blind, randomised, placebo-controlled trial. Lancet 2020;395 (10240):1855–63.

[78] Konstantinopoulos PA, Cheng SC, Wahner Hendrickson AE, Penson RT, Schumer ST, Doyle LA, et al. Berzosertib plus gemcitabine versus gemcitabine alone in platinum-resistant high-grade serous ovarian cancer: a multicentre, open-label, randomised, phase 2 trial. Lancet Oncol 2020;21(7): 957–68.

[79] Kanda S, Ohe Y, Goto Y, Horinouchi H, Fujiwara Y, Nokihara H, et al. Five-year safety and efficacy data from a phase Ib study of nivolumab and chemotherapy in advanced non-small-cell lung cancer. Cancer Sci 2020;111(6):1933–42.

[80] Takahashi K, Osaka Y, Ota Y, Watanabe T, Iwasaki K, Tachibana S, et al. Phase II study of docetaxel, cisplatin, and 5-fluorouracil chemoradiotherapy for unresectable esophageal cancer. Anticancer Res 2020;40(5):2827–32.

[81] Tassinari D, Cherubini C, Tamburini E, Drudi F, Papi M, Fantini M, et al. Antimetabolites in the treatment of advanced pleural mesothelioma: a network meta-analysis of randomized clinical trials. J Chemother 2017;29(6):365–71.

[82] Shimada K, Yamaguchi M, Atsuta Y, Matsue K, Sato K, Kusumoto S, et al. Rituximab, cyclophosphamide, doxorubicin, vincristine, and prednisolone combined with high-dose methotrexate plus intrathecal chemotherapy for newly diagnosed intravascular large B-cell lymphoma (PRIMEUR-IVL): a multicentre, single-arm, phase 2 trial. Lancet Oncol 2020;21(4):593–602.

[83] Branvall E, Ekberg S, Eloranta S, Wasterlid T, Birmann BM, Smedby KE. Statin use is associated with improved survival in multiple myeloma: a Swedish population-based study of 4315 patients. Am J Hematol 2020;95(6):652–61.

[84] Saleh R, Elkord E. Acquired resistance to cancer immunotherapy: role of tumor-mediated immuno-suppression. Semin Cancer Biol 2020;65:13–27.

[85] Vinay DS, Ryan EP, Pawelec G, Talib WH, Stagg J, Elkord E, et al. Immune evasion in cancer: mechanistic basis and therapeutic strategies. Semin Cancer Biol 2015;35(Suppl.):S185–98.

[86] Franklin RA, Liao W, Sarkar A, Kim MV, Bivona MR, Liu K, et al. The cellular and molecular origin of tumor-associated macrophages. Science 2014;344(6186):921–5.

[87] Freemerman AJ, Johnson AR, Sacks GN, Milner JJ, Kirk EL, Troester MA, et al. Metabolic reprogramming of macrophages: glucose transporter 1 (GLUT1)-mediated glucose metabolism drives a proinflammatory phenotype. J Biol Chem 2014;289(11):7884–96.

[88] Schmid MC, Avraamides CJ, Dippold HC, Franco I, Foubert P, Ellies LG, et al. Receptor tyrosine kinases and TLR/IL1Rs unexpectedly activate myeloid cell PI3kgamma, a single convergent point promoting tumor inflammation and progression. Cancer Cell 2011;19(6):715–27.

[89] Mantovani A, Sica A, Sozzani S, Allavena P, Vecchi A, Locati M. The chemokine system in diverse forms of macrophage activation and polarization. Trends Immunol 2004;25(12):677–86.

[90] Colegio OR, Chu NQ, Szabo AL, Chu T, Rhebergen AM, Jairam V, et al. Functional polarization of tumour-associated macrophages by tumour-derived lactic acid. Nature 2014;513(7519):559–63.

[91] Ferrante CJ, Pinhal-Enfield G, Elson G, Cronstein BN, Hasko G, Outram S, et al. The adenosine-dependent angiogenic switch of macrophages to an M2-like phenotype is independent of interleukin-4 receptor alpha (IL-4Ralpha) signaling. Inflammation 2013;36(4):921–31.

[92] Henze AT, Mazzone M. The impact of hypoxia on tumor-associated macrophages. J Clin Invest 2016;126(10):3672–9.

[93] Bosca L, Gonzalez-Ramos S, Prieto P, Fernandez-Velasco M, Mojena M, Martin-Sanz P, et al. Metabolic signatures linked to macrophage polarization: from glucose metabolism to oxidative phosphorylation. Biochem Soc Trans 2015;43(4):740–4.

[94] Wu L, Zhang X, Zheng L, Zhao H, Yan G, Zhang Q, et al. RIPK3 orchestrates fatty acid metabolism in tumor-associated macrophages and Hepatocarcinogenesis. Cancer Immunol Res 2020;8 (5):710–21.

[95] Galdiero MR, Bonavita E, Barajon I, Garlanda C, Mantovani A, Jaillon S. Tumor associated macrophages and neutrophils in cancer. Immunobiology 2013;218(11):1402–10.

[96] Galdiero MR, Garlanda C, Jaillon S, Marone G, Mantovani A. Tumor associated macrophages and neutrophils in tumor progression. J Cell Physiol 2013;228(7):1404–12.

[97] Walmsley SR, Print C, Farahi N, Peyssonnaux C, Johnson RS, Cramer T, et al. Hypoxia-induced neutrophil survival is mediated by HIF-1alpha-dependent NF-kappaB activity. J Exp Med 2005;201 (1):105–15.

[98] Fridlender ZG, Sun J, Kim S, Kapoor V, Cheng G, Ling L, et al. Polarization of tumor-associated neutrophil phenotype by TGF-beta: "N1" versus "N2" TAN. Cancer Cell 2009;16(3):183–94.

[99] Keppel MP, Saucier N, Mah AY, Vogel TP, Cooper MA. Activation-specific metabolic requirements for NK cell IFN-gamma production. J Immunol 2015;194(4):1954–62.

[100] Velasquez SY, Killian D, Schulte J, Sticht C, Thiel M, Lindner HA. Short term hypoxia synergizes with interleukin 15 priming in driving glycolytic gene transcription and supports human natural killer cell activities. J Biol Chem 2016;291(25):12960–77.

[101] Cong J, Wang X, Zheng X, Wang D, Fu B, Sun R, et al. Dysfunction of natural killer cells by FBP1-induced inhibition of glycolysis during lung cancer progression. Cell Metab 2018;28(2):243–55. e5.

[102] Lamas B, Vergnaud-Gauduchon J, Goncalves-Mendes N, Perche O, Rossary A, Vasson MP, et al. Altered functions of natural killer cells in response to L-arginine availability. Cell Immunol 2012;280(2):182–90.

[103] Harmon C, Robinson MW, Hand F, Almuaili D, Mentor K, Houlihan DD, et al. Lactate-mediated acidification of tumor microenvironment induces apoptosis of liver-resident NK cells in colorectal liver metastasis. Cancer Immunol Res 2019;7(2):335–46.

[104] Brand A, Singer K, Koehl GE, Kolitzus M, Schoenhammer G, Thiel A, et al. LDHA-associated lactic acid production blunts tumor Immunosurveillance by T and NK cells. Cell Metab 2016;24 (5):657–71.

[105] Young A, Ngiow SF, Gao Y, Patch AM, Barkauskas DS, Messaoudene M, et al. A2AR adenosine signaling suppresses natural killer cell maturation in the tumor microenvironment. Cancer Res 2018;78(4):1003–16.

[106] Dong H, Bullock TN. Metabolic influences that regulate dendritic cell function in tumors. Front Immunol 2014;5:24.

[107] Nasi A, Rethi B. Disarmed by density: a glycolytic break for immunostimulatory dendritic cells? Onco Targets Ther 2013;2(12), e26744.

[108] Krawczyk CM, Holowka T, Sun J, Blagih J, Amiel E, DeBerardinis RJ, et al. Toll-like receptor-induced changes in glycolytic metabolism regulate dendritic cell activation. Blood 2010;115(23):4742–9.

[109] Krystel-Whittemore M, Dileepan KN, Wood JG. Mast cell: a multi-functional master cell. Front Immunol 2015;6:620.

[110] Sekar Y, Moon TC, Slupsky CM, Befus AD. Protein tyrosine nitration of aldolase in mast cells: a plausible pathway in nitric oxide-mediated regulation of mast cell function. J Immunol 2010;185(1):578–87.

[111] Morse MA, Hall JR, Plate JM. Countering tumor-induced immunosuppression during immunotherapy for pancreatic cancer. Expert Opin Biol Ther 2009;9(3):331–9.

[112] Corzo CA, Condamine T, Lu L, Cotter MJ, Youn JI, Cheng P, et al. HIF-1alpha regulates function and differentiation of myeloid-derived suppressor cells in the tumor microenvironment. J Exp Med 2010;207(11):2439–53.

[113] Noman MZ, Desantis G, Janji B, Hasmim M, Karray S, Dessen P, et al. PD-L1 is a novel direct target of HIF-1alpha, and its blockade under hypoxia enhanced MDSC-mediated T cell activation. J Exp Med 2014;211(5):781–90.

[114] Waickman AT, Powell JD. mTOR, metabolism, and the regulation of T-cell differentiation and function. Immunol Rev 2012;249(1):43–58.

[115] Chen JL, Lucas JE, Schroeder T, Mori S, Wu J, Nevins J, et al. The genomic analysis of lactic acidosis and acidosis response in human cancers. PLoS Genet 2008;4(12), e1000293.

[116] Duvel K, Yecies JL, Menon S, Raman P, Lipovsky AI, Souza AL, et al. Activation of a metabolic gene regulatory network downstream of mTOR complex 1. Mol Cell 2010;39(2):171–83.

[117] Fischer K, Hoffmann P, Voelkl S, Meidenbauer N, Ammer J, Edinger M, et al. Inhibitory effect of tumor cell-derived lactic acid on human T cells. Blood 2007;109(9):3812–9.

[118] Fernández-Ramos AA, Poindessous V, Marchetti-Laurent C, Pallet N, Loriot MA. The effect of immunosuppressive molecules on T-cell metabolic reprogramming. Biochimie 2016;127:23–36.

[119] Blagih J, Coulombe F, Vincent EE, Dupuy F, Galicia-Vazquez G, Yurchenko E, et al. The energy sensor AMPK regulates T cell metabolic adaptation and effector responses in vivo. Immunity 2015;42(1):41–54.

[120] Shen H, Shi LZ. Metabolic regulation of TH17 cells. Mol Immunol 2019;109:81–7.

[121] Shi LZ, Wang R, Huang G, Vogel P, Neale G, Green DR, et al. HIF1alpha-dependent glycolytic pathway orchestrates a metabolic checkpoint for the differentiation of TH17 and Treg cells. J Exp Med 2011;208(7):1367–76.

[122] Caro-Maldonado A, Wang R, Nichols AG, Kuraoka M, Milasta S, Sun LD, et al. Metabolic reprogramming is required for antibody production that is suppressed in anergic but exaggerated in chronically BAFF-exposed B cells. J Immunol 2014;192(8):3626–36.

[123] Jellusova J, Rickert RC. The PI3K pathway in B cell metabolism. Crit Rev Biochem Mol Biol 2016;51(5):359–78.

[124] Condeelis J, Pollard JW. Macrophages: obligate partners for tumor cell migration, invasion, and metastasis. Cell 2006;124(2):263–6.

[125] Qian BZ, Pollard JW. Macrophage diversity enhances tumor progression and metastasis. Cell 2010;141(1):39–51.

[126] Ojalvo LS, King W, Cox D, Pollard JW. High-density gene expression analysis of tumor-associated macrophages from mouse mammary tumors. Am J Pathol 2009;174(3):1048–64.

[127] Aras S, Zaidi MR. TAMeless traitors: macrophages in cancer progression and metastasis. Br J Cancer 2017;117(11):1583–91.

[128] Mantovani A, Sica A. Macrophages, innate immunity and cancer: balance, tolerance, and diversity. Curr Opin Immunol 2010;22(2):231–7.

[129] Fridman WH, Zitvogel L, Sautès-Fridman C, Kroemer G. The immune contexture in cancer prognosis and treatment. Nat Rev Clin Oncol 2017;14(12):717–34.

[130] Zhou D, Huang C, Lin Z, Zhan S, Kong L, Fang C, et al. Macrophage polarization and function with emphasis on the evolving roles of coordinated regulation of cellular signaling pathways. Cell Signal 2014;26(2):192–7.

[131] Chang M, Hamilton JA, Scholz GM, Elsegood CL. Glycolytic control of adjuvant-induced macrophage survival: role of PI3K, MEK1/2, and Bcl-2. J Leukoc Biol 2009;85(6):947–56.

[132] Cheng SC, Quintin J, Cramer RA, Shepardson KM, Saeed S, Kumar V, et al. mTOR- and HIF-1alpha-mediated aerobic glycolysis as metabolic basis for trained immunity. Science 2014;345 (6204):1250684.

[133] Rivera LB, Meyronet D, Hervieu V, Frederick MJ, Bergsland E, Bergers G. Intratumoral myeloid cells regulate responsiveness and resistance to antiangiogenic therapy. Cell Rep 2015;11(4):577–91.

[134] Semenza GL, Roth PH, Fang HM, Wang GL. Transcriptional regulation of genes encoding glycolytic enzymes by hypoxia-inducible factor 1. J Biol Chem 1994;269(38):23757–63.

[135] Tannahill GM, Curtis AM, Adamik J, Palsson-McDermott EM, McGettrick AF, Goel G, et al. Succinate is an inflammatory signal that induces IL-1beta through HIF-1alpha. Nature 2013;496 (7444):238–42.

[136] Biswas SK, Mantovani A. Macrophage plasticity and interaction with lymphocyte subsets: cancer as a paradigm. Nat Immunol 2010;11(10):889–96.

[137] Iles KE, Forman HJ. Macrophage signaling and respiratory burst. Immunol Res 2002;26 (1–3):95–105.

[138] Mu X, Shi W, Xu Y, Xu C, Zhao T, Geng B, et al. Tumor-derived lactate induces M2 macrophage polarization via the activation of the ERK/STAT3 signaling pathway in breast cancer. Cell Cycle 2018;17(4):428–38.

[139] Lewis CE, De Palma M, Naldini L. Tie2-expressing monocytes and tumor angiogenesis: regulation by hypoxia and angiopoietin-2. Cancer Res 2007;67(18):8429–32.

[140] Lemke G, Lu Q. Macrophage regulation by Tyro 3 family receptors. Curr Opin Immunol 2003;15 (1):31–6.

[141] Pello OM, De Pizzol M, Mirolo M, Soucek L, Zammataro L, Amabile A, et al. Role of c-MYC in alternative activation of human macrophages and tumor-associated macrophage biology. Blood 2012;119(2):411–21.

[142] Palsson-McDermott EM, Curtis AM, Goel G, Lauterbach MA, Sheedy FJ, Gleeson LE, et al. Pyruvate kinase M2 regulates Hif-1alpha activity and IL-1beta induction and is a critical determinant of the Warburg effect in LPS-activated macrophages. Cell Metab 2015;21(1):65–80.

[143] Odegaard JI, Chawla A. Alternative macrophage activation and metabolism. Annu Rev Pathol 2011;6:275–97.

[144] Menegaut L, Thomas C, Lagrost L, Masson D. Fatty acid metabolism in macrophages: a target in cardio-metabolic diseases. Curr Opin Lipidol 2017;28(1):19–26.

[145] Korbecki J, Bajdak-Rusinek K. The effect of palmitic acid on inflammatory response in macrophages: an overview of molecular mechanisms. Inflamm Res 2019;68(11):915–32.

[146] Huang SC, Everts B, Ivanova Y, O'Sullivan D, Nascimento M, Smith AM, et al. Cell-intrinsic lysosomal lipolysis is essential for alternative activation of macrophages. Nat Immunol 2014;15(9):846–55.

[147] Li Y, Cai L, Wang H, Wu P, Gu W, Chen Y, et al. Pleiotropic regulation of macrophage polarization and tumorigenesis by formyl peptide receptor-2. Oncogene 2011;30(36):3887–99.

[148] Soliman E, Elhassanny A, Malur A, McPeek M, Bell A, Leffler N, et al. Impaired mitochondrial function of alveolar macrophages in carbon nanotube-induced chronic pulmonary granulomatous disease. Toxicology 2020;445, 152598.

[149] Pourcet B, Feig JE, Vengrenyuk Y, Hobbs AJ, Kepka-Lenhart D, Garabedian MJ, et al. LXRalpha regulates macrophage arginase 1 through PU.1 and interferon regulatory factor 8. Circ Res 2011;109 (5):492–501.

[150] Nath N, Kashfi K. Tumor associated macrophages and 'NO'. Biochem Pharmacol 2020;176, 113899.

[151] Pourcet B, Pineda-Torra I. Transcriptional regulation of macrophage arginase 1 expression and its role in atherosclerosis. Trends Cardiovasc Med 2013;23(5):143–52.

[152] Bronte V, Zanovello P. Regulation of immune responses by L-arginine metabolism. Nat Rev Immunol 2005;5(8):641–54.

[153] Geiger R, Rieckmann JC, Wolf T, Basso C, Feng Y, Fuhrer T, et al. L-Arginine modulates T cell metabolism and enhances survival and anti-tumor activity. Cell 2016;167(3):829–42. e13.

[154] Jha AK, Huang SC, Sergushichev A, Lampropoulou V, Ivanova Y, Loginicheva E, et al. Network integration of parallel metabolic and transcriptional data reveals metabolic modules that regulate macrophage polarization. Immunity 2015;42(3):419–30.

[155] Palmieri EM, Menga A, Martin-Perez R, Quinto A, Riera-Domingo C, De Tullio G, et al. Pharmacologic or genetic targeting of glutamine synthetase skews macrophages toward an M1-like phenotype and inhibits tumor metastasis. Cell Rep 2017;20(7):1654–66.

[156] Liu M, Chen Y, Wang S, Zhou H, Feng D, Wei J, et al. Alpha-ketoglutarate modulates macrophage polarization through regulation of PPARgamma transcription and mTORC1/p70S6K pathway to ameliorate ALI/ARDS. Shock 2020;53(1):103–13.

[157] Gabrilovich DI. Myeloid-derived suppressor cells. Cancer Immunol Res 2017;5(1):3–8.

[158] Nagaraj S, Gupta K, Pisarev V, Kinarsky L, Sherman S, Kang L, et al. Altered recognition of antigen is a mechanism of CD8[+] T cell tolerance in cancer. Nat Med 2007;13(7):828–35.

[159] Tartour E, Pere H, Maillere B, Terme M, Merillon N, Taieb J, et al. Angiogenesis and immunity: a bidirectional link potentially relevant for the monitoring of antiangiogenic therapy and the development of novel therapeutic combination with immunotherapy. Cancer Metastasis Rev 2011;30 (1):83–95.

[160] Casella I, Feccia T, Chelucci C, Samoggia P, Castelli G, Guerriero R, et al. Autocrine-paracrine VEGF loops potentiate the maturation of megakaryocytic precursors through Flt1 receptor. Blood 2003;101(4):1316–23.

[161] Shojaei F, Wu X, Qu X, Kowanetz M, Yu L, Tan M, et al. G-CSF-initiated myeloid cell mobilization and angiogenesis mediate tumor refractoriness to anti-VEGF therapy in mouse models. Proc Natl Acad Sci U S A 2009;106(16):6742–7.

[162] Condamine T, Mastio J, Gabrilovich DI. Transcriptional regulation of myeloid-derived suppressor cells. J Leukoc Biol 2015;98(6):913–22.

[163] Condamine T, Kumar V, Ramachandran IR, Youn JI, Celis E, Finnberg N, et al. ER stress regulates myeloid-derived suppressor cell fate through TRAIL-R-mediated apoptosis. J Clin Invest 2014;124 (6):2626–39.

[164] Lee BR, Chang SY, Hong EH, Kwon BE, Kim HM, Kim YJ, et al. Elevated endoplasmic reticulum stress reinforced immunosuppression in the tumor microenvironment via myeloid-derived suppressor cells. Oncotarget 2014;5(23):12331–45.

[165] Thevenot PT, Sierra RA, Raber PL, Al-Khami AA, Trillo-Tinoco J, Zarreii P, et al. The stress-response sensor chop regulates the function and accumulation of myeloid-derived suppressor cells in tumors. Immunity 2014;41(3):389–401.

[166] Wang Y, Jia A, Bi Y, Wang Y, Liu G. Metabolic regulation of myeloid-derived suppressor cell function in cancer. Cells 2020;9(4):1011.

[167] Hossain F, Al-Khami AA, Wyczechowska D, Hernandez C, Zheng L, Reiss K, et al. Inhibition of fatty acid oxidation modulates immunosuppressive functions of myeloid-derived suppressor cells and enhances cancer therapies. Cancer Immunol Res 2015;3(11):1236–47.

[168] Jian SL, Chen WW, Su YC, Su YW, Chuang TH, Hsu SC, et al. Glycolysis regulates the expansion of myeloid-derived suppressor cells in tumor-bearing hosts through prevention of ROS-mediated apoptosis. Cell Death Dis 2017;8(5), e2779.

[169] Husain Z, Huang Y, Seth P, Sukhatme VP. Tumor-derived lactate modifies antitumor immune response: effect on myeloid-derived suppressor cells and NK cells. J Immunol 2013;191(3):1486–95.

[170] Cimen BC, Elzey BD, Crist SA, Ellies LG, Ratliff TL. Expression of cationic amino acid transporter 2 is required for myeloid-derived suppressor cell-mediated control of T cell immunity. J Immunol 2015;195(11):5237–50.

[171] Hu C, Pang B, Lin G, Zhen Y, Yi H. Energy metabolism manipulates the fate and function of tumour myeloid-derived suppressor cells. Br J Cancer 2020;122(1):23–9.

[172] Szefel J, Danielak A, Kruszewski WJ. Metabolic pathways of L-arginine and therapeutic consequences in tumors. Adv Med Sci 2019;64(1):104–10.

[173] Zoso A, Mazza EM, Bicciato S, Mandruzzato S, Bronte V, Serafini P, et al. Human fibrocytic myeloid-derived suppressor cells express IDO and promote tolerance via Treg-cell expansion. Eur J Immunol 2014;44(11):3307–19.

[174] Hammami I, Chen J, Bronte V, DeCrescenzo G, Jolicoeur M. L-Glutamine is a key parameter in the immunosuppression phenomenon. Biochem Biophys Res Commun 2012;425(4):724–9.

[175] Al-Khami AA, Zheng L, Del VL, Hossain F, Wyczechowska D, Zabaleta J, et al. Exogenous lipid uptake induces metabolic and functional reprogramming of tumor-associated myeloid-derived suppressor cells. Onco Targets Ther 2017;6(10), e1344804.

[176] Wu AA, Drake V, Huang HS, Chiu S, Zheng L. Reprogramming the tumor microenvironment: tumor-induced immunosuppressive factors paralyze T cells. Onco Targets Ther 2015;4(7), e1016700.

[177] Crook KR, Jin M, Weeks MF, Rampersad RR, Baldi RM, Glekas AS, et al. Myeloid-derived suppressor cells regulate T cell and B cell responses during autoimmune disease. J Leukoc Biol 2015;97(3):573–82.

[178] Obermajer N, Muthuswamy R, Odunsi K, Edwards RP, Kalinski P. PGE(2)-induced CXCL12 production and CXCR4 expression controls the accumulation of human MDSCs in ovarian cancer environment. Cancer Res 2011;71(24):7463–70.

[179] Veglia F, Tyurin VA, Blasi M, De Leo A, Kossenkov AV, Donthireddy L, et al. Fatty acid transport protein 2 reprograms neutrophils in cancer. Nature 2019;569(7754):73–8.

[180] Saravia J, Chapman NM, Chi H. Helper T cell differentiation. Cell Mol Immunol 2019;16(7):634–43.

[181] Liao W, Lin JX, Leonard WJ. Interleukin-2 at the crossroads of effector responses, tolerance, and immunotherapy. Immunity 2013;38(1):13–25.

[182] Saito S, Nakashima A, Shima T, Ito M. Th1/Th2/Th17 and regulatory T-cell paradigm in pregnancy. Am J Reprod Immunol 2010;63(6):601–10.

[183] Zhang N, Bevan MJ. CD8(+) T cells: foot soldiers of the immune system. Immunity 2011;35(2):161–8.

[184] Siska PJ, Rathmell JC. T cell metabolic fitness in antitumor immunity. Trends Immunol 2015;36(4):257–64.

[185] MacIver NJ, Michalek RD, Rathmell JC. Metabolic regulation of T lymphocytes. Annu Rev Immunol 2013;31:259–83.

[186] Maciolek JA, Pasternak JA, Wilson HL. Metabolism of activated T lymphocytes. Curr Opin Immunol 2014;27:60–74.

[187] Michalek RD, Gerriets VA, Jacobs SR, Macintyre AN, MacIver NJ, Mason EF, et al. Cutting edge: distinct glycolytic and lipid oxidative metabolic programs are essential for effector and regulatory CD4 + T cell subsets. J Immunol 2011;186(6):3299–303.

[188] Lochner M, Berod L, Sparwasser T. Fatty acid metabolism in the regulation of T cell function. Trends Immunol 2015;36(2):81–91.

[189] Beier UH, Angelin A, Akimova T, Wang L, Liu Y, Xiao H, et al. Essential role of mitochondrial energy metabolism in Foxp3(+) T-regulatory cell function and allograft survival. FASEB J 2015;29(6):2315–26.

[190] Menk AV, Scharping NE, Moreci RS, Zeng X, Guy C, Salvatore S, et al. Early TCR Signaling induces rapid aerobic glycolysis enabling distinct acute T cell effector functions. Cell Rep 2018;22(6):1509–21.

[191] Sena LA, Li S, Jairaman A, Prakriya M, Ezponda T, Hildeman DA, et al. Mitochondria are required for antigen-specific T cell activation through reactive oxygen species signaling. Immunity 2013;38(2):225–36.

[192] Leone RD, Powell JD. Metabolism of immune cells in cancer. Nat Rev Cancer 2020;20(9):516–31.

[193] Frauwirth KA, Riley JL, Harris MH, Parry RV, Rathmell JC, Plas DR, et al. The CD28 signaling pathway regulates glucose metabolism. Immunity 2002;16(6):769–77.

[194] Wang R, Dillon CP, Shi LZ, Milasta S, Carter R, Finkelstein D, et al. The transcription factor Myc controls metabolic reprogramming upon T lymphocyte activation. Immunity 2011;35(6):871–82.

[195] Finlay DK, Rosenzweig E, Sinclair LV, Feijoo-Carnero C, Hukelmann JL, Rolf J, et al. PDK1 regulation of mTOR and hypoxia-inducible factor 1 integrate metabolism and migration of CD8 + T cells. J Exp Med 2012;209(13):2441–53.

[196] Osthus RC, Shim H, Kim S, Li Q, Reddy R, Mukherjee M, et al. Deregulation of glucose transporter 1 and glycolytic gene expression by c-Myc. J Biol Chem 2000;275(29):21797–800.

[197] Chang CH, Curtis JD, Maggi LJ, Faubert B, Villarino AV, O'Sullivan D, et al. Posttranscriptional control of T cell effector function by aerobic glycolysis. Cell 2013;153(6):1239–51.

[198] Kidani Y, Elsaesser H, Hock MB, Vergnes L, Williams KJ, Argus JP, et al. Sterol regulatory element-binding proteins are essential for the metabolic programming of effector T cells and adaptive immunity. Nat Immunol 2013;14(5):489–99.

[199] Swamy M, Pathak S, Grzes KM, Damerow S, Sinclair LV, van Aalten DM, et al. Glucose and glutamine fuel protein O-GlcNAcylation to control T cell self-renewal and malignancy. Nat Immunol 2016;17(6):712–20.

[200] Yang W, Bai Y, Xiong Y, Zhang J, Chen S, Zheng X, et al. Potentiating the antitumour response of CD8(+) T cells by modulating cholesterol metabolism. Nature 2016;531(7596):651–5.

[201] Nii T, Segawa H, Taketani Y, Tani Y, Ohkido M, Kishida S, et al. Molecular events involved in up-regulating human Na +-independent neutral amino acid transporter LAT1 during T-cell activation. Biochem J 2001;358(Pt 3):693–704.

[202] Carr EL, Kelman A, Wu GS, Gopaul R, Senkevitch E, Aghvanyan A, et al. Glutamine uptake and metabolism are coordinately regulated by ERK/MAPK during T lymphocyte activation. J Immunol 2010;185(2):1037–44.

[203] Nakaya M, Xiao Y, Zhou X, Chang JH, Chang M, Cheng X, et al. Inflammatory T cell responses rely on amino acid transporter ASCT2 facilitation of glutamine uptake and mTORC1 kinase activation. Immunity 2014;40(5):692–705.

[204] Ren W, Liu G, Yin J, Tan B, Wu G, Bazer FW, et al. Amino-acid transporters in T-cell activation and differentiation. Cell Death Dis 2017;8(3), e2655.

[205] Ferrer CM, Sodi VL, Reginato MJ. O-GlcNAcylation in cancer biology: linking metabolism and Signaling. J Mol Biol 2016;428(16):3282–94.

[206] Ravez S, Corbet C, Spillier Q, Dutu A, Robin AD, Mullarky E, et al. Alpha-ketothioamide derivatives: a promising tool to interrogate phosphoglycerate dehydrogenase (PHGDH). J Med Chem 2017;60(4):1591–7.

[207] Vander HM, Lunt SY, Dayton TL, Fiske BP, Israelsen WJ, Mattaini KR, et al. Metabolic pathway alterations that support cell proliferation. Cold Spring Harb Symp Quant Biol 2011;76:325–34.

[208] DeBerardinis RJ, Lum JJ, Hatzivassiliou G, Thompson CB. The biology of cancer: metabolic reprogramming fuels cell growth and proliferation. Cell Metab 2008;7(1):11–20.

[209] Chiarugi P, Cirri P. Metabolic exchanges within tumor microenvironment. Cancer Lett 2016;380 (1):272–80.

[210] Gottfried E, Kunz-Schughart LA, Ebner S, Mueller-Klieser W, Hoves S, Andreesen R, et al. Tumor-derived lactic acid modulates dendritic cell activation and antigen expression. Blood 2006;107 (5):2013–21.

[211] Bader JE, Voss K, Rathmell JC. Targeting metabolism to improve the tumor microenvironment for cancer immunotherapy. Mol Cell 2020;78(6):1019–33.

[212] Zhang D, Tang Z, Huang H, Zhou G, Cui C, Weng Y, et al. Metabolic regulation of gene expression by histone lactylation. Nature 2019;574(7779):575–80.

[213] Rabinowitz JD, Enerback S. Lactate: the ugly duckling of energy metabolism. Nat Metab 2020;2 (7):566–71.

[214] Hui S, Ghergurovich JM, Morscher RJ, Jang C, Teng X, Lu W, et al. Glucose feeds the TCA cycle via circulating lactate. Nature 2017;551(7678):115–8.

[215] Dhup S, Dadhich RK, Porporato PE, Sonveaux P. Multiple biological activities of lactic acid in cancer: influences on tumor growth, angiogenesis and metastasis. Curr Pharm Des 2012;18(10):1319–30.

[216] Cairns RA, Harris IS, Mak TW. Regulation of cancer cell metabolism. Nat Rev Cancer 2011;11 (2):85–95.

[217] Garcia-Heredia JM, Carnero A. Decoding Warburg's hypothesis: tumor-related mutations in the mitochondrial respiratory chain. Oncotarget 2015;6(39):41582–99.

[218] Gatenby RA, Gawlinski ET. A reaction-diffusion model of cancer invasion. Cancer Res 1996;56 (24):5745–53.

[219] Khatib-Massalha E, Bhattacharya S, Massalha H, Biram A, Golan K, Kollet O, et al. Lactate released by inflammatory bone marrow neutrophils induces their mobilization via endothelial GPR81 signaling. Nat Commun 2020;11(1):3547.

[220] Wieman HL, Wofford JA, Rathmell JC. Cytokine stimulation promotes glucose uptake via phosphatidylinositol-3 kinase/Akt regulation of Glut1 activity and trafficking. Mol Biol Cell 2007;18(4):1437–46.

[221] Zhou D, Jiang L, Jin L, Yao Y, Wang P, Zhu X. Glucose transporter-1 cooperating with AKT signaling promote gastric cancer progression. Cancer Manage Res 2020;12:4151–60.

[222] Chen B, Tang H, Liu X, Liu P, Yang L, Xie X, et al. miR-22 as a prognostic factor targets glucose transporter protein type 1 in breast cancer. Cancer Lett 2015;356(2 Pt B):410–7.

[223] Gonzalez-Menendez P, Hevia D, Alonso-Arias R, Alvarez-Artime A, Rodriguez-Garcia A, Kinet S, et al. GLUT1 protects prostate cancer cells from glucose deprivation-induced oxidative stress. Redox Biol 2018;17:112–27.

[224] Wu Q, Ba-Alawi W, Deblois G, Cruickshank J, Duan S, Lima-Fernandes E, et al. GLUT1 inhibition blocks growth of RB1-positive triple negative breast cancer. Nat Commun 2020;11(1):4205.

[225] Goos JA, de Cuba EM, Coupé VM, Diosdado B, Delis-Van DP, Karga C, et al. Glucose transporter 1 (SLC2A1) and vascular endothelial growth factor a (VEGFA) predict survival after resection of colorectal cancer liver metastasis. Ann Surg 2016;263(1):138–45.

[226] Anderson M, Marayati R, Moffitt R, Yeh JJ. Hexokinase 2 promotes tumor growth and metastasis by regulating lactate production in pancreatic cancer. Oncotarget 2017;8(34):56081–94.

[227] Wang L, Xiong H, Wu F, Zhang Y, Wang J, Zhao L, et al. Hexokinase 2-mediated Warburg effect is required for PTEN- and p53-deficiency-driven prostate cancer growth. Cell Rep 2014;8(5):1461–74.

[228] Li Z, Li X, Wu S, Xue M, Chen W. Long non-coding RNA UCA1 promotes glycolysis by upregulating hexokinase 2 through the mTOR-STAT3/microRNA143 pathway. Cancer Sci 2014;105(8):951–5.

[229] Agnihotri S, Mansouri S, Burrell K, Li M, Mamatjan Y, Liu J, et al. Ketoconazole and posaconazole selectively target HK2-expressing glioblastoma cells. Clin Cancer Res 2019;25(2):844–55.

[230] Xu S, Zhou T, Doh HM, Trinh KR, Catapang A, Lee JT, et al. An HK2 antisense oligonucleotide induces synthetic lethality in HK1(−)HK2(+) multiple myeloma. Cancer Res 2019;79(10):2748–60.

[231] Chesney J. 6-Phosphofructo-2-kinase/fructose-2,6-bisphosphatase and tumor cell glycolysis. Curr Opin Clin Nutr Metab Care 2006;9(5):535–9.

[232] Hasawi NA, Alkandari MF, Luqmani YA. Phosphofructokinase: a mediator of glycolytic flux in cancer progression. Crit Rev Oncol Hematol 2014;92(3):312–21.

[233] Clem B, Telang S, Clem A, Yalcin A, Meier J, Simmons A, et al. Small-molecule inhibition of 6-phosphofructo-2-kinase activity suppresses glycolytic flux and tumor growth. Mol Cancer Ther 2008;7(1):110–20.

[234] Wong N, Ojo D, Yan J, Tang D. PKM2 contributes to cancer metabolism. Cancer Lett 2015;356(2 Pt A):184–91.

[235] Christofk HR, Vander HM, Harris MH, Ramanathan A, Gerszten RE, Wei R, et al. The M2 splice isoform of pyruvate kinase is important for cancer metabolism and tumour growth. Nature 2008;452 (7184):230–3.

[236] Le A, Cooper CR, Gouw AM, Dinavahi R, Maitra A, Deck LM, et al. Inhibition of lactate dehydrogenase A induces oxidative stress and inhibits tumor progression. Proc Natl Acad Sci U S A 2010;107(5):2037–42.

[237] Liu X, Yang Z, Chen Z, Chen R, Zhao D, Zhou Y, et al. Effects of the suppression of lactate dehydrogenase A on the growth and invasion of human gastric cancer cells. Oncol Rep 2015;33 (1):157–62.

[238] Rellinger EJ, Craig BT, Alvarez AL, Dusek HL, Kim KW, Qiao J, et al. FX11 inhibits aerobic glycolysis and growth of neuroblastoma cells. Surgery 2017;161(3):747–52.

[239] Le A, Rajeshkumar NV, Maitra A, Dang CV. Conceptual framework for cutting the pancreatic cancer fuel supply. Clin Cancer Res 2012;18(16):4285–90.

[240] Hassell KN. Histone deacetylases and their inhibitors in cancer epigenetics. Diseases 2019;7(4):57.

[241] Glozak MA, Seto E. Histone deacetylases and cancer. Oncogene 2007;26(37):5420–32.

[242] Wagner VP, Martins MD, Castilho RM. Histones acetylation and cancer stem cells (CSCs). Methods Mol Biol 2018;1692:179–93.

[243] Chae YC, Vaira V, Caino MC, Tang HY, Seo JH, Kossenkov AV, et al. Mitochondrial Akt regulation of hypoxic tumor reprogramming. Cancer Cell 2016;30(2):257–72.

[244] Firth JD, Ebert BL, Ratcliffe PJ. Hypoxic regulation of lactate dehydrogenase A. Interaction between hypoxia-inducible factor 1 and cAMP response elements. J Biol Chem 1995;270(36):21021–7.

[245] Kim JW, Tchernyshyov I, Semenza GL, Dang CV. HIF-1-mediated expression of pyruvate dehydrogenase kinase: a metabolic switch required for cellular adaptation to hypoxia. Cell Metab 2006;3 (3):177–85.

[246] Pollard PJ, Briere JJ, Alam NA, Barwell J, Barclay E, Wortham NC, et al. Accumulation of Krebs cycle intermediates and over-expression of HIF1alpha in tumours which result from germline FH and SDH mutations. Hum Mol Genet 2005;14(15):2231–9.

[247] Lu H, Dalgard CL, Mohyeldin A, McFate T, Tait AS, Verma A. Reversible inactivation of HIF-1 prolyl hydroxylases allows cell metabolism to control basal HIF-1. J Biol Chem 2005;280 (51):41928–39.

[248] Nogueira V, Hay N. Molecular pathways: reactive oxygen species homeostasis in cancer cells and implications for cancer therapy. Clin Cancer Res 2013;19(16):4309–14.

[249] De Preter G, Neveu MA, Danhier P, Brisson L, Payen VL, Porporato PE, et al. Inhibition of the pentose phosphate pathway by dichloroacetate unravels a missing link between aerobic glycolysis and cancer cell proliferation. Oncotarget 2016;7(3):2910–20.

[250] Jin L, Zhou Y. Crucial role of the pentose phosphate pathway in malignant tumors. Oncol Lett 2019;17(5):4213–21.

[251] Jiang P, Du W, Wang X, Mancuso A, Gao X, Wu M, et al. p53 regulates biosynthesis through direct inactivation of glucose-6-phosphate dehydrogenase. Nat Cell Biol 2011;13(3):310–6.

[252] Hong X, Song R, Song H, Zheng T, Wang J, Liang Y, et al. PTEN antagonises Tcl1/hnRNPK-mediated G6PD pre-mRNA splicing which contributes to hepatocarcinogenesis. Gut 2014;63 (10):1635–47.

[253] Jiang P, Du W, Wu M. Regulation of the pentose phosphate pathway in cancer. Protein Cell 2014;5 (8):592–602.

[254] Patra KC, Hay N. The pentose phosphate pathway and cancer. Trends Biochem Sci 2014;39 (8):347–54.

[255] Ferrer CM, Lynch TP, Sodi VL, Falcone JN, Schwab LP, Peacock DL, et al. O-GlcNAcylation regulates cancer metabolism and survival stress signaling via regulation of the HIF-1 pathway. Mol Cell 2014;54(5):820–31.

[256] Jozwiak P, Forma E, Brys M, Krzeslak A. O-GlcNAcylation and metabolic reprograming in cancer. Front Endocrinol (Lausanne) 2014;5:145.

[257] DeBerardinis RJ. Serine metabolism: some tumors take the road less traveled. Cell Metab 2011;14 (3):285–6.

[258] Ma X, Li B, Liu J, Fu Y, Luo Y. Phosphoglycerate dehydrogenase promotes pancreatic cancer development by interacting with eIF4A1 and eIF4E. J Exp Clin Cancer Res 2019;38(1):66.

[259] Fan J, Teng X, Liu L, Mattaini KR, Looper RE, Vander HM, et al. Human phosphoglycerate dehydrogenase produces the oncometabolite D-2-hydroxyglutarate. ACS Chem Biol 2015;10(2):510–6.

[260] Zhao X, Fu J, Du J, Xu W. The role of D-3-phosphoglycerate dehydrogenase in cancer. Int J Biol Sci 2020;16(9):1495–506.

[261] Liu J, Guo S, Li Q, Yang L, Xia Z, Zhang L, et al. Phosphoglycerate dehydrogenase induces glioma cells proliferation and invasion by stabilizing forkhead box M1. J Neuro-Oncol 2013;111(3):245–55.

[262] Jia XQ, Zhang S, Zhu HJ, Wang W, Zhu JH, Wang XD, et al. Increased expression of PHGDH and prognostic significance in colorectal cancer. Transl Oncol 2016;9(3):191–6.

[263] Gromova I, Gromov P, Honma N, Kumar S, Rimm D, Talman ML, et al. High level PHGDH expression in breast is predominantly associated with keratin 5-positive cell lineage independently of malignancy. Mol Oncol 2015;9(8):1636–54.

[264] Mullarky E, Mattaini KR, Vander HM, Cantley LC, Locasale JW. PHGDH amplification and altered glucose metabolism in human melanoma. Pigment Cell Melanoma Res 2011;24(6):1112–5.

[265] Ackerman D, Simon MC. Hypoxia, lipids, and cancer: surviving the harsh tumor microenvironment. Trends Cell Biol 2014;24(8):472–8.

[266] Swierczynski J, Hebanowska A, Sledzinski T. Role of abnormal lipid metabolism in development, progression, diagnosis and therapy of pancreatic cancer. World J Gastroenterol 2014;20(9):2279–303.

[267] Kuemmerle NB, Rysman E, Lombardo PS, Flanagan AJ, Lipe BC, Wells WA, et al. Lipoprotein lipase links dietary fat to solid tumor cell proliferation. Mol Cancer Ther 2011;10(3):427–36.

[268] Catalina-Rodriguez O, Kolukula VK, Tomita Y, Preet A, Palmieri F, Wellstein A, et al. The mitochondrial citrate transporter, CIC, is essential for mitochondrial homeostasis. Oncotarget 2012;3(10):1220–35.

[269] Ozkaya AB, Ak H, Atay S, Aydin HH. Targeting mitochondrial citrate transport in breast cancer cell lines. Anti Cancer Agents Med Chem 2015;15(3):374–81.

[270] Zaidi N, Swinnen JV, Smans K. ATP-citrate lyase: a key player in cancer metabolism. Cancer Res 2012;72(15):3709–14.

[271] Hanai J, Doro N, Sasaki AT, Kobayashi S, Cantley LC, Seth P, et al. Inhibition of lung cancer growth: ATP citrate lyase knockdown and statin treatment leads to dual blockade of mitogen-activated protein kinase (MAPK) and phosphatidylinositol-3-kinase (PI3K)/AKT pathways. J Cell Physiol 2012;227(4):1709–20.

[272] Zhou Y, Bollu LR, Tozzi F, Ye X, Bhattacharya R, Gao G, et al. ATP citrate lyase mediates resistance of colorectal cancer cells to SN38. Mol Cancer Ther 2013;12(12):2782–91.

[273] Beckner ME, Fellows-Mayle W, Zhang Z, Agostino NR, Kant JA, Day BW, et al. Identification of ATP citrate lyase as a positive regulator of glycolytic function in glioblastomas. Int J Cancer 2010;126(10):2282–95.

[274] Wang Y, Wang Y, Shen L, Pang Y, Qiao Z, Liu P. Prognostic and therapeutic implications of increased ATP citrate lyase expression in human epithelial ovarian cancer. Oncol Rep 2012;27(4):1156–62.

[275] Schug ZT, Peck B, Jones DT, Zhang Q, Grosskurth S, Alam IS, et al. Acetyl-CoA synthetase 2 promotes acetate utilization and maintains cancer cell growth under metabolic stress. Cancer Cell 2015;27(1):57–71.

[276] Bourbeau MP, Bartberger MD. Recent advances in the development of acetyl-CoA carboxylase (ACC) inhibitors for the treatment of metabolic disease. J Med Chem 2015;58(2):525–36.

[277] Ye B, Yin L, Wang Q, Xu C. ACC1 is overexpressed in liver cancers and contributes to the proliferation of human hepatoma Hep G2 cells and the rat liver cell line BRL 3A. Mol Med Rep 2019;19(5):3431–40.

[278] Rios GM, Steinbauer B, Srivastava K, Singhal M, Mattijssen F, Maida A, et al. Acetyl-CoA carboxylase 1-dependent protein acetylation controls breast cancer metastasis and recurrence. Cell Metab 2017;26(6):842–55. e5.

[279] Yang JH, Kim NH, Yun JS, Cho ES, Cha YH, Cho SB, et al. Snail augments fatty acid oxidation by suppression of mitochondrial ACC2 during cancer progression. Life Sci Alliance 2020;3(7):e202000683.

[280] Luo J, Hong Y, Lu Y, Qiu S, Chaganty BK, Zhang L, et al. Acetyl-CoA carboxylase rewires cancer metabolism to allow cancer cells to survive inhibition of the Warburg effect by cetuximab. Cancer Lett 2017;384:39–49.

[281] Menendez JA, Lupu R. Fatty acid synthase and the lipogenic phenotype in cancer pathogenesis. Nat Rev Cancer 2007;7(10):763–77.

[282] Menendez JA, Lupu R. Fatty acid synthase (FASN) as a therapeutic target in breast cancer. Expert Opin Ther Targets 2017;21(11):1001–16.

[283] Lu T, Sun L, Wang Z, Zhang Y, He Z, Xu C. Fatty acid synthase enhances colorectal cancer cell proliferation and metastasis via regulating AMPK/mTOR pathway. Onco Targets Ther 2019;12:3339–47.

[284] Hopperton KE, Duncan RE, Bazinet RP, Archer MC. Fatty acid synthase plays a role in cancer metabolism beyond providing fatty acids for phospholipid synthesis or sustaining elevations in glycolytic activity. Exp Cell Res 2014;320(2):302–10.

[285] Nomura DK, Long JZ, Niessen S, Hoover HS, Ng SW, Cravatt BF. Monoacylglycerol lipase regulates a fatty acid network that promotes cancer pathogenesis. Cell 2010;140(1):49–61.

[286] Mason P, Liang B, Li L, Fremgen T, Murphy E, Quinn A, et al. SCD1 inhibition causes cancer cell death by depleting mono-unsaturated fatty acids. PLoS One 2012;7(3), e33823.

[287] Young RM, Ackerman D, Quinn ZL, Mancuso A, Gruber M, Liu L, et al. Dysregulated mTORC1 renders cells critically dependent on desaturated lipids for survival under tumor-like stress. Genes Dev 2013;27(10):1115–31.

[288] Gu Z, Shan K, Chen H, Chen YQ. N-3 polyunsaturated fatty acids and their role in cancer chemoprevention. Curr Pharmacol Rep 2015;1(5):283–94.

[289] Chushi L, Wei W, Kangkang X, Yongzeng F, Ning X, Xiaolei C. HMGCR is up-regulated in gastric cancer and promotes the growth and migration of the cancer cells. 587(1). Gene; 2016. p. 42–7.

[290] Ding X, Zhang W, Li S, Yang H. The role of cholesterol metabolism in cancer. Am J Cancer Res 2019;9(2):219–27.

[291] Xiao X, Tang JJ, Peng C, Wang Y, Fu L, Qiu ZP, et al. Cholesterol modification of smoothened is required for hedgehog signaling. Mol Cell 2017;66(1):154–62. e10.

[292] Kuzu OF, Noory MA, Robertson GP. The role of cholesterol in cancer. Cancer Res 2016;76 (8):2063–70.

[293] Zhang X, Song Y, Feng M, Zhou X, Lu Y, Gao L, et al. Thyroid-stimulating hormone decreases HMG-CoA reductase phosphorylation via AMP-activated protein kinase in the liver. J Lipid Res 2015;56(5):963–71.

[294] Freed-Pastor WA, Mizuno H, Zhao X, Langerød A, Moon SH, Rodriguez-Barrueco R, et al. Mutant p53 disrupts mammary tissue architecture via the mevalonate pathway. Cell 2012;148(1–2):244–58.

[295] Beckwitt CH, Brufsky A, Oltvai ZN, Wells A. Statin drugs to reduce breast cancer recurrence and mortality. Breast Cancer Res 2018;20(1):144.

[296] Alfaqih MA, Allott EH, Hamilton RJ, Freeman MR, Freedland SJ. The current evidence on statin use and prostate cancer prevention: are we there yet? Nat Rev Urol 2017;14(2):107–19.

[297] Zhang Y, Liang M, Sun C, Qu G, Shi T, Min M, et al. Statin use and risk of pancreatic cancer: an updated meta-analysis of 26 studies. Pancreas 2019;48(2):142–50.

[298] Ibáñez-Sanz G, Guinó E, Pontes C, Quijada-Manuitt MÁ, de la Peña-Negro LC, Aragón M, et al. Statin use and the risk of colorectal cancer in a population-based electronic health records study. Sci Rep 2019;9(1):13560.

[299] Bjarnadottir O, Romero Q, Bendahl PO, Jirström K, Rydén L, Loman N, et al. Targeting HMG-CoA reductase with statins in a window-of-opportunity breast cancer trial. Breast Cancer Res Treat 2013;138(2):499–508.

[300] Zhang XL, Geng J, Zhang XP, Peng B, Che JP, Yan Y, et al. Statin use and risk of bladder cancer: a meta-analysis. Cancer Causes Control 2013;24(4):769–76.

[301] Goldstein MR, Mascitelli L, Pezzetta F. Do statins prevent or promote cancer? Curr Oncol 2008;15 (2):76–7.

[302] Ma X, Bi E, Lu Y, Su P, Huang C, Liu L, et al. Cholesterol induces CD8(+) T cell exhaustion in the tumor microenvironment. Cell Metab 2019;30(1):143–56. e5.

[303] Ma X, Bi E, Huang C, Lu Y, Xue G, Guo X, et al. Cholesterol negatively regulates IL-9-producing CD8(+) T cell differentiation and antitumor activity. J Exp Med 2018;215(6):1555–69.

[304] Beltowski J, Semczuk A. Liver X receptor (LXR) and the reproductive system—a potential novel target for therapeutic intervention. Pharmacol Rep 2010;62(1):15–27.

[305] Nelson ER, Wardell SE, Jasper JS, Park S, Suchindran S, Howe MK, et al. 27-Hydroxycholesterol links hypercholesterolemia and breast cancer pathophysiology. Science 2013;342(6162):1094–8.

[306] Revilla G, Pons MP, Baila-Rueda L, Garcia-Leon A, Santos D, Cenarro A, et al. Cholesterol and 27-hydroxycholesterol promote thyroid carcinoma aggressiveness. Sci Rep 2019;9(1):10260.

[307] Guo F, Hong W, Yang M, Xu D, Bai Q, Li X, et al. Upregulation of 24(R/S),25-epoxycholesterol and 27-hydroxycholesterol suppresses the proliferation and migration of gastric cancer cells. Biochem Biophys Res Commun 2018;504(4):892–8.

[308] Wang S, Yao Y, Rao C, Zheng G, Chen W. 25-HC decreases the sensitivity of human gastric cancer cells to 5-fluorouracil and promotes cells invasion via the TLR2/NF-kappaB signaling pathway. Int J Oncol 2019;54(3):966–80.

[309] Chen L, Zhang L, Xian G, Lv Y, Lin Y, Wang Y. 25-Hydroxycholesterol promotes migration and invasion of lung adenocarcinoma cells. Biochem Biophys Res Commun 2017;484(4):857–63.

[310] Cyster JG, Dang EV, Reboldi A, Yi T. 25-Hydroxycholesterols in innate and adaptive immunity. Nat Rev Immunol 2014;14(11):731–43.

[311] Hensley CT, Wasti AT, DeBerardinis RJ. Glutamine and cancer: cell biology, physiology, and clinical opportunities. J Clin Invest 2013;123(9):3678–84.

[312] Watford M. Glutamine and glutamate: nonessential or essential amino acids? Anim Nutr 2015;1 (3):119–22.

[313] Masisi BK, El AR, Alfarsi L, Rakha EA, Green AR, Craze ML. The role of glutaminase in cancer. Histopathology 2020;76(4):498–508.

[314] Mates JM, Segura JA, Martin-Rufian M, Campos-Sandoval JA, Alonso FJ, Marquez J. Glutaminase isoenzymes as key regulators in metabolic and oxidative stress against cancer. Curr Mol Med 2013;13 (4):514–34.

[315] Gao P, Tchernyshyov I, Chang TC, Lee YS, Kita K, Ochi T, et al. C-Myc suppression of miR-23a/b enhances mitochondrial glutaminase expression and glutamine metabolism. Nature 2009;458 (7239):762–5.

[316] Corn PG, Ricci MS, Scata KA, Arsham AM, Simon MC, Dicker DT, et al. Mxi1 is induced by hypoxia in a HIF-1-dependent manner and protects cells from c-Myc-induced apoptosis. Cancer Biol Ther 2005;4(11):1285–94.

[317] Gordan JD, Bertout JA, Hu CJ, Diehl JA, Simon MC. HIF-2alpha promotes hypoxic cell proliferation by enhancing c-myc transcriptional activity. Cancer Cell 2007;11(4):335–47.

[318] Metallo CM, Gameiro PA, Bell EL, Mattaini KR, Yang J, Hiller K, et al. Reductive glutamine metabolism by IDH1 mediates lipogenesis under hypoxia. Nature 2011;481(7381):380–4.

[319] Seltzer MJ, Bennett BD, Joshi AD, Gao P, Thomas AG, Ferraris DV, et al. Inhibition of glutaminase preferentially slows growth of glioma cells with mutant IDH1. Cancer Res 2010;70(22):8981–7.

[320] Emadi A, Jun SA, Tsukamoto T, Fathi AT, Minden MD, Dang CV. Inhibition of glutaminase selectively suppresses the growth of primary acute myeloid leukemia cells with IDH mutations. Exp Hematol 2014;42(4):247–51.

[321] Katt WP, Lukey MJ, Cerione RA. A tale of two glutaminases: homologous enzymes with distinct roles in tumorigenesis. Future Med Chem 2017;9(2):223–43.

[322] Hu W, Zhang C, Wu R, Sun Y, Levine A, Feng Z. Glutaminase 2, a novel p53 target gene regulating energy metabolism and antioxidant function. Proc Natl Acad Sci U S A 2010;107(16):7455–60.

[323] Suzuki S, Tanaka T, Poyurovsky MV, Nagano H, Mayama T, Ohkubo S, et al. Phosphate-activated glutaminase (GLS2), a p53-inducible regulator of glutamine metabolism and reactive oxygen species. Proc Natl Acad Sci U S A 2010;107(16):7461–6.

[324] Nabe S, Yamada T, Suzuki J, Toriyama K, Yasuoka T, Kuwahara M, et al. Reinforce the antitumor activity of CD8(+) T cells via glutamine restriction. Cancer Sci 2018;109(12):3737–50.

[325] Kuo MT, Savaraj N, Feun LG. Targeted cellular metabolism for cancer chemotherapy with recombinant arginine-degrading enzymes. Oncotarget 2010;1(4):246–51.

[326] Scott L, Lamb J, Smith S, Wheatley DN. Single amino acid (arginine) deprivation: rapid and selective death of cultured transformed and malignant cells. Br J Cancer 2000;83(6):800–10.

[327] Feun LG, Kuo MT, Savaraj N. Arginine deprivation in cancer therapy. Curr Opin Clin Nutr Metab Care 2015;18(1):78–82.

[328] You J, Chen W, Chen J, Zheng Q, Dong J, Zhu Y. The oncogenic role of ARG1 in progression and metastasis of hepatocellular carcinoma. Biomed Res Int 2018;2018, 2109865.

[329] Takahashi H, Sakakura K, Kudo T, Toyoda M, Kaira K, Oyama T, et al. Cancer-associated fibroblasts promote an immunosuppressive microenvironment through the induction and accumulation of protumoral macrophages. Oncotarget 2017;8(5):8633–47.

[330] Monu NR, Frey AB. Myeloid-derived suppressor cells and anti-tumor T cells: a complex relationship. Immunol Investig 2012;41(6–7):595–613.

[331] Pesce JT, Ramalingam TR, Mentink-Kane MM, Wilson MS, El KK, Smith AM, et al. Arginase-1-expressing macrophages suppress Th2 cytokine-driven inflammation and fibrosis. PLoS Pathog 2009;5(4), e1000371.

[332] Mondanelli G, Bianchi R, Pallotta MT, Orabona C, Albini E, Iacono A, et al. A relay pathway between arginine and tryptophan metabolism confers immunosuppressive properties on dendritic cells. Immunity 2017;46(2):233–44.

[333] Mondanelli G, Iacono A, Allegrucci M, Puccetti P, Grohmann U. Immunoregulatory interplay between arginine and tryptophan metabolism in health and disease. Front Immunol 2019;10:1565.

[334] Rodriguez PC, Quiceno DG, Zabaleta J, Ortiz B, Zea AH, Piazuelo MB, et al. Arginase I production in the tumor microenvironment by mature myeloid cells inhibits T-cell receptor expression and antigen-specific T-cell responses. Cancer Res 2004;64(16):5839–49.

[335] Munn DH, Shafizadeh E, Attwood JT, Bondarev I, Pashine A, Mellor AL. Inhibition of T cell proliferation by macrophage tryptophan catabolism. J Exp Med 1999;189(9):1363–72.

[336] Liu M, Wang X, Wang L, Ma X, Gong Z, Zhang S, et al. Targeting the IDO1 pathway in cancer: from bench to bedside. J Hematol Oncol 2018;11(1):100.

[337] Liu X, Shin N, Koblish HK, Yang G, Wang Q, Wang K, et al. Selective inhibition of IDO1 effectively regulates mediators of antitumor immunity. Blood 2010;115(17):3520–30.

[338] Liu H, Shen Z, Wang Z, Wang X, Zhang H, Qin J, et al. Increased expression of IDO associates with poor postoperative clinical outcome of patients with gastric adenocarcinoma. Sci Rep 2016;6:21319.

[339] Li R, Wei F, Yu J, Li H, Ren X, Hao X. IDO inhibits T-cell function through suppressing Vav1 expression and activation. Cancer Biol Ther 2009;8(14):1402–8.

[340] Munn DH, Mellor AL. Indoleamine 2,3-dioxygenase and tumor-induced tolerance. J Clin Invest 2007;117(5):1147–54.

[341] Shi J, Chen C, Ju R, Wang Q, Li J, Guo L, et al. Carboxyamidotriazole combined with IDO1-Kyn-AhR pathway inhibitors profoundly enhances cancer immunotherapy. J Immunother Cancer 2019;7(1):246.

[342] Mezrich JD, Fechner JH, Zhang X, Johnson BP, Burlingham WJ, Bradfield CA. An interaction between kynurenine and the aryl hydrocarbon receptor can generate regulatory T cells. J Immunol 2010;185(6):3190–8.

[343] Castro-Portuguez R, Sutphin GL. Kynurenine pathway, NAD(+) synthesis, and mitochondrial function: targeting tryptophan metabolism to promote longevity and healthspan. Exp Gerontol 2020;132:110841.

[344] Luckheeram RV, Zhou R, Verma AD, Xia B. CD4(+)T cells: differentiation and functions. Clin Dev Immunol 2012;2012:925135.

[345] Harkcom WT, Ghosh AK, Sung MS, Matov A, Brown KD, Giannakakou P, et al. NAD + and SIRT3 control microtubule dynamics and reduce susceptibility to antimicrotubule agents. Proc Natl Acad Sci U S A 2014;111(24):E2443–52.

[346] Nikiforov MA, Chandriani S, O'Connell B, Petrenko O, Kotenko I, Beavis A, et al. A functional screen for Myc-responsive genes reveals serine hydroxymethyltransferase, a major source of the one-carbon unit for cell metabolism. Mol Cell Biol 2002;22(16):5793–800.

[347] Maddocks O, Athineos D, Cheung EC, Lee P, Zhang T, van den Broek N, et al. Modulating the therapeutic response of tumours to dietary serine and glycine starvation. Nature 2017;544 (7650):372–6.

[348] Amelio I, Cutruzzola F, Antonov A, Agostini M, Melino G. Serine and glycine metabolism in cancer. Trends Biochem Sci 2014;39(4):191–8.

[349] Ma EH, Bantug G, Griss T, Condotta S, Johnson RM, Samborska B, et al. Serine is an essential metabolite for effector T cell expansion. Cell Metab 2017;25(2):345–57.

[350] Sas K, Szabo E, Vecsei L. Mitochondria, oxidative stress and the kynurenine system, with a focus on ageing and neuroprotection. Molecules 2018;23(1):191.

[351] Nesuashvili L, Hadley SH, Bahia PK, Taylor-Clark TE. Sensory nerve terminal mitochondrial dysfunction activates airway sensory nerves via transient receptor potential (TRP) channels. Mol Pharmacol 2013;83(5):1007–19.

[352] Houten SM, Violante S, Ventura FV, Wanders RJ. The biochemistry and physiology of mitochondrial fatty acid beta-oxidation and its genetic disorders. Annu Rev Physiol 2016;78:23–44.

[353] Marchi S, Giorgi C, Suski JM, Agnoletto C, Bononi A, Bonora M, et al. Mitochondria-ros crosstalk in the control of cell death and aging. J Signal Transduct 2012;2012(329635).

[354] Youle RJ, Karbowski M. Mitochondrial fission in apoptosis. Nat Rev Mol Cell Biol 2005;6 (8):657–63.

[355] Bravo-Sagua R, Parra V, López-Crisosto C, Díaz P, Quest AF, Lavandero S. Calcium transport and signaling in mitochondria. Compr Physiol 2017;7(2):623–34.

[356] Peron M, Meneghetti G, Dinarello A, Martorano L, Betto RM, Facchinello N, et al. Mitochondrial STAT3 regulates proliferation of tissue stem cells. bioRxiv 2020. 2020.07.17.208264.

[357] Mitra K. Mitochondrial fission-fusion as an emerging key regulator of cell proliferation and differentiation. BioEssays 2013;35(11):955–64.

[358] Denisenko TV, Gorbunova AS, Zhivotovsky B. Mitochondrial involvement in migration, invasion and metastasis. Front Cell Dev Biol 2019;7:355.

[359] Ishikawa K, Takenaga K, Akimoto M, Koshikawa N, Yamaguchi A, Imanishi H, et al. ROS-generating mitochondrial DNA mutations can regulate tumor cell metastasis. Science 2008;320 (5876):661–4.

[360] Droge W. Free radicals in the physiological control of cell function. Physiol Rev 2002;82(1):47–95.

[361] Hamanaka RB, Chandel NS. Mitochondrial reactive oxygen species regulate cellular signaling and dictate biological outcomes. Trends Biochem Sci 2010;35(9):505–13.

[362] Li N, Ragheb K, Lawler G, Sturgis J, Rajwa B, Melendez JA, et al. Mitochondrial complex I inhibitor rotenone induces apoptosis through enhancing mitochondrial reactive oxygen species production. J Biol Chem 2003;278(10):8516–25.

[363] Sena LA, Chandel NS. Physiological roles of mitochondrial reactive oxygen species. Mol Cell 2012;48 (2):158–67.

[364] Rashida GJ, Wu R, Wang R. Metabolic reprogramming in modulating T cell reactive oxygen species generation and antioxidant capacity. Front Immunol 2018;9:1075.

[365] Kaminski MM, Roth D, Krammer PH, Gulow K. Mitochondria as oxidative signaling organelles in T-cell activation: physiological role and pathological implications. Arch Immunol Ther Exp (Warsz) 2013;61(5):367–84.

[366] Jang KJ, Mano H, Aoki K, Hayashi T, Muto A, Nambu Y, et al. Mitochondrial function provides instructive signals for activation-induced B-cell fates. Nat Commun 2015;6:6750.

[367] He C, Ryan AJ, Murthy S, Carter AB. Accelerated development of pulmonary fibrosis via Cu, Zn-superoxide dismutase-induced alternative activation of macrophages. J Biol Chem 2013;288 (28):20745–57.

[368] Nakahira K, Haspel JA, Rathinam VA, Lee SJ, Dolinay T, Lam HC, et al. Autophagy proteins regulate innate immune responses by inhibiting the release of mitochondrial DNA mediated by the NALP3 inflammasome. Nat Immunol 2011;12(3):222–30.

[369] Lasithiotaki I, Tsitoura E, Samara KD, Trachalaki A, Charalambous I, Tzanakis N, et al. NLRP3/Caspase-1 inflammasome activation is decreased in alveolar macrophages in patients with lung cancer. PLoS One 2018;13(10), e0205242.

[370] Dostert C, Petrilli V, Van Bruggen R, Steele C, Mossman BT, Tschopp J. Innate immune activation through Nalp3 inflammasome sensing of asbestos and silica. Science 2008;320(5876):674–7.

[371] Brydges SD, Broderick L, McGeough MD, Pena CA, Mueller JL, Hoffman HM. Divergence of IL-1, IL-18, and cell death in NLRP3 inflammasomopathies. J Clin Invest 2013;123(11):4695–705.

[372] Zhou H, Ivanov VN, Lien YC, Davidson M, Hei TK. Mitochondrial function and nuclear factor-kappaB-mediated signaling in radiation-induced bystander effects. Cancer Res 2008;68(7):2233–40.

[373] Li S, Yan T, Yang JQ, Oberley TD, Oberley LW. The role of cellular glutathione peroxidase redox regulation in the suppression of tumor cell growth by manganese superoxide dismutase. Cancer Res 2000;60(14):3927–39.

[374] Quijano C, Trujillo M, Castro L, Trostchansky A. Interplay between oxidant species and energy metabolism. Redox Biol 2016;8:28–42.

[375] Sivitz WI, Yorek MA. Mitochondrial dysfunction in diabetes: from molecular mechanisms to functional significance and therapeutic opportunities. Antioxid Redox Signal 2010;12(4):537–77.

[376] De Simone R, Ajmone-Cat MA, Pandolfi M, Bernardo A, De Nuccio C, Minghetti L, et al. The mitochondrial uncoupling protein-2 is a master regulator of both M1 and M2 microglial responses. J Neurochem 2015;135(1):147–56.

[377] Miwa S, Jow H, Baty K, Johnson A, Czapiewski R, Saretzki G, et al. Low abundance of the matrix arm of complex I in mitochondria predicts longevity in mice. Nat Commun 2014;5:3837.

[378] Mimaki M, Wang X, McKenzie M, Thorburn DR, Ryan MT. Understanding mitochondrial complex I assembly in health and disease. Biochim Biophys Acta 2012;1817(6):851–62.

[379] Ko JH, Yoon SO, Lee HJ, Oh JY. Rapamycin regulates macrophage activation by inhibiting NLRP3 inflammasome-p38 MAPK-NFkappaB pathways in autophagy- and p62-dependent manners. Oncotarget 2017;8(25):40817–31.

[380] Sies H. Strategies of antioxidant defense. Eur J Biochem 1993;215(2):213–9.

[381] Kirichok Y, Krapivinsky G, Clapham DE. The mitochondrial calcium uniporter is a highly selective ion channel. Nature 2004;427(6972):360–4.

[382] Núñez L, Valero RA, Senovilla L, Sanz-Blasco S, García-Sancho J, Villalobos C. Cell proliferation depends on mitochondrial Ca^{2+} uptake: inhibition by salicylate. J Physiol 2006;571(Pt 1):57–73.

[383] Giorgi C, Romagnoli A, Pinton P, Rizzuto R. Ca^{2+} signaling, mitochondria and cell death. Curr Mol Med 2008;8(2):119–30.

[384] Narendra DP, Jin SM, Tanaka A, Suen DF, Gautier CA, Shen J, et al. PINK1 is selectively stabilized on impaired mitochondria to activate Parkin. PLoS Biol 2010;8(1), e1000298.

[385] Botbol Y, Guerrero-Ros I, Macian F. Key roles of autophagy in regulating T-cell function. Eur J Immunol 2016;46(6):1326–34.

[386] Pua HH, Guo J, Komatsu M, He YW. Autophagy is essential for mitochondrial clearance in mature T lymphocytes. J Immunol 2009;182(7):4046–55.

[387] Willinger T, Flavell RA. Canonical autophagy dependent on the class III phosphoinositide-3 kinase Vps34 is required for naive T-cell homeostasis. Proc Natl Acad Sci U S A 2012;109(22):8670–5.

[388] Alissafi T, Kalafati L, Lazari M, Filia A, Kloukina I, Manifava M, et al. Mitochondrial oxidative damage underlies regulatory T cell defects in autoimmunity. Cell Metab 2020;32(4):591–604. e7.

[389] Wang N, Yuan J, Karim MR, Zhong P, Sun YP, Zhang HY, et al. Effects of mitophagy on regulatory T cell function in patients with myasthenia gravis. Front Neurol 2020;11:238.

[390] Shimada K, Crother TR, Karlin J, Dagvadorj J, Chiba N, Chen S, et al. Oxidized mitochondrial DNA activates the NLRP3 inflammasome during apoptosis. Immunity 2012;36(3):401–14.

[391] Huang LS, Hong Z, Wu W, Xiong S, Zhong M, Gao X, et al. mtDNA activates cGAS signaling and suppresses the YAP-mediated endothelial cell proliferation program to promote inflammatory injury. Immunity 2020;52(3):475–86. e5.

[392] Guo Y, Chi X, Wang Y, Heng BC, Wei Y, Zhang X, et al. Mitochondria transfer enhances proliferation, migration, and osteogenic differentiation of bone marrow mesenchymal stem cell and promotes bone defect healing. Stem Cell Res Ther 2020;11(1):245.

[393] D'Souza AD, Parikh N, Kaech SM, Shadel GS. Convergence of multiple signaling pathways is required to coordinately up-regulate mtDNA and mitochondrial biogenesis during T cell activation. Mitochondrion 2007;7(6):374–85.

[394] Desdin-Micó G, Soto-Heredero G, Mittelbrunn M. Mitochondrial activity in T cells. Mitochondrion 2018;41:51–7.

[395] Vyas S, Zaganjor E, Haigis MC. Mitochondria and cancer. Cell 2016;166(3):555–66.

[396] Zapico SC, Ubelaker DH. mtDNA mutations and their role in aging, diseases and forensic sciences. Aging Dis 2013;4(6):364–80.

[397] van Gisbergen MW, Voets AM, Starmans MH, de Coo IF, Yadak R, Hoffmann RF, et al. How do changes in the mtDNA and mitochondrial dysfunction influence cancer and cancer therapy? Challenges, opportunities and models. Mutat Res Rev Mutat Res 2015;764:16–30.

[398] Wallace DC. Mitochondria and cancer. Nat Rev Cancer 2012;12(10):685–98.

[399] De Luise M, Girolimetti G, Okere B, Porcelli AM, Kurelac I, Gasparre G. Molecular and metabolic features of oncocytomas: seeking the blueprints of indolent cancers. Biochim Biophys Acta Bioenerg 2017;1858(8):591–601.

[400] Mambo E, Chatterjee A, Xing M, Tallini G, Haugen BR, Yeung SC, et al. Tumor-specific changes in mtDNA content in human cancer. Int J Cancer 2005;116(6):920–4.

[401] Guha M, Srinivasan S, Ruthel G, Kashina AK, Carstens RP, Mendoza A, et al. Mitochondrial retrograde signaling induces epithelial-mesenchymal transition and generates breast cancer stem cells. Oncogene 2014;33(45):5238–50.

[402] Xu Y, Lu S. Transforming growth factor-beta1-induced epithelial to mesenchymal transition increases mitochondrial content in the A549 non-small cell lung cancer cell line. Mol Med Rep 2015;11(1):417–21.

[403] Kerkhofs M, Bittremieux M, Morciano G, Giorgi C, Pinton P, Parys JB, et al. Emerging molecular mechanisms in chemotherapy: Ca(2+) signaling at the mitochondria-associated endoplasmic reticulum membranes. Cell Death Dis 2018;9(3):334.

[404] Eisner V, Picard M, Hajnoczky G. Mitochondrial dynamics in adaptive and maladaptive cellular stress responses. Nat Cell Biol 2018;20(7):755–65.

[405] Desai SP, Bhatia SN, Toner M, Irimia D. Mitochondrial localization and the persistent migration of epithelial cancer cells. Biophys J 2013;104(9):2077–88.

[406] Zhao J, Zhang J, Yu M, Xie Y, Huang Y, Wolff DW, et al. Mitochondrial dynamics regulates migration and invasion of breast cancer cells. Oncogene 2013;32(40):4814–24.

[407] Wang Y, Subramanian M, Yurdagul AJ, Barbosa-Lorenzi VC, Cai B, de Juan-Sanz J, et al. Mitochondrial fission promotes the continued clearance of apoptotic cells by macrophages. Cell 2017;171 (2):331–45. e22.

[408] Buck MD, O'Sullivan D, Klein GR, Curtis JD, Chang CH, Sanin DE, et al. Mitochondrial dynamics controls T cell fate through metabolic programming. Cell 2016;166(1):63–76.

[409] Baixauli F, Martin-Cófreces NB, Morlino G, Carrasco YR, Calabia-Linares C, Veiga E, et al. The mitochondrial fission factor dynamin-related protein 1 modulates T-cell receptor signalling at the immune synapse. EMBO J 2011;30(7):1238–50.

[410] Abarca-Rojano E, Muñiz-Hernández S, Moreno-Altamirano MM, Mondragón-Flores R, Enriquez-Rincón F, Sánchez-García FJ. Re-organization of mitochondria at the NK cell immune synapse. Immunol Lett 2009;122(1):18–25.

[411] Kang YJ, Bang BR, Han KH, Hong L, Shim EJ, Ma J, et al. Regulation of NKT cell-mediated immune responses to tumours and liver inflammation by mitochondrial PGAM5-Drp1 signalling. Nat Commun 2015;6:8371.

[412] Valentine G, Prince A, Aagaard KM. The neonatal microbiome and metagenomics: what do we know and what is the future? NeoReviews 2019;20(5):e258–71.

[413] Laville E, Perrier J, Bejar N, Maresca M, Esque J, Tauzin AS, et al. Investigating host microbiota relationships through functional metagenomics. Front Microbiol 2019;10:1286.

[414] Frankel AE, Deshmukh S, Reddy A, Lightcap J, Hayes M, McClellan S, et al. Cancer immune checkpoint inhibitor therapy and the gut microbiota. Integr Cancer Ther 2019;18, 1534735419846379.

[415] Fessler J, Matson V, Gajewski TF. Exploring the emerging role of the microbiome in cancer immunotherapy. J Immunother Cancer 2019;7(1):108.

[416] Kaiser J. Gut microbes shape response to cancer immunotherapy. Science 2017;358(6363):573.

[417] Bain CC, Mowat AM. Macrophages in intestinal homeostasis and inflammation. Immunol Rev 2014;260(1):102–17.

[418] Sánchez-Alcoholado L, Ramos-Molina B, Otero A, Laborda-Illanes A, Ordonez R, Medina JA, et al. The role of the gut microbiome in colorectal cancer development and therapy response. Cancers (Basel) 2020;12(6):1406.

[419] Ponziani FR, Bhoori S, Castelli C, Putignani L, Rivoltini L, Del CF, et al. Hepatocellular carcinoma is associated with gut microbiota profile and inflammation in nonalcoholic fatty liver disease. Hepatology 2019;69(1):107–20.

[420] Gupta H, Youn GS, Shin MJ, Suk KT. Role of gut microbiota in Hepatocarcinogenesis. Microorganisms 2019;7(5):121.

[421] Meng C, Bai C, Brown TD, Hood LE, Tian Q. Human gut microbiota and gastrointestinal cancer. Genomics Proteomics Bioinformatics 2018;16(1):33–49.

[422] Li D, He R, Hou G, Ming W, Fan T, Chen L, et al. Characterization of the esophageal microbiota and prediction of the metabolic pathways involved in esophageal cancer. Front Cell Infect Microbiol 2020;10:268.

[423] Parida S, Sharma D. The microbiome-estrogen connection and breast cancer risk. Cells 2019;8 (12):1642.

[424] Vicente-Dueñas C, Janssen S, Oldenburg M, Auer F, González-Herrero I, Casado-García A, et al. An intact gut microbiome protects genetically predisposed mice against leukemia. Blood 2020;136 (18):2003–17.

[425] Li Y, Tinoco R, Elmen L, Segota I, Xian Y, Fujita Y, et al. Gut microbiota dependent anti-tumor immunity restricts melanoma growth in Rnf5(−/−) mice. Nat Commun 2019;10(1):1492.

[426] Gensollen T, Iyer SS, Kasper DL, Blumberg RS. How colonization by microbiota in early life shapes the immune system. Science 2016;352(6285):539–44.

[427] Olszak T, An D, Zeissig S, Vera MP, Richter J, Franke A, et al. Microbial exposure during early life has persistent effects on natural killer T cell function. Science 2012;336(6080):489–93.

[428] An D, Oh SF, Olszak T, Neves JF, Avci FY, Erturk-Hasdemir D, et al. Sphingolipids from a symbiotic microbe regulate homeostasis of host intestinal natural killer T cells. Cell 2014;156(1–2):123–33.

[429] Crabbe PA, Nash DR, Bazin H, Eyssen H, Heremans JF. Immunohistochemical observations on lymphoid tissues from conventional and germ-free mice. Lab Investig 1970;22(5):448–57.

[430] Cahenzli J, Koller Y, Wyss M, Geuking MB, McCoy KD. Intestinal microbial diversity during early-life colonization shapes long-term IgE levels. Cell Host Microbe 2013;14(5):559–70.

[431] Gollwitzer ES, Saglani S, Trompette A, Yadava K, Sherburn R, McCoy KD, et al. Lung microbiota promotes tolerance to allergens in neonates via PD-L1. Nat Med 2014;20(6):642–7.

[432] Li R, Zhou R, Wang H, Li W, Pan M, Yao X, et al. Gut microbiota-stimulated cathepsin K secretion mediates TLR4-dependent M2 macrophage polarization and promotes tumor metastasis in colorectal cancer. Cell Death Differ 2019;26(11):2447–63.

[433] Sivan A, Corrales L, Hubert N, Williams JB, Aquino-Michaels K, Earley ZM, et al. Commensal Bifidobacterium promotes antitumor immunity and facilitates anti-PD-L1 efficacy. Science 2015;350(6264):1084–9.

[434] Yokoyama MT, Carlson JR. Microbial metabolites of tryptophan in the intestinal tract with special reference to skatole. Am J Clin Nutr 1979;32(1):173–8.

[435] Rooks MG, Garrett WS. Gut microbiota, metabolites and host immunity. Nat Rev Immunol 2016;16 (6):341–52.

[436] Rothhammer V, Borucki DM, Tjon EC, Takenaka MC, Chao CC, Ardura-Fabregat A, et al. Microglial control of astrocytes in response to microbial metabolites. Nature 2018;557(7707):724–8.

[437] Kaufmann E, Sanz J, Dunn JL, Khan N, Mendonca LE, Pacis A, et al. BCG educates hematopoietic stem cells to generate protective innate immunity against tuberculosis. Cell 2018;172(1–2):176–90. e19.

[438] Hersh EM, Gutterman JU, Mavligit GM. BCG as adjuvant immunotherapy for neoplasia. Annu Rev Med 1977;28:489–515.

[439] Dejea CM, Fathi P, Craig JM, Boleij A, Taddese R, Geis AL, et al. Patients with familial adenomatous polyposis harbor colonic biofilms containing tumorigenic bacteria. Science 2018;359(6375):592–7.

[440] Meisel M, Hinterleitner R, Pacis A, Chen L, Earley ZM, Mayassi T, et al. Microbial signals drive pre-leukaemic myeloproliferation in a Tet2-deficient host. Nature 2018;557(7706):580–4.

[441] Muller PA, Matheis F, Schneeberger M, Kerner Z, Jové V, Mucida D. Microbiota-modulated CART (+) enteric neurons autonomously regulate blood glucose. Science 2020;370(6514):314–21.

[442] Geller LT, Barzily-Rokni M, Danino T, Jonas OH, Shental N, Nejman D, et al. Potential role of intratumor bacteria in mediating tumor resistance to the chemotherapeutic drug gemcitabine. Science 2017;357(6356):1156–60.

[443] Rothschild D, Weissbrod O, Barkan E, Kurilshikov A, Korem T, Zeevi D, et al. Environment dominates over host genetics in shaping human gut microbiota. Nature 2018;555(7695):210–5.

[444] Imhann F, Vich VA, Bonder MJ, Lopez MA, Koonen D, Fu J, et al. The influence of proton pump inhibitors and other commonly used medication on the gut microbiota. Gut Microbes 2017;8 (4):351–8.

[445] Ouyang J, Isnard S, Lin J, Fombuena B, Marette A, Routy B, et al. Metformin effect on gut microbiota: insights for HIV-related inflammation. AIDS Res Ther 2020;17(1):10.

[446] Tropini C, Moss EL, Merrill BD, Ng KM, Higginbottom SK, Casavant EP, et al. Transient osmotic perturbation causes long-term alteration to the gut microbiota. Cell 2018;173(7):1742–54. e17.

[447] Olson CA, Vuong HE, Yano JM, Liang QY, Nusbaum DJ, Hsiao EY. The gut microbiota mediates the anti-seizure effects of the ketogenic diet. Cell 2018;173(7):1728–41. e13.

[448] Zhao L, Zhang F, Ding X, Wu G, Lam YY, Wang X, et al. Gut bacteria selectively promoted by dietary fibers alleviate type 2 diabetes. Science 2018;359(6380):1151–6.

[449] Zhu W, Winter MG, Byndloss MX, Spiga L, Duerkop BA, Hughes ER, et al. Precision editing of the gut microbiota ameliorates colitis. Nature 2018;553(7687):208–11.

[450] Zhernakova A, Kurilshikov A, Bonder MJ, Tigchelaar EF, Schirmer M, Vatanen T, et al. Population-based metagenomics analysis reveals markers for gut microbiome composition and diversity. Science 2016;352(6285):565–9.

[451] Chen D, Wu J, Jin D, Wang B, Cao H. Fecal microbiota transplantation in cancer management: current status and perspectives. Int J Cancer 2019;145(8):2021–31.

[452] Li SS, Zhu A, Benes V, Costea PI, Hercog R, Hildebrand F, et al. Durable coexistence of donor and recipient strains after fecal microbiota transplantation. Science 2016;352(6285):586–9.

[453] Surawicz CM, Brandt LJ, Binion DG, Ananthakrishnan AN, Curry SR, Gilligan PH, et al. Guidelines for diagnosis, treatment, and prevention of *Clostridium difficile* infections. Am J Gastroenterol 2013;108 (4):478–98 [quiz 499].

[454] Konturek PC, Haziri D, Brzozowski T, Hess T, Heyman S, Kwiecien S, et al. Emerging role of fecal microbiota therapy in the treatment of gastrointestinal and extra-gastrointestinal diseases. J Physiol Pharmacol 2015;66(4):483–91.

[455] Rosshart SP, Vassallo BG, Angeletti D, Hutchinson DS, Morgan AP, Takeda K, et al. Wild mouse gut microbiota promotes host fitness and improves disease resistance. Cell 2017;171(5):1015–28. e13.

[456] Zhou D, Pan Q, Shen F, Cao HX, Ding WJ, Chen YW, et al. Total fecal microbiota transplantation alleviates high-fat diet-induced steatohepatitis in mice via beneficial regulation of gut microbiota. Sci Rep 2017;7(1):1529.

[457] de Groot P, Nikolic T, Pellegrini S, Sordi V, Imangaliyev S, Rampanelli E, et al. Faecal microbiota transplantation halts progression of human new-onset type 1 diabetes in a randomised controlled trial. Gut 2020. gutjnl-2020-322630.

[458] Mullard A. Oncologists tap the microbiome in bid to improve immunotherapy outcomes. Nat Rev Drug Discov 2018;17(3):153–5.

[459] Thomas RM, Gharaibeh RZ, Gauthier J, Beveridge M, Pope JL, Guijarro MV, et al. Intestinal microbiota enhances pancreatic carcinogenesis in preclinical models. Carcinogenesis 2018;39(8):1068–78.

[460] Elstrom RL, Bauer DE, Buzzai M, Karnauskas R, Harris MH, Plas DR, et al. Akt stimulates aerobic glycolysis in cancer cells. Cancer Res 2004;64(11):3892–9.

[461] Robey IF, Lien AD, Welsh SJ, Baggett BK, Gillies RJ. Hypoxia-inducible factor-1alpha and the glycolytic phenotype in tumors. Neoplasia 2005;7(4):324–30.

[462] Csibi A, Blenis J. Appetite for destruction: the inhibition of glycolysis as a therapy for tuberous sclerosis complex-related tumors. BMC Biol 2011;9:69.

[463] Jiang X, Kenerson H, Aicher L, Miyaoka R, Eary J, Bissler J, et al. The tuberous sclerosis complex regulates trafficking of glucose transporters and glucose uptake. Am J Pathol 2008;172(6):1748–56.

[464] Hu M, Chen X, Ma L, Ma Y, Li Y, Song H, et al. AMPK inhibition suppresses the malignant phenotype of pancreatic cancer cells in part by attenuating aerobic glycolysis. J Cancer 2019;10 (8):1870–8.

[465] Faubert B, Boily G, Izreig S, Griss T, Samborska B, Dong Z, et al. AMPK is a negative regulator of the Warburg effect and suppresses tumor growth in vivo. Cell Metab 2013;17(1):113–24.

[466] Hussien R, Brooks GA. Mitochondrial and plasma membrane lactate transporter and lactate dehydrogenase isoform expression in breast cancer cell lines. Physiol Genomics 2011;43(5):255–64.

[467] Choi SY, Collins CC, Gout PW, Wang Y. Cancer-generated lactic acid: a regulatory, immunosuppressive metabolite? J Pathol 2013;230(4):350–5.

[468] Wang C, Ma J, Zhang N, Yang Q, Jin Y, Wang Y. The acetyl-CoA carboxylase enzyme: a target for cancer therapy? Expert Rev Anticancer Ther 2015;15(6):667–76.

[469] Godin-Ethier J, Hanafi LA, Piccirillo CA, Lapointe R. Indoleamine 2,3-dioxygenase expression in human cancers: clinical and immunologic perspectives. Clin Cancer Res 2011;17(22):6985–91.

[470] Whiteside TL. The role of immune cells in the tumor microenvironment. Cancer Treat Res 2006;130:103–24.

[471] Gajewski TF, Schreiber H, Fu YX. Innate and adaptive immune cells in the tumor microenvironment. Nat Immunol 2013;14(10):1014–22.

[472] Noman MZ, Hasmim M, Lequeux A, Xiao M, Duhem C, Chouaib S, et al. Improving cancer immunotherapy by targeting the hypoxic tumor microenvironment: new opportunities and challenges. Cell 2019;8(9). https://doi.org/10.3390/cells8091083.

[473] Semenza GL. Targeting HIF-1 for cancer therapy. Nat Rev Cancer 2003;3(10):721–32.

[474] Barsoum IB, Smallwood CA, Siemens DR, Graham CH. A mechanism of hypoxia-mediated escape from adaptive immunity in cancer cells. Cancer Res 2014;74(3):665–74.

[475] Bollinger T, Gies S, Naujoks J, Feldhoff L, Bollinger A, Solbach W, et al. HIF-1alpha- and hypoxia-dependent immune responses in human CD4 + CD25high T cells and T helper 17 cells. J Leukoc Biol 2014;96(2):305–12.

[476] Palazón A, Martínez-Forero I, Teijeira A, Morales-Kastresana A, Alfaro C, Sanmamed MF, et al. The HIF-1alpha hypoxia response in tumor-infiltrating T lymphocytes induces functional CD137 (4-1BB) for immunotherapy. Cancer Discov 2012;2(7):608–23.

[477] Nowak EC, Lines JL, Varn FS, Deng J, Sarde A, Mabaera R, et al. Immunoregulatory functions of VISTA. Immunol Rev 2017;276(1):66–79.

[478] Deng J, Li J, Sarde A, Lines JL, Lee YC, Qian DC, et al. Hypoxia-induced VISTA promotes the suppressive function of myeloid-derived suppressor cells in the tumor microenvironment. Cancer Immunol Res 2019;7(7):1079–90.

[479] Zhang H, Lu H, Xiang L, Bullen JW, Zhang C, Samanta D, et al. HIF-1 regulates CD47 expression in breast cancer cells to promote evasion of phagocytosis and maintenance of cancer stem cells. Proc Natl Acad Sci U S A 2015;112(45):E6215–23.

[480] Mauge L, Terme M, Tartour E, Helley D. Control of the adaptive immune response by tumor vasculature. Front Oncol 2014;4:61.

[481] Kohlhapp FJ, Mitra AK, Lengyel E, Peter ME. MicroRNAs as mediators and communicators between cancer cells and the tumor microenvironment. Oncogene 2015;34(48):5857–68.

[482] Pio R, Ajona D, Lambris JD. Complement inhibition in cancer therapy. Semin Immunol 2013;25 (1):54–64.

CHAPTER SEVEN

Epigenetic programming of the immune responses in cancer

Abbey A. Saadey[a,*]**, Amir Yousif**[a,*]**, and Hazem E. Ghoneim**[a,b]

[a]Department of Microbial Infection and Immunity, College of Medicine, The Ohio State University, Columbus, OH, United States
[b]The James Comprehensive Cancer Center, Pelotonia Institute for Immuno-Oncology, The Ohio State University, Columbus, OH, United States

Contents

1. Introduction

Immune surveillance is an important host-defense strategy against spontaneously arising tumor cells or invading pathogens. In cancer, the function of the immune system is to recognize and eliminate transformed cells before they evolve into tumors, and then control the tumors that do form. However, evolving tumors employ several strategies to

---

*Equal contribution.

Cancer Immunology and Immunotherapy
https://doi.org/10.1016/B978-0-12-823397-9.00007-7

successfully evade immune detection, and/or suppress antitumor immune responses. The intricate interplay between cancer and the host's immune system involves a process coined as "cancer immunoediting" consisting of three phases: (a) tumor elimination by immune surveillance; (b) equilibrium between tumor growth and host's antitumor immune responses; and (c) tumor escape from host defenses [1, 2]. Progressive phenotypic and functional changes occur within tumor-infiltrating immune cells and evolving cancer cells throughout these phases. Such changes eventually establish an immunosuppressive tumor microenvironment (TME) conducive to cancer progression (as discussed in Chapters 2, 5, and 6). Enormous efforts have been focused on identifying extrinsic and cell-intrinsic mechanisms that interfere with immune responses within TME, such as poor antigen presentation, metabolic competition, immunosuppressive cytokines, impaired T-cell infiltration, and T-cell dysfunction [3]. Targeting molecular switches that mediate tumor-induced immune evasion or immune dysfunction has provided insights into developing novel cancer immunotherapies.

Over the past two decades, several advances in understanding the interactions between cancer and the immune system have resulted in breakthrough discoveries for treating cancer. Most of the successful immunotherapeutic approaches function by exploiting the selective killing power of cytotoxic CD8 + T-cells. Among T-cell-based immunotherapies is immune checkpoint blockade (ICB), developed to block either surface inhibitory receptors on dysfunctional T-cells or their ligands (e.g., PD-1, CTLA-4, PD-L1). ICB therapy has shown remarkable efficacy for treating multiple types of previously refractory cancers (e.g., metastatic melanoma, nonsmall cell lung carcinoma—NSCLC, hepatocellular carcinoma, gastric cancer, Merkel cell cancer) [4]. Recently, chimeric antigen receptor (CAR) T-cells have also emerged as a successful immunotherapy that employs genetic engineering of patient-derived T-cells to redirect their specificity against selected antigens expressed on target tumor cells. CD19-targeted CAR T-cells were the first CAR T-cell therapy approved by the FDA for treating relapsed or refractory B-cell acute lymphoblastic leukemia in both children and adults [5]. Despite the unprecedented clinical benefits achieved by T-cell immunotherapies, many patients remain nonresponsive to ICB [4], while CAR T-cells have poor in vivo persistence in leukemic patients and limited efficacy for treating solid cancers [5, 6]. A better understanding of cell-intrinsic mechanisms that regulate molecular and cellular barriers to cancer immunotherapy will provide insights into novel treatment strategies that effectively harness our immune system to eradicate tumors.

Transcriptional plasticity is an essential feature of the immune system to functionally adapt its responses against a broad spectrum of pathogens or cancers. These functional changes in responding immune cells are regulated downstream microenvironmental cues through cell-intrinsic mechanisms. Epigenetic programming serves as a fundamental cell-intrinsic mechanism that regulates cell's transcriptional output and controls cell's identity and fate commitment [7, 8]. Epigenetic mechanisms involve covalent structural modifications in the chromatin that modulate chromatin accessibility and gene expression programs without altering DNA sequences. To regulate gene expression programs, cell's epigenome engages a diverse array of mechanisms, such as DNA methylation, posttranslational histone modifications, chromatin remodeling, and noncoding RNAs, with differences in nature, impact, and stability features [7, 8]. Aberrant regulation of epigenetic programming may disrupt cellular differentiation and functional state, causing significant human diseases, as seen in including cancer, obesity, diabetes, lupus, asthma, and several neurological disorders [7, 9–16]. Despite the significant role of epigenetic mechanisms in regulating immune cell functions, studies have mainly focused on the role of epigenetic dysregulation within tumor cells. This was, in part, due to the broad heterogeneity across immune cell types. In addition, epigenetic profiling of different subsets of tumor-infiltrating immune cells was technically challenging, given their small cell numbers. However, recent advances in deep sequencing technologies have enabled better investigation of the changes in transcriptional and epigenetic programs within diverse types of cancer-associated immune cells (Box 1).

BOX 1 Advances in molecular profiling.

Bisulfite sequencing:

When studying the functions of DNA methylation, it is important to profile methylated levels at individual CpGs in the genome to analyze the changes during cell differentiation, development, and disease. Several reliable methods have been developed, including whole-genome or locus-specific bisulfite-sequencing [17], where sodium bisulfite treatment of single-stranded DNA leads to the deamination of cytosines, converting them to uracil. However, the 5-methylcytosines are not susceptible to bisulfite conversion, thereby protecting methylated cytosines. Methylated CpGs are then located by comparing treated and untreated sequences and can be further analyzed with next-generation sequencing at specific genomic loci or on a whole-genome scale. It is noteworthy that BS-seq does not distinguish between 5mC and 5hmC modifications, therefore, more assays were recently developed for global profiling 5-hydroxymethylation marks at cytosines in mammalian genomes, such as Tet-assisted bisulfite sequencing (TAB-Seq) [18].

Chromatin immunoprecipitation (ChIP):

Chromatin immunoprecipitation is a commonly used method for studying protein-DNA interactions [19, 20]. This method is used to identify locations in the genome bound by specific proteins at a particular gene sequence. Identifying the genetic targets of DNA-binding proteins and the mechanism of protein-DNA interaction is vital to understanding cellular functions such as gene transcription, epigenetic repression, signal transduction, chromosome segregation, and DNA replication. It is a valuable tool for studying histone modifications associated with a gene promoter, the binding of transcription factors, or other DNA-binding proteins at a specific locus where healthy levels can be compared with diseased states.

Cross-linked protein-DNA complexes in chromatin can be captured by antibodies directed against the protein of interest. The captured DNA is then purified and can be used for various

Continued

BOX 1 Advances in molecular profiling—cont'd

applications. Using ChIP in combination with microarray detection (ChIP-on-chip) and sequencing technology (ChIP-seq) allows for the mapping of genome-wide patterns of histone modifications that may lead to a greater understanding of the interplay between different modifications during transcriptional regulation.

ATAC-Seq:

Assay for transposase accessible chromatin with sequencing (ATAC-seq) is a technique for mapping chromatin accessibility both on a single-cell and genome-wide level [21, 22]. This technology has enabled the assessment of open chromatin landscape within rare immune cell populations, such as antigen-specific CD8+ T-cells. Chromatin accessibility is measured using hyperactive Tn5 transposase, which inserts into accessible regions of chromatin while simultaneously fragmenting the open DNA and ligating sequencing primers, tagging open DNA sites. The sequenced DNA can then be used to determine chromatin regions of increased accessibility, where this openness can be a proxy for inferring the level of simplicity for a transcription factor to bind to an accessible region. Data analysis, thereafter, allows for transcription factor binding sites and nucleosome mapping. ATAC-seq data is more advantageous in providing new insights into gene regulation and expression when combined with other epigenetic analyses, such as DNA methylation, lending a more comprehensive view.

Hi-C assay:

Pinpointing the structural and functional relationships of the genome, from a three-dimensional (3D) perspective, is a meaningful foundational question in biology. Chromosome conformation capture (3C)-derived techniques, notably the high-throughput chromosome conformation capture (Hi-C), an extension of 3C [23, 24], has helped outline the 3D genome structure. In a genome-wide manner, Hi-C can identify distant interactions, and can in theory also be used to detect chromatin loops across the genome. Hi-C combines formaldehyde-crosslinking of chromatin with restriction enzyme digestion, DNA ligation, where a biotin-labeled nucleotide is integrated at the ligation junction of each fragment, thereby making way for the purification of the DNA ligation junctions.

Direct analysis via high-throughput DNA-sequencing detects neighboring loci genome-wide. This analysis highlights several features of genomic structural composition, such as the separation of chromatin into distinct active and inactive compartments, and the folding of DNA into self-associating domains and loops [25]. As a result, 3D-folding of chromatin compartmentalizes the genome and allows for closer spatial interactions of distant genetic features, such as promoters and enhancers. Therefore, visualizing the relationship between chromosome organization and genome activity will further improve our understanding of genomic and epigenomic processes.

Epigenetic programming plays an instrumental role in dictating the fate of activated immune cells. It does not only orchestrate dynamic transcriptional changes downstream of immune signaling pathways but can also maintain a history of these changes within experienced "memory" immune cells. Such transcriptional memory is an important feature of the adaptive immune system that is required for establishing long-term protection against prior infections or tumors. Understanding the molecular mechanisms that regulate the acquisition and maintenance of this epigenetic memory is crucial to develop effective vaccines and long-term protection by immunotherapies. Recent studies have advanced our understanding of how epigenetics regulates immune cell differentiation and function, with exciting insights for developing novel epigenetic therapies to boost antitumor immunity [26–28]. In this chapter, we will summarize epigenetic modifications employed within immune cells, and review our current understanding of how epigenetic mechanisms regulate the development of immune responses during cancer. We will also highlight the critical role of de novo epigenetic programming in restraining the responsiveness of antitumor T-cells to cancer immunotherapy. Additionally, we will discuss how targeting epigenetic mechanisms within immune cells has great potential to overcome the molecular and cellular barriers to cancer immunotherapy.

2. Epigenetic mechanisms

The term *epigenetics* was classically used to define heritable changes in phenotype without changing the genotype by Waddington in 1942 [29, 30]. Recently, epigenetics more broadly describes changes in chromatin biology that affect gene expression states without altering the DNA sequence itself [8]. Epigenetic changes are regular and natural events that can be influenced by several external factors such as age, environment, lifestyle, and disease. Through different epigenetic marks, gene expression can be selectively activated, silenced, or fine-tuned at the transcriptional and/or posttranscriptional levels to control cell growth, proliferation, differentiation, and death. In addition, during developmental processes, these modifications play a critical role in how genetically identical cells can terminally differentiate into different cell types (skin cells, liver cells, brain cells, etc.) [8, 31]. Mounting evidence has also highlighted epigenetic changes as valuable diagnostic biomarkers for disease risk or progression as these changes may preface a variety of disorders and diseases [32]. Recent advances in technologies for assessing various epigenetic marks have pushed forward the investigation of mechanisms underlying chromatin stability, gene regulation, and transcriptional plasticity during cancer [8] (Box 1). With a better understanding of epigenetics, new avenues have emerged for developing novel epigenetic therapies for cancer and an increasing number of diseases.

2.1 DNA methylation and demethylation

DNA methylation is a genetically programmed modification that involves the addition of a methyl group onto the $5'$ position of a pyrimidine ring on cytosines mostly within the CpG dinucleotide [33, 34]. A high frequency of CpG dinucleotides tend to cluster in regions known as CpG islands. These regions, commonly found in gene promoters, are constitutively hypomethylated in normal cells, while the main body of a gene is typically hypermethylated. The same pattern is observed in CpG island shores, regions of lower CpG density upstream of CpG islands [35]. Generally speaking, DNA methylation has been functionally linked to transcriptional repression. Early studies show that *in vitro* methylated DNA is transcriptionally inactive when transfected into cultured mammalian cells [36, 37], with evidence that methylation is involved in gene-silencing events critical for mammalian development (*e.g.*, genomic imprinting, X-chromosome inactivation, silencing repetitive DNA sequences, and preserving chromatin stability) [35]. Recent advances in multi-layered epigenomic and transcriptomic analyses have now allowed for the simultaneous measurement of DNA methylation, chromatin accessibility, and gene expression levels, showing that changes in DNA methylation and chromatin accessibility or gene expression are negatively correlated, particularly at the gene regulatory regions [38].

A highly conserved machinery controls the acquisition and maintenance of DNA methylation programs. This concerted machinery includes (a) Writer enzymes—DNA methyltransferases (DNMT1, DNMT3A, and DNMT3B), which are responsible for both maintenance (DNMT1) and de novo methylation (DNMT3A/3B) [39–41]; (b) Readers of the CpG methylation marks, such as methyl-CpG-binding domain proteins (MBDs) that recognize and bind preferentially to methylated CpG sites; and (c) Erasers that catalyze the oxidation of methyl groups on CpG sites resulting in their active removal, such as ten-eleven translocase enzymes (TET1, TET2, and TET3) (Fig. 1). MBDs mediate transcriptional silencing after DNA methylation by recruiting transcriptional repressor complexes, and/or by steric hindrance to transcription factor binding caused by the methyl groups on CpGs at gene regulatory regions [33].

Following the erasure of DNA methylation after egg fertilization, brand-new DNA methylation occurs during embryonic development by Dnmt3a and Dnmt3b enzymes that are guided by specific DNA-binding proteins [40]. DNA methylation patterns temporarily break down during DNA replication resulting in hemimethylated DNA where the parent strand is methylated and the nascent strand is unmethylated [42]. These patterns, however, can be faithfully maintained and inherited in daughter cells after cell division [43]. A multidomain protein UHRF1 (ubiquitin-like with PHD and ring finger domains 1) recruits DNMT1 to the hemimethylated DNA and together facilitate remethylation after replication [44, 45]. This allows for the methylation patterns of the precursor cell to be maintained onto the newly synthesized DNA strands, and as a result, the methylation across the genome is stable through multiple rounds of cell division. Several studies have examined the importance of DNMT1, through its mutation in differentiated and embryonic stem cells from both mice and humans [46–48]. In mice, *Dnmt1* deficiency is embryonically lethal as cells lose almost all genomic DNA methylation marks. Mutations in mouse embryonic stem cells are more tolerated but their differentiation induces cell death. Similarly, deletion of DNMT1 is lethal to embryonic stem or differentiated cells in humans [49]. These studies highlight the importance of DNMT1 in preserving the heritability of DNA methylation programs, which constitutes a fundamental mechanism for epigenetic memory and genomic stability.

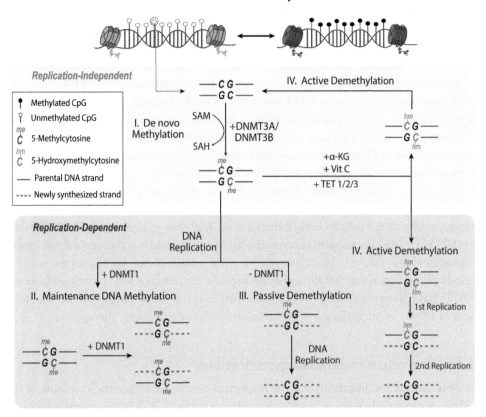

Fig. 1 Mechanisms of DNA methylation. I. De novo methylation, DNA methyltransferase 3A (DNMT3A) and DNMT3B add new methylation marks to unmodified cytosines, resulting in 5-methylcytosine (*red bases*). II. Maintenance methylation, during DNA replication, DNMT1 preferentially methylates semi-conservative DNA to maintain established methylation patterns. III. Passive demethylation, occurs in the absence of maintenance DNA methylation activity within the newly replicated DNA resulting in unmethylated DNA. IV. Active demethylation is directed by ten-eleven translocation (TET) proteins in either a replication-dependent or -independent manner. TET proteins mediate oxidation of 5-methylcytosine (5mC, *red bases*) to 5-hydroxymethylcytosine (5hmC, *green bases*).

Natural reprogramming of DNA methylation patterns can ensue throughout one's lifetime. These reprogramming events occur in response to various extrinsic or cell-intrinsic signals, such as the reported reprogramming events during normal embryonic development, oncogenic transformation, or cellular aging. DNA demethylation or hydroxymethylation can naturally occur to allow for "epigenetic reprogramming." Through either an active or passive demethylation process, methylated CpGs convert back to an unmethylated state that is often coupled with transcriptional reactivation. Active DNA demethylation refers to an active enzymatic process involving TET1, TET2, and TET3

DNA hydroxylases that oxidize 5-methylcytosine (5mC), converting it into 5-hydroxymethylcytosine (5hmC). Additional oxidations catalyzed by TET enzymes result in 5-formylcytosine (5fC) and 5-carboxylcytosine (5caC) [50, 51], which can be removed efficiently by thymine DNA glycosylase-mediated base excision repair [18]. In the context of immune cells, stimulated T-cells were shown to actively demethylate the IL-2 promoter-enhancer region in the absence of DNA replication [52]. TET enzymatic activity may also induce loss of DNA methylation marks through passive demethylation, due to DNMT1 having a reduced affinity to oxidized methyl groups during DNA replication [53], leading to a lack of maintenance methylation during cell division [51].

Dysregulation in DNA methylation can potentially play a role in driving cancer survival and progression. This is strikingly evident when, for example, the promoter regions of tumor suppressor genes are silenced through hypermethylation [54, 55]. The reversibility of DNA methylation presents with the logical approach to target DNMTs as a means of reprogramming tumor cells to induce their arrest. With this goal in mind, DNMT inhibitors (DNMTi) (*e.g.,* decitabine, azacytidine) are being studied as a therapeutic for various malignancies and are, so far, approved to treat some blood malignancies, such as myelodysplastic syndrome (MDS) and acute myeloid leukemia (AML) [56–58]. However, DNMTi have shown extensive toxicity, so further research on more specific, less toxic targets of DNA (de)methylation is an important area for epigenetic therapeutics.

2.2 Histone posttranslational modifications

Covalent histone modifications play an important role as an epigenetic mechanism that regulates temporal changes in gene expression programs. The basic repeating unit of chromatin is the nucleosome, which consists of DNA ($\sim$147 bp long) coiled around a histone octamer (a pair of four histone proteins-H2A, H2B, H3, H4) [59]. Nucleosomes are joined together by a linker histone known as H1. The "histone tails" are flexible and unstructured regions that protrude from the surface of the globular domain. These tails are important to nucleosome variability, as many of the amino acid residues, mostly on the N-terminus side of H3 and H4, are subject to extensive posttranslational chemical modifications including acetylation, methylation, phosphorylation, ubiquitination, and SUMOylation [60]. The diversity of histone posttranslational modifications (PTMs) accentuates the epigenetic complexity, which regulate how tightly DNA compacts around the nucleosome (Fig. 2B).

Similar to DNA (de)methylation, specific enzymes mediate the addition ("writers") or removal ("erasers") of histone PTMs (*e.g.,* acetyl, phosphoryl, and methyl groups). For example, the acetylation of key lysine residues of H3 and H4 is "written" by enzymes known as histone acetyltransferases (HATs) known to play a key role in transcriptional activation. The positively charged lysine residues at histone tails interact with the negatively charged phosphates of DNA, facilitating tight binding, and rendering the DNA largely inaccessible to transcription factors [61]. When specific lysine residues

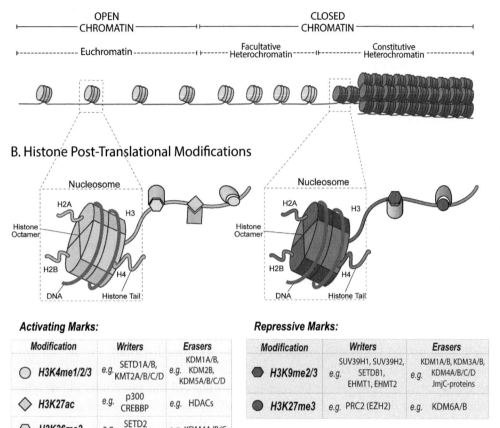

A. Chromatin Remodeling

B. Histone Post-Translational Modifications

Activating Marks:

Modification	Writers	Erasers
○ H3K4me1/2/3	e.g. SETD1A/B, KMT2A/B/C/D	e.g. KDM1A/B, KDM2B, KDM5A/B/C/D
◇ H3K27ac	e.g. p300 CREBBP	e.g. HDACs
⬡ H3K36me3	e.g. SETD2 NSD1	e.g. KDM4A/B/C

Repressive Marks:

Modification	Writers	Erasers
⬢ H3K9me2/3	e.g. SUV39H1, SUV39H2, SETDB1, EHMT1, EHMT2	e.g. KDM1A/B, KDM3A/B, KDM4A/B/C/D JmjC-proteins
● H3K27me3	e.g. PRC2 (EZH2)	e.g. KDM6A/B

Fig. 2 Chromatin remodeling and histone posttranslational modifications. (A) Chromatin remodeling: chromatin is organized into open or closed states. Euchromatin (open state; *green*) is accessible to DNA binding proteins and may be either transcriptionally active or inactive depending on which proteins are bound. Heterochromatin (closed state) is condensed and transcriptionally silenced. Heterochromatin is further classified into constitutive (*red*) and facultative heterochromatin (*yellow*) based on the level of accessibility. (B) Histone posttranslational modifications (PTMs): DNA is wrapped around nucleosomes that consist of two copies each of the histone proteins H2A, H2B, H3 and H4. The N-terminal tails of histone proteins extend from the nucleosome allowing for posttranslational modifications. Each chromatin state has specific histone posttranslational modifications that are associated with it—Note, this is not an exhaustive list of all H3 posttranslational modifications. Lysine (K) are subject to (de)methylation (me) and/or (de)acetylation (ac) by writer/eraser enzymes—examples depicted in tables [61]. Modifications: H3K4me1/2/3 (histone H3 lysine 4 mono/di/tri-methylation); H3K27ac (histone H3 lysine 27 acetylation); H3K4me3 (histone H3 lysine 4 trimethylation); H3K9me2/3 (histone H3 lysine 9 mono/di/tri-methylation); H3K27me3 (histone H3 lysine 27 trimethylation). Writers: SETD1A/B, histone-lysine *N*-methyltransferase; SETDB1, histone-lysine *N*-methyltransferase SETDB1; SUV39H1/2, histone-lysine *N*-methyltransferase SUV39H1/2; KMT, histone-lysine *N*-methyltransferase (2A/B/C/D); p300, histone acetyltransferase p300; CREBBP, CREB-binding protein; EHMT1/2, histone-lysine *N*-methyltransferase EHMT1/2; PRC2, Polycomb repressive complex 2. Readers: KDM, lysine-specific demethylase (1A/B, 2B, 3A/B, 4A/B/C/D, 5A/B/C/D, 6A/B); HDACs, histone deacetylases; JmjC, Jumonji domain-containing C proteins.

are acetylated, the positive charges are neutralized to weaken the interactions between histone tails and the DNA, thereby allowing the nucleosome to have more flexibility and a looser structure. After nucleosome shifting and chromatin decondensation, gene regulatory regions may become accessible to transcription factors, activating gene expression [62]. Alternatively, the action of histone deacetylases (HDACs) "erases" the acetylation and therefore represses transcription.

Histone methylation is another modification in which histone methyltransferases (HMTs) add methyl groups to histone tails, commonly on specific lysine or arginine residues in H3 and H4, which may result in an activating or repressing effect on transcription. In contrast, histone demethylases mediate the removal of histone methyl groups and are classified into two families: Lysine-specific demethylase 1 (LSD1) and Jumonji C (JmjC) domain-containing demethylases [61, 63–65]. Various effector proteins or "histone readers" such as bromodomains (*e.g.,* BRD2, BRD3, BRD4) and chromodomains (*e.g.,* HP1, CHD1, Pc), recognize and bind to histone PTMs, indirectly regulating gene expression. Both bromodomains and chromodomains, motifs that recognize acetylated histone residues [66–68] and methylated lysine residues on histone tails [61–63], respectively, facilitate downstream chromatin remodeling events.

Intriguingly, certain chromatin regions have histone PTMs that are bivalent, marked with seemingly antagonistic modifications. This chromatin can consist of both an inactivating mark, H3K27me3 (histone H3 lysine 27 trimethylation), and an activating mark, H3K4me3 (histone H3 lysine 4 trimethylation) [72]. These bivalent modifications are usually seen at promoters of key genes regulating cell differentiation and development. Owing to the presence of the H3K27me3 repressive marks, these genes are not expressed. However, during differentiation when the repressive marks are removed, genes become poised for rapid activation as the H3K4me3 permissive mark is already present [61].

Multiple protein complexes may coordinate changes in histone PTMs and the chromatin compaction. One example is the Polycomb complex group (PcG) that has been shown to play a pivotal role in gene repression. Both main classes of PcG proteins, polycomb repressor complexes PRC1 and PRC2 [73], can bind to specific promoters repressing their activity in a concerted manner. While PRC2 is responsible for the trimethylation of the H3K27 residues through its subunit EZH2, a histone methyltransferase (writer) [74], PRC1 complex binds to H3K27me3 through the chromodomain of CBX7 [75], resulting in chromatin compaction and gene silencing [76].

2.3 Chromatin remodeling and the "epigenetic code"

Chromatin is a highly dynamic structure, subject to remodeling into an open or closed state, as it interacts with relevant proteins and signals from upstream pathways.

Euchromatin (open state) is accessible to DNA-binding proteins and may be either transcriptionally active or inactive depending on which proteins are bound (transcriptional activators or repressors). Conversely, heterochromatin (closed state) is highly compact, condensed, and transcriptionally silenced. This condensed conformation protects the underlying DNA regions from damage or from being accessed and exploited by eager transcription machinery. Heterochromatin is further classified into constitutive and facultative heterochromatin (Fig. 2A). Facultative heterochromatin often constitutes genomic regions involving genes that must be silenced during development. The main characteristic of facultative heterochromatin is its flexibility to either adopt an open or closed conformation in a context-dependent manner. In contrast, constitutive heterochromatin is deemed to be in genomic regions that is usually void of genes and is considered more condensed and invariable than facultative heterochromatin [77–79]. Global epigenetic features of heterochromatin consisting of histone hypermethylation and hypoacetylation results in chromatin compaction.

The findings that histone marks help orchestrate interactions among writers, erasers, and readers to regulate dynamic changes in gene expression programs led to the concept known as the "histone code theory." In addition, accurate histone PTMs are critical to achieve proper chromatin remodeling. However, this "epigenetic code" underlying histone PTMs has been recently challenged [80]. For example, methylation of both DNA and histones together leads to repression of transcription. In this coordinated process, multiple levels of epigenetic control are activated to enforce gene silencing. Proteins that can recognize and bind to methylated DNA can also combine with the proteins engaged in the deacetylation of histones. Therefore, when DNA is methylated, nearby histones are also deacetylated, resulting in heightened inhibitory effects on transcription. Similarly, DNA demethylation does not attract histone deacetylating enzymes, thus permitting histone residues to remain acetylated, keeping transcription active. Furthermore, most studies that consider histone marks an epigenetic code were largely based on correlative analysis without showing a causal relationship. Recent studies have indicated that histone modifications previously associated with gene expression do not directly cause transcriptional activation. Additionally, many histone-modifying proteins retain noncatalytic functions that prevail over their enzymatic roles [81]. Overall, these findings have challenged the instructive function of histone modifications for transcriptional regulation [80]. A broader concept of an "epigenetic code" involving complex interactions of DNA methylation, the histone code, and other epigenetic modifications is now more recognized for regulating transcriptional programs [82].

2.4 Noncoding RNA

With advanced high-throughput RNA-sequencing technologies, studies have shown that more than 70% of the human genome is transcribed into RNA molecules, including messenger RNAs (mRNA) that can be translated into proteins, and nonprotein-coding RNAs [83]. Noncoding RNAs (ncRNAs) comprise heterogeneous groups of house-keeping (*e.g.,* ribosomal RNAs, transfer RNAs) and regulatory RNA molecules with various biological functions, such as chromatin remodeling, transcriptional and posttranscriptional regulation. Based on their length, regulatory ncRNAs can be classified into small ncRNAs; microRNAs (miRNAs); and long noncoding (lnc) RNAs that have over 200 nucleotide-length. Growing evidence has indicated that some ncRNAs play important roles in gene regulation through interactions with the epigenetic machinery. They can recruit and interact with histone-modifying complexes or modulate the activity of DNA methyltransferases, influencing heterochromatin formation and gene silencing. For example, X-inactive specific transcript (Xist) is a lncRNA [84] that accumulates the entire length of one female X-chromosome. This results in the recruitment of Polycomb group complexes (PcG), such as PRC2, triggering extensive histone methylation and transcriptional silencing across the X chromosome. In contrast, through a complementary nucleotide sequence, microRNAs and small interfering RNAs (siRNAs) bind to a specific target messenger RNA, inducing cleavage, degradation, or inhibition of mRNA translation into proteins [85, 86]. In addition to targeting mRNA for degradation or translational silencing, small noncoding RNAs can also assemble into a separate RNA-induced transcriptional silencing (RITS) complex that uses the small ncRNA to direct the complex to the transcription site of a target mRNA, silencing gene expression [87].

3. Overview of cancer-immunity cycle

Cancer is a complex progressive disease that involves continuous crosstalks among cancerous cells and tumor-infiltrating immune and stromal cells. A puzzling question in the field of immuno-oncology is how tumors avoid destruction despite the heterogeneous populations of immune cells observed in many human tumors [88]. Although the concept of immune surveillance against pathogens has been early accepted, the role of the immune system in cancer progression *versus* eradication remained controversial in the last century. Extensive efforts over the last three decades have revealed multiple aspects of the cancer-immunity cycle and how the dual host-protective and tumor-promoting actions of the immune system shape cancer disease outcomes [89].

During early stages of tumor development, neo-antigens expressed by transformed cells and danger signals released from dying cells at the tumor site trigger the first phase of antitumor immune responses [3]. At this phase, coordinated groups of innate and adaptive tumor-suppressing immune cells are activated and recruited to the tumor site. If properly stimulated and accumulated within the arising tumor, immune cells will eliminate the most immunogenic cancerous cells. This antitumor response is mainly mediated by cytotoxic CD8+ T-cells, helper CD4+ T-cells, natural killer (NK) cells, and inflammatory macrophages [3]. However, less immunogenic tumor cell variants may not be completely eliminated, as the tumor progresses into an equilibrium phase where tumor outgrowth is kept under control by the host antitumor immune response. In this case, the tumor may remain clinically unapparent as long as the tumor-specific immunity remains intact in the host. In some cases, the tumor may eventually evolve to escape the immune response by undergoing immunoediting or developing immunosuppressive TME. This late phase is characterized by accumulation of immunosuppressive cells within the TME, such as regulatory CD4+ T-cells (Tregs), myeloid-derived suppressor cells (MDSCs), and anti-inflammatory macrophages [3]. In this section, we will highlight key transcriptional programs and functional changes within tumor-suppressing and tumor-promoting immune cell populations in the context of cancer development. We will then discuss how epigenetic programming plays a central role in imprinting these transcriptional programs within immune cells in response to the dynamic evolution of the TME.

3.1 Antitumor immune cells

3.1.1 CD8+ T-cells

CD8+ T-cells are central mediators of the adaptive antitumor immunity. They are highly skilled at effectively surveilling and detecting tumors by recognizing specific antigens displayed on the surface of the tumor cells on major histocompatibility complex class-1 (MHC-1) molecules. First, upon priming and activation in the draining lymph nodes, naïve CD8+ T-cells differentiate into effector cytotoxic T-lymphocytes (CTLs) which migrate to the TME. Infiltrating CTLs directly kill immunogenic tumor cells [90, 91] via the release of granules containing granzymes and perforin, or by FasL-mediated apoptosis [92–96].

However, malignant tumors exploit tolerance pathways that regulate and avert uncontrolled effector T-cell immune responses that cause host immunopathology and/or autoimmunity [97]. Under persistent antigen stimulation and chronic inflammation present within the TME, CTLs progressively lose effector function, developing T-cell exhaustion [98–102]. This environment leads to a sustained overexpression of inhibitory receptors dampening tumor-killing activity, a hallmark of exhausted T-cells (Tex) [103]. Further transcriptional and epigenetic studies have revealed that T cell exhaustion is a distinct differentiation state of antigen-specific CD8+ T-cells that develop during cancer and chronic infections, exhibiting features unique from other T-cell states, such as naïve, effector, or memory.

3.1.2 T_H1 CD4+ T-cells

Depending on their microenvironment, CD4+ "helper" T-cells display ample phenotypic plasticity as they differentiate into multiple different subsets (*e.g.,* T_H1, T_H2, T_H17). T_H1 CD4+ T-cells "help" enhance the effector function of other immune cells [104,

105]. Upon activation and recognition of antigens presented on MHC class II molecules, naïve CD4+ T-cells secrete IL-2 to activate cytotoxic CD8+ T-cells, inducing their antitumor effector functions. CD4+ T-cells can also indirectly help CD8+ T-cell response by activating proinflammatory dendritic cells (DCs) [106]. Signals from DCs, notably IL-12, promote T_H1 CD4+ T-cells to produce proinflammatory cytokines, such as IFNγ and TNFα, that allow for direct antitumor effects [107].

3.1.3 Natural killer cells

"Natural killer" or NK cells act as an antitumor immune defense, particularly when tumor cells evade T-cell recognition. However, NK cells are closely tied to adaptive immunity. Similar to effector CD8+ T-cells, NK cells induce natural cytotoxicity against primary tumor cells, protecting the host from further metastatic growth. NK cell's effector function depends on the surveillance of inhibitory and stimulatory signals from target cells, including tumors and virus-infected cells [3, 108]. NK cell inhibitory receptors recognize tumor cells that lack MHC-I and exert their killing effects through the release of perforin and granzyme B, which induces apoptotic cell death [109]. They can also recognize antibodies coating target tumor cells to induce cell death through antibody-dependent cellular cytotoxicity (ADCC) [110].

In addition to their cytotoxic activity against tumor cells, NK cells modulate the adaptive antitumor immune response through the production of IFNγ that promotes CD4+ T-cell differentiation toward T_H1 polarization [111, 112]. However, within the TME, chronic inflammation and persistent exposure of NK cells via NKG2D receptor to ligands expressed by tumor cells results in dysfunction of NK cells. Furthermore, inhibitory cytokines (*e.g.,* IL-10 and TGF-β), produced by tumors and other tumor-promoting immune cells, create a suppressive microenvironment favoring tumor cells that subvert NK-cell control. These "selected" tumor cells downregulate their expression of NKG2D ligands, impairing NK cell recognition [3, 113, 114].

3.2 Tumor-promoting immune cells

3.2.1 Regulatory CD4+ T-cells

CD4+ regulatory T-cells (Tregs) are important mediators of peripheral immune tolerance through their suppressive effects on the activation and proliferation of CD4+ and CD8+ T-cells [115]. Yet, Tregs may also drive antitumor immune dysfunction through multiple mechanisms including: (a) secretion of anti-inflammatory cytokines (*e.g.,* TGF-β, IL-10); (b) inhibition of DC-derived costimulatory signals; (c) IL-2 deprivation in the TME; and (d) direct killing of effector antitumor immune cells [116–118]. Indeed, the accumulation of Tregs within the TME has been strongly correlated with tumor progression in mice and humans [119, 120].

3.2.2 Tumor-associated macrophages and myeloid-derived suppressor cells

Similar to Tregs, the accumulation of tumor-associated macrophages (TAMs) promotes the immunosuppressive TME. Tumor-derived chemokines and signals such as CCL2, VEGF, and M-CSF induce the recruitment of immune cells and promote TAM polarization into an M2-like phenotype [121]. TAMs may directly or indirectly contribute to T-cell dysfunction through inhibitory receptor engagement and/or secreted anti-inflammatory cytokines [122]. In addition to tumor-promoting macrophages, myeloid-derived suppressor cells (MDSCs) represent a heterogeneous group of myeloid progenitor cells that can also accumulate within the TME and inhibit T-cell function [123] through: (a) secretion of TGF-β and IL-10 [124]; (b) depletion of amino acids (arginine/cysteine) that are essential for appropriate T-cell function [125]; (c) promoting the immunosuppressive activity of Tregs [126]; and/or (d) reduced ability to present tumor antigens.

Both TAMs and MDSCs can induce multiple common ways of immunosuppression. MDSCs and M2 macrophages secret high levels of IL-10 and TGF-β [122, 123, 127]; express PD-L1 to bind the inhibitory receptor PD-1 on CD8+ T-cells inducing T-cell exhaustion; and can directly inhibit CD8+ T-cells cytotoxicity by producing high levels of amino acid–degrading enzymes, such as arginase 1 and indoleamine-2,3-dioxygenase (IDO) [128, 129]. In addition, elevated levels of reactive oxygen species (ROS) in the TME are produced by MDSCs, TAMs, and other cells causing oxidative stress when interacting with T-cells [130]. Collectively, any or a combination of these repressive actions occurring against antitumor T-cells in the TME contributes to the immunosuppression conditions favoring tumor survival, development, and immune evasion, ultimately promoting the exhaustion of CD8+ T-cells [99, 131, 132].

4. Epigenetic regulation of immune cells in the tumor microenvironment

As innate and adaptive immune cells become activated by the presence of a pathogen or tumor, they undergo epigenetic alterations while mounting their effector functions. Each step of immune cell differentiation following its activation may involve drastic phenotypic and functional changes. This activation-induced transcriptional reprogramming is primarily driven by epigenetic modifications in the chromatin that impact interactions between cell-type-specific transcription factors and their target genes. The activation state of immune cells is regulated by the nature, magnitude, and/or duration of extrinsic signals. Various epigenetic mechanisms within the responding cells integrate these signals to calibrate the cell's transcriptional activity and possibly establish memory of the microenvironmental cues, which may dictate cell-fate commitment.

The chronic inflammatory nature of TME in many human cancers shares similarities with the inflammatory microenvironment during chronic virus infections [88, 133].

Both diseases are characterized by prolonged exposure to either virus-derived or tumor-derived antigens. Given these similarities, most of our understanding of the epigenetic regulation of antitumor immune responses was extrapolated from animal and human studies of chronic virus infections [98]. The lymphocytic choriomeningitis virus (LCMV) infection model in mice has been a highly valuable and tractable experimental tool that allows controlled analysis of innate and adaptive immune cell activation, particularly CD8 + T-cells, under settings of acute *versus* chronic antigen stimulation [134]. Many discoveries based on this model system have significantly impacted our understanding of human immunology. Examples include the discovery of MHC restriction [135], the identification of memory and exhausted antiviral T-cell immune responses [136–138], and development of anti-PD1 checkpoint blockade therapy [103]. Yet, it remains largely unknown how the unique features of TME (*e.g.,* hypoxia, nutrient depletion, acidic pH, inflammatory signals) contribute to the epigenetic programming of dysfunctional antitumor immune cells.

4.1 Epigenetic regulation of CD8+ T-cells

Studies of functional CD8 + T-cells generated during acute infections have provided insights into understanding the transcriptional and epigenetic regulation of T-cell dysfunction during chronic infections or cancer. We will highlight how epigenetic changes regulate distinct differentiation pathways of CD8 + T-cells during acute *versus* chronic antigen stimulations.

4.1.1 During acute antigen exposure

During an acute virus infection, antivirus CD8 + T-cell responses can effectively clear all virus-infected cells, resulting in a quick recovery of the host. During the first wave of the innate immune response, professional antigen-presenting cells, such as DCs, migrate from the infection site to the draining lymph nodes to stimulate the more selective, adaptive immune responses. To initiate antivirus CD8 + T-cell immune response, naïve virus-specific CD8 + T-cells must first engage their T-cell receptors (TCR) with the MHC I-cognate peptide complex expressed on the surface of DCs. Following proper activation and co-stimulation, naïve T-cell undergoes clonal expansion accompanied by extensive changes in transcriptional and epigenetic programs, generating what are known as cytotoxic T lymphocytes. These effector T cells quickly migrate to the inflamed infection site, where they exhibit selective highly cytotoxic activity against virus-infected target cells. The majority (> 90%) of effector CD8 + T-cells are terminally differentiated (defined as terminal effectors—TE), as they undergo cell death upon successful clearance of the antigen source. In contrast, a minor subset of effector CD8 + T-cells, defined as memory precursors (MP), continues differentiation generating distinct populations of memory CD8 + T-cells. These antigen-experienced T-cells persist in a quiescent state while maintaining a transcriptional memory of the effector program.

Thereby, memory T-cells sustain a poised state to quickly recall their effector function and clear the pathogen more effectively than naïve T-cells upon future antigen re-exposures [139–142]. These unique features of memory T-cells are mainly mediated by a stable epigenetic memory acquired during CD8+ T-cell differentiation.

The phenotypic transitions CD8+ T-cells undergo during an acute infection are driven by epigenetic changes impacting the gene expression pattern of each state. Studies examining changes in DNA methylation and histone modifications highlighted the significance of epigenetic mechanisms during different stages of T-cell differentiation (Table 1). During naïve-to-effector transition, CD8+ T-cells undergo extensive remodeling in DNA methylation programs, histone modifications, and the open chromatin landscape, which can be categorized into two major patterns: (a) On-Off changes in genes encoding naïve or memory-related molecules that are downregulated in effector T-cells, such as naïve T-cell transcription factors (*e.g.*, *Tcf7*, *Lef1*, *Bach2*, *Id3*, *Foxp1*) and lymphoid tissue homing markers and pro-survival molecules (*e.g.*, *Ccr7*, *Sell*, *Il7ra*); (b) Off-On changes in genes encoding effector function programs that are acquired in stimulated CD8+ T-cells, such as *Ifnγ*, *Tbx21*, *Eomes*, *Id2*, *Myc*, *Gzmb*, *Prf1*, and *Tnfα* (Table 1). Generally speaking, the On-Off epigenetic changes involve de novo DNA methylation and/or increased deposition of repressive histone modifications (*e.g.*, H3K27me3). These changes may be coupled with reduced chromatin accessibility at gene regulatory regions, such as promoters and enhancers. In contrast, Off-On epigenetic programming can be enforced by DNA demethylation, increased deposition of permissive histone PTMs (*e.g.*, H3K4me1/3, H3K27Ac), and/or loss of repressive histone marks, coupled with gain of chromatin accessibility [139, 140, 146, 158]. This global epigenetic remodeling enforces the effector differentiation of acutely stimulated CD8+ T-cells.

Memory T-cell subsets are defined by their anatomic residency features into three major subsets: (a) central memory T-cells (T_{CM}) that mainly reside in lymphoid tissues; (b) effector memory T-cells (T_{EM}) that circulate outside the lymphoid organs in the blood and peripheral tissues; and (c) tissue-resident memory T-cells that localize in mucosal tissues. Another memory subset was recently identified as stem cell-like memory CD8+ T-cells (T_{SCM}) that retain high degrees of stemness and self-renewal capacity [159, 160]. These memory T-cell subsets exhibit some transcriptional differences (*e.g.*, tissue homing ability, proliferation potential, metabolic features). Yet, they share common characteristics, including their ability to self-renew and rapidly recall their effector function upon a second exposure to the same antigen [161].

Several differentiation models have been proposed to describe the origin of memory CD8+ T-cells. We will briefly discuss our current understanding of CD8+ T-cell differentiation in the context of two models based on recent epigenetic and transcriptional studies. In the first model that we described below as "Bifurcated Progressive Differentiation" model, memory CD8+ T-cells arise from

Table 1 Transcriptional and epigenetic changes in key genes regulating CD8+ T-cell differentiation: dynamic changes in gene expression and epigenetic programs of key genes expressed by CD8+ T-lymphocytes at each differentiation stage: naïve (N), effector (E), memory (M) and exhausted (Exh).

Gene	Gene Expression (*upon TCR stimulation)				DNA Methylation ↑ - Hypermethylated ↓ - Hypomethylated ↑↓ Varied in cell subsets				Chromatin Accessibility				Histone PTMs (+) Activating (+) Repressive		
Transcription factors	N	E	M	Exh	N	E	M	Exh	N	E	M	Exh	N	E	M
Tbx21 (Tbet)					↑	↓	↑↓	↑	Closed	Open	Open	Intermediate	+	+	+
Tcf7 (Tcf1)					↓	↑	↑↓	↑	Open	Closed	Open	Closed	+	+	+
Eomes					↑	↑↓	↑↓	↓	Closed	Open	Intermediate	Open	+	++	+
Batf					↑	↓	↑↓	↓	Closed	Open	Intermediate	Open	++	+	+
Tox					↑↓	↑↓	↑	↓	Open	Closed	Intermediate	Open	++	+	+
Effector Cytokines	N	E	M	Exh	N	E	M	Exh	N	E	M	Exh	N	E	M
IFNγ		*			↑	↓	↓	↑	Closed	Open	Open	Closed	+	+	+
TNFα		*			↑	↓	↓	↑	Closed	Open	Open	Intermediate	++	+	+
IL-2		*			↑	↓	↓	↑	Closed	Open	Open	Closed	++	++	+
Gzmb					↑	↓	↓	↓↑	Closed	Open	Open	Intermediate	+	+	+
Prf1					↑	↓	↓	↓↑	Closed	Open	Open	Closed	++	+	+
Inhibitory Receptors	N	E	M	Exh	N	E	M	Exh	N	E	M	Exh	N	E	M
Pdcd1 (PD-1)					↑	↓	↑	↓	Closed	Intermediate	Closed	Open	++	+	++?
CTLA-4					↑	↓	↑	↓	Closed	Open	Intermediate	Open	+	+	+
Lag3					↑	↓	↑↓	↓	Closed	Open	Closed	Intermediate	+	+	+
Havcr2 (Tim3)					↑	↓	↑↓	↓	Closed	Open	Intermediate	Open	+	+	++?
Lymphoid Tissue Homing	N	E	M (TCM)	Exh	N	E	M	Exh	N	E	M	Exh	N	E	M
Ccr7					↓	↑	↑↓	↑	Open	Closed	Intermediate	Closed	+	+	+
Sell (CD62L)					↓	↑	↑↓	↑	Open	Closed	Intermediate	Closed	+	+	+
Chemokine Receptors	N	E	M (TCM)	Exh	N	E	M	Exh	N	E	M	Exh	N	E	M
Cxcr5					↑	↑↓	↑	↑↓	Closed	Open	Closed	Intermediate	++?	+	+
Cx3cr1					↑	↑↓	↑↓	↑	Closed	Open	Open	Intermediate	+	+	+
Ccr5					↑	↑↓	↑↓	↑	Closed	Open	Open	Intermediate	+	+	+
Cytokine Receptors	N	E	M	Exh	N	E	M	Exh	N	E	M	Exh	N	E	M
IL7r					↓	↑	↓	↑	Open	Intermediate	Open	Closed	+	+	+
IL2ra					↑	↓	↓	↑	Closed	Open	Open	Closed	+	+	+

Changes in gene expression was determined using RNA-seq data and protein expression analyses upon TCR stimulation [143–145]. Direction of line indicates upregulation, downregulation, or maintained expression during naïve–effector (*black*), effector–memory (*green*), or effector–exhausted (*red*) T-cell transitions. Changes in DNA methylation were measured using published epigenomic datasets [139, 146–149]. *Red upward arrow* represents hypermethylation, *blue downward arrow* represents hypomethylation, *dual arrows* represent varied methylation of CpGs in the gene locus. Changes in chromatin accessibility were assessed from ATAC-seq data [150–153], while changes in histone posttranslational modifications (PTM) data were examined from published ChIP-seq datasets [154–157]. *Blue symbols* represent activating PTM marks; *red symbols* refer to repressive PTM marks. Addition of histone PTM represented by (+), removal of mark (−).

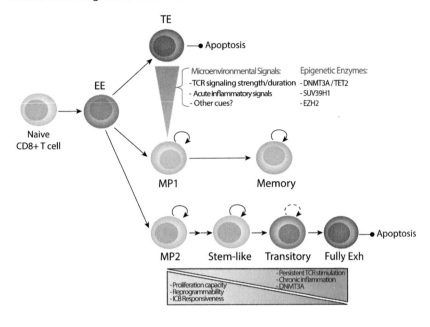

A- Bifurcated Progressive Differentiation Model

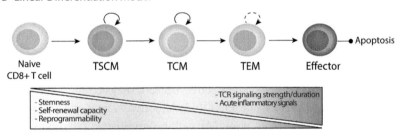

B- Linear Differentiation Model

Fig. 3 Models of CD8+ T-cell differentiation. (A) Bifurcated progressive differentiation model: EE, early effector; TE, terminal effector; MP, memory precursor CD8+ T-cells developed during acute (MP1) or chronic infections and cancer (MP2). (B) Linear differentiation model: T_{SCM}, stem cell memory T-cells; T_{CM}, central memory T-cells; T_{EM}, effector memory T-cells; T_{EFF}, effector CD8+ T-cells.

a minor subset of effector cells called MPs (Fig. 3A). When the pathogen is cleared, TCR stimulation diminishes, resulting in cell death of the TE subset. However, the MP subset survives and generates memory cells. During effector-to-memory transition, MP cells undergo further epigenetic modifications concomitant with re-expression of memory-associated genes, while maintaining the epigenetic programs of effector genes.

The above-described model has been recently supported by several molecular studies in mice and humans. In an elegant human study, the authors did *in vivo* deuterium-based labeling of antigen-specific human CD8+ T-cells after live yellow fever virus (YFV) vaccination [141]. This method enabled longitudinal analyses of transcriptional and epigenetic changes within blood-circulating YFV-specific CD8+ T-cells over 10 years following vaccination. This prolonged investigation has demonstrated that: (a) human memory CD8+ T-cells arise from proliferating effector CD8+ T-cells; (b) memory T-cells retain an epigenetically poised, DNA demethylated state at the effector function genes (*e.g.*, *GZMB, PRF1, IFNγ*); and (c) memory T-cells turn back their epigenetic clock to re-express naïve/memory-associated genes (*e.g.*, *IL7R, CCR7*) that were initially silenced during the effector stage of T-cell differentiation [141]. These patterns of transcriptional and epigenetic changes during the effector-to-memory transition were also shown to be driven by epigenetic dedifferentiation in mice after acute LCMV infection [142].

An alternative model of CD8+ T-cell differentiation describes a "Linear Differentiation" of naïve-to-memory cells, in which memory CD8+ T-cells arise from naïve T-cells and prior to effector T-cell differentiation (Fig. 3B). This model places memory T-cells as an intermediate step along T cell differentiation pathway (Naïve → T_{SCM} → T_{CM} → T_{EM} → Effector). This model has been mainly based on the observed phenotypic and transcriptional similarities between memory cells and both naïve and effector cells, and the possibility that memory cells are masked during the acute phase of the immune response by the massive expansion of effector CD8+ T-cells [27]. Indeed, global DNA methylation and histone PTMs profiling revealed that human T_{SCM} cells are epigenetically closer to and resemble naïve CD8+ T-cells [159, 160, 162]. Although this model is plausible, it cannot explain the recently discovered role of de novo DNA methylation in regulating memory CD8+ T-cell differentiation. DNA methylation analysis of effector CD8+ T-cell subsets (MP and TE) demonstrated that MP cells do not maintain the hypomethylated state of naïve/memory-related genes. Instead, they acquire DNA methylation programs at these genes, while undergoing DNA demethylation at effector-related genes. After resting in antigen-free environment, MP cells progressively undergo DNA demethylation at the epigenetically silenced naïve genes, allowing their re-expression in the developed memory T-cells [142, 147]. In addition, studies have demonstrated that conditional deletion of Dnmt3a in activated T-cells inhibits epigenetic silencing of naïve-associated genes in effector CD8+ T-cells, allowing faster development of T_{CM} cells [142, 147]. These findings suggest that memory T-cells arise after effector T cell differentiation, and that de novo DNA methylation plays an important role in regulating the progressive dedifferentiation of MP cells into memory T-cells.

Although resting memory CD8+ T-cells no longer express high levels of effector molecules, epigenetic analysis of the effector loci show that they remain accessible to transcriptional factors. The promoter regions at some effector loci may gain DNA methylation or repressive histone PTMs during the memory state [163]. Yet, upon TCR stimulation these epigenetic marks are erased faster in memory than in naïve CD8+ T-cells. These findings highlight the unique maintenance of the recall effector function in memory T-cells (Table 1).

Investigating the epigenetic regulation of functional memory CD8+ T-cell differentiation has provided important insights for understanding how epigenetics may regulate antitumor T-cell immunity. In the following section, we will discuss the implications of recent epigenetic studies of dysfunctional CD8+ T-cells while addressing the following questions:

(a) Do CD8+ T-cells undergo different epigenetic programming under settings of prolonged antigen stimulation and inflammation?

(b) Do epigenetic mechanisms regulate the biology of T-cell dysfunction during cancer or chronic infections?

(c) Can we manipulate epigenetic programs in exhausted CD8+ T-cells to enhance their antitumor activity and response to immunotherapies?

4.1.2 During chronic antigen exposure

The ultimate outcome of CD8+ T-cell differentiation is impacted by the strength, amount, and/or duration of antigen stimulation and inflammatory signals in the milieu. While MP1 CD8+ T-cells differentiate into memory cells following antigen clearance in acute infections, a similar subset—MP2 cells—progressively lose effector and memory potential under prolonged antigen stimulation during cancer or chronic infections (Fig. 3). These dysfunctional or exhausted T-cells exhibit unique features, such as poor production of effector cytokines (*e.g.*, IL-2, TNFα, IFNγ), metabolic dysregulation, sustained upregulation of surface inhibitory receptors (*e.g.*, PD-1, CTLA-4, TIM-3), impaired homeostatic proliferation, and reduced cytokine-dependent survival [98]. Transcriptional studies have revealed remarkable heterogeneity within exhausted T-cells linked to the progressive nature of T-cell-exhaustion development. These heterogenous exhausted T-cell populations can be divided into three main subsets: (a) Partially exhausted or "stem-like" progenitor T-cells (phenotypically defined as PD-1int TCF1+ CXCR5+ LY108+ TIM-3− CD101 −) that retain effector function, metabolic fitness, and higher proliferation capacity; (b) Transitory cytolytic T-cells (PD-1+ TBET+ CXCR5− CX3CR1+ TIM-3+ CD101 −) that exhibit higher cytotoxic activity and retain proliferation potential; and (c) Fully exhausted T-cells (PD-1high TOX+ CXCR5− CX3CR1low TIM-3+ CD101 +) that show signs of severe exhaustion with complete loss of effector cytokine production and proliferative potential, in addition to metabolic dysfunction [27, 143, 164, 165]. Importantly, stem-like progenitor and transitory subsets of exhausted T-cells provide proliferative burst after ICB therapy, while fully exhausted cells remain refractory to rejuvenation [143, 164, 166]. As such, terminal T-cell exhaustion remains a major barrier to T-cell-based cancer immunotherapies [4, 26, 98].

We and others have recently shown that exhausted T-cells acquire unique DNA methylation programs and open chromatin landscape during chronic virus infection or cancer [146, 150–153]. Epigenetic profiling studies have revealed distinct chromatin accessibility changes coupled to the development of T-cell exhaustion [150–153]. Similar to the extensive chromatin remodeling observed during naïve-to-effector CD8+ T-cell transition in acute infections, CD8+ T-cells undergo massive reprogramming of their open chromatin landscape during the early stage of tumors or chronic infections. Effector function-related genes become accessible in early generated antitumor CD8+ T-cells coupled with transcriptional activation. In contrast, naïve-related programs are down-regulated and lose chromatin accessibility (examples are shown in Table 1). Despite these general similarities in effector cells during acute and chronic stimulations, early tumor-infiltrating CD8+ T-cells (TILs) exhibits many unique open chromatin peaks [151]. For example, enhancer peaks in the *IFNγ* locus, that become more accessible during effector differentiation, were found inaccessible in TILs. Interestingly, a unique peak ∼23 kb upstream to the *Pdcd1* gene promoter (encoding PD-1) was found open in mouse tumor-infiltrating or chronically LCMV-stimulated CD8+ T-cells [150–153]. This finding suggests a regulatory role of this region in enforcing the upregulation of PD-1 on exhausted T-cells.

As chronically stimulated CD8+ T-cells become dysfunctional, they progressively divert from the epigenetic route of memory differentiation (Fig. 3A, Table 1). During persistent antigen exposure, exhausted T-cells lose chromatin accessibility in many genes regulating effector function (*e.g.*, *IFNγ*, *Tbx21*—encoding Tbet, *Myc*) while maintaining the inaccessibility of naïve-associated genes (*e.g.*, *Sell, Tcf7, Lef1, Ccr7*). Counterintuitively, they also continue to gain more open chromatin peaks during the effector-to-exhausted transition [150–153]. These open chromatin patterns are linked to several features underlying terminal exhaustion, including: (a) heightened expression of inhibitory receptors is coupled with more open peaks in their corresponding genes (*e.g.*, *Pdcd1, Havcr2*—encoding Tim-3, *Lag3*); (b) increased expression of terminal differentiation and cell-death-related genes (*e.g.*, *Batf, Eomes, Fasl*); and (c) reduced metabolic fitness-associated genes [167]. Interestingly, the HMG-box Tox transcription factor was recently discovered to be uniquely upregulated in MP2 cells and required for the generation and maintenance of partially exhausted T-cells in chronic infections or tumors [168–172]. Further epigenetic analysis of the *Tox* locus has revealed distinct open chromatin peaks and DNA demethylation events in Tex cells [168]. Importantly, genetic deletion of Tox initially enhanced effector function in CD8+ T-cells, but eventually induced a massive quantitative loss of Tex cells, implying that targeting TOX may not be an effective therapeutic approach, given its counteractive functions in exhaustion and survival. These findings suggest that changes in open chromatin landscape have a regulatory function in exhausted T-cells. Yet, it is not clear how this epigenetic landscape is maintained during exhaustion, and whether specific changes in chromatin accessibility impact the biology of exhausted T-cells. It also remains unknown whether the reported chromatin accessibility changes are causal in the development of T-cell exhaustion or merely coupled to transcriptional reprogramming of chronically stimulated CD8+ T-cells.

DNA methylation plays a fundamental role in establishing the epigenetic landscape of dysfunctional T-cell populations during chronic stimulation. We discovered that Dnmt3a-mediated de novo DNA methylation enforces silencing of effector and memory/stemness-related genes and promotes terminal T-cell exhaustion [146]. Our work has shown that conditional deletion of Dnmt3a (*Dnmt3a* cKO) in activated CD8 + T-cells blocked their progression toward the full-exhaustion state. Furthermore, Dnmt3a-deficient antigen-specific CD8 + T-cells were exhaustion-resistant during the CD4 + T-cell-helpless chronic LCMV infection—an animal model of severe T-cell exhaustion [146]. At the mechanistic level, antigen-specific cKO CD8 + T-cells maintained higher quantities while retaining their capacity to recall effector cytokines despite prolonged TCR stimulation. In addition, Dnmt3a-deficient CD8 + T-cells continued to express higher levels of stemness-associated transcription factors, such as Tcf1 and Lef1. These data established de novo DNA methylation programming as a novel, cell-intrinsic mechanism that is causally linked to T-cell exhaustion.

Further examination of de novo DNA methylation and its role in T-cell dysfunction has shown that during the effector-to-exhausted transition, virus-specific T-cells gain ~1200 new methylation programs including in genes related to effector function, such as *IFNγ* [146]. Alternatively, memory T-cells developed after acute infection maintain demethylation at the *IFNγ* locus, thereby quickly expressing IFNγ upon re-stimulation. DNA methylation gained at loci related to cell-cycle genes, such as *Myc*, may underlie the limited the proliferative capacity of exhausted T-cells. Additionally, Tcf1 is critical for the self-renewal capacity of CD8 + T-cells. In contrast to memory T-cells that demethylate the *Tcf7* locus during effector-to-memory differentiation, exhausted T-cells maintain the de novo methylation program acquired during the effector stage of the immune response [146]. As observed during chronic virus infections, de novo DNA methylation also contributes to the dysfunctional phenotype of PD-1 + TILs. Tumor-specific CD8 + T-cells isolated from cancer models in mice acquire similar methylation marks in loci such as *IFNγ*, *Myc,* and *Tcf7*, leading to decreased expression of markers associated with effector function, proliferation, and memory differentiation [146].

Similar to the distinct open chromatin features observed at inhibitory receptors, a unique DNA demethylation state was identified at regulatory regions of the *PDCD1* locus in human and mouse CD8 + T-cells during chronic HIV or LCMV infections, respectively [148, 173, 174]. This unmethylated program was reported to be imprinted early during chronic infection in effector CD8 + T-cells [174]. The *PDCD1* locus contains several CpG sites that remain primarily methylated in naïve T-cells, then undergo demethylation following strong TCR activation, allowing PD-1 expression. During acute infections, the locus becomes remethylated as CD8 + T-cells entering the memory phase downregulate PD-1. However, during chronic virus infections, exhausted T-cells maintain the unmethylated state of the *PDCD1* locus [148]. *Pdcd1* demethylation and consequent PD-1 expression is strongly correlated to the TCR signaling strength, with higher and more prolonged viral loads linked to more demethylation and higher accessibility at the *Pdcd1* locus. Even with diminishing viral load, the region remains unmethylated in exhausted T-cells, allowing them to rapidly re-express PD-1 upon future TCR stimulation [148]. Therefore, chronic stimulation has a profound impact on the epigenetic programming of CD8+ T-cells, driving the exhaustion state rather than memory differentiation.

At the level of chromatin organization, special AT-rich sequence-binding protein-1 (SATB1) has an important role in regulating chromatin looping and reorganization during T-cell differentiation [175–177]. SATB1 was found to recruit nucleosome remodeling deacetylase complex to genomic regions, and preferentially target closed chromatin by accessing specific nucleosome-embedded motifs [178]. It orchestrates essential spatiotemporal chromatin remodeling during T-cell development and helper CD4+ T-cell differentiation [143, 163, 164]. Recently, SATB1 was reported to control PD-1 expression through binding to regulatory regions of the *Pdcd1* locus [177]. This interaction results in the removal of the permissive histone acetyl marks, limiting PD-1 expression on early activated T-cells. However, within the TGF-β-rich TME, SATB1 is epigenetically silenced in TILs, unleashing its repressive effect on PD-1 expression. Furthermore, phosphorylated Smad2/3 downstream TGF-β signaling binds to *Pdcd1* enhancers promoting PD-1 upregulation [177]. These findings indicate that multiple epigenetic mechanisms crosstalk to enforce the dysfunctional state of CD8+ T-cells during cancer.

4.2 Epigenetic regulation of NK cells

Alike CD8+ T-cells, antitumor activity of NK cells is regulated by epigenetic programming. IFNγ is an important effector molecule that mediates NK cell function. Notably, its expression was shown to be epigenetically regulated in activated NK cells by histone PTMs that allow binding of Tbet transcription factor to a conserved noncoding sequence (CNS) region upstream of the *IFNγ* locus. This CNS incorporates histone acetylation marks that enhance chromatin accessibility of the *IFNγ* locus in both T- and NK cells [179]. Additionally, IL-2 was reported to enhance histone acetylation marks at a distal region promoting IFNγ expression in NK cells [180]. These studies suggest that, in response to upstream signals, NK cell-effector function is intrinsically regulated at the epigenetic level. In contrast, under chronic stimulatory conditions, NK cells have shown limited cytotoxic activity, reduced ability to produce cytokines, and increased expression of inhibitory receptors during chronic infections [181, 182] and cancer [183, 184]. Such NK cell dysfunction has been also linked to epigenetic remodeling. Chronic infection of human cytomegalovirus (hCMV) is known to drive expansion of NK cells expressing the NKG2C activating receptor in infected patients. Genome-wide DNA methylation analysis of NKG2C+ NK cells from hCMV-infected donors has revealed that many loci undergo epigenetic changes associated with reduced effector activity following persistent stimulation with NKG2C activating antibodies [182]. Such changes include demethylation of *LAG3* and *PDCD1* promoter regions, allowing upregulation of these inhibitory receptors, while increased DNA methylation at key chromatin regulators, such as SATB1, was also observed. Further challenge of the chronically stimulated NK cells with hCMV epitope-expressing endothelial cells induced lower levels of IFNγ expression, suggesting a link between epigenetic remodeling and impaired effector function in NK cells [181].

NK cells have also displayed a dysfunctional signature in cancer patients. Tumor-infiltrating NK cells in patients with NSCLC have shown reduced expression of several NK cell-activating receptors compared to blood-circulating NK cells, including NKp30, NKp80, and ILT2 [184]. These intratumoral NK cells have also reduced degranulation, measured by CD107a expression, and impaired IFNγ production [184]. Furthermore, a cytolytic subset within peripheral blood-circulating NKs ($CD56^{dim}$) from renal cell carcinoma patients were found to express high levels of PD-1 [183]. These observations suggest a therapeutic potential of PD-1 blockade therapy to reinvigorate dysfunctional NK cells during cancer.

Given the role of epigenetic remodeling in regulating NK cell function, currently available nonselective epigenetic therapies may alter the effector function of NK cells. Indeed, treatment of NK cells isolated from NSCLC patients with an HDAC inhibitor known as valproic acid (VPA) was found to limit NK cell cytolytic activity. Notably, VPA treatment impaired NK cell degranulation and IFNγ production, along with increased PD-1-PD-L1-mediated apoptosis of NK cells [185]. DNA hypermethylation and increased suppressive histone methylation marks were also observed in the promoter region of the NKG2D activating receptor gene in VPA-treated NK cells [185]. Together, these findings suggest that epigenetic alterations in NK cells during chronic infections or cancer may imprint their limited ability to clear target cells and contribute to the tumor persistence.

4.3 Epigenetic regulation of Treg cells

Regulatory CD4+ T-cells suppress the antitumor functions of CD8+ T-cells and NK cells, allowing uncontrolled tumor growth. Epigenetic regulation is known to control the differentiation of naïve CD4+ T-cells into Tregs, specifically by impacting the expression of the Treg lineage-specifying transcription factor Foxp3 and other Treg-associated genes [186]. During Treg activation, CD28 co-stimulation induces the expression of EZH2—H3K27 methyltransferase—that adds repressive methylation marks to other lineage-specific genes while allowing for the maintenance of the Treg identity [187]. In contrast, conditional deletion of EZH2 in mouse Tregs impaired their ability to prevent autoimmunity, suggesting the importance of this histone-modifying pathway in normal Treg function [187, 188].

TGF-β signaling has a central role in the epigenetic programming of Tregs within the TME. Through the SMAD2/3 intracellular signaling pathway, tumor-derived TGF-β induces the expression of SMYD3—H3K4 histone methyltransferase—in inducible Treg cells (iTregs). Expression of SMYD3 is associated with activating Foxp3 expression driving iTreg differentiation [189]. Treg survival in the TME is further promoted by tumor-secreted microRNAs [190]. In several human and mouse cancers, miR-214 was found to be upregulated in both tumor tissue and peripheral blood. This microRNA mechanistically silences the expression of the PTEN tumor-suppressor gene, resulting in Treg expansion. Furthermore, mouse Lewis lung carcinoma (LLC) tumor-derived miR-214 was associated with increased IL-10 production by Tregs and enhanced tumor growth [190], suggesting an epigenetic influence on Treg activity. These data highlight the significant impact of epigenetic regulation on the development and immunosuppressive function of Treg cells within the TME.

4.4 Epigenetic regulation of MDSCs

MDSCs are also involved in suppressing antitumor immune responses by soluble mediators and/or promoting the generation of Tregs and TAMs [191]. The development and suppressive function of MDSCs were shown to be epigenetically regulated within the TME. A reduced expression of retinoblastoma gene *RB1* is related to MDSC accumulation and is correlated with increased tumor growth. In contrast to hematopoietic stem cells and mature DCs that normally express RB1, ChIP-seq analysis of MDSCs from tumor-bearing mice has shown an increased binding of the HDAC2 enzyme to the *RB1* promoter region, inducing its downregulation in MDSCs. However, siRNA silencing of HDAC2 restored *RB1* expression, suggesting the critical role of HDAC2 in MDSC development [192]. This evidence indicates that MDSC differentiation may be driven by epigenetic modifications. Additionally, tumor-derived TGF-β induces upregulation of several microRNAs in MDSCs supporting their accumulation within the TME. In particular, miR-494 has been shown to reduce the expression of the PTEN gene and enhance the Akt survival pathway in MDSCs, while promoting CXCR4-mediated chemotaxis of MDSCs [193]. A recent study has also reported an interesting role of Tet2 enzyme in promoting antitumor immunity through an indirect control of MDSCs. The authors found that Tet2 deficiency enhanced IL-6 upregulation after tumor challenge, which further stimulated the expansion of granulocytic MDSCs [194]. These findings demonstrate that various epigenetic mechanisms regulate the development, survival, and accumulation of MDSCs in tumors.

4.5 Epigenetic regulation of TAMs

The development of TAMs has been shown to be epigenetically regulated downstream IL-4-signaling. IL-4-driven STAT6 activation induces upregulation of Jumonji domain-containing 3 (Jmjd3) histone demethylase, which removes the repressive H3K27 methylation marks from the marker genes of tumor-promoting M2-macrophages. Subsequently, expressing these marker genes drives the epigenetic reprogramming of tumor-infiltrating macrophages into an M2-like state [195]. Another example of epigenetic regulation of TAMs was linked to the decreased expression of MHC-II on the surface of TAMs, compared to inflammatory macrophages. MHC-II downregulation acts as a tumor-promoting mechanism by reducing the presentation of tumor-associated antigens to antitumor T-cells. Decoy receptor 3 (DcR3), a soluble receptor in the TNF receptor family known to be expressed by tumor cells, has shown an epigenetic regulatory effect that reduces MHC-II expression on TAMs. DcR3 promotes downstream signaling in TAMs that induces histone deacetylation of the MHC-II master transactivator CIITA gene promoter, resulting in downregulation of MHC-II and subsequently reduced tumor antigen-presentation [196].

While tumor-reactive T-lymphocytes and NK cells are subject to epigenetic changes that limit their antitumor activity, the immunosuppressive functions of Tregs, MDSCs, and TAMs are additionally driven by epigenetic regulation influenced by the TME. The considerable role of epigenetic modifications on transcriptional programming of the immune cells associated with tumor progression or antitumor immune responses, poses as an important target for boosting cancer immunotherapies.

5. Epigenetic barriers to cancer immunotherapy

Cancer immunotherapies aim to boost the quantity and/or quality of antitumor immune cells, while overcoming the immunosuppressive TME. Understanding the molecular mechanisms regulating antitumor immunity is imperative to develop effective immunotherapeutic strategies. Owing to the association between the sustained upregulation of surface inhibitory molecules and decreased effector function in tumor-reactive T-cells, ICB therapy has emerged to block these suppressive signaling pathways. Yet, ICB therapy has shown limited clinical success in ∼70%–80% of patients with cancer [4]. Transcriptional studies of mouse and human Tex cells have indicated that ICB reinvigorates only the partially exhausted subsets of antitumor CD8+ T-cells. Furthermore, after showing an initial boost following ICB treatment, the rejuvenated CD8+ T-cells eventually progress toward a terminally exhausted state under persistent antigen stimulation. Given the substantial impact of epigenetic modifications on antitumor CD8+ T-cell functions, acquired epigenetic programs in Tex cells remain a major cell-intrinsic barrier to immunotherapies. Interestingly, we and others have found a minimal change in exhaustion-associated DNA methylation and chromatin accessibility programs after ICB treatment [146, 152]. Such stable nature of exhaustion-associated epigenetic programs may explain the observed transient response to ICB treatment. It also highlights the need to develop novel approaches for targeting T-cell epigenetic programs to enhance the effectiveness of ICB therapy.

IL-2 cytokine therapy was an early immunotherapeutic approach to boost antitumor immune responses, which was approved by the FDA for treating metastatic renal carcinoma and metastatic melanoma [197]. However, its clinical application remains relatively restricted due to its nonselective immunomodulatory effects (*e.g.*, induced preferential expansion of Tregs) and the high toxicity of high-dose IL-2 treatment [197]. On the other hand, Tex cells display poor responsiveness to multiple cytokines that normally maintain homeostatic proliferation and survival (*e.g.*, IL-7, IL-15) or activation of CD8+ T-cells (*e.g.*, IL-2, IL-12, IL-21) due to downregulation of their receptors. Therefore, Tex cells mainly depend on continuous antigen stimulation to maintain their cell proliferation [198–200]. Animal studies of chronic LCMV infection showed that IL-2 therapy boosts PD-L1 blockade-mediated reinvigoration of Tex cells [201]. This finding was further supported in an animal model of cancer, in conjunction with tumor-antigen-targeted antibody and T-cell vaccine [202]. Nonetheless, the effect of IL-2 on fully exhausted CD8+ T-cells remains relatively limited due to the epigenetic silencing of cytokine receptors (Table 1). IL-15 therapy is another cytokine that was used to boost antiviral or antitumor CD8+ T-cell immunity. Interestingly, IL-15 treatment in simian immunodeficiency virus (SIV)-infected macaques enhanced the viral loads despite significantly increasing the quantities of NK cells and SIV-specific CD8+ T-cells [203, 204]. A recent study has shown that *in vitro* IL-15 treatment was able to rejuvenate effector function of early activated antitumor CD8+ T-cells, rather than the late fully exhausted TILs [151]. These findings emphasize the strong impact of the epigenetic state of Tex cells on their response to immunotherapies, including cytokine therapy. Modulating this epigenetic state in Tex cells will be of great interest to develop effective combined immunotherapies for treating cancer.

Despite the remarkable success of adoptive T-cell (ACT) therapy for treating leukemia, it remains ineffective in solid tumors. The poor efficacy of ACT in solid tumors has been linked to several challenges, including: (a) poor *in vivo* survival of adoptively transferred T-cells; (b) limited T-cell infiltration into tumors; (c) CAR T-cell dysfunction within the TME [6]. These barriers have been also identified for the endogenous antitumor CD8+ T-cell immune responses, implying that similar biological mechanisms regulate both endogenous and adoptively transferred antitumor T-cells. Given the mounting evidence of the substantial influence of epigenetic programming on CD8+ T-cell differentiation and function, it is likely that the epigenetic features of CAR T-cells play a pivotal role in regulating the efficacy of ACT. Indeed, a recent human study has reported that *TET2* disruption, randomly occurred in a clone of CAR T-cells due to lentiviral vector insertion, resulted in complete remission in a patient with chronic lymphocytic leukemia [205]. Interestingly, TET2-disrupted CAR T-cells displayed T_{CM}-like phenotype with enhanced tumor-killing activity and long-lived memory potential [205].

A similar phenotype was reported in mouse *Tet2*-deficient CD8+ T-cells during acute virus infection [206], further supporting an overlapping role of epigenetic regulation in the biology of endogenous and CAR T-cells. These findings suggest that epigenetic regulation has significant impact on the efficacy and overall response of CAR T-cell therapy, and manipulating key epigenetic factors in T-cells has potential to improve current immunotherapies.

6. Epigenetic reprogramming of immune cells in the tumor microenvironment

Developing new approaches to epigenetically reprogram dysfunctional immune cells toward a functional state may offer a novel strategy to overcome cell-intrinsic obstacles to cancer immunotherapies. Currently available options for epigenetic therapies include inhibitors of DNA methylase and histone deacetylase enzymes. Both groups of enzymes have been associated with silencing effector and memory programs and enforcing terminal differentiation and full exhaustion. Treatment of activated mouse CD8+ T-cells during chronic virus infection and cancer models with a DNA demethylating agent, known as decitabine, can interfere with the acquisition of de novo methylation programs associated with the development of the exhaustion phenotype. We found that short-term, low-dose decitabine treatment has a synergistic effect on anti-PD-L1-mediated rejuvenation of Tex cells [146]. This sequential treatment strategy resulted in an enhanced ability of ICB-refractory Tex cells to produce effector molecules and undergo proliferative response after anti-PD-L1 treatment in animal models of chronic infection or cancer. As a genetic proof-of-concept, we found that CD8+ T-cells lacking Dnmt3a-mediated de novo methylation responded more efficiently to ICB, with striking capacity to proliferate and secrete IFNγ and IL-2 effector cytokines [146]. These findings suggest that targeting exhaustion-driving epigenetic mechanisms can synergize with ICB therapy. However, DNMT inhibitors act in a nonspecific manner and thus may result in toxicity to other host cells. In addition, DNMT3A mutations are common in blood malignancies, many of which are linked to pre-leukemic dedifferentiation [207, 208]. Thus, one must be cautious about targeting all DNMT3A-mediated epigenetic programs, since this may be advantageous for tumor progression. Further research on more specific approaches for targeting DNA methylation in immune cells will be of great interest for therapeutic manipulation of T-cell differentiation.

Epigenetic therapies may also modulate the transcriptional programs of immunosuppressive immune cells within the TME, such as TAMs, or tumor cells. In a recent study using a mouse model of ovarian cancer, combined DNMTi and α-difluoromethylornithine—an ornithine decarboxylase inhibitor—treatment induced depletion of M2-like macrophages, while enriching M1-like proinflammatory macrophages within the TME. The epigenetic modulation of TAMs subsequently enhanced tumor control through the recruitment of antitumor T-cells and NK cells [209], suggesting that the potential effects of epigenetic therapies need to be carefully assessed in a broader immune context during cancer.

On the other hand, targeting the epigenetic dysregulation in tumor cells has been an active research area for treating cancer over the last two decades. Some epigenetic therapeutic interventions have proven successful in reprogramming cancer-associated epigenetic abnormalities by: (a) inducing differentiation of transformed cells into nonmalignant cells; (b) targeting self-renewal ability of cancer-stem-like cells; (c) inhibiting tumor metastasis; and/or (d) sensitizing tumor cells for better immune detection and antitumor activity by reversing the epigenetic silencing of antigen presentation genes or inducing interferon responses [28]. Among the FDA-approved epigenetic therapies are DNMTi, such as azacytidine and decitabine, for treating MDS or AML, and HDACi (*e.g.*, entinostat, vorinostat, panobinostat) for treating advanced-stage breast cancer, cutaneous or peripheral T-cell lymphoma, or relapsed multiple myeloma [28]. Combinations of these treatments in addition to the current portfolio of chemotherapeutic agents or immunotherapies are currently investigated in preclinical animal models and several clinical trials. However, off-target effects remain one of the major limitations of the current epigenetic therapies, making it crucial to better understand how these epigenetic abnormalities are acquired or maintained in tumor cells to design more selective interventions.

Designing new therapeutic approaches that directly or indirectly target epigenetic mechanisms regulating the dysfunction of antitumor T- and NK cells, as well as tumor immune evasion, can profoundly improve clinical responses, especially when used with current immunotherapies. Deeper understanding of the epigenetic mechanisms driving normal immune cell differentiation *versus* dysfunction in cancer would reveal a wide array of potential targets to epigenetically reprogram immune cells toward a functional state. Epigenetic intervention of dysfunctional antitumor immune cells can therefore lead to improved rejuvenation following immunotherapeutic treatment, and eventually bring about memory immune cell populations that provide robust long-lived protection against tumors.

References

[1] Mittal D, Gubin MM, Schreiber RD, Smyth MJ. New insights into cancer immunoediting and its three component phases—elimination, equilibrium and escape. Curr Opin Immunol 2014;27:16–25.
[2] Burnet FM. The concept of immunological surveillance. Prog Exp Tumor Res 1970;13:1–27.
[3] Vesely MD, Kershaw MH, Schreiber RD, Smyth MJ. Natural innate and adaptive immunity to cancer. Annu Rev Immunol 2011;29:235–71.
[4] Sharma P, Allison JP. Dissecting the mechanisms of immune checkpoint therapy. Nat Rev Immunol 2020;20(2):75–6.
[5] Guedan S, Ruella M, June CH. Emerging cellular therapies for cancer. Annu Rev Immunol 2019;37:145–71.
[6] Wagner J, Wickman E, DeRenzo C, Gottschalk S. CAR T cell therapy for solid tumors: bright future or dark reality? Mol Ther 2020;28(11):2320–39.

[7] Kelly TK, De Carvalho DD, Jones PA. Epigenetic modifications as therapeutic targets. Nat Biotechnol 2010;28(10):1069–78.

[8] Cavalli G, Heard E. Advances in epigenetics link genetics to the environment and disease. Nature 2019;571(7766):489–99.

[9] Fouse SD, Costello JF. Epigenetics of neurological cancers. Future Oncol 2009;5(10):1615–29.

[10] Jones PA, Baylin SB. The epigenomics of cancer. Cell 2007;128(4):683–92.

[11] Villeneuve LM, Natarajan R. The role of epigenetics in the pathology of diabetic complications. Am J Physiol Ren Physiol 2010;299(1):F14–25.

[12] Javierre BM, Fernandez AF, Richter J, Al-Shahrour F, Martin-Subero JI, Rodriguez-Ubreva J, et al. Changes in the pattern of DNA methylation associate with twin discordance in systemic lupus erythematosus. Genome Res 2010;20(2):170–9.

[13] Adcock IM, Ito K, Barnes PJ. Histone deacetylation: an important mechanism in inflammatory lung diseases. COPD 2005;2(4):445–55.

[14] Egger G, Liang G, Aparicio A, Jones PA. Epigenetics in human disease and prospects for epigenetic therapy. Nature 2004;429(6990):457–63.

[15] Urdinguio RG, Sanchez-Mut JV, Esteller M. Epigenetic mechanisms in neurological diseases: genes, syndromes, and therapies. Lancet Neurol 2009;8(11):1056–72.

[16] Feng J, Fan G. The role of DNA methylation in the central nervous system and neuropsychiatric disorders. Int Rev Neurobiol 2009;89:67–84.

[17] Frommer M, McDonald LE, Millar DS, Collis CM, Watt F, Grigg GW, et al. A genomic sequencing protocol that yields a positive display of 5-methylcytosine residues in individual DNA strands. Proc Natl Acad Sci USA 1992;89(5):1827–31.

[18] Yu M, Hon GC, Szulwach KE, Song C-X, Zhang L, Kim A, et al. Base-resolution analysis of 5-hydroxymethylcytosine in the mammalian genome. Cell 2012;149(6):1368–80.

[19] Solomon MJ, Larsen PL, Varshavsky A. Mapping protein-DNA interactions in vivo with formaldehyde: evidence that histone H4 is retained on a highly transcribed gene. Cell 1988;53(6):937–47.

[20] Park PJ. ChIP-seq: advantages and challenges of a maturing technology. Nat Rev Genet 2009;10(10):669–80.

[21] Buenrostro JD, Giresi PG, Zaba LC, Chang HY, Greenleaf WJ. Transposition of native chromatin for fast and sensitive epigenomic profiling of open chromatin, DNA-binding proteins and nucleosome position. Nat Methods 2013;10(12):1213–8.

[22] Chen X, Miragaia RJ, Natarajan KN, Teichmann SA. A rapid and robust method for single cell chromatin accessibility profiling. Nat Commun 2018;9(1):5345.

[23] Lieberman-Aiden E, van Berkum NL, Williams L, Imakaev M, Ragoczy T, Telling A, et al. Comprehensive mapping of long-range interactions reveals folding principles of the human genome. Science 2009;326(5950):289–93.

[24] Dekker J, Rippe K, Dekker M, Kleckner N. Capturing chromosome conformation. Science 2002;295(5558):1306–11.

[25] Rao SSP, Huntley MH, Durand NC, Stamenova EK, Bochkov ID, Robinson JT, et al. A 3D map of the human genome at kilobase resolution reveals principles of chromatin looping. Cell 2014;159(7):1665–80.

[26] Ghoneim HE, Zamora AE, Thomas PG, Youngblood BA. Cell-intrinsic barriers of T cell-based immunotherapy. Trends Mol Med 2016;22(12):1000–11.

[27] Henning AN, Roychoudhuri R, Restifo NP. Epigenetic control of CD8 + T cell differentiation. Nat Rev Immunol 2018;18(5):340–56.

[28] Topper MJ, Vaz M, Marrone KA, Brahmer JR, Baylin SB. The emerging role of epigenetic therapeutics in immuno-oncology. Nat Rev Clin Oncol 2020;17(2):75–90.

[29] Waddington CH. Canalization of development and the inheritance of acquired characters. Nature 1942;150(3811):563–5.

[30] Waddington CH. The epigenotype. 1942. Int J Epidemiol 2011;41(1):10–3.

[31] Goldberg AD, Allis CD, Bernstein E. Epigenetics: a landscape takes shape. Cell 2007;128(4):635–8.

[32] Feinberg AP. Phenotypic plasticity and the epigenetics of human disease. Nature 2007;447(7143):433–40.

[33] Razin A, Riggs AD. DNA methylation and gene function. Science 1980;210(4470):604–10.

[34] Bird A, Taggart M, Frommer M, Miller OJ, Macleod D. A fraction of the mouse genome that is derived from islands of nonmethylated, CpG-rich DNA. Cell 1985;40(1):91–9.

[35] Deaton AM, Bird A. CpG islands and the regulation of transcription. Genes Dev 2011;25 (10):1010–22.

[36] Vardimon L, Kressmann A, Cedar H, Maechler M, Doerfler W. Expression of a cloned adenovirus gene is inhibited by in vitro methylation. Proc Natl Acad Sci USA 1982;79(4):1073–7.

[37] Stein R, Razin A, Cedar H. In vitro methylation of the hamster adenine phosphoribosyltransferase gene inhibits its expression in mouse L cells. Proc Natl Acad Sci USA 1982;79(11):3418–22.

[38] Clark SJ, Argelaguet R, Kapourani C-A, Stubbs TM, Lee HJ, Alda-Catalinas C, et al. scNMT-seq enables joint profiling of chromatin accessibility DNA methylation and transcription in single cells. Nat Commun 2018;9(1):781.

[39] Jones PA. Functions of DNA methylation: islands, start sites, gene bodies and beyond. Nat Rev Genet 2012;13(7):484–92.

[40] Seisenberger S, Peat JR, Hore TA, Santos F, Dean W, Reik W. Reprogramming DNA methylation in the mammalian life cycle: building and breaking epigenetic barriers. Philos Trans R Soc Lond B Biol Sci 2013;368(1609). Available from: https://www.ncbi.nlm.nih.gov/pmc/articles/PMC3539359/.

[41] Okano M, Bell DW, Haber DA, Li E. DNA methyltransferases Dnmt3a and Dnmt3b are essential for de novo methylation and mammalian development. Cell 1999;99(3):247–57.

[42] Sharif J, Koseki H. Hemimethylation: DNA's lasting odd couple. Science 2018;359(6380):1102–3.

[43] Wigler M, Levy D, Perucho M. The somatic replication of DNA methylation. Cell 1981;24 (1):33–40.

[44] Xie S, Qian C. The growing complexity of UHRF1-mediated maintenance DNA methylation. Genes (Basel) 2018;9(12):600–11.

[45] Bostick M, Kim JK, Estève P-O, Clark A, Pradhan S, Jacobsen SE. UHRF1 plays a role in maintaining DNA methylation in mammalian cells. Science 2007;317(5845):1760–4.

[46] Lei H, Oh SP, Okano M, Juttermann R, Goss KA, Jaenisch R, et al. De novo DNA cytosine methyltransferase activities in mouse embryonic stem cells. Development 1996;122(10):3195–205.

[47] Anon. Targeted mutation of the DNA methyltransferase gene results in embryonic lethality. Cell 1992;69(6):915–26.

[48] Liao J, Karnik R, Gu H, Ziller MJ, Clement K, Tsankov AM, et al. Targeted disruption of DNMT1, DNMT3A and DNMT3B in human embryonic stem cells. Nat Genet 2015;47(5):469–78.

[49] Brown KD, Robertson KD. DNMT1 knockout delivers a strong blow to genome stability and cell viability. Nat Genet 2007;39(3):289–90.

[50] Ito S, Shen L, Dai Q, Wu SC, Collins LB, Swenberg JA, et al. Tet proteins can convert 5-methylcytosine to 5-formylcytosine and 5-carboxylcytosine. Science 2011;333(6047):1300–3.

[51] Kohli RM, Zhang Y. TET enzymes, TDG and the dynamics of DNA demethylation. Nature 2013;502(7472):472–9.

[52] Bruniquel D, Schwartz RH. Selective, stable demethylation of the interleukin-2 gene enhances transcription by an active process. Nat Immunol 2003;4(3):235–40.

[53] Valinluck V, Sowers LC. Endogenous cytosine damage products alter the site selectivity of human DNA maintenance methyltransferase DNMT1. Cancer Res 2007;67(3):946–50.

[54] He W, Li X, Xu S, Ai J, Gong Y, Gregg JL, et al. Aberrant methylation and loss of CADM2 tumor suppressor expression is associated with human renal cell carcinoma tumor progression. Biochem Biophys Res Commun 2013;435(4):526–32.

[55] Xia L, Huang W, Bellani M, Seidman MM, Wu K, Fan D, et al. CHD4 has oncogenic functions in initiating and maintaining epigenetic suppression of multiple tumor suppressor genes. Cancer Cell 2017;31(5). 653–668.e7.

[56] Bohl SR, Bullinger L, Rücker FG. Epigenetic therapy: azacytidine and decitabine in acute myeloid leukemia. Expert Rev Hematol 2018;11(5):361–71.

[57] Döhner H, Estey E, Grimwade D, Amadori S, Appelbaum FR, Büchner T, et al. Diagnosis and management of AML in adults: 2017 ELN recommendations from an international expert panel. Blood 2017;129(4):424–47.

[58] Gnyszka A, Jastrzebski Z, Flis S. DNA methyltransferase inhibitors and their emerging role in epigenetic therapy of cancer. Anticancer Res 2013;33(8):2989–96.

[59] Luger K, Mäder AW, Richmond RK, Sargent DF, Richmond TJ. Crystal structure of the nucleosome core particle at 2.8 Å resolution. Nature 1997;389(6648):251–60.

[60] Huang H, Sabari BR, Garcia BA, Allis CD, Zhao Y. SnapShot: histone modifications. Cell 2014;159 (2). 458–458.e1.

[61] Hyun K, Jeon J, Park K, Kim J. Writing, erasing and reading histone lysine methylations. Exp Mol Med 2017;49(4):e324.

[62] Huang H, Lin S, Garcia BA, Zhao Y. Quantitative proteomic analysis of histone modifications. Chem Rev 2015;115(6):2376–418.

[63] Kooistra SM, Helin K. Molecular mechanisms and potential functions of histone demethylases. Nat Rev Mol Cell Biol 2012;13(5):297–311.

[64] D'Oto A, Tian Q-W, Davidoff AM, Yang J. Histone demethylases and their roles in cancer epigenetics. J Med Oncol Ther 2016;1(2):34–40.

[65] Shi Y, Whetstine JR. Dynamic regulation of histone lysine methylation by demethylases. Mol Cell 2007;25(1):1–14.

[66] Dhalluin C, Carlson JE, Zeng L, He C, Aggarwal AK, Zhou M-M, et al. Structure and ligand of a histone acetyltransferase bromodomain. Nature 1999;399(6735):491–6.

[67] Jacobson RH, Ladurner AG, King DS, Tjian R. Structure and function of a human TAFII250 double bromodomain module. Science 2000;288(5470):1422–5.

[68] Fujisawa T, Filippakopoulos P. Functions of bromodomain-containing proteins and their roles in homeostasis and cancer. Nat Rev Mol Cell Biol 2017;18(4):246–62.

[69] Bannister AJ, Zegerman P, Partridge JF, Miska EA, Thomas JO, Allshire RC, et al. Selective recognition of methylated lysine 9 on histone H3 by the HP1 chromo domain. Nature 2001;410 (6824):120–4.

[70] Lachner M, O'Carroll D, Rea S, Mechtler K, Jenuwein T. Methylation of histone H3 lysine 9 creates a binding site for HP1 proteins. Nature 2001;410(6824):116–20.

[71] Nakayama J, Rice JC, Strahl BD, Allis CD, Grewal SI. Role of histone H3 lysine 9 methylation in epigenetic control of heterochromatin assembly. Science 2001;292(5514):110–3.

[72] Bannister AJ, Falcão AM, Castelo-Branco G. Chapter 2: Histone modifications and histone variants in pluripotency and differentiation. In: Göndör A, editor. Chromatin regulation and dynamics. Boston: Academic Press; 2017. p. 35–64. Available from: http://www.sciencedirect.com/science/article/pii/B9780128033951000022.

[73] Schuettengruber B, Bourbon H-M, Di Croce L, Cavalli G. Genome regulation by polycomb and trithorax: 70 years and counting. Cell 2017;171(1):34–57.

[74] Schuettengruber B, Cavalli G. Recruitment of polycomb group complexes and their role in the dynamic regulation of cell fate choice. Development 2009;136(21):3531–42.

[75] Qin S, Li L, Min J. Chapter 3: The chromodomain of polycomb: methylation reader and beyond. In: Pirrotta V, editor. Polycomb group proteins. Academic Press; 2017. p. 33–56. Available from: http://www.sciencedirect.com/science/article/pii/B9780128097373000039.

[76] Eskeland R, Leeb M, Grimes GR, Kress C, Boyle S, Sproul D, et al. Ring1B compacts chromatin structure and represses gene expression independent of histone ubiquitination. Mol Cell 2010;38 (3):452–64.

[77] Trojer P, Reinberg D. Facultative heterochromatin: is there a distinctive molecular signature? Mol Cell 2007;28(1):1–13.

[78] Saksouk N, Simboeck E, Déjardin J. Constitutive heterochromatin formation and transcription in mammals. Epigenetics Chromatin 2015;8(1):3.

[79] Żylicz JJ, Heard E. Molecular mechanisms of facultative heterochromatin formation: an X-chromosome perspective. Annu Rev Biochem 2020;89(1):255–82.

[80] Morgan MAJ, Shilatifard A. Reevaluating the roles of histone-modifying enzymes and their associated chromatin modifications in transcriptional regulation. Nat Genet 2020;52(12):1271–81.

[81] Aubert Y, Egolf S, Capell BC. The unexpected non-catalytic roles of histone modifiers in development and disease. Trends Genet 2019;35(9):645–57.

[82] Jenuwein T, Allis CD. Translating the histone code. Science 2001;293(5532):1074–80.

[83] Angrand P-O, Vennin C, Le Bourhis X, Adriaenssens E. The role of long non-coding RNAs in genome formatting and expression. Front Genet 2015;6. Available from: https://www.ncbi.nlm.nih.gov/pmc/articles/PMC4413816/.

[84] Penny GD, Kay GF, Sheardown SA, Rastan S, Brockdorff N. Requirement for Xist in X chromosome inactivation. Nature 1996;379(6561):131–7.

[85] Peschansky VJ, Wahlestedt C. Non-coding RNAs as direct and indirect modulators of epigenetic regulation. Epigenetics 2014;9(1):3–12.

[86] Morris KV, Mattick JS. The rise of regulatory RNA. Nat Rev Genet 2014;15(6):423–37.

[87] Moazed D. Small RNAs in transcriptional gene silencing and genome defence. Nature 2009;457 (7228):413–20.

[88] Coussens LM, Zitvogel L, Palucka AK. Neutralizing tumor-promoting chronic inflammation: a magic bullet? Science 2013;339(6117):286–91.

[89] Palucka AK, Coussens LM. The basis of oncoimmunology. Cell 2016;164(6):1233–47.

[90] Russell JH, Ley TJ. Lymphocyte-mediated cytotoxicity. Annu Rev Immunol 2002;20(1):323–70.

[91] Savage PA, Leventhal DS, Malchow S. Shaping the repertoire of tumor-infiltrating effector and regulatory T cells. Immunol Rev 2014;259(1):245–58.

[92] Gordy C, He Y-W. Endocytosis by target cells: an essential means for perforin- and granzyme-mediated killing. Cell Mol Immunol 2012;9(1):5–6.

[93] Dvorak AM, Galli SJ, Marcum JA, Nabel G, der Simonian H, Goldin J, et al. Cloned mouse cells with natural killer function and cloned suppressor T cells express ultrastructural and biochemical features not shared by cloned inducer T cells. J Exp Med 1983;157(3):843–61.

[94] Masson D, Tschopp J. Isolation of a lytic, pore-forming protein (perforin) from cytolytic T-lymphocytes. J Biol Chem 1985;260(16):9069–72.

[95] Nagata S, Golstein P. The Fas death factor. Science 1995;267(5203):1449–56.

[96] Ashkenazi A, Dixit VM. Death receptors: signaling and modulation. Science 1998;281(5381):1305–8.

[97] Vodnala SK, Eil R, Kishton RJ, Sukumar M, Yamamoto TN, Ha N-H, et al. T cell stemness and dysfunction in tumors are triggered by a common mechanism. Science 2019;363(6434). Available from: https://science.sciencemag.org/content/363/6434/eaau0135.

[98] McLane LM, Abdel-Hakeem MS, Wherry EJ. CD8 T cell exhaustion during chronic viral infection and cancer. Annu Rev Immunol 2019;37(1):457–95.

[99] Zhang Z, Liu S, Zhang B, Qiao L, Zhang Y, Zhang Y. T cell dysfunction and exhaustion in cancer. Front Cell Dev Biol 2020;8. Available from: https://www.ncbi.nlm.nih.gov/pmc/articles/PMC7027373/.

[100] Blank CU, Haining WN, Held W, Hogan PG, Kallies A, Lugli E, et al. Defining 'T cell exhaustion. Nat Rev Immunol 2019;19(11):665–74.

[101] Wherry EJ. T cell exhaustion. Nat Immunol 2011;12(6):492–9.

[102] Angelosanto JM, Blackburn SD, Crawford A, Wherry EJ. Progressive loss of memory T cell potential and commitment to exhaustion during chronic viral infection. J Virol 2012;86(15):8161–70.

[103] Barber DL, Wherry EJ, Masopust D, Zhu B, Allison JP, Sharpe AH, et al. Restoring function in exhausted CD8 T cells during chronic viral infection. Nature 2006;439(7077):682–7.

[104] Keene J-A, Forman J. Helper activity is required for the in vivo generation of cytotoxic T lymphocytes. J Exp Med 1982;155(3):768–82.

[105] Bennett SRM, Carbone FR, Karamalis F, Miller JFAP, Heath WR. Induction of a CD8+ cytotoxic T lymphocyte response by cross-priming requires cognate CD4+ T cell help. J Exp Med 1997;186 (1):65–70.

[106] Hor JL, Whitney PG, Zaid A, Brooks AG, Heath WR, Mueller SN. Spatiotemporally distinct interactions with dendritic cell subsets facilitates CD4+ and CD8+ T cell activation to localized viral infection. Immunity 2015;43(3):554–65.

[107] Vacaflores A, Chapman NM, Harty JT, Richer MJ, Houtman JCD. Exposure of human CD4 T cells to IL-12 results in enhanced TCR-induced cytokine production, altered TCR signaling, and increased oxidative metabolism. PLoS One 2016;11(6), e0157175.

[108] Waldhauer I, Steinle A. NK cells and cancer immunosurveillance. Oncogene 2008;27(45):5932–43.

[109] Zamai L, Ahmad M, Bennett IM, Azzoni L, Alnemri ES, Perussia B. Natural killer (NK) cell-mediated cytotoxicity: differential use of TRAIL and Fas ligand by immature and mature primary human NK cells. J Exp Med 1998;188(12):2375–80.

[110] Orange JS. Natural killer cell deficiency. J Allergy Clin Immunol 2013;132(3):515–25.

[111] Moretta A, Marcenaro E, Sivori S, Chiesa MD, Vitale M, Moretta L. Early liaisons between cells of the innate immune system in inflamed peripheral tissues. Trends Immunol 2005;26(12):668–75.

[112] Martín-Fontecha A, Thomsen LL, Brett S, Gerard C, Lipp M, Lanzavecchia A, et al. Induced recruitment of NK cells to lymph nodes provides IFN-γ for TH1 priming. Nat Immunol 2004;5 (12):1260–5.

[113] Bryceson YT, Ljunggren H-G. Tumor cell recognition by the NK cell activating receptor NKG2D. Eur J Immunol 2008;38(11):2957–61.

[114] Sungur CM, Murphy WJ. Positive and negative regulation by NK cells in cancer. Crit Rev Oncog 2014;19:57–66.

[115] Veiga-Parga T, Sehrawat S, Rouse BT. Role of regulatory T cells during virus infection. Immunol Rev 2013;255(1):182–96.

[116] Tinoco R, Alcalde V, Yang Y, Sauer K, Zuniga EI. Cell-intrinsic transforming growth factor-β signaling mediates virus-specific CD8 + T cell deletion and viral persistence in vivo. Immunity 2009;31 (1):145–57.

[117] Sharma S, Stolina M, Lin Y, Gardner B, Miller PW, Kronenberg M, et al. T cell-derived IL-10 promotes lung cancer growth by suppressing both T cell and APC function. J Immunol 1999;163 (9):5020–8.

[118] Sawant DV, Yano H, Chikina M, Zhang Q, Liao M, Liu C, et al. Adaptive plasticity of IL-10 + and IL-35 + T reg cells cooperatively promotes tumor T cell exhaustion. Nat Immunol 2019;20 (6):724–35.

[119] Penaloza-MacMaster P, Kamphorst AO, Wieland A, Araki K, Iyer SS, West EE, et al. Interplay between regulatory T cells and PD-1 in modulating T cell exhaustion and viral control during chronic LCMV infection. J Exp Med 2014;211(9):1905–18.

[120] Li C, Jiang P, Wei S, Xu X, Wang J. Regulatory T cells in tumor microenvironment: new mechanisms, potential therapeutic strategies and future prospects. Mol Cancer 2020;19(1):116.

[121] Chen Y, Zhang S, Wang Q, Zhang X. Tumor-recruited M2 macrophages promote gastric and breast cancer metastasis via M2 macrophage-secreted CHI3L1 protein. J Hematol Oncol 2017;10(1):36.

[122] Pathria P, Louis TL, Varner JA. Targeting tumor-associated macrophages in cancer. Trends Immunol 2019;40(4):310–27.

[123] Ostrand-Rosenberg S. Myeloid-derived suppressor cells: more mechanisms for inhibiting antitumor immunity. Cancer Immunol Immunother 2010;59(10):1593–600.

[124] Li H, Han Y, Guo Q, Zhang M, Cao X. Cancer-expanded myeloid-derived suppressor cells induce anergy of NK cells through membrane-bound TGF-β1. J Immunol 2009;182:240–9.

[125] Srivatsava MK, et al. Myeloid-derived suppressor cells inhibit T-cell activation by depleting cystine and cysteine. Cancer Res 2010;70(1):68–77.

[126] Huang B, Pan P-Y, Li Q, Sato AI, Levy DE, Bromberg J, et al. Gr-1 + CD115 + immature myeloid suppressor cells mediate the development of tumor-induced T regulatory cells and T-cell anergy in tumor-bearing host. Cancer Res 2006;66(2):1123–31.

[127] Chen Y, Song Y, Du W, Gong L, Chang H, Zou Z. Tumor-associated macrophages: an accomplice in solid tumor progression. J Biomed Sci 2019;26(1):78.

[128] Colegio OR, Chu N-Q, Szabo AL, Chu T, Rhebergen AM, Jairam V, et al. Functional polarization of tumour-associated macrophages by tumour-derived lactic acid. Nature 2014;513(7519):559–63.

[129] Zhao Q, Kuang D-M, Wu Y, Xiao X, Li X-F, Li T-J, et al. Activated CD69 + T cells foster immune privilege by regulating IDO expression in tumor-associated macrophages. J Immunol 2012;188 (3):1117–24.

[130] Chen X, Song M, Zhang B, Zhang Y. Reactive oxygen species regulate T cell immune response in the tumor microenvironment. Oxid Med Cell Longev 2016;2016:1580967.

[131] Wherry EJ, Kurachi M. Molecular and cellular insights into T cell exhaustion. Nat Rev Immunol 2015;15(8):486–99.

[132] Jiang Y, Li Y, Zhu B. T-cell exhaustion in the tumor microenvironment. Cell Death Dis 2015;6(6): e1792.

[133] Snell LM, McGaha TL, Brooks DG. Type I interferon in chronic virus infection and cancer. Trends Immunol 2017;38(8):542–57.

[134] Zehn D, Wherry EJ. Immune memory and exhaustion: clinically relevant lessons from the LCMV model. In: Schoenberger SP, Katsikis PD, Pulendran B, editors. Crossroads between innate and adaptive immunity V. Advances in experimental medicine and biology. Cham: Springer International Publishing; 2015. p. 137–52.

[135] Zinkernagel RM, Doherty PC. Restriction of in vitro T cell-mediated cytotoxicity in lymphocytic choriomeningitis within a syngeneic or semiallogeneic system. Nature 1974;248(5450):701–2.

[136] Gallimore A, Glithero A, Godkin A, Tissot AC, Plückthun A, Elliott T, et al. Induction and exhaustion of lymphocytic choriomeningitis virus-specific cytotoxic T lymphocytes visualized using soluble tetrameric major histocompatibility complex class I-peptide complexes. J Exp Med 1998;187 (9):1383–93.

[137] Zajac AJ, Blattman JN, Murali-Krishna K, Sourdive DJD, Suresh M, Altman JD, et al. Viral immune evasion due to persistence of activated T cells without effector function. J Exp Med 1998;188 (12):2205–13.

[138] Butz EA, Bevan MJ. Massive expansion of antigen-specific CD8 + T cells during an acute virus infection. Immunity 1998;8(2):167–75.

[139] Scharer CD, Barwick BG, Youngblood BA, Ahmed R, Boss JM. Global DNA methylation remodeling accompanies CD8 T cell effector function. J Immunol 2013;191(6):3419–29.

[140] Russ BE, Olshanksy M, Smallwood HS, Li J, Denton AE, Prier JE, et al. Distinct epigenetic signatures delineate transcriptional programs during virus-specific CD8(+) T cell differentiation. Immunity 2014;41(5):853–65.

[141] Akondy RS, Fitch M, Edupuganti S, Yang S, Kissick HT, Li KW, et al. Origin and differentiation of human memory CD8 T cells after vaccination. Nature 2017;552(7685):362–7.

[142] Youngblood B, Hale JS, Kissick HT, Ahn E, Xu X, Wieland A, et al. Effector CD8 T cells dedifferentiate into long-lived memory cells. Nature 2017;552(7685):404–9.

[143] Hudson WH, Gensheimer J, Hashimoto M, Wieland A, Valanparambil RM, Li P, et al. Proliferating transitory T cells with an effector-like transcriptional signature emerge from PD-1 + stem-like CD8 + T cells during chronic infection. Immunity 2019;51(6). 1043–1058.e4.

[144] Pereira RM, Hogan PG, Rao A, Martinez GJ. Transcriptional and epigenetic regulation of T cell hyporesponsiveness. J Leukoc Biol 2017;102(3):601–15.

[145] Hudson WH, Prokhnevska N, Gensheimer J, Akondy R, McGuire DJ, Ahmed R, et al. Expression of novel long noncoding RNAs defines virus-specific effector and memory CD8 + T cells. Nat Commun 2019;10(1):196.

[146] Ghoneim HE, Fan Y, Moustaki A, Abdelsamed HA, Dash P, Dogra P, et al. De novo epigenetic programs inhibit PD-1 blockade-mediated T cell rejuvenation. Cell 2017;170(1). 142–157.e19.

[147] Ladle BH, Li K-P, Phillips MJ, Pucsek AB, Haile A, Powell JD, et al. De novo DNA methylation by DNA methyltransferase 3a controls early effector CD8 + T-cell fate decisions following activation. Proc Natl Acad Sci USA 2016;113(38):10631–6.

[148] Youngblood B, Oestreich KJ, Ha S-J, Duraiswamy J, Akondy RS, West EE, et al. Chronic virus infection enforces demethylation of the locus that encodes PD-1 in antigen-specific CD8(+) T cells. Immunity 2011;35(3):400–12.

[149] Yang R, Cheng S, Luo N, Gao R, Yu K, Kang B, et al. Distinct epigenetic features of tumor-reactive CD8 + T cells in colorectal cancer patients revealed by genome-wide DNA methylation analysis. Genome Biol 2019;21(1):2.

[150] Sen DR, Kaminski J, Barnitz RA, Kurachi M, Gerdemann U, Yates KB, et al. The epigenetic landscape of T cell exhaustion. Science 2016;354(6316):1165–9.

[151] Philip M, Fairchild L, Sun L, Horste EL, Camara S, Shakiba M, et al. Chromatin states define tumour-specific T cell dysfunction and reprogramming. Nature 2017;545(7655):452–6.

[152] Pauken KE, Sammons MA, Odorizzi PM, Manne S, Godec J, Khan O, et al. Epigenetic stability of exhausted T cells limits durability of reinvigoration by PD-1 blockade. Science 2016;354 (6316):1160–5.

[153] Scott-Browne JP, López-Moyado IF, Trifari S, Wong V, Chavez L, Rao A, et al. Dynamic changes in chromatin accessibility occur in CD8+ T cells responding to viral infection. Immunity 2016;45(6):1327–40.

[154] Araki Y, Fann M, Wersto R, Weng N-P. Histone acetylation facilitates rapid and robust memory CD8 T cell response through differential expression of effector molecules (eomesodermin and its targets: perforin and granzyme B). J Immunol 2008;180(12):8102–8.

[155] Manna S, Kim JK, Baugé C, Cam M, Zhao Y, Shetty J, et al. Histone H3 lysine 27 demethylases Jmjd3 and Utx are required for T-cell differentiation. Nat Commun 2015;6:8152.

[156] Araki Y, Wang Z, Zang C, Wood WH, Schones D, Cui K, et al. Genome-wide analysis of histone methylation reveals chromatin state-based regulation of gene transcription and function of memory CD8+ T cells. Immunity 2009;30(6):912–25.

[157] He B, Xing S, Chen C, Gao P, Teng L, Shan Q, et al. CD8+ T cells utilize highly dynamic enhancer repertoires and regulatory circuitry in response to infections. Immunity 2016;45(6):1341–54.

[158] Gray SM, Amezquita RA, Guan T, Kleinstein SH, Kaech SM. Polycomb repressive complex 2-mediated chromatin repression guides effector CD8+ T cell terminal differentiation and loss of multipotency. Immunity 2017;46(4):596–608.

[159] Gattinoni L, Lugli E, Ji Y, Pos Z, Paulos CM, Quigley MF, et al. A human memory T cell subset with stem cell-like properties. Nat Med 2011;17(10):1290–7.

[160] Roychoudhuri R, Lefebvre F, Honda M, Pan L, Ji Y, Klebanoff CA, et al. Transcriptional profiles reveal a stepwise developmental program of memory CD8(+) T cell differentiation. Vaccine 2015;33(7):914–23.

[161] Omilusik KD, Goldrath AW. Remembering to remember: T cell memory maintenance and plasticity. Curr Opin Immunol 2019;58:89–97.

[162] Abdelsamed HA, Moustaki A, Fan Y, Dogra P, Ghoneim HE, Zebley CC, et al. Human memory CD8 T cell effector potential is epigenetically preserved during in vivo homeostasis. J Exp Med 2017;214(6):1593–606.

[163] Kersh EN, Fitzpatrick DR, Murali-Krishna K, Shires J, Speck SH, Boss JM, et al. Rapid demethylation of the IFN-gamma gene occurs in memory but not naive CD8 T cells. J Immunol 1950;176. Available from: https://pubmed.ncbi.nlm.nih.gov/16547244/.

[164] Beltra J-C, Manne S, Abdel-Hakeem MS, Kurachi M, Giles JR, Chen Z, et al. Developmental relationships of four exhausted CD8+ T cell subsets reveals underlying transcriptional and epigenetic landscape control mechanisms. Immunity 2020;52(5). 825–841.e8.

[165] Zander R, Schauder D, Xin G, Nguyen C, Wu X, Zajac A, et al. CD4+ T cell help is required for the formation of a cytolytic CD8+ T cell subset that protects against chronic infection and cancer. Immunity 2019;51(6). 1028–1042.e4.

[166] Im SJ, Hashimoto M, Gerner MY, Lee J, Kissick HT, Burger MC, et al. Defining CD8+ T cells that provide the proliferative burst after PD-1 therapy. Nature 2016;537(7620):417–21.

[167] Yu Y-R, Imrichova H, Wang H, Chao T, Xiao Z, Gao M, et al. Disturbed mitochondrial dynamics in CD8+ TILs reinforce T cell exhaustion. Nat Immunol 2020;21(12):1540–51.

[168] Alfei F, Kanev K, Hofmann M, Wu M, Ghoneim HE, Roelli P, et al. TOX reinforces the phenotype and longevity of exhausted T cells in chronic viral infection. Nature 2019;571(7764):265–9.

[169] Khan O, Giles JR, McDonald S, Manne S, Ngiow SF, Patel KP, et al. TOX transcriptionally and epigenetically programs CD8+ T cell exhaustion. Nature 2019;571(7764):211–8.

[170] Yao C, Sun H-W, Lacey NE, Ji Y, Moseman EA, Shih H-Y, et al. Single-cell RNA-seq reveals TOX as a key regulator of CD8+ T cell persistence in chronic infection. Nat Immunol 2019;20(7):890–901.

[171] Scott AC, Dündar F, Zumbo P, Chandran SS, Klebanoff CA, Shakiba M, et al. TOX is a critical regulator of tumour-specific T cell differentiation. Nature 2019;571(7764):270–4.

[172] Seo H, Chen J, González-Avalos E, Samaniego-Castruita D, Das A, Wang YH, et al. TOX and TOX2 transcription factors cooperate with NR4A transcription factors to impose CD8+ T cell exhaustion. Proc Natl Acad Sci USA 2019;116(25):12410–5.

[173] Youngblood B, Noto A, Porichis F, Akondy RS, Ndhlovu ZM, Austin JW, et al. Cutting edge: prolonged exposure to HIV reinforces a poised epigenetic program for PD-1 expression in virus-specific CD8 T cells. J Immunol 2013;191(2):540–4.

[174] Ahn E, Youngblood B, Lee J, Lee J, Sarkar S, Ahmed R. Demethylation of the PD-1 promoter is imprinted during the effector phase of CD8 T cell exhaustion. J Virol 2016;90(19):8934–46.

[175] Alvarez JD, Yasui DH, Niida H, Joh T, Loh DY, Kohwi-Shigematsu T. The MAR-binding protein SATB1 orchestrates temporal and spatial expression of multiple genes during T-cell development. Genes Dev 2000;14(5):521–35.

[176] Cai S, Lee CC, Kohwi-Shigematsu T. SATB1 packages densely looped, transcriptionally active chromatin for coordinated expression of cytokine genes. Nat Genet 2006;38(11):1278–88.

[177] Stephen TL, Payne KK, Chaurio RA, Allegrezza MJ, Zhu H, Perez-Sanz J, et al. SATB1 expression governs epigenetic repression of PD-1 in tumor-reactive T cells. Immunity 2017;46(1):51–64.

[178] Ghosh RP, Shi Q, Yang L, Reddick MP, Nikitina T, Zhurkin VB, et al. Satb1 integrates DNA binding site geometry and torsional stress to differentially target nucleosome-dense regions. Nat Commun 2019;10(1):3221.

[179] Hatton RD, Harrington LE, Luther RJ, Wakefield T, Janowski KM, Oliver JR, et al. A distal conserved sequence element controls Ifng gene expression by T cells and NK cells. Immunity 2006;25 (5):717–29.

[180] Bream JH, Hodge DL, Gonsky R, Spolski R, Leonard WJ, Krebs S, et al. A distal region in the interferon-gamma gene is a site of epigenetic remodeling and transcriptional regulation by interleukin-2. J Biol Chem 2004;279(39):41249–57.

[181] Merino A, Zhang B, Dougherty P, Luo X, Wang J, Blazar BR, et al. Chronic stimulation drives human NK cell dysfunction and epigenetic reprograming. J Clin Invest 2019;129(9):3770–85.

[182] Schlums H, Cichocki F, Tesi B, Theorell J, Beziat V, Holmes TD, et al. Cytomegalovirus infection drives adaptive epigenetic diversification of NK cells with altered signaling and effector function. Immunity 2015;42(3):443–56.

[183] MacFarlane AW, Jillab M, Plimack ER, Hudes GR, Uzzo RG, Litwin S, et al. PD-1 expression on peripheral blood cells increases with stage in renal cell carcinoma patients and is rapidly reduced after surgical tumor resection. Cancer Immunol Res 2014;2(4):320–31.

[184] Platonova S, Cherfils-Vicini J, Damotte D, Crozet L, Vieillard V, Validire P, et al. Profound coordinated alterations of intratumoral NK cell phenotype and function in lung carcinoma. Cancer Res 2011;71(16):5412–22.

[185] Shi X, Li M, Cui M, Niu C, Xu J, Zhou L, et al. Epigenetic suppression of the antitumor cytotoxicity of NK cells by histone deacetylase inhibitor valproic acid. Am J Cancer Res 2016;6(3):600–14.

[186] Ohkura N, Kitagawa Y, Sakaguchi S. Development and maintenance of regulatory T cells. Immunity 2013;38(3):414–23.

[187] DuPage M, Chopra G, Quiros J, Rosenthal WL, Morar MM, Holohan D, et al. The chromatin-modifying enzyme Ezh2 is critical for the maintenance of regulatory T cell identity after activation. Immunity 2015;42(2):227–38.

[188] Arvey A, van der Veeken J, Samstein RM, Feng Y, Stamatoyannopoulos JA, Rudensky AY. Inflammation-induced repression of chromatin bound by the transcription factor Foxp3 in regulatory T cells. Nat Immunol 2014;15(6):580–7.

[189] Nagata DEDA, Ting H-A, Cavassani KA, Schaller MA, Mukherjee S, Ptaschinski C, et al. Epigenetic control of Foxp3 by SMYD3 H3K4 histone methyltransferase controls iTreg development and regulates pathogenic T-cell responses during pulmonary viral infection. Mucosal Immunol 2015;8 (5):1131–43.

[190] Yin Y, Cai X, Chen X, Liang H, Zhang Y, Li J, et al. Tumor-secreted miR-214 induces regulatory T cells: a major link between immune evasion and tumor growth. Cell Res 2014;24(10):1164–80.

[191] Liu M, Zhou J, Chen Z, Cheng AS-L. Understanding the epigenetic regulation of tumours and their microenvironments: opportunities and problems for epigenetic therapy. J Pathol 2017;241(1):10–24.

[192] Youn J-I, Kumar V, Collazo M, Nefedova Y, Condamine T, Cheng P, et al. Epigenetic silencing of retinoblastoma gene regulates pathologic differentiation of myeloid cells in cancer. Nat Immunol 2013;14(3):211–20.

[193] Liu Y, Lai L, Chen Q, Song Y, Xu S, Ma F, et al. MicroRNA-494 is required for the accumulation and functions of tumor-expanded myeloid-derived suppressor cells via targeting of PTEN. J Immunol 2012;188(11):5500–10.

[194] Li S, Feng J, Wu F, Cai J, Zhang X, Wang H, et al. TET2 promotes anti-tumor immunity by governing G-MDSCs and CD8+ T-cell numbers. EMBO Rep 2020;21(10), e49425.

[195] Ishii M, Wen H, Corsa CAS, Liu T, Coelho AL, Allen RM, et al. Epigenetic regulation of the alternatively activated macrophage phenotype. Blood 2009;114(15):3244–54.

[196] Chang Y-C, Chen T-C, Lee C-T, Yang C-Y, Wang H-W, Wang C-C, et al. Epigenetic control of MHC class II expression in tumor-associated macrophages by decoy receptor 3. Blood 2008;111 (10):5054–63.

[197] Jiang T, Zhou C, Ren S. Role of IL-2 in cancer immunotherapy. Oncoimmunology 2016;5(6). Available from: https://www.ncbi.nlm.nih.gov/pmc/articles/PMC4938354/.

[198] Ingram JT, Yi JS, Zajac AJ. Exhausted CD8 T cells downregulate the IL-18 receptor and become unresponsive to inflammatory cytokines and bacterial co-infections. PLoS Pathog 2011;7(9), e1002273.

[199] Shin H, Blackburn SD, Blattman JN, Wherry EJ. Viral antigen and extensive division maintain virus-specific CD8 T cells during chronic infection. J Exp Med 2007;204(4):941–9.

[200] Wherry EJ, Barber DL, Kaech SM, Blattman JN, Ahmed R. Antigen-independent memory CD8 T cells do not develop during chronic viral infection. Proc Natl Acad Sci USA 2004;101(45):16004–9.

[201] West EE, Jin H-T, Rasheed A-U, Penaloza-MacMaster P, Ha S-J, Tan WG, et al. PD-L1 blockade synergizes with IL-2 therapy in reinvigorating exhausted T cells. J Clin Invest 2013;123(6):2604–15.

[202] Moynihan KD, Opel CF, Szeto GL, Tzeng A, Zhu EF, Engreitz JM, et al. Eradication of large established tumors in mice by combination immunotherapy that engages innate and adaptive immune responses. Nat Med 2016;22(12):1402–10.

[203] Mueller YM, Petrovas C, Bojczuk PM, Dimitriou ID, Beer B, Silvera P, et al. Interleukin-15 increases effector memory CD8+ t cells and NK cells in simian immunodeficiency virus-infected macaques. J Virol 2005;79(8):4877–85.

[204] Mueller YM, Do DH, Altork SR, Artlett CM, Gracely EJ, Katsetos CD, et al. IL-15 treatment during acute simian immunodeficiency virus (SIV) infection increases viral set point and accelerates disease progression despite the induction of stronger SIV-specific CD8+ T cell responses. J Immunol 2008;180(1):350–60.

[205] Fraietta JA, Nobles CL, Sammons MA, Lundh S, Carty SA, Reich TJ, et al. Disruption of TET2 promotes the therapeutic efficacy of CD19-targeted T cells. Nature 2018;558(7709):307–12.

[206] Carty SA, Gohil M, Banks LB, Cotton RM, Johnson ME, Stelekati E, et al. The loss of TET2 promotes CD8+ T cell memory differentiation. J Immunol 2018;200(1):82–91.

[207] Brunetti L, Gundry MC, Goodell MA. DNMT3A in leukemia. Cold Spring Harb Perspect Med 2017;7(2), a030320.

[208] Mayle A, Yang L, Rodriguez B, Zhou T, Chang E, Curry CV, et al. Dnmt3a loss predisposes murine hematopoietic stem cells to malignant transformation. Blood 2015;125(4):629–38.

[209] Travers M, Brown SM, Dunworth M, Holbert CE, Wiehagen KR, Bachman KE, et al. DFMO and 5-azacytidine increase M1 macrophages in the tumor microenvironment of murine ovarian cancer. Cancer Res 2019;79(13):3445–54.

CHAPTER EIGHT

Cellular therapeutics in immuno-oncology

Gulzar Ahmad[a] and Mansoor M. Amiji[b]
[a]Repertoire Immune Medicines, Cambridge, MA, United States
[b]School of Pharmacy, Bouve College of Health Sciences, Northeastern University, Boston, MA, United States

Contents

1. Introduction

Cancer is usually treated with different therapeutic modalities including chemotherapy, radiation, surgery, molecular-targeted therapy, and endocrine therapy [1, 2]. During past few years, extensive research and advancement in cellular therapy has resulted in a paradigm shift in cancer treatment and is showing mild safety and encouraging activity against different types of cancers [2]. Among these therapies, immunotherapy can be in the form of

Cancer Immunology and Immunotherapy
https://doi.org/10.1016/B978-0-12-823397-9.00008-9

cytokines, vaccinations, monoclonal antibodies (mAbs), immune check point inhibitors, and adoptive cell transfer. In case of adoptive cell therapy, a robust immune response is generated against a specific tumor type through manipulation of specific immune cells *ex vivo*. Three main immune cells under study now a days are T cell, NK cell, and macrophages. Up till now, most of the work has been done in adoptive T cell therapy field, and T cell therapy is showing promising results against different types of cancers [3]. Other immune cell type-based therapies under study because of their additive benefits. Current T cell therapies can be applied for the treatment of cancer through three different approaches. In the first approach, antitumor T cells are isolated from primary tumor tissue of the patient, expand these tumor infiltrating lymphocytes (TIL) *ex vivo*, and put back into the patient to treat cancer. However, because cancer patients undergone different treatment modalities, and during this treatment regimen, these patients lose TIL at every step of the way which result in insufficient TIL for treatment of these patients. This is one of the reasons this method is not fruitful to all types of cancer patients and has shown efficacy only in those patients whose tumor has higher number of TILs to start with [4]. In the second approach, synthetic T cell receptor (sTCR) is genetically transferred to T cells, these T cells are expanded in the laboratory and infused into patient [5]. However, because of its high selectivity, this method has limited success. Third and most successful immunotherapy approach till today is chimeric antigen receptor. We highlighted the CAR cell therapy for T cells, NK cells, and macrophages in this chapter.

2. Chimeric antigen receptor (CAR)

Chimeric antigen receptor (CAR) is the most commonly used approach for generating tumor target-specific immune cell through genetic modification of these cells. CAR has been introduced into different types of immune cells including T cell, NK cells, and macrophages, and these CAR transduced immune cells are named as CAR-T, CAR-NK, and CAR-M, respectively. The concept of CAR was first introduced by Gross and others in 1989, when they fused the antibody-binding domain of Fab with the TCR signaling domain CD3ζ and called it T body [6]. Chimeric antigen receptor typically consists of three main regions. First, CAR region is an extracellular ligand binding domain of a single chain variable fragment of the antibody (scFv) [2, 3]. This region is responsible for recognizing specific target antigen independent of the peptide-HLA complex. Second region of a CAR construct contains a transmembrane domain and a hinge/extracellular spacer [2, 3]. This region links other two components of a CAR and enhance overall flexibility, stability, and dimerization of the construct. Third region is a cytoplasmic signaling domain and costimulatory molecules [2, 3]. This region is responsible for transmission of signal, and most of the CAR construct modification is done in this region to further improve signal transmission capacity of a specific CAR [7] (Fig. 1).

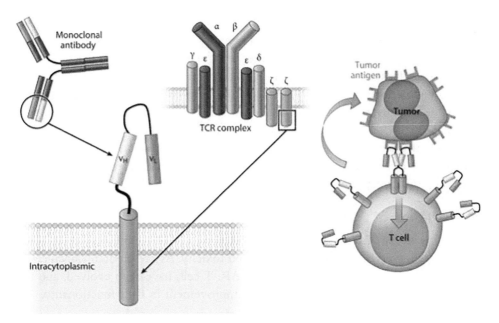

Fig. 1 The basic structure of chimeric antigen receptors (CARs). The most common CARs combine the extracellular antigen-recognition site of a monoclonal antibody and the intracellular domains of a T cell receptor (TCR) complex molecule. Clustering of CARs induced by antigen binding on the surface of tumor cells initiates signal transduction that leads to T cell activation and killing of tumor cells. *(Figure reused with permission from the authors: Ramos CA, Heslop HE, Brenner MK. CAR-T cell therapy for lymphoma. Annu Rev Med 2016;67:165–83. https://doi.org/10.1146/annurev-med-051914-021702.)*

3. CAR-T cells

CAR-T therapy is one of the most studied approaches and has performed better against hematological malignancies than solid tumors.

CAR-T cells are using genetically modifying with synthetic immune receptors. These synthetic immune receptors redirect T cells through recognition of specific antigens expressed on tumor cells to eradicate them [2, 3]. There are several advantages of CAR-T cell therapy compared to other immune cell therapies. For example, CAR-T cells can be generated from nonspecifically activated polyclonal T cells which make isolation and amplification of these cells less cumbersome compared to natural tumor specific CD4$^+$ and CD8$^+$ T cells [2, 3]. Also, CAR-T cells are not bound to recognize an extracellular antigen through MHC-dependent manner. This means that these CAR-T cells can recognize even those tumor cells which has either reduced HLA expression or antigen processing. Therefore, CAR-T cells are more effective against cancer because both degree of antigen processing and HLA expression level are one of the main determinants of immunological escape of the tumor [8]. Along with above advantages, CAR-T cells has the ability to actively and specifically home to tumor microenvironment, where they can persist and expand after recognizing the tumor-associated antigens (TAA). These advantages make CAR-T cell therapy more promising than other cellular immune therapies especially mAbs by producing long-lasting response against specific tumor type [9]. Lastly, CAR-T cells have the potential to cross blood brain barrier, which makes these highly desired therapies against metastasized tumor that have entered into central nervous system [9]. However, systemic delivery of these cells to central nervous system can also cause some adverse effects that need to be considered while recommending this therapy.

3.1 The design of CAR-T

Since the discovery of CARs, scientists are working to further improve their efficacy against different types of tumors and reduce their toxicity. As a result of this advancement in CAR-T design, different generations of CAR-T cells have been evolved, and these CARs are categorized according to order of improvement in their functionality.

3.1.1 First-generation CARs

This generation of CAR constructs consists of an extracellular scFv-binding domain for binding to a targeted tumor-associated antigen (TAA) and CD3ζ containing cytoplasmic domain. In this generation of CAR molecule, first signal was triggered due to binding to scFv to TAA and second signal which is required for T cell activation was provided through CD3ζ containing cytoplasmic domain or the Fc receptor c [2, 3, 9].

3.1.2 Second-generation CARs

To further improve CARs, in second-generation CAR construct, instead of two-step T cell activation, one step activation was introduced. This one step activation of CAR molecules was due to combining of CD3ζ containing intracellular domain with a costimulatory molecule either CD38 or 4-1BB. This modification resulted in simultaneous induction of both signal 1 and 2 upon binding to targeted tumor-associated antigen [2, 3, 9]. Although this generation of CARs does not outperform the antitumor efficacy in preclinical models compared to first generation CARs, it does improve the persistence of CAR-T cells when 4-1BB was introduced as costimulatory molecule and enhanced the proliferation of CAR-T cells through CD28 combining approach [2, 3, 9].

3.1.3 Third-generation CARs

Instead of single costimulatory molecule, in third-generation CAR constructs, along with intracellular CD3ζ containing domain, two costimulatory domains were introduced. It means after binding to targeted tumor-associated antigen, one signal comes from CD3ζ intracellular domain and two signals from costimulatory molecules. In most

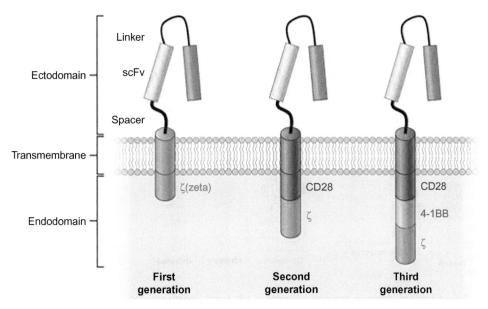

Fig. 2 Three generations of chimeric antigen receptors (CARs). First-generation CARs include an extracellular domain (ectodomain), usually derived from a single-chain variable fragment (scFv), composed of the antigen-binding regions of both heavy and light chains of a monoclonal antibody; a transmembrane domain; and an intracellular domain (endodomain) with a cell-signaling component derived from the T cell receptor, usually the ζ chain. Most subsequent CARs have followed this same structural pattern, with incorporation of one (second-generation CARs) or more (third-generation CARs) accessory or costimulatory signaling components, such as CD28, CD137 (4-1BB), and CD134 (OX40). These additional costimulatory endodomains improve T cell activation and proliferation, and thus may promote killing of target tumor cells. *(Figure reused with permission from the authors: Ramos CA, Heslop HE, Brenner MK. CAR-T cell therapy for lymphoma. Annu Rev Med 2016;67:165–83. https://doi.org/10.1146/annurev-med-051914-021702.)*

of the third-generation CAR molecules, these costimulatory molecules are either combination of intracellular domains of 4-1BB and CD28 or OX40 and CD28 [2, 3, 7–9] (Fig. 2). Till now, very few clinical trials have used third-generation CARs because it is still under investigation whether third-generation CARs are better than second-generation ones.

3.1.4 Fourth-generation CARs
To further improve the efficacy and safety profile of CAR molecules, in fourth-generation CARs along with traditional third-generation CAR designed construct were decorated with a universal killing cytokine-like IL-12 [2, 3, 9]. The purpose of this strategy was to modulate the tumor microenvironment after T cells transduced with such CAR home to the tumor. Inside the tumor, accumulated cytokine from the CAR

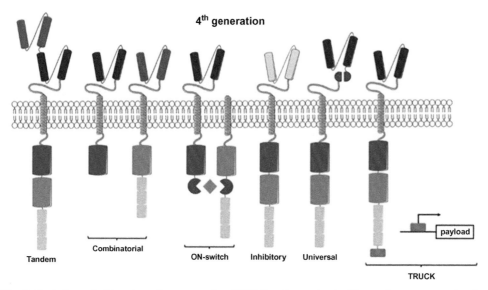

Fig. 3 Evolution of chimeric antigen receptor (CAR)-T cell constructs. The main components of a chimeric antigen receptor and its evolution over time. Examples of fourth-generation CAR construct organized by subgroups. *scFv*, single-chain variable fragment; *TM*, transmembrane; *TRUCK*, T cells redirected for universal cytokine killing. *(Figure reused with permission from the authors: Pfefferle A, Huntington ND. You have got a fast CAR: chimeric antigen receptor NK cells in cancer therapy. Cancers (Basel) 2020;12(3):706. https://doi.org/10.3390/cancers12030706.)*

can trigger an innate immune response against tumor–associated antigen possessing tumor cells and hence keep the tumor under control for longer period of time [10] (Fig. 3).

3.1.5 Further strategies for improving the design of CAR molecules

Different next-generation designs are under investigation for improving the antitumor efficacy and safety profile of existing CARs. One of such design is named "Armored CARs," to combat the immunosuppressive tumor microenvironment [3, 11]. In Armored CAR molecule, a dominant negative form of immunosuppressive receptor or immune check point inhibitor or a cytokine is coexpressed along with traditional CAR molecule in T cells to enhance their efficacy, primarily in solid tumor where tumor microenvironment is usually more immunosuppressive [3, 10, 11]. Second such design is called "Split receptor CARs" in which along with traditional antigen-binding receptor, another receptor is included which is responsive only in the presence of exogenous inducers. Split receptor CAR design, along with improving the outcome of the CAR molecule, also give control over the dosage, location, and timing of these CAR-transduced T cells [3, 12]. Third is called "Off-switch CARs," these CAR constructs use a suicide gene that can destroy T cells transduced with such CAR constructs in the event of adverse outcome from these cells. This adverse response can be either in the form of oncogenic transformation or excessive toxicity because of these cells. For example, proteolytic domain of Cas9 which is linked to human KF506-binding protein allow conditional dimerization only in the presence of rapamycin [13]. Fourth design is named as "Tagged CARs," as the name suggests, these CARs are marked with a specific tag that can be recognized by a specific monoclonal antibody and cleared from the body in case of adverse outcome from these CARs [14]. Fifth design is called "Dual CARs" that express two types of CARs that recognize two different types of targets, and recognition of both the targets is required for activation of these CARs [15]. Then there are TanCARs and iCARs, both of which are designed to restrict the response of T cells with such CAR constructs when needed.

3.2 Production of CAR-T cell

T cells are isolated and purified from patient blood by using one of the well-established protocols [16, 17]. These isolated T cells are then cryopreserved in special bags to be shipped in frozen form to a manufacturing facility where T cells can be processed. Special care is taken to preserve the quality of T cells during all these steps of freezing and thawing [18, 19]. After arrival at designated center, these T cells are thawed and enriched usually through the use of different microbeads, including, CD3, CD4, CD8, and CD62L [19, 20]. After the enrichment of T cells, these cells are activated by using different monoclonal antibodies, such as anti-CD3 and anti-CD28, which are usually coated on cell-sized microbeads and by the use of different cytokines including IL-2, IL-7, and IL-15. This T cell activation approach results in proliferation of both memory and effector T cells [21]. Expanded T cells are now ready to be transduced with potential tumor-associated target antigen. Foreign gene can be transduced into the expanded T cells through both viral and nonviral systems. Among viral systems, lentiviral has some advantages compared to others because they can infect both dividing and nondividing cells. Along with this, lentiviral vectors are capable of incorporating larger transgenes into the target T cells compared to retroviral vectors [16, 17, 22]. Even though viral vectors reduce the CAR production time compared to nonviral systems, viral vectors can cause some adverse events of insertional mutagenesis.

Because of their comparative safety of insertional mutagenesis, nonviral systems, including mRNA transfection, and transposon systems are also being used in CAR production [17, 23]. After the transfer of foreign gene, these transduced cells are activated and expanded usually using anti-CD3 and anti-CD28 beads. However, the use of such microbeads is prone to aggregation, and this limitation of these beads can reduce the final CAR-T functional cells production. To overcome this shortcoming, tetrameric CD3/CD28 antibody complex has been introduced that can improve the number of functional CAR-T cells in final CAR-T therapy product [16, 22, 23].

Even though this tetrameric antibody complex is useful, it can expand a special subset of immune cells and can alter the ratio of CD4 to CD8 and can also expand some untransduced T cells. After the expansion of transduced CAR-T cells, the survival of these cells is critical that need to at least 90% post thaw following 10 months of cryopreservation. After passing through all these critical parameters, best lots of CAR-T cell products are finally used for human clinical application in both hematological malignancies and in solid tumors.

3.3 Use of CAR-T cell for cancer treatment

Treatment status of both hematological malignancies and solid tumors is looking promising but still need further work to cure cancer. Historically for the treatment of cancer, surgery, radiation, chemotherapy, and some hormonal therapies have been the first choice as therapy [1]. However, recently, immunotherapy has emerged as paradigm shift in this regard. Among immunotherapy options, check point inhibitors have been introduced as a drug of choice against selective cancers, especially against immune responsive tumors. After the success of check point inhibitors, different CARs have shown some promising results.

3.3.1 CAR-T cell for hematological malignancies

CAR-T cell therapy has shown some promising results against hematological malignancies. This better response of CAR-T cells against blood cancer is partially because of best selection of potential target antigen. CAR-T is one of the most important immunotherapies currently in use and different targets are under investigation [24]. CD19 is an ideal target antigen, especially in the case of relapsed and/or refractor B-cell acute lymphoblastic leukemia (B-ALL) in some non-Hodgkin lymphomas [25]. CD19 is ubiquitously expressed on a broad range of differentiated B cells, ranging from pro-B cells to memory B cells. But is not expressed on most of the essential cell types, including hematopoietic stem cells. This cell type specific expression of potential target antigen reduces the risk of "on target-off tumor" toxicity. The efficacy of CD19-targeted CAR-T cell therapy ranges from 70% to 90% in both pediatric and adult patients, across different treatment centers in USA and UK [25]. This variable response across different treatment centers is because of different factors. Some of these factors are choice of an approach being used during design of a specific CAR, technology being used to transfer foreign gene into CAR-T cells, dose of the CAR-T infused cells, use of myeloablative agent, and effect of particular design in a clinical trial. For example, the use of electroporation for gene transfer instead of viral vector can cause exhaustion of T cells even before even they are infused into patients. Use of a specific costimulatory domain 4-1BB or CD28 can also result in variable efficacy response. Use of immune-depletion reagent before the start of CAR-T treatment can also affect the persistence and expansion of adopted CAR-T cells. Fludarabine and cyclophosphamide-based preconditioning each has its own unique mode of depleting immune cells. So far, the use of preconditioning chemotherapy both fludarabine and cyclophosphamide has been reported to increase the persistence of infused CAR-T cells and has shown better response in clinical trials. In clinical trials, where fludarabine was used as preconditioning chemotherapy, infused cells showed 100 times greater expansion of infused CAR-T cells and were detectable even after 1 month in preconditioned treated patients compared to nontreated patients [25, 26]. Along with their engraftment benefits, myeloablative immunodepletion has also been useful agent among those patients who need an allogeneic hematopoietic stem cell transplantation (alloHSCT). Some other factors, including ratio of $CD4^+$ to $CD8^+$ in CAR-T cell product, can also affect the final outcome of CAR therapy and might contribute to final output. In the case of B-ALL, age of the patient has also been reported to be an important variable in CAR-T response. In B-ALL patients, adult patients have been reported to have higher rate of relapse and mortality compared to young ones, and this efficacious disparity among young and adult patients is because of different number of immune cells, immune-senescence, and functional status of the thymus between two age groups [25, 26]. Along with above factors, variability in tumor type/volume and pre-treatment history of the patient are also important factors for predicting the response output and lifelong memory phenotype of the infused CAR-T cells [25, 26]. Currently, different laboratories are working toward discovering the contribution of above factors either alone or in combination with preconditioning chemotherapy in rate of CAR-T cells persistence both at preclinical and clinical settings. CAR-T therapy does not show same rate of success among all hematological malignances. For example, compared to B-ALL success, the CAR-T infused perform modestly against chronic lymphocytic leukemia (CLL) [27]. Different factors have been attributed to this poor response in the case of CLL, including inherit dysfunctional nature of effect T cells from CLL patients, preferred rate of migration, and penetration of infused CAR-T cell in lymph nodes compared to bone marrow, higher population of immune-suppressive cells and immune suppression because of upregulation of checkpoint inhibitory receptors [28]. Even though CD19 has been proven to be an excellent therapeutic target, additional potential targets for CAR-T cells against hematological malignancies have been identified and are being validated. These potential CAR-T cell therapy targets are showing promising preclinical results, and few of them are even making their way to clinic. Some of these CAR-T targets are CD30, ROR1 (inactive tyrosine-protein kinase transmembrane receptor) and Ig kappa light chain for different B-cell malignancies [29]. B-cell maturation antigen (BCMA), CD138, CS1, and CD38 for multiple myeloma [30]. For the treatment of AML, CD33 and CD123 have been validated and are partially characterized [31].

3.3.2 CAR-T cell for solid tumors

Historically, CAR-T cell therapy has been reported to be more successful against hematological malignancies than solid tumor. This biased disparity in success of CAR-T cells against solid tumors has been due to some significant challenges. Some of the these challenges from solid tumors include highly immune-suppressive tumor microenvironment, difficulty in selection of antigen, which is usually because of higher rate of antigen heterogeneity compared to hematological malignancies, and finally problematic "on-target/off-tumor" toxicity because of expression of CAR-T therapy targets in some essential organs along with tumor [32]. CAR-T cell therapy against various potential targets for different solid tumors are either undergoing preclinical validation or being characterized in clinical trials. For example, in case of brain tumor, IL13Ra2, EGFRvIII, and HER2 are well studied potential targets for CAR-T therapy, while in case of breast cancer, HER2, c-Met, and MUC1 are undergoing different preclinical studies [32]. In this section, we will summarize the current status of CAR-T therapy against these potential targets and will present the future of other CAR-T therapy candidates in solid tumors. In case of brain tumor, these potential targets (IL13Ra2, EGFR variant III, and HER2) were selected because of their comparatively higher expression in tumor compared to normal brain. Tumor-restricted expression of target antigens is very important for all solid tumors, but in case of brain tumor this is of upmost importance because "on-target/off-tumor" toxicity can even lead to mortality of the patient in this tumor. MAGE-A3 is a well-known example of on-target/off-tumor toxicity that unfortunately results in inflammatory response and lethal destruction of neurons [33]. MAGE-A3-related toxicity force the clinicians to emphasize the selection of potential target antigen, the preparative regimens of these targets, and the dose of CAR-T cells during infusion into patients. In brain tumor, first ever, CAR-T therapy target was IL13Ra2, which is a high affinity IL-13 receptor overexpressed in brain tumor but is not expressed in normal brain tissue and is a prognostic indicator of poor survival. This CAR-T therapy target has been reported to be expressed by both differentiated and stem cell-like glioma cells, rendering both these cell populations to be susceptible to CAR-T therapy [34]. First, the generation of CAR-T cells against IL13Ra2 showed efficacy in both *in vitro* and preclinical studies. However, further improved second-generation CAR-T therapy against this glioblastoma target after inclusion of 4-1BB as costimulatory domain enhanced the antitumor efficacy up to 10 times [35]. This increased anti-glioblastoma activity was confirmed through increased tumor necrosis by MRI and other imaging techniques, reduction in IL13Ra2+ tumor cells, and detection of infused CAR-T cells away from site of cell injection. Second well studied potential glioblastoma target is EGFRvIII.

Epidermal growth factor receptor (EGFR) is a well-known member of a family of receptor tyrosine kinases. In case of glioblastoma, this receptor has been either mutated or genetically amplified in about 50% of cases, render it a potential antitumor target. Mutated version of EGFR, named as variant III, has been reported to have tumor-restricted expression and does not found in normal brain. This tumor-specific presence of this EGFR variant III makes it an ideal candidate for CAR-T therapy [36]. In both preclinical studies and clinical trials, CAR-T immunotherapy against EGFRvIII, at lower dose, was safe but lead to pulmonary edema at highest dose. Even though infusion of these CAR-T cells through intravenously after lymphodepletion showed access to brain. However, neither persistence of these cells in the brain for long time nor antitumor efficacy of these infused cells was impressive. The impaired efficacy of the infused CAR-T infused cells has partially been due to highly immunosuppressive tumor environment. This environment is caused due to higher number of T regulatory and other immunosuppressive cells. Upregulation of the proteins is involved in the immunosuppressive phenotype of the tumor including IDO1, IL-10, and PD-L1 check points. This upregulation emphasizes the use of immune checkpoint inhibitors combinations in addition to CAR-T cell therapy. Although the expression of EGFRvIII is tumor-restricted, but it is unstable throughout the course of the disease, making reoccurrence of brain tumor a likelihood during treatment through antigen escape [36, 37]. Another well-studied anti-glioblastoma target for CAR-T immunotherapy is HER2, which has also been studied in the case of breast cancer.

HER2 is another member of receptor tyrosine kinase. Unless EGFR, activating mutation in this receptor is less common but its overexpression has been reported most commonly in breast cancer and some other cancers. In case of glioblastoma, it is overexpressed in 80% of tumors. Overexpression of HER2 has been quite often result in malignant transformation [38]. CAR-T therapy against HER2 tumors has been studied both in preclinical and clinical settings. Anti-HER2 CAR-T showed remarkable results *in vitro* and output of these CAR-T cells *in vivo* was also impressive. Even in phase 1 dose escalation clinical trials, anti-HER2 CAR-T cell dose well tolerated [39]. Other than above anti-brain tumor CAR-T targets, Chondroitin sulfate proteoglycan 4 (CSPG4), CD47, CD133 and Erythropoietin-producing hepatocellular carcinoma (EphA2) are also under investigation evaluation as potential anti-glioblastoma CAR-T targets.

Similar to brain tumor, CAR-T immunotherapy is also used for breast cancer. For breast cancer, along with HER2, MUC1 and c-MET are potential antitumor targets for CAR-T cell therapy. Other solid tumors for which CARs are being tried are prostate cancer, colorectal cancer, ovarian cancer and metastatic renal cell carcinoma [38–40]. CAR-T cell immune therapy need further work to be applicable to solid tumors because of various challenges.

3.4 Limitations of CAR-T cell therapy for cancer treatment

CAR-T cell therapy has been emerged as a promising cancer treatment over the past few years and is preforming good against hematological malignancies, but its efficacy against different solid tumors still need further work. The weak performance of this therapy is due to different challenges that this therapy has to face against solid tumors. These challenges range all the way from rarity of tumor-restricted target antigen, toxicity, immuno-suppressive tumor microenvironment, and many others. Each of these limitations is elaborated below.

3.4.1 Status of target antigen

Currently, various potential antigens have been targeted for both hematological malignancies and solid tumors in different clinical trials. For CAR-T cell therapy, tumor-restricted expression of the target antigen is of optimum importance because CARs consisting of scFv domains are only able to access external antigens unlike naïve TCR (T cell receptors) which can target internal antigen after their processing and presentation through MHC molecules to T cells. This confines the accessibility of CAR-T cell therapy to limited antigens compromise their capability. In most of cases, CARs can target only one antigen at a time, but now a days with the advancement in the designing approach of CARs, multiple antigens can be targeted with single CAR molecule. For example, "split signal CARs" can target two antigens at a time and let T cell activation only in those tumors which express multiple antigens [3, 11]. Similarly, tandem CARs can also recognize multiple antigens. Therefore, CARs targeting multiple antigens would be instrumental against cancer immune escape phenomenon which had proven a mode of resistance against previous cancer treatment options including chemotherapy and can become to be a dilemma for cell-based therapies including CAR-T cell therapy. Along with targeting multiple antigens, targeting tumor stroma would be helpful to overcome immune-escape challenge. Because tumor stroma by secreting different cytokines and growth factors and by providing nutrient supplements for tumor growth are important facilitator to tumor immune–suppression environment [3, 9, 11].

3.4.2 Persistence of infused CAR-T cells

Higher rate of persistence of infused cells in peripheral blood would increase the chances of access of these cells into tumor bed. Different factors can effect *in vivo* persistence of CAR-T cells, including molecular designing of CARs, preconditioning of patients with myeloablating chemotherapy, use of specific techniques, and conditions for the culture of these CAR-T-transduced cells, mode of foreign gene integration, phenotype and quality of T cells obtained from patient, ratio of different subsets of immune cells, quality of the final CAR-T cell product, and response of the host patient against infused cells. For example, inclusion of 4–1BB as costimulatory domain in the design of CAR has been reported to enhance the persistence of infused cells. Infused CAR-T cells with memory phenotype persist longer in patients than exhausted, differentiated, and activated T cells. Preconditioning of the patient before infusion of CAR-T cells has also proven helpful in increasing their persistence. Use of different cytokines supplemented with media during culture also improve the phenotype of CAR-T cells. Instead of using as supplement, CAR-T can also be designed to secret these cytokines for improving the persistence and overall health of the infused cells and avoiding systemic toxicity at the same time [11, 12]. Persistence of infused cells can also be improved by blocking the signals that compromise the efficacy of the infused CAR-T cell product.

3.4.3 Homing of infused CAR-T cells

A critical parameter for the success of infused CAR-T cells is their ability to bind its target antigen on tumor cells which is only possible if these cells can migrate to the tumor microenvironment. One of the reasons for the success of CAR-T cell therapy against hematological malignancies, especially against B-ALL, is to the ease of the infused CAR-T cells to bind their potential targets in the presence of different costimulatory receptor ligands. However, because of the complicated nature of the solid tumor, homing of infused CAR-T cells inside the tumor is very challenging. Chemokines play an important role in migration of lymphocytes including infused cells; however, chemokine system is very complex because it is involved in the secretion of both lymphocyte migration facilitator and their regulatory chemokines. Either the external supplementation or genetic modification of CAR-T cells to secrete lymphocyte migration facilitating chemokines would be helpful in enhancing the homing these infused CAR-T cells inside the tumor [3].

3.4.4 Immunosuppressive microenvironment of the tumor

Success of any treatment modality and especially immunotherapy dependents on the microenvironment of the tumor. Solid tumor bed consists of plenty of tumor growth facilitating nutrients, population of different immune suppressive cells that prevent anti-tumor effect of immune system, different immunosuppressive cytokines, and higher expression of different coinhibitory receptors. Immunosuppressive cells inside the tumor bed include tumor-associated macrophages, immature myeloid cells, and regulatory T cells. CAR-T cells perform their antitumor efficacy after overcoming all these immunosuppressive factors [11, 12]. During this antitumor activity, cytotoxic function of these infused CAR-T cells can either become suppressed or their ability to engage with target antigen is impaired and that results in their compromised antitumor efficacy. Furthermore, lack of nutrients for these infused CAR-T cells, oxidative stress, hypoxic conditions, and acidic pH of the tumor bed can limit their proliferation and survival. Along with above factors, different upregulation of different immune checkpoint inhibitors including programmed death 1 (PD1), cytotoxic T-lymphocyte-associated protein 4 (CTLA-4), Tigit, Tim-3, and Lag-3 has been reported in tumor microenvironment that can impair the effector function of transferred CAR-T cells [9, 11, 12]. All of above factors either directly or indirectly have been involved in compromising the efficacy of infused CAR-T cells in case of solid tumors. This immune-suppressive tumor microenvironment can be copped by implementing different approaches, by designing CARs with dominant negative form of TGFβ, to avoid apoptosis, by knocking down of Fas/Fas ligand, or by overexpressing survival genes including BCL-XL and by including of different cytokines, such as IL-12 and IL-15, to improve the tumor microenvironment. Urging to further improve not only the design of these CARs but also try combinatorial therapy approach to get maximum benefit from this emerging immunotherapy [11, 12].

3.4.5 Toxicity because of CAR-T cells

Although CAR-T cell therapy is showing some promising results in case of hematological malignancies, there are some toxicity concerns. This toxicity is primarily caused due to "on-target/off-tumor" effect because in most of cases, along with tumor potential target, is also expressed on normal tissues. Therefore, the choice of ideal CAR-T target whose expression is only restricted to tumor is necessary to ameliorate this toxicity. Along with on-target/off-tumor toxicity, incorporation of different costimulatory domains in CAR design has also been reported to cause toxicities. Different types of CAR-T-related toxicities have been reported, among them cytokine-release-syndrome (CRS), macrophage activation syndrome (MAS), and neurological related toxicities are well studied. CRS is caused because of higher activation of T cells and macrophages, which result in excessive secretion of different cytokines including interferon-gamma (IFN-γ), interleukin (IL-1), interleukin IL-10, granulocyte macrophage colony-stimulating factor (GM-CSF) [41]. Although some of the cytokines are also required for antitumor activity, but excess of these cytokines can result in life-threating condition. That is why CAR-T cell therapy is considered as "double-edge-sword." Even though exact cause of neurotoxicity is still unknown, but this toxicity is usually has been linked to either higher level of different cytokines or presence of infused CAR-T cells in spinal fluids [42]. Neurotoxicity can be in the form of expressive aphasia, delirium, cerebral edema, confusion, or seizures. Different approaches are under consideration to ameliorate this toxicity from infused CAR-T cells. For example, along with selection of ideal tumor target, infusion of CAR-T cells with transient expression, optimization of dose of these CAR-T cells, genetic modification of CAR-T cells expressing either suicide or safety switch along with CARs for self-regulation in case of over-toxicity either through over expansion or of infused CAR-T cells, and excessive secretion of different cytokines and use of antibodies against most notorious cytokines like IL-6 would be helpful against above toxicities [41, 42]. Not all the toxicities are related to CAR-T cell therapy, some of these toxicities can be due to patient characteristics and treatment regimen. For example, patients with higher tumor burden at the time of treatment initiation, higher level of angiopoietin-2, thrombocytopenia, and von Willebrand factor are at greater risk of developing these toxicities [41, 42].

3.5 Strategies for improving CAR-T cells

Different approaches are under investigation to further improve the antitumor efficacy and safety profile of infused CAR-T cells. The toxicity from CAR-T cells can be lessened by two different strategies; either by decreasing the signal strength of infused CAR-T cells or by lowering the level of harmful cytokine production. Both above safety milestones can be obtained by engineering novel CAR constructs by incorporating "safety switch" [41, 42]. In order to improve the persistence of infused CAR-T cells, culture protocols that will preserve the phenotype of infused CAR-T cells because higher percentage of infused cells in naïve or central memory phenotype have been reported to improve their persistence. The infused CAR-T cells can also be kept in less differentiated phenotypes either by reducing the number of days in culture or use of different cytokines cocktails [41, 42]. The efficacy of these infused CAR-T cells can also be improved either by deleting a gene that suppress the activity of CAR-T cells or overexpress a gene that facilitate the expansion and persistence of these cells. Along with genetic manipulation of CAR constructs, supplementation with different chemokines can also improve the homing of infused CAR-T cells. Safety of the CAR-T product can also be improved through regional or local delivery of these cells and also by reducing the number of cells. "Tandem CARs" approach has also been beneficial not only in controlling the tumor but also in improving the safety of transferred CAR-T cells. For improving the efficacy, the immunosuppressive tumor microenvironment needs to be overcame either by the use of check point inhibitor combination or use of dominant negative mutant of immuno-suppressive molecules [11, 12].

4. CAR-NK cells

Natural killer (NK) cells are primary defense tool against external invaders and perform their role through innate immune system. NK cells were discovered in 1975 from two different research laboratories [43]. Along with their defensive role, NK cells are also an important potential effector in immune oncology. These cells have the ability to target different tumors without human leukocyte antigen (HLA) or antigen presensitization. Unlike T cells, activation of NK cells is controlled through integration of signals responsible for stimulation and inhibition. These cells get activated when the signals originated from activating receptor such as natural killer group 2, member D (NKG2D), and CD16 are dominant than signal from inhibitory receptors. NK cells were considered important after their role in malignancy was reported in NK-deficient patients. These patients did not show increased rate of malignancy but become susceptible to different viral pathogens [44]. Later on, similar findings were also reported from mice studies where impaired NK activity make them vulnerable to different pathogens. Currently, much of the research work on chimeric antigen receptor (CAR) has been focused on T cells because of their promising results against hematological malignancies [3, 4]. Two CAR-T cell products have also obtained approval from Food and Drug Administration (FDA) for human use because of their promising results against hematological malignancies [4, 5]. However, toxicity issues related to CAR-T product are big concern, which has been discussed in detail in previous section of this chapter. NK cells are being considered as an alternate to CAR-T cell therapy because of some advantages compared to CAR-T cell product.

NK cells have better safety profile, there is almost no induction of graft *versus* host disease (GvHD) from transplantation of allogeneic NK cells and the ease of NK cell availability because these cells can be obtained from different sources, including umbilical cord blood (UCB), human embryonic stem cell (hESC), peripheral blood (PB), human induced pluripotent stem cell (hiPSC), and NK cell lines [45].

4.1 The design of CAR-NK

Design of chimeric antigen receptor is very important to get maximum benefit from cells transduces with CAR constructs. Similar to CAR-T molecules, design of CAR-NK is typically based on CAR-T designing except some unique activation and signaling features of NK cells. CAR-NK construct consists of an antigen recognition domain and therefore can be either single-chain variable fragment (scFv) or natural killer group 2 member D (NKG2D) [43, 45]. Second, it contains a transmembrane domain and a signaling domain that provide signals for the activation of NK cells. Instead of adding scFv as an antigen recognizing domain, by including NKG2D, CAR-NK can recognize multiple ligands, including UL-16, major histocompatibility complex class I chain-related A and also MICB which is expressed on different types of tumor cells. Unlike scFv domain-containing CARs, NKG2D CAR along with recognizing around 90% of different human tumor types can also target immunosuppressive cells including regulatory T cells (Tregs) and myeloid derived suppressor cells (MDSCs), which express NKG2D [43, 45]. Similar to CAR-T constructs, CAR-NK has different generations. In case of first-generation CAR-Ns, the signaling domain is either CD3ζ or DNAX–activation protein 12 (DAP12). Each of these domains has their own value. DAP12 domain might activate NK cells better than CD3ζ signaling domain and CD3ζ domain works better as signaling domain compared to DAP12. This generation of CAR-NK has shown promising results against different malignancies especially, CD19, HER2, and CD20 target antigens expressing tumors [45, 46]. Second-generation CAR-NK, in addition to standard CD3ζ signaling domain, has an additional either CD28 or 4-1BB signaling domain. Both of these additional signaling domains have their pros and cons. For example, when 4-1BB is used as signaling domain, it improves the lysis ability of CAR-NK against target cells. Usually, second-generation CAR-NK are more effective than first-generation CAR-NK. Third-generation CAR-NK have two costimulatory domains added to it to further improve the efficacy of this chimeric antigen receptor construct [47]. Along with improving the activation of CAR-NK cells, there is a dire need to target immunosuppressive tumor microenvironment, and for this purpose, fourth-generation constructs have the ability to secrete different immune-modulating cytokines including IL-12, which can improve response of the infused CAR-NK cells (Fig. 3). Further strategies are under study to improve the design of CAR-NK.

4.2 Production of CAR-NK cell

For CAR-NK cell production, NK cells can either be obtained through autologous or allogeneic approach from bone marrow, products of apheresis, umbilical cord blood cells (UCB), peripheral blood monocytes (PBMCs), induces pluripotent stem cells (iPSCs), and human embryonic stem cells (hESCs) [48] (Fig. 4). For autologous source of NK cells, primary NK cells are mostly isolated from blood PBMCs from cancer patients who has gone through different chemotherapy treatment regimens [43, 46]. Cells obtained from such patients have lesser yield and the obtained NK cells also has compromised functionality including their cell killing ability. Also, autologous NK cells can have variability from patient to patient in downstream application like efficiency to transduce CAR genes, activation state of infused CAR-NK cells that can affect the overall quality, and composition of the final product [43, 46]. Autologous infused CAR-NK cells have shorter life of span compared to CAR-T cell products. The shorter

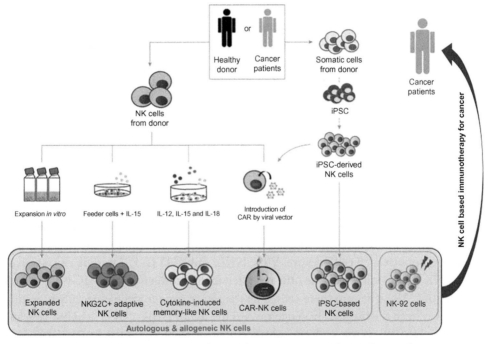

Fig. 4 Methods of NK cell-based immunotherapy for cancer. Types of NK cells used for cancer immunotherapy. Allogeneic NK cells and autologous NK cells, manipulated with cytokines (IL-12, IL-15, and IL-18) and CAR, resulted in various types of NK cells for cancer immunotherapy. Somatic cells from the donor could be used as functional iPSC-based NK cells. Irradiated NK-92 cell line also served as an important resource for NK cell-based cancer therapy. *(Figure reused with permission from the authors: Shin MH, Kim J, Lim SA, Kim J, Kim SJ, Lee KM. NK cell-based immunotherapies in cancer. Immune Netw 2020;20(2):e14. https://doi.org/10.4110/in.2020.20.e14.)*

life span of these cells reduces the toxicity concerns because of reduced off-target effects. However, this short life span of infused CAR-NK cells also compromises their ability to regress tumor. Whereas allogenic NK cells are not only cost effective but it can also be safely infused into a HLA-mismatched recipient patient [45, 46]. For allogenic NK cell production, plenty of NK cells can be obtained even from peripheral human blood because buffy coat layer of a healthy human PBMC consists of 10%–15% CD3-negative and CD56-positive NK cells which results into 10–20 million NK cells from one buffy coat layer. Also, immobilized apheresis can have 5%–15% NK cells. Allogenic NK cells are purified from PBMCs through standard protocol. During which CD3+ cells are depleted first and then enriched from CD3-cells with CD56 using standard immuno-magnetic bead separation technique [43, 46]. Other than allogenic NK cells can also be obtained and expanded from UCB *in vitro* through a two-step differentiation process by using a defined cocktail of different cytokines. NK cells obtained through this way can be expanded up to 15,000-fold with 100% purity and possess powerful functional ability. Similar to UCB, derivation of NK cells for clinical scale from iPSCs and hESCs is done through two-stage culture protocol by using a specially defined media that use recombinant protein-based components [43, 46].

Even though the primary NK cells are good source to start with, but limited expansion potential and life span of these primary NK cells make their use in clinic less favorable. Because unlike T cells, stable human NK cell lines are available that can be expanded *in vitro* continuously without above limitations. One of such cell lines is NK-92, isolated from human non-Hodgkin lymphoma patients [49]. This cell line express NK cell activating receptors including NKp46, NKp30, and NKG2D which can target many different cancer types. NK-92 cell line possesses many characteristics of activated primary NK cells such as high level of cytotoxic enzyme granzyme A and B but lacks most of the inhibitory receptors. Because of these characteristics, NK-92 cell line has been introduced into human clinical trials and is showing promising results toward safety and efficacy as CAR-NK product [46, 49]. In addition to NK-92, there are other immortal NK cell lines that are under investigation for their use in CAR-NK application [46, 49]. Some of the most popular such cell lines are, YTS, NKL, and KHYG-1. After the expansion of primary NK cells or stable cell lines, the next step in the production of CAR-NK product is transfer of CAR gene into these cells. The transfer of foreign gene especially in resting NK cells has been the biggest challenge toward the use of CAR-NK cells as human cancer therapy [46, 49]. Now with the advancement of science, similar to standard CAR immune cells, foreign gene can be transferred into NK through both viral and nonviral vectors. In case of viral CAR transduction approach, both retrovirus and lentiviral particles have been applied that lead to stable integration of transferred gene into cell. This stable integration later on can cause adverse effects because of integration of insertional mutations into transduced NK cells and also cause toxicity because of continuous activation of infused CAR-NK cells into patient. Even though retroviral transduction has not only better efficacy compared to lentivirus but also has higher rate of insertional mutagenesis in primary patient-derived PBMC [44, 45, 49]. Hence, lentiviral particles transduction in human primary PBMCs needs further improvement. However, in case of cord blood-derived NK cells, transduction efficiency of lentivirus is improved [50]. Because of toxicity and insertional mutagenesis concerns from viral vector approach, nonviral mode of transfer of foreign gene into primary human-derived PBMCs is also getting attention. Naked DNA usually do not integrate into host cell genome, these have low immunogenicity and are least expansive compared to other approaches. Among the nonviral approaches, sleeping beauty (SB) transposon system is most popular because it has added features of viral vector. SB system can stably express the transferred foreign gene into transduced NK cells. Sleeping beauty transposon system has shown good results against non-Hodgkin's lymphoma and acute lymphoblastic leukemia patients [51]. In contrast to primary NK cells either from PMBCs or UBCs, the transfer of foreign into immortal cell line like NK-92 is comparatively more efficient. Hence, both viral and nonviral systems work for transfer of CAR-NK into these cell lines [49, 50].

After isolation, expansion, transduction of NK cells with different generations of CARs, cryopreservation and transportation of CAR-NK cells to the clinic are very critical. Because after freeze/thaw, the viability of this cell product has been impaired in most of the cases and it need critical thinking to maintain not only the optimum viability but also the functionality of these infused cells.

4.3 Use of CAR-NK cell for cancer treatment

Recently, CAR-NK cells are making their way to clinic because of their added advantages compared to CAR-T therapy. During this early on application of CAR-NK therapy for human cancers, CAR-NK cells are being used against hematological malignancies, and very few clinical trials are being conducted with solid tumors [49–51]. Most of the clinical studies with CAR-NK are in early phases of clinical trials and are assessing safety, maximum tolerated dose and in some trials relative tumor regression efficacy [50, 51]. For example, a CAR of NK cells, obtained from umbilical cord blood, engineered with CD19-CD28-zeta-2A-iCasp9-IL-15 is investigating safety and efficacy in cohort of patients with either refractory or replaced CD19-positive B-cell lymphoma. For this particular clinical trial, NK cells were derived from umbilical cord blood and transduced with CAR which has iCasp9/IL-15 suicide genes incorporated into the design [49–51]. These genes are meant to secrete IL-15 on one hand to facilitate the survival of the infused cells and inducible caspase-9 (iCasp9), a suicide gene to eliminate the CAR-NK cells upon their over activation on other hand. Another similar type of clinical trial is assessing the safety and relative efficacy of CB-NK cells in special B-cell lymphoma malignancies which were earlier treated either with stem cell transplantation or with higher dose of chemotherapy. Other clinical trials against CD22, CD19, and combined CD19/CD22 potential targets are also under investigation [50, 51]. In case of solid tumors, NKG2D ligand-targeted CAR-NK cells are assessing safety and efficacy, in ovarian cancer study, infused cells are targeting anti-mesothelin and in case of prostate cancer, the potential target for CAR-NK therapy is anti–Prostate Specific Membrane Antigen (anti-PSMA) [50, 51]. All of these above human clinical trials are in their initial stages. Some other clinical trials are also exploring the possibility of CAR-NK use and any safety concerns from this therapy. These CAR-NK are mostly transduced in NK-92 cell lines, hence named as CAR-NK92. CAR-NK92 are evaluating safety against potential targets of CD33, CD7, MUC-1, and HER-2 in solid tumors [49–51]. So far, CD33-directed CAR-NK92 cells have not shown any major adverse effects in relapsed refractory AML patients, indicating that CAR-NK therapy is a safer option than CAR-T cell therapy.

Most of work for CAR-NK is done *in vitro* and in preclinical models. For example, preclinical studies of CAR-NK cells have shown very promising results against different types of cancers. CAR-NK92 cells, redirected against epidermal growth factor receptor (EGFR), showed significant killing of glioblastoma cells *in vitro* [49–51]. Also, CAR-NK92 cells against GD2 had strong killing effect in neuroblastoma. Similar preclinical results have been reported against potential targets from hepatocellular carcinoma, osteosarcoma, prostate, breast, and pancreatic cancers [52].

In the case of acute lymphoblastic leukemia, CAR-NK, redirected against CD5, kills different T cell lymphoma and leukemia cell lines *in vitro* and regress tumor in T-ALL xenograft mouse. CD19-redirected CAR-NK has shown success against acute myeloid leukemia (AML), primary chronic lymphocytic leukemia (CLL), and B-cell lymphoma and leukemia [53]. In case of multiple myeloma (MM), CAR-NK92, redirected against glycoprotein CD2 subset 1 (CS1), not only enhanced IFN-γ secretion but also inhibited their cell growth and prolonged the survival of mice which were bearing multiple myeloma tumor [53]. Along with above targets studied for both solid tumors and hematological malignancies, various other potential targets against different types of cancers are under investigation under preclinical settings, including CD20, CD138, CD3, CD123 in hematological malignancies and GD2, EpCAM, mutant EGFRvIII, WTI, and ROR-1 in solid tumors [52, 53].

4.4 Advantages of CAR-NK cell for cancer treatment

Even though, CAR-T cell therapy is showing promising results against some hematological malignancies. Its application against cancer is slowed because of some limitations, especially toxicity issues. These CAR-related concerns made scientists to explore other immune cell types for future CAR therapies. NK cells have been reported as a possible alternate to T cells because of their superior CAR-driving capabilities than T cells due to following advantages.

First, CAR–NK cells have been proven to be safer than CAR–T cell therapy through different clinical trials. Especially allogeneic NK cells are well tolerated and do not lead to any serious toxicity and GVHD response. It means autologous cells are not the only option for CAR–NK cell therapy. One of the deleterious effects of CAR–T is longer persistence of CAR–T cells that leads to on-target/off-tumor effect. During this longer persistence, infused T cells cause severe long-lasting B-cell deficiency because of attack on normal mature and or progenitor B cells [54]. Since CAR–NK has limited lifespan in circulation, it results in comparatively few on-target/off-tumor effects. CAR–NK cells are also safer because of difference in the type of cytokines being produced from these cells compared to CAR–T cells. Activated NK cells usually produce GM-CSF and IFN-γ cytokines, whereas the CAR–T cells give rise to the excessive production of different pro-inflammatory cytokines like TNF-α, IL-6, and IL-1, which result in severe toxicity [53, 54].

Second, CAR–T cells need a sensitization phase to get activated, during this phase, an unprimed T cells interact with antigen-presenting cells to become activated. This activated T lymphocytes then become effecter T cells that kill the target tumor cell [53, 54]. However, NK cell activation neither require sensitization nor matching of human leukocyte antigen. Instead of this activation of NK cells is controlled by the integration of stimulatory and inhibitory signals. It means, in addition to killing target cell through CAR-specific mechanism, NK cells can also kill the tumor cells by recognizing different ligands through diverse activating receptors, including NKG2D, NKp44, DNAM-1, NKp30, and NKp46 [53, 54]. These activating NK receptors mostly recognize those ligands that are expressed on tumor cells either due to long-term treatment or because of stress induction. Along with receptor activation death of tumor cells, NK cells can also kill the tumor cells through antibody-dependent cell-mediated cytotoxicity (ADCC) [53, 54]. It means unlike CAR–T cells that can only kill CAR-redirected tumor cells, CAR–NK cells along with killing tumor cells expressing tumor-associated antigen can also kill those tumor cells which only express ligands for NK cell receptors [53, 54].

Last and final advantage of using CAR–NK is the easy and abundance of access. NK cells can be obtained from multiple sources, including hESCs, PBMC, iPCSs, UCB and even in the form of different immortal cell lines, best studied cell lines among these is NK-92 which was originated from a 50-year old man with malignant non-Hodgkin's lymphoma [53, 54]. The beneficial aspect of these cell lines is their homogenous cell population that can be easily expanded for clinical applications under good manufacturing practice standards (GMP).

4.5 Limitation of CAR-NK cell for cancer treatment

Even though CAR–NK are much safer and showing promising results in preclinical models. Still, there are a lot of challenges that need to be addressed to make this adoptive immune cell therapy more successful. One of the limitations of CAR-NK cells is the design of CAR being adopted from CAR-T cells [53, 54]. This CAR design still need improvement to be used in CAR-NK cells because the distance between the CAR-NK cell surface and location of CAR binding epitopes, affect their antigen binding ability and hence their activation [55]. So, in order to get better activation and antitumor efficacy and improved safety, design of CAR in CAR-NK cells need further improvements. Second drawback of CAR-NK cell therapy is reduced transduction efficiency of lentiviral vector and increased induction of insertional mutation by retroviral vector in primary NK cells [54, 55]. Even though nonviral vectors have overcome some of these limitations, but they have their own challenges. For example, mRNA transfection has better rate of foreign gene transfer but since transferred gene does not incorporate into NK cell gene, it keeps the expression of transfected gene for few days and then lost that result in their impaired antitumor efficacy [54, 55]. Even though, immortal cell lines do not have any issue of transduction or transfection but since these cell lines for example NK-92 has been derived from a cancer patient who had NK lymphoma [53, 54]. Hence, there is still risk of tumor engraftment after CAR-NK cells are infused into the patient. Along with this, CAR-NK cells are well known for having very low persistence *in vivo*. Even though, low persistence is considered good thing that make CAR-NK cell therapy safer than CAR-T cells, but this low *in vivo* persistence can also affect the antitumor activity of the final product of CAR-NK cells [53–55]. Similar to CAR-T cells, CAR-NK also has difficulty to penetrate into the tumor bed due to some of the immunosuppression factors including TGFβ. Final CAR-NK cell product also very prone to freeze/thaw cycle that can not only decrease the percentage of live CAR transduced NK cells but also impact their functionality [56]. In case of use of allogenic NK cells for making CAR-NK cell therapy product, some of these allogenic cells might be contaminated with T cells that can result into either some sort of lymphoproliferative disorders or GvHD [55, 56]. In order to avoid these disorders, purity of allogenic NK cells is of utmost importance. Therefore, for purifying NK cells, a clinical grade protocol should be followed that assure maximum purity of NK cells without presence of any of the myeloid-derived suppressive or regulatory T cells along with other immune cells.

4.6 Strategies for improving CAR-NK cells

For further improving CAR-NK cell product, we can follow similar strategies that had been proven fruitful in case of CAR-T cells. For example, in order to improve the safety of CAR-NK cells, antigen selection is very important. Tumor restricted antigen selection will not only reduce on-target/off-tumor toxicity but will also improve the antitumor efficacy of infused CAR-NK cells. For example, MUC1 along with tumor cells is also expressed on normal cells with variable expression but an aberrantly glycosylated form of MUC1 (Tn-MUC1) is only expressed on tumor cells. Therefore, CAR-NK transduced cells CAR that are specific against Tn-MUC1 would be safer and more efficacious against this tumor-associated antigen [57]. To restrict, activity of CAR-T cells to tumor only, significant progress has been made in case of CAR-T cell therapy and these can be implemented to CAR-NK cells with some modifications. For example, tumor microenvironment in case of most of the tumors is deprived of adequate amount of oxygen, so by designing CARs that only work in hypoxic environment will limit the activity of infused CAR-NK cells to the tumor area without causing any on-target/off-tumor toxicity. In case of CAR-T cells, an oxygen sensitive CAR has already been designed by incorporating an oxygen level responsive subdomain of HIF1a gene that only function in tumor area [58]. Another similar example is use of protease peptides in masked CAR design. In this strategy, masking peptide block the protease-sensitive linker and antigen binding site interaction until it gets inside the tumor. In the tumor bed, after becoming active protease cleave the linker and set free CAR to function against potential tumor-associated antigen target. Along with above approaches, combinatorial antigen recognition strategy involving *syn*-Notch system has also be instrumental in lessening the safety concerns in case of CAR-T cell therapy. In this system, CAR molecule work against two target antigens at the same time and CAR-NK work only on those target cell that express both synNotch and CAR ligands. If one of these two ligands are not expressed, CAR will not work. This double targeted approach makes CAR activity more specific and reduce toxicity issues [59]. In order to implement all these strategies into CAR-NK, further characterization is required. For example, study of dose/response between different ligands and their respective receptors need further validation. Also, the specificity, strength and duration of synNotch system along with CAR expression kinetics need further investigations [59].

Antitumor efficacy of CAR-NK cells can be improved by removing or silencing the inhibitory receptors during the design of CARs for NK cells, we can make them productive against tumors. After entrance into the tumor bed, next big hurdle for CAR-NK cells is the immune-suppressive environment of the tumor which not has only higher expression of different ligands like NKG2A, TIM3, and PD-1 to name few but also higher number of different immune-suppressive cells including MDSCs and Tregs [60]. This immune-suppressive tumor microenvironment not only compromise the functionality but also shorten the persistence of the infused CAR-NK cells. Therefore, during design of CAR, silencing of NK cell inhibitory receptors should be incorporated.

In addition to above strategies for improving efficacy and safety of CAR-NK cells, strategies explained for CAR-T cells in the first portion of this chapter can also be explored. Since persistence of CAR-NK cells is very low *in vivo*, incorporation of different cytokine secreting genes into the CAR-NK design would prolong their stay inside the body and might improve antitumor efficacy [60]. CRISPR/Cas9 technique has been used to knock-in certain CARs at a constant locus of T cell receptor that not only result into better internalization but also delay the exhaustion of effector T cells and keep T cell in memory phenotype for longer. Similarly, CRISPR/Cas9 strategy can also be implemented for CAR-NK cells to improve their antitumor efficacy and safety profile.

5. CAR-M cells

Macrophages are part of innate immune system that act as one of the first defense systems of the body against foreign invaders. These cells are meant to remove the old and damaged cells along with cellular debris from the body. Macrophages are found throughout the body and get rid of the threatening and old cells through a process called phagocytosis. These immune cells are usually the first cells to make their way inside the tumor where in some cases, these make up to 50% of the total tumor mass [61]. These are the only immune cells that can make their way inside the tumor because rest of immune cells cannot get inside tumor by themselves and need some sort of support to get there [61]. However, instead of being useful in eating tumor, these are being exploited and abused in facilitating tumor growth and has been reported to be associated with worse prognosis across a range of tumor types. Also, tumor cells have a specialized signaling mechanisms and ability to disguise on molecular basis that macrophages cannot engulf them. Rather tumor cells use them for protection through recruitment of other immunosuppressive cells including myeloid suppressive and T regulatory cells [61]. Different institutions across the globe are trying to understand why macrophages are unable to attack the tumor cells and how these innate immune soldiers can be tamed to attack cancer. In recent times, scientists are coming up with different strategies to restore the natural talent of macrophages and also stop their induction into tumor-associated macrophages (TAMs). Macrophages have the ability to penetrate inside the tumor and phagocytose the cells other than the tumor cells because of is expression of an "eat me not" signal called CD47 on tumor cells [62]. Macrophages express SIRPa on their surface that recognize CD47 on other cells and stop attacking them. CD47 is expressed on all the normal cells of the body and that is how normal system of the body protects itself from macrophages by distinguishing self from nonself cells. From different studies, it has been proven that CD47 is overexpressed in various types of cancers compared to their normal counterparts [62]. This is true for both hematological malignancies and solid tumors. For example, acute myeloid leukemia (AML) stem cells have higher level of CD47 compared to their normal cells. Also, different solid tumors including glioblastoma, melanoma, ovarian cancer, osteosarcoma, and squamous cell carcinoma of head and neck express higher level of CD47 as compared to their normal surrounding tissues [62]. Since macrophages play an important role in CD47/SIRPa blockage, it makes perfect sense to exploit the field of synthetic biology and cell therapy to engineer CAR macrophages to target CD47 overexpressing tumor cells [63]. Since past few years, different institutions and companies are working on CAR-macrophages that would be able to sense specific tumor-associated antigens including CD47 and will respond against it either through secretion of different cytokines or through activation of process of phagocytosis [62, 63].

5.1 The design of CAR-M

The design of CAR-macrophages has also been based on other CARs, primarily CAR-T constructs with minor changes. Similar to CAR-T, CAR-M molecule contains the extracellular single chain antibody variable fragment (scFv), responsible for recognizing the specific cell surface antigen [3–5]. For example, for B-cell-related malignancies, CAR-M will be designed with extracellular scFv against CD19, which is highly expressed in these cancers. CAR-M also contains a transmembrane domain-like CD8 which works for CAR-T also [3–5]. Since, mode of action of CAR-M is different than CAR-T, CAR-M molecule needs cytoplasmic domain that can promote phagocytosis. Therefore, unlike CAR-T, CAR-M molecule usually either has Megf10 or the common gamma subunit of Fc receptor (FcR-γ) [62, 63]. Even though CD3ζ domain has also shown good results in preclinical studies of CAR-M, CAR-M macrophages have anti-tumor efficiency similar to endogenous cell system. CAR-M cells containing either Megf10 or FcR-γ were able to engulf not only target-coated beads but also live cells, emphasizing the workability of this system [62, 63]. Furthermore, by creating a "tandem CAR-M," scientists were able to enhance the full cell engulfment of these transduced macrophages. Therefore, by assembling further motifs, phagocyting efficiency of these CAR-Ms can be further improved. Overall, CAR-M is a successful strategy through which CAR-M can be directed against a specific target expressing tumor cells to get rid to cancer. To further refine the output of these CARs, along with other modification, synthetic Notch (SynNotch) receptor approach from Lim *et al.*, can be applied [59].

5.2 Production of CAR-M cell

For using macrophages in CAR-M therapy, these are taken from peripheral blood monocytes from the patients and then engineered *ex vivo* after their expansion. Macrophages are generally resistant to infection by the standard viral vectors like retrovirus and lentivirus which are mostly used in cell and gene therapy field [62, 63]. Recently, lentivirus has been successful in transducing these macrophages. Once the foreign gene engineered CAR-Ms are transduced into macrophages, these not only expressed the CAR-M but also transform themselves into more inflammatory cell types [62, 63]. This change in their phenotype make these macrophages not only to attack the tumor cells and kill themselves but also stimulate other immune cells to attack the tumor cells that leads to a broader immune response. Stronger immune attack means better control of the tumor. After decades of research, scientists are now able to transduce macrophages with these CARs up to some extend but still need further investigation to make it work in most efficient manner [62, 63].

5.3 Use of CAR-M cell for cancer treatment

CAR-M macrophages are very early in their phase of application as cellular therapy. Most of the work currently has been done either *in vitro* human samples or some preclinical tumor models. First proposed clinical trial of CAR-M from a company named Carisma has been delayed due to coronavirus pandemic. This purposed trial is with an anti-HER2 CAR-M as monotherapy in HER2 overexpressing metastatic primarily solid tumors which includes lung, breast, esophageal, ovarian, and gastric cancers [61]. However, there are different clinical trials going on in different phases by using mono or bispecific antibodies against different targets but mostly against CD47 and some trials against SIRPa across the globe [62, 63]. So far, main concern from these antibody trails is on–target/off–tumor toxicity issue due the expression of CD47 on normal tissues also. SIRPa approach is expected to reduce the toxicity but still too early to speculate [62, 63]. Some other targets are also under investigation like CSF1R and TIE2. Even though these also have safety concerns, but scientists are trying to restrict their action against tumor cells to make them safer and more efficacious. Along with these old targets, some potential new targets are also emerging including CD24.

6. Conclusions

Cellular therapies are emerging as best therapeutic option not only hematological malignancies but also for solid tumors. Adoptive cell therapies like tumor infiltrating lymphocytes are moving slowly but chimeric antigen receptor is performing successfully especially CAR-T cell therapy against hematological malignancies. However, rate of success of CAR-T cell therapy for solid tumor is bit slower, primarily due to lack of best antigen target selection, severe toxicity issues, and hard to get inside the tumor bed, etc. Because of the slow success rate of CAR-T cell therapy, scientists are trying other immune cell as a potential delivery agent for these genetically engineered CARs in the patient. After T cells, NK cells have got better repute because of their comparatively safer approach, but CAR-NK also need further validation and improvement in their final CAR-NK product to make success in solid tumors. Lastly, since one of the factors for impaired efficacy of both CAR-T and CAR-NK is their lack of penetration into the tumor bed. Macrophages are the immune cells that can get inside the solid tumor, and in most cases make up to 50% to the tumor mass, making them a desirable candidate to be considered for chimeric antigen cell therapy. Different companies are exploring their options for using macrophages for CAR, but it is still in very naïve state and will take some time to prove itself as established cellular therapy. Overall, cellular therapy is showing promising results, but their slow pace is an issue. Therefore, combination of these ongoing cellular therapies with some other immunotherapy agents will be right approach to fill the gap until we get a cellular therapy as single agent drug.

References

[1] Siegel RL, Miller KD, Jemal A. Cancer statistics, 2017. CA Cancer J Clin 2017;67(1):7–30.

[2] Wang J, Zhou P. New approaches in CAR-T cell immunotherapy for breast cancer. Adv Exp Med Biol 2017;1026:371–81. https://doi.org/10.1007/978-981-10-6020-5_17.

[3] Cheadle EJ, et al. CAR T cells: driving the road from the laboratory to the clinic. Immunol Rev 2014;257(1):91–106.

[4] Oble DA, et al. Focus on TILs: prognostic significance of tumor infiltrating lymphocytes in human melanoma. Cancer Immun 2009;9:3.

[5] Ankri C, et al. Human T cells engineered to express a programmed death 1/28 costimulatory retargeting molecule display enhanced antitumor activity. J Immunol 2013;191:4121–9. 24026081.

[6] Gross G, Waks T, Eshhar Z. Expression of immunoglobulin-T-cell receptor chimeric molecules as functional receptors with antibody-type specificity. Proc Natl Acad Sci USA 1989;86(24):10024–8. https://doi.org/10.1073/pnas.86.24.10024.

[7] Ramos CA, Heslop HE, Brenner MK. CAR-T cell therapy for lymphoma. Annu Rev Med 2016;67:165–83. https://doi.org/10.1146/annurev-med-051914-021702.

[8] Singh R, Paterson Y. Immunoediting sculpts tumor epitopes during immunotherapy. Cancer Res 2007;67(5):1887–92.

[9] Sun M, et al. Construction and evaluation of a novel humanized HER2-specific chimeric receptor. Breast Cancer Res 2014;16(3):R61.

[10] Pfefferle A, Huntington ND. You have got a fast CAR: chimeric antigen receptor NK cells in cancer therapy. Cancers (Basel) 2020;12(3):706. https://doi.org/10.3390/cancers12030706.

[11] Jin C, Yu D, Essand M. Prospects to improve chimeric antigen receptor T-cell therapy for solid tumors. Immunotherapy 2016;8:1355–61.

[12] Wu C-Y, Roybal KT, Puchner EM, et al. Remote control of therapeutic T cells through a small molecule-gated chimeric receptor. Science 2015;350, aab4077.

[13] Straathof KC, Pule MA, Yotnda P, et al. An inducible caspase 9 safety switch for T-cell therapy. Blood 2005;105:4247–54.

[14] Philip B, Kokalaki E, Mekkaoui L, et al. A highly compact epitope-based marker/suicide gene for easier and safer T-cell therapy. Blood 2014;124:1277–87.

[15] Kloss CC, Condomines M, Cartellieri M, et al. Combinatorial antigen recognition with balanced signaling promotes selective tumor eradication by engineered T cells. Nat Biotechnol 2013;31:71–5.

[16] Piscopo NJ, Mueller KP, Das A, et al. Bioengineering solutions for manufacturing challenges in CAR T cells. Biotechnol J 2018;13. https://doi.org/10.1002/biot.201700095.

[17] Poorebrahim M, Sadeghi S, Fakhr E, et al. Production of CAR T-cells by GMP-grade lentiviral vectors: latest advances and future prospects. Crit Rev Clin Lab Sci 2019;56(6):393–419. https://doi.org/10.1080/10408363.2019.1633512.

[18] Stacey GN, Merten OW. Host cells and cell banking. Methods Mol Biol 2011;737:45–88.

[19] Levine B. Performance-enhancing drugs: design and production of redirected chimeric antigen receptor (CAR) T cells. Cancer Gene Ther 2015;22:79–84.

[20] Hollyman D, Stefanski J, Przybylowski M, et al. Manufacturing validation of biologically functional T cells targeted to CD19 antigen for autologous adoptive cell therapy. J Immunother 2009;32:169–80.

[21] Ghassemi S, Bedoya F, Nunez-Cruz S, et al. Shortened T cell culture with IL-7 and IL-15 provides the most potent chimeric antigen receptor (CAR)-modified T cells for adoptive immunotherapy. J Immunol 2016;196:214.

[22] Vormittag P, Gunn R, Ghorashian S, et al. A guide to manufacturing CAR T cell therapies. Curr Opin Biotechnol 2018;53:164–81.

[23] Ghassemi S, Nunez-Cruz S, O'Connor RS, et al. Reducing ex vivo culture improves the antileukemic activity of chimeric antigen receptor (CAR)-T cells. Cancer Immunol Res 2018;6:1100–9.

[24] Xin Yu J, Hubbard-Lucey VM, Tang J. The global pipeline of cell therapies for cancer. Nat Rev Drug Discov 2019;18(11):821–2. https://doi.org/10.1038/d41573-019-00090-z.

[25] Lee DW, et al. T cells expressing CD19 chimeric antigen receptors for acute lymphoblastic leukaemia in children and young adults: a phase 1 dose-escalation trial. Lancet 2014;385:517–28. 25319501.

[26] Park JH, et al. CD19-targeted 19-28z CAR modified autologous T cells induce high rates of complete remission and durable responses in adult patients with relapsed, refractory B-cell ALL. Blood 2014;124:382.

[27] Porter DL, et al. Chimeric antigen receptor T cells persist and induce sustained remissions in relapsed refractory chronic lymphocytic leukemia. Sci Transl Med 2015;7. 303ra139.

[28] Jitschin R, et al. CLL-cells induce IDOhi CD14+HLA-DRlo myeloid-derived suppressor cells that inhibit T-cell responses and promote TRegs. Blood 2014;124:750–60. 24850760.

[29] Hudecek M, et al. The B-cell tumor-associated antigen ROR1 can be targeted with T cells modified to express a ROR1-specific chimeric antigen receptor. Blood 2010;116:4532–41. 20702778.

[30] Drent E, et al. CD38 chimeric antigen receptor engineered T cells as therapeutic tools for multiple myeloma. Blood 2014;124:4759.

[31] Mardiros A, et al. T cells expressing CD123-specific chimeric antigen receptors exhibit specific cytolytic effector functions and antitumor effects against human acute myeloid leukemia. Blood 2013;122:3138–48. 24030378.

[32] Marusyk A, Polyak K. Tumor heterogeneity: causes and consequences. Biochim Biophys Acta 2010;1805:105–17. 19931353.

[33] Morgan RA, Chinnasamy N, Abate-Daga D, et al. Cancer regression and neurological toxicity following anti-MAGE-A3 TCR gene therapy. J Immunother 2013;36(2):133–51.

[34] Jarboe JS, Johnson KR, Choi Y, Lonser RR, Park JK. Expression of interleukin-13 receptor alpha2 in glioblastoma multiforme: implications for targeted therapies. Cancer Res 2007;67(17):7983–6.

[35] Brown CE, Aguilar B, Starr R, et al. Optimization of IL13Ralpha2-targeted chimeric antigen receptor T cells for improved anti-tumor efficacy against glioblastoma. Mol Ther 2018;26(1):31–44.

[36] O'Rourke DM, Nasrallah MP, Desai A, et al. A single dose of peripherally infused EGFRvIII-directed CAR T cells mediates antigen loss and induces adaptive resistance in patients with recurrent glioblastoma. Sci Transl Med 2017;9(399), eaaa0984.

[37] Hicks MJ, Chiuchiolo MJ, Ballon D, et al. Anti-epidermal growth factor receptor gene therapy for glioblastoma. PLoS One 2016;11(10), e0162978.

[38] Hynes NE, MacDonald G. ErbB receptors and signaling pathways in cancer. Curr Opin Cell Biol 2009;21:177–84. https://doi.org/10.1016/j.ceb.2008.12.010.

[39] Ahmed N, Salsman VS, Kew Y, Shaffer D, Powell S, Zhang YJ, et al. HER2-specific T cells target primary glioblastoma stem cells and induce regression of autologous experimental tumors. Clin Cancer Res 2010;16:474–85. https://doi.org/10.1158/1078-0432.CCR-09-1322.

[40] Lamers CH, et al. Treatment of metastatic renal cell carcinoma with CAIX CAR-engineered T cells: clinical evaluation and management of on-target toxicity. Mol Ther 2013;21(4):904–12.

[41] Maude SL, Barrett D, Teachey DT, et al. Managing cytokine release syndrome associated with novel T cell-engaging therapies. Cancer J 2014;20:119–22.

[42] Davila ML, Riviere I, Wang X, et al. Efficacy and toxicity management of 19-28z CAR T cell therapy in B cell acute lymphoblastic leukemia. Sci Transl Med 2014;6. 224ra25.

[43] Herberman RB, Nunn ME, Holden HT, Lavrin DH. Natural cytotoxic reactivity of mouse lymphoid cells against syngeneic and allogeneic tumors. II. Characterization of effector cells. Int J Cancer 1975;16:230–9.

[44] Biron CA, Byron KS, Sullivan JL. Severe herpesvirus infections in an adolescent without natural killer cells. N Engl J Med 1989;320:1731–5.

[45] Ni Z, Knorr DA, Bendzick L, et al. Expression of chimeric receptor CD4zeta by natural killer cells derived from human pluripotent stem cells improves in vitro activity but does not enhance suppression of HIV infection in vivo. Stem Cells 2014;32:1021–31.

[46] Hermanson DL, Kaufman DS. Utilizing chimeric antigen receptors to direct natural killer cell activity. Front Immunol 2015;6:195.

[47] Hu Y, Tian ZG, Zhang C. Chimeric antigen receptor (CAR)-transduced natural killer cells in tumor immunotherapy. Acta Pharmacol Sin 2018;39(2):167–76. https://doi.org/10.1038/aps.2017.125.

[48] Shin MH, Kim J, Lim SA, Kim J, Kim SJ, Lee KM. NK cell-based immunotherapies in cancer. Immune Netw 2020;20(2). https://doi.org/10.4110/in.2020.20.e14, e14.

[49] Gong JH, Maki G, Klingemann HG. Characterization of a human cell line (NK-92) with phenotypical and functional characteristics of activated natural killer cells. Leukemia 1994;8:652–8.

[50] Boissel L, Betancur M, Lu W, Wels WS, Marino T, Van Etten RA, et al. Comparison of mRNA and lentiviral based transfection of natural killer cells with chimeric antigen receptors recognizing lymphoid antigens. Leuk Lymphoma 2012;53:958–65.

[51] Monjezi R, Miskey C, Gogishvili T, Schleef M, Schmeer M, Einsele H, et al. Enhanced CAR T-cell engineering using non-viral sleeping beauty transposition from minicircle vectors. Leukemia 2017;31:186–94.

[52] Chang YH, Connolly J, Shimasaki N, et al. A chimeric receptor with NKG2D specificity enhances natural killer cell activation and killing of tumor cells. Cancer Res 2013;73:1777–86.

[53] Oelsner S, Friede ME, Zhang C, Wagner J, Badura S, Bader P, Ullrich E, Ottmann OG, Klingemann H, Tonn T, et al. Continuously expanding CAR NK-92 cells display selective cytotoxicity against B-cell leukemia and lymphoma. Cytotherapy 2017;19:235–49.

[54] Kalos M, Levine BL, Porter DL, Katz S, Grupp SA, Bagg A, et al. T cells with chimeric antigen receptors have potent antitumor effects and can establish memory in patients with advanced leukemia. Sci Transl Med 2011;3. 95ra73.

[55] Zhang C, Oberoi P, Oelsner S, et al. Chimeric antigen receptor-engineered NK-92 cells: an off-the-shelf cellular therapeutic for targeted elimination of cancer cells and induction of protective antitumor immunity. Front Immunol 2017;8:533.

[56] van Ostaijen-ten Dam MM, Prins HJ, Boerman GH, et al. Preparation of cytokine-activated NK cells for use in adoptive cell therapy in cancer patients: protocol optimization and therapeutic potential. J Immunother 2016;39:90–100.

[57] Posey Jr AD, Schwab RD, Boesteanu AC, Steentoft C, Mandel U, Engels B, et al. Engineered CAR T cells targeting the cancer-associated Tn-glycoform of the membrane mucin MUC1 control adenocarcinoma. Immunity 2016;44:1444–54.

[58] Juillerat A, Marechal A, Filhol JM, Valogne Y, Valton J, Duclert A, et al. An oxygen sensitive self-decision making engineered CAR T-cell. Sci Rep 2017;7:39833.

[59] Morsut L, Roybal KT, Xiong X, Gordley RM, Coyle SM, Thomson M, et al. Engineering customized cell sensing and response behaviors using synthetic notch receptors. Cell 2016;164:780–91.

[60] Zhang S, Ke X, Zeng S, Wu M, Lou J, Wu L, et al. Analysis of CD8 + Treg cells in patients with ovarian cancer: a possible mechanism for immune impairment. Cell Mol Immunol 2015;12:580–91.

[61] Klichinsky M, Ruella M, Shestova O, et al. Human chimeric antigen receptor macrophages for cancer immunotherapy. Nat Biotechnol 2020. https://doi.org/10.1038/s41587-020-0462-y.

[62] Majeti R, Chao MP, Alizadeh AA, et al. CD47 is an adverse prognostic factor and therapeutic antibody target on human acute myeloid leukemia stem cells. Cell 2009;138(2):286–99.

[63] Morrissey MA, Williamson AP, Steinbach AM, et al. Chimeric antigen receptors that trigger phagocytosis. eLife 2018;7, e36688.

> CHAPTER NINE

T-cell engaging bispecific antibody therapy

Patty A. Culp, Jeremiah D. Degenhardt, Danielle E. Dettling, and Chad May
Maverick Therapeutics, Brisbane, CA, United States

Contents

Cancer Immunology and Immunotherapy
https://doi.org/10.1016/B978-0-12-823397-9.00002-8

1. Inherently active T-cell engagers

1.1 Overview

Since the discovery of T-cells as potent mediators of cytotoxic activity against foreign or infected cells, scientists have sought to harness that cytotoxic power to kill cells in a targeted manner, with the goal of circumventing both the antigen specificity and major histocompatibility (MHC)-dependence of that function. Initial studies demonstrated that antibodies against either the α/β chains or CD3 of the T-cell receptor (TCR) complex were able to trigger T-cell activation and killing of the cells to which those antibodies were bound, either through covalent attachment of the antibody directly to the target cell surface or by creating hetero-conjugates comprising an anti-T-cell antibody and a second antibody recognizing an antigen on the surface of the target cell [1–4].

With the realization that T-cells could be redirected to exert cytotoxicity against target cells in an MHC-independent manner, scientists began to generate therapeutics to kill malignant tumor cells using this mechanism. This first section will focus on the evolution and molecular diversity of inherently active T-cell engagers (TCEs), which are designed to bind simultaneously to T-cells and a cell-surface antigen expressed on malignant cells. The characteristics of the platforms discussed in this section are summarized in Table 1; several good reviews of bispecific molecules contain illustrations of the molecular design of these platforms [5–7]. The simultaneous binding of the bispecific to T-cells and tumor cells creates a cytolytic synapse that results in activation of the T-cell, which then releases perforins and granzymes to kill the tumor cell. Inherently active T-cell engagers bind but do not activate, or minimally activate, T-cells until the other arm of the bispecific co-engages a tumor cell antigen. This characteristic prevents robust systemic activation of circulating T-cells upon administration.

Most bispecific molecules that redirect T-cell cytotoxicity engage T-cells through the CD3 epsilon chain (CD3ε), a key signaling component of the TCR complex expressed on CD8 and CD4 α/β T-cells, γ/δ T-cells, and NKT cells. Several other bispecific molecules have been designed to avoid engaging a subset of α/β T-cells, regulatory T-cells, and instead engage the proinflammatory γ/δ T-cells. T-cells that express γ/δ chains may be activated either through engagement of their TCR or through other activating receptors that are also expressed on NK cells, including CD16, NKp30, and NKG2D [8, 9]. Like α/β T-cells, activated γ/δ T-cells upregulate CD25, release the proinflammatory cytokines IFNγ and TNFα, and exert cytotoxicity through the release of perforins and granzymes.

The initial bispecific molecules that demonstrated the potential for targeted T-cell cytotoxicity used hetero-conjugates of two separate antibodies covalently joined through crosslinking. This strategy of generating bispecifics is a cumbersome process and limitations in technology at that time did not provide sufficient consistency in the process to

Table 1 Inherently active T-cell engagers.

Platform	MW (kDa)	T-cell binding	TAA binding	HLE moiety	Fc arrangement	HC/LC[a] association	Developed by
BiTE	50	Monovalent	Monovalent	None	—	scFv	Amgen
DART	50	Monovalent	Monovalent	None	—	Coiled-coil	MacroGenics
TandAb	106	Bivalent	Bivalent	None	—	n/a	Affimed
sdAb × sdAb	26	Monovalent	Monovalent	None	—	n/a	VU University Med Center, Amsterdam
Tribody	100	Monovalent	Bivalent	None	—	Single Fab/scFv	Christian–Albrechts University of Kiel
Small mol. conjugate	25	Monovalent	Small molecule	None	—	n/a	Calibr, Scripps Research
ImmTAC	75	Monovalent	Monovalent	None	—	scFv	Immunocore
HLE BiTE	100	Monovalent	Monovalent	Fc	scFc	scFv	Amgen
DART–Fc	110	Monovalent	Monovalent	Fc	Heterodimer	scFv	MacroGenics
ADAPTIR	150	Bivalent	Bivalent	Fc	Homodimer	scFv	Aptevo
TriTAC	100	Monovalent	Monovalent	Anti-HSA	—	scFv/sdAb	Harpoon
IgM	>800	Monovalent	Decavalent	HSA	—	Fab/scFv	IGM Biosciences
XPAT	140	Monovalent	Monovalent	XTEN	—	scFv	Amunix
Triomab	150	Monovalent	Monovalent	Fc	Rat/mouse heterodimer	Species-specific	Fresenius Biotech/TRION Pharma
Duobody	150	Monovalent	Monovalent	Fc	Heterodimer	Preassociated	Genmab
Biclonics, ART-Ig, Veloci-Bi	150	Monovalent	Monovalent	Fc	Heterodimer	Common LC	Merus, Chugai, Regeneron
XmAb, BEAT	125 or 150	Monovalent	Monovalent or Bivalent	Fc	Heterodimer	Single Fab/scFv	Xencor, Glenmark
Uniab	125	Monovalent	Monovalent or Bivalent	Fc	Heterodimer	Single Fab/sdAb	Teneobio
CrossMAb	150 or 175	Monovalent	Monovalent or Bivalent	Fc	Heterodimer	Swapped CH1/CL	Roche/Genentech
IgG-scFv	200	Bivalent	Bivalent	Fc	Homodimer	Fab2 + scFv2	Memorial Sloan-Kettering

[a]HC, heavy chain; LC, light chain.

enable development of a therapeutic using this approach. Thus, other methods to generate bispecific molecules were explored. Multiple molecular designs for T-cell engager bispecifics were subsequently produced, two of which became platforms for therapeutic bispecifics, the single-chain bispecific and the Triomab [10–12]. These molecules represent two very different strategies for designing bispecific molecules. The Triomab is a hybrid monoclonal antibody, expressed from a quadroma, which itself is formed by the fusion of two hybridoma cell lines producing antibodies against different antigens. Triomabs will be discussed in more detail later. In contrast, the single-chain bispecific contains just the binding sites of two different antibodies and is generated through protein engineering. This molecular design is the origin of the BiTE molecule, which has become a key bispecific T-cell engager platform.

1.2 Short half-life TCEs

1.2.1 BiTEs

BiTE (bispecific T-cell engaging) molecules contain two tandem binding units, each of which is comprised of a single-chain Fv (scFv) containing the variable heavy (Vh) and variable light (Vl) sequences derived from a standard antibody. The Vh and Vl subunits and the two scFvs are linked together with artificial linkers that provide flexibility for proper folding. One scFv of the BiTE binds CD3ε and the other binds a tumor-associated antigen (TAA). Within the BiTE platform, the scFvs are generally positioned in the TAA-CD3 order, the CD3 scFv is fixed in the Vh/Vl orientation, as are the linker sequences and lengths. For each new BiTE molecule, the orientation of the TAA Vh and Vl is determined. BiTE molecules are approximately 50 kDa and fairly compact. The small size is considered to be an advantage for oncology therapeutics, as the high interstitial pressure of tumors may limit tumor penetration of large or bulky molecules [13–15]. One challenge of the BiTE molecules is that the scFvs are held together only through electrostatic and hydrophobic interactions, allowing for considerable "breathing" in the molecules. Thus, domain exchange readily occurs between scFvs in separate molecules, creating aggregates and stability issues for the molecule, but these issues can be at least partly ameliorated by including additional disulfide bonds in the scFvs of the BiTE molecule [5, 16]. Owing to their small size and lack of half-life extension moiety, BiTE molecules have been shown to exhibit short half-lives in humans, approximately 1–2 h, requiring continuous intravenous infusion to maintain therapeutic exposures [17, 18].

BiTE molecules have been generated that target a range of tumor associated antigens (TAAs), including CD19 (blinatumomab), EpCAM (MT110/solitomab), CEA (AMG 211), PSMA (AMG 212/BAY2010112/pasotuxizumab), BCMA (AMG 420/BI 836909), CD33 (AMG 330), the EGFRvIII variant (AMG 596), EGFR, EphA2, B7H6, and TrfR [19–29]. These molecules direct T-cell cytotoxicity *in vitro* on TAA-expressing tumor cell lines and on primary human samples at low picomolar or subpicomolar concentrations. Moreover, BiTEs promote cytotoxicity by both CD8+

and CD4+ T-cells in an MHC-independent manner and can eradicate established tumors in mouse models at doses as low as 50 μg/kg [19–21, 30, 31]. The potent preclinical activity of blinotumomab translated to profound efficacy in patients with B-cell malignancies and was approved in 2014 for the treatment of B-cell acute lymphoblastic leukemia (B-ALL) and at this time is the only T-cell engaging bispecific to have been approved for systemic administration [6].

1.2.2 DARTs

Another bispecific T-cell engager only containing Vh and Vl regions is the DART (dual-affinity retargeting) protein. This molecule is a diabody, formed from two separate chains, one of which contains the Vl of anti-CD3 and the Vh of anti-TAA, and the other that contains the Vl of anti-TAA and Vh of anti-CD3. Within each chain, a 5 amino acid linker between the Vh and Vl prevents intramolecular scFv formation, such that intermolecular domain interactions are favored. Pairs of oppositely charged coiled-coil domains are included in the two chains to promote heterodimerization and minimize the formation of nonfunctional homodimers [32]. In addition, the dimer is stabilized by a disulfide bond between the C termini of the two chains. DART molecules are similar in size to BiTEs, approximately 50–60 kDa, providing the potential advantage of better penetration into tumors. However, because the size is below the renal clearance threshold and these molecules lack half-life extension domains, DARTs, such as BiTEs, are expected to have short half-lives in humans.

Multiple DART molecules have been described, the first being a CD3xCD19 bispecific [33]. This molecule promoted potent T-cell cytotoxicity activity *in vitro* at subpicomolar concentrations on all cell lines tested. In this same study, a TCRxCD19 DART, which engages T-cells through the α/β chains of the TCR produced very similar results, supporting a second option for designing T-cell engagers. In a head-to-head comparison between the CD3xCD19 DART and a BiTE with the same Vh and Vl sequences, the DART was approximately 10-fold more potent than the BiTE. A second DART molecule, MGD006, a CD3xCD123 bispecific for acute myelogenous leukemia, demonstrated potent activity against AML cell lines and primary AML blasts [34, 35].

1.2.3 TandAbs

Tandem diabody molecules, or TandAbs, are tetravalent bispecific molecules and are homodimers of tandemly arrayed Vh and Vl sequences. Each chain of a T-cell engaging TandAb contains the Vh and Vl for both CD3 and TAA and short intervening linkers to restrict intramolecular scFv formation. Thus, functional antibody combining sites are formed by a Vh on one chain and a Vl on the opposite chain. These molecules are approximately twice the size of BiTEs and DARTs, approximately 106 kDa, which is above the threshold for renal clearance, providing these molecules with much longer half-lives in humans, 1–3 days [36].

T-cell engaging TandAbs targeting different TAA, including CD19 (AFM11), CD33 (AMV564), and EGFRvIII, have been described [37–39]. Each of these TandAb molecules exhibits potent activity *in vitro*, directing the T-cell-dependent lysis of TAA-expressing cell lines at low picomolar or high femtomolar concentrations. Despite being highly potent *in vitro*, the TandAbs have shown unexpectedly modest activity in mouse tumor models. Tested in prophylactic models, AFM11 completely prevented tumor growth only at the highest dose tested, 5 mg/kg, and the CD3xEGFRvIII TandAb slowed the growth of tumors by approximately 50% when dosed at 0.5 mg/kg [37, 39]. Administration of AMV564 at 1 mg/kg into mice with established tumors resulted in approximately 50% tumor growth inhibition [38].

As expected from their bivalency to both CD3 and TAA, TandAbs exhibit avid binding to their target antigens [37, 39]. In a direct comparison, the CD3xCD19 TandAb was found to be approximately 10-fold more potent at inducing cytotoxicity *in vitro* than BiTE molecules containing the same Vh and Vl sequences [37]. In an earlier study, however, the CD3xCD19 BiTE, MT103 or blinotumomab, was found to promote cytotoxicity at 600–1000 × more potently than a TandAb molecule [40]. This discrepancy may be due to differences in the binding domains used in the two studies, as binding domains may be more active in some structures than in others.

1.2.4 sdAb × sdAb bispecific

One of the smallest bispecific T-cell-engaging molecules thus far generated is comprised of two tandem sdAbs (single-domain antibodies, also called VHHs, nanobodies) [41]. sdAbs, the variable regions derived from heavy-chain only antibodies (usually camelid in origin), have multiple attractive features from a protein engineering perspective, including the lack of a light chain, which avoids the Vh/Vl pairing problem, their relatively good stability, and their small size, which likely allows for better penetration into tissues [42]. The small size of this bispecific molecule, approximately 26 kDa, would be expected to have rapid renal clearance and thus a short half-life *in vivo*.

The bispecific described by de Bruin *et al.*, 7D12-5GS-6H4, targets γ/δ T-cells and EGFR. The γ/δ T-cell binding domain binds an invariant epitope within the Vδ2 chain of Vg9Vδ2 cells, the most common γ/δ T-cell subset in the blood [43]. 7D12-5GS-6H4 was shown to promote cytotoxicity of γ/δ T-cells against EGFR-expressing tumor cells *in vitro* at subnanomolar concentrations. This molecule appears to be less potent than the BiTE and DART molecules described previously, but whether that is due to the difference between targeting α/β and γ/δ effector cells or whether the potency can be improved by optimizing the binding domains remains to be determined.

1.2.5 Tribody

The tribody design, initially described in 2000, is a Fab-scFv$_2$ design, consisting of a Fab against one target and scFvs against a second target attached to the C-terminus of the Ch and Cl chains of the Fab [44]. Thus, the tribody is monovalent for the Fab target and bivalent for the scFv target. In the tribody platform, the two scFvs bind the TAA and the single Fab binds the effector cell. A molecule that contains two binding sites to the TAA provides additional binding strength to the tumor cell, through avidity.

The tribody design has been adopted as the scaffold for several bispecific molecules, including $(CD20)_2xCD16$, $(HER2)_2xV\gamma9$, and $(HER2)_2xCD16$, all of which have the ability to promote cytotoxicity by γ/δ T-cells [45–47]. The molecular weight of the tribody is approximately 100 kDa, rendering it too large for first-pass renal clearance. Given the lack of a half-life extension moiety, these molecules are expected to have circulating half-lives shorter than that of an antibody, but substantially greater than a BiTE.

In *in vitro* assays, the $(HER2)_2xV\gamma9$ bispecific was able to induce lysis of HER2 expressing cell lines by γ/δ T-cells isolated from the blood of healthy volunteers or pancreatic cancer patients [46]. The molecule was not assessed in a range of concentrations, so its potency cannot be estimated, but the $(HER2)_2xV\gamma9$ bispecific was shown to be more active in this assay than a HER2xCD3 single-chain bispecific. In contrast, in assays using γ/δ T-cells isolated from ascites or tumors (TILs) on autologous tumor cells, the HER2xCD3 bispecific was more active than the $(HER2)_2xV\gamma9$ bispecific, suggesting differences between the functional capacity of circulating γ/δ T-cells and tumor-resident γ/δ T-cells [48].

1.2.6 Small molecule targeting TAA

A unique approach to generating T-cell-engaging bispecifics is by conjugating the CD3-binding region to a small molecule specific for the TAA. One such conjugate, CCW702, engages CD3 through an scFv, to which is attached a synthetic small molecule specific for prostate-specific membrane antigen (PSMA), resulting in a molecule with picomolar potency in inducing T-cell cytotoxicity *in vitro* [49]. As expected from its small size, the half-life of this molecule is relatively short, approximately 5–6 h, but inhibited the growth of established tumors in mouse models at 1 mg/kg.

1.2.7 Tumor cell targeting via recombinant T-cell receptors

Recombinant T-cell receptors (TCRs) are comprised of the α and β chains derived from naturally occurring T-cell receptors that recognize specific peptides in the context of major histocompatibility complex (MHC). The ImmTAC platform (Immune mobilizing monoclonal T-cell receptors against cancer) is a class of T-cell engaging molecules that contain a recombinant TCR, whereby the α and β heterodimer is stabilized via a disulfide bond, and a CD3 scFv fused to the TCR β chain [50]. With a molecular weight of approximately 75 kDa, the ImmTACs are above the renal clearance threshold and have a circulating half-life of approximately 6–8 h.

A major advantage of using recombinant TCRs to target TAAs is the potential to vastly broaden the repertoire of TAAs that may be targeted with a T-cell engager.

While antibodies can bind only to cell-surface proteins, which comprise a small fraction of the proteins expressed by a given cell, TCRs bind to peptides derived from any expressed protein within the cell [51]. This increase in the pool of candidate TAAs has the potential to develop T-cell engagers against targets with more cancer-restricted expression, resulting in increased potencies and/or reduced toxicities. Additionally, TCRs can engage cells expressing a relatively small number of peptide antigen-positive molecules, fewer than 100 per cell [52].

However, there are several challenges with using recombinant TCRs in a T-cell engager, the first being that successful engagement relies on expression of MHC by tumor cells. Downregulation of MHC is frequently observed in tumors as a means of escape from T-cell-mediated immune destruction, potentially rendering such cells also resistant to T-cell engagers using TCRs [53, 54]. Another challenge is that TCR recognition of the MHC:peptide complex is MHC allele-specific, thus limiting the number of patients that could potentially benefit from such a therapy.

While the binding of the native TCR to the MHC:peptide complex is often a low affinity interaction, improving TCR affinities is a straightforward process and has resulted in potent ImmTAC molecules [52]. Multiple ImmTAC molecules have been described, including tebantafusp, which targets gp100 and is specific for HLA-A2; ImmTAC-NYE, which binds both NY-ESO-1 and LAGE-2 and is specific for HLA-A2; ImmTAC-MAGE, targeting MAGE-A3 and is specific for HLA-A1; and ImmTAC-MEL, which targets Melan-A/MART-1 and is specific for HLA-A2. These molecules exhibit low-to-mid picomolar potency in *in vitro* cytotoxicity assays, and ImmTAC-MAGE largely prevented the growth of established tumors when dosed at 0.1 mg/kg [52, 55].

1.3 TCEs containing half-life extension moieties

Many of the molecules described previously, due to their small size and lack of half-life extension moieties exhibit, or are expected to have, relatively short half-lives in humans. For highly potent molecules, a short half-life may be beneficial in the event of serious adverse events, as the serum concentrations of such molecules would rapidly decrease, which should allow for a faster resolution of the adverse event. However, a short half-life introduces the challenge of maintaining the drug exposures required for efficacy; thus, very frequent or continuous dosing is required, providing a convenience issue for patients. Several half-life extended (HLE) bispecific T-cell engagers have been developed, including Fc-fusions, fusions to albumin or to albumin-binding regions, fusions to unstructured protein domains, and engineered monoclonal antibodies.

1.3.1 Fc-fusion proteins

Fusing the Fc (CH2 and CH3) portion of an antibody to a small protein is a common method to provide half-life extension through the ability of the Fc both to enlarge the molecule and to allow pH–dependent binding to neonatal Fc receptor (FcRn), which engages the IgG salvage pathway to promote recycling [56–58]. However, attaching the naturally dimeric Fc to a molecule for which monovalent binding is desired provides its own challenges. Several engineering strategies have been employed to generate monovalent bispecific Fc fusion proteins.

For the single-chain bispecific BiTE molecules, the Fc is attached to the C-terminus of the bispecific scFv sequences by way of two Fc regions in tandem, forming a single-chain Fc (scFc). A number of such HLE BiTEs have been generated with specificities to a range of TAA, including CD19 (AMG 562), PSMA (AMG 160), FLT3 (AMG 427), CD33 (AMG 673), BCMA (AMG 701), DLL3 (AMG 757), MUC17 (AMG 199), and Claudin 18.2 (AMG 910) [59]. These HLE BiTEs exhibit potent T-cell-directed cytotoxicity *in vitro* and in mouse models *in vivo*, though in some cases their potencies are somewhat reduced compared to standard BiTE molecules lacking HLE. These HLE BiTEs are larger in size, approximately 100 kDa, and as expected, they have been shown to have prolonged circulation in cynomolgus monkeys, from 5 to 7 days [60–62].

To provide half-life extension for DART molecules, two different strategies for creating Fc-fusions have been employed. One strategy created DART-Fc proteins comprised of three separate chains: one chain comprising the TAA Vl, CD3 Vh, and CH2/CH3 of Fc, a second chain containing CD3 Vl and TAA Vh; and a third chain containing the Fc CH2/CH3 sequences [63, 64]. These three chains are stabilized by the coiled-coil sequences and the disulfide bond at the C-terminus of the DART chains, and Fc heterodimerization and stability is driven by the use of knob-into-hole technology and disulfide bonds in the Fc. The knob-into-hole technology, first described in 1996, has subsequently been optimized and has become a standard method for creating heterodimeric Fc's [65]. In a second strategy to generate DART-Fc proteins, each of the standard DART molecule chains was attached at its C-terminus to an Fc chain, in which knob-into-holes were used to favor heterodimerization, creating the LP-DART molecule [66]. Both of these strategies produce DART Fc fusion proteins of approximately 110 kDa, and as expected, exhibit a prolonged half-life of 4–7 days in cynomolgus monkeys or human FcRn-transgenic mice [63, 64, 66]. Multiple Fc-containing DART molecules have exhibited potent, subpicomolar activity *in vitro* and eradicated established tumors in mice at 0.5 mg/kg [63, 64, 66, 67].

Another molecular design that uses the CH2 and CH3 domains of the Fc as a scaffold onto which antibody binding sites can be attached is the scFv–Fc–scFv format. This format is a homodimer and when constructed as a bispecific, is bivalent for both antigens. Several T-cell-engaging bispecific molecules of this configuration, called ADAPTIR proteins, have been formed by attaching the CD3 scFvs to the C-terminus of CH3 and the TAA scFvs to the N-terminus of CH2 [68, 69]. APVO414 (MOR209/ES414), which targets CD3 and PSMA, induced T-cell cytotoxicity of PSMA-expressing tumor cells at low picomolar concentrations *in vitro* [68]. In a direct comparison with a single-chain bispecific scFv molecule encoded by the sequences from BAY2010112, ES414 was approximately 30-fold more potent than the single-chain bispecific scFv. APVO436, a CD3xCD123 bispecific, promoted T-cell cytotoxicity of both AML cell lines and primary AML samples *in vitro* at low picomolar concentrations and was found to have similar potency as a bispecific diabody encoded by sequences from MGD006. APVO436 exhibited activity in a mouse model of AML, slowing the growth of established tumors when dosed at 150 μg/kg.

1.3.2 Half-life extension through albumin binding

Like antibodies, human serum albumin (HSA) maintains a relatively long circulating half-life in humans by recycling via FcRn [70]. This property of HSA has prompted the development of antibody-based molecules that contain an HSA-binding region, a strategy that has successfully conferred the longer half-life of HSA to the protein bound to it [71, 72].

The TriTAC is a T-cell engaging bispecific platform that includes an HSA-binding region to provide half-life extension [73]. The TriTAC comprises three binding units in series: a TAA-binding sdAb, an HSA-binding sdAb, and a CD3 scFv, yielding a molecule of approximately 52 kDa. In the absence of albumin binding, such a molecule is below the renal clearance threshold and would be expected to have a relatively short half-life. In cynomolgus monkeys, the half-lives of the TriTACs have been found to be 3–5 days, supporting the ability of the molecules to bind albumin *in vivo* [74–76].

TriTAC molecules targeting EGFR, MSLN (HPN536), PSMA (HPN424), BCMA (HPN217), and DLL3 (HPN328) have been described, and these molecules promote T-cell cytotoxicity at picomolar concentrations [73–77]. In mouse models, TriTAC treatment results in regression of established tumors at 0.1–0.5 mg/kg [73, 74, 76].

1.3.3 Fusions to albumin

A related mechanism that co-opts the long half-life of albumin is through generation of recombinant fusions with albumin itself. A T-cell engaging molecule that incorporates HSA is IGM-2323, a CD3xCD20 IgM-based molecule [78]. In this molecule, CD20 binding is achieved by a pentamer of IgM molecules, resulting in considerable avidity against the TAA. Monovalent CD3 engagement is achieved with a single scFv linked to one end of the J chain, a separate protein sequence required for pentamer assembly. HSA is linked to the other end of J chain, and the resulting molecule is very large, more than 800 kDa. IGM-2323 exhibits low picomolar potency in T-cell cytotoxicity assays.

1.3.4 XTEN fusions

An additional strategy for promoting half-life extension is through attachment of bulking moieties, such as PEG or unstructured protein sequences (XTEN) [79, 80]. Bulking agents provide additional mass and substantial increases in the volume of the molecule, which minimizes renal clearance. Moreover, such bulking moieties can provide better solubility and stability to the protein. T-cell-engaging bispecific utilizing such a strategy are the ProTIA (Protease Triggered Immune Activators) and XPAT (XTENylated Protease-Activated T-cell Engager) platforms [81, 82]. The ProTIA and XPAT molecules are comprised of a single-chain bispecific scFv molecule targeting CD3 and TAA, attached to an XTEN polypeptides. In ProTIA molecules, the XTEN sequence is attached downstream of the anti-CD3 scFv, while in the XPAT molecules, both the N and C termini of the bispecific scFv are attached to XTEN sequences. ProTIA and XPAT molecules targeting EpCAM, HER2, and EGFR have been described and induce T-cell cytotoxicity *in vitro* at picomolar concentrations and regress established tumors in mice at 1–2 mg/kg. In cynomolgus monkeys, HER2-XPAT exhibits an increase in half-life over the cognate non-XTEN single-chain bispecific molecule, allowing for weekly administration compared to continuous infusion.

1.4 Monoclonal antibody-derived TCEs

1.4.1 Triomab

The monoclonal antibody (mAb) is a conceptually appealing structure from which to generate bispecifics due to its natural bivalency. However, forcing the formation of heterodimeric heavy chains and ensuring the appropriate association of the light chains creates challenges in designing bispecifics, challenges that were initially resolved by creating the Triomab, a rat/mouse hybrid monoclonal antibody [12]. The rat/mouse heterodimeric heavy chain molecule was readily purified away from the rat/rat and mouse/mouse homodimeric molecules using pH gradients, and within the molecules, the heavy and light chains naturally paired in a species-specific manner. Several T-cell-engaging Triomabs have been described, targeting EpCAM (catumaxomab), GD2 (TRBs07/ektomab), HER2 (ertumaxomab), and CD20 (FBTA05), which induced tumor cell cytotoxicity by T-cells *in vitro* at picomolar concentrations [83–86]. In clinical studies, the toxicity of catumaxomab limited dosing to 5 µg when administered systemically; thus, the molecule was evaluated for local administration and was approved in 2009 for intraperitoneal administration to patients with malignant ascites [87–90]. However, as expected from the rat and mouse origins of the Triomab, the majority of patients developed antirat or antimouse antibodies after catumaxomab administration, which likely limited the efficacy of this molecule [90].

1.4.2 Duobody

The Duobody takes advantage of a process that human IgG4 antibodies naturally perform, that of Fab-arm exchange, whereby two bivalent monospecific antibodies exchange half molecules (one heavy chain + one light chain) to form two monovalent bispecific antibodies [91]. This process is facilitated by residues unique to IgG4 constant regions in the upper hinge (S228) and lower CH3 domain (R409). Introducing R409K and F405L mutations into IgG1 molecules allows for efficient Fab-arm exchange of this isotype in the presence of mild reducing agents [92]. Duobodies have been generated from either modified IgG1 antibodies or from IgG4 antibodies via this controlled Fab-arm exchange method, a process that is reported yield up to 95% of the desired product [92–96]. Multiple Duobody molecules, targeting different TAA, are currently in development, including GEN3013 (CD3xCD20), JNJ-63709178 (CD3xCD123), GEN1044 (CD3x5T4), JNJ-64407564 (CD3xGPRC5D), and JNJ-64007957/teclistamab (CD3xBCMA) [93–96]. These molecules induce T-cell cytotoxicity *in vitro* at subnanomolar to subpicomolar concentrations and can eradicate established tumors at doses as low as 0.5 mg/kg.

The Duobody platform represents a method to generate heterodimeric antibodies post-production. However, other protein engineering strategies have focused on driving heterodimerization of heavy chains during antibody expression. These modifications have introduced bulky and/or charged amino acids in the CH3 domain, the region that drives Fc dimerization [5, 97]. Successful heterodimerization of heavy chains is a key step towards generating a bispecific molecule; however, in a molecule containing two heavy chains with different antigen specificities, either heavy chain can associate at random with either of their two light chains, yielding a large fraction of molecules with at least one nonfunctional binding site. Using different protein engineering approaches to address the "light chain problem", a range of T-cell-engaging IgG-based bispecific molecules have been developed.

1.4.3 Common light chain

The most straightforward means of solving the light chain association problem in a molecule containing a heterodimeric heavy chain is to use a common light chain, whereby only the heavy chain variable domain contributes to the binding specificity of the antibody. Multiple groups have employed the use of common light chains to build T-cell-engaging bispecifics. In one platform using a common light chain, heterodimerization is driven by electrostatic interactions in the CH3 domain, resulting in >95% purity of the desired product [98, 99]. A T-cell-engaging bispecific molecule based on this platform, MCLA-117, targets CD3 and CLEC12A for AML. This molecule induced T-cell lysis of CLEC12A-expressing cell lines at subnanomolar concentrations *in vitro* and was able to induce lysis of primary AML blasts.

In a second mAb-based platform using a common light chain, charged residues have been introduced into the CH3 to favor heterodimerization of the heavy chains, as well as additional modifications to facilitate purification of the heterodimer away from the homodimers and other impurities [100, 101]. A T-cell engaging molecule based on this platform, ERY974, targeting CD3 and GPC3, exhibited subnanomolar potency *in vitro* and eradicated established tumors in mouse models at 1 mg/kg [101].

In a third platform using a common light chain, heavy chain heterodimers are purified away from homodimers via differential Protein A binding. This platform takes advantage of the lack of Protein A binding by IgG3 antibodies, a characteristic that is determined by two amino acids in the CH3 domain [102]. Co-expression of a heavy chain competent for Protein A binding and a heavy chain containing the CH3 IgG3 modifications yields heterodimers that can be purified from the homodimers via a pH gradient, with a yield of 40%–50% [103]. Using this platform, several bispecific T-cell engagers have been generated: REGN1979/odronextamab, which targets CD3 and CD20, REGN5458, a CD3xBCMA bispecific, and REGN4018, targeting CD3 and MUC16 [103–106]. All three of these molecules have demonstrated potent activity in inducing T-cell cytotoxicity of TAA-expressing cells *in vitro*, at picomolar concentrations. REGN1979 and REGN4018 have also been effective in eradicating established tumors, as large as 500–900 mm^3 in mice, at 0.5 mg/kg. In an *in vitro* cytotoxicity assay, REGN5458 compared favorably with a chimeric antigen receptor (CAR)-T-cells expressing a BCMA-targeted scFv derived from REGN5458.

1.4.4 Fab + scFv arms

Several groups have solved the multiple light chain problem by engineering molecules that contain a standard Fab in one arm and an scFv in the other arm. Thus, the Fab arm light chain can associate only with the Fab heavy chain, as the scFv heavy chain is covalently linked to its light chain. In these molecules, CD3 binding is achieved through the Fab, while the scFv engages the TAA. In the XmAb platform, molecules containing either a single scFv or two tandem scFvs are in development. As described earlier, molecules designed to bind bivalently to TAA provides increased binding to the tumor cell; in contrast, maintaining monovalent CD3 binding prevents CD3 crosslinking and TAA-independent T-cell activation. XmAb molecules have been described that target a range of TAAs, including CD38 (AMG 424) [107], STEAP1 (AMG 509) [108], ENPP (XmAb30819) [109], SSTR2 (XmAb18087/tidutamab) [110], CD20 (XmAb13676/plamotanab) [111], and CD123 (XmAb14045/vibecotamab) [112, 113]. These molecules have demonstrated the ability to promote T-cell cytotoxicity of TAA-expressing cells *in vitro* at picomolar concentrations.

A related platform, BEAT, is also comprised of a heterodimeric Fc, with one arm binding CD3 via a Fab and the other arm binding TAA via an scFv. In the BEAT platform, the CH3 region of the Fc has been modified to contain residues derived from the α/β TCR to promote Fc heterodimerization [114]. Several such T-cell-engaging molecules are in development, targeting the TAA's CD38 (ISB 1342/GBR1342) and HER2 (ISB 1302/GBR1302). These molecules induce T-cell cytotoxicity of TAA-expressing cells *in vitro* at picomolar concentrations [115, 116],

1.4.5 Fab + sdAb arms

Another means of eliminating the light chain pairing problem in a molecule containing a heterodimeric heavy chain is to substitute the Fab structures in one arm of the mAb with one or more sdAbs, which do not need to pair with light chains, as described previously. A T-cell-engaging bispecific based on this strategy incorporates knob–into–hole technology to favor heavy chain heterodimerization. One arm binds CD3 through a standard Fab, while TAA binding is achieved by two tandem sdAbs, attached to the second heavy chain just above the hinge region [117]. TNB-383B, a CD3xBCMA bispecific, is bivalent for BCMA binding, but monovalent for CD3. This molecule was selected for its relatively low affinity for CD3 (30 nM), which was reflected in its modest potency *in vitro*, inducing T-cell cytotoxicity of BCMA-expressing tumor cells at subnanomolar concentrations. However, in a mouse model, TNB-383B was very active, substantially inhibiting the growth of established tumors at doses as low as 0.5 µg/kg.

1.4.6 CrossMAb

In the CrossMAb platform, the Fc heterodimer is formed by knob–into–hole technology, and one of the Fab arms has been modified such that the CH1 and CL constant regions are swapped, forcing correct association of the two light chains with their respective heavy chains [118, 119]. A variation on this design contains a second TAA-binding Fab attached to the N–terminus of the CD3-binding Fab, creating a molecule that is bivalent for TAA binding, but monovalent for CD3 binding, referred to as a 2:1 T-cell bispecific (TCB). Several such CrossMAbs are currently in development, targeting CEA (RG7902/cibisatamab), BCMA (EM801), and CD20 (RG6026/glofitamab) [120–122]. In *in vitro* T-cell cytotoxicity assays, these molecules range from low nanomolar to low picomolar potencies, and regress established tumors at 0.5 mg/kg or lower.

1.4.7 IgG-(scFv)₂

Engineering a second binding specificity as an scFv onto a standard mAb avoids both the Fc heterodimerization and light chain pairing challenges. In one such platform, anti-CD3 scFvs are attached to the C termini of the light chains of an anti-TAA mAb, rendering the molecule bivalent for both CD3 and TAA binding [123]. A GD2xCD3 bispecific built on this platform, hu3F8-BsAb, exhibited femtomolar activity in T-cell cytotoxicity assays *in vitro*, and regressed established tumors at 2 mg/kg [124].

2. The challenges of targeting solid tumor indications

A significant attribute of TCEs is the highly potent T-cell-mediated killing and the low level of target co-engagement required for activation and subsequent cytotoxic activity. As measured by B-cell depletion *in vitro*, the relative potency of the anti-CD19

TCE blinatumomab is close to 100,000-fold increase over the anti-CD20 mAb rituximab [125]. Unlike standard IgG mAbs, which rely on the much less potent Fc domain-mediated antibody-dependent cellular cytotoxicity (ADCC) to induce effector cell killing, the level of surface antigen required for potent T-cell killing mediated by TCEs has been reported at fewer than 100 copies of antigen/cell [52]. This is similar to what has been reported for T-cell activation in response to peptide/MHC complexes on the surface of antigen presenting cell, where as few as 10 TCRs engaging peptide/MHC complexes is sufficient for activation [126, 127]. Because TCEs have the potential to be so highly potent, when left unchecked they run the risk of inducing harmful systemic and localized inflammatory responses.

2.1 Toxicities associated with TCEs

Neurotoxicity has been reported in a subset of patients with hematological B-cell malignancies that receive TCEs, such as blinatumomab [128]. However, to-date this dose limiting toxicity (DLT) has not been observed in patients receiving TCE targeting solid tumor antigens. In the context of solid tumor indications TCEs have generally demonstrated two types of toxicities, the first is cytokine release syndrome (CRS), and the second is on-target/off-tumor normal tissue damage.

2.1.1 Cytokine release syndrome

Cytokine release syndrome (CRS) is a systemic inflammatory response that may be initiated by a variety of factors, including T-cell therapies. While mild cases can present as flu-like illness, more severe responses may lead to of life threatening cardiovascular, pulmonary, and renal involvement [129]. CRS has been reported as a DLT in preclinical safety studies where the TCEs cross-react to nonhuman primate targets, as well as in patients [29, 130, 131]. While concentrations of several circulating cytokine are increased in CRS after TCE treatment, IL-6 is believed to be the cytokine contributing most significantly to DLT, as anti-IL-6 blocking monoclonal antibodies (mAbs) have been shown to rapidly reverse life-threatening CRS in patients administered TCEs [132]. Besides inhibition of IL-6, corticosteroids are also often used to treat CRS [133, 134] or are given prophylactically prior to dosing to avoid DLTs, with minimal impact on T-cell proliferation or killing [135]. CRS has been associated with the maximum serum concentration (Cmax) of the TCEs upon the conclusion of dosing [136]. For TCEs dosed intravenously (IV), the Cmax is anticipated to be highest upon completion of administration. Therefore, cytokine release often peaks within hours after administration of the TCE. Alternate methods of administration, such as subcutaneous (SC) dosing lessens cytokine release and delays the transient spike due to the difference in the pharmacokinetics of exposure via this route [131].

Methods to ameliorate CRS and allow efficacious doses of TCEs to be reached have been reported. They are often referred to as "lead-in" or "priming" dose strategies, and are based on observations that cytokine release is mitigated by repeat TCE dosing. This stepwise dose escalation approach begins with an initial lower dose that induces a low-to-moderate cytokine increase but does not induce CRS, followed by administration of higher maintenance doses. This allows for maintenance dose concentrations to be reached that would otherwise not be tolerated if given as the starting dose. This strategy has been pursued for several programs to reduce TCE-mediated CRS and achieve higher maximum tolerated doses in both the preclinical and clinical settings [130, 131, 137].

2.1.2 On-target/off-tumor tissue damage

On-target/off-tumor normal tissue damage, a second DLT observed with TCEs, is generally dependent on-target expression in healthy organs and is associated with the duration of TCE exposure over time [136]. Due to this, DLTs associated with cytokine release may be observed before normal tissue destruction is detected, especially when the TCE is administered by IV infusion over a short time, as the maximum TCE concentration will be driving the CRS response. In the case of TCEs against CD19 or CD20, which target B-cell related malignancies, the sustained depletion of peripheral healthy B-cells that also express the target is generally well tolerated by patients. However, in the case of TCEs designed to treat solid tumor indications, even low levels of target expressed on normal tissues may be sufficient to damage healthy organs and severely restrict the dose concentrations that patients can tolerate, subsequently reducing the opportunity for antitumor response.

For example, a TCE targeting EGFR was tested in cynomolgus monkeys. Because both the anti-EGFR and anti-CD3 binding domains were species cross-reactive, this was deemed a relevant safety species, as the authors demonstrate equivalent TCE-mediated T-cell killing of human and nonhuman primate (NHP) EGFR expressing cells. While at low concentrations the TCE was well tolerated, the administration of higher concentrations resulted in severe liver and kidney associated toxicities that required the animals be euthanized [29]. Histopathological analysis indicated signs of toxicity consistent with low levels of EGFR expression in those tissues. Toxicities were also reported in a Phase I study using a TCE against EpCAM. Results demonstrated what appeared to be off-tumor/on-target-related DLTs that limited dose escalation and prevented a therapeutic window from being achieved [130]. Unlike the anti-EGFR TCE noted previously, the anti-EpCAM TCE was not selected for cross reactivity to the orthologous NHP EpCAM protein. Had TCEs been designed that bind equivalently to both human and NHP antigens, EpCAM and CD3, they may have been able to identify the DLTs associated with the program before moving it into the much more costly phase 1 trial. Cross reactivity to the NHP protein, along with a good understanding of differential target expression across the two species, is important for predicting tolerability in patients and more safely advancing programs into the clinic.

2.2 Identifying tumor-specific targets

Due to the potency of TCEs, which are able to induce cytotoxic T-cell activity with a relatively small number of CD3 co-engagements [52], low level target antigen expression on normal tissues is still sufficient to initiate a destructive T-cell response. These low levels of antigen may be below the limit of detection by the most frequently utilized method of immunohistochemistry, which is lacking in sensitivity. It has been noted that TCEs that target antigens expressed on hematological cancers are often tolerated at efficacious doses, even when they deplete healthy cells that express the same antigen. This has been demonstrated by blinatumomab, which is clinically approved to treat B-cell malignancies, since patients can tolerate a transient reduction of their normal B-cells [17]. A number of TCEs targeting different antigens that are being developed to treat multiple myeloma have also demonstrated depletion of healthy immune cell lineages expressing the antigen, and appear to be tolerated [121, 138, 139].

However, for inherently active TCEs targeting solid tumor indications, antigen expression in vital organs is much less likely to be tolerated. Therefore, the selection of antigens has generally focused on targets thought to be differentially expressed on tumor relative to normal healthy tissue to achieve a therapeutic window. Still, this presents a significant challenge to the highly potent TCEs to safely treat solid tumor indications, as even low level antigen expression in normal tissues can lead to the DLTs at concentrations below those needed to achieve sustained antitumor activity. For example, development of catumaxomab, the EpCAM-targeted TCE was terminated after demonstrating unresolvable DLTs consistent with target expression in the liver, with none of the patients on treatment demonstrating an objective response [140]. And as noted previously, the EpCAM-targeted BiTE demonstrated apparent on-target/off-tumor-mediated DLTs and was also terminated [130]. More recently, a DART targeting the differentially expressed B7H3 protein [141] was placed on partial clinical hold due to adverse events [142]. Later, its development was discontinued along with another solid tumor targeting DART direct toward GPA33 [143], a target that has also been reported to demonstrate differential tumor expression relative to normal tissue [144].

While rare, others have also taken advantage of mutations in protein sequences that are uniquely expressed on the cell surface of tumors. Including a constitutively activated splice variant of EGFR that has lost exons 2–7. This truncated form is called EGFR variant III isoform (EGFRvIII), a tumor-specific variant that has generated broad interest for use in targeted therapies, including CD3 engagement [26, 39]. EGFRvIII is found in high prevalence in glioblastoma multiforme, with up to 25%–33% prevalence in patients, and often presents together with wildtype EGFR [145].

The high potency of TCEs, along with the limited number of unique tumor-specific targets, reduces their therapeutic window and limits their potential for success in the clinic. This is especially true in solid tumor indications, where unlike B-cell malignancies, even a very low level of normal tissue expression of the target may not be tolerated. The combined data suggests that novel approaches are needed to increase the therapeutic window and allow these therapies to realize their full potential, especially in solid tumor indications.

3. Next-generation TCEs
3.1 Affinity tuned TCEs

For the treatment of solid tumor indications, many groups are developing novel therapeutics to reduce the potential for toxicity and increase tumor-specific activity, allowing improved patient response. One strategy among the next-generation of TCEs includes tuning the binding affinity to CD3 to decouple cytokine release from T-cell-mediated target-cell killing. This approach has demonstrated significant decreases in cytokine expression, while maintaining *in vivo* efficacy. A reduction in CD3 binding affinity provided a selectivity window of 1000-fold towards high HER2-expressing cells [117]. Additionally, a high affinity anti-CLL-1 TCE demonstrated comparable *in vivo* efficacy when paired with a panel of anti-CD3 binding domains that included a range of affinities. However, after testing in NHPs, it was observed that only the lower affinity anti-CD3 binding domains were well tolerated and able to deplete the tumor cells. [146]. The reduced binding affinity to CD3 also resulted in higher exposures of the TCE in NHPs relative to the higher affinity anti-CD3 binding domains, suggesting CD3 expressed on T-cells had less of a sink effect with inclusion of the lower affinity binding domain. Modification of the CD3 binding affinity was also performed in the context of a CD38 T-cell engager after an initial lead demonstrated nontolerated cytokine release in nonhuman primate studies. Comparing leads with modified CD3 binding affinity demonstrated that depletion of B-cells could be maintained, while decreasing the total amount of cytokine release elicited [107]. While modification of CD3 binding affinity may reduce cytokine-release-associated toxicities, it may not overcome TCE-mediated on-target/off-tumor toxicity, as they remain inherently active and do not provide a mechanism to distinguish between tumor and normal tissues. One question that remains unanswered with this approach is what impact will it will have on T-cell infiltration into the tumor, since T-cell recruitment is induced by proinflammatory cytokine and chemokine release [147].

Another affinity tuning approach being pursued to increase the selectivity of TCEs to the tumor and improve the therapeutic window is directed towards preferentially binding cells where antigen is expressed at higher density. This approach incorporates two-low affinity tumor antigen binding domains into the TCE to direct it to preferentially engage tumor cells that express relatively high levels of the antigen compared to normal cells [120, 148–151]. Modification of the affinity has effects on *in vitro* potency, exposure and *in vivo* activity, the results of which are specific to the design of the TCE. Using the 2:1 TCB design described earlier, one report demonstrates affinity tuned anti-CEA binding domains allow for preferential targeting of tumor cells expressing approximately 10,000 cell surface CEA antigen/cell or greater; with limited activity on cells expressing lower levels of the antigen that is closer to the normal CEA expression range [120]. Additionally, using similar 2:1 TCB formats, two TCEs were affinity tuned to selectively deplete either HER2 or PSMA-overexpressing cells, while avoiding cells with lower expression [150, 151] Together demonstrating the success of this strategy in the preclinical setting. Still, this approach comes with its own potential pitfalls. Although weakening the affinity towards the target antigen may reduce TCE binding to healthy cells that express lower levels of target antigen, it may also allow for tumor cells expressing low levels of antigen to escape.

These combined data support the approach of tuning the affinity of the binding domains to CD3 and the target antigen to improve safety and tumor selectivity, and to potentially improve the therapeutic window.

3.2 TCEs targeting tumor-specific peptides

The use of TCRs to recognize tumor neoantigens presented in the context of HLA is a strategy being utilized by several groups to circumvent the expression of many tumor antigen targets on normal tissues. Instead of antibody derived binding domains, these TCEs incorporate recombinant TCRs that are selected to target HLA complexed tumor-specific peptide sequences that are uniquely presented on the surface of tumor cells. The recombinant TCR binds to the tumor-specific HLA/peptide complex on the cancer cell while co-engaging CD3 on the T-cell. This interaction with the TCE initiates effector-cell-mediated killing of the tumor, while sparing normal healthy tissues where the peptide is expected to be absent. TCRs typically interact relatively weakly with HLA/peptide complexes, a result of thymic selection to minimize autoreactivity by depleting high affinity T-cells during their development [152]. Therefore, to improve potency recombinant TCRs are often engineered and selected for high affinity binding to the specific HLA/peptide complex of interest [52]. While these TCEs are inherently active buy design, the peptides sequences being targeted are often unique to the tumor, through mutations or are members of the well-studied cancer/testis antigens [153], and are unlikely to be expressed in normal tissues. Still, through the process of increasing the recombinant TCR's affinity for the HLA/peptide complex, cross-reactivity of the TCR to other HLA/peptide complexes is possible, and may impact safety [154].

The previously described ImmTAC platform from Immunocore is being developed for use in both solid tumor and hematological indications. ImmTACs bind HLA-A*0201-positive tumors that present the peptide sequence of interest along with CD3 on T-cells. Initial preclinical data from ImmTACs studies targeting gp100, MAGE-A3, MART-1, LAGE-1, and NY-ESO-1 demonstrate high sensitivity and potency, eliminating cells expressing as low as 2–10 copies of pMHC per cell [52, 55]. A significant burden to this approach is that the TCEs are restricted to a specific subset of patients that express the relevant HLA allele, limiting the availability to a select number of patients for a given therapeutic. An added complication is that because these therapeutics are designed to target peptides in the context of an HLA, toxicology assessment is not feasible in NHP, requiring the development and use of *in vitro* and *ex vivo* studies to assess safety before progressing into patients [155].

While some have pursued recombinant TCRs paired with anti-CD3 binding domains, others have engineered TCR mimic antibody binding domains that recognize and bind strongly to tumor-specific HLA/peptide complexes and have paired them with anti-CD3 binding domains. These TCEs have demonstrated preclinical activity on several tumors that present the HLA/peptide complex, including a splice variant of Survivin [156] and the intracellular oncoprotein WT1 [157]. Like recombinant TCRs, engineering high affinity binding domains to HLA/peptide complexes is challenging due to the small size of the peptide being presented, which limits binding epitope availability and increases the chances of cross reactivity to the HLA alone, regardless of what peptide is being presented in the complex [158].

3.3 Conditionally active TCEs

Conditionally active TCEs take advantage of the tumor's unique biology to increase tumor specificity and improve safety, thereby increasing the therapeutic window. It is well established that the tumor microenvironment (TME) demonstrates increased protease activity relative to healthy tissues [159]. Several groups have designed conditionally active TCEs that are dosed as prodrugs that are unable to co-engage T-cells and tumor cells, and are only converted to active TCEs upon entering the TME. The conversion step from prodrug to active TCE typically requires a proteolytic cleavage event directed towards an amino acid linker that tethers a masking domain that inhibits the binding domains from engaging the tumor antigen and/or CD3 on the T-cell. Only after proteolytic cleavage can the TCE specifically target the tumor. Many different protease cleavable linkers have been incorporated to into the different TCE designs to take advantage of the various protease families that have been reported to be active in tumors, including serine proteases and matrix metalloproteinases [160–162].

CytomX has utilized their Probody design over multiple therapeutic programs. The Probody is based on an antibody format [162], with peptide masks fused to the end of each antigen binding domain through a protease cleavable linker that blocks binding to the antigen. The Probody design has been used to engineer a conditionally active TCE format that prohibits both target antigen binding and T-cell engagement via CD3 prior to protease-mediated activation. Because it is based on an IgG mAb design, the Fc domain provides HLE in both its inactive prodrug and active TCE states.

The conditionally active anti–EGFR x CD3 Probody demonstrated potent *in vitro* activity and tumor regressions in preclinical studies, as well as an improved safety window in NHP studies when compared with an inherently active positive control TCE [163].

The Probody format was followed by AMUNIXs XPAT (XTENylated Protease-Activated T-Cell Engager) design that utilizes their XTEN masking technology to block both CD3 and tumor antigen binding. The XTEN masks are fused to the TCE by protease cleavable linkers, and once cleaved by active proteases the XTEN masks are released. The TCE can then co-engage the tumor antigen of interest along with CD3 on T-cells. Addition of the XTENs increases the molecular weight (MW) of the TCE and provides HLE after administration, which reduces the need for more frequent dosing in patients. Because the active PAT form has a lower MW after the loss of the XTEN masks, it is able to be more rapidly cleared, a design mechanism purported to minimize exposure and subsequent toxicity to healthy target-expressing tissues. Like the Probody design, the conditionally active anti–EGFR × CD3 and anti–HER2 XPATs demonstrated potent *in vitro* activity and tumor regressions in preclinical studies, as well as an improved safety windows in NHP studies when compared with an inherently active positive control TCEs [81].

Maverick Therapeutics has also developed the conditionally active TCE COBRA (Conditional Bispecific Redirected Activation) platform [161, 164]. The COBRA design includes two sdAbs that can bind tumor antigen upon administration, but only engages CD3 on T-cells following a protease-mediated activation event. In its prodrug state the formation of an active CD3 binding domain is constrained by its unique folded design. A protease cleavable linker separates the active COBRA fragment from the inactivating fragment and an antihuman serum albumin (HSA) sdAb at the end of the molecule, which gives the intact prodrug increased half-life after administration. Once the linker is proteolytically cleaved, the inactive fragment is released and the active fragment that is bound to the cell surface of the tumor dimerizes with a second active fragment forming a CD3 binding domain, allowing for COBRA co-engagement of tumor antigen and CD3 on T-cells. Due to the four-tumor antigen binding sdAbs in the active COBRA dimer, tetrameric binding results in high avidity to the tumor antigen expressed on the cell surface. When the inactive fragment is released after protease cleavage the anti-HSA sdAb is released with it, thereby removing the HLE domain from the active COBRA fragment. As demonstrated with the anti-EGFR COBRA MVC-101 [164], this allows for the rapid clearance of active COBRA molecules should they escape the TME. Recent preclinical studies report MVC-101 is conditionally active, highly potent *in vitro*, and demonstrates regression of established solid tumors in preclinical efficacy studies at relatively low doses. [161, 164].

More recently, Roche reported on their conditionally active TCE based on the 2:1 TCB design. They describe protease-activated antifolate receptor 1 (FOLR1) and antimesothelin (MSLN) TCBs that fuses anti-idiotypic anti-CD3 masks to the anti-CD3 Fab through a protease-cleavable linker [160]. As with the other conditionally active designs described earlier, they report conditionally active and highly potent TCEs that demonstrate efficacy against established solid tumor models.

While it is still relatively early for the conditionally active TCEs, a number of them are moving into clinical trials, as other novel conditionally active formats are being disclosed. Over the next few years, we may know if these conditionally active designs are able to take advantage of the unique TME and preferentially target tumors over normal tissues that express the same antigens. Potentially allowing for TCEs to more effective and safer in patients with solid tumors indications.

4. Potential mechanisms of resistance to TCEs

Blinotumomab has proven to be successful in patients with B-ALL, being the first and only T-cell engaging bispecific to be approved for systemic administration. While several other inherently active T-cell engagers have shown promise in hematologic malignancies, attaining responses with T-cell engagers has proven to be more challenging in patients with solid tumors. The potential of inherently active T-cell engagers in solid tumors has often been limited by toxicities, as discussed earlier in this chapter. In addition, multiple resistance mechanisms are also thought to limit the activity of inherently active TCEs in solid tumors, the first being the high interstitial pressure and poorly organized vasculature in solid tumors, which limits the ability of therapeutics to penetrate tumors [13–15]. This challenge is common to all therapeutics, especially biologics, and multiple groups are developing and evaluating strategies to facilitate tumor penetration of therapeutics [165–167].

An additional potential resistance mechanism to inherent TCE in solid tumors is a lack of T-cells within the tumor tissue or exclusion of T-cells from the tumor core, the region within the tumor tissue comprised of tumor cells. In general, "cold" tumors, those containing few T-cells or where T-cells are limited to stromal regions, correlate with worse prognosis for patients than "hot" tumors, where T-cells are located within the tumor cores [168]. Because T-cell engagers rely on the simultaneous binding of T-cells and tumor cells, a lack of T-cell infiltration into tumor cell regions necessarily limits their activity. Strategies to convert cold tumors to hot tumors is an area of intense investigation, which if successful, would not only benefit patients overall, but could also increase the sensitivity of tumors to TCEs [169, 170].

Engagement of checkpoint inhibitory receptors on T-cells with their cognate ligands on tumor cells provides an additional potential mechanism of resistance to T-cell engagers. T-cells migrating into tumor tissues become activated in an antigen-specific manner, which is followed by upregulation of inhibitory receptors on the T-cells as part of a normal negative feedback mechanism to limit their activity [171, 172].

In conjunction, the tumor cells upregulate the ligands for such receptors, and engagement of the ligand/receptor pair provides a potent resistance mechanism against destruction of the tumor cells by T-cells [172]. One of the most well-known of such inhibitory ligand/receptor pairs is PD-L1/PD1; disrupting the interaction between PD1 on the T-cells and PD-L1 on the tumor cells has been shown to enhance the potency of T-cell engagers *in vitro* [173]. Overcoming such a resistance mechanism to T-cell engagers in patients may be addressed via combination therapy strategies, as discussed later in this chapter.

Tumor heterogeneity is also a potential mechanism of resistance to targeted therapies in solid tumors, including T-cell engagers [174, 175] Within a given tumor, populations of cells either not expressing or substantially reduced expression a given TAA, enables those cells to escape destruction by T-cells via a TCE. Heterogeneity may be due to pre-existing small populations of tumor cells with varying levels of target expression or may arise via antigen loss by tumors that had initially been uniform. In either case, treatment with a cytotoxic therapy selects for and allows the expansion of tumor cells that lacks the TAA, and tumor heterogeneity is a challenge for attaining durable responses with T-cell engaging therapies.

One strategy to overcome the tumor heterogeneity challenge is to target two tumor antigens simultaneously. In the case of a low-frequency event such as antigen loss or downmodulation, the likelihood that a single tumor cell will downmodulate two separate antigens simultaneously would be an exceedingly rare event. Thus, T-cell engager molecules with the ability to target multiple TAA's may have an advantage in creating durable responses. Several T-cell engager molecules, in their current configurations, are bivalent for TAA binding, including TandAbs, 2:1 CrossmAbs, ADAPTIRs, Tribodies, and COBRAs; thus, these molecules can be readily modified as TCEs that bind two different TAA's. However, additional molecular designs can likely accommodate an additional TAA binding site to form dual-targeting T-cell engagers to address the issue of tumor heterogeneity.

5. The limitations of preclinical *in vivo* efficacy models

While all *in vivo* mouse tumor efficacy models have limitations that impact their translatability into the clinic, some have more limitations than others. In the context of TCEs it is important to understand what the limitations of each model are when designing a study, and how those limitations may impact the interpretation of the data. For this overview we will focus on five generally employed efficacy model designs, discuss their pros and cons in relationship to solid tumor indications, and where relevant, point out recent modifications of these models that have been reported, which may improve their read-outs (see Table 2). One limitation that is relevant to all the efficacy models is the lack of cross reactivity between anti-CD3 binding domains to human and

Table 2 Pros and cons of preclinical mouse efficacy models to study solid tumor indications.

Tumor models	Pros	Cons
Co-mix Model	• Low threshold to observe TCE mediated activity	• Does not measure solid tumor regressions • Lacks T-cell recruitment and infiltration • Lacks mouse stromal cell incorporation • Unrealistic E:T starting ratios • Often uses daily dosing regimen
Humanized PBMC Engraftment Model	• Can measure TCE mediated tumor regressions • Requires T-cell infiltration • Incorporates mouse stromal cells • More realistic E:T starting ratios • Allows for less frequent dosing regimens	• Systemic xenoreactive T-cells mediate GVHD (use of MHC class I and II KO strains may overcome GVHD) • Limited window for experimentation due to GVHD
Adoptive Human T-cell Transfer Model	• Can measure TCE mediated tumor regressions • Requires T-cell infiltration • Incorporates mouse stromal cells • More realistic E:T starting ratios • Allows for less frequent dosing regimens	• May underestimate efficacy due to limited number of T-cells administered • May limit efficacy with TCEs that have mouse cross reactive binding domains and normal lung expression of the target
Humanized HSC Engraftment Model	• Can measure TCE mediated tumor regressions • Requires T-cell infiltration • Incorporates mouse stromal cells • More realistic E:T starting ratios • Allows for less frequent dosing regimens	• Impact of donor-to-donor variability in studies with large cohorts • Per mouse cost is high

| Humanized CD3 Transgenic Models | • Can measure TCE mediated tumor regressions
• Requires T-cell infiltration
• Incorporates mouse stromal cells
• More realistic E:T starting ratios
• Allows for less frequent dosing regimens | • Limited commercial availability
• Less TCE mediated activity (less likely to demonstrate regressions)
• Limited dosing window due to ADA
• Limited syngeneic tumor models
• May need to engineer syngeneic tumor models to express the human antigen |

murine CD3. Owing to low sequence homology between the species, antibody binding domains have not been identified that can engage both human and murine CD3. This has resulted in the need for most preclinical efficacy models to implant both human tumor cells and human T-cells into immunodeficient mice, as the T-cells are the actual cytotoxic agent being induced by the TCEs and are required for activity. As described in the following section, there are several ways preclinical models have been designed to include human T-cells, and these differences contribute strongly to how one can interpret the results.

5.1 Co-mix Model

The first tumor efficacy model, generally referred to as the Co-mix (or Ad-mix) Model, involves mixing a set ratio of isolated human T-cells or human peripheral blood monocytes (PBMCs) with human tumor cells (cultured cells or single cell suspensions from freshly isolated tumor tissue). This mixture of tumor and immune cells is then implanted subcutaneously (SC) in the flanks of immunodeficient mice. The mice then typically receive their first dose of the TCE within 24 h after implantation, and this is followed by successive daily doses of the TCE for generally for 4–5 days [33, 64, 66, 176, 177]. This tumor model is the least rigorous of the five we describe here and is generally considered to be a low-bar model to demonstrate TCE activity against a tumor expressing the relevant target antigen. However, along with the ease of measuring TCE activity in this model comes several limitations.

One limitation of the Co-mix Model is that it does not measure tumor regressions, but rather inhibition of tumor out-growth. Inhibition of tumor outgrowth is achieved with significantly lower doses compared to concentrations required to demonstrate regressions of established tumors in mice. For example, when the same TCE against P-cadherin (PF-00671008) was tested in both the Co-mix Model and the Adoptive Human T-cell Transfer Model (an established tumor model described as follows), the dose required to inhibit tumor out-growth in the Co-mix Model [66] was 500-fold less than the dose required to demonstrate complete regressions of established tumors using the Adoptive Human T-cell Transfer Model [67] when testing on the same tumor xenograft. Additionally, therapeutic candidates that demonstrate tumor inhibition, but do not demonstrate regression of established solid tumors *in vivo* are generally considered to be less likely to translate to patient responses in the clinic, thereby limiting the utility of the data collected from Co-mix Model.

Other limitations of the Co-mix Model are due to the nature of premixing the T-cells and tumor cells before implanting them, since in this model the T-cells are not required to home in on and infiltrate the tumor before observing activity. It is well established that the tumor microenvironment (TME) may restrict both T-cell recruitment and infiltration into the tumor [147]. Chemokine attracted T-cells must navigate the irregular tumor vasculature and overcome various barriers before entering the tumor site, including tumor endothelial cells and pericytes [178–183]. The absence of tumor vessels in the Co-mix Model neglects their influence on T-cell trafficking into the tumor.

Additionally, due to the tumor and immune cell mixture being implanted in conjunction with the rapid commencement of dosing, the tumors have not had time to become established prior to treatment, which severely restricts the contribution of the various stromal cell subtypes in the make-up of tumor architecture. Tumor stromal cells are known to contribute significantly to the milieu of the TME and have direct influence on T-cell activity. Once infiltrated, the stromal cells act to impede the recruited T-cell effector functions. The stromal cell types that negatively impact T-cell activity include cancer associated fibroblasts [184], mesenchymal stem cells [185], tumor associated macrophages [186], myeloid derived suppresser cells [187], tumor associated neutrophils [188], and immature dendritic cells [189]. The lack of tumor stromal cell contribution in the Co-mix Model becomes even more of a limitation in the context of testing conditionally active TCEs that take advantage of increases in proteases activity in the TME to differentiate between normal and tumor tissues that express the target antigen. Because stromal cell populations are known to contribute significantly to the protease activity measured in tumors [190], their absence in the Co-mix Model in the context of conditionally active TCEs may alter the level of protease expression and activation, as well as the specific proteases that are expressed and active in the TME.

Another limitation of the Co-mix Model is the ratio of T-cell effector (E) cells to tumor target (T) cells implanted in mice is not physiologically relevant compared to what is observed in patient tumors. Patient tumors have been reported to contain a broad range of CD3+ T-cell infiltrates, averaging anywhere from 1% to 10% depending on the tumor type [191]. The E:T ratios that are generally tested *in vivo* using the Co-mix Model are in the 5:1 to 10:1 range, which is much greater than the actual T-cell density in patient tumors. One example of how disproportionate this model can be is demonstrated by the requirement of a 1000:1 E:T ratio to completely inhibit tumor outgrowth in the Co-mix Model, while lower E:T ratios only delayed tumor outgrowth [176]. The lop-sided starting number of effector T-cells relative to the target tumor cells used in the Co-mix Model limits its usefulness in determining relevant *in vivo* activity of TCEs.

And one last notable limitation of the Co-mix Model is related to the more recent use of HLE TCE designs. In this model the mice typically receive successive daily doses of the TCE for a set number of days very soon after effector and target cell implantation. Because early TCE designs often did not include domains that extend their half-lives *in vivo*, it was reasonable to dose them daily due to their short exposure times. However, most current TCE designs directed towards solid tumor indications now include half-life extension domains, yet daily dosing is still reported when using this model [64, 66, 77, 192, 193]. When these HLE designs are given daily in the Co-mix Model, it is expected that each subsequent administration will result in accumulation of the circulating TCE concentration. Therefore, exposure of the TCE in the blood may be much higher relative to the reported daily dose administered. This may mislead one to believe the lower dose was efficacious, without realizing the exposure was significantly higher due to the frequent dosing regimen with a HLE TCE.

5.2 Humanized PBMC Engraftment Model

The next tumor efficacy model, generally referred to as the Humanized PBMC Engraftment Model is an established tumor model. Unlike the Co-mix Model, this model requires T-cell recruitment and infiltration for efficacy, the tumors incorporate mouse derived stromal cells, it has a more physiologically relevant E:T ratio, it can be dosed less frequently over a longer duration, and importantly it can be used to measure tumor regressions, as opposed to tumor outgrowth. Initially this model involves implanting a set number of human PBMCs intraperitoneally (IP) into the peritoneal cavity of immunodeficient mice, generally between 5×10^6 and 2×10^7 PBMCs [194]. The T-cells in the immune cell mixture then expand systemically overtime due to their xenoreactivity against the host mouse tissues, and the mice subsequently develop graft *versus* host disease (GVHD). As the T-cells are engrafting in the mice, tumor cells are implanted SC into the mice and given time to grow and become established, often to a volume that is palpable, in the range of 100–300 mm^3, before the start of treatment.

This brings us to the limitations of this model. As the xenoreactive T-cells continually expand they are actively cytotoxic against mouse tissues resulting in the mice eventually becoming moribund [195]. Because the health of the mice declines overtime due to GVHD, it restricts the window allowed for experimentation after the tumors have become established, often limited to around 2–3 weeks. Therefore, a good understanding of the T-cell engraftment rates in the strain of mice being used, along with the individual tumor growth rates, are critical for a well-designed study. The timing of the PBMC implantation relative to the tumor cell implantation will vary from study-to-study, depending on the growth rate of tumor cell line selected for the xenograft relative to the engraftment rate of the donor human T-cells. To maximize the duration of the study, the start of dosing is often early in the engraftment phase that results in a high variability in the relative percent of human T-cell engraftment between the mice. If early time points in the study are being sampled, this variability in engraftment levels may result in the data being more difficult to interpret, so the authors recommend prior to treatment to not only randomize the mice into groups based on the volume of their tumors, but also on the percent level of T-cell engraftment in the peripheral blood at the time of dose initiation.

Owing to the systemic poly-activation of human T-cells in the PBMC Engraftment Model, it is more difficult to analyze their responses precisely. To overcome this limitation, recently developed strains of immunodeficient mice have been generated that knock-out (KO) both mouse major histocompatibility (MHC) class I and II genes [195, 196]. The loss of MHC class I and II in mouse tissues significantly reduces the level of GVHD mediated by the xenoreactive human T-cells. These PBMC engrafted MHC class I and II KO mice survive significantly longer (>100 days), while the human T-cells maintain stable levels of engraftment and retain their function overtime [195]. It is worth noting that the rate of T-cell engraftment is slower and the total percent of T-cells engrafted is lower compared to mice that express MHC class I and/or II, which may make efficacy data generated by the parental and double KO mouse strains more difficult to directly compare.

5.3 Adoptive Human T-cell Transfer Model

Another established tumor model, generally referred to as the Adoptive Human T-cell Transfer Model, is like the Human PBMC Engraftment Model in that it requires T-cell recruitment and infiltration to measure efficacy. Also, the tumors incorporate mouse derived stromal cells and it utilizes a more relevant E:T ratio. This model can be dosed less frequently over a longer duration as well, and can be used to measure tumor regressions. This model begins with SC human tumor cell implantation into immunodeficient mice, which are then given time to grow and become established. While the tumors grow to volumes similar to those noted earlier, human T-cells are expanded in culture in parallel for around 10 days. Once the tumors are established, mice receive an IV injection of human T-cells, generally around $1–2 \times 10^6$, which is then followed by the first dose of the TCE within 24 h [67]. From the authors' experience, if administering fewer T-cells the study-to-study variability becomes more apparent, and if given more T-cells there is an increased likelihood that mice in the study will develop GVHD, which typically is not observed in this model if given the sufficient number of T-cells. Owing to the relatively low number of T-cells administered, the E:T ratio is likely underestimating the efficacious dose when testing TCEs, since the T-cells and not the TCE are delivering the cytotoxic response. The authors consider this one of the most rigorous models, since at the time of the first dose only a small percentage of the T-cells have reached the established tumors. When treated with TCEs those effector T-cells that reside early in the tumor expand, and as has been reported, recruitment and infiltration of additional peripheral T-cells may take between 5 and 8 days to peak [197]. We have observed, as have others [198], that adoptively transferred T-cells are primarily retained in the lungs after IV administration. When testing TCEs that cross react to mouse antigens expressed in the lung (even at very low levels), it is the authors' experience that this can lead to a phenomenon where the implanted tumors do not respond using this model. Direct comparison of several mouse cross-reactive and mouse noncross-reactive TCEs to the same target antigen have demonstrated this observation (unpublished data). And in similar studies where the target antigen is not expressed on normal lungs, the mouse cross reactive TCEs are as efficacious as the nonmouse cross reactive TCEs (unpublished data). While the authors haven't performed the studies necessary to fully demonstrate the direct mechanism, we speculate that low level human T-cell activation in the lungs mediated by TCEs bound to the target, induce cytokine and chemokine release, which results in the T-cells being retained and unable to be recruited to the site of the tumor. It is also worth noting that when those same mouse cross-reactive TCEs are tested in the Human PBMC Engraftment Model they are as efficacious as the nonmouse cross reactive TCEs. This is presumably because the human T-cells are both higher in number and better distributed at the time of dosing in the mice due to GVHD and are better able to be recruited to the site of the tumor.

5.4 Humanized HSC Engraftment Model

Efficacy with TCEs on established human tumors has also been studied preclinically in the Humanized Hematopoietic Stem Cell (HSC) Engrafted Model [199]. In this model the immunodeficient mice are myeloablated and humanized by adoptive transfer of human CD34+ HSCs. The mice are then housed for several weeks to allow for multilineage human immune cell development, including T-cell maturation. This model has been validated for the study of immuno-oncology therapeutics [200], as the human T-cells mediate inflammatory responses against tumors with no donor cell immune reactivity towards the host. This model allows for long-term studies to be performed on mice to better measure the durability of antitumor responses. A recent study testing TCEs in this model reported complete tumor regressions with no sign of recurrence in half the animals treated 10 weeks after dosing ceased [200]. However, also noted in the report, donor-to-donor variability is a factor as one donor demonstrated a robust TCE-mediated tumor response, while a second donor did not respond. Since a single donor is only able to engraft a limited number of mice, moderate-to-larger study designs require more than one HSC donor. This variability should be considered closely when choosing this model as it typically is the most expensive option.

5.5 Humanized CD3 Transgenic Mouse Model

All the murine models described previously reconstitute T-cell-mediated immune responses to varying levels, but none of these methods produce fully immunocompetent mice. These efficacy models are valuable to different degrees to evaluate the activity of TCEs in an *in vivo* setting. Still, the ability to test TCE efficacy preclinically in an established mouse tumor model with a fully developed immune system may increase our understanding of their therapeutic potential. Which brings us to the last *in vivo* efficacy model we will review in this chapter, the Humanized CD3 Transgenic Mouse Model. Owing to the low sequence homology, an agonist antibody that is cross-reactive to human and mouse CD3 has not been identified. To side-step this several transgenic mouse models have been engineered to express human CD3 on mouse T-cells and have been tested in the context of TCE-mediated antitumor activity. Of the three different CD3 chains contained within the TCR complex (δ, ε, and γ) most agonist antibodies have been generated are against the ε chain. A variety of strategies have been pursued to produce mice that express relevant levels of human CD3ε on mouse T-cells for use in TCE studies.

One approach has been to express human CD3ε on mouse T-cells, while maintaining endogenous mouse CD3ε, CD3δ, and CD3γ chain expression. Several groups that have taken this route report transient growth inhibition of established tumors, yet have not demonstrated complete regressions using this design [201, 202]. This suggests adding human CD3ε expression alone may not be sufficient to fully engage the mouse T-cell response. A more recent and complex engineering approach knocked out all three mouse chains and replaced them with human CD3ε, δ and γ expression. The T-cells in these transgenic mice matured and were able to respond effectively against established tumors and demonstrated tumor regressions using TCEs [203].

As noted earlier, having a fully developed immune system may be beneficial in understanding TCE-mediated cytolytic activity. Still, these transgenic mice also have limitations that lessens our ability to fully exploit their systemic immune response. Because these mice are immunocompetent, prolonged administration of human or humanized TCE proteins often leads to the development of antidrug antibodies (ADAs). ADA response are typically observed 10 days after the initial dose of human or humanized protein. The ADA response neutralizes the TCE, thereby reducing its exposure in the tumor and its ability to act against it. And in severe cases, the ADA response may induce anaphylaxis and death through an immune complex-mediated inflammatory response that includes cytokine release and complement activation [204]. Another limitation of the transgenic CD3 model is the number and variety of syngeneic mouse tumor models that are readily accessible. Generally, two strains of immunocompetent mice are used in oncology to measure efficacy against established tumor, C57Bl/6 and BALB/C. For each strain of mice a number of tumors have been induced through various methods, but compared to human xenograft models, which are most often derived from patient samples, a relatively low number of mouse derived tumors are available, and those that are available only cover a narrow segment of tumor types [205]. Owing to this, the mouse syngeneic tumors that are accessible often do not express the tumor antigen of interest, and if they do the TCE does not cross react to the mouse sequence. In this case, one must engineer the mouse tumor cell line to express the human tumor antigen and then implant it into the mice. However, because these mice have a functional immune system, the MHC class I presented "foreign" peptide sequence derived from expressing the human tumor antigen is likely to be identified through immune surveillance and contribute to the inhibition of tumor growth independent of TCE treatment [206].

TCEs that have been tested in several of these solid tumor efficacy models are now, or soon will be, progressing in clinical trials. As more clinical data becomes available, we will develop a better sense of which preclinical tumor models may better predicted response in patients, as well as what doses are required to achieve them.

6. TCEs targeting solid tumors in the clinic

As the previous sections have noted, the development of clinically useful TCEs for solid tumor indications have met several challenges due to both toxicity and efficacy. As noted in the first section, protein engineers have also spent a substantial amount of time working on designs to overcome these issues.

The first TCE to enter the clinical market was catumaxomab in 2009. Developed by Fresenius Biotech and Trion Pharma (and later acquired by Neovii), catumaxomab is a trifunctional bispecific with binding domains specific for human EpCAM and CD3ε as well as a functional IgG Fc domain with binding to Fcγ receptors. It was approved by the EMA for intraperitoneal infusion for the treatment of malignant ascites due to epithelial carcinomas [90]. While catumaxomab was withdrawn from the market in 2017 due to commercial reasons, its approval kicked off the search for additional bispecific T-cell engagers for the treatment of cancer. In 2011, Micromet entered into a development collaboration with Amgen to develop two solid tumor targeting molecules on an entirely new platform called Bispecific T-cell Engagers (BiTEs). Amgen would go on to acquire Micromet in 2012 and continue to develop the BiTE technology.

In December 2014, Amgen had its first successful approval on the platform with blinatumomab being approved for the treatment of Philadelphia chromosome-negative precursor B-cell acute lymphoblastic leukemia (B-cell ALL). While Amgen has had less success with solid tumor targeting BiTEs, it has still secured its position as the leader in development of T-cell-engaging bispecifics; currently with 11 clinical BiTE programs and additional collaborations with Xencor on the XmAB technology. However, since the approval of blinatumomab, there have been no additional approvals of T-cell engagers in either the hematological or solid tumor space and follow-on trials with additional targets have been slow to come. Instead, as described previously, research has mainly focused on modifying the existing platforms and developing new platforms to overcome the hurdles encountered trying to bring this potent modality to solid tumors.

Over the past decade of development, more than a dozen additional pharmaceutical and biotechnology companies entered the T-cell engaging bispecific space with multiple new platforms. By 2018 this research had begun to mature and as of October 2020 there are 28 active clinical trials of bispecific T-cell engagers and 29 completed or terminated trials in solid tumor indications. The trials span a range of tumor targets including those that are broadly expressed, narrowly expressed, and tumor-specific as well as different platforms, and indications. Most of the trials (43) were either phase 1 or phase 1/phase 2 trials, showing the increased rate of development within this modality over the past 2 years (Table 3) and also the challenges still being encountered. The following sections

Table 3 Ongoing and completed clinical trials of bispecific T-cell engaging therapeutics.

Company	Target	Molecular identity	Clinical status	Type	Combinations	Clinical trials
Amgen	PSMA	AMG160	Ph1	HLE	Pembrolizumab (anti-PD-1); Etanercept (TNF inhibitor)	NCT03792841
	PSMA	AMG212/ BAY 2010112	Ph1	BITE		NCT01723475
	EGFR VIII	AMG596	Ph1	BITE	AMG 404 (anti-PD-1)	NCT03296696
	DLL3	AMG757	Ph1	HLE	Pembrolizumab (anti-PD-1)	NCT03319940
	MUC17	AMG199	Ph1	HLE		NCT04117958
	CLDN18	AMG910	Ph1	HLE		NCT04260191
	STEAP1	AMG509	Ph1	XmAb		NCT04221542
	CEA	AMG211/ MEDI565	Discontinued	BITE		NCT01284231; NCT02291614
	EpCAM	AMG110/ MT110	Discontinued	BITE		NCT00635596; NCT02806960
Immunocore	gp100	IMCgp100 (tebentafusp)	Pivotal	ImmTAC	Durvalumab (anti-PD-1); tremelimumab (CTLA-4)	NCT02889861; NCT03070392; NCT01211262; NCT01209676; NCT02570308; NCT02535078
	MAG-A4	IMC-C103C	Ph1/2	ImmTAC	Atezolizumab (anti-PD-L1)	NCT03973333
	PRAME	IMC-F106C	Ph1/2	ImmTAC	Anti-PD-(L)1	NCT04262466
	NY-ESO-1	GSK01/ IMCnyeso	Ph1/2	ImmTAC		NCT03515551
Roche (Genentech; Chugai)	GPC3	ERY974	Completed Ph1	TRAB		NCT02748837
	CEA	Cibisatamab	Ph1	TCB	Pretreatment with Obinutuzumab (CD20), combo with Atezolizumab (PD-L1) and Tocilizumab (IL6)	NCT02650713; NCT02324257; NCT03337698; NCT03866239
	HER2	RG6194/ BTRC4017A	Ph1	TDB		NCT03448042

Continued

Table 3 Ongoing and completed clinical trials of bispecific T-cell engaging therapeutics—cont'd

Company	Target	Molecular identity	Clinical status	Type	Combinations	Clinical trials
Harpoon	PSMA	HPN424	Ph1/2	TriTAC		NCT03577028
	MSLN	HPN536	Ph1/2	TriTAC		NCT03872206
Pfizer	GUCY2c	PF-07062119	Ph1	DART	Anti-PD-1; anti-VEGF	NCT04171141
	CDH3	PF-06671008	Discontinued	DART		NCT02659631
MacroGenics	B7H3	MGD009	Discontinued	DART	MGA012 (PD-1)	NCT03406949; NCT02628535
	gpA33	MGD007	Discontinued	DART	MGA012 (PD-1)	NCT03531632; NCT02248805
Neovii/ Fresenius	EpCAM	Catumaxomab	Withdrawn[a]	Triomab		NCT00377429; NCT00326885; NCT01065246; NCT00464893; NCT01320020; NCT00822809; NCT00563836; NCT00836654; NCT00352833
	HER2	Ertumaxomab	Discontinued	Triomab		NCT01569412; NCT00351858; NCT00522457; NCT00452140
Glenmark/ Ichnos	HER2	GBR1302	Ph1	BEAT		NCT02829372; NCT03983395
Xencor	SSTR2	XmAb18087	Ph1	XmAB		NCT03411915
Aptevo	PSMA	APVO414/ ES414	Discontinued	ADAPTIR		NCT02262910
Regeneron	MUC16	REGN4018	Ph1/2	Fc mutant IgG	Cemiplimab (anti-PD-1)	NCT03564340
Amphivena	CD33	AMV564	Ph1	ReSTORE		NCT04128423[b]
Genmab	5T4	GEN1044	Ph1	DuoBody		NCT04424641
GEMoaB	PSCA	GEM3PSCA	Ph1	ATAC		NCT03927573
University Hospital Tübingen	PSMA	CC-1	Ph1			NCT04104607; NCT04496674

[a]Only listing trials run by Neovii and Fresenius. LintonPharm Co., Ltd. is resuming clinical trials on this compound with a Ph3 study planned this year (2020).
[b]Only listing trials in solid tumor indications.

summarize the current state of clinical development for bispecific TCE platforms in solid tumor indications.

6.1 Broadly expressed tumor targets

Bispecific T-cell engagers targeting broadly expressed tumor antigens such as EpCAM, EGFR, B7H3, and CDH3 have shown limited clinical success. To date, these programs have all been terminated either before entering the clinic or during phase 1 clinical development. While most companies have cited business decisions as the driver for canceling the clinical programs, the limited published data has demonstrated that tolerability and a limited therapeutic window may have contributed to the lack of success for some programs [130]. The majority of broadly expressed tumor antigens are also broadly expressed on many healthy tissues. Even though these targets may be more highly expressed in tumors (*e.g.*, EGFR), the level of expression on the healthy tissues is still sufficient to lead to unacceptable toxicities for this potent modality as discussed in earlier sections.

Solitomab (AMG110/MT110) was the second BiTE program to enter clinical development and the first systemically dosed bispecific TCE for a solid tumor indication. Similar to catumaxomab, solitomab targets EpCAM and CD3ε. EpCAM is a type I transmembrane glycoprotein, with broad normal tissue expression, which plays a role in multiple cellular processes including proliferation, migration, adhesion, differentiation and signaling [207]. In normal tissues it is primarily found on the basolateral surfaces and within tight junctions [207]. EpCAM has also been shown to be frequently over-expressed in a wide range of epithelial cancers making it an attractive target for cancer diagnostics and therapeutics. The overexpression and cellular localization in normal tissues were cited as factors in selecting EpCAM as a target for the BiTE platform [130]. In the phase 1 trial, the starting dose of solitomab was set at 1 μg/day, flat dosing, based on the *in vitro* Minimal Anticipated Biological Effect Level (MABEL). Given the short half-life of the BiTE platform ($\sim$4.5 h in this case), solitomab was administered via a portable IV pump as a continuous IV infusion (cIV). Pharmacodynamic assessments confirmed activity consistent with the proposed mechanism of action of this TCE. The primary observations included a prolonged reduction in peripheral T-cell counts due to T-cell redistribution into tissues and increases in peripheral cytokines consistent with T-cell activation and inflammation. The most common dose-limiting toxicities (DLTs) observed were grade 3–4 liver enzyme increases and diarrhea. The liver enzyme increases were observed early after the start of dosing and resolved under continued administration. Given this observation, both single and multiple-step lead-in dosing schedules, referred to in the trial as "run-in" doses, were evaluated. Under the single-step lead-in dosing schedules, patients received a lower initial dose (3 μg/day) for 1 week, followed by an increased dose for subsequent weeks. Using this strategy, a maximum tolerated dose (MTD) of 24 μg/day was achieved. Multiple-step lead-in dosing schedules were also evaluated. Again, primarily starting with 3 µg/day for the first week, patient dosing was increased to 12 µg/day for 1–3 weeks before an additional dose increase was implemented. Using this multi-step strategy, the administered dose was expanded to as high as 96 µg/day. However, even with the lead-in dosing schedules and an addition of dexamethasone during the first 3 days of each dosing cycle, doses above 24 µg/day resulted in unacceptable toxicities with DLTs observed in three of seven patients receiving 48 µg/day and two of seven patients receiving 96 µg/day. A dose level of 24 µg/day was considered inadequate, as a dose of at least 48 µg/day was required for concentrations of solitomab comparable to the *in vitro* EC90 values. Patient response data also showed that stable disease was the best overall response with a single unconfirmed partial response suggesting the achievable dose level was inadequate for robust tumor response [130].

Two other programs that have more recently been terminated are MGD009, a B7H3 targeting DART from MacroGenics, and PF-06671008, a CDH3 (P-cadherin) targeting Fc-DART from Pfizer in collaboration with MacroGenics. While limited data is available for both programs, and both programs were officially terminated for business decisions, there is evidence that both programs observed toxicities. MGD009 was placed on a partial clinical hold for hepatic adverse events (transient elevations of transaminase) in late 2018 and removed from hold in early 2019 [142, 208]. No additional data on dose levels reached or toxicity is currently available. The PF-06671008 phase 1 clinical trial was terminated in early 2019 and study results were posted in early 2020 [209]. Study results show that PF-06671008 was administered by weekly IV infusion starting at 1.5 ng/kg and the highest dose tested was 400 ng/kg. Lead-in dosing was attempted in this study as well, with a single patient receiving a priming dose of 200 ng/kg followed by a maintenance dose of 300 ng/kg. In addition, two patients were treated using subcutaneous (SC) injection at 200 ng/kg. While grading information was not provided for the adverse events observed, SAEs related to cytokine release syndrome (CRS) were reported in most patients starting in the 50 ng/kg dose cohort. Both lead-in dosing and SC dosing cohorts reported no SAEs but did show AEs related to CRS. The dose levels achieved were also well below the preclinical efficacious dose of 1 µg/kg required for inhibition of tumor outgrowth [66] and the 500 µg/kg needed to for tumor regressions in established tumor models [67].

6.2 Tumor-specific mutations

To circumvent the cytokine release syndrome and on-target/off-tumor toxicity observed in the broadly expressed targets, other groups have employed an additional strategy, the identification of tumor-specific antigens. The main source of these antigens is cancer-specific mutations of surface proteins. As discussed earlier, EGFRvIII mutations occur frequently in glioblastoma where EGFR amplification and overexpression are also common [145]. AMG 596 is a BiTE molecule under development by Amgen that specifically targets EGFRvIII. Amgen initiated a phase 1a/1b clinical trial in April 2018 to treat glioblastoma patients whose tumors tested positive for EGFRvIII. The trial is anticipated to complete in March 2021. Interim data published in November 2019 on 15 patients, suggest that AMG 596 may be tolerated, although 50% of patients experienced serious AEs. The early data also showed hints of response. In the eight patients with sufficient follow-up to evaluate response, there was one patient with a partial response (PR) and two patients with stable disease (SD) [210]. No additional data is currently available for this molecule.

6.3 Tumor targets with restricted expression

Given the initial findings with targets showing more broad normal tissue expression and the paucity of relevant tumor-specific targets in many solid tumor indications, many groups have focused on identifying targets with either more specific tumor expression or a greater differential in expression between tumor and normal tissue. An extremely popular target in this category in PSMA. Currently there are three ongoing and two completed clinical trials targeting PSMA with a TCE. PSMA, also known as FOLH1, is a type II membrane protein expressed on all prostate tissue with increased expression on prostate carcinomas. PSMA has limited expression in other normal tissues. The earliest T-cell engaging bispecific targeting PSMA to enter the clinic was pasotuxizumab (AMG212/BAY2010112). Originally developed by Micromet on the BiTE platform, pasotuxizumab entered a phase 1 clinical trial in November 2012 for the treatment of metastatic castration-resistant prostate cancer (mCRPC). The phase 1 trial had two arms, the first with SC dosing and the second with cIV dosing. No data has been reported on the SC cohort, but an update was given on the cIV cohort at ASCO 2019. The trial was terminated early due to the sponsor changing from Bayer to Amgen, and so an MTD was not reached, but 16 patients were treated starting at 5 μg/day up to 80 μg/day. The best overall response was SD in three patients and noncomplete response/nonprogressive disease in three patients. Cytokine release syndrome was observed in 1 of 4 patients at the 40 μg/day dose and 2 of 2 patients at the 80 μg/day dose level. Grade 3 or greater AEs were seen in all dose levels and in 63% of patients overall with the most common AEs being fever and chills. Long-term reductions in circulating Prostate-Specific Antigen (PSA) levels were observed in two patients. One of these was a patient in the 40 μg/day group treated for 14 months, and the second was a patient in the 80 μg/day group treated for 19.2 months [211].

After terminating the pasotuxizumab trial in September 2018, Amgen began a trial of AMG 160, an HLE PSMA targeting BiTE in February 2019. The half-life extension allows AMG 160 to be given every 2 weeks by IV infusion rather than the cIV dosing required for the first-generation BiTEs. Interim results from the phase 1 trial were presented at ESMO 2020. Six dose levels ranging from 0.003 mg up to 0.9 mg were tested and an MTD had not been reached. CRS was the most common AE occurring in 90% of patients with 25% experiencing grade 3 CRS. Out of 15 patients with tumors evaluable by RECIST criteria, there were two patients with a confirmed PR, one with an unconfirmed PR, and eight with stable disease. Overall, 68% of patients showed reductions in PSA levels with 34% having decreases greater than 50% [212]. Dose escalation is continuing for this molecule.

In 2014, APVO414/ES414, built on the ADAPTIR platform entered a phase 1 clinical trial. Limited data was shared on the phase 1 trial of APVO414, however, a corporate press release from Aptevo in August 2017 revealed that patients were developing high-titer ADAs to the platform and Aptevo was shifting to a cIV infusion [213]. The company later discontinued development of the APVO414 to focus on next-generation platforms [214].

Following these trials, the next company to enter the clinic was Harpoon Therapeutics with HPN424 in 2018. HPN424 is built on a unique HLE platform called the TriTAC. While the phase 1 trial for HPN424 is still ongoing, interim data was presented at ASCO 2020. The data show that Harpoon started dosing HPN424 weekly at a MABEL dose of 1.4 ng/kg IV. At the time of the presentation, 44 patients had been treated with 96 ng/kg being the highest dose reached. Grade 3+ CRS was observed in two patients at 24 ng/kg prompting the start of dexamethasone pretreatment in subsequent groups. Multiple regimens of dexamethasone treatment have been evaluated across the dose levels. A single DLT (grade 3 lipase increase) was observed at 96 ng/kg [215]. While the doses of HPN424 are well below the 2 μg/kg dose levels required to see tumor outgrowth inhibition in mouse xenograft models [75], two patients did show greater than 50% decrease in PSA levels, and one patient with decreased PSA has continued on treatment for 84 weeks with two intra-patient dose escalations. Circulating tumor cell (CTC) analysis also showed CTC decreases of greater than 50% in 10 patients [215]. Only a single patient had developed ADA for HPN424. Dose escalation is ongoing for this molecule.

The most recent PSMA-targeted bispecific TCE to enter the clinic is CC-1, built on the IgGsc platform [216]. Developed by the University Hospital Tübingen, two phase 1 clinical trials have been initiated for this candidate. The first of two trials started recruiting in September of 2019. No data was available for these trials at the time of writing.

In addition to PSMA, some other narrowly expressed targets that are being evaluated for TCE bispecifics are DLL3 and CLDN18.2. Amgen has begun phase 1 trials utilizing their HLE BiTE format for both targets. DLL3, the Delta Like Canonical Notch Ligand 3 is an inhibitory member of the delta ligand family of genes. It shows extremely limited normal tissue expression but is selectively expressed in high-grade neuroendocrine tumors such as small cell lung cancer (SCLC) [217, 218]. Within SCLC upward of 80% of tumors are positive for DLL3 by IHC [218]. AMG 757 is an HLE BiTE molecule targeting the DLL3 ligand for treatment or patients with SCLC. Amgen started a large phase 1 trial in 2017 and is exploring both monotherapy and combination therapy with pembrolizumab. The trial is scheduled to complete in 2022. No updates have been given on AMG 757 to date. CLDN18 is a member of the claudin superfamily of genes consists of 18 homologous genes in humans that are integral components of cellular tight junctions [219]. Normal tissue expression of CLDN18 is restricted to gastric epithelium but it is expressed on a large fraction of gastric and pancreatic adenocarcinomas. AMG 910 is an HLE BiTE targeting the second isoform of CLDN18, CLDN18.2. A phase 1 dose escalation/expansion trial for AMG 910 was begun in June 2020 for patients with CLDN18.2 positive gastric cancers. The targeted primary completion date is 2024. Currently no updates have been given for this trial.

6.4 TCR based TCE platforms

Currently, the most advanced TCE clinical programs in solid tumors are based on the Immunocore ImmTAC platform. Rather than targeting a cell surface protein, this platform uses a recombinant TCR to target HLA presenting tumor-associated peptides. The lead program, tebentafusp, targets an HLA-A*0201 complexed with a peptide from the melanoma-associated antigen gp100 [220]. gp100, also known as the premelanosome protein PMEL, is a type-1 transmembrane glycoprotein primarily expressed in melanosomes, the melanin-producing organelles of melanocytes. gp100 has is highly expressed in all stages of malignant melanoma, with lower expression in normal melanocytes and limited expression in other healthy tissues. The TCR in tebentafusp targets HLA complexed with a specific peptide, $gp100_{280-288}$. The HLA targeting of tebentafusp limits its use to individuals carrying the HLA-A*0201 haplotype. Initial human testing of tebentafusp began in 2010 with a Ph0 intratumoral dosing study, the goal of which was to evaluate the single dose pharmacodynamics and a Ph1 study in advanced metastatic melanoma. Results of the Ph1 study showed promising clinical activity. The safety profile was acceptable with the most common grade 3/4 adverse events being manageable skin related toxicities and were likely on target. Less than 5% of patients experienced grade 3 or greater cytokine release syndrome [220]. Out the 47 patients considered evaluable, there were four partial responses and 14 patients with stable disease. However, within the subset of patients with uveal melanoma ($N = 14$) there were two partial responses and eight stable disease with a disease control rate of 57%, showing encouraging results in this hard-to-treat tumor type [220, 221]. Given the promising results of this phase 1 trial in uveal melanoma and the lack of other options for these patients, a phase 1/phase 2 trial focused only on metastatic uveal melanoma was initiated [220, 222]. This phase 1/phase 2 trial was still ongoing at the time of writing with an expected readout in October 2020. Results from the phase 1 arm of the trial showed consistent response with the initial phase 1 study, with 3 of 17 patients achieving a partial response and 11 of 17 with disease control for more than 16 weeks [220]. An additional phase 1b/phase 2 study evaluating the combination of tebentafusp with durvalumab (anti-PD-L1) and/or tremelimumab (anti-CTLA-4) in cutaneous melanoma patients was also initiated.

Immunocore has begun clinical development of three additional molecules utilizing their ImmTAC platform: GSK01, IMC-C103C, and IMC-F106C. All three programs are targeting HLA-A2 antigens complexed with peptides from genes that are members of the Cancer Testis family of genes. GSK01 (also known as IMCnyeso), targets NY-ESO-1 (cancer/testis antigen 1B; CTAG1B) gene, which is expressed in a wide variety of solid tumor indications. IMC-C103C targets MAGE-A4 and IMC-F106C targets PRAME, both of which are highly expressed in melanoma. For all three of these programs, Immunocore has begun phase 1/phase 2 clinical trials in target-selected patient populations. GSK01 is being tested only as a monotherapy, while both IMC-C103C and IMC-F106C and be tested alone and in combination with checkpoint inhibitors. The GSK01 trial is scheduled to readout at the end of 2020, while IMC-C103C will read out in 2021 and IMC-F106C in 2022. Currently no interim data has been presented on any of these three programs [223].

6.5 Increased tumor selectivity through affinity tuning

In order to improve tumor selectivity and try to reduce normal tissue binding and CRS, some groups have developed constructs that take advantage of a normal property of monospecific antibodies that has been engineered out of many bispecific antibodies. In a standard antibody, binding from the two identical arms provides additional strength through process known as avidity. Avidity not only improves the apparent affinity of binders through having multiple interactions, but it can also improve specificity to highly expressed targets by allowing the binding of multiple target molecules at once. By developing bispecific TCE with two binding domains for the tumor targets and a single domain for CD3ε, these molecules can more specifically target proteins that are over expressed on tumors relative to normal tissues. Roche and Xencor have both developed similar 2:1 binding domain platforms.

The Roche platform, referred to as 2:1 TCB, has been utilized in cibisatamab, a CEA targeted bispecific T-cell engager. Clinical development of cibisatamab began in 2014 and to date, Roche has run two phase 1 trials, one monotherapy and one in combination with atezolizumab (an anti-PD-L1 monoclonal antibody). An interim update was given at the 2017 ASCO conference. The initial data was intriguing with some partial responses, stable disease and metabolic partial responses in microsatellite stable colorectal cancer, an indication with limited response to checkpoint inhibitors [224]. The toxicity profile was acceptable with the most common AEs being pyrexia and infusion related reactions. However, grade 4 and 5 DLTs were observed at the 600 μg dose level. Additional data has been slow to come and at AACR 2019 in a podium talk by Dominik Ruttinger it was discussed that many patients were developing ADA. In May 2019 Roche started a third phase 1 trial of cibisatamab, this time with patients receiving a 2-week predosing of obinutuzumab, an anti-CD20 monoclonal antibody. This combination therapy may be an attempt to reduce the occurrence of ADA by first depleting the B-cell population in these patients. No updates have been given on this combination trial.

Using the Xencor XmAb 2 + 1 platform, Amgen has begun a phase 1 trial of AMG 509 targeting STEAP1. Six transmembrane epithelial antigen of the prostate 1 (STEAP1) is a cell surface antigen predominantly expressed in prostate cancer [225]. The study, which started in March of 2020, is a dose escalation/expansion in metastatic castrate resistant prostate cancer. No updates are currently available for this program.

Owing to the issues with ADA for cibisatamab and the recent start of the XmAb programs, it remains to be seen whether these 2:1 TCE platforms will improve the selectivity and thereby the therapeutic window of these molecules.

6.6 Half-life extension

The earliest bispecific T-cell engaging molecules were built on platforms such as the BiTE and DART where the small size ~50 kDa and lack of Fc domains to aid in cellular recycling resulted in therapeutics with extremely short half-life of only a few hours. Owing to the short half-life and infusion related reactions at high doses, many of these programs required low-dose continuous infusion. This dosing regimen did improve tolerability but cIV dosing is inconvenient for patients with solid tumors. Owing to these issues, many companies have begun exploring HLE platforms as noted earlier in the chapter. Multiple half-life extension methods have been utilized in clinical programs with the most common being the addition of Fc domains either as fusion proteins or being modifications of more standard antibody platforms. These Fc containing bispecific T-cell engagers have half-lives on the order of 1 week or greater allowing for weekly and even bi-weekly dosing. In addition, some recent entries in the clinic have utilized binding domains specific to HSA. Early clinical data shows that the HSA binding domains provide half-life of 2–4 days [215], much shorter than the Fc domains but still consistent with weekly dosing. Given the early stages of most clinical trials utilizing HLE platforms, it remains to be seen which platforms and what pharmacokinetic properties best fit this modality.

6.7 Dealing with ADA

ADAs are a common issue for antibody and biologic therapeutics [226, 227]. During the history of antibody therapeutic development, new methods for the discovery, modification, and production of more human-like proteins, from chimeric, to humanized, to fully human, have helped reduce the occurrence and severity of ADA response.

However, even for fully human antibody therapeutics, patients can still develop ADA [228]. The presence of ADA can have a variety of impacts on both the therapeutic and the patients including affecting drug disposition and PK, neutralization of the biologic, and hypersensitivity reactions such as infusion related reactions and inflammatory reactions [226]. While the same processes of humanization used for monoclonal antibodies are applied to bispecific antibodies, due to the additional protein modifications of these platforms, they are inherently less natural looking, which could contribute to the development of ADA. As pointed out in previous sections, the development of ADA has led to the termination of multiple therapeutic candidates [213, 229]. As more TCE platforms enter later phase clinical development, it will become clear if ADA is larger issue with TCE bispecific molecules than it has been for monoclonal antibodies. If it is, it may become necessary to develop additional strategies to combat the development of ADA similar to the Roche cibisatamab trial.

6.8 Combination therapies

Activation of T-cells through binding and engagement of CD3ε by TCEs causes a series of downstream events. The activated T-cells release the content of their cytotoxic granules (granzyme and perforin) triggering apoptosis of the cells in close proximity. In addition, the T-cells turn on a set of pathways leading to proliferation. The T-cells also release various cytokines and chemokines creating a cascade of immunologic events in nearby cells. One of these cytokines is interferon gamma (IFNγ). The release of IFNγ triggers an upregulation of PD-L1 in nearby cells and binding of PD-L1 by PD-1 on T-cells causes an inhibition of TCR-mediated activation. The result is a negative feedback loop leading to the suppression of T-cell cytotoxic response. This process has been demonstrated *in vitro* through the use of bispecific TCEs and the upregulation of PD-L1 has also been shown to be a mechanism of resistance to blinatumomab in the clinic [230]. Given this built in mechanism of resistance, the combination of check-point inhibitors targeting the PD-1/PD-L1 interaction with TCEs is a rational combination. Many current clinical studies of TCEs have begun testing these combinations in the early phases. Currently, there is not sufficient data to determine if the combination therapies will significantly improve response rates or outcomes, but the initial data does show some hints. Interim data from initial phase 1 trials with cibisatamab showed a 5% partial response rate in the monotherapy arm and a 20% partial response rate in combination with atezolizumab. The current phase 1b/2 trial is moving forward only with the combination therapy. As more clinical studies complete over the next 2 years, the extent of the benefit of PD-1/PD-L1 combinations will emerge.

6.9 Next-generation TCEs

Given the toxicities seen from inherently active T-cell engagers targeting solid tumor targets, much work has focused on developing novel conditionally active platforms to reduce this toxicity. Three companies are currently working to bring the concept of prodrug therapeutics to biologics to reduce normal tissue toxicity, while maintaining the potency of bispecific T-cell engagers. CytomX, Amunix, and Maverick Therapeutics have all revealed that they are working on protease activatable T-cell engagers that are close to being tested in the clinic. The three platforms utilize different masking mechanisms and antibody designs but are theoretically quite similar. All three platforms are designed to be inert until masking domains are cleaved off by proteases, which are known to be overactive in tumor microenvironments.

The Amunix platform, XPAT, is built on a dual-scFv platform similar to a BiTE but adds protease cleavable polypeptide sequences known as Pro-XTEN to mask both the CD3 and tumor targeting domains via steric hinderance and also increase the half-life of the dual-scFv. Once in the tumor microenvironment, the peptides are cleaved off by the active proteases revealing a dual-scFv that is capable of binding tumors and engaging T-cells. The lead program, AMX-818, targeting HER2 is currently in IND enabling studies. Preclinical safety studies showed that the AMX-818 has >1000-fold higher tolerated Cmax than the unmasked HER2-PAT [81].

CytomX is also preparing to enter the clinic with its first CD3 engaging bispecific Probody, CX-904. The Probody platform is based on a standard Fc-containing antibody design but the binding domains are specifically masked via a protease cleavable masking peptide. The Probody platform already has some clinical data to validate the approach with CX-072, an anti-PD-L1 monoclonal Probody. Data from CX-072 has shown that the Probody is stable in circulation, with increased cleavage within the tumor microenvironment [231]. CX-904, being co-developed by CytomX and Amgen, is an EGFR-targeted bispecific Probody currently in preclinical development with a target IND date of late 2021. Based on preclinical data, the tolerated exposure of the Probody is ~10,000-fold higher than the unmasked counterpart [163].

Finally, Maverick Therapeutics has two molecules built on their COBRA platform that are scheduled to begin phase 1/phase 2 clinical trials in 2021 [232]. The COBRA platform is a completely unique bispecific T-cell engager design. Combining traditional antibody Vh and Vl domains with sdAbs, and a unique inactivation through mutated Vh and Vl domains, the COBRA structure combines aspects of the BiTE, DART, and TriTAC formats into a protease cleavable prodrug [161, 164]. MVC-101, the lead program, targets the validated tumor antigen EGFR, while its second program in development, MVC-280, targets B7H3 (CD276), a widely expressed target that is a member of the B7 family of proteins that includes CD86 and PD-L1. Preclinical models demonstrate these conditionally active TCEs have a therapeutic index 30–100 times greater than that of a standard inherently active T-cell engager [232].

Given the early stages of these conditionally activated prodrug platforms, it remains to be seen if the concept will prove out in clinical studies. If successful, these platforms could significantly improve the safety of these potent tumor-killing modality within solid tumor indications and significantly expand their utility in the clinic.

References

[1] Kranz DM, Tonegawa S, Eisen HN. Attachment of an anti-receptor antibody to non-target cells renders them susceptible to lysis by a clone of cytotoxic T lymphocytes. Proc Natl Acad Sci U S A 1984;81:7922–6.

[2] Liu MA, et al. Heteroantibody duplexes target cells for lysis by cytotoxic T lymphocytes. Proc Natl Acad Sci U S A 1985;82:8648–52.

[3] Perez P, et al. Specific targeting of cytotoxic T cells by anti-T3 linked to anti-target cell antibody. Nature 1985;316:354–6.

[4] Staerz UD, Kanagawa O, Bevan MJ. Hybrid antibodies can target sites for attack by T cells. Nature 1985;314:628–31.

[5] Brinkmann U, Kontermann RE. The making of bispecific antibodies. MAbs 2017;9:182–212.

[6] Suurs FV, et al. A review of bispecific antibodies and antibody constructs in oncology and clinical challenges. Pharmacol Ther 2019;201:103–19.

[7] Kontermann RE, Brinkmann U. Bispecific antibodies. Drug Discov Today 2015;20:838–47.

[8] Correia DV, Lopes A, Silva-Santos B. Tumor cell recognition by γδ T lymphocytes: T-cell receptor vs. NK-cell receptors. Oncoimmunology 2013;2, e22892.

[9] Liu Y, Zhang C. The role of human γδ T cells in anti-tumor immunity and their potential for cancer immunotherapy. Cell 2020;9.

[10] Gruber M, et al. Efficient tumor cell lysis mediated by a bispecific single chain antibody expressed in Escherichia coli. J Immunol 1994;152:5368–74.

[11] Mack M, Riethmüller G, Kufer P. A small bispecific antibody construct expressed as a functional single-chain molecule with high tumor cell cytotoxicity. Proc Natl Acad Sci U S A 1995;92:7021–5.

[12] Lindhofer H, et al. Preferential species-restricted heavy/light chain pairing in rat/mouse quadromas. Implications for a single-step purification of bispecific antibodies. J Immunol 1995;155:219–25.

[13] Jain RK. Physiological barriers to delivery of monoclonal antibodies and other macromolecules in tumors. Cancer Res 1990;50:814s–9s.

[14] Minchinton AI, Tannock IF. Drug penetration in solid tumours. Nat Rev Cancer 2006;6:583–92.

[15] Thurber GM, Schmidt MM, Wittrup KD. Antibody tumor penetration: transport opposed by systemic and antigen-mediated clearance. Adv Drug Deliv Rev 2008;60:1421–34.

[16] Bruenke J, et al. Effective lysis of lymphoma cells with a stabilised bispecific single-chain Fv antibody against CD19 and FcgammaRIII (CD16). Br J Haematol 2005;130:218–28.

[17] Klinger M, et al. Immunopharmacologic response of patients with B-lineage acute lymphoblastic leukemia to continuous infusion of T cell-engaging CD19/CD3-bispecific BiTE antibody blinatumomab. Blood 2012;119:6226–33.

[18] Lee KJ, et al. Clinical use of blinatumomab for B-cell acute lymphoblastic leukemia in adults. Ther Clin Risk Manag 2016;12:1301–10.

[19] Löffler A, et al. A recombinant bispecific single-chain antibody, CD19 x CD3, induces rapid and high lymphoma-directed cytotoxicity by unstimulated T lymphocytes. Blood 2000;95:2098–103.

[20] Brischwein K, et al. MT110: a novel bispecific single-chain antibody construct with high efficacy in eradicating established tumors. Mol Immunol 2006;43:1129–43.

[21] Friedrich M, et al. Regression of human prostate cancer xenografts in mice by AMG 212/BAY2010112, a novel PSMA/CD3-bispecific BiTE antibody cross-reactive with non-human primate antigens. Mol Cancer Ther 2012;11:2664–73.

[22] Friedrich M, et al. Preclinical characterization of AMG 330, a CD3/CD33-bispecific T-cell-engaging antibody with potential for treatment of acute myelogenous leukemia. Mol Cancer Ther 2014;13:1549–57.

[23] Hammond SA, et al. Selective targeting and potent control of tumor growth using an EphA2/CD3-Bispecific single-chain antibody construct. Cancer Res 2007;67:3927–35.

[24] Osada T, et al. Metastatic colorectal cancer cells from patients previously treated with chemotherapy are sensitive to T-cell killing mediated by CEA/CD3-bispecific T-cell-engaging BiTE antibody. Br J Cancer 2010;102:124–33.

[25] Oberst MD, et al. CEA/CD3 bispecific antibody MEDI-565/AMG 211 activation of T cells and subsequent killing of human tumors is independent of mutations commonly found in colorectal adenocarcinomas. MAbs 2014;6:1571–84.

[26] Choi BD, et al. Systemic administration of a bispecific antibody targeting EGFRvIII successfully treats intracerebral glioma. Proc Natl Acad Sci U S A 2013;110:270–5.

[27] Fu M, et al. Therapeutic bispecific T-cell engager antibody targeting the transferrin receptor. Front Immunol 2019;10:1396.

[28] Wu MR, et al. B7H6-specific bispecific T cell engagers Lead to tumor elimination and host antitumor immunity. J Immunol 2015;194:5305–11.

[29] Lutterbuese R, et al. T cell-engaging BiTE antibodies specific for EGFR potently eliminate KRAS- and BRAF-mutated colorectal cancer cells. Proc Natl Acad Sci 2010;107:12605–10.

[30] Schlereth B, et al. Eradication of tumors from a human colon cancer cell line and from ovarian cancer metastases in immunodeficient mice by a single-chain Ep-CAM-/CD3-bispecific antibody construct. Cancer Res 2005;65:2882–9.

[31] Offner S, et al. Induction of regular cytolytic T cell synapses by bispecific single-chain antibody constructs on MHC class I-negative tumor cells. Mol Immunol 2006;43:763–71.

[32] Johnson S, et al. Effector cell recruitment with novel Fv-based dual-affinity re-targeting protein leads to potent tumor cytolysis and in vivo B-cell depletion. J Mol Biol 2010;399:436–49.

[33] Moore PA, et al. Application of dual affinity retargeting molecules to achieve optimal redirected T-cell killing of B-cell lymphoma. Blood 2011;117:4542–51.

[34] Chichili GR, et al. A CD3xCD123 bispecific DART for redirecting host T cells to myelogenous leukemia: preclinical activity and safety in nonhuman primates. Sci Transl Med 2015;7:289ra82.

[35] Al-Hussaini M, et al. Targeting CD123 in acute myeloid leukemia using a T-cell-directed dual-affinity retargeting platform. Blood 2016;127:122–31.

[36] Westervelt P, et al. Phase 1 first-in-human trial of AMV564, a bivalent bispecific (2:2) CD33/CD3 T-cell engager, in patients with relapsed/refractory acute myeloid leukemia (AML). Blood 2019;134:834.

[37] Reusch U, et al. A tetravalent bispecific TandAb (CD19/CD3), AFM11, efficiently recruits T cells for the potent lysis of CD19(+) tumor cells. MAbs 2015;7:584–604.

[38] Reusch U, et al. Characterization of CD33/CD3 tetravalent bispecific tandem diabodies (TandAbs) for the treatment of acute myeloid leukemia. Clin Cancer Res 2016;22:5829–38.

[39] Ellwanger K, et al. Highly specific and effective targeting of EGFRvIII-positive tumors with TandAb antibodies. Front Oncol 2017;7:100.

[40] Mølhøj M, et al. CD19-/CD3-bispecific antibody of the BiTE class is far superior to tandem diabody with respect to redirected tumor cell lysis. Mol Immunol 2007;44:1935–43.

[41] de Bruin RCG, et al. A bispecific nanobody approach to leverage the potent and widely applicable tumor cytolytic capacity of Vγ9Vδ2-T cells. Onco Targets Ther 2017;7, e1375641.

[42] Kijanka M, et al. Nanobody-based cancer therapy of solid tumors. Nanomedicine 2015;10:161–74.

[43] Joalland N, Scotet E. Emerging challenges of preclinical models of anti-tumor immunotherapeutic strategies utilizing Vγ9Vδ2 T cells. Front Immunol 2020;11:992.

[44] Schoonjans R, et al. Fab chains as an efficient heterodimerization scaffold for the production of recombinant bispecific and trispecific antibody derivatives. J Immunol 2000;165:7050–7.

[45] Glorius P, et al. The novel tribody [(CD20)(2)xCD16] efficiently triggers effector cell-mediated lysis of malignant B cells. Leukemia 2013;27:190–201.

[46] Oberg HH, et al. Novel bispecific antibodies increase γδ T-cell cytotoxicity against pancreatic cancer cells. Cancer Res 2014;74:1349–60.

[47] Oberg HH, et al. Tribody [(HER2)2xCD16] is more effective than trastuzumab in enhancing γδ T cell and natural killer cell cytotoxicity against HER2-expressing cancer cells. Front Immunol 2018;9:814.

[48] Oberg HH, et al. Bispecific antibodies enhance tumor-infiltrating T cell cytotoxicity against autologous HER-2-expressing high-grade ovarian tumors. J Leukoc Biol 2020;107:1081–95.

[49] Kim CH, et al. Bispecific small molecule-antibody conjugate targeting prostate cancer. Proc Natl Acad Sci U S A 2013;110:17796–801.

[50] Lowe KL, et al. Novel TCR-based biologics: mobilising T cells to warm 'cold' tumours. Cancer Treat Rev 2019;77:35–43.

[51] Blum JS, Wearsch PA, Cresswell P. Pathways of antigen processing. Annu Rev Immunol 2013;31:443–73.

[52] Liddy N, et al. Monoclonal TCR-redirected tumor cell killing. Nat Med 2012;18:980–7.

[53] Bodmer WF, et al. Tumor escape from immune response by variation in HLA expression and other mechanisms. Ann N Y Acad Sci 1993;690:42–9.

[54] Garrido F, Aptsiauri N. Cancer immune escape: MHC expression in primary tumours versus metastases. Immunology 2019;158:255–66.

[55] McCormack E, et al. Bi-specific TCR-anti CD3 redirected T-cell targeting of NY-ESO-1- and LAGE-1-positive tumors. Cancer Immunol Immunother 2013;62:773–85.

[56] Junghans RP, Anderson CL. The protection receptor for IgG catabolism is the beta2-microglobulin-containing neonatal intestinal transport receptor. Proc Natl Acad Sci U S A 1996;93:5512–6.

[57] Roopenian DC, et al. The MHC class I-like IgG receptor controls perinatal IgG transport, IgG homeostasis, and fate of IgG-fc-coupled drugs. J Immunol 2003;170:3528–33.

[58] Kuo TT, Aveson VG. Neonatal Fc receptor and IgG-based therapeutics. MAbs 2011;3:422–30.

[59] Einsele H, et al. The BiTE (bispecific T-cell engager) platform: development and future potential of a targeted immuno-oncology therapy across tumor types. Cancer 2020;126:3192–201.

[60] Goyos A, et al. Generation of half-life extended anti-BCMA Bite® antibody construct compatible with once-weekly dosing for treatment of multiple myeloma (MM). Blood 2017;130:5389.

[61] Lorenczewski G, et al. Generation of a half-life extended anti-CD19 BiTE® antibody construct compatible with once-weekly dosing for treatment of CD19-positive malignancies. Blood 2017;130:2815.

[62] Bailis J, et al. Preclinical evaluation of AMG 160, a next-generation bispecific T cell engager (BiTE) targeting the prostate-specific membrane antigen PSMA for metastatic castration-resistant prostate cancer (mCRPC). J Clin Oncol 2019;37:301.

[63] Liu L, et al. MGD011, A CD19 x CD3 dual-affinity retargeting bi-specific molecule incorporating extended circulating half-life for the treatment of B-cell malignancies. Clin Cancer Res 2017;23:1506–18.

[64] Moore PA, et al. Development of MGD007, a gpA33 x CD3-bispecific DART protein for T-cell immunotherapy of metastatic colorectal Cancer. Mol Cancer Ther 2018;17:1761–72.

[65] Atwell S, et al. Stable heterodimers from remodeling the domain interface of a homodimer using a phage display library. J Mol Biol 1997;270:26–35.

[66] Root AR, et al. Development of PF-06671008, a highly potent anti-P-cadherin/anti-CD3 bispecific DART molecule with extended half-life for the treatment of cancer. Antibodies (Basel, Switzerland) 2016;5:6.

[67] Fisher TS, et al. A CD3-bispecific molecule targeting P-cadherin demonstrates T cell-mediated regression of established solid tumors in mice. Cancer Immunol Immunother 2018;67(2):247–59.

[68] Hernandez-Hoyos G, et al. MOR209/ES414, a novel bispecific antibody targeting PSMA for the treatment of metastatic castration-resistant prostate cancer. Mol Cancer Ther 2016;15:2155–65.

[69] Comeau MR, et al. APVO436, a bispecific anti-CD123 x anti-CD3 ADAPTIR™ molecule for redirected T-cell cytotoxicity, induces potent T-cell activation, proliferation and cytotoxicity with limited cytokine release. Cancer Res 2018;78:1786.

[70] Chaudhury C, et al. The major histocompatibility complex-related fc receptor for IgG (FcRn) binds albumin and prolongs its lifespan. J Exp Med 2003;197:315–22.

[71] O'Connor-Semmes RL, et al. GSK2374697, a novel albumin-binding domain antibody (AlbudAb), extends systemic exposure of exendin-4: first study in humans- -PK/PD and safety. Clin Pharmacol Ther 2014;96:704–12.

[72] Hoefman S, et al. Pre-clinical intravenous serum pharmacokinetics of albumin binding and non-half-life extended nanobodies®. Antibodies 2015;4:141–56.

[73] Austin RJ, et al. TriTACs, a novel class of T cell-engaging protein constructs designed for the treatment of solid tumors. Mol Cancer Ther 2020;20:109–20.

[74] Law CL, et al. Preclinical and nonclinical characterization of HPN217: a tri-specific T cell activating construct (TriTAC) targeting B cell maturation antigen (BCMA) for the treatment of multiple myeloma. Blood 2018;132:3225.

[75] Lemon B, et al. HPN424, a half-life extended, PSMA/CD3-specific TriTAC for the treatment of metastatic prostate cancer. Cancer Res 2018;78:1773.

[76] Aaron WH, et al. HPN328: an anti-DLL3 T cell engager for treatment of small cell lung cancer. Mol Cancer Ther 2019;18:C033.

[77] Austin R, et al. HPN536, a T cell-engaging, mesothelin/CD3-specific TriTAC for the treatment of solid tumors. Cancer Res 2018;78:1781.

[78] Schellenberger V, et al. A recombinant polypeptide extends the in vivo half-life of peptides and proteins in a tunable manner. Nat Biotechnol 2009;27:1186–90.

[79] Schellenberger V. AMX-268 an EpCAM-targeted T cell engager with best-in-class therapeutic index. In: Protein and Antibody Engineering Summit; 2018.

[80] AlQahtani AD, et al. Strategies for the production of long-acting therapeutics and efficient drug delivery for cancer treatment. Biomed Pharmacother 2019;113:108750.

[81] Cattaruzza F, et al. HER2-XPAT and EGFR-XPAT: pro-drug T-cell engagers (TCEs) engineered to address on-target, off-tumor toxicity with potent efficacy in vitro and in vivo and large safety margins in NHP. Cancer Res 2020;80:3376.

[82] Sim B-C. AMX-168, a long-acting, tumor protease-sensitive bispecific precursor for the treatment of solid malignancies. In: Proceedings for the American association for cancer research annual meeting 2017; 2017. p. 3638.

[83] Zeidler R, et al. Simultaneous activation of T cells and accessory cells by a new class of intact bispecific antibody results in efficient tumor cell killing. J Immunol 1999;163:1246–52.

[84] Ruf P, et al. Two new trifunctional antibodies for the therapy of human malignant melanoma. Int J Cancer 2004;108:725–32.

[85] Kiewe P, et al. Phase I trial of the trifunctional anti-HER2 x anti-CD3 antibody ertumaxomab in metastatic breast cancer. Clin Cancer Res 2006;12:3085–91.

[86] Stanglmaier M, et al. Bi20 (FBTA05), a novel trifunctional bispecific antibody (anti-CD20 x anti-CD3), mediates efficient killing of B-cell lymphoma cells even with very low CD20 expression levels. Int J Cancer 2008;123:1181–9.

[87] Sebastian M, et al. Treatment of non-small cell lung cancer patients with the trifunctional monoclonal antibody catumaxomab (anti-EpCAM x anti-CD3): a phase I study. Cancer Immunol Immunother 2007;56:1637–44.

[88] Burges A, et al. Effective relief of malignant ascites in patients with advanced ovarian cancer by a trifunctional anti-EpCAM x anti-CD3 antibody: a phase I/II study. Clin Cancer Res 2007;13:3899–905.

[89] Seimetz D, Lindhofer H, Bokemeyer C. Development and approval of the trifunctional antibody catumaxomab (anti-EpCAM x anti-CD3) as a targeted cancer immunotherapy. Cancer Treat Rev 2010;36:458–67.

[90] Linke R, Klein A, Seimetz D. Catumaxomab: clinical development and future directions. MAbs 2010;2:129–36.

[91] van der Neut Kolfschoten M, et al. Anti-inflammatory activity of human IgG4 antibodies by dynamic Fab arm exchange. Science 2007;317:1554–7.

[92] Labrijn AF, et al. Efficient generation of stable bispecific IgG1 by controlled Fab-arm exchange. Proc Natl Acad Sci U S A 2013;110:5145–50.

[93] Gaudet F, et al. Development of a CD123xCD3 bispecific antibody (JNJ-63709178) for the treatment of acute myeloid leukemia (AML). Blood 2016;128:2824.

[94] Kemper K, et al. DuoBody®-CD3x5T4 shows potent preclinical anti-tumor activity in vitro and in vivo in a range of cancer indications. J Immunother Cancer 2019;7(Suppl 1):283.

[95] Engelberts PJ, et al. DuoBody-CD3xCD20 induces potent T-cell-mediated killing of malignant B cells in preclinical models and provides opportunities for subcutaneous dosing. EBioMedicine 2020;52:102625.

[96] Pillarisetti K, et al. A T-cell-redirecting bispecific G-protein-coupled receptor class 5 member D x CD3 antibody to treat multiple myeloma. Blood 2020;135:1232–43.

[97] Chen S, et al. Immunoglobulin gamma-like therapeutic bispecific antibody formats for tumor therapy. J Immunol Res 2019;2019:4516041.

[98] De Nardis C, et al. A new approach for generating bispecific antibodies based on a common light chain format and the stable architecture of human immunoglobulin G1. J Biol Chem 2017;292:14706–17.

[99] van Loo PF, et al. MCLA-117, a CLEC12AxCD3 bispecific antibody targeting a leukaemic stem cell antigen, induces T cell-mediated AML blast lysis. Expert Opin Biol Ther 2019;19:721–33.

[100] Sampei Z, et al. Identification and multidimensional optimization of an asymmetric bispecific IgG antibody mimicking the function of factor VIII cofactor activity. PLoS One 2013;8, e57479.

[101] Ishiguro T, et al. An anti-glypican 3/CD3 bispecific T cell-redirecting antibody for treatment of solid tumors. Sci Transl Med 2017;9:eaal4291.

[102] Jendeberg L, et al. Engineering of Fc(1) and Fc(3) from human immunoglobulin G to analyse subclass specificity for staphylococcal protein a. J Immunol Methods 1997;201:25–34.

[103] Smith EJ, et al. A novel, native-format bispecific antibody triggering T-cell killing of B-cells is robustly active in mouse tumor models and cynomolgus monkeys. Sci Rep 2015;5:17943.

[104] Varghese B, et al. A novel CD20xCD3 bispecific fully human antibody induces potent anti-tumor effects against B cell lymphoma in mice. Blood 2014;124:4501.

[105] Dilillo DJ, et al. REGN5458, a bispecific BCMAxCD3 T cell engaging antibody, demonstrates robust in vitro and in vivo anti-tumor efficacy in multiple myeloma models, comparable to that of BCMA CAR T cells. Blood 2018;132:1944.

[106] Crawford A, et al. A mucin 16 bispecific T cell-engaging antibody for the treatment of ovarian cancer. Sci Transl Med 2019;11:eaau7534.

[107] Zuch de Zafra CL, et al. Targeting multiple myeloma with AMG 424, a novel anti-CD38/CD3 bispecific T-cell-recruiting antibody optimized for cytotoxicity and cytokine release. Clin Cancer Res 2019;25:3921–33.

[108] Kelly WK, et al. Phase I study of AMG 509, a STEAP1 x CD3 T cell-recruiting XmAb 2 + 1 immune therapy, in patients with metastatic castration-resistant prostate cancer (mCRPC). J Clin Oncol 2020;38:TPS5589.

[109] Nisthal A, et al. Abstract 2286: XmAb30819, an XmAb®2 + 1 ENPP3 x CD3 bispecific antibody for RCC, demonstrates safety and efficacy in in vivo preclinical studies; 2020. p. 2286.

[110] Lee SH, et al. Anti-SSTR2 × anti-CD3 bispecific antibody induces potent killing of human tumor cells in vitro and in mice, and stimulates target-dependent T cell activation in monkeys: a potential immunotherapy for neuroendocrine tumors. Cancer Res 2017;77:3633.

[111] Chu SY, et al. Immunotherapy with long-lived anti-CD20 × anti-CD3 bispecific antibodies stimulates potent T cell-mediated killing of human B cell lines and of circulating and lymphoid B cells in monkeys: a potential therapy for B cell lymphomas and leukemias. Blood 2014;124:3111.

[112] Chu SY, et al. Immunotherapy with long-lived anti-CD123 × anti-CD3 bispecific antibodies stimulates potent T cell-mediated killing of human AML cell lines and of CD123+ cells in monkeys: a potential therapy for acute myelogenous Leukemia. Blood 2014;124:2316.

[113] Ravandi F, et al. Complete responses in relapsed/refractory acute myeloid leukemia (AML) patients on a weekly dosing schedule of XmAb14045, a CD123 x CD3 T-cell-engaging bispecific antibody: initial results of a phase 1 study. Blood 2018;132:763.

[114] Skegro D, et al. Immunoglobulin domain interface exchange as a platform technology for the generation of Fc heterodimers and bispecific antibodies. J Biol Chem 2017;292:9745–59.

[115] Back J. GBR1302-BEAT bispecific antibody targeting CD3 and HER2 demonstrates a higher anti-tumor potential than current HER2-targeting therapies. Ann Oncol 2016;27(Suppl 8). VIII2.

[116] Richter JR, et al. Phase 1, multicenter, open-label study of single-agent bispecific antibody T-cell engager GBR 1342 in relapsed/refractory multiple myeloma. J Clin Oncol 2018;36:TPS81.

[117] Trinklein ND, et al. Efficient tumor killing and minimal cytokine release with novel T-cell agonist bispecific antibodies. MAbs 2019;11:639–52.

[118] Schaefer W, et al. Immunoglobulin domain crossover as a generic approach for the production of bispecific IgG antibodies. Proc Natl Acad Sci U S A 2011;108:11187–92.

[119] Klein C, Schaefer W, Regula JT. The use of CrossMAb technology for the generation of bi- and multispecific antibodies. MAbs 2016;8:1010–20.

[120] Bacac M, et al. A novel carcinoembryonic antigen T-cell bispecific antibody (CEA TCB) for the treatment of solid tumors. Clin Cancer Res 2016;22:3286–97.

[121] Seckinger A, et al. Target expression, generation, preclinical activity, and pharmacokinetics of the BCMA-T cell bispecific antibody EM801 for multiple myeloma treatment. Cancer Cell 2017;31:396–410.

[122] Bacac M, et al. CD20-TCB with obinutuzumab pretreatment as next-generation treatment of hematologic malignancies. Clin Cancer Res 2018;24:4785–97.

[123] Orcutt KD, et al. A modular IgG-scFv bispecific antibody topology. Protein Eng Des Sel 2010;23:221–8.

[124] Xu H, et al. Retargeting T cells to GD2 pentasaccharide on human tumors using bispecific humanized antibody. Cancer Immunol Res 2015;3:266–77.

[125] Dreier T, et al. Extremely potent, rapid and costimulation-independent cytotoxic T-cell response against lymphoma cells catalyzed by a single-chain bispecific antibody. Int J Cancer 2002;100:690–7.

[126] Irvine DJ, et al. Direct observation of ligand recognition by T cells. Nature 2002;419:845–9.

[127] Manz BN, et al. T-cell triggering thresholds are modulated by the number of antigen within individual T-cell receptor clusters. Proc Natl Acad Sci U S A 2011;108:9089–94.

[128] Klinger M, et al. Adhesion of T cells to endothelial cells facilitates blinatumomab-associated neurologic adverse events. Cancer Res 2020;80:91–101.

[129] Shimabukuro-Vornhagen A, et al. Cytokine release syndrome. J Immunother Cancer 2018;6:56.

[130] Kebenko M, et al. A multicenter phase 1 study of solitomab (MT110, AMG 110), a bispecific EpCAM/CD3 T-cell engager (BiTE®) antibody construct, in patients with refractory solid tumors. Onco Targets Ther 2018;7, e1450710.

[131] Chen X, et al. A modeling framework to characterize cytokine release upon T-cell-engaging bispecific antibody treatment: methodology and opportunities. Clin Transl Sci 2019;12:600–8.

[132] Maude SL, et al. Managing cytokine release syndrome associated with novel T cell-engaging therapies. Cancer J 2014;20:119–22.

[133] Grupp SA, et al. Chimeric antigen receptor-modified T cells for acute lymphoid leukemia. N Engl J Med 2013;368:1509–18.

[134] Brentjens RJ, et al. CD19-targeted T cells rapidly induce molecular remissions in adults with chemotherapy-refractory acute lymphoblastic leukemia. Sci Transl Med 2013;5:177ra38.

[135] Brandl C, et al. The effect of dexamethasone on polyclonal T cell activation and redirected target cell lysis as induced by a CD19/CD3-bispecific single-chain antibody construct. Cancer Immunol Immunother 2007;56:1551–63.

[136] Kamperschroer C, et al. Summary of a workshop on preclinical and translational safety assessment of CD3 bispecifics. J Immunotoxicol 2020;17:67–85.

[137] Jacobs K, et al. Lead-in dose optimization to mitigate cytokine release syndrome in AML and MDS patients treated with flotetuzumab, a CD123 x CD3 Dart® molecule for T-cell redirected therapy. Blood 2017;130(Suppl. 1):3856.

[138] Panowski SH, et al. Preclinical efficacy and safety comparison of CD3 bispecific and ADC modalities targeting BCMA for the treatment of multiple myeloma. Mol Cancer Ther 2019;18:2008–20.

[139] Li J, et al. Membrane-proximal epitope facilitates efficient T cell synapse formation by anti-FcRH5/CD3 and is a requirement for myeloma cell killing. Cancer Cell 2017;31:383–95.

[140] Mau-Sørensen M, et al. A phase I trial of intravenous catumaxomab: a bispecific monoclonal antibody targeting EpCAM and the T cell coreceptor CD3. Cancer Chemother Pharmacol 2015;75:1065–73.

[141] Tolcher AW, et al. Phase 1, first-in-human, open label, dose escalation ctudy of MGD009, a humanized B7-H3 x CD3 dual-affinity re-targeting (DART) protein in patients with B7-H3-expressing neoplasms or B7-H3 expressing tumor vasculature. J Clin Oncol 2016;34:TPS3105.

[142] Anon. MGD009 partial clinical hold. [cited 2020 October 23]; Available from http://ir.macrogenics.com/news-releases/news-release-details/macrogenics-announces-partial-clinical-hold-mgd009-phase-1.

[143] MacroGenics, Inc. MacroGenics outlines corporate priorities for 2020 [Accessed on 23 October 2020]. Available from http://ir.macrogenics.com/news-releases/news-release-details/macrogenics-outlines-corporate-priorities-2020.

[144] Ackerman ME, et al. A33 antigen displays persistent surface expression. Cancer Immunol Immunother 2008;57:1017–27.

[145] An Z, et al. Epidermal growth factor receptor and EGFRvIII in glioblastoma: signaling pathways and targeted therapies. Oncogene 2018;37:1561–75.

[146] Leong SR, et al. An anti-CD3/anti-CLL-1 bispecific antibody for the treatment of acute myeloid leukemia. Blood 2017;129:609–18.

[147] Lanitis E, et al. Mechanisms regulating T-cell infiltration and activity in solid tumors. Ann Oncol 2017;28:xii18–32.

[148] Mazor Y, et al. Enhanced tumor-targeting selectivity by modulating bispecific antibody binding affinity and format valence. Sci Rep 2017;7.

[149] Diego E. Bispecific T-cell engagers: towards understanding variables influencing the in vitro potency and tumor selectivity and their modulation to enhance their efficacy and safety. Methods 2019;154:102–17.

[150] Slaga D, et al. Avidity-based binding to HER2 results in selective killing of HER2-overexpressing cells by anti-HER2/CD3. Sci Transl Med 2018;10.

[151] Nisthal A, et al. Abstract 5663: affinity tuned XmAb®2 + 1 PSMA x CD3 bispecific antibodies demonstrate selective activity in prostate cancer models; 2020. p. 5663.

[152] Viret C, Janeway CA. MHC and T cell development. Rev Immunogenet 1999;1:91–104.

[153] Gibbs ZA, Whitehurst AW. Emerging contributions of cancer/testis antigens to neoplastic behaviors. Trends Cancer 2018;4:701–12.

[154] Cameron BJ, et al. Identification of a Titin-derived HLA-A1-presented peptide as a cross-reactive target for engineered MAGE A3-directed T cells. Sci Transl Med 2013;5:197ra103.

[155] Harper J, et al. An approved in vitro approach to preclinical safety and efficacy evaluation of engineered T cell receptor anti-CD3 bispecific (ImmTAC) molecules. PLoS One 2018;13, e0205491.

[156] Kurosawa N, et al. High throughput development of TCR-mimic antibody that targets survivin-2B80-88/HLA-A*A24 and its application in a bispecific T-cell engager. Sci Rep 2019;9:9827.

[157] Dao T, et al. Therapeutic bispecific T-cell engager antibody targeting the intracellular oncoprotein WT1. Nat Biotechnol 2015;33:1079–86.

[158] Holland CJ, et al. Specificity of bispecific T cell receptors and antibodies targeting peptide-HLA. J Clin Investig 2020;130:2673–88.

[159] Dudani JS, Warren AD, Bhatia SN. Harnessing protease activity to improve cancer care. Annu Rev Cancer Biol 2018;2:353–76.

[160] Geiger M, et al. Protease-activation using anti-idiotypic masks enables tumor specificity of a folate receptor 1-T cell bispecific antibody. Nat Commun 2020;11:3196.

[161] Panchal A, et al. COBRA: a highly potent conditionally active T cell engager engineered for the treatment of solid tumors. MAbs 2020;**12**:1792130.

[162] Desnoyers LR, et al. Tumor-specific activation of an EGFR-targeting probody enhances therapeutic index. Sci Transl Med 2013;5:207ra144.

[163] Boustany LM, et al. Abstract A164: EGFR-CD3 bispecific Probody therapeutic induces tumor regressions and increases maximum tolerated dose >60-fold in preclinical studies; 2018. p. A164.

[164] Dettling D, et al. Abstract 557: COBRA: a novel conditionally active bispecific antibody that regresses established solid tumors in mice; 2019. p. 557.

[165] Rosenblum D, et al. Progress and challenges towards targeted delivery of cancer therapeutics. Nat Commun 2018;9.

[166] Saggar JK, et al. The tumor microenvironment and strategies to improve drug distribution. Front Oncol 2013;3:154.

[167] Marcucci F, et al. Improving drug uptake and penetration into Tumors: current and forthcoming opportunities. Front Oncol 2013;3:161.

[168] Fridman WH, et al. The immune contexture in human tumours: impact on clinical outcome. Nat Rev Cancer 2012;12:298–306.

[169] Gajewski TF, et al. Cancer immunotherapy targets based on understanding the T cell-inflamed versus non-T cell-inflamed tumor microenvironment. Adv Exp Med Biol 2017;1036:19–31.

[170] Duan Q, et al. Turning cold into hot: firing up the tumor microenvironment. Trends Cancer 2020;6:605–18.

[171] Baitsch L, et al. Extended co-expression of inhibitory receptors by human CD8 T-cells depending on differentiation, antigen-specificity and anatomical localization. PLoS One 2012;7, e30852.

[172] Pardoll DM. The blockade of immune checkpoints in cancer immunotherapy. Nat Rev Cancer 2012;12:252–64.

[173] Krupka C, et al. Blockade of the PD-1/PD-L1 axis augments lysis of AML cells by the CD33/CD3 BiTE antibody construct AMG 330: reversing a T-cell-induced immune escape mechanism. Leukemia 2015;30:484–91.

[174] Yuan Y. Spatial heterogeneity in the tumor microenvironment. Cold Spring Harb Perspect Med 2016;6:a026583.

[175] Ramón YCS, et al. Clinical implications of intratumor heterogeneity: challenges and opportunities. J Mol Med (Berl) 2020;98:161–77.

[176] Dreier T, et al. T cell costimulus-independent and very efficacious inhibition of tumor growth in mice bearing subcutaneous or leukemic human B cell lymphoma xenografts by a CD19-/CD3- bispecific single-chain antibody construct. J Immunol 2003;170:4397–402.

[177] Amann M, et al. Therapeutic window of an EpCAM/CD3-specific BiTE antibody in mice is determined by a subpopulation of EpCAM-expressing lymphocytes that is absent in humans. Cancer Immunol Immunother 2008;58:95–109.

[178] Castermans K, Griffioen AW. Tumor blood vessels, a difficult hurdle for infiltrating leukocytes. Biochim Biophys Acta 2007;1776:160–74.

[179] Griffioen AW, et al. Tumor angiogenesis is accompanied by a decreased inflammatory response of tumor-associated endothelium. Blood 1996;88:667–73.

[180] Griffioen AW, et al. Endothelial intercellular adhesion molecule-1 expression is suppressed in human malignancies: the role of angiogenic factors. Cancer Res 1996;56:1111–7.

[181] Bouzin C, et al. Effects of vascular endothelial growth factor on the lymphocyte-endothelium interactions: identification of caveolin-1 and nitric oxide as control points of endothelial cell anergy. J Immunol 2007;178:1505–11.

[182] Buckanovich RJ, et al. Endothelin B receptor mediates the endothelial barrier to T cell homing to tumors and disables immune therapy. Nat Med 2008;14:28–36.

[183] Delfortrie S, et al. Egfl7 promotes tumor escape from immunity by repressing endothelial cell activation. Cancer Res 2011;71:7176–86.

[184] De Jaeghere EA, Denys HG, De Wever O. Fibroblasts fuel immune escape in the tumor microenvironment. Trends Cancer 2019;5:704–23.

[185] Razmkhah M, Abtahi S, Ghaderi A. Mesenchymal stem cells, immune cells and tumor cells crosstalk: a sinister triangle in the tumor microenvironment. Curr Stem Cell Res Ther 2019;14:43–51.

[186] Noy R, Pollard JW. Tumor-associated macrophages: from mechanisms to therapy. Immunity 2014;41:49–61.

[187] Yang Y, et al. Myeloid-derived suppressor cells in tumors: from mechanisms to antigen specificity and microenvironmental regulation. Front Immunol 2020;11:1371.

[188] Masucci MT, Minopoli M, Carriero MV. Tumor associated neutrophils. their role in tumorigenesis, metastasis, prognosis and therapy. Front Oncol 2019;9:1146.

[189] Veglia F, Gabrilovich DI. Dendritic cells in cancer: the role revisited. Curr Opin Immunol 2017;45:43–51.

[190] DeClerck YA, et al. Proteases, extracellular matrix, and cancer: a workshop of the path B study section. Am J Pathol 2004;164:1131–9.

[191] Holl EK, et al. Examining peripheral and tumor cellular immunome in patients with Cancer. Front Immunol 2019;10:1767.

[192] Wesche H, et al. Abstract 3814: TriTACs are novel T cell-engaging therapeutic proteins optimized for the treatment of solid tumors and for long serum half-life; 2018. p. 3814.

[193] Lin SJ, et al. ProTriTAC: a protease-activatable T cell engager platform that links half-life extension to functional masking and expands therapeutic window to enable targeting of broadly expressed tumor antigens. In: The society for immunotherapy of cancer 33rd annual meeting; 2018. Abstract P608.

[194] Morillon YMI, et al. Model for investigation of immune and anti-tumor effects mediated by the bifunctional immunotherapeutic Bintrafusp alfa. Front Oncol 2020;10:549.

[195] Brehm MA, et al. Lack of acute xenogeneic graft-versus-host disease, but retention of T-cell function following engraftment of human peripheral blood mononuclear cells in NSG mice deficient in MHC class I and II expression. FASEB J 2018;33:3137–51.

[196] Yaguchi T, et al. Human PBMC-transferred murine MHC class I/II-deficient NOG mice enable long-term evaluation of human immune responses. Cell Mol Immunol 2018;15:953–62.

[197] Gupta VR, et al. Molecular imaging reveals biodistribution of P-cadherin LP-DART bispecific and trafficking of adoptively transferred T cells in mouse xenograft model. Oncotarget 2020;11:1344–57.

[198] Visioni A, et al. Intra-arterial versus intravenous adoptive cell therapy in a mouse tumor model. J Immunother 2018;41:313–8.

[199] Shultz LD, et al. Human lymphoid and myeloid cell development in NOD/LtSz-scid IL2R gamma null mice engrafted with mobilized human hemopoietic stem cells. J Immunol 2005;174:6477–89.

[200] Yao L-C, et al. Abstract 5186: human stem cell humanized mouse model for in vivo evaluation of bispecific antibody efficacy and drug-induced cytokine release; 2020. p. 5186.

[201] Yang M, et al. Abstract 5669: establishment of a human CD3ε transgenic mouse model to assess antitumor efficacy of human T-cell-redirecting bispecific antibodies; 2018. p. 5669.

[202] Junttila TT, et al. Antitumor efficacy of a bispecific antibody that targets HER2 and activates T cells. Cancer Res 2014;74:5561–71.

[203] Ueda O, et al. Entire CD3ε, δ, and γ humanized mouse to evaluate human CD3-mediated therapeutics. Sci Rep 2017;7.

[204] Murphy JT, et al. Anaphylaxis caused by repetitive doses of a GITR agonist monoclonal antibody in mice. Blood 2014;123:2172–80.

[205] Olson B, et al. Mouse models for cancer immunotherapy research. Cancer Discov 2018;8:1358–65.

[206] Estin CD, et al. Transfected mouse melanoma lines that express various levels of human melanoma-associated antigen p97. J Natl Cancer Inst 1989;81:445–8.

[207] Schnell U, Kuipers J, Giepmans BNG. EpCAM proteolysis: new fragments with distinct functions? Biosci Rep 2013;33.

[208] Anon. MGD009 clinical hold release. [cited 2020 October 23]; Available from: http://ir.macrogenics.com/news-releases/news-release-details/macrogenics-announces-removal-partial-clinical-hold-mgd009.

[209] Anon. PF-06671008 clinical trial results. [cited 2020 October 23]; Available from: https://clinicaltrials.gov/ct2/show/results/NCT02659631.

[210] Rosenthal MA, et al. ATIM-49 (LTBK-01). AMG 596, a novel anti-EGFRVIII bispecific T cell engager (BITE®) molecule for the treatment of glioblastoma (GBM): planned interim analysis in recurrent GBM (RGBM). Neuro-Oncology 2019;21:vi283.

[211] Hummel H-D, et al. Phase 1 study of pasotuxizumab (BAY 2010112), a PSMA-targeting bispecific T cell engager (BiTE) immunotherapy for metastatic castration-resistant prostate cancer (mCRPC). J Clin Oncol 2019;37:5034.

[212] Tran B, et al. Phase I study of AMG 160, a half-life extended bispecific T-cell engager (HLE BiTE) immune therapy targeting prostate-specific membrane antigen (PSMA), in patients with metastatic castration-resistant prostate cancer (mCRPC). J Clin Oncol 2020;38:TPS261.

[213] Anon. Aptevo ES414 press release. [cited 2020 October 23]; Available from: https://www.globenewswire.com/news-release/2017/08/31/1106420/0/en/Aptevo-Therapeutics-and-MorphoSys-End-Joint-Development-and-Commercialization-Agreement-for-MOR209-ES414.htm.

[214] Anon. Aptevo Q3 2018 report. Available from https://aptevotherapeutics.gcs-web.com/news-releases/news-release-details/aptevo-therapeutics-reports-third-quarter-2018-financial-results.

[215] Bendell JC, et al. First-in-human phase I study of HPN424, a tri-specific half-life extended PSMA-targeting T-cell engager in patients with metastatic castration-resistant prostate cancer (mCRPC). J Clin Oncol 2020;38:5552.

[216] Anon. Universitätsklinikum Tübingen press release. [cited 2020 October 23]; Available from: https://dktk.dkfz.de/en/about-us/news/bispecific-antibody-be-tested-treatment-prostate-cancer.

[217] Sharma SK, et al. Noninvasive interrogation of DLL3 expression in metastatic small cell lung cancer. Cancer Res 2017;77:3931–41.

[218] Owen DH, et al. DLL3: an emerging target in small cell lung cancer. J Hematol Oncol 2019;12:61.

[219] Heiskala M, Peterson PA, Yang Y. The roles of claudin superfamily proteins in paracellular transport. Traffic 2001;2:92–8.

[220] Damato BE, et al. Tebentafusp: T cell redirection for the treatment of metastatic uveal melanoma. Cancer 2019;11:971.

[221] Sato T, et al. Redirected T cell lysis in patients with metastatic uveal melanoma with gp100-directed TCR IMCgp100: overall survival findings. J Clin Oncol 2018;36:9521.

[222] Middleton MR, et al. Safety, pharmacokinetics and efficacy of IMCgp100, a first-in-class soluble TCR-antiCD3 bispecific t cell redirector with solid tumour activity: results from the FIH study in melanoma. J Clin Oncol 2016;34:3016.

[223] Anon. Immunocore pipeline. [cited 2020 October 23]; Available from: https://www.immunocore.com/our-science/pipeline.

[224] Tabernero J, et al. Phase Ia and Ib studies of the novel carcinoembryonic antigen (CEA) T-cell bispecific (CEA CD3 TCB) antibody as a single agent and in combination with atezolizumab: preliminary efficacy and safety in patients with metastatic colorectal cancer (mCRC). J Clin Oncol 2017;35:3002.

[225] Anon. Amgen pipeline. [cited 2020 October 23]; Available from: https://www.amgenpipeline.com.

[226] Krishna M, Nadler SG. Immunogenicity to biotherapeutics—the role of anti-drug immune complexes. Front Immunol 2016;7.

[227] Davda J, et al. Immunogenicity of immunomodulatory, antibody-based, oncology therapeutics. J Immunother Cancer 2019;7.

[228] Harding FA, et al. The immunogenicity of humanized and fully human antibodies. MAbs 2014;2:256.

[229] Moek KL, et al. Phase I study of AMG 211/MEDI-565 administered as continuous intravenous infusion (cIV) for relapsed/refractory gastrointestinal (GI) adenocarcinoma. Ann Oncol 2018;29.

[230] Köhnke T, et al. Increase of PD-L1 expressing B-precursor ALL cells in a patient resistant to the CD19/CD3-bispecific T cell engager antibody blinatumomab. J Hematol Oncol 2015;8:111.

[231] Lyman S, et al. Evidence of intratumoral localization, activation, and immunomodulatory effect of CX-072, a probody therapeutic targeting PD-L1, in a phase I/II trial. J Clin Oncol 2020;38:3108.

[232] Anon. Maverick Therapeutics press release. [cited 2020 October 26]; Available from: https://www.mavericktx.com/press/pipeline-updates-conditionally-active-t-cell-engaging-cobra-platform/.

CHAPTER TEN

Role of microbiome in cancer immunotherapy

Edda Russo[a], Federico Boem[b], and Amedeo Amedei[a]
[a]Department of Experimental and Clinical Medicine, University of Florence, Florence, Italy
[b]Department of Literature and Philosophy, University of Florence, Florence, Italy

Contents

1. Introduction

Cancer is a primary cause of death globally, with one in seven deaths worldwide [1, 2] and lung, breast, colorectal, and prostate cancers are the four types of malignancy most commonly diagnosed [1, 2]. In addition, tumor global incidence is rising rapidly, due to aging and population growth; but, the improvement of therapies, new progresses in early diagnosis, and advances in cancer prevention have significantly increased the number of saved lives yearly with respect to the past [3]. For a long time, surgery, chemotherapy, and radiotherapy were the most common types of cancer treatment available, employed either alone or in different combinations. Nevertheless, while much progress has been made in malignancy care, despite therapy, a large percentage of patients still died, implying that more effective targets for tumor treatment are needed [1]. Consequently, two major

Cancer Immunology and Immunotherapy
https://doi.org/10.1016/B978-0-12-823397-9.00010-7

revolutions in cancer research and treatment are fulfilling the need for targeted treatments, and the first is based on the identification of actionable genetic alterations in oncogene-driven cancers. The second one now under way focuses on significant advancements in cancer immunotherapy, a treatment that activates or suppresses the host immune system, becoming a crucial therapeutic option, being the first choice in many cases. The role of the immune system on cancer control has been a subject of discussion for many years, but the recent opinion is that malignancy is a genetic disorder that occurs through the acquisition of multiple mutations by somatic cells, causing the barriers that normally take somatic cells under control to be overcome (see the review in Ref. [4]).

Commonly, the host immune system is involved in tumor prevention with many roles, such as (i) prevention of the setting of an inflammatory environment pro-tumorigenesis; (ii) suppression virus-induced tumors contrasting the viral infections; and (iii) elimination of tumor cells that express ligands for receptors active on innate immune cells and tumor antigens recognized by the adaptive immune cells [5].

Conversely, tumor cells, due to their genetic instability, can evade immunosurveillance, possibly tracing the path of cancer initiation. In this context, a targeted immunotherapy turns as a groundbreaking anticancer approach, capable to enhance the host anticancer immune response, and, concomitantly, improve to "hit" recurrence mechanisms and tumor resistance [6]. Currently, immunotherapy has become of huge interest to clinicians, researchers, and pharmaceutical companies, because of its promise for revolutionizing tumor therapy treating various cancer types, with reduced risk of adverse events compared to the conventional therapies [7]. Because this approach is so promising, *Science* named "cancer immunotherapy" "Breakthrough of the Year" in 2013. Nowadays, a range of tumor immunotherapy approaches has demonstrated effectiveness in many patients, involving immune checkpoint blockers, cell-based treatments, monoclonal antibodies, and cancer vaccines [3, 8].

As previously mentioned, the comprehension of the human immune system role toward cancer onset and progression has improved exponentially. Nevertheless, an intricate network of dynamics involving host and environmental factors can influence the individual immune response. Between them, the microbiota has currently acquired growing consideration because of its emerging role as a modulator of the immune response (as reviewed in Ref. [9]). Indeed, multiple lines of evidence provide the notion that the gut microbiota (GM) is able to shape the immune system and indirectly modulate the progression of cancer through the immune system in several organs. Therefore, a positive alteration of the GM composition and metabolic activity might represent a significant and innovative approach to reduce the risk of carcinogenesis and tumor progression. The elucidation of the relationship between the immune system and other systems, such as the gut microbiome, in different diseases has generated important results, giving rise to significant clinical applications. In particular, numerous reports analyzed the discrepancy between immunotherapy-resistant and immunotherapy-sensitive cohorts, proving that specific GM signatures were strictly related to therapy effect. Additionally, further evidence supported interventional GM modulation as an effective treatment to enhance efficacy and relieve resistance during immunotherapy. So, the rising knowledge of the GM relevance in health and pathological status and identification of the host-microbe mutualism (at both metabolic and immunological levels) become crucial for a good understanding of several pathologies, principally tumor. All of these discoveries can be useful for several immune-mediated pathologies, but overall, the success reached with some kind of immunotherapies in current years has enlightened new methods to investigate, but above all manipulate the immune system for host advantage. Currently, discovering new ways to selectively control the GM to prevent carcinogenesis and tumor progression represents a stimulating challenge. In the near future, high-quality interventional human studies and mechanistic experimental reports might provide the scientific premise for the clinical application of GM modulating factors (*e.g.*, probiotics, prebiotics, or fecal microbiota transplant—FMT) in tumor treatment and other multifactorial human illness.

Our chapter introduces cancer immunotherapy strategies, cancer immunology, and explores the emerging role of gut microbiota in the immunotherapy landscape. Furthermore, we will highlight the potential therapeutic utility of directly manipulating commensal microbiota as an approach to enhance the efficacy of cancer immunotherapy. Finally, starting from an original theoretical point of view, we will discuss how the GM will reshape our approach to cancer immunotherapy, introducing the concept of "holobiont."

2. Main immunotherapy approaches: From the past to nowadays

Immunotherapy typically consists of harnessing the host immune system to produce an anticancer response, which is often maintained after the end of treatment, thus indicating a role for immune system modulation and alteration [10].

The practice of artificially improving the immune system is not a recent development, but interestingly, the very beginning of "sensu lato" immunotherapy may be traced back to China, around the third-century BC [11], with variola minor virus inoculation to prevent the disease of smallpox [12]. In 1718, Lady Montague also reported this habit in the Ottoman Empire, and in 1796, Edward Jenner showed protective immunity against smallpox by inoculation with the common cowpox virus [11]. This event was largely known as the age start of vaccines that undoubtedly changed modern medicine and saved millions of lives around the world. The "sensu stricto" immunotherapy, on the other hand, focuses on the treatment of cancer and reflects the more new era of immunotherapy history [8]. As early as the late 1800s, when William B. Coley, known as the father of tumor immunotherapy, observed cases in which neoplasia disappeared with the contraction of *erysipelas*, a microbial infection, so interest in cancer immunology and the use of the immune system as a method to eliminate malignant cells was obvious. In particular, Dr. Coley arranged a mixture of killed *Serratia marcescens* and *Streptococcus pyogenes*, known as "Coley's toxins," based on his experiments, and administered them to subjects who had various forms of inoperable tumors. Coley's toxins have been effective in curing some tumor types, particularly sarcomas, and have been used for several decades as a cancer therapy [13]. Another important phase in tumor immunotherapy took place in 1957, when E. Donnall Thomas studied, with some success, the stem cell transplantation, administering bone marrow from healthy people to subjects affected by advanced leukemia [13]. Subsequently, Steven A. Rosenberg delivered activated immune cells and cytokines to melanoma patients in the late 1980s. Throughout the 20th century, cancer therapies involving the immune system were created as most researchers hypothesized that the immune system was involved in tumor recognition and removal and considered immunodeficiency as the cause of cancer growth [3]. There have been many developments in understanding how cancer is recognized, removed, and prevented by the immune system over the past few decades [7]. In particular, developments in molecular and tumor biology have modified cancer-treatment paradigms dramatically over the past 15 years [14]. Finally, in 2018, Tasuku Honjo and James P. Allison were awarded the Nobel Prize for the discovery of cancer treatment via the suppression of negative immune regulation.

Several forms of immunotherapy are currently used in cancer treatment, but the most widely used approach is to modulate immune checkpoints (IC), especially programmed cell–death receptor-1 (PD-1) or programmed cell–death ligand-1 (PD-L1) and cytotoxic lymphocyte-associated T-4 (CTLA-4) molecules. Although new inhibitory and stimulatory pathways (*e.g.*, T-cell immunoglobulin/TIM-3, lymphocyte activation gene/LAG-3, V-domain Ig T-cell activation suppressor/VISTA, inducible costimulator/ICOS, OX40, 4-1BB) have also emerged as targets [15]. The immune checkpoint inhibitors (ICI) are one of the main current anticancer treatment [16] represented by immunomodulatory monoclonal antibodies (mAbs) that block IC receptors. On the surface of cancer cells, IC molecules interact with $CD8^+$ T-cells, allowing cancer to evade from immune system cells [17]. ICIs suppress the associations between IC molecules and inhibitory T-cellular receptors. These act as immune response brakes, resulting in sustained antitumor responses [13]. Nowadays, ICIs are emerging as potential therapies for different cancers and in some patients, they induce a remarkable, long-term, and well-tolerated therapy reaction that offers term survival benefits [18]. Immunomodulatory mAbs that block IC receptors, such as CTLA-4 and PD-1, or IC ligands, such as PD-L1 ligand, can avoid checkpoint ligand/receptor interactions. Actually, multiple proteins programmed for PD-1 and PD-L1 in the therapy of hematological and solid malignancies have been approved by the Food and Drug Administration (FDA), and are able to provide substantial clinical benefits in the treatment of several tumors. The action mode for ICIs, however, suggests a real change in oncology, rather than being targeted at killing cancer cells, as is the case for radiation and chemotherapy therapies, they target cancer-induced immunosuppression. In particular, CTLA-4 was the first

IC receptor to be clinically targeted. It is expressed only on T-cell surface, where the primary role is to control the early-stage activation amplitude of T-cells [8]. The FDA approved the anti-CTLA-4 ipilimumab for the cure of melanoma in 2011, representing the beginning of a new age for tumor immunotherapy. While several studies indicate that it interferes with kinase signals caused by CD28 and T-cell receptors (TCR), the precise molecular mechanisms by which CTLA-4 inhibits T-cell activation remain under study. In addition, the process by which CTLA-4 amplifies Tregs' (regulatory T-cells) inhibitory activity is unclear, but it has been shown that ICIs greatly decrease Tregs' capacity to regulate both the anticancer immune response and autoimmunity [8].

In addition, a potential immunotherapy target has also risen as the PD-1 receptor present on activated T-cells. The PD-1 primary function is to restrict the activity of T-cells in peripheral tissues to counteract autoimmunity during the response to inflammatory infection [13]. The binding of PD-1 on $CD8^+$ T-cells to its ligands, PD-L1 or PD-L2, contributes to apoptosis as well as reduced T-cell proliferation and development of cytokines. PD-1 is strongly expressed on Tregs; the binding to its ligand promotes the Treg proliferation, allowing the $CD4^+$ and $CD8^+$ T-cell effector functions to be silenced. Unfortunately, some immune checkpoint treatments drawbacks are of concern, like a heterogeneous response where certain patients receive a maximum response, while others never do. In addition, due to alternate immune escape mechanisms, tumor recurrence can happen; unfortunately, optimal biomarkers to predict reaction and toxicity are missing. The occurrence of autoimmune-like reactions and new adverse events and the expense associated with this treatment were other big challenges. Thus, methods that depart from controlling immune checkpoints are under study as alternatives.

Regarding another current immunotherapy approach, the CAR-T was first mentioned by Zelig Eshhar et al. in 1993 and was among the most promising immunotherapy methodologies [19]. CARs was inserted ex vivo into autologous T-cells through genetic modification to increase their function and specificity against antigens present on the surface of the malignant cells [13]. Thus, as it binds to the antigen/MHC complex [13], the TCR is modified such that its antigen-binding component is conjugated to an artificial signaling molecule that gives activation of T-cells. The enlarged CAR T-cells are injected back into the patient after alteration, where they can precisely attack and kill cancerous cells [20]. In this way, CAR signaling can completely replace endogenous TCR, allowing a strong, fast, and non–HLA-restricted targeted cytotoxic immune response. As a result, adoptive cell therapy approaches focused on CAR are resistant to cancer escape mechanisms resulting from depletion of the HLA molecule [20]. With respect to the therapy of patients with relapsed or refractory B-cell cancers, such as acute and persistent lymphocytic leukemia [7], a significant number of CARs targeting different tumors have been produced and have demonstrated promising results. However, targeting solid tumors with CAR T-cells has only produced moderate outcomes. The absence of sufficiently specific tumor surface antigens is currently the main weakness of CAR therapy [20]. Efforts to target specific antigens such as CAIX, HER2/neu, and CD33 have been made, but this has resulted in substantial toxicity and healthy tissue injury. As a result, the search for improved targets is, however, ongoing. Signaling domains involved in T-cell activation include CAR-T cells such as CD28, CD3, ICOS, 4-1BB, and OX40. A second-generation CAR links an antigen-binding domain to the T-cell signaling domain coupled to the intracellular signaling domain of a costimulatory molecule, *e.g.*, CD28. CAR-T cell constructs can include several signaling domains that favor the longevity and activation of CAR-T cells, thereby allowing the effector functions of MHC-independent CAR T-cells to be enforced [21–23]. The FDA approved CAR T-cells guided against CD19 in August 2017 for the treatment of acute lymphoblastic leukemia and then axicabtagene ciloleucel in October 2017 for the treatment of R/R large B-cell lymphoma [22–24]. As reported earlier, although promising clinical activities of CAR T-cells have been documented in hematological malignancies, many barriers to the successful application of CAR T-cells in solid tumors need to be addressed [25]. In detail, the ineffective trafficking of CAR T-cells to cancer sites, the absence of ideal cancer-specific antigens, the tumor immune-suppressive microenvironment, and the possibility of acquiring on-target/off-tumor toxicities arising from an attack on host cells expressing the intended tumor antigen are some of these hurdles. In cancer immunotherapy, future research to enhance specificity and safety is likely to take CAR T-cell therapy into the central stage [26].

In addition, another type of immunotherapy therapy, such as *monoclonal antibodies (mAbs)*, has been a significant treatment for numerous cancers, including breast, lymphoma, and colorectal cancer (CRC) malignancies over the past 20 years [3]. The mAbs are artificial versions of large proteins generated by a particular clone of B-cells that have a special specificity of antigen that enables them to bind to epitopes on or in the plasma of the cancer cell [13]. Radioactive contaminants, chemotherapy medications, or toxins are mixed with conjugated mAbs so that they can serve as a vehicle for guiding these agents into cancer cells.

Furthermore, as immune modulators, compounds such as cytokines have also been used, such as interferons, chemokines, interleukins, and growth factors [13]. Such molecules guide the proliferation, activation, migration, and suppression of leukocytes [20]. Some specific cytokines can specifically strengthen or inhibit the reaction of T-cells to cancer cells.

Finally, cancer vaccines are used to stimulate the host secretion of antibodies that attack the tumor's peptides or antigens. Cancer antigens are also introduced into the microenvironment of the systemic circulation and the tumor. One purpose of cancer vaccinations is to promote the attack and eradication of cancer cells by the immune system [14]. To this end, cancer vaccines include entire cancer cells, cancer cell sections, or purified antigens that strengthen the immune response against cancer cells. Tumor vaccines may be peptide-based, immune cell-based, DC-based, or tumor cell-based. The findings of therapeutic vaccine against existing tumors proved to be only marginally favorable, with clinical advantage for cancer patients, despite encouraging clinical outcomes achieved with immune checkpoint blockade and CAR therapy. However, Melief and colleagues, as the explanations for the failure of cancer eradication [27], suggested the suboptimal vaccine formulation and the existence of an immunosuppressive cancer microenvironment.

Nevertheless, as some of them are resistant to therapy, not all cancer patients may undergo immunotherapy [28]. There has been an increasing attention in understanding the immune properties of the tumor microenvironment in recent years; in particular, the characterization of tumor infiltrating lymphocytes (TILs) has an important role in reacting to inhibition of the checkpoint [29].

3. A new emerging actor in cancer immunotherapy response: The human microbiome

As known, the *microbiota* represents the communities of microorganisms living in coexistence with their hosts, while the *microbiome* refers to the collection of genomes from all the microorganisms in the environment. Regarding its complex composition, the human microflora encompasses the bacterial microbiome, the virome (eukaryotic and bacteriophages viruses), the archaeal microbiome, the meiofauna (unicellular protozoa and helminth worms), and the mycobiome (fungi) [30]. The human microflora represents a diverse and intricate ecosystem living at portals of entry on all epithelial barriers, such as skin and mucosa (from the gastrointestinal, to respiratory and urogenital tract), acquired after delivery through vertical transmission and then modeled by environmental exposure throughout life.

In general, the composition of the human microbiota is strictly personal, but the diversity in the structure among the body sites is greater than it is between individuals. However, it is possible to term a bacterial community "core" indicating conserved communities of a healthy microbiota that is commonly present within different body sites. Notably, the gastrointestinal tract is the organ that contains the larger fraction of microorganisms producing molecules that can be consumed as nutrients, making it a favored place for colonization.

To date, although there have been over 50 bacterial phyla described, only 2 of them dominate the human GM: the *Firmicutes and Bacteroidetes,* whereas *Actinobacteria, Proteobacteria, Fusobacteria, Cyanobacteria,* and *Verrucomicrobia* appear in minor proportion [31]. Interestingly, about 70% of the human microbiota is composed of microorganisms that cannot be cultivated by current microbiological techniques. Indeed, the traditional culture-based methodologies capture less than 30% of our microbiota [32]. However, today, genomic next-generation sequencing (NGS) analysis has been crucial to analyze the bacterial microbiota profile and the metagenome, because these techniques can give more information about the microflora impact in host metabolic reaction, cancer progression, and inflammation [33, 34].

During this last decade, we assisted to a plethora of studies relating the host microbiota with normal physiology, such as food digestion and immune system stimulation [35]. In fact, the human microbiome is necessary to body physiology, as it could generate a huge quantity of molecules/metabolites able to beneficially cooperate with the host. Additionally, it plays an important role in the life-long programming of acquired and innate immune responses; moreover, the microflora fine-tunes the delicate equilibrium between infection, inflammation, and tolerance of commensal antigens and food (Fig. 1) [36–38]. Concerning GM, the communities of microorganisms are divided from the internal gut milieu by a monolayer of epithelial cells, which acts as a chemical and physical barrier, and simultaneously maintain the crosstalk between the immune host system and the external environment. Additionally, the epithelial surfaces have developed defensive actions to counteract microbiota infiltration as adaptive (T- and B-lymphocytes) and innate (macrophages and dendritic cells) immune responses. It should be noted that almost 80% of the immunologically active cells belong to the immune system associated with the mucosa. The majority of these cells are located in the gastrointestinal system, where immunogenic factors, such as components of the microbial flora and food, are at the highest concentration compared to the other body districts.

Generally, the commensal bacterial flora is symbiotic, but it can cause a pathological state after translocation through the mucosa or in specific situations such as immunodeficiency.

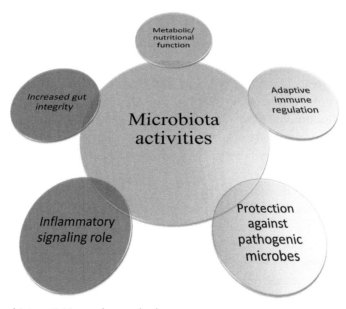

Fig. 1 Gut microbiota activities on human body.

On the other hand, more evidence suggests a relationship between disturbances in the homeostasis of microbiota communities (dysbiosis) and pathologic conditions [39–41]. In addition, as previously reported, data of different studies indicate a link between malignancy and the commensal microbiota [42], including current compelling evidence regarding the GM role in modulating the immune system and thus the reaction to tumor immunotherapy [43–49].

4. Crosstalk between the microbiome and immunity in cancer

4.1 The direct role of microbiome in carcinogenesis and progression

As mentioned, both at epithelial barriers and inside sterile tissue, the microflora has been shown to be associated with the starting and progression of different cancer types. In fact, numerous experiments in bacteria-associated germ-free models have shown signs of cancer-promoting properties of the microbiota on sporadic and genetically caused tumors in many organs, including breast, skin, liver, intestine, and lungs [50–52]. Furthermore, it has been shown that the commensal ecosystems living in the mucosae influence both local and distant tumor initiation. Indeed, by delivering a toxic metabolite or an oncogenic substance, or indirectly by inducing inflammation or immunosuppression, microbes can directly act as cancer-transforming agents. There is convincing evidence for the microbiome influence in the tumors' development, with several experiments stressing the association with the immunity axis (Fig. 2).

In 1975, the GM was linked for the first time to intestinal tumor development [51]. Afterward, Vannucci and colleagues demonstrated that germ-free rats, compared with animals having a similar genotype and a normal microflora, develop smaller neoplasia, as randomly as after carcinogenesis induced by chemical products [52]. In comparison, germfree mice exhibit reduced tumor development and fewer oncogenic mutations in

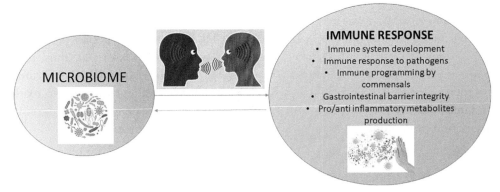

Fig. 2 Crosstalk between microbiome and immune response in cancer.

colitis-associated cancer and adenomatous polyposis coli (APC)-related colorectal cancer [53]. Furthermore, intestinal microbiota depletion by antibiotics in mice decreased the occurrence of cancer in the colon and liver [50], as does the eradication of particular pathogens.

Besides, multiple theories were proposed about the active action of the microbiome in cancer initiation; for instance, Tjalsma and colleagues [54] first proposed the *bacterial driver-passenger* paradigm to explain the microbial role in the CRC development.

Different indigenous intestinal bacteria, called "driver bacteria," will generate DNA damage according to this model and drive genome instability to induce the initial stages of tumor development. As a result, with either tumor-promoting or tumor-suppressing properties (bacterial passengers), the bacterial drivers are eventually substituted by commensal gut bacteria. This model suggests that as a function of the growing tumor, disease development induces changes in the microenvironment, resulting in a reformed selective strain on the microbial population.

In 2011, Sears and Pardoll [55] proposed that *alpha bugs* (some microbiome elements with unusual virulence characteristics), such as enterotoxigenic *Bacteroides fragilis* (ETBF), are specifically pro-oncogenic and capable of remolding the immune response of the mucosal and colonic microbial-removing organisms, thereby promoting cancer development.

In addition, the *keystone pathogen* assumption also explains the significance of the human microbiota in host health and disease. Indeed, in the ecological literature, the word "keystone" has been used to describe organisms whose impacts on their populations are excessively high compared to their abundance and who are considered to constitute the "keystone" of the system of the environment. This assumption suggests that by remodeling a typically benign microbiome into a dysbiotic one, certain low abundance microbial pathogens can orchestrate inflammatory diseases [56].

Finally, it was shown that inflammatory responses elicited by bacteria could increase cancer progression [57]. Some bacteria can cause changes in the mucosa permeability, enabling the migration of bacteria and bacterial toxins (such as lipopolysaccharide). The inflammatory process leads to cancer growth, development, and treatment, but it is uncertain if commensal bacteria affect inflammation in the sterile tumor microenvironment.

A growing number of studies has demonstrated the inflammation impact in developing settings that can fundamentally change local immune responses and, thus, tissue homeostasis. In specific, inflammatory mediators [such as interleukin (IL)-1 and tumor necrosis factor-alpha (TNF-alpha) prostaglandin-2 derivatives and IL-8, nitric oxide] and inflammatory pathway molecules are well known to be involved in the progressive association of immune cells and tissue cells undergoing transformation [58].

4.2 The role of adaptive immune response in cancer progression

The structure of adaptive immunity has been developed by vertebrates to contrast pathogens that avoid or defeat innate immune response. The elements of the adaptive immune system are generally inactivated; however, when they are triggered by the presence of dangerous agents (*e.g.*, infectious, cancer cells, etc.), they develop effective pathways to neutralize or remove the pathogens. Any of these reactions, however, may be protumorigenic. There are two kinds of adaptive immune responses: humoral immunity, mediated by B-cell-generated antibodies, and cell-mediated immunity, mediated by T-cells, separated into two subsets: $CD8^+$ (CTL) and $CD4^+$ or T helper (Th). The comparticipation of both T-cells' subsets [4] is necessary for an efficient antitumor immune response.

In preclinical models and in cancer patients, the functions of T helper cells in antitumor immunity have currently been widely elucidated. For priming tumor-specific CTL and for the secondary expansion and memory of $CD8^+$ T-cells, the $CD4^+$ T-cells are essential [59]. However, the discovery of regulatory T-cells (Treg) and Th17 cells not only modified the Th cell differentiation classical Th1/Th2 paradigm, but also greatly changed the notion of the Th role of in antitumor immunity [60]. Additionally, tumor-infiltrating Tregs have been shown to cause an immunosuppressive microenvironment, inhibit successful antitumor immunity [61, 62], and become a significant barrier to the effectiveness of tumor immunotherapy [63].

Moreover, the functional role of human Th17 cells in tumor immunity is already debated because varied cancer forms have been demonstrated for both protumor and antitumor effects [64–66]. In addition to defending against cancer cells and pathogens, for the formation of diverse bacterial populations in the gut, acquired immunity is crucial. Microbiome-driven carcinogenesis pathways vary considerably across organs, and the microbiome may modulate the growth of tumors by direct and indirect behavior. Microbial products can directly or indirectly promote tumor establishment, suggesting that the bacteria themselves are unable to encourage cancer development unless they interfere with the immune system. Eventually, it is also probable that defects in particular immune response processes cause the expansion of some bacteria activating the protumor immune response.

4.3 Antibacterial-specific immune response and tumor starting

Inflammatory and immune responses originating from the host are an essential driving factor of the microbial population structure and may lead to dysbiosis when altered. Any microbiota members modify the adaptive immune response and, in essence, foster the cancer development. The Th17 cells are one of the key factors by which the microbiome will indirectly stimulate tumorigenesis.

In order to facilitate homeostasis, the microbiota actively regulate intestinal T-cell responses. Th17 cells, but complex compensatory mechanisms are needed to control the Th17 cells in turn.

In physiological conditions, the commensal bacteria stimulate IL-1β production in the mammalian intestine to sustain a basal concentration of Th17 cells under in the lamina propria [67]. However, substantial numbers of naive Th cells differentiate into Th17 in response to pathogenic extracellular bacterial or fungal infections under the control of IL-23, IL-1β, IL-6, or TGF-β on the mucosal surfaces of the respiratory and enteric tract [68]. Moreover, if some pathways are altered, the Th17 cells could become pathogenic, causing chronic inflammation. When stimulated with TGF-β and IL-6, the transcription factor RORγt (retinoic acid receptor-related orphan receptor gamma t) and Th17-specific cytokines (such as IL-17 and IL-22) are upregulated by the antigen-activated Th cells [69].

Usually, CD4^{+} T-cells that express RORγt increase promote microbicidal protein secretion and the establishment of close junctions, which contributes to the intestinal epithelium's barrier function, but in the same, may have a protumorigenic effect.

Compelling data, as mentioned earlier, indicates that Th17 cells and their cytokines are highly related to tumor development, but the functional Th17 contribution in cancer is still unclear, as it seems to have both procancerogenic and anticancerogenic activities, and the quality of the responses seem to depend on the nature of cancer [70]. Moreover, IL-17F, IL-21, IL-22, interferon (IFN)-γ and granulocyte-macrophage colony-stimulating factor (GM-CSF) can be secreted by Th17 cells in addition to IL-17A [71]. Initially, Th17 reactions, and in particular the IL-17 activity itself, were thought to stimulate cancer development, angiogenesis, and invasion [72]. The Th17 cells are capable of fostering the growth of colorectal cancer in a mouse model, caused by colon inflammation [73]. Genetically predisposed mouse (APCmin/+) studies crossed with IL-17A deficient mice suggest a dramatic deficiency in intestinal tumorigenesis [74]. Furthermore, APCmin/+ mice that cannot respond to IL-17 have been shown to produce fewer tumors in the colon [70]. In addition, the function of Th17 cells in humans has been studied in subjects with numerous forms of neoplasia, including prostate and ovarian cancer, and several others [75–84]. Most of these experiments have studied peripheral blood Th17 cells, but Th17 cells could be stimulated or enrolled into the cancer microenvironment [85].

Studies of the enterotoxigenic *B. fragilis* provide more clear evidence for the function of bacterially induced tumor growth through Th17 cells. This colonic human bacterium secretes *B. fragilis* (BFT) toxin that triggers inflammatory diarrhea in humans. Several mouse models, predisposed to develop intestinal cancer, indicate that between colonization of *B. fragilis* and nontoxigenic *B. fragilis*, only the first induce colitis and triggers colonic tumors [73].

Especially *B. fragilis* causes selective colonic signal transducer and transcription-3 (STAT3) activation with colitis characterized by a selective Th17 response. The blockade of IL-17 and IL-23 receptor using antibodies, is able to inhibit *B. fragilis*-induced colitis, neoplasia formation, and colonic hyperplasia.

These findings suggest a mechanism of inflammation-induced cancer based on STAT3 and Th17 by a typical human commensal bacterium, offering a new mechanistic insight into human colon carcinogenesis. Moreover, the Th17 response activates the expansion of neutrophil cells needed for the clearance of invading bacteria upon contact with particular bacteria [86].

For defense against mucosal pathogens such as *Klebsiella pneumoniae* and *Salmonella typhimurium*, the Th17 reaction is essential. Mice with Th17 cytokine deficiency exhibit significant pathology during infection with *Salmonella* or *C. rodentium*, showing enhanced bacterial translocation into lymph nodes [87]. Segmented filamentous bacteria (SFB), belonging to nonculturable *Clostridia*-related species, and flagellin-positive bacteria are other species of bacteria capable of activating Th17. In the host epithelial membrane, these bacteria interact with the epithelial cells, inducing systemic inflammation mediated by the release of IL-17 and IL-22, which is likely to facilitate the growth of intestinal cancer. Recently, SFB has been shown to help the generation of Th17 cells [88] and robust Th17 responses have been identified only in animals colonized by SFB [89]. In order to boost Th17 and CTL-cell responses against cancer cells, some researchers were inspired to approach the microbiome study based on such interesting results [90]. Th17 cells, as previously mentioned, express other cytokines such as IL-22, which in murine models [91] and human studies [92] has been related to gastrointestinal tumorigenesis. Inflammasome-regulated IL-22BP (IL-22 R) modulates tumorigenesis in the intestine and in cancer of the human colon [93]. In addition, IL-22 is linked to human colorectal cancer growth through activation of STAT3. Several extracolonic cancers, such as hepatocellular carcinoma and nonsmall cell lung cancer [94], have previously demonstrated a procarcinogenic role for IL-22 through the STAT3 pathway. In addition, IL-22 can induce the development of inducible nitric oxide synthase (iNOS) and procarcinogenic nitric oxygen species in human colon carcinoma cell lines in combination with IFN-γ [95] (Fig. 3).

Finally, in reaction to numerous microbiome molecules, such as flagellin [96], the cytokine IL-23 is produced by myeloid cells. IL-23 is capable of fostering Th17-type responses characterized by the activation of IL-17 and IL-22 cytokines [97]. Finally, in human colon adenocarcinoma, IL-23 was found to increase and encourage tumor growth through the activation of proinflammatory responses [98].

5. The "microbiome-immunity axis" influences the effectiveness of cancer immunotherapy

In addition to the GM role in the host immune system, present persuasive evidence links the microbiota signatures of individual patients with the favorable outcome of tumor immunotherapy in cancer vaccinations [7], ICIs [47] and CAR-T [99] (as reviewed in Ref. [100]).

The first suggestion of the immunotherapy function of GM came from studies of total-body irradiation, which increased the effectiveness of the transmission of tumor-specific T-cells by translocating the Toll-like 4L receptor (TLR4L) lipopolysaccharide

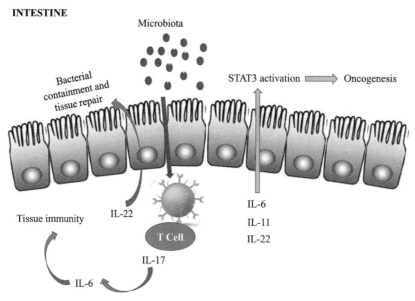

Fig. 3 Specific immune response against gut microbiome contribute to oncogenesis through STAT 3 activation.

from the intestinal lumen to secondary lymphoid organs [101]. Soon after, experiments in mouse models demonstrated the presence of a crosstalk between various microbial communities and ICIs [48,49,102], suggesting that GM modulation could increase the effectiveness of CTLA-4 and PD-1 checkpoint blockade. In particular, mice with melanoma and separate GM profiles displayed a distinct immune-mediated antitumor response; moreover, the authors found that mice had tumor-specific T-cell responses and CTL tumor cell aggregation. Interestingly, it was also documented that this distinct response [48] was reset by cohousing or fecal microbiota transplantation. Notably, FMT using feces from patients (who proved clinical response to ICIs) transferred a "responder" phenotype to recipient mice, whereas "nonresponder" patient stools tended to confer nonresponsiveness to recipients [45, 46] (Fig. 4). Such evidence emphasizes that the nonresponder/responder conditions of recipient mice were induced via FMT from the donor GM structure. In reconstituted mouse models, novel microbial species were found that conferred enhanced immune-mediated cancer regulation. Vètizou and colleagues, for example, studied multiple patients with neoplasia treated with antibiotic medication at the beginning of CTLA-4 therapy [49]. They found that the genera *Bacteroides* and *Burkholderia* were linked to GM's antitumor action, hypothesizing that, in response to *Bacteroides* and *Burkholderia*, IL-12 released by innate immune cells can activate the T-cells. In addition to this, the GM was transferred directly in germfree mice or indirectly, by giving those feces enriched in *Bacteroides* of Ipilimumab-treated patients.

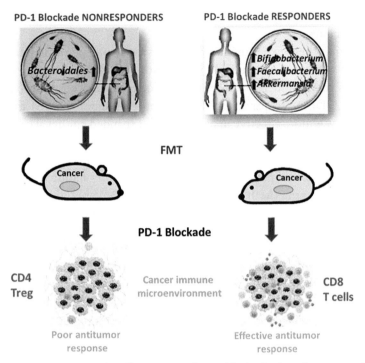

Fig. 4 Gut microbiota impact on the effectiveness of PD-1 blockade. Specific gut microbiota profiles correlate with response to PD-1 blockade in tumor patients. Fecal Microbiota Transplantation (FMT) from responders into mice improves responses to anti-PD-1 treatment and correlates with increased anticancer CD8$^+$ cells in the tumor environment. Mice receiving FMT from nonresponders' patients did not respond to anti-PD-1 therapy, and tumor microenvironment is enriched of immune suppressive CD4$^+$ Tregs.

The response to the checkpoint inhibitors was enhanced in all cases [49]. *Faecalibacterium* [103] and *Verrucomicrobiaceae* [47] were also associated with the antitumor GM activity in lung cancer patients treated with nivolumab or pembrolizumab PD-1 inhibitors. While *Enterococcus faecium, Collinsella aerofaciens*, and *Bifidobacterium longum* have also been enriched in patients with a positive response to PD-1 blockade in another study. When the fecal samples of responders' patients were transferred to germfree mice, a slower tumor evolution and enhanced therapeutic effects were detected, compared to mice receiving nonresponders' fecal samples. Furthermore, a decrease in Tregs and increased CTL in the cancer microenvironment were documented [46]. In any case, where the GM was missing or manipulated, clinical effectiveness was curtailed. These initial experiments on mouse models show a key impact of gut microbiota in the anticancer treatment by ICIs, which encouraged therapeutic efforts to evaluate the GM role in human ICIs-based therapies.

Corroborating these experimental findings, several independent retrospective analyses of human cohorts of metastatic lung, prostate, and bladder cancer patients revealed the deleterious role of different classes of antibiotics taken around the introduction of mAbs to PD1/PDL-1. In particular, Routy and colleagues showed that anti-PD-1/PD-L1 therapy was more effective in patients with expanded epithelial cancer not treated with antibiotics, compared with the therapy effect of those who received antibiotics. They prove that anomalous GM signatures [47] may be linked to primary resistance to ICIs. This evidence emphasizes that the administration of antibiotics will destabilize the GM, thus weakening the ICIs blockade response. Notably, the study of fecal specimens obtained at the time of diagnosis revealed that GM populations enriched in *Alistipes* and *Akkermansia* were present and altered by patients who reacted positively to PD-1 blockade.

The sequencing of baseline stool samples from patients treated with antibodies against PD-1-based therapies may have been the most provocative evidence showing the role of commensal microbiota in the therapeutic effectiveness of cancer immunotherapy. Based on their microbial metagenomic fingerprint, recent developments in sequencing technology have strengthened our ability to stratify patients [47–49]. There are, however, differences between the numerous scientific studies involving patients with different genetic and dietary patterns, clinical trials performed in different geographical locations in the United States or Europe, and different types of tumors, including lung cancer, cancer of the renal cells, and melanoma.

Furthermore, germfree mice were treated with FMT before the anti-PD-1 therapy. The oral supplementation of *Akkermansia muciniphila* restored the weakened immune response of germfree mice treated with FMT from nonresponders. It was demonstrated that these bacteria stimulated PD-1 blockade in an IL-12-dependent way by increasing the enrollment of $CXCR3^+CCR9^+$ $CD4^+$ T-cells into mouse tumor beds [47]. Gopalakrishnan *et al.* studied the intestinal and oral microbiota in patients with metastatic melanoma under anti-PD-1 treatment. Patients whose intestinal specimens were enriched with *Clostridiales, Faecalibacterium*, or *Ruminococcaceae* displayed a protected cytokine reaction to anti-PD-1 therapy and more peripheral blood effector T-cells ($CD8^+$ and $CD4^+$); on the other hand, patients with *Bacteroidales* enrichment showed higher levels of Tregs with a decreased cytokine reaction [45]. In general, immune profiling indicated an improved antitumor and systemic immunity in responding melanoma subjects with a favorable GM, as well as in germfree mice receiving FMT from responding subjects. In a recent study, patients with nonsmall cell lung cancer, undergoing Nivolumab treatment, were observed. A greater diversity of fecal microbiota was detected before and after treatment with anti-PD-1. Compared with the lower diversity of microbiota, this disparity was correlated with extended "progression-free survival time" [104]. In comparison, patients responding to treatment with Nivolumab displayed a higher baseline variety of gut microbiota, which retained a stable structure during treatment [104]. Finally, a report of Tanoue and colleagues has recognized 11 healthy human-related bacterial strains that stimulate CTL cells and are involved in inhibiting colorectal cancer evolution if used with ICIs [105]. Nevertheless, GM can induce toxicity to the ICIs. This effect was first observed in animal models and then in humans [42,43,106]. Within the *Ruminococcaceae* family and the *Firmicutes* phylum, the GM clades associated with both ICIs reaction and toxicity have been determined [43, 45]. On the other hand, within the *Bacteroidales* order, the GM clades missing reaction to ICIs have been found, but a greater richness in these taxa typically reduces the toxicity frequency [43,106].

Retrospective analyses of patients treated with second-line therapies for FDA-approved indications need to be confirmed in future prospective clinical trials and validated in other treatment settings (*e.g.*, first-line or second-line immunotherapy and additional forms of cancer). In addition, diet and supplements might have an effect on the patient's responsivity to anticancer immunotherapy, probably due to the modulation of their GM [107]. Analysis of fecal samples from 113 patients with melanoma under anti-PD-1/PD-L1 therapy indicated that the favorable reaction to treatment could be associated with higher microbial diversity. Participants offered details on their eating habits and use of probiotic supplements, but only 40% of them reported taking the supplements successfully. In addition, researchers observed that high intake of fibers (including fruits, vegetables, and whole grains) was positively related in previous studies to the existence of microorganisms associated with a strong anti-PD-1 treatment response [107]. On the contrary, a high sugar and processed meat intake has been adversely related to these microbial florae. In particular, an enrichment of the *Ruminococcaceae* family was associated with better therapy response, while *Bacteroidales* bacteria were associated with poorer response. In addition, patients on a high-fiber diet were around five times more likely (respect to patients on a low-fiber diet) to respond to anti-PD-1 therapy (Pembrolizumab or Nivolumab). Surprisingly, probiotic supplement use was associated with less diversity of gut bacteria [107]. There has been considerable interest in deciphering microbiome signatures based on these published papers in order to conceive of the most effective microbial consortia for therapeutic use in combination with immunotherapy, but this would definitely be an iterative method with several questions about both the host and the administration of the microbial products.

A very recent study identifies a previously unknown microbial metabolite immune pathway activated by immunotherapy that may be exploited to develop microbial-based adjuvant therapies [108]. In this study, three bacterial species—*Bifidobacterium pseudolongum*, *Lactobacillus johnsonii*, and *Olsenella* species significantly enhanced efficacy of immune checkpoint inhibitors in four mouse models of cancer. In particular, *B. pseudolongum* modulated enhanced immunotherapy response through production of the metabolite inosine. Decreased gut barrier function induced by immunotherapy increased systemic translocation of inosine and activated antitumor T-cells. The effect of inosine was dependent on T-cell expression of the adenosine A2A receptor and required costimulation [108].

6. Microbiota shaping: From molecular immunotherapy to "eco-immunotherapy"

One way to improve immunotherapeutic interventions is constituted by microbial manipulation.

In this respect, diverse, non-necessarily mutually exclusive alternatives, are possible. Interventions aimed at shaping microbiota and modify its activity can be both direct and indirect interventions. For instance, both acting on nutrition and lifestyle habits are indirect modes to functionally influence the microbiota. Unfortunately, these approaches normally take time to be efficacious. Moreover, diet and healthy routines can promote desirable conditions, but might not be suitable as therapies in case of acute clinical situations. Another problem of these approaches is that the administration of probiotics and prebiotics can definitely be positive for a favorable immune modulation (for instance, in colorectal cancer patients) [109] but these treatments are still too difficult to standardize and control due to the high number of variables in place. FMT on the other hand, has shown to be a more direct and effective type of intervention, displaying appreciable and durable outcomes in some well-known pathological conditions (such as *Clostridium difficile* infections) [110]. Interestingly, more recently FMT has been used to treat, or to complementary improve standard treatments, concerning diverse pathologies such as autoimmune [111,112] or neurodegenerative diseases [113].

A criticism with these forms of approaches is that they still lack a detailed mechanistic understanding. Therefore, it has been argued for a new venue, called "oncomicrobiotics," consisting of a better comprehension of bacterial capacities and metabolic products to develop tailored drugs. These approaches promise, in theory, to be particularly effective due to what they have been labeled as "microbiota-conditioning strategies" [114]. This kind of intervention will necessarily rely on big-data science, to discern and evaluate millions of distinct microorganismal outcomes that might affect distinct molecular pathways.

Another possibility could be genetically modifying microbiota itself [115,116], implying the creation and use of genetically modified microbial strains. In this sense, the technology of CRISPR-Cas systems is a very suitable tool for various forms of genetic editing [117–119].

First, "additive interventions," are ways to generate modified microbial strains to foster probiotic effects. Alternatively, "subtractive therapies," can try to mimic antibiotic functions also via specifically designed bacteriophages (to counteract the well-known phenomenon of antibiotic resistance and, also, the nonspecificity). Accordingly, engineered bacteriophages will serve a new, more tailored and specific, way to affect the microbial functionality and to govern their number and composition. Third, "modulatory therapies" will influence bacterial genic expression (also by controlling their number), always through designed phages. Lastly, live microbial by-products might serve as immunostimulants to improve immunotherapeutic responses.

As a result, it is now clear that GM manipulation constitutes a fruitful tool to improve therapeutic outcomes [45–47]. Research has also begun to shed light on several possible mechanisms by which the microbiota could influence these crucial factors in therapies.

These proposed mechanisms encompass several possibilities. First, there is the direct interplay of microbial metabolic activities (*e.g.*, via PAMPs—pathogen-associated molecular patterns) with antigen-presenting cells (APCs) and other innate effectors (via pattern-recognition receptors—PRRs—like Toll-like receptors—TLRs). As a result, these interaction mechanisms can stimulate an adaptive immune response. Second, microbiota can induce the production of different cytokines, thus potentially affecting also areas that are distant from the source of microbial activities. Because of that, researchers are now developing approaches and strategies to better identify these signatures to design specific favorable microbial populations to enhance immunotherapeutic interventions [120].

The complexity of these mechanisms is such that even small differences can lead to extremely different results. For instance, an *in vitro* study [121] has shown that distinct strains of *Enterococcus faecalis*, co-cultured with colorectal cancer cell lines, display antitumor features. The researchers tested several strains and their metabolic products on diverse CRC cell lines. Surprisingly, four out of eight strains exhibited antiproliferative features on two cell lines, while the others did not have effect.

These aspects indicate also that the genetic background of the holobiont is only one element of the picture. Bacterial species, populations, modes of interactions, environmental conditions, and cancer types (considering also, tumor stage, interheterogeneity, and intraheterogeneity) are fundamental as well [122,123].

Thus, by reinvigorating the need of an ecological take on the issue, these outcomes reveal that the only presence of a given bacterial genus as such is not enough to condition the therapy efficacy. This means that rather than focusing just on the microbial taxa and phylogeny, relationships among species and individuals within the holobiont (see the following) need to be urgently considered. Accordingly, more effective and precise treatments will be implemented when different layers of interventions will be more integrated and comprehensively addressed.

Finally, it is essential to consider that microbiota can also generate unwanted and undesirable outcomes. In fact, the alteration of certain relations within bacterial communities, like physiological disruption under some conditions, can favor tumor development rather than enhancing immunotherapy. Because of that, future immunotherapeutic strategies should also concentrate on clearer comprehension of microbial activities and features, thus fostering certain kind of functions and restraining others [42].

Fascinatingly, all these findings suggest diverse new ways to improve immunotherapy. On one side, a more refined, ecologically oriented, understanding of microbiota's functions and its interplay with the immune system is definitely a venue that promises to increase the identification of specific bio-signatures valuable for both diagnosis and prognosis. On the other side, the result of such a change probably urges the need of a revision of the traditional conceptualization. As a matter of fact, a proficuous improvement of traditional immunotherapy will require microbiota shaping through modalities and procedures that exceed current strategies.

Indeed, all these approaches are definitely interesting and worthy to be pursued. However, the lack of clear and deep understanding of the interactions among bacteria in diverse contexts and the complex, largely unknown, relationship of these activities with human physiology suggests caution.

As a matter of fact, ecologists praise for attention when manipulations are at stake. Because of that, some scholars proposed to adopt the so-called "holobiont-perspective" to tackle this issue. Indeed, a holobiont can be defined as the assemblage of diverse life forms (both eukaryotes and prokaryotes), engaging in a tangled network of dynamical relationships (thus not necessarily all beneficial and susceptible to change over time) that all together participate to the generation of a "functional whole." Each holobiont is somehow unique and can highly differ in its assemblage, functionality, and ecological relationships. Thus, what we usually refer to as "human individuals" can be seen as holobionts, composed of the "host" and its "microbiota." Concerning health, this means that the same subject or "person," through time, can display extremely distinct features compared with previous conditions, an aspect that makes quite difficult to define what a "healthy microbiota" is in general.

It is also important to note that the systemic-ecological network of microbial populations is, generally speaking, functionally redundant. Analogous properties can be obtained by the activities of distinct species and genera. In the end, healthy conditions might depend on bio-geographical and social factors as well as various contexts. Accordingly, from a therapeutic point of view, it is possible to argue in favor of the need to also change the conceptual perspective aside with methodological innovations.

First, as previously reported, microbiota has to be conceived in a broader sense, including not only bacteria but also viruses, archaea, fungi, and others. The interactions of all these actors deserve rigorous investigations as they are the "nodes" of a system that is the holobiont, which in turns becomes the main actor in the study of the whole immune response. A better knowledge of species and genera involved is necessary but not sufficient. Indeed, since the holobiont is a network, it cannot be studied just by recomposing its elements investigated separately.

That is why it is arguable to put more efforts for the establishment of new areas of inquiry (like, as already done in other sectors of scientific investigation, the ecological evolutionary developmental biology (Eco-Evo-Devo)) [124]. These new research fields are the result of the integration of diverse research agendas, methodologies, and kinds of expertise (spanning from molecular embryology to population ecology), which might show huge implications on health definition and potential therapeutic interventions [110].

7. How the holobiont perspective can reshape our approach to cancer immunotherapy

As previously argued, immunotherapy constitutes a great promise for biomedical research. When compared with other forms of therapeutic interventions, immunotherapy is perceived to be a fundamental alternative in tumor treatment, given its potential to both affect the neoplasia and its microenvironment. This systemic feature is not accidental. The immunotherapy efficacy lies in the fact that it makes use of components and functions of the patient's immune system in order to tackle both tumor initiation and progression. Despite the fact that in many cases this approach has proved to be extraordinarily effective, in other situations, as previously reported, patients still do not respond to it in a positive manner. This diversity of responses has been a source of perplexity for many researchers and clinicians, making it difficult even to present evidence and data that could provide a unitary form of explanation.

The discovery that GM strongly influences the immune system is not only fundamental from an empirical point of view but also suggests the need for a rethinking at the theoretical/conceptual level of some fundamental biological notions. In other words, this perspective indicates the urgency of changing and updating the current ideas concerning *symbiosis*, which piercingly discriminates the host from its "associates" microorganisms [125,126]. Accordingly, as already mentioned, what is macroscopically categorized as a single organism should be rather intended as the assemblage of diverse life forms (both eukaryotes and prokaryotes), engaging in a tangled network of dynamical relationships (thus not necessarily all beneficial and susceptible to change over time) that all together participate to the generation of a "functional whole," *i.e.*, the *holobiont* [125,127,128]. Consequently, what is traditionally defined as *the human immune system* is rather a wider and more complex one, including microbial activities. In addition, given its interaction with both the nervous and the endocrine systems, the extended immune system plays the role to surveil and guarantee both homeostasis and the activities of the entire "functional whole."

This is not just speculation. As previously mentioned, recent findings have shown that microbiota is essential for shaping the immune system prerogatives in the gut, also contributing to the maintenance of the local homeostasis [129]. An important case to clarify this fact is constituted by interaction of microbes with the production of the mucus layer covering epithelial cells in the intestine [130]. A very important function of mucus is to preserve the gut epithelium from damages of the digestive secretions. Nevertheless, the mucus layer is also a crucial component for bacterial colonization. In order to colonize the gut, several microbial species establish an intense crosstalk with the mucus layer, forming a real ecological niche. From a physiological point of view, we usually associate the function of mucus with the protection of the human organism. On the other hand, as already mentioned, mucus also has a fundamental ecological function. However, such a feature crucially affects the health of the entire system. In fact, by adopting the holobiont notion, it is possible to reconcile these two perspectives within a common frame. Accordingly, it becomes clear how the way of describing functions and mechanisms of already known physiological processes can drastically change, offering new possibilities for investigation and, above all, new approaches to therapeutic interventions. Interestingly, it has been shown that microbial activity can have an impact way beyond the locality of the ecological niche in which microorganisms reside. Indeed, many studies have demonstrated how microbiota contributes to a variety of physiological functions. In particular, the microbiota extends the effects of its activities through the close relationship with the immune system, thus impacting not only locally but globally, with possible effects on distant areas of the organism and even on a systemic level [103]. Because of that, it has been suggested that microbiota plays an essential role in shaping the type and the degree of the entire immune response, now responding to an infection, now maintaining a condition of low inflammation [103]. This complex scenario can be partly explained due to the magnitude of microbiota itself. Indeed, microbial cells, just in the gut (thus not considering other areas) surpass (or are equal to) the total number of host human cells [129,131]. Indeed, growing evidence suggests that the number and variety of associated microorganisms, their capabilities and plasticity, systemically influences the entire physiology of the functional whole [103,126,132]. On this aspect, it has been shown that microbial metabolites, either directly or due to the crosstalk with the immune system, may affect different physiological functions and mechanisms throughout the body, as in the case of the so-called gut-brain axis [133,134]. This expression refers not to any anatomical structure but rather to a functional relationship between distant structures that is quite stable. This functional relationship indicates that the crosstalk between the human part and the microbial in the holobiont embodies a critical feature for sustaining global health conditions [135] Comprehensively, this scenario suggests an almost dialectic situation. On one side, the microbiota shapes the activity of the immune system. On the other side, the immune system itself has evolved to interplay with associated microorganisms and their ecological structure [131]. So, it should not seem strange to start considering the microbiota as a proper functional part of the immune system [127]. Because of that, numerous researches have begun to consider the microbiota in the study of diseases. As already explained, the observations made in the last 20 years show that the intestinal microbiota contributes to the general well-being of the holobiont. This means that aberrant activities of the microbiota component can play an important role, like physiological disruptions, in the etiology of various disorders, including obesity, diabetes, metabolic diseases, and, of course, autoimmune diseases and cancer [136,137]. Espousing this perspective, evidence of these phenomena can reveal how complex networks and growing microbial communities, in their activities, are related, in different ways, to various cancers [114,120]. A paradigmatic case is constituted by *Helicobacter pylori*, which definitely represents a crucial risk factor for gastric cancers. Indeed, recent studies have demonstrated how microbial communities in *H. pylori*-positive patients often present an increase of *Proteobacteria*, *Spirochaetes*, and *Acidobacteria*, and a decrease of *Actinobacteria*, *Bacteroidetes*, and *Firmicutes*. However, this shows that it is not just the single species or strain to be considered. On the contrary, the causal net involved in pathogenesis highly relies on the ecological and systemic dimension [138–140]. Thus, this scenario could support the development of new approaches to cancer therapies, especially cancer immunotherapies, focused on the understanding of gut microbiota dynamics and its manipulation. In these aspects, recent studies [45, 47–49] have highlighted how the microbiota is a key factor in determining the efficacy of anti-PD-L1 and anti-CTLA-4 immunotherapeutic interventions.

Moreover, as extensively mentioned earlier, it has been observed that ecological relationship among microbial populations was a crucial aspect for effective therapy. These results have proved to be greater when, by administering fecal transplantation, germfree mice positively responded to therapy that was previously inefficacious. Last, but not the least, the presence of certain species reduced complications and side effects, lowering the frequency of subclinical colitis.

Summing up, increasing evidence from these studies indicates that diverse individuals, presenting different responses to therapeutic interventions, seem to display what has been labeled as "microbial signature." Like other determinants (*e.g.*, genetic background), diverse signatures have been associated with more or less favorable health conditions, with crucial consequences on both the local and systemic immune response.

8. Conclusions

As a conclusion, it should not be seen as weird to praise the creation and development of an extended immunotherapy, or *eco-immunology*. This research field will, necessarily, integrate views, methods, and criteria of diverse disciplines. On the other hand, assimilating such diverse, however intriguing, ways of doing research may not be so easy and definitely run the risk of becoming sterile or superficial. Therefore, a more pluralistic and integrated approach will require a shared theoretical framework already devoted, by internal constitution, to the complexity consideration both for the treatment of diverse levels of description and explanation, and in the capacity to deal with different kinds of data in a unitary way. This is therefore a metatheoretical reason for supporting the need for an ecological perspective in immunology. In fact, ecology, due to its attention to relations, processes, different scales of description and variety of data processed, is perhaps the most suitable discipline to promote this type of integration in a successful manner.

Such integration will demand great efforts, given that any experimental science has the specific manipulation criteria of the experimental systems. The criteria that are tools to determine, within a discipline, whether a discovery is valuable or not, differ from one branch of knowledge. Because of that, new types of experimental procedures and new standards to evaluate the experimental results should necessarily be designed. Such a perspective is not just a mere speculation. Ecology has already begun to address the problem of global experimental manipulations [141]. In fact, the comprehension of the holobiont as the privileged biological unit to study the immune response urges the need of an approach capable of encompassing the "whole eco-immune system." This is crucial to provide a more combined and harmonized understanding of the whole responses to external impacts, such as diet, lifestyle, genetic components, social factors, and so on. Indeed, ecology can provide a good lesson. Researchers working on tropical ecosystems routinely face the challenge of studying these complex objects. Experimental manipulations aware of the systemic reverberations on the entire whole are essential owing to the necessity to provide suggestions concerning the properties of the entire system itself along with its dynamics in responding to internal and external stimuli. Similarly, to a tropical forest as "whole," the holobiont can be seen as an assemblage of biomes. Thus, its functions cannot be fully addressed without a global look. Thus, ecological immunotherapeutic interventions should necessarily mimic the impact of causal factors on the entire "ecosystem." Eco-immunotherapy of cancer is, nowadays, only a theoretical suggestion. However, the proposal is very concrete. In fact, as reported in the following, the ecologists have made the study of complex dynamics their object of research. It is not foolish to think that (in the light of the growing evidence on the integrated perspective that sees the holobiont as a privileged unit in the organization of living entities) including experts, such as population, theoretical, or microbial ecologists, in the study and design of new therapeutic approaches and protocols is something feasible in the future. Science is founded (or should have been founded) on its own methodological rigor, but its progress also arises from the ability to create and invent new ways of looking at phenomena, far beyond the boundaries created by the very same methods, updating them, overcoming them, and even waiting to be overcome in the future.

Acknowledgment

The research was funded with a grant from the Foundation "Ente Cassa di Risparmio di Firenze."

Consent for publication

Not applicable.

Conflict of Interest

The author declares no conflict of interest, financial or otherwise.

References

[1] Tartari F, Santoni M, Burattini L, Mazzanti P, Onofri A, Berardi R. Economic sustainability of anti-PD-1 agents nivolumab and pembrolizumab in cancer patients: recent insights and future challenges. Cancer Treat Rev 2016;48:20–4. https://doi.org/10.1016/j.ctrv.2016.06.002.

[2] Zugazagoitia J, Guedes C, Ponce S, Ferrer I, Molina-Pinelo S, Paz-Ares L. Current challenges in Cancer treatment. Clin Ther 2016;38(7):1551–66. https://doi.org/10.1016/j.clinthera.2016.03.026.

[3] Weiner LM. Cancer immunology for the clinician. Clin Adv Hematol Oncol 2015;13(5):299–306.

[4] Schreiber RD, Old LJ, Smyth MJ. Cancer immunoediting: integrating immunity's roles in cancer suppression and promotion. Science 2011;331(6024):1565–70. https://doi.org/10.1126/science.1203486.

[5] Thorsson V, Gibbs DL, Brown SD, Wolf D, Bortone DS, Ou Yang TH, Shmulevich I. The immune landscape of cancer. Immunity 2018;48(4):812–30. e814 https://doi.org/10.1016/j.immuni.2018.03.023.

[6] Emens LA, Ascierto PA, Darcy PK, Demaria S, Eggermont AMM, Redmond WL, Marincola FM. Cancer immunotherapy: opportunities and challenges in the rapidly evolving clinical landscape. Eur J Cancer 2017;81:116–29. https://doi.org/10.1016/j.ejca.2017.01.035.

[7] Yang Y. Cancer immunotherapy: harnessing the immune system to battle cancer. J Clin Invest 2015;125(9):3335–7. https://doi.org/10.1172/JCI83871.

[8] Pardoll D. Cancer and the immune system: basic concepts and targets for intervention. Semin Oncol 2015;42(4):523–38. https://doi.org/10.1053/j.seminoncol.2015.05.003.

[9] Russo E, Taddei A, Ringressi MN, Ricci F, Amedei A. The interplay between the microbiome and the adaptive immune response in cancer development. Ther Adv Gastroenterol 2016;9(4):594–605. https://doi.org/10.1177/1756283X16635082.

[10] Khalil DN, Smith EL, Brentjens RJ, Wolchok JD. The future of cancer treatment: immuno-modulation, CARs and combination immunotherapy. Nat Rev Clin Oncol 2016;13(5):273–90. https://doi.org/10.1038/nrclinonc.2016.25.

[11] Decker WK, da Silva RF, Sanabria MH, Angelo LS, Guimaraes F, Burt BM, Paust S. Cancer immunotherapy: historical perspective of a clinical revolution and emerging preclinical animal models. Front Immunol 2017;8:829. https://doi.org/10.3389/fimmu.2017.00829.

[12] Lombard M, Pastoret PP, Moulin AM. A brief history of vaccines and vaccination. Rev Sci Tech 2007;26(1):29–48. https://doi.org/10.20506/rst.26.1.1724.

[13] Alatrash G, Jakher H, Stafford PD, Mittendorf EA. Cancer immunotherapies, their safety and toxicity. Expert Opin Drug Saf 2013;12(5):631–45. https://doi.org/10.1517/14740338.2013.795944.

[14] Karlitepe A, Ozalp O, Avci CB. New approaches for cancer immunotherapy. Tumour Biol 2015;36 (6):4075–8. https://doi.org/10.1007/s13277-015-3491-2.

[15] Sharma P, Allison JP. The future of immune checkpoint therapy. Science 2015;348(6230):56–61. https://doi.org/10.1126/science.aaa8172.

[16] Robert C, Thomas L, Bondarenko I, O'Day S, Weber J, Garbe C, Wolchok JD. Ipilimumab plus dacarbazine for previously untreated metastatic melanoma. N Engl J Med 2011;364(26):2517–26. https://doi.org/10.1056/NEJMoa1104621.

[17] Borghaei H, Paz-Ares L, Horn L, Spigel DR, Steins M, Ready NE, Brahmer JR. Nivolumab versus docetaxel in advanced nonsquamous non-small-cell lung cancer. N Engl J Med 2015;373 (17):1627–39. https://doi.org/10.1056/NEJMoa1507643.

[18] Helissey C, Vicier C, Champiat S. The development of immunotherapy in older adults: new treatments, new toxicities? J Geriatr Oncol 2016;7(5):325–33. https://doi.org/10.1016/j.jgo.2016.05.007.

[19] Grupp SA, Kalos M, Barrett D, Aplenc R, Porter DL, Rheingold SR, June CH. Chimeric antigen receptor-modified T cells for acute lymphoid leukemia. N Engl J Med 2013;368(16):1509–18. https://doi.org/10.1056/NEJMoa1215134.

[20] Klener Jr P, Otahal P, Lateckova L, Klener P. Immunotherapy approaches in cancer treatment. Curr Pharm Biotechnol 2015;16(9):771–81. https://doi.org/10.2174/1389201016666150619114554.

[21] Abid MB. The revving up of CARs. Gene Ther 2018;25(3):162. https://doi.org/10.1038/s41434-018-0015-x.

[22] June CH, Sadelain M. Chimeric antigen receptor therapy. N Engl J Med 2018;379(1):64–73. https://doi.org/10.1056/NEJMra1706169.

[23] Li S, Zhang J, Wang M, Fu G, Li Y, Pei L, Qian C. Treatment of acute lymphoblastic leukaemia with the second generation of CD19 CAR-T containing either CD28 or 4-1BB. Br J Haematol 2018;181 (3):360–71. https://doi.org/10.1111/bjh.15195.

[24] Neelapu SS, Locke FL, Bartlett NL, Lekakis LJ, Miklos DB, Jacobson CA, Go WY. Axicabtagene Ciloleucel CAR T-cell therapy in refractory large B-cell lymphoma. N Engl J Med 2017;377 (26):2531–44. https://doi.org/10.1056/NEJMoa1707447.

[25] Yu S, Li A, Liu Q, Li T, Yuan X, Han X, Wu K. Chimeric antigen receptor T cells: a novel therapy for solid tumors. J Hematol Oncol 2017;10(1):78. https://doi.org/10.1186/s13045-017-0444-9.

[26] Liu B, Song Y, Liu D. Clinical trials of CAR-T cells in China. J Hematol Oncol 2017;10(1):166. https://doi.org/10.1186/s13045-017-0535-7.

[27] Melief CJ, van Hall T, Arens R, Ossendorp F, van der Burg SH. Therapeutic cancer vaccines. J Clin Invest 2015;125(9):3401–12. https://doi.org/10.1172/JCI80009.

[28] Topalian SL, Hodi FS, Brahmer JR, Gettinger SN, Smith DC, McDermott DF, Sznol M. Safety, activity, and immune correlates of anti-PD-1 antibody in cancer. N Engl J Med 2012;366 (26):2443–54. https://doi.org/10.1056/NEJMoa1200690.

[29] Cesano A, Warren S. Bringing the next generation of immuno-oncology biomarkers to the clinic. Biomedicines 2018;6(1). https://doi.org/10.3390/biomedicines6010014.

[30] Stappenbeck TS, Virgin HW. Accounting for reciprocal host-microbiome interactions in experimental science. Nature 2016;534(7606):191–9. https://doi.org/10.1038/nature18285.

[31] Eckburg PB, Bik EM, Bernstein CN, Purdom E, Dethlefsen L, Sargent M, Relman DA. Diversity of the human intestinal microbial flora. Science 2005;308(5728):1635–8. https://doi.org/10.1126/science.1110591.

[32] Fraher MH, O'Toole PW, Quigley EM. Techniques used to characterize the gut microbiota: a guide for the clinician. Nat Rev Gastroenterol Hepatol 2012;9(6):312–22. https://doi.org/10.1038/nrgastro.2012.44.

[33] Human Microbiome Project Consortium. Structure, function and diversity of the healthy human microbiome. Nature 2012;486(7402):207–14. https://doi.org/10.1038/nature11234.

[34] Kau AL, Ahern PP, Griffin NW, Goodman AL, Gordon JI. Human nutrition, the gut microbiome and the immune system. Nature 2011;474(7351):327–36. https://doi.org/10.1038/nature10213.

[35] Ackerman J. The ultimate social network. Sci Am 2012;306(6):36–43. https://doi.org/10.1038/scientificamerican0612-36.

[36] Belkaid Y, Naik S. Compartmentalized and systemic control of tissue immunity by commensals. Nat Immunol 2013;14(7):646–53. https://doi.org/10.1038/ni.2604.

[37] de Vos WM, de Vos EA. Role of the intestinal microbiome in health and disease: from correlation to causation. Nutr Rev 2012;70(Suppl 1):S45–56. https://doi.org/10.1111/j.1753-4887.2012.00505.x.

[38] Sender R, Fuchs S, Milo R. Revised estimates for the number of human and bacteria cells in the body. PLoS Biol 2016;14(8). https://doi.org/10.1371/journal.pbio.1002533, e1002533.

[39] Czesnikiewicz-Guzik M, Muller DN. Scientists on the spot: salt, the microbiome, and cardiovascular diseases. Cardiovasc Res 2018;114(10):e72–3. https://doi.org/10.1093/cvr/cvy171.

[40] Jangi S, Gandhi R, Cox LM, Li N, von Glehn F, Yan R, Weiner HL. Alterations of the human gut microbiome in multiple sclerosis. Nat Commun 2016;7:12015. https://doi.org/10.1038/ncomms12015.

[41] Kim D, Zeng MY, Nunez G. The interplay between host immune cells and gut microbiota in chronic inflammatory diseases. Exp Mol Med 2017;49(5). https://doi.org/10.1038/emm.2017.24, e339.

[42] Zitvogel L, Daillere R, Roberti MP, Routy B, Kroemer G. Anticancer effects of the microbiome and its products. Nat Rev Microbiol 2017;15(8):465–78. https://doi.org/10.1038/nrmicro.2017.44.

[43] Frankel AE, Coughlin LA, Kim J, Froehlich TW, Xie Y, Frenkel EP, Koh AY. Metagenomic shotgun sequencing and unbiased metabolomic profiling identify specific human gut microbiota and metabolites associated with immune checkpoint therapy efficacy in melanoma patients. Neoplasia 2017;19 (10):848–55. https://doi.org/10.1016/j.neo.2017.08.004.

[44] Geller LT, Barzily-Rokni M, Danino T, Jonas OH, Shental N, Nejman D, Straussman R. Potential role of intratumor bacteria in mediating tumor resistance to the chemotherapeutic drug gemcitabine. Science 2017;357(6356):1156–60. https://doi.org/10.1126/science.aah5043.

[45] Gopalakrishnan V, Spencer CN, Nezi L, Reuben A, Andrews MC, Karpinets TV, Wargo JA. Gut microbiome modulates response to anti-PD-1 immunotherapy in melanoma patients. Science 2018;359(6371):97–103. https://doi.org/10.1126/science.aan4236.

[46] Matson V, Fessler J, Bao R, Chongsuwat T, Zha Y, Alegre ML, Gajewski TF. The commensal microbiome is associated with anti-PD-1 efficacy in metastatic melanoma patients. Science 2018;359 (6371):104–8. https://doi.org/10.1126/science.aao3290.

[47] Routy B, Le Chatelier E, Derosa L, Duong CPM, Alou MT, Daillere R, Zitvogel L. Gut microbiome influences efficacy of PD-1-based immunotherapy against epithelial tumors. Science 2018;359 (6371):91–7. https://doi.org/10.1126/science.aan3706.

[48] Sivan A, Corrales L, Hubert N, Williams JB, Aquino-Michaels K, Earley ZM, Gajewski TF. Commensal bifidobacterium promotes antitumor immunity and facilitates anti-PD-L1 efficacy. Science 2015;350(6264):1084–9. https://doi.org/10.1126/science.aac4255.

[49] Vètizou M, Pitt JM, Daillere R, Lepage P, Waldschmitt N, Flament C, Zitvogel L. Anticancer immunotherapy by CTLA-4 blockade relies on the gut microbiota. Science 2015;350(6264):1079–84. https://doi.org/10.1126/science.aad1329.

[50] Dapito DH, Mencin A, Gwak GY, Pradere JP, Jang MK, Mederacke I, Schwabe RF. Promotion of hepatocellular carcinoma by the intestinal microbiota and TLR4. Cancer Cell 2012;21(4):504–16. https://doi.org/10.1016/j.ccr.2012.02.007.

[51] Reddy BS, Narisawa T, Maronpot R, Weisburger JH, Wynder EL. Animal models for the study of dietary factors and cancer of the large bowel. Cancer Res 1975;35(11 Pt. 2):3421–6.

[52] Vannucci L, Stepankova R, Kozakova H, Fiserova A, Rossmann P, Tlaskalova-Hogenova H. Colorectal carcinogenesis in germ-free and conventionally reared rats: different intestinal environments affect the systemic immunity. Int J Oncol 2008;32(3):609–17.

[53] Rakoff-Nahoum S, Medzhitov R. Role of toll-like receptors in tissue repair and tumorigenesis. Biochemistry (Mosc) 2008;73(5):555–61. https://doi.org/10.1134/s0006297908050088.

[54] Tjalsma H, Boleij A, Marchesi JR, Dutilh BE. A bacterial driver-passenger model for colorectal cancer: beyond the usual suspects. Nat Rev Microbiol 2012;10(8):575–82. https://doi.org/10.1038/nrmicro2819.

[55] Sears CL, Pardoll DM. Perspective: alpha-bugs, their microbial partners, and the link to colon cancer. J Infect Dis 2011;203(3):306–11. https://doi.org/10.1093/jinfdis/jiq061.

[56] Hajishengallis G, Darveau RP, Curtis MA. The keystone-pathogen hypothesis. Nat Rev Microbiol 2012;10(10):717–25. https://doi.org/10.1038/nrmicro2873.

[57] Fukata M, Abreu MT. Role of Toll-like receptors in gastrointestinal malignancies. Oncogene 2008;27(2):234–43. https://doi.org/10.1038/sj.onc.1210908.

[58] Mantovani A, Allavena P, Sica A, Balkwill F. Cancer-related inflammation. Nature 2008;454 (7203):436–44. https://doi.org/10.1038/nature07205.

[59] Janssen EM, Lemmens EE, Wolfe T, Christen U, von Herrath MG, Schoenberger SP. CD4+ T cells are required for secondary expansion and memory in CD8+ T lymphocytes. Nature 2003;421 (6925):852–6. https://doi.org/10.1038/nature01441.

[60] Wang X, Yang Y, Moore DR, Nimmo SL, Lightfoot SA, Huycke MM. 4-hydroxy-2-nonenal mediates genotoxicity and bystander effects caused by enterococcus faecalis-infected macrophages. Gastroenterology 2012;142(3):543–51. e547 https://doi.org/10.1053/j.gastro.2011.11.020.

[61] Niccolai E, Cappello P, Taddei A, Ricci F, D'Elios MM, Benagiano M, Amedei A. Peripheral ENO1-specific T cells mirror the intratumoral immune response and their presence is a potential

prognostic factor for pancreatic adenocarcinoma. Int J Oncol 2016;49(1):393–401. https://doi.org/10.3892/ijo.2016.3524.

[62] Niccolai E, Ricci F, Russo E, Nannini G, Emmi G, Taddei A, Amedei A. The different functional distribution of "not effector" T cells (Treg/Tnull) in colorectal cancer. Front Immunol 2017;8:1900. https://doi.org/10.3389/fimmu.2017.01900.

[63] Curiel TJ. Regulatory T cells and treatment of cancer. Curr Opin Immunol 2008;20(2):241–6. https://doi.org/10.1016/j.coi.2008.04.008.

[64] Amedei A, Munari F, Bella CD, Niccolai E, Benagiano M, Bencini L, D'Elios MM. Helicobacter pylori secreted peptidyl prolyl cis, trans-isomerase drives Th17 inflammation in gastric adenocarcinoma. Intern Emerg Med 2014;9(3):303–9. https://doi.org/10.1007/s11739-012-0867-9.

[65] Wilke CM, Kryczek I, Wei S, Zhao E, Wu K, Wang G, Zou W. Th17 cells in cancer: help or hindrance? Carcinogenesis 2011;32(5):643–9. https://doi.org/10.1093/carcin/bgr019.

[66] Amedei A, Niccolai E, Benagiano M, Della Bella C, Cianchi F, Bechi P, et al. Ex vivo analysis of pancreatic cancer-infiltrating T lymphocytes reveals that ENO-specific Tregs accumulate in tumor tissue and inhibit Th1/Th17 effector cell functions. Cancer Immunol Immunother 2013;62(7):1249–60. https://doi.org/10.1007/s00262-013-1429-3.

[67] Shaw MH, Kamada N, Kim YG, Nunez G. Microbiota-induced IL-1beta, but not IL-6, is critical for the development of steady-state TH17 cells in the intestine. J Exp Med 2012;209(2):251–8. https://doi.org/10.1084/jem.20111703.

[68] Ouyang W, Kolls JK, Zheng Y. The biological functions of T helper 17 cell effector cytokines in inflammation. Immunity 2008;28(4):454–67. https://doi.org/10.1016/j.immuni.2008.03.004.

[69] Korn T, Bettelli E, Oukka M, Kuchroo VK. IL-17 and Th17 cells. Annu Rev Immunol 2009;27:485–517. https://doi.org/10.1146/annurev.immunol.021908.132710.

[70] Grivennikov SI, Wang K, Mucida D, Stewart CA, Schnabl B, Jauch D, Karin M. Adenoma-linked barrier defects and microbial products drive IL-23/IL-17-mediated tumour growth. Nature 2012;491(7423):254–8. https://doi.org/10.1038/nature11465.

[71] Zheng Y, Valdez PA, Danilenko DM, Hu Y, Sa SM, Gong Q, Ouyang W. Interleukin-22 mediates early host defense against attaching and effacing bacterial pathogens. Nat Med 2008;14(3):282–9. https://doi.org/10.1038/nm1720.

[72] Numasaki M, Fukushi J, Ono M, Narula SK, Zavodny PJ, Kudo T, Lotze MT. Interleukin-17 promotes angiogenesis and tumor growth. Blood 2003;101(7):2620–7. https://doi.org/10.1182/blood-2002-05-1461.

[73] Wu S, Rhee KJ, Albesiano E, Rabizadeh S, Wu X, Yen HR, Sears CL. A human colonic commensal promotes colon tumorigenesis via activation of T helper type 17 T cell responses. Nat Med 2009;15(9):1016–22. https://doi.org/10.1038/nm.2015.

[74] Chae WJ, Gibson TF, Zelterman D, Hao L, Henegariu O, Bothwell AL. Ablation of IL-17A abrogates progression of spontaneous intestinal tumorigenesis. Proc Natl Acad Sci USA 2010;107(12):5540–4. https://doi.org/10.1073/pnas.0912675107.

[75] Charles KA, Kulbe H, Soper R, Escorcio-Correia M, Lawrence T, Schultheis A, Hagemann T. The tumor-promoting actions of TNF-alpha involve TNFR1 and IL-17 in ovarian cancer in mice and humans. J Clin Invest 2009;119(10):3011–23. https://doi.org/10.1172/JCI39065.

[76] Derhovanessian E, Adams V, Hahnel K, Groeger A, Pandha H, Ward S, Pawelec G. Pretreatment frequency of circulating IL-17+ CD4+ T-cells, but not Tregs, correlates with clinical response to whole-cell vaccination in prostate cancer patients. Int J Cancer 2009;125(6):1372–9. https://doi.org/10.1002/ijc.24497.

[77] Dhodapkar KM, Barbuto S, Matthews P, Kukreja A, Mazumder A, Vesole D, Dhodapkar MV. Dendritic cells mediate the induction of polyfunctional human IL17-producing cells (Th17-1 cells) enriched in the bone marrow of patients with myeloma. Blood 2008;112(7):2878–85. https://doi.org/10.1182/blood-2008-03-143222.

[78] Horlock C, Stott B, Dyson PJ, Morishita M, Coombes RC, Savage P, Stebbing J. The effects of trastuzumab on the CD4+CD25+FoxP3+ and CD4+IL17A+ T-cell axis in patients with breast cancer. Br J Cancer 2009;100(7):1061–7. https://doi.org/10.1038/sj.bjc.6604963.

[79] Inozume T, Hanada K, Wang QJ, Yang JC. IL-17 secreted by tumor reactive T cells induces IL-8 release by human renal cancer cells. J Immunother 2009;32(2):109–17. https://doi.org/10.1097/CJI.0b013e31819302da.

[80] Koyama K, Kagamu H, Miura S, Hiura T, Miyabayashi T, Itoh R, Gejyo F. Reciprocal CD4+ T-cell balance of effector CD62Llow CD4+ and CD62LhighCD25+ CD4+ regulatory T cells in small cell lung cancer reflects disease stage. Clin Cancer Res 2008;14(21):6770–9. https://doi.org/10.1158/1078-0432.CCR-08-1156.

[81] Wang W, Edington HD, Rao UN, Jukic DM, Radfar A, Wang H, Kirkwood JM. Effects of high-dose IFNalpha2b on regional lymph node metastases of human melanoma: modulation of STAT5, FOXP3, and IL-17. Clin Cancer Res 2008;14(24):8314–20. https://doi.org/10.1158/1078-0432.CCR-08-0705.

[82] Yang ZZ, Novak AJ, Ziesmer SC, Witzig TE, Ansell SM. Malignant B cells skew the balance of regulatory T cells and TH17 cells in B-cell non-Hodgkin's lymphoma. Cancer Res 2009;69(13):5522–30. https://doi.org/10.1158/0008-5472.CAN-09-0266.

[83] Zhang B, Rong G, Wei H, Zhang M, Bi J, Ma L, Fang G. The prevalence of Th17 cells in patients with gastric cancer. Biochem Biophys Res Commun 2008;374(3):533–7. https://doi.org/10.1016/j.bbrc.2008.07.060.

[84] Zhang JP, Yan J, Xu J, Pang XH, Chen MS, Li L, Zheng L. Increased intratumoral IL-17-producing cells correlate with poor survival in hepatocellular carcinoma patients. J Hepatol 2009;50(5):980–9. https://doi.org/10.1016/j.jhep.2008.12.033.

[85] Kryczek I, Banerjee M, Cheng P, Vatan L, Szeliga W, Wei S, Zou W. Phenotype, distribution, generation, and functional and clinical relevance of Th17 cells in the human tumor environments. Blood 2009;114(6):1141–9. https://doi.org/10.1182/blood-2009-03-208249.

[86] Blaschitz C, Raffatellu M. Th17 cytokines and the gut mucosal barrier. J Clin Immunol 2010;30(2):196–203. https://doi.org/10.1007/s10875-010-9368-7.

[87] Raffatellu M, Santos RL, Verhoeven DE, George MD, Wilson RP, Winter SE, Baumler AJ. Simian immunodeficiency virus-induced mucosal interleukin-17 deficiency promotes Salmonella dissemination from the gut. Nat Med 2008;14(4):421–8. https://doi.org/10.1038/nm1743.

[88] Suzuki K, Meek B, Doi Y, Muramatsu M, Chiba T, Honjo T, Fagarasan S. Aberrant expansion of segmented filamentous bacteria in IgA-deficient gut. Proc Natl Acad Sci USA 2004;101(7):1981–6. https://doi.org/10.1073/pnas.0307317101.

[89] Gaboriau-Routhiau V, Rakotobe S, Lecuyer E, Mulder I, Lan A, Bridonneau C, Cerf-Bensussan N. The key role of segmented filamentous bacteria in the coordinated maturation of gut helper T cell responses. Immunity 2009;31(4):677–89. https://doi.org/10.1016/j.immuni.2009.08.020.

[90] Gajewski TF, Schreiber H, Fu YX. Innate and adaptive immune cells in the tumor microenvironment. Nat Immunol 2013;14(10):1014–22. https://doi.org/10.1038/ni.2703.

[91] Kirchberger S, Royston DJ, Boulard O, Thornton E, Franchini F, Szabady RL, Powrie F. Innate lymphoid cells sustain colon cancer through production of interleukin-22 in a mouse model. J Exp Med 2013;210(5):917–31. https://doi.org/10.1084/jem.20122308.

[92] Niccolai E, Taddei A, Ricci F, Rolla S, D'Elios MM, Benagiano M, Amedei A. Intra-tumoral IFN-gamma-producing Th22 cells correlate with TNM staging and the worst outcomes in pancreatic cancer. Clin Sci (Lond) 2016;130(4):247–58. https://doi.org/10.1042/CS20150437.

[93] Jiang R, Wang H, Deng L, Hou J, Shi R, Yao M, Sun B. IL-22 is related to development of human colon cancer by activation of STAT3. BMC Cancer 2013;13:59. https://doi.org/10.1186/1471-2407-13-59.

[94] Zhang W, Chen Y, Wei H, Zheng C, Sun R, Zhang J, Tian Z. Antiapoptotic activity of autocrine interleukin-22 and therapeutic effects of interleukin-22-small interfering RNA on human lung cancer xenografts. Clin Cancer Res 2008;14(20):6432–9. https://doi.org/10.1158/1078-0432.CCR-07-4401.

[95] Ziesche E, Bachmann M, Kleinert H, Pfeilschifter J, Muhl H. The interleukin-22/STAT3 pathway potentiates expression of inducible nitric-oxide synthase in human colon carcinoma cells. J Biol Chem 2007;282(22):16006–15. https://doi.org/10.1074/jbc.M611040200.

[96] Kinnebrew MA, Buffie CG, Diehl GE, Zenewicz LA, Leiner I, Hohl TM, Pamer EG. Interleukin 23 production by intestinal CD103(+)CD11b(+) dendritic cells in response to bacterial flagellin enhances mucosal innate immune defense. Immunity 2012;36(2):276–87. https://doi.org/10.1016/j.immuni.2011.12.011.

[97] Liang SC, Tan XY, Luxenberg DP, Karim R, Dunussi-Joannopoulos K, Collins M, Fouser LA. Interleukin (IL)-22 and IL-17 are coexpressed by Th17 cells and cooperatively enhance expression of antimicrobial peptides. J Exp Med 2006;203(10):2271–9. https://doi.org/10.1084/jem.20061308.

[98] Langowski JL, Zhang X, Wu L, Mattson JD, Chen T, Smith K, Oft M. IL-23 promotes tumour incidence and growth. Nature 2006;442(7101):461–5. https://doi.org/10.1038/nature04808.

[99] Abid MB, Shah NN, Maatman TC, Hari PN. Gut microbiome and CAR-T therapy. Exp Hematol Oncol 2019;8:31. https://doi.org/10.1186/s40164-019-0155-8.

[100] Russo E, Nannini G, Dinu M, Pagliai G, Sofi F, Amedei A. Exploring the food-gut axis in immunotherapy response of cancer patients. World J Gastroenterol 2020;26(33):4919–32. https://doi.org/10.3748/wjg.v26.i33.4919.

[101] Paulos CM, Wrzesinski C, Kaiser A, Hinrichs CS, Chieppa M, Cassard L, Restifo NP. Microbial translocation augments the function of adoptively transferred self/tumor-specific CD8+ T cells via TLR4 signaling. J Clin Invest 2007;117(8):2197–204. https://doi.org/10.1172/JCI32205.

[102] Taur Y, Jenq RR, Perales MA, Littmann ER, Morjaria S, Ling L, Pamer EG. The effects of intestinal tract bacterial diversity on mortality following allogeneic hematopoietic stem cell transplantation. Blood 2014;124(7):1174–82. https://doi.org/10.1182/blood-2014-02-554725.

[103] Gopalakrishnan V, Helmink BA, Spencer CN, Reuben A, Wargo JA. The influence of the gut microbiome on cancer, immunity, and cancer immunotherapy. Cancer Cell 2018;33(4):570–80. https://doi.org/10.1016/j.ccell.2018.03.015.

[104] Jin Y, Dong H, Xia L, Yang Y, Zhu Y, Shen Y, Lu S. The diversity of gut microbiome is associated with favorable responses to anti-programmed death 1 immunotherapy in Chinese patients with NSCLC. J Thorac Oncol 2019;14(8):1378–89. https://doi.org/10.1016/j.jtho.2019.04.007.

[105] Tanoue T, Morita S, Plichta DR, Skelly AN, Suda W, Sugiura Y, Honda K. A defined commensal consortium elicits CD8 T cells and anti-cancer immunity. Nature 2019;565(7741):600–5. https://doi.org/10.1038/s41586-019-0878-z.

[106] Chaput N, Lepage P, Coutzac C, Soularue E, Le Roux K, Monot C, Carbonnel F. Baseline gut microbiota predicts clinical response and colitis in metastatic melanoma patients treated with ipilimumab. Ann Oncol 2017;28(6):1368–79. https://doi.org/10.1093/annonc/mdx108.

[107] Spencer CN, Gopalakrishnan V, McQuade J, Andrews MC, Helmink B, Khan MAW, Sirmans E, Haydu L, Cogdill A, Burton E, Amaria R, Patel S, Glitza I, Davies M, Posada E, Hwu W-J, Diab A, Nelson K, Tawbi H, Wong M, Jenq RR, Cohen L, Daniel-MacDougall C, Wargo JA. The gut microbiome (GM) and immunotherapy response are influenced by host lifestyle factors. In: AACR Annual Meeting 2019 Online Proceedings and Itinerary Planner Home; 2019.

[108] Mager LF, Burkhard R, Pett N, Cooke NCA, Brown K, Ramay H, McCoy KD. Microbiome-derived inosine modulates response to checkpoint inhibitor immunotherapy. Science 2020;369(6510):1481–9. https://doi.org/10.1126/science.abc3421.

[109] De Almeida CV, de Camargo MR, Russo E, Amedei A. Role of diet and gut microbiota on colorectal cancer immunomodulation. World J Gastroenterol 2019;25(2):151–62. https://doi.org/10.3748/wjg.v25.i2.151.

[110] Blaser MJ. The past and future biology of the human microbiome in an age of extinctions. Cell 2018;172(6):1173–7. https://doi.org/10.1016/j.cell.2018.02.040.

[111] Balakrishnan B, Taneja V. Microbial modulation of the gut microbiome for treating autoimmune diseases. Expert Rev Gastroenterol Hepatol 2018;12(10):985–96. https://doi.org/10.1080/17474124.2018.1517044.

[112] Makkawi S, Camara-Lemarroy C, Metz L. Fecal microbiota transplantation associated with 10 years of stability in a patient with SPMS. Neurol Neuroimmunol Neuroinflamm 2018;5(4). https://doi.org/10.1212/NXI.0000000000000459, e459.

[113] Mandrioli J, Amedei A, Cammarota G, Niccolai E, Zucchi E, D'Amico R, Masucci L. FETR-ALS study protocol: a randomized clinical trial of fecal microbiota transplantation in amyotrophic lateral sclerosis. Front Neurol 2019;10:1021. https://doi.org/10.3389/fneur.2019.01021.

[114] Pitt JM, Vetizou M, Waldschmitt N, Kroemer G, Chamaillard M, Boneca IG, Zitvogel L. Fine-tuning cancer immunotherapy: optimizing the gut microbiome. Cancer Res 2016;76(16):4602–7. https://doi.org/10.1158/0008-5472.CAN-16-0448.

[115] Fuentes S, de Vos WM. How to manipulate the microbiota: fecal microbiota transplantation. Adv Exp Med Biol 2016;902:143–53. https://doi.org/10.1007/978-3-319-31248-4_10.

[116] Landry BP, Tabor JJ. Engineering diagnostic and therapeutic gut bacteria. Microbiol Spectr 2017;5(5). https://doi.org/10.1128/microbiolspec.BAD-0020-2017.

[117] Bober JR, Beisel CL, Nair NU. Synthetic biology approaches to engineer probiotics and members of the human microbiota for biomedical applications. Annu Rev Biomed Eng 2018;20:277–300. https://doi.org/10.1146/annurev-bioeng-062117-121019.

[118] Gibson SB, Green SI, Liu CG, Salazar KC, Clark JR, Terwilliger AL, Ramig RF. Constructing and characterizing bacteriophage libraries for phage therapy of human infections. Front Microbiol 2019;10:2537. https://doi.org/10.3389/fmicb.2019.02537.

[119] Ramachandran G, Bikard D. Editing the microbiome the CRISPR way. Philos Trans R Soc Lond Ser B Biol Sci 2019;374(1772):20180103. https://doi.org/10.1098/rstb.2018.0103.

[120] Helmink BA, Khan MAW, Hermann A, Gopalakrishnan V, Wargo JA. The microbiome, cancer, and cancer therapy. Nat Med 2019;25(3):377–88. https://doi.org/10.1038/s41591-019-0377-7.

[121] De Almeida CV, Lulli M, di Pilato V, Schiavone N, Russo E, Nannini G, Amedei A. Differential responses of colorectal cancer cell lines to enterococcus faecalis' strains isolated from healthy donors and colorectal cancer patients. J Clin Med 2019;8(3). https://doi.org/10.3390/jcm8030388.

[122] McGranahan N, Swanton C. Clonal heterogeneity and tumor evolution: past, present, and the future. Cell 2017;168(4):613–28. https://doi.org/10.1016/j.cell.2017.01.018.

[123] Prasetyanti PR, Medema JP. Intra-tumor heterogeneity from a cancer stem cell perspective. Mol Cancer 2017;16(1):41. https://doi.org/10.1186/s12943-017-0600-4.

[124] Gilbert SF, Bosch TC, Ledon-Rettig C. Eco-evo-devo: developmental symbiosis and developmental plasticity as evolutionary agents. Nat Rev Genet 2015;16(10):611–22. https://doi.org/10.1038/nrg3982.

[125] Bordenstein SR, Theis KR. Host biology in light of the microbiome: ten principles of holobionts and hologenomes. PLoS Biol 2015;13(8). https://doi.org/10.1371/journal.pbio.1002226, e1002226.

[126] Eberl G. A new vision of immunity: homeostasis of the superorganism. Mucosal Immunol 2010;3(5):450–60. https://doi.org/10.1038/mi.2010.20.

[127] Amedei A, Boem F. I've gut a feeling: microbiota impacting the conceptual and experimental perspectives of personalized medicine. Int J Mol Sci 2018;19(12). https://doi.org/10.3390/ijms19123756.

[128] Suarez J, Stencel A. A part-dependent account of biological individuality: why holobionts are individuals and ecosystems simultaneously. Biol Rev Camb Philos Soc 2020;95(5):1308–24. https://doi.org/10.1111/brv.12610.

[129] Belkaid Y, Harrison OJ. Homeostatic immunity and the microbiota. Immunity 2017;46(4):562–76. https://doi.org/10.1016/j.immuni.2017.04.008.

[130] Deplancke B, Gaskins HR. Microbial modulation of innate defense: goblet cells and the intestinal mucus layer. Am J Clin Nutr 2001;73(6):1131S–41S. https://doi.org/10.1093/ajcn/73.6.1131S.

[131] Belkaid Y, Hand TW. Role of the microbiota in immunity and inflammation. Cell 2014;157(1):121–41. https://doi.org/10.1016/j.cell.2014.03.011.

[132] Schroeder BO, Backhed F. Signals from the gut microbiota to distant organs in physiology and disease. Nat Med 2016;22(10):1079–89. https://doi.org/10.1038/nm.4185.

[133] Johnson KV, Foster KR. Why does the microbiome affect behaviour? Nat Rev Microbiol 2018;16(10):647–55. https://doi.org/10.1038/s41579-018-0014-3.

[134] Nicholson JK, Holmes E, Kinross J, Burcelin R, Gibson G, Jia W, Pettersson S. Host-gut microbiota metabolic interactions. Science 2012;336(6086):1262–7. https://doi.org/10.1126/science.1223813.

[135] Varade J, Magadan S, Gonzalez-Fernandez A. Human immunology and immunotherapy: main achievements and challenges. Cell Mol Immunol 2020. https://doi.org/10.1038/s41423-020-00530-6.

[136] Fan Y, Pedersen O. Gut microbiota in human metabolic health and disease. Nat Rev Microbiol 2020. https://doi.org/10.1038/s41579-020-0433-9.

[137] Niccolai E, Boem F, Emmi G, Amedei A. The link "cancer and autoimmune diseases" in the light of microbiota: evidence of a potential culprit. Immunol Lett 2020;222:12–28. https://doi.org/10.1016/j.imlet.2020.03.001.

[138] Bik EM, Eckburg PB, Gill SR, Nelson KE, Purdom EA, Francois F, Relman DA. Molecular analysis of the bacterial microbiota in the human stomach. Proc Natl Acad Sci USA 2006;103(3):732–7. https://doi.org/10.1073/pnas.0506655103.

[139] Iizasa H, Ishihara S, Richardo T, Kanehiro Y, Yoshiyama H. Dysbiotic infection in the stomach. World J Gastroenterol 2015;21(40):11450–7. https://doi.org/10.3748/wjg.v21.i40.11450.

[140] Maldonado-Contreras A, Goldfarb KC, Godoy-Vitorino F, Karaoz U, Contreras M, Blaser MJ, Dominguez-Bello MG. Structure of the human gastric bacterial community in relation to helicobacter pylori status. ISME J 2011;5(4):574–9. https://doi.org/10.1038/ismej.2010.149.

[141] Fayle TM, Turner EC, Basset Y, Ewers RM, Reynolds G, Novotny V. Whole-ecosystem experimental manipulations of tropical forests. Trends Ecol Evol 2015;30(6):334–46. https://doi.org/10.1016/j.tree.2015.03.010.

CHAPTER ELEVEN

STING pathway and modulation for cancer immunotherapy

Ting Su[*], Nadia Tasnim Ahmed[*], Shurong Zhou[*], Xiang Liu[*], and Guizhi Zhu

Department of Pharmaceutics and Center for Pharmaceutical Engineering and Sciences, School of Pharmacy, Virginia Commonwealth University, Richmond, VA, United States

Contents

1. Introduction

Various pattern recognition receptors (PRRs) in the mammalian innate immune system serve as one of the first defense lines against pathogen infections by detecting pathogen-associated molecular patterns (PAMPs) and damaged-associated molecular patterns (DAMPs). These PRRs include toll-like receptors (TLRs), NOD-like receptors (NLRs), (RIG-1)-like receptors (RLRs), C-type lectin receptors (CLRs), as well as cyclic GMP–AMP synthase (cGAS)-stimulator of IFN genes (STING). While TLRs detect lipid-based ligands or nucleic acids in the forms of double-stranded RNA, single-stranded RNA, or unmethylated CpG, cGAS and STING are cytosolic nucleic acid sensors [1] that detect cytosolic dsDNA by cGAS and cyclic dinucleotides (CDNs) by STING. STING was first reported to be a critical mediator for the transcription of innate immune genes upon the stimulation of some invading bacteria, DNA viruses, or transfected DNA [2–4]. Encoded by *Tmem173*, STING is a protein with 379 amino acids in human cells and 378 amino acids in murine cells [2]. STING is

[*] These authors contributed equally.

Cancer Immunology and Immunotherapy
https://doi.org/10.1016/B978-0-12-823397-9.00011-9

expressed in various cell types, including T-cells, macrophages, and dendritic cells (DCs) that are central in the regulation of immune homeostasis. STING is located on the membrane of endoplasmic reticulum (ER) and consists of several functional domains [5]. The N-terminal domain of STING consists of several transmembrane helices that anchor STING to the ER [6–8]. The C-terminal domain of STING resides in cell cytosol as a V-shape dimer and consists of a subunit for binding of CDNs and phosphorylation site for TANK-binding kinase 1 (TBK1) that is essential for downstream signaling [8–10]. STING activation is finely regulated with other proteins to maintain homeostasis [11]. Ca^{2+} sensor stromal interaction molecule 1 (STIM1) is associated with STING to retain it in the ER membrane [12]. The deficiency in STIM1 can spontaneously activate the STING pathway and strongly enhance the expression of type I interferons after viral infection in both murine and human cells. Recently, a STING stabilizer protein-Tollip (toll-interacting protein) was identified to be a critical cofactor that can prevent STING degradation through direct interaction with STING [13]. It has been reported that the resting-state of STING protein is strictly regulated by a constant tug-of-war between "stabilizer" Tollip and "degrader" IRE1α lysosome. Tollip deficiency renders STING protein unstable in immune cells, leading to severely dampened STING signaling capacity.

2. cGAS–STING signaling pathway in cancer

2.1 STING activation and signaling

STING can be activated directly by natural CDNs, such as cyclic di-GMP(c-di-GMP) and cyclic di-AMP (c-di-AMP) that can be secreted by bacteria. Some small molecular STING agonists are shown in Table 1. Cytosolic DNA species (dsDNA or RNA: DNA hybrids) sensing by STING pathway requires additional proteins such as dsDNA-sensing cytosolic cGAS [14–16]. In the presence of dsDNA, cGAS uses intracellular ATP and GTP as substrates to catalyze the production of 2′3'-cyclic GMP–AMP (cGAMP) [15]. cGAMP can then bind with STING, leading to a conformational change and the polymerization of STING [6, 17]. This polymerization in ER results in the relocation of STING to the Golgi. Then, STING migrates to perinuclear Golgi to complete the palmitoylation process, which is essential for STING activation [18]. Activated STING then recruits TANK-binding kinase 1 (TBK1), which phosphorylates and activates the transcription factor interferon (IFN) regulatory factor 3 (IRF3) and nuclear factor-κB (NF-κB). These transcription factors then traffic into the nucleus to induce type I IFN-related gene transcription [19].

Besides driving innate immune responses via DNA sensing, STING also plays an important role in restricting RNA virus replication [4, 20–22]. It has been demonstrated that STING can be activated and its expression is upregulated during RNA virus infection [23]. Cells lacking STING are vulnerable to diverse RNA virus infections, and the replication of RNA

Table 1 Examples of small molecular STING agonists.

		Ref.
Nucleotidyl small molecular STING agonists	c-di-GMP	[73]
	3'3'-cGAMP	[74]
	2'3'-cGAMP	[75]
	ADU-S100 (ML RR-S2 CDA)	[76]
	Cyclic adenosine monophosphate-inosine monophosphates (cAIMPs)	[74]
Nonnucleotidyl small molecular STING agonists	Murine STING agonists	
	5,6-Dimethylxanthenone-4-acetic acid (DMXAA)	[77]
	Flavone-8-acetic acid (FAA)	[78]
	2,7-Bis(2-diethylamino ethoxy)fluoren-9-one (tilorone)	[79]
	10-Carboxymethyl-9-acridanone (CMA)	[80]
	Human STING agonists	
	DiABZIs	[81]
	Dispiro diketopiperzine (DSDP)	[82]
	Benzo[b][1, 4]thiazine-6-carboxamide (G10)	[83]
	a-Mangostin (19, a-MG)	[84]
	Benzamide (BNBC) and its analogues	[8]
	Bicyclic benzamides	[85]
	Benzothiophene derivatives	[86]

viruses is enhanced [2, 24, 25]. STING-mediated antiviral responses in RNA viruses might have distinct mechanisms compared with DNA viruses [26]. This process involves a membrane fusion that stimulates interferon production in a STING-dependent but cGAS-independent manner during influenza A virus infection [27]. Influenza A virus can trigger the leakage of mitochondrial DNA (mtDNA) translocate into the cytosol [28], and it has been found that intracellular mtDNA leakage into the cytosol also engages the cGAS–STING axis [29]. In addition, STING restricts RNA viruses in a transcription-independent manner and acts as an important restriction factor to prevent viral protein synthesis [30]. However, due to contrast observations in STING regulating RNA virus-mediated type I IFN and cytokine production, the role of STING in influencing host defense during RNA virus infection still awaits further investigation [25, 30, 31]. Also, alternative isoforms of STING are involved in counteracting RNA virus infection in different ways for type I IFN signaling pathway or RNA virus-induced antiviral signaling transduction [32].

Some synthetic biomaterials also have shown the ability to activate the cGAS–STING signaling pathway. A library of ultra-pH-sensitive copolymers with various different tertiary amines was tested as STING-activating nanovaccines for tumor immunotherapy, and one of the polymers, PC7A, with six-ring structure was identified as the lead for cancer immunotherapy [33]. In mice, the PC7A polymers activated antigen-presenting cells (APCs) in the draining lymph nodes and stimulated type I IFN production in a STING-dependent manner. In another example, an ionizable lipid-like material with ring structure also showed STING-mediated adjuvant effect, which promoted mRNA delivery and antitumor efficacy due to the enhanced antigen presentation [34]. Furthermore, cationic polymer chitosan promoted dendritic cell maturation and enhance cytotoxic T-lymphocytes responses through a type I IFN responses, which dramatically improved the antitumor immunity [35]. Mechanistically, the adjuvant-like ability of chitosan was from damaging mitochondria, followed by enhancing mitochondrial reactive oxygen species production and releasing mtDNA into the cytosol to activate cGAS [36]. When antigens with chitosan were loaded into microneedles, it elicited higher and more sustained specific antibody responses and greatly enhanced antigen immunogenicity [37].

Beyond eliciting type I IFN responses, STING also mediates IFN-independent functions. Mice defective for STING-mediated IFN response still resist against HSV-1 infection to the extent that is close to that in wild-type mice, suggesting the antiviral activity through an IFN-independent manner [38]. In addition to the induction of inflammatory cytokines, cGAMP-induced autophagy by STING pathway through a mechanism independent of TBK1 activation. This process is important for the clearance of DNA and viruses in the cytosol [39, 40]. Bacterial-induced cell–cell fusion leads to the activation of autophagy and corresponding cell death, which is independent of type I IFN production although through cGAS–STING signaling pathway [41]. STING activation can impact the maintenance of cell proliferation and genomic stability. STING depletion conferred a shorter doubling time compared with wild-type, implicating STING has cell-intrinsic functions in the regulation of the cell cycle [42]. In another example, upon STING activation, the reduction of proliferation led to the apoptosis of T-cells and B-cells [43–45].

With abundant studies on STING activation for innate immune response, aberrant STING activation due to usual cytosolic DNA or STING gain-of-function mutations may result in autoimmune diseases. For example, under normal conditions, self-DNA from necrotic or inappropriately apoptotic cells will be strictly regulated through distinct regulatory mechanisms, such as DNA turnover, subcellular DNA compartmentalization, and the existence of a dynamic threshold of cGAS and STING activation also helps to restrict the trigger for cGAS–STING pathway [46]. However, if the cells fail to distinguish between self and foreign nucleic acid or eliminate self-DNA wrongly, it may trigger aberrant immune responses inadvertently [47], which could be a critical factor in facilitating inflammatory disease [48]. For instance, acute and chronic *in vivo* mitochondrial stress leads to a STING-meditated type I interferon response in Parkinson's disease mouse model, which can be rescued by deletion of STING, suggesting that STING gets involved in the possible mechanism for inflammation-driven diseases [49]. Moreover, STING gain-of-function mutations cause autoimmunity and immunodeficiency in mice and STING-associated vasculopathy with onset in infancy (SAVI) in humans [50]. Some mutants can cause critical constitutive activation of STING and thereby induce an inflammatory disease [51]. Mouse models show that heterozygous STING N153S mice with SAVI-associated mutation will develop immunodeficiency [52], causing systemic inflammation, lung disease as well as T-cell cytopenia [53, 54]. Recent study showed that this mutation will also disrupt lymph node organogenesis and innate lymphoid cell development [55]. STING-N154S disrupts calcium homeostasis, which chronically activates ER stress and primes T-cell death [56].

2.2 STING activation in cancer immunotherapy

cGAS–STING pathway acts as essential immune surveillance mediators by mediating type-I interferon responses and the control of cellular inflammatory responses [57]. Inflammation is a double-edged sword in cancer development, and chronic inflammation can drive carcinogenesis and promotes tumors growth [58, 59]. Therefore, STING-induced chronic inflammation may aggravate inflammation-aggravated cancer [19]. On the other hand, strong inflammation can have antitumor activities [60]. Specifically, therapy-induced inflammation developed in response to various anticancer therapies can activate the host immune system, which can promote antitumor immune responses [58, 61]. cGAS–STING pathway may also modulate the tumor immune microenvironment by ameliorating tumor immunosuppression, which is a promising therapeutic strategy to improve cancer immunotherapy [62–64]. For example, colorectal cancer patients with higher STING expression have increased intratumoral CD8$^+$ T-cell infiltration and less frequent lymphovascular invasion in the early stage, both of which can promote antitumor immune responses and tumor therapeutic efficacy [65].

STING agonists such as cGAMP or CDNs have the potential as potent immuno-stimulant adjuvants for cancer immunotherapy. Currently, there have been a number of developmental small molecular STING agonists that bind to STING directly. Intratumoral administration of STING agonists can trigger type I interferon-driven inflammation, which can regulate the tumor microenvironment for stimulating dendritic cell activation and tumor antigen presentation to prime antitumor T-cells [66–69]. In melanoma mouse models, intratumoral injection of cGAMP can potently enhance CD8$^+$ T-responses in the tumor microenvironment in a STING-dependent manner [67]. Further investigation reveals that tumor endothelial cells, not dendritic cells, are the principal IFN-β producers in response to STING activation in tumors [67]. The systemic CD8$^+$ T-cell antitumor immunity can control the growth of both the treated and contralateral tumor. Intratumoral administration of CDNs, such as cGAMP, can also accumulate potent macrophages into tumor site via a STING-dependent mechanism [70, 71]. These macrophages are potent antitumor effector cells showing the production of TNFα and high expression levels of Cxcl10, Cxcl11, Ifnb1, and IFN-induced molecules. When combining low-dose checkpoint modulators together with CDG through intratumoral injection, approximately 50% of mice showed curative abscopal immunity in nontargeted organs, which greatly depends on the infiltration of systemic tumor-specific T-cells, reduction of the suppressive myeloid polarization, and downregulation of the M2 marker on tumor-associated macrophages [72].

Understanding the expression levels of STING may help to predict prognostic outcomes and the cancer therapeutic efficacy. Tumor cells actively inhibit STING to avoid the DNA-sensing machinery, which can effectively ablate downstream IFN signaling [87]. Moreover, many cancer cell lines such as human breast cancer cell lines (MCF-7, T47-D, and MDA-MB-231), human ovarian cancer lines (A1847, A2780, and ES2), and several cancerous melanoma cell lines (G361, MeWo, SK-MEL-5, SK-MEL-2, SK-MEL-28, and WM115) are shown to have defect STING expression compared with nontumorigenic cells [88, 89]. These defects cause the concomitant loss of cytosolic DNA signaling and cytokine production in cancer cells. For example, in patients with gastric cancer, STING expression is remarkably decreased in tumor tissues compared to nontumor tissues, which reduces patients' survival and is linked to a poor prognosis of gastric cancer [90]. For some tumor types, the suppressed STING pathway may render cancer cells more susceptible to oncolytic virus treatment [91, 92]. In SK-MEL-5 and SK-MEL-2 melanoma cell lines, low levels of STING are associated with increased sensitivity to oncolytic herpes simplex virus lysis treatments [93]. STING deficiency alters cytokine release induced by oncolytic virus infection compared normal tumor cells, such as less TNFα production but significantly increased interleukin 1 beta (IL-1β) release. Another study observed that silenced STING expression could be a major contributor to Merkel cell carcinoma immune escape; therefore, reintroducing STING functionality may be a viable path to alter this immune resistance [94]. Also, in nasopharyngeal carcinoma (NPC), STING levels in tumor tissues are lower than in adjacent nontumor tissues, correlated with the accumulation of myeloid-derived suppressor cell (MDSC) in the tumor site and poor prognosis of NPC patients [95].

Several other therapies have shown an impact on the STING signaling pathway to induce antitumor immunity during cancer therapies that directly target cancer cells [96, 97]. For example, ionizing radiation can induce tumor cell apoptosis, which functions as an effective immunogenic tumor vaccine and induce an effective antitumor response. STING protein, but not MyD88, is required in DCs for the IFN-β induction and type I IFN-dependent antitumor effects, which is essential for radiation-induced adaptive immune responses [98]. It has been shown that the mitochondrial DNA (mtDNA) triggering STING pathway is important for the efficient antitumor response based on the irradiated cancer cell vaccine, while TLR9 and IL-1β signaling are not [99]. Induced by ionizing radiation, cancer cells, rather than host cells, are the dominant producers to efficiently produce and export cGAMP. More extracellular cGAMP acts as a danger signal sensed by host STING then lead to infiltration of DCs and cytotoxic T-cell activation [100]. Several chemotherapeutic drugs may lead to genomic DNA damage and other cellular stresses that can induce the production of cytosolic DNA [101].

Chemotherapeutic drugs such as cisplatin and hydroxyurea are S-phase DNA damage drugs. When treating breast tumors deficient in DNA repair with these drugs, the induction of increased cytosolic DNA is associated with chemokine expression and a cause of inflammatory microenvironment. Importantly, STING knockdown reduces chemokine expression, reinforcing the importance of STING in the activation of immune response [102]. The same phenomenon was also observed in Topotecan (TPT)-treated tumors. TPT inhibits tumor growth by triggering the secretion of exosomes containing immuno-stimulatory DNA, which promotes DC maturation and $CD8^+$ T-cell activation in a STING-dependent antitumor immunity [103]. Poly(ADP-ribose) polymerase (PARP) inhibition drugs can suppress repair of DNA lesions and induce DNA damage response (DDR) [104]. These PARP inhibitors have been tested clinically and approved for the treatment of breast cancer and ovarian cancer. The mechanisms underlying the therapeutic effects of PARP inhibitors greatly rely on the generation of cytosolic dsDNA and micronuclei in tumor cells. Thus, STING pathway activation has shown the potential to elicit antitumor immune responses for tumor immunotherapy [105–107]. Cisplatin can increase both cGAS and STING protein levels in orthotopic ovarian tumors along with boosting antigen presentation and accumulation of tumor-infiltrating lymphocytes [108]. Chemotherapeutic drug paclitaxel treatment induces unstable nuclear membrane of micronuclei for recruitment of cGAS/STING activation, which activates type I IFN and TNFα to trigger a proapoptotic secretome induced apoptotic priming signals [109]. These new findings shift traditional anticancer drugs or therapies for a comprehensive understanding of immunostimulatory functions and profound developments for new applications.

3. Targeting the STING pathway for cancer immunotherapy

The cGAS–STING pathway can be activated by tumor-derived DNA, such as DNA from dead tumor cells, which can be micronuclei, cytoplasmic chromatin fragments, and free telomeric DNA, leading to induction of cell senescence, inflammation, and antitumor immunity followed by divergent effects on tumorigenesis [68, 110–115]. Structural studies have shown that upon binding of dsDNA, conformational changes persist on Cgas, which promotes its catalytic activity for cGAMP production. cGAMP binds to a small pocket of the STING dimer and promotes its translocation [15, 46, 75, 116, 117]. Through serial phosphorylation events of STING, TANK-binding kinase 1 (TBK1) and interferon regulatory transcription factor 3 (IRF3) are then recruited and activated from the ER via Golgi apparatus to perinuclear microsomes [118, 119]. In response to cytosolic dsDNA, STING also activates NF-κB in a TBK1-dependent manner, which in collaboration with IRF3 facilitates dsDNA-induced gene expression of type I IFNs [120, 121].

$CD8^+$ T-cells play a critical role in antitumor immunity that is elicited via the cross-presentation of tumor antigens by DCs. Type I IFN production by DCs is important for priming of CD8$^+$ T-cells against immunogenic tumors [122, 123]. In the tumor micro-environment, STING modulates IFN-β expression in DCs, which is correlated with the DC uptake of tumor-derived DNA [68]. CD8$^+$ T-cell activation and tumor rejection remain defective in STING deficient mice. Thus, host cGAS–STING pathway activation and type I IFN induction in DCs fosters cross-presentation of tumor antigens to trigger T-cells for tumor control. Activation of intrinsic cGAS–STING pathway by cytosolic DNA plays a role in cellular senescence, which is a prominent tumor suppression mechanism associated with loss of nuclear lamina protein lamin B1 and recognition of aberrant cytosolic chromatin fragments by cGAS [112, 114]. Consequently, production of type I IFNs and senescence-associated secretory phenotype (SASP) factors is facilitated by the cGAS–STING pathway to promote senescence [110, 114].

3.1 STING activation for combination therapy

STING agonists can work as adjuvants in combination therapy using chemotherapeutic drugs and radiation. Genotoxic effects caused by radiation and chemotherapeutic agents induce the formation of micronuclei and cytoplasmic chromatin fragments and therefore activate cGAS–STING pathway (Fig. 1) [98, 110–112, 114, 115]. Consistently, the anti-tumor effects of radiation are promoted by synergistic radiation therapy and intratumoral administration of cGAMP with enhanced T-cell responses in mice in a STING-dependent manner [98]. Combination treatment of fluorouracil (5-FU) along with cGAMP not only improves the antitumor activity of 5-FU but also reduces the tissue toxicity associated with it [124]. In a similar manner, combination treatment of cyclic di-guanylate (c-di-GMP) with a Listeria monocytogenes-based vaccine containing the tumor-associated antigen MAGE-b (Listeria-Mage-b) eradicated metastases and halted tumor growth in a metastatic breast cancer model [125]. Besides combination with chemotherapy or radiotherapy, STING agonists can enhance the tumor therapeutic efficacy of immune checkpoint blockers. Immune checkpoint pathways help tumor cells to escape the host immune recognition. Cancer cells that express immune checkpoint PD-L1 inhibit T-cell functions via binding with the receptor PD-1 on T-cells [126]. CTLA-4 is another immune checkpoint that induces immunological tolerance toward tumor cells [127]. Various immune checkpoint blockers, such as anti-PD-1, anti-PD-L1, and anti-CTLA-4 antibodies, are used to promote antitumor response by blocking immune checkpoints. However, immune checkpoint blockade treatment has suboptimal therapeutic efficacy in "cold" tumors, which are minimally infiltrated by immune cells, as shown in the majority of cancer patients. To address this issue, cancer vaccines such as STINGVAX that is formulated based on granulocyte–macrophage colony-stimulating factor (GM-CSF) with synthetic or bacterial CDNs, have been studied to promote tumor infiltration of CD8$^+$ T-cells in the tumor microenvironment,

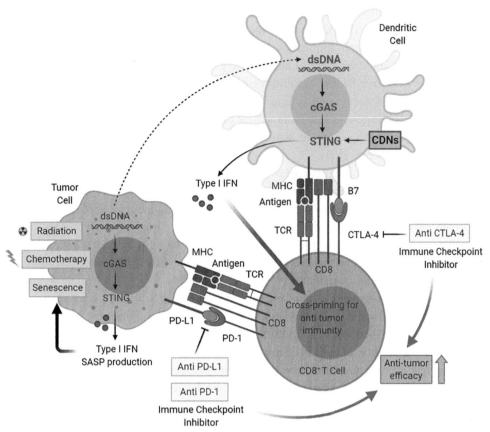

Fig. 1 cGAS–STING pathway in tumor suppression and combination tumor immunotherapy. Cytoplasmic chromatin DNA initiates the cGAS–STING pathway for type I IFN responses and senescence-associated secretory phenotype (SASP) production inducing cellular senescence. Uptake of exogenous DNA by DCs may activate cGAS–STING pathway and produce type I IFNs, which promotes antitumor T-cell responses. Immune checkpoints such as PD-L1, PD-1, and CTLA-4 are negatively associated with antitumor T-cell responses. Ongoing studies have been exploring the therapeutic efficacy of the combination of STING agonists and immune checkpoint inhibitors, radiation therapy, and chemotherapy. *(Figure was created with Biorender.com.)*

turning "cold" tumors into "hot" ones, whereby promoting the therapeutic efficacy of immune checkpoint blockers [128].

4. Drug delivery systems for STING agonists

Although STING agonists demonstrated potent antitumor effects, in current clinical testing, the administration of STING agonists is often limited to intratumoral routes in part due to by their intrinsic properties and the resulting poor distant delivery efficiency. The ability to deliver STING agonists to distant tissues (*e.g.*, tumor, lymphoid tissues) can broaden their application in cancer immunotherapy. First, small molecule STING agonists are vulnerable to enzyme degradation [129] in the physiological environment. Second, free STING agonists displayed limited uptake and retention in distant tissues due to rapid dissemination into the blood [130]. Development of drug delivery systems for STING agonists can address these problems by protecting them from enzyme degradation, prolonging their circulation time in the body, and improving the intracellular delivery efficiency [131].

Nanoparticles and microparticles are under investigation for distant delivery of STING agonists. Nanoparticles with diameters less than 200 nm are typically considered relatively easy to be drained to lymph nodes [132]. Larger–size nanoparticles are drained relatively slowly to lymph nodes but have prolonged retention in lymph nodes. Microparticles tend to remain in the injection sites [33]. Particles with different sizes can be used in different applications via various injection strategies to fulfill the specific requirements in anticancer immune drug delivery.

For example, polymeric vectors and lipid nanoparticles have been studied as delivery systems for STING agonists [133]. Cationic polymers have long been used in nucleic acid delivery because of their low immunogenicity, stability in circulation, and feasibility of multiple functional modifications [134]. Similar to nucleic acid, STING agonists such as CDNs are negatively charged and need to be released into cytoplasm, and therefore, cationic polymers have been studied to improve CDNs cytosolic delivery. Worth to mention, these cationic vectors often have short half-lives in physiological conditions and will be hydrolytically degraded, which reduces their cytotoxicity caused by the cationic charges [135]. For instance, CDNs encapsulated in cationic polymer poly(beta–amino ester) (PBAEs) showed a significantly higher cellular uptake in comparison to free CDNs, resulting in robust immune response at much lower drug concentration. When combined with immune checkpoint inhibitors, intratumoral injection of PBAE nanoparticles remarkably increased the antitumor activity [136]. Most recently, synthetic poly(ethylene glycol)-*block*-[(2-(diethylamino)ethylmethacrylate)-*co*-(butylmethacrylate)-*co*-(pyridyl disulfide ethyl methacrylate)] (PEG-DBP) copolymers not only enhanced the CDNs encapsulation efficiency when formulating but also enabled the release of the STING agonists in endosomal pH environment, compared to the polymers without crosslinking disulfide bonds (Fig. 2). Furthermore, these nanoparticle systems are able to improve the therapeutic efficacy of CDNs by triggering type I IFN responses and enhancing tumoral immunogenicity via both intratumoral and intravenous injection, which are promising for the treatment of nonaccessible tumors using STING agonists [137].

In addition to particular drug delivery systems, macromaterials, such as hydrogels have also been investigated for the delivery of STING agonists in cancer immunotherapy. One example is submicron–sized linear poly-ethyleneimine (LPEI)/hyaluronic acid (HA)

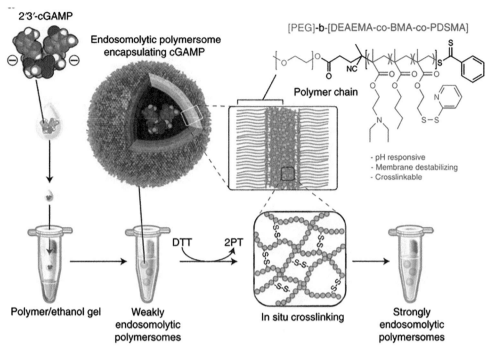

Fig. 2 Schematic illustration of endosomolytic polymersomes production using PEG-DBP copolymers to formulate 2′3′-cGAMP. *(Reprinted from Louttit C, Park KS, Moon JJ. Bioinspired nucleic acid structures for immune modulation. Biomaterials 2019. https://doi.org/10.1016/j.biomaterials.2019.119287.)*

hydrogels (Fig. 3) [138]. The cell uptake tracking of the HA hydrogel showed 0.5 μm particles targeted phagocytic macrophages to effectively induce cytokine at low concentrations of STING agonists while alleviating off-target side effects.

In another study, the pendant hydroxyl group in homopolysaccharide is converted into an acetal group to form acetalated dextran (Ace-DEX). With diameters around 1.5 μm, these particles trafficked to lymph nodes along with antigen-presenting cells without subsequent systemic dissemination, which enables the formulation to induce long-term immunity at a 100-fold lower concentration of unformulated CDNs [139]. Owing to the anionic proteins in the interstitial fluid, negative surface of the microparticles also facilitates lymph node draining via charge repulsion effect. Besides, the formulation can be processed in relatively large batches using electrospray with high CDN encapsulation efficiency as well as greater monodispersity [140]. Controlled release is another desired property of drug delivery system. Lasting release of the encapsulated drug results in drug accumulation in tumor site, which could overcome drug resistance [141]. Besides, prolonged drug release will protect patients from multiple injections, which can cause long-period injection pains and poor adherence [142]. A type of STINGel based on

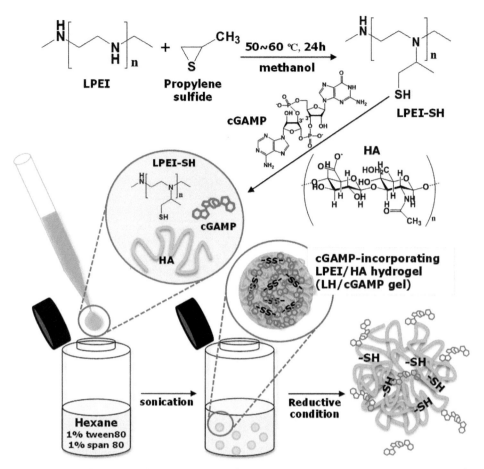

Fig. 3 Schematic illustration of the preparation of cGAMP-loaded LPEI/HA hydrogels. *(Reprinted from Lee E, Jang H-E, Kang YY, Kim J, Ahn J-H, Mok H. Submicron-sized hydrogels incorporating cyclic dinucleotides for selective delivery and elevated cytokine release in macrophages. Acta Biomater 2016;29:271–281. https://doi.org/10.1016/j.actbio.2015.10.025.)*

electrostatic interactions between negative CDNs and positive peptide nanofibers demonstrated significantly prolonged drug release time in comparison to the collagen gels [143]. More recently, cubic polylactic-*co*-glycolic acid (PLGA) microparticles with CDNs loaded inside the enclosed cavity have extended the CDN release up to 2 weeks. A single injection of these microparticles demonstrated comparable antitumor efficacy relative to multiple injections of free CDNs in various tumor models [142]. This system could be optimized by reducing particle wall thickness to enhance drug loading efficiency as well as by increasing particle layers to realize in pulses release.

Lipid nanoparticles are another widely used drug delivery system for nucleic acids [144]. Owing to their amphiphilicity nature, lipids spontaneously form bilayers when

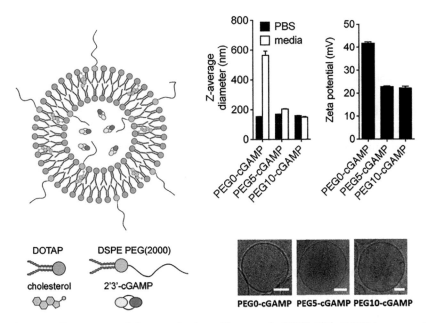

Fig. 4 Schematic illustration and characterization of liposomal cGAMP. 2'3'-cGAMP is encapsulated in cationic liposome, which is synthesized from 1,2-dioleoyl-3-trimethylammonium-propane (DOTAP) and cholesterol. The liposome is partially modified by PEG-containing lipid [1,2-distearoyl-sn-gly-cero-3-phosphoethanolamine-N-[methoxy(polyethylene glycol)-2000] (DSPE-PEG(2000))] (Scale bar: 50 nm). *(Reprinted and reorganized from Koshy ST, Cheung AS, Gu L, Graveline AR, Mooney DJ. Liposomal delivery enhances immune activation by STING agonists for cancer immunotherapy. Adv Biosyst 2017;1 (1–2). https://doi.org/10.1002/adbi.201600013.)*

dispersing in an aqueous medium to create an interior aqueous core that can encapsulate hydrophilic molecules such as CDNs (Fig. 4). Liposomes loaded with STING agonists have been studied to be administrated by inhalation. Inhaled phosphatidylserine coated liposomes encapsulating cGAMP were rapidly disseminated in the lung and endocytosed by antigen-presenting cells. When synergized with radiotherapy, such system induced robust anticancer immunity and greatly prolonged the survival times of mice with lung metastases and re-challenged tumors [62]. Subcutaneous injection of liposomes can also lead to their uptake by antigen-presenting cells in draining lymph nodes. For instance, subcutaneous injection of CDNs encapsulated in 150 nm partially PEGylated phospha-tidylcholine liposomes enabled much more STING agonists accumulation in lymph nodes than free CDNs [130].

Previously used lipids are permanently positively charged and these cationic lipids are used to load anionic nucleic acid via electrostatic interaction for nucleic acid drug delivery, but these cationic polymers might be associated with long-term toxicity [145, 146]. To address this issue, ionizable lipids are developed with optimal pKa values that allow

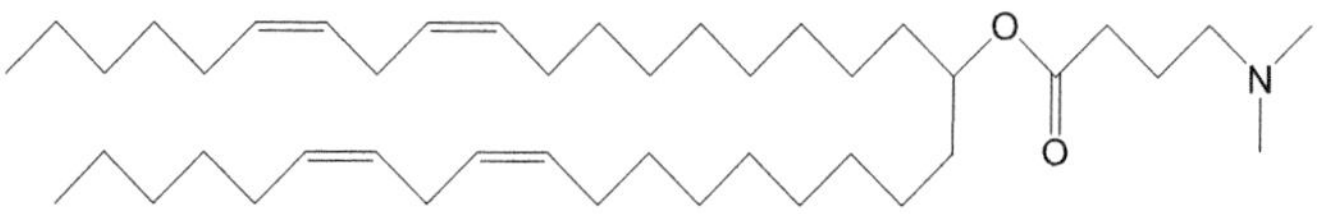

Fig. 5 Chemical structure of ionizable lipid ᴅ-Lin-MC3-DMA [145].

these lipids to be conditionally positively charged only in acidic environments such as that in the endolysosome. Ionizable lipids are composed of three parts: the amine head group, the linker group, and the hydrophobic tails (Fig. 5). The lipids are positively charged in acidic environment such as endosomal environment (pH < pKa), which enable them to disrupt endosomal membrane to release the loaded drug. The electrostatic charge of these lipids will be converted to be neutral at pH 7.4 (pH > pKa), which reduce their toxicity relative to cationic lipids under the physiological environment [73].

5. Summary and outlook

The cGAS–STING signaling pathway is involved in many types of cancer, infection diseases, autoimmune diseases, as well as senescence. Cancer immunotherapy via the modulation of the cGAS–STING pathway has been explored by, for example, using STING agonists. Based on the promising preclinical evidence for these therapeutic strategies thus far, it is critical to consolidate clinically translatable therapeutics as well as their formulations, including drug delivery systems, given the unique physicochemical properties of some of these therapeutics such as CDNs.

References

[1] Liu Z, Han C, Fu Y-X. Targeting innate sensing in the tumor microenvironment to improve immunotherapy. Cell Mol Immunol 2020;17(1):13–26. https://doi.org/10.1038/s41423-019-0341-y.
[2] Ishikawa H, Barber GN. STING is an endoplasmic reticulum adaptor that facilitates innate immune signalling. Nature 2008;455(7213):674–8. https://doi.org/10.1038/nature07317.
[3] Sun W, Li Y, Chen L, et al. ERIS, an endoplasmic reticulum IFN stimulator, activates innate immune signaling through dimerization. Proc Natl Acad Sci 2009;106(21):8653–8. https://doi.org/10.1073/pnas.0900850106.
[4] Zhong B, Yang Y, Li S, et al. The adaptor protein MITA links virus-sensing receptors to IRF3 transcription factor activation. Immunity 2008;29(4):538–50. https://doi.org/10.1016/j.immuni.2008.09.003.
[5] de Oliveira Mann CC, Orzalli MH, King DS, Kagan JC, Lee ASY, Kranzusch PJ. Modular architecture of the STING C-terminal tail allows interferon and NF-κB signaling adaptation. Cell Rep 2019;27(4):1165–1175.e5. https://doi.org/10.1016/j.celrep.2019.03.098.
[6] Shang G, Zhang C, Chen ZJ, Bai X-C, Zhang X. Cryo-EM structures of STING reveal its mechanism of activation by cyclic GMP-AMP. Nature 2019;567(7748):389–93. https://doi.org/10.1038/s41586-019-0998-5.
[7] Yin Q, Tian Y, Kabaleeswaran V, et al. Cyclic di-GMP sensing via the innate immune signaling protein STING. Mol Cell 2012;46(6):735–45. https://doi.org/10.1016/j.molcel.2012.05.029.
[8] Zhang C, Shang G, Gui X, Zhang X, Bai X-C, Chen ZJ. Structural basis of STING binding with and phosphorylation by TBK1. Nature 2019;567(7748):394–8. https://doi.org/10.1038/s41586-019-1000-2.

[9] Zhao B, Du F, Xu P, et al. A conserved PLPLRT/SD motif of STING mediates the recruitment and activation of TBK1. Nature 2019. https://doi.org/10.1038/s41586-019-1228-x.

[10] Shu C, Yi G, Watts T, Kao CC, Li P. Structure of STING bound to cyclic di-GMP reveals the mechanism of cyclic dinucleotide recognition by the immune system. Nat Struct Mol Biol 2012;19 (7):722–4. https://doi.org/10.1038/nsmb.2331.

[11] Landman SL, Ressing ME, van der Veen AG. Balancing STING in antimicrobial defense and autoinflammation. Cytokine Growth Factor Rev 2020;55:1–14. https://doi.org/10.1016/j. cytogfr.2020.06.004.

[12] Srikanth S, Woo JS, Wu B, et al. The Ca2+ sensor STIM1 regulates the type I interferon response by retaining the signaling adaptor STING at the endoplasmic reticulum. Nat Immunol 2019;20 (2):152–62. https://doi.org/10.1038/s41590-018-0287-8.

[13] Pokatayev V, Yang K, Tu X, et al. Homeostatic regulation of STING protein at the resting state by stabilizer TOLLIP. Nat Immunol 2020;21(2):158–67. https://doi.org/10.1038/s41590-019-0569-9.

[14] Sun L, Wu J, Du F, Chen X, Chen ZJ. Cyclic GMP-AMP synthase is a cytosolic DNA sensor that activates the type-I interferon pathway. Science 2013;339(6121). https://doi.org/10.1126/ science.1232458.

[15] Wu J, Sun L, Chen X, et al. Cyclic GMP-AMP is an endogenous second messenger in innate immune signaling by cytosolic DNA. Science 2013;339(6121):826–30. https://doi.org/10.1126/ science.1229963.

[16] Mankan AK, Schmidt T, Chauhan D, et al. Cytosolic RNA:DNA hybrids activate the cGAS–STING axis. EMBO J 2014;33(24):2937–46. https://doi.org/10.15252/embj.201488726.

[17] Ergun SL, Fernandez D, Weiss TM, Li L. STING polymer structure reveals mechanisms for activation, hyperactivation, and inhibition. Cell 2019;178(2):290–301.e10. https://doi.org/10.1016/j. cell.2019.05.036.

[18] Mukai K, Konno H, Akiba T, et al. Activation of STING requires palmitoylation at the Golgi. Nat Commun 2016;7(1):11932. https://doi.org/10.1038/ncomms11932.

[19] Barber GN. STING: infection, inflammation and cancer. Nat Rev Immunol 2015;15(12):760–70. https://doi.org/10.1038/nri3921.

[20] Burdette DL, Vance RE. STING and the innate immune response to nucleic acids in the cytosol. Nat Immunol 2013;14(1):19–26. https://doi.org/10.1038/ni.2491.

[21] Lee J, Ghonime MG, Wang R, Cassady KA. The antiviral apparatus: STING and oncolytic virus restriction. Mol Ther Oncolytics 2019;13:7–13. https://doi.org/10.1016/j.omto.2019.02.002.

[22] Ahn J, Barber GN. STING signaling and host defense against microbial infection. Exp Mol Med 2019;51(12):1–10. https://doi.org/10.1038/s12276-019-0333-0.

[23] Maringer K, Fernandez-Sesma A. Message in a bottle: lessons learned from antagonism of STING signalling during RNA virus infection. Cytokine Growth Factor Rev 2014;25(6):669–79. https:// doi.org/10.1016/j.cytogfr.2014.08.004.

[24] Wuertz KM, Treuting PM, Hemann EA, et al. STING is required for host defense against neuropathological West Nile virus infection. PLoS Pathog 2019;15(8). https://doi.org/10.1371/journal. ppat.1007899, e1007899.

[25] Ishikawa H, Ma Z, Barber GN. STING regulates intracellular DNA-mediated, type I interferon-dependent innate immunity. Nature 2009;461(7265):788–92. https://doi.org/10.1038/nature08476.

[26] Ni G, Ma Z, Damania B. cGAS and STING: at the intersection of DNA and RNA virus-sensing networks. PLoS Pathog 2018;14(8). https://doi.org/10.1371/journal.ppat.1007148, e1007148.

[27] Holm CK, Rahbek SH, Gad HH, et al. Influenza A virus targets a cGAS-independent STING pathway that controls enveloped RNA viruses. Nat Commun 2016;7. https://doi.org/10.1038/ ncomms10680.

[28] Moriyama M, Koshiba T, Ichinohe T. Influenza A virus M2 protein triggers mitochondrial DNA-mediated antiviral immune responses. Nat Commun 2019;10(1):4624. https://doi.org/10.1038/ s41467-019-12632-5.

[29] West AP, Shadel GS. Mitochondrial DNA in innate immune responses and inflammatory pathology. Nat Rev Immunol 2017;17(6):363–75. https://doi.org/10.1038/nri.2017.21.

[30] Franz KM, Neidermyer WJ, Tan Y-J, Whelan SPJ, Kagan JC. STING-dependent translation inhibition restricts RNA virus replication. Proc Natl Acad Sci U S A 2018;115(9):E2058–67. https:// doi.org/10.1073/pnas.1716937115.

[31] Schoggins JW, MacDuff DA, Imanaka N, et al. Pan-viral specificity of IFN-induced genes reveals new roles for cGAS in innate immunity. Nature 2014;505(7485):691–5. https://doi.org/10.1038/nature12862.

[32] Xu L, Yu D, Peng L, et al. An alternative splicing of tupaia STING modulated anti-RNA virus responses by targeting MDA5-LGP2 and IRF3. J Immunol 2020;204(12):3191–204. https://doi.org/10.4049/jimmunol.1901320.

[33] Luo M, Wang H, Wang Z, et al. A STING-activating nanovaccine for cancer immunotherapy. Nat Nanotechnol 2017;12(7):648–54. https://doi.org/10.1038/nnano.2017.52.

[34] Miao L, Li L, Huang Y, et al. Delivery of mRNA vaccines with heterocyclic lipids increases antitumor efficacy by STING-mediated immune cell activation. Nat Biotechnol 2019;37(10):1174–85. https://doi.org/10.1038/s41587-019-0247-3.

[35] Shi G-N, Zhang C-N, Xu R, et al. Enhanced antitumor immunity by targeting dendritic cells with tumor cell lysate-loaded chitosan nanoparticles vaccine. Biomaterials 2017;113:191–202. https://doi.org/10.1016/j.biomaterials.2016.10.047.

[36] Carroll EC, Jin L, Mori A, et al. The vaccine adjuvant chitosan promotes cellular immunity via DNA sensor cGAS-STING-dependent induction of type I interferons. Immunity 2016;44(3):597–608. https://doi.org/10.1016/j.immuni.2016.02.004.

[37] Mc C, Ky L, Mh L, Cw L. Enhancing immunogenicity of antigens through sustained intradermal delivery using chitosan microneedles with a patch-dissolvable design. Acta Biomater 2017;65:66–75. https://doi.org/10.1016/j.actbio.2017.11.004.

[38] Wu J, Dobbs N, Yang K, Yan N. Interferon-independent activities of mammalian STING mediate antiviral response and tumor immune evasion. Immunity 2020;53(1):115–126.e5. https://doi.org/10.1016/j.immuni.2020.06.009.

[39] Gui X, Yang H, Li T, et al. Autophagy induction via STING trafficking is a primordial function of the cGAS pathway. Nature 2019;567(7747):262–6. https://doi.org/10.1038/s41586-019-1006-9.

[40] Watson RO, Manzanillo PS, Cox JS. Extracellular M. tuberculosis DNA targets bacteria for autophagy by activating the host DNA-sensing pathway. Cell 2012;150(4):803–15. https://doi.org/10.1016/j.cell.2012.06.040.

[41] Ku JWK, Chen Y, Lim BJW, Gasser S, Crasta KC, Gan Y-H. Bacterial-induced cell fusion is a danger signal triggering cGAS-STING pathway via micronuclei formation. Proc Natl Acad Sci U S A 2020;117(27):15923–34. https://doi.org/10.1073/pnas.2006908117.

[42] Ranoa DRE, Widau RC, Mallon S, et al. STING promotes homeostasis via regulation of cell proliferation and chromosomal stability. Cancer Res 2019;79(7):1465–79. https://doi.org/10.1158/0008-5472.CAN-18-1972.

[43] Tang C-HA, Zundell JA, Ranatunga S, et al. Agonist-mediated activation of STING induces apoptosis in malignant B cells. Cancer Res 2016;76(8):2137–52. https://doi.org/10.1158/0008-5472.CAN-15-1885.

[44] Gulen MF, Koch U, Haag SM, et al. Signalling strength determines proapoptotic functions of STING. Nat Commun 2017;8(1):427. https://doi.org/10.1038/s41467-017-00573-w.

[45] Cerboni S, Jeremiah N, Gentili M, et al. Intrinsic antiproliferative activity of the innate sensor STING in T lymphocytes. J Exp Med 2017;214(6):1769–85. https://doi.org/10.1084/jem.20161674.

[46] Ablasser A, Goldeck M, Cavlar T, et al. cGAS produces a 2′-5′-linked cyclic dinucleotide second messenger that activates STING. Nature 2013;498(7454):380–4. https://doi.org/10.1038/nature12306.

[47] Gao D, Li T, Li X-D, et al. Activation of cyclic GMP-AMP synthase by self-DNA causes autoimmune diseases. Proc Natl Acad Sci U S A 2015;112(42):E5699–705. https://doi.org/10.1073/pnas.1516465112.

[48] Motwani M, Pesiridis S, Fitzgerald KA. DNA sensing by the cGAS-STING pathway in health and disease. Nat Rev Genet 2019;20(11):657–74. https://doi.org/10.1038/s41576-019-0151-1.

[49] Sliter DA, Martinez J, Hao L, et al. Parkin and PINK1 mitigate STING-induced inflammation. Nature 2018;561(7722):258–62. https://doi.org/10.1038/s41586-018-0448-9.

[50] Liu Y, Jesus AA, Marrero B, et al. Activated STING in a vascular and pulmonary syndrome. N Engl J Med 2014;371(6):507–18. https://doi.org/10.1056/NEJMoa1312625.

[51] Jeremiah N, Neven B, Gentili M, et al. Inherited STING-activating mutation underlies a familial inflammatory syndrome with lupus-like manifestations. J Clin Invest 2014;124(12):5516–20. https://doi.org/10.1172/JCI79100.

[52] Bennion BG, Ingle H, Ai TL, et al. A human gain-of-function STING mutation causes immunodeficiency and gammaherpesvirus-induced pulmonary fibrosis in mice. J Virol 2019;93(4). https://doi.org/10.1128/JVI.01806-18.

[53] Luksch H, Stinson WA, Platt DJ, et al. STING-associated lung disease in mice relies on T cells but not type I interferon. J Allergy Clin Immunol 2019;144(1):254–266.e8. https://doi.org/10.1016/j.jaci.2019.01.044.

[54] Warner JD, Irizarry-Caro RA, Bennion BG, et al. STING-associated vasculopathy develops independently of IRF3 in mice. J Exp Med 2017;214(11):3279–92. https://doi.org/10.1084/jem.20171351.

[55] Bennion BG, Croft CA, Ai TL, et al. STING gain-of-function disrupts lymph node organogenesis and innate lymphoid cell development in mice. Cell Rep 2020;31(11):107771. https://doi.org/10.1016/j.celrep.2020.107771.

[56] Wu J, Chen Y-J, Dobbs N, et al. STING-mediated disruption of calcium homeostasis chronically activates ER stress and primes T cell death. J Exp Med 2019;216(4):867–83. https://doi.org/10.1084/jem.20182192.

[57] Wan D, Jiang W, Hao J. Research advances in how the cGAS-STING pathway controls the cellular inflammatory response. Front Immunol 2020;11. https://doi.org/10.3389/fimmu.2020.00615.

[58] Greten FR, Grivennikov SI. Inflammation and cancer: triggers, mechanisms, and consequences. Immunity 2019;51(1):27–41. https://doi.org/10.1016/j.immuni.2019.06.025.

[59] Aller M-A, Arias A, Arias J-I, Arias J. Carcinogenesis: the cancer cell–mast cell connection. Inflamm Res 2019;68(2):103–16. https://doi.org/10.1007/s00011-018-1201-4.

[60] Abdolvahab MH, Darvishi B, Zarei M, Majidzadeh-A K, Farahmand L. Interferons: role in cancer therapy. Immunotherapy 2020;12(11):833–55. https://doi.org/10.2217/imt-2019-0217.

[61] Shalapour S, Lin X-J, Bastian IN, et al. Inflammation-induced IgA+ cells dismantle anti-liver cancer immunity. Nature 2017;551(7680):340–5. https://doi.org/10.1038/nature24302.

[62] Liu Y, Crowe WN, Wang L, et al. An inhalable nanoparticulate STING agonist synergizes with radiotherapy to confer long-term control of lung metastases. Nat Commun 2019;10(1):5108. https://doi.org/10.1038/s41467-019-13094-5.

[63] Jing W, McAllister D, Vonderhaar EP, et al. STING agonist inflames the pancreatic cancer immune microenvironment and reduces tumor burden in mouse models. J Immunother Cancer 2019;7(1):115. https://doi.org/10.1186/s40425-019-0573-5.

[64] Harabuchi S, Kosaka A, Yajima Y, et al. Intratumoral STING activations overcome negative impact of cisplatin on antitumor immunity by inflaming tumor microenvironment in squamous cell carcinoma. Biochem Biophys Res Commun 2020;522(2):408–14. https://doi.org/10.1016/j.bbrc.2019.11.107.

[65] Chon HJ, Kim H, Noh JH, et al. STING signaling is a potential immunotherapeutic target in colorectal cancer. J Cancer 2019;10(20):4932–8. https://doi.org/10.7150/jca.32806.

[66] Corrales L, Glickman LH, McWhirter SM, et al. Direct activation of STING in the tumor microenvironment leads to potent and systemic tumor regression and immunity. Cell Rep 2015;11(7):1018–30. https://doi.org/10.1016/j.celrep.2015.04.031.

[67] Demaria O, Gassart AD, Coso S, et al. STING activation of tumor endothelial cells initiates spontaneous and therapeutic antitumor immunity. Proc Natl Acad Sci 2015;112(50):15408–13. https://doi.org/10.1073/pnas.1512832112.

[68] Woo S-R, Fuertes MB, Corrales L, et al. STING-dependent cytosolic DNA sensing mediates innate immune recognition of immunogenic tumors. Immunity 2014;41(5):830–42. https://doi.org/10.1016/j.immuni.2014.10.017.

[69] Wang X, Cao R, Zhang H, et al. The anti-influenza virus drug, arbidol is an efficient inhibitor of SARS-CoV-2 in vitro. Cell Discov 2020;6(1):1–5. https://doi.org/10.1038/s41421-020-0169-8.

[70] Ohkuri T, Kosaka A, Ishibashi K, et al. Intratumoral administration of cGAMP transiently accumulates potent macrophages for antitumor immunity at a mouse tumor site. Cancer Immunol Immunother 2017;66(6):705–16. https://doi.org/10.1007/s00262-017-1975-1.

[71] Cheng N, Watkins-Schulz R, Junkins RD, et al. A nanoparticle-incorporated STING activator enhances antitumor immunity in PD-L1-insensitive models of triple-negative breast cancer. JCI Insight 2018;3(22). https://doi.org/10.1172/jci.insight.120638.

[72] Ager CR, Reilley MJ, Nicholas C, Bartkowiak T, Jaiswal AR, Curran MA. Intratumoral STING activation with T-cell checkpoint modulation generates systemic antitumor immunity. Cancer Immunol Res 2017;5(8):676–84. https://doi.org/10.1158/2326-6066.CIR-17-0049.

[73] Burdette DL, Monroe KM, Sotelo-Troha K, et al. STING is a direct innate immune sensor of cyclic-di-GMP. Nature 2011;478(7370):515–8. https://doi.org/10.1038/nature10429.

[74] Lioux T, Mauny M-A, Lamoureux A, et al. Design, synthesis, and biological evaluation of novel cyclic adenosine-inosine monophosphate (cAIMP) analogs that activate stimulator of interferon genes (STING). J Med Chem 2016;59(22):10253–67. https://doi.org/10.1021/acs.jmedchem.6b01300.

[75] Zhang X, Shi H, Wu J, et al. Cyclic GMP-AMP containing mixed phosphodiester linkages is an endogenous high-affinity ligand for STING. Mol Cell 2013;51(2):226–35. https://doi.org/10.1016/j.molcel.2013.05.022.

[76] Gajewski TF. The next hurdle in cancer immunotherapy: overcoming the non-T cell-inflamed tumor microenvironment. Semin Oncol 2015;42(4):663–71. https://doi.org/10.1053/j.seminoncol.2015.05.011.

[77] Prantner D, Perkins DJ, Lai W, et al. 5,6-Dimethylxanthenone-4-acetic acid (DMXAA) activates stimulator of interferon gene (STING)-dependent innate immune pathways and is regulated by mitochondrial membrane potential. J Biol Chem 2012;287(47):39776–88. https://doi.org/10.1074/jbc.M112.382986.

[78] Gong J, Chehrazi-Raffle A, Reddi S, Salgia R. Development of PD-1 and PD-L1 inhibitors as a form of cancer immunotherapy: a comprehensive review of registration trials and future considerations. J Immunother Cancer 2018;6(1):8. https://doi.org/10.1186/s40425-018-0316-z.

[79] Tang J, Yu JX, Hubbard-Lucey VM, Neftelinov ST, Hodge JP, Lin Y. Trial watch: the clinical trial landscape for PD1/PDL1 immune checkpoint inhibitors. Nat Rev Drug Discov 2018;17(12):854–5. https://doi.org/10.1038/nrd.2018.210.

[80] Cavlar T, Deimling T, Ablasser A, Hopfner K-P, Hornung V. Species-specific detection of the antiviral small-molecule compound CMA by STING. EMBO J 2013;32(10):1440–50. https://doi.org/10.1038/emboj.2013.86.

[81] Ramanjulu JM, Pesiridis GS, Yang J, et al. Design of amidobenzimidazole STING receptor agonists with systemic activity. Nature 2018;564(7736):439–43. https://doi.org/10.1038/s41586-018-0705-y.

[82] Liu B, Tang L, Zhang X, et al. A cell-based high throughput screening assay for the discovery of cGAS-STING pathway agonists. Antivir Res 2017;147:37–46. https://doi.org/10.1016/j.antiviral.2017.10.001.

[83] Sali TM, Pryke KM, Abraham J, et al. Characterization of a novel human-specific STING agonist that elicits antiviral activity against emerging alphaviruses. PLoS Pathog 2015;11(12). https://doi.org/10.1371/journal.ppat.1005324, e1005324.

[84] Zhang Y, Sun Z, Pei J, et al. Identification of α-mangostin as an agonist of human STING. ChemMedChem 2018;13(19):2057–64. https://doi.org/10.1002/cmdc.201800481.

[85] Banerjee M, Middya S, Basu S, et al. Small molecule modulators of human sting. Published online December 27, 2018. Accessed January 22, 2021 https://patentscope.wipo.int/search/en/detail.jsf?docId=WO2018234808; 2018.

[86] Andreeva L, Hiller B, Kostrewa D, et al. cGAS senses long and HMGB/TFAM-bound U-turn DNA by forming protein-DNA ladders. Nature 2017;549(7672):394–8. https://doi.org/10.1038/nature23890.

[87] Kitajima S, Ivanova E, Guo S, et al. Suppression of STING associated with LKB1 loss in KRAS-driven lung cancer. Cancer Discov 2019;9(1):34–45. https://doi.org/10.1158/2159-8290.CD-18-0689.

[88] de Queiroz NMGP, Xia T, Konno H, Barber GN. Ovarian cancer cells commonly exhibit defective STING signaling which affects sensitivity to viral oncolysis. Mol Cancer Res 2019;17(4):974–86. https://doi.org/10.1158/1541-7786.MCR-18-0504.

[89] Berger G, Marloye M, Lawler SE. Pharmacological modulation of the STING pathway for cancer immunotherapy. Trends Mol Med 2019;25(5):412–27. https://doi.org/10.1016/j.molmed.2019.02.007.

[90] Song S, Peng P, Tang Z, et al. Decreased expression of STING predicts poor prognosis in patients with gastric cancer. Sci Rep 2017;7(1):39858. https://doi.org/10.1038/srep39858.

[91] Xia T, Konno H, Barber GN. Recurrent loss of STING signaling in melanoma correlates with susceptibility to viral oncolysis. Cancer Res 2016;76(22):6747–59. https://doi.org/10.1158/0008-5472. CAN-16-1404.

[92] Xia T, Konno H, Ahn J, Barber GN. Deregulation of STING signaling in colorectal carcinoma constrains DNA damage responses and correlates with tumorigenesis. Cell Rep 2016;14(2):282–97. https://doi.org/10.1016/j.celrep.2015.12.029.

[93] Bommareddy PK, Zloza A, Rabkin SD, Kaufman HL. Oncolytic virus immunotherapy induces immunogenic cell death and overcomes STING deficiency in melanoma. Oncoimmunology 2019;8(7). https://doi.org/10.1080/2162402X.2019.1591875.

[94] Liu W, Kim GB, Krump NA, Zhou Y, Riley JL, You J. Selective reactivation of STING signaling to target Merkel cell carcinoma. Proc Natl Acad Sci U S A 2020;117(24):13730–9. https://doi.org/10.1073/pnas.1919690117.

[95] Zhang C-X, Ye S-B, Ni J-J, et al. STING signaling remodels the tumor microenvironment by antagonizing myeloid-derived suppressor cell expansion. Cell Death Differ 2019;26(11):2314–28. https://doi.org/10.1038/s41418-019-0302-0.

[96] Yum S, Li M, Frankel AE, Chen ZJ. Roles of the cGAS-STING pathway in cancer immunosurveillance and immunotherapy. Annu Rev Cancer Biol 2019;3(1):323–44. https://doi.org/10.1146/annurev-cancerbio-030518-055636.

[97] Yum S, Li M, Chen ZJ. Old dogs, new trick: classic cancer therapies activate cGAS. Cell Res 2020;30 (8):639–48. https://doi.org/10.1038/s41422-020-0346-1.

[98] Deng L, Liang H, Xu M, et al. STING-dependent cytosolic DNA sensing promotes radiation-induced type I interferon-dependent antitumor immunity in immunogenic tumors. Immunity 2014;41(5):843–52. https://doi.org/10.1016/j.immuni.2014.10.019.

[99] Fang C, Mo F, Liu L, et al. Oxidized mitochondrial DNA sensing by STING signaling promotes the antitumor effect of an irradiated immunogenic cancer cell vaccine. Cell Mol Immunol 2020;1–13. https://doi.org/10.1038/s41423-020-0456-1. Published online May 12.

[100] Carozza JA, Böhnert V, Nguyen KC, et al. Extracellular 2′3′-cGAMP is an immunotransmitter produced by cancer cells and regulated by ENPP1. bioRxiv 2019;539312. https://doi.org/10.1101/539312. Published online October 7.

[101] Hong C, Tijhuis AE, Foijer F. The cGAS paradox: contrasting roles for cGAS-STING pathway in chromosomal instability. Cell 2019;8(10). https://doi.org/10.3390/cells8101228.

[102] Parkes EE, Walker SM, Taggart LE, et al. Activation of STING-dependent innate immune signaling by S-phase-specific DNA damage in breast cancer. J Natl Cancer Inst 2017;109(1). https://doi.org/10.1093/jnci/djw199.

[103] Kitai Y, Kawasaki T, Sueyoshi T, et al. DNA-containing exosomes derived from cancer cells treated with topotecan activate a STING-dependent pathway and reinforce antitumor immunity. J Immunol 2017;198(4):1649–59. https://doi.org/10.4049/jimmunol.1601694.

[104] Lord CJ, Ashworth A. PARP inhibitors: synthetic lethality in the clinic. Science 2017;355 (6330):1152–8. https://doi.org/10.1126/science.aam7344.

[105] Ding L, Kim H-J, Wang Q, et al. PARP inhibition elicits STING-dependent antitumor immunity in Brca1-deficient ovarian cancer. Cell Rep 2018;25(11):2972–2980.e5. https://doi.org/10.1016/j. celrep.2018.11.054.

[106] Sen T, Rodriguez BL, Chen L, et al. Targeting DNA damage response promotes antitumor immunity through STING-mediated T-cell activation in small cell lung cancer. Cancer Discov 2019;9 (5):646–61. https://doi.org/10.1158/2159-8290.CD-18-1020.

[107] Pantelidou C, Sonzogni O, De Oliveria TM, et al. PARP inhibitor efficacy depends on CD8 + T-cell recruitment via intratumoral STING pathway activation in BRCA-deficient models of triple-negative breast cancer. Cancer Discov 2019;9(6):722–37. https://doi.org/10.1158/2159-8290. CD-18-1218.

[108] Grabosch S, Bulatovic M, Zeng F, et al. Cisplatin-induced immune modulation in ovarian cancer mouse models with distinct inflammation profiles. Oncogene 2019;38(13):2380–93. https://doi.org/10.1038/s41388-018-0581-9.

[109] Lohard S, Bourgeois N, Maillet L, et al. STING-dependent paracriny shapes apoptotic priming of breast tumors in response to anti-mitotic treatment. Nat Commun 2020;11(1):259. https://doi.org/10.1038/s41467-019-13689-y.

[110] Yang H, Wang H, Ren J, Chen Q, Chen ZJ. cGAS is essential for cellular senescence. Proc Natl Acad Sci 2017;114(23):E4612–20. https://doi.org/10.1073/pnas.1705499114.

[111] Mackenzie KJ, Carroll P, Martin C-A, et al. cGAS surveillance of micronuclei links genome instability to innate immunity. Nature 2017;548(7668):461–5. https://doi.org/10.1038/nature23449.

[112] Dou Z, Ghosh K, Vizioli MG, et al. Cytoplasmic chromatin triggers inflammation in senescence and cancer. Nature 2017;550(7676):402–6. https://doi.org/10.1038/nature24050.

[113] Chen Y-A, Shen Y-L, Hsia H-Y, Tiang Y-P, Sung T-L, Chen L-Y. Extrachromosomal telomere repeat DNA is linked to ALT development via cGAS-STING DNA sensing pathway. Nat Struct Mol Biol 2017;24(12):1124–31. https://doi.org/10.1038/nsmb.3498.

[114] Glück S, Guey B, Gulen MF, et al. Innate immune sensing of cytosolic chromatin fragments through cGAS promotes senescence. Nat Cell Biol 2017;19(9):1061–70. https://doi.org/10.1038/ncb3586.

[115] Harding SM, Benci JL, Irianto J, Discher DE, Minn AJ, Greenberg RA. Mitotic progression following DNA damage enables pattern recognition within micronuclei. Nature 2017;548(7668):466–70. https://doi.org/10.1038/nature23470.

[116] Gao P, Ascano M, Wu Y, et al. Cyclic [G(2′,5′)pA(3′,5′)p] is the metazoan second messenger produced by DNA-activated cyclic GMP-AMP synthase. Cell 2013;153(5):1094–107. https://doi.org/10.1016/j.cell.2013.04.046.

[117] Gao P, Ascano M, Zillinger T, et al. Structure-function analysis of STING activation by c[G(2′,5′)pA(3′,5′)p] and targeting by antiviral DMXAA. Cell 2013;154(4):748–62. https://doi.org/10.1016/j.cell.2013.07.023.

[118] Liu S, Cai X, Wu J, et al. Phosphorylation of innate immune adaptor proteins MAVS, STING, and TRIF induces IRF3 activation. Science 2015;347(6227):aaa2630. https://doi.org/10.1126/science.aaa2630.

[119] Tanaka Y, Chen ZJ. STING specifies IRF3 phosphorylation by TBK1 in the cytosolic DNA signaling pathway. Sci Signal 2012;5(214):ra20. https://doi.org/10.1126/scisignal.2002521.

[120] Chen Q, Boire A, Jin X, et al. Carcinoma-astrocyte gap junctions promote brain metastasis by cGAMP transfer. Nature 2016;533(7604):493–8. https://doi.org/10.1038/nature18268.

[121] Li T, Chen ZJ. The cGAS-cGAMP-STING pathway connects DNA damage to inflammation, senescence, and cancer. J Exp Med 2018;215(5):1287–99. https://doi.org/10.1084/jem.20180139.

[122] Fuertes MB, Kacha AK, Kline J, et al. Host type I IFN signals are required for antitumor CD8 + T cell responses through CD8{alpha}+ dendritic cells. J Exp Med 2011;208(10):2005–16. https://doi.org/10.1084/jem.20101159.

[123] Diamond MS, Kinder M, Matsushita H, et al. Type I interferon is selectively required by dendritic cells for immune rejection of tumors. J Exp Med 2011;208(10):1989–2003. https://doi.org/10.1084/jem.20101158.

[124] Li T, Cheng H, Yuan H, et al. Antitumor activity of cGAMP via stimulation of cGAS-cGAMP-STING-IRF3 mediated innate immune response. Sci Rep 2016;6:19049. https://doi.org/10.1038/srep19049.

[125] Chandra D, Quispe-Tintaya W, Jahangir A, et al. STING ligand c-di-GMP improves cancer vaccination against metastatic breast cancer. Cancer Immunol Res 2014;2(9):901–10. https://doi.org/10.1158/2326-6066.CIR-13-0123.

[126] Iwai Y, Hamanishi J, Chamoto K, Honjo T. Cancer immunotherapies targeting the PD-1 signaling pathway. J Biomed Sci 2017;24(1):26. https://doi.org/10.1186/s12929-017-0329-9.

[127] Callahan MK, Wolchok JD, Allison JP. Anti-CTLA-4 antibody therapy: immune monitoring during clinical development of a novel immunotherapy. Semin Oncol 2010;37(5):473–84. https://doi.org/10.1053/j.seminoncol.2010.09.001.

[128] Wang H, Hu S, Chen X, et al. cGAS is essential for the antitumor effect of immune checkpoint blockade. Proc Natl Acad Sci U S A 2017;114(7):1637–42. https://doi.org/10.1073/pnas.1621363114.

[129] Kato K, Nishimasu H, Oikawa D, et al. Structural insights into cGAMP degradation by Ecto-nucleotide pyrophosphatase phosphodiesterase 1. Nat Commun 2018;9(1):4424. https://doi.org/10.1038/s41467-018-06922-7.

[130] Hanson MC, Crespo MP, Abraham W, et al. Nanoparticulate STING agonists are potent lymph node-targeted vaccine adjuvants. J Clin Invest 2015;125(6):2532–46. https://doi.org/10.1172/JCI79915.

[131] Su T, Zhang Y, Valerie K, Wang X-Y, Lin S, Zhu G. STING activation in cancer immunotherapy. Theranostics 2019;9(25):7759–71. https://doi.org/10.7150/thno.37574.

[132] Nishioka Y, Yoshino H. Lymphatic targeting with nanoparticulate system. Adv Drug Deliv Rev 2001;47(1):55–64. https://doi.org/10.1016/S0169-409X(00)00121-6.

[133] Sung YK, Kim SW. Recent advances in polymeric drug delivery systems. Biomater Res 2020;24(1):12. https://doi.org/10.1186/s40824-020-00190-7.

[134] Liu Z, Zhang Z, Zhou C, Jiao Y. Hydrophobic modifications of cationic polymers for gene delivery. Prog Polym Sci 2010;35(9):1144–62. https://doi.org/10.1016/j.progpolymsci.2010.04.007.

[135] Sunshine JC, Peng DY, Green JJ. Uptake and transfection with polymeric nanoparticles are dependent on polymer end-group structure, but largely independent of nanoparticle physical and chemical properties. Mol Pharm 2012;9(11):3375–83. https://doi.org/10.1021/mp3004176.

[136] Wilson DR, Sen R, Sunshine JC, Pardoll DM, Green JJ, Kim YJ. Biodegradable STING agonist nanoparticles for enhanced cancer immunotherapy. Nanomed Nanotechnol Biol Med 2018;14(2):237–46. https://doi.org/10.1016/j.nano.2017.10.013.

[137] Shae D, Becker KW, Christov P, et al. Endosomolytic polymersomes increase the activity of cyclic dinucleotide STING agonists to enhance cancer immunotherapy. Nat Nanotechnol 2019;14(3):269–78. https://doi.org/10.1038/s41565-018-0342-5.

[138] Lee E, Jang H-E, Kang YY, Kim J, Ahn J-H, Mok H. Submicron-sized hydrogels incorporating cyclic dinucleotides for selective delivery and elevated cytokine release in macrophages. Acta Biomater 2016;29:271–81. https://doi.org/10.1016/j.actbio.2015.10.025.

[139] Junkins RD, Gallovic MD, Johnson BM, et al. A robust microparticle platform for a STING-targeted adjuvant that enhances both humoral and cellular immunity during vaccination. J Control Release 2018;270:1–13. https://doi.org/10.1016/j.jconrel.2017.11.030.

[140] Watkins-Schulz R, Tiet P, Gallovic MD, et al. A microparticle platform for STING-targeted immunotherapy enhances natural killer cell- and CD8 + T cell-mediated antitumor immunity. Biomaterials 2019;205:94–105. https://doi.org/10.1016/j.biomaterials.2019.03.011.

[141] Markman JL, Rekechenetskiy A, Holler E, Ljubimova JY. Nanomedicine therapeutic approaches to overcome cancer drug resistance. Adv Drug Deliv Rev 2013;65(13–14):1866–79. https://doi.org/10.1016/j.addr.2013.09.019.

[142] Lu X, Miao L, Gao W, et al. Engineered PLGA microparticles for long-term, pulsatile release of STING agonist for cancer immunotherapy. Sci Transl Med 2020;12(556). https://doi.org/10.1126/scitranslmed.aaz6606.

[143] Leach DG, Dharmaraj N, Piotrowski SL, et al. STINGel: controlled release of a cyclic dinucleotide for enhanced cancer immunotherapy. Biomaterials 2018;163:67–75. https://doi.org/10.1016/j.biomaterials.2018.01.035.

[144] Coelho T, Adams D, Silva A, et al. Safety and efficacy of RNAi therapy for transthyretin amyloidosis. N Engl J Med 2013;369(9):819–29. https://doi.org/10.1056/NEJMoa1208760.

[145] Hafez IM, Maurer N, Cullis PR. On the mechanism whereby cationic lipids promote intracellular delivery of polynucleic acids. Gene Ther 2001;8(15):1188–96. https://doi.org/10.1038/sj.gt.3301506.

[146] Mendonça MCP, Radaic A, Garcia-Fossa F, da Cruz-Höfling MA, Vinolo MAR, de Jesus MB. The in vivo toxicological profile of cationic solid lipid nanoparticles. Drug Deliv Transl Res 2020;10(1):34–42. https://doi.org/10.1007/s13346-019-00657-8.

CHAPTER TWELVE

Oncolytic viruses in immunotherapy

Ilse Hernandez-Aguirre[a,b] and Kevin A. Cassady[c,d]

[a]Medical Scientist Training Program, The Ohio State University Wexner Medical Center, Columbus, OH, United States
[b]The Research Institute at Nationwide Children's Hospital, Center for Childhood Cancer and Blood Diseases, The Ohio State University College of Medicine, Columbus, OH, United States
[c]Department of Pediatrics, Division of Pediatric Infectious Diseases, Nationwide Children's Hospital, Columbus, OH, United States
[d]The Research Institute at Nationwide Children's Hospital, Center for Childhood Cancer and Blood Diseases, Division of Pediatric Infectious Diseases, The Ohio State University College of Medicine, Pelotonia Institute for Immuno-Oncology, Columbus, OH, United States

Contents

Cancer Immunology and Immunotherapy
https://doi.org/10.1016/B978-0-12-823397-9.00012-0

1. Introduction—Oncolytic viroimmunotherapy

Virotherapy involves using therapeutic viruses as a conditionally replicating agent or as a gene expression platform to induce direct cytotoxic or immune-related antitumor activity against malignant tumors. Both experimental and clinical evidence suggest that oncolytic virus (OV) lytic damage, tumor microenvironment changes, release of tumor-associated antigens, and the resultant immune-mediated activity (both innate and adaptive) contribute to OV antitumor activity. In some cases, OV therapy produces a durable antitumor response and likely involves the generation of tumor antigen-specific adaptive immune cell populations that persist after virus replication and gene expression has completed. OVs have evolved from early conservatively designed attenuated vectors focused on safety to now more complex engineered OVs. These newer generation vectors combine enhanced selective viral replication within tumor cells using cancer conditional gene expression or entry with immunomodulatory gene expression to enhance the antitumor activity of the recruited immune cells. In this way OV therapy is evolving to both enhance direct cytoreductive activity with gene therapy related activity to further modulate the antiviral immune cells recruited to the tumor to enhance their antitumor activity.

2. A brief history of oncolytic viruses

Case reports of probable virus infection-associated tumor remissions were described as early as the mid-1800s; however, the potential significance of naturally occurring virus infections upon cancer became a more active pursuit at the turn of the 20th century. While attempting to understand leukemia, physicians noted that in some cases, an infection with a virus led to transient leukemia regression [1]. In 1904, Dr. George Dock documented the case of a leukemia patient whose acute myeloid leukemia (AML) initially regressed following what is now considered a likely influenza infection. He described a reduction in the number of abnormal leukocytes and in her splenomegaly. Months later, however, her spleen and leukocytes rapidly increased and she eventually succumbed to her cancer. He goes on to note that such cases force us to learn not only about pathology, but about something that could potentially be of therapeutic value.

It was not until the 1950s that oncolytic viruses were named and studied as such. Defined as oncolytic in that they lyse tumors, viruses were investigated for their ability to induce tumor cell death and reduce tumor growth [2]. Scientists investigated Russian encephalitis virus's effect on mouse sarcomas and Egypt Virus's activity on a human carcinoma in rat models and used evolutionary pressure (serial passage) to increase oncolytic activity [3, 4]. Others continued to explore viruses as cancer therapy. In 1954, investigators implanted RPL-12 chicken lymphoma into chickens' pectoral muscles and superimposed infection with St. Louis Encephalitis virus. Although they noted tumor regression, virus infection did not appear to directly lyse the tumor cells. This suggested that while St Louis Encephalitis virus infection contributed to the response, an indirect mechanism was responsible for the anticancer activity [5]. Their results suggested that virus infection stimulated the host immune response and rendered the tumor cells more susceptible to phagocytosis and the antitumor activity was limited to tumors in periphery but did not occur when tumors were implanted in immune privileged sites [5].

Virotherapy studies in the 1960s focused on delineating the consequences of tumor lysis and explored the possibility of a virus administered therapy rather than the effects of chance infections. Researchers continued to test different viruses in different tumor models as a therapy. In 1963, they provided evidence that the immune response contributed to virotherapy. While they continued to demonstrate tumor lysis and regression, they also suggested the possibility for antitumor immunity [6]. The virotherapy treated mouse survivors were resistant to tumor growth upon rechallenge months after initial treatment. The studies indicated that immunoglobulins were integral for this antitumor immunity and created tumor homogenates for use as an anticancer vaccine [7]. They recognized that tumor homogenates infected with an oncolytic strain of influenza were immunogenic and protected mice from Ehrlich ascites tumor, whereas uninfected tumor homogenates were not effective, even when *in vitro*-cultured virus was added. They theorized that, since influenza viruses were known to incorporate antigens derived from lysed cells, the oncolytic influenza virus similarly incorporated tumor antigens, turning them into potent antigens that could be recognized in conjunction with the viral antigens.

Despite these advances, toxicities associated with naturally occurring viruses limited virotherapy's development. It was not until recombinant techniques had advanced sufficiently and viral gene functions were better characterized and understood that virotherapy became more attractive as a potential therapeutic approach in the 1980s and 1990s. There was also an appreciation as to how retroviruses could promote tumor growth and encoded potential oncogenic activity. Investigators entertained the possibilities of using therapeutic retroviruses to counter these oncogenes, induce apoptosis, or to deliver new antigens to improve immune cell recognition [8, 9]. Early studies examined the use of virus-based gene expression in combination with antiviral therapy to mediate tumor cell damage [10–13]. These results produced some activity; however, it was with the advent of conditional viral replication competent vectors that oncolytic virotherapy began to show clear antitumor responses in select patients. Eventually viral vectors were studied as potential vectors for gene expression in cancer [14]. Researchers also began to delve into combination therapies, such as HSV vectors in combination with ganciclovir to treat disseminated brain tumors [15]. In parallel with these initial efforts, researchers continued to identify viral genes and their role in viral infection and pathogenesis. As scientific discoveries improved our understanding of viral gene functions and their role in regulating cellular functions and contribution to viral infection, genetic construction of oncolytic viruses with improved antitumor activity occurred. An example of this is the discovery of the HSV-1 neurovirulence genes that permitted construction of an oncolytic virus safe for treating brain tumors. Researchers identified mutations that were dispensable for HSV replication in human cell culture [16], but that rendered the virus less neurovirulent in mouse models [17, 18]. Specifically, deletion of gene $\gamma_1 34.5$ did not affect the replication of HSV, and actually led to cell death of neuroblastoma cells [19]. Further studies identified that the deletion of $\gamma_1 34.5$ gene enabled viral evasion of PKR and translational arrest [20] and led to the speculation that $\gamma_1 34.5$ genetically engineered HSV could be used for treatment of experimental brain tumors based on successful animal experiments [21]. Further studies unraveled how the $\gamma_1 34.5$ gene product regulated other important cellular functions including Beclin-1-mediated autophagy [22], TBK1-mediated IRF3 signaling, and interferon (IFN) β production in some infected cells [23]. Similarly, other investigators found that certain mutations of adenoviral E1A and E1B rendered the virus unable to form E1A/Rb complexes [24]; they used this knowledge to specifically target cancer cells that have dysfunctional Rb pathways [25, 26]. Fueyo *et al.* created a tumor selective adenovirus with a 24 bp deletion in E1A, termed Delta-24, which lysed human glioma cells but not nonmalignant cells [27]. Delta-24 was further modified to make Delta-24-RGD by added RGD-4C peptide to increase binding to integrins due to the fact that Delta-24 was unable to efficiently infect many highly malignant tumors due to their lack of coxsackie-adenovirus-receptor (CAR) expression [28]. In 1977, Hashiro *et al.* found reovirus to have preferential cytotoxicity in certain tumor cell lines and spontaneously transformed cell lines, whereas normal cell lines were resistant [29]. Upon further investigation, reovirus was shown to selectively replicate in cells with increased epidermal growth factor receptor (EGFR) or Ras GTPase/mitogen-activated protein kinase (RAS/MAPK) activity [30–32]. Eventually, reovirus was administered to tumor-bearing mice to investigate whether this virus could be used for therapy [33]. Those studies led to Reolysin, a reovirus, translated to clinical trials [34]. In parallel, investigators developed recombinant vaccinia viruses (VV) for tumor antigen expression in preclinical cancer vaccine studies since the late 1980s [35–37], and clinical trials in the late 1990s [38]. The possibilities of engineering therapeutic viruses capable of tumor associated replication while remaining attenuated and with restricted replication in non-malignant cells began to emerge.

Over the last three decades, efforts have focused on engineering viruses that efficiently target cancer cell infection and lysis and modulate the virus recruited immune cell response. G207 is an aneurovirulent virus restricted to actively replicating cells due to its inability to efficiently synthesize nucleotides because of a $U_L 39$ viral ribonucleotide reductase mutation [39]. This conditional replication lysed tumor cells (oncolysis) and improved survival in rodent studies. In parallel, researchers in the United Kingdom were developing 1716 an aneurovirulent oHSV for tumor therapy [40]. These viruses further moved into clinical trials to test their safety in humans [41, 42]. In studies in the United

Kingdom and the United States, investigators showed that engineered oncolytic viruses designed to conditionally replicate in tumors were safe in humans. These conservatively designed viruses focused on safety but demonstrated clinical and radiographic evidence of antitumor response in 49% (17 of 35) of evaluable patients. The results of these trials showed that an oHSV (1) could be safely injected into the brains of MG patients, (2) could be safely injected repeatedly, and (3) could be combined with adjunctive therapy for an improved response [43–48]. As several clinical trials ensued throughout the world, two unarmed OVs were approved for clinical use in the mid-2000s. An unmodified echovirus (Rigivir) with some naturally occurring or intrinsic tumor selectivity was approved in Latvia [49]. China also approved the use of an engineered oncolytic adenovirus (H101) containing an E1B deletion, in combination with chemotherapy to treat squamous cell cancer of head and neck or esophagus cancer [50–52]. However, efforts then moved toward improving the replication, oncolytic activity, and immunomodulatory effects of the virus therapy. A group of researchers further engineered an ICP34.5 mutated oncolytic virus with another deletion of ICP34.5 and a deletion of ICP47 to create a virus with enhanced oncolytic effects, while potentiating the antitumor immunity by inserting the GM-CSF gene [53]. Termed T-VEC (talimogene laherpaverec), this virus became the first Food and Drug Administration (FDA) approved virotherapy. Owing to the cost of cGMP production of clinical grade virus and regulatory safety studies required for these biologics, clinical translation has lagged and many of the current clinical trials involve vectors developed decades earlier that are just now entering clinical trial. More advanced viral vectors are in preclinical studies and as OV therapy gains both commercial and therapeutic traction the time to clinical translation is anticipated to accelerate.

Oncolytic viral therapy has evolved from an initial observation of tumor regression following an inter-current viral infection to now meticulously engineered therapeutic approach. Although initial efforts focused on improving virus related lytic activity, scientists now use the viral genome as a gene expression platform to modulate the immune response during virus infection of the tumor. As other immunotherapies mature and enter the clinic, combination therapies involving oncolytic virus infection and sequential administration of these immune active agents are anticipated.

3. Overview of oncolytic virotherapy

Although some viruses are inherently oncolytic (*e.g.*, nonhuman viruses naturally restricted to replication in tumor cells), most viruses developed for clinical use, including naturally oncolytic viruses, have undergone some genetic engineering. These modifications further attenuate the virus, improve tumor cell specificity (restricting viral entry, gene expression, replication or egress to cancer cells) to make them safe for clinical use, or increase their therapeutic efficacy (cytokine, chemokine, or immunologically active gene expression). An overview of the stages of OV delivery, their direct and indirect immune-mediated activity and opportunities for immunotherapeutic combination therapies are summarized in Fig. 1 [54]. Virus delivery can involve direct injection, convection-enhanced delivery (CED), systemic administration, or carrier cell delivery. Currently for solid tumors most studies use direct intratumoral OV delivery as a rapid method of distributing the virus into the primary tumor. There is great interest in systemic delivery as it would target both primary and metastatic tumors that virus delivery to the tumor is limited in due to poor penetration and immune neutralization [55, 56]. Following delivery, infection, gene expression and, in the case of oncolytic vectors, replication, cell lysis and cell-to-cell spread occurs. For other replication incompetent or stable gene expression platforms this is comprised of selective gene expression by the virus. Both approaches involve conditional or tumor-specific viral gene expression, tumor targeting based upon entry receptor preference, and selective replication in the cancerous cell followed by lysis and spread. Current viral design in most instances attempts to maintain the attenuation of earlier vectors while improving the tumor associated viral gene expression, replication, and spread. This balance between viral replication and safety, however, is an ongoing issue for OV therapy. Virus infection/gene expression stimulates an antiviral immune response. This begins first with intrinsic antiviral immune activity (IFN, chemokine, ISG, DAMP/PAMP activity) from the infected cell leading to innate (dendritic PMN, NK) and adaptive (T and B) immune cell recruitment and further antiviral activity. Viruses are often designed to modify this immune-

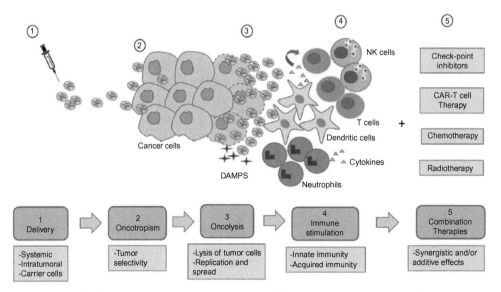

Fig. 1 Overview of oncolytic virus antitumor activity and therapeutic approaches to enhance immune activity.

mediated activity to enhance the immune-mediated antitumor effect (in the case of oncolytic vectors) or to prolong viral based gene expression (in the case of replication defective viruses). Current efforts rely on enhancing viral tumor cells lysis and through viral-based immunostimulatory/immunomodulatory gene expression to enhance cellular activity (*e.g.*, GM–CSF, IL-2, IL12, IL15, CD40L, IFN b) T-cell engagers, immune checkpoint inhibitors) or combining OV treatment with other immunostimulatory (RT) or adoptive cell therapies.

4. Oncolytic virus safety and efficacy

As a biological therapy developed from pathogens, oncolytic viral therapy has received appropriate regulatory scrutiny to ensure its safety in patients. Oncolytic viruses must balance efficient replication and gene expression within the tumor environment with attenuation in noncancerous cells. It is this dynamic balance between efficient viral activity in the cancer and safety throughout the noncancerous tissue that has driven virus design. While tumor specificity is important for virus safety, additional criteria are also required. OVs are evaluated for their transmission potential beyond the treated patient (human and environmental), undesirable side effects, genetic stability and the potential for secondary or escape mutations during replication, and the effect of prior immunity on viral replication activity [57]. Early engineered OVs were conservatively designed, often encoding multiple mutations that target critical viral functions to ensure safety. These OVs were remarkably effective in a subset of patients despite their limited gene expression/replication and are still in use. Over the last decades, efforts have focused on improving their therapeutic efficacy. One concern is that the redundant genetic mutations used to attenuate and ensure safety restrict efficient viral gene expression, replication, and therapeutic efficacy in infected tumor cells. This has led to newer generation OVs with improved conditional gene expression and replication efficiency. Other efforts have focused on arming viruses with nonviral genes to further improve OV therapeutic activity.

4.1 Delivery, uptake, and cell-to-cell spread

Solid tumors can prevent systemically delivered virus from physically reaching the tumor; they limit efficient viral replication and cell-to-cell spread within it. Furthermore, for many of the OV based therapies, pre-existing neutralizing antibodies may exist that further limit viral entry. Most virotherapies involve highly attenuated vectors and direct injection into the tumor provides a dose advantage that may facilitate viral replication. Direct injection also can reduce the effects of neutralizing antibodies and other antiviral proteins and induces a more robust inflammatory cytokine/chemokine response in the treated tumor. Additionally, solid tumors are composed of intricate lymphatic and blood vessels that rarely mimic normal vascularity [58]. This complex vascularity, along with a dense extracellular matrix (ECM), creates an environment with interstitial hypertension. This impairs viral infiltration due to the size of viruses. Although still small and considered nanoparticles, they are big enough that they are hindered by the available space in the matrix as well as eliminated pressure gradients, due to the high interstitial fluid pressure in the tumor. This means viruses may cause only local effects in these tumors. CED is one approach to overcome this. By using high pressure, the OV is disseminated throughout the interstitial space of the tumor, improving tumor cell delivery. Viral particle size, receptor access, and pressure dynamics within the tumor, however, influence the efficacy of this approach, and what works for one virus may not work for another. Another approach to normalize the tumor interstitial matrix is by degrading the collagen and gly-cosaminoglycan content in the ECM by expressing bacterial collagenase or matrix metalloproteinases (MMPs) from the virus [59, 60]. This approach has been used with HSV and ADV to modify the ECM and improve viral spread using proteins (relaxin and Hyaluronidase) that modify angiogenesis [61, 62]. Relaxin upregulates MMPs to degrade ECM, therefore oncolytic adenoviruses expressing it successfully increased viral spread in multiple tumor models [63]. When armed with hyaluronidase, such as VCN-01, adenoviruses decreased hyaluronic acid amounts in tumor ECM, leading to increased survival in murine glioma models [62]. On the other hand, oncolytic adenoviruses also have been engineered to carry inhibitors to MMPs, such as tissue inhibitor of metalloproteinases (TIMPs). In doing so, these viruses can then inhibit the degradation of tumor ECM, leading to inhibition of tumor proliferation, migration, and angiogenesis. An oncolytic adenovirus expressing TIMP2, increased viral replication in ovarian tumors but distribution of the virus, usually stopped by the ECM, was not analyzed [64]. Oncolytic adenoviruses have also been armed with genes expressing antiangiogenic factors. By having a virus express soluble vascular endothelial growth factor (VEGF) receptor 3 (FP3), which reduces cancer cell VEGF expression, researchers saw increased oncolytic activity in lung cancer xenograft models [65]. Similarly, oncolytic adenoviruses have also been engineered to express soluble Flt1, a portion of the VEGF-1 receptor, or soluble Delta-like 4 (*DII4*), which prevents Notch signaling from maturing endothelium, and these reduced vascular formation in breast cancer models [66].

4.2 Cell entry

Improving cancer cell uptake of virus is one method for enhancing tumor specificity. Viruses are obligate intracellular pathogens and must deliver their genetic material across the cell membrane. Viruses have evolved to employ essential cell mechanisms and important cell surface receptors to achieve this end [67]. Many viruses undergo a two–stage attachment to the host cell (although exceptions exist, Vaccinia virus, for example, uses receptor-independent modes for viral entry). The initial low-affinity attachment is followed by high affinity interactions with more specific receptors. Viral glycoprotein binding can involve meta-stable protein conformations that upon binding to the cellular receptor results in conformational changes that approximates the virus closer to the cell surface. This facilitates entry and activation of signaling pathways that allow direct fusion or endocytic uptake in some cases. Oncolytic viruses can specifically target cancer cells due to high expression of certain receptors in tumor cells. For example, cancer cells sometimes over express intracellular adhesion molecule-1 (ICAM-1) and decay accelerating factor (DAF), both of which are receptors for coxsackie virus A21 [68]. Another example is echovirus type 1, which preferentially enters ovarian cancer cells because they overexpress the I domain of integrin α2β1 [69]. Further examples include polioviruses, which can infect tumor cells expressing CD155, which is abundant on many cell types [70, 71]. However, if viruses use a receptor that is commonly found throughout the body, there is possibility for off-target effects. Therefore, it is best to target cancer cells with tumor-specific receptors or using intracellular approaches.

Many oncolytic adenoviruses are Serotype 3 based and use the native coxsackie-adenovirus receptor (CAR) to bind to target cells, however CAR is vastly distributed throughout the body and is poorly expressed or absent on some malignant cells. This favors off-target activity and reduces oncolytic activity in some tumors [72, 73]. By re-engineering the ADV entry molecule (e.g., inclusion of RGD motifs) or by using receptors from other serotypes that recognize receptors enriched in tumors (serotype switching) investigators can enhance cancer cell uptake of virus and improve antitumor activity. Other, oncolytic viruses (e.g., HSV) have broad cell tropism and utilize a diverse set of receptors and modes of cell entry. Investigators have attempted to improve tumor-specific re-targeting of these viruses using single-chain antibodies against tumor-specific proteins to enhance oHSV infection specificity, for example for HER2/neu positive tumors [74].

5. Viral gene expression, replication and oncolysis

5.1 Intracellular events important for viral replication

An alternative strategy to receptor targeting involves genetically modifying viruses to improve their replication in cancer cells while suppressing replication efficiency in non-malignant cells. Viruses often require an efficient replication cycle and cell-to-cell spread to sustain infection in the host: oncolytic viruses are no different. The efficiency of viral entry, uncoating, gene expression, replication, capsid assembly, maturation, and release, are integral to oncolytic virus cancer cell lysis. Host antiviral response mechanisms that limit each of these steps have evolved that not only restrict naturally occurring viruses but also OV replication efficiency. Identification of biological differences in these pathways between malignant and nonmalignant cells has been instrumental in improving OV design and enhancing conditional virus replication in tumor cells and selective cancer cell lysis. By eliminating viral genes that are dispensable in cancer cells but that are important for replication in nonmalignant cells (*e.g.*, viral genes involved in biosynthetic or nucleic acid production, Ribonucleotide reductase, evading RB/p53 regulated cell-cycle changes, or IFN-related host antiviral activity) this both attenuates the virus and provides selective replication advantage for the virus in malignant cells. A variation on this approach involves the selective expression of these genes using tumor-specific promoters [75]. Examples of these two common cellular pathways that are aberrant in cancer cells and some of the OV modifications used to exploit these pathways to improve conditional virus activity are summarized below and in Fig. 2 [76].

5.2 RB/p53 and p16/RB related cell-cycle changes

Viruses require a high biosynthetic environment to generate the large amounts of nucleic acids and amino acids necessary for their replication. Accordingly, many viruses shift the infected cell cycle into a highly synthetic phase through a series of virus regulated CDK dependent phosphorylation events in the infected cell. These unregulated cell-cycle shifts and virus-stimulated cell signaling responses can induce apoptosis in the infected cells, therefore viruses often encode genes that prevent apoptosis [77]. Examples of viral proteins that directly target Rb family members for inactivation to create a more hospitable environment for viral replication include oncoproteins E7 (from human papillomavirus), E1A (from adenovirus), and the large T (tumor) antigen (from simian virus 40) [78]. In addition to creating a biosynthetic environment in the host cell, viruses also encode genes (ribonucleotide reductase, dihydrofolate reductase, thymidine kinase, thymidylate synthetase) that augment host cellular production and allow additional bulk nucleotide synthesis important for efficient viral gene expression and replication. Cancer cells also require a robust biosynthetic and proliferative environment to undergo unregulated DNA replication and resist apoptosis and therefore frequently contain mutations in these same pathways similar to those targeted by the viral genes. It is therefore possible to engineer and eliminate these important genes from oncolytic viruses because the cancer cells complement the viral gene loss. In the nonmalignant cell, the virus is attenuated and cannot shift cell cycle, inhibit apoptosis, or recreate the biosynthetic environment necessary for efficient viral replication. However, in the cancer cell, because these mutations are often present, this provides an environment that supports viral replication. Many ADV vectors (*e.g.*, ONYX-15 and H101) utilize this principal and contain E1A and or E1B gene mutations [79–81]. As shown in Fig. 2A and B, the adenoviral E1A and E1B proteins are important for viral evasion of p53 related growth arrest and apoptotic response. Their absence halts viral replication and induces apoptosis in nonmalignant cells [82]. Domains of E1A are also involved in E2F related transcriptional changes in the cell [83–85]. These viral proteins become dispensable in cancer cells with p16 or RB functional loss and benefit from these cells' highly proliferative biosynthetic environment. In addition to directly regulating the p53 axis, gene modifications can also take advantage of

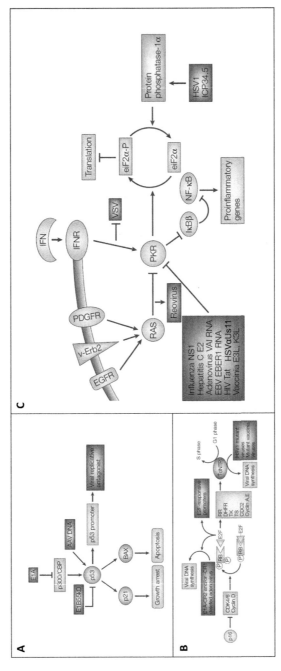

Fig. 2 Common cellular pathways mutated in cancers that permit selective viral replication and conditional oncolytic activity.

the increased biosynthetic activity in highly proliferative cancer cells. Vaccinia and oHSV vectors have also been designed with ribonucleotide reductase or thymidine kinase mutations. This prevents the virus from efficiently synthesizing nucleic acids or ribose moieties necessary for nucleotide production and replication, providing viral dependence on the cancer cell environment for replication [86–90].

5.3 Host antiviral response (PRR, PKR, IFN signaling)

OV specificity and conditional replication in tumor cells can also be achieved by exploiting differences in intrinsic antiviral pathways that often occur in malignant cells. Pattern recognition receptors (PRRs) are present in cells and monitor the extracellular and cytosolic environment for foreign molecules with pathogen associated molecular patterns (PAMPs) suggestive of infection or aberrant genetic expression in the cell. These PRRs upon binding "non-self" molecules (dsRNA, LPS, CPG DNA 5′ uncapped RNA, Diacyl and Triacyl lipopeptides) activate signaling pathways and transcriptional changes that initiate IFN, chemokine, cytokine, and IFN-dependent gene production both within the infected and neighboring cells. While some cancer cells (*e.g.*, inflammation-based tumors and carcinomas) harness these pathways to stimulate proliferative responses, their transcriptional products can also restrict cancer cell growth through both direct (tumor cell inhibition) and indirect (immune-mediated) effects. Consequently, malignant cells often eliminate important components of these pathways or suppress the IFN-mediated response and these often lead to IFN signaling defects in the cancer cells [91]. In addition to direct IFN signaling mutations, cancer cells can also suppress or mutate interferon stimulated genes that limit cancer growth. Examples of this include chromosome 9P21 deletions in glioblastomas that frequently eliminate the type I IFN genes, IFN α and β genes, and also the PRR response loss in many cancers (*e.g.*, STING cGAS, PKR) [92–97]. Other cancers modify signaling pathways (*e.g.*, PTEN, IRF3 signaling, Ras/MAPK) to decrease the deleterious effects of these pathways on tumor cell growth [31,98–101]. Viruses encode genes that counteract these PRR pathways and the IFN response to enable it to synthesize necessary gene products and replicate in the infected cell. Mutating or deleting genes instrumental in countering these host antiviral response pathways attenuates the virus allowing its safe use. The antiviral pathways restrict the virus in nonmalignant cells, whereas malignant cells can selectively support viral replication. Although each virus is unique, they are all obligate intracellular pathogens and rely on host cellular machinery for viral gene expression [102]. Protein kinase R is an IFN stimulated dsRNA activated cytoplasmic protein that regulates protein translation in infected cells or cells with limited amino acid stores. When nonmalignant cells are infected with virus, they secrete IFN, which inhibits protein translation in the infected cell and primes neighboring cells, thus suppressing efficient viral replication and cell-to-cell spread. Oncolytic viruses are frequently engineered to mutate or eliminate viral genes that counteract the IFN or IFN-related gene products that restrict viral replication. These mutations attenuate the virus and restrict efficient replication in nonmalignant cells where these pathways are intact. Because these pathways are frequently dysfunctional in cancerous cells this allows these attenuated viruses to selectively replicate in malignant cells. For example, matrix (M) protein from vesicular stomatitis virus (VSV) can inhibit the translation of IFN in the affected cell. Mutations in the VSV M protein, suppress viral replication in nonmalignant cells providing conditional replication in tumor cells [103]. HSVs principal neurovirulence gene $\gamma_1 34.5$ encodes a protein (ICP34.5) that counteracts multiple IFN-related functions in the infected cell including PKR-mediated translational arrest [18,20,22,23,104,105]. ICP34.5 binds and redirects the host protein phosphatase-1alpha to dephosphorylate eukaryotic initiation factor 2 (eIF-2) allowing efficient protein translation in infected cells. This HSV gene is frequently deleted in oncolytic HSVs and attenuates the virus. Cancer cells in some cases have PKR deletional mutations or contain Ras mutations or high MAPK signaling activity which suppresses PKR autophosphorylation and activity [106]. This increases translational activity in the cancer cell and selectively complements the mutations in some oncolytic viruses improving their gene expression and replication (*e.g.*, VSV, reovirus, HSV $\Delta\gamma_1 34.5$). However, not all cancer cells contain these mutations [107,108] and even in cancer cells with increased Ras/MAPK activity viral replication efficiency remains less than that of a wild-type HSV in some tumor cells (up to $10,000 \times$ lower). Some recent OV efforts have focused on selectively improving protein translation and viral replication in cancer cells [107,109–115]. In addition to $\gamma_1 34.5$ mutations, other viral gene mutations improve HSV tumor selective replication.

Tumor cell specificity can also be achieved by transcriptionally regulating important viral genes rather than deleting them from the OV genome. Selectively restricting viral gene expression to target cells allows the virus to replicate more efficiently in the cancer while remaining attenuated in nonmalignant cells. This can be achieved using tumor-specific promoters. Depending on the target cancer of interest, investigators have used nestin (glioma), PSA (prostate), AFP (Hepatocellular), estrogen (breast) and survivin promoters to regulate viral genes or cytotoxic genes to regulate OV activity in tumors [111,116–125]. While this provides selective expression in these cancers, the disadvantage is that these OVs have a narrow specificity and cannot be repurposed for other cancers that lack these tumor-specific promoters. Modifications to the OV (promoter modifications) require repeat cGMP production, biotoxicology studies, and regulatory approval. Another way of achieving transcriptional regulation is to incorporate miRNA recognition into critical viral genes. By placing essential genes under miRNA, which are expressed in healthy cells but not in cancer cells, viruses cannot replicate in healthy cells expressing the miRNA [126]. Researchers engineered four copies of the miR-124 recognition site, which is expressed in neuron but not glioblastoma cells, and placed it on the HSV gene ICP4, which is an essential gene that increases transcription of other viral genes. Because ICP4 is an essential gene, this limits ICP4 and subsequent HSV gene transcription in neurons thus limiting viral gene expression and replication to tumor cells and thus protecting healthy tissue [127].

6. Immune-mediated antitumor activity

While direct lysis of cancer cells remains an integral component of virotherapy, the virus-induced immune response is increasingly recognized as a major contributor to OV activity and thought to be the principal mechanism responsible for durable antitumor responses. The antiviral immune response was historically viewed as an impediment to OV therapy, by restricting oncolytic virus replication and cell-to-cell spread. In 1998, investigators engineered oncolytic HSV to express cytokines mIL-4/mIL10 to reduce immune related oncolysis in an effort to prolong viral replication. This was compared to oHSV that encodes IL-12, alone or in combination with CCL2 as a cytokine/chemokine therapy, to enhance cytotoxic and T-cell activity in the tumor [128–131]. The pre-clinical results showed that immunostimulatory cytokine therapy improved outcome whereas oHSV expressing Th2-related cytokines eliminated any OV survival advantage over saline therapy. Evidence now shows replicating virus and gene expression is instrumental in attracting immune cells to the tumor microenvironment and inducing the immune-mediated antitumor response as illustrated in Fig. 3 [132]. Recent data also suggests that immune-mediated cell lysis could promote early OV release and spread [133]. In addition to innate immune cell effectors that initially lyse tumor cells and alter the cytokine profile within the OV treated microenvironment (Fig. 3A), this immune cell lysis also plays an important role in the release of tumor associated antigens that are then available for processing and presentation to infiltrating adaptive immune cells (Fig. 3B–D) [134,135].

Increasingly, virotherapy is being used as a gene expression platform to express nonviral genes to improve immunotherapeutic activity. For some viruses this follows a more traditional gene therapeutic approach involving a nonlytic engineered virus (retroviral, lentivirus, adeno-associated virus [AAV], and nonreplicating adenovirus) to conditionally express genes. In addition, conditionally replication competent oncolytic viruses that express immune-modulatory genes are in pre-clinical and clinical studies. To enhance myeloid population in the tumor, investigators inserted GM-CSF [46,136]. To further promote antitumor activity using cell-mediated cytotoxicity via NK cells and CD8 T-cells against tumor, IL-15 was inserted into conditionally replicating HSV alone or with IL-15 receptor alpha [137,138].

6.1 OV immunomodulatory gene expression

Bioactive cytokines and immunomodulatory gene products that enhance the antitumor cellular response are increasingly being engineered into the viral genome to regulate

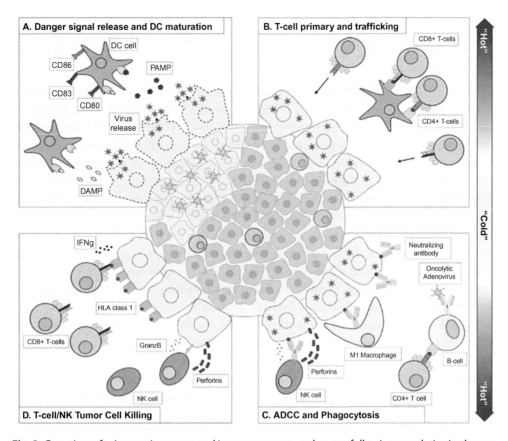

Fig. 3 Overview of microenvironment and immune response changes following oncolytic virotherapy. (A) Initial virus infection produces pathogen associated molecular pattern and danger-associated molecular pattern signals. (B) This enhances antigen presentation of newly uncovered tumor antigen to effector T-cells. (C) OV therapy can enhance antigen-dependent cellular cytotoxicity and phagocytosis of tumor cells and (D) cytotoxic (CD8 and NK) cell killing of the tumor.

immune cells recruited during virotherapy. These viruses expressed immunomodulatory genes are being used to overcome immunosuppressive factors in the tumor environment or to enhance or prolong antitumor immune functions during virotherapy. Viral designs include GM-CSF, IL-2, IL-12, IL-15 as well as CD40L are very common practices to make a tumor more accessible to immune cells. By adding the expression of GM-CSF, the adenovirus JX-594 destroys cancer cells while also stimulating the immune system [139]. By arming a virus with IFNβ expression, the virus would be more selective for cancer cells by increasing the anticancer effect and enhancing virus inactivation in normal tissues [140]. By including the expression of IL-2 and IL-12, the virus can further activate the immune system [141,142]. By inducing CD40L expression in tumor cells, the virus can enhance both cellular and humoral immunity and also improve memory response against further challenge with the same tumor [143]. By inserting a 15-prostaglandin dehydrogenase expression cassette, oHSV can express this tumor suppressor protein that can degrade PGE2, which promotes tumor growth [144]. By targeting PGE2, viruses can help overcome immunosuppression [145].

OVs can also be engineered to express chemokines to stimulate T-cell extravasation into the tumor parenchyma, overcoming the hurdle that is the immunosuppressive tumor microenvironment. These chemokines include CXCL9, CXCL10, and CXCL11 [146], and can induce a strong antitumor response by attracting T-cells to tumors. An oncolytic vaccinia virus expressing CCL5-enhanced therapeutic effects, which correlated with increased presence of the virus and increased levels of infiltrating lymphocytes in the tumor [109]. Others have reported that virus infection alone saturates the CXCL9–11 chemokine response and that viral expression (e.g., CXCL9 from VSV) did not enhance T-cell trafficking despite producing a steep chemokine gradient [147].

Viruses can also be used to induce the expression of tumor-specific antibodies to enhances cellular immune engagement [146]. Yu et al. engineered a virus that secretes a bispecific T-cell engager, consisting of two single-chain variable fragments for CD3 and the tumor cell surface antigen Epha2. This redirected T-cells to the tumor and induced T-cell activation, as well as the expected oncolysis seen with unmodified virus [148]. This engineered virus not only killed infected tumor cells also but enhanced bystander effect of uninfected tumor cells.

Researchers hypothesized that they could combine the tumor lysis of oncolytic viruses with further immunostimulatory activity by combining checkpoint inhibitors. Although there are multiple clinical trials testing the therapy of checkpoint blockade administered with oncolytic viruses separately, researchers are also creating oncolytic viruses that express checkpoint inhibitors in the local infected tumor microenvironment. By having virus expressed antibodies *in situ*, scientists can avoid adverse effects of systemic antibody therapy and ensure that antibody therapy reaches the tumor. Engeland et al. created an attenuated oncolytic measles virus that also encoded antibodies against CTLA4 and PD-1, separately [149]. They observed delayed tumor progression and overall increased survival rates in mouse models. To avoid systemic effects of CTLA4 administration and increase CTLA4 at the tumor site, investigators engineered an oncolytic adenovirus expressing complete human mAb against CTLA4 [150]. They were able to see higher antibody amounts at the tumor, low systemic levels, and T-cells from patients were stimulated. Jiang et al. took their oncolytic adenovirus backbone and engineered it to express murine OX40 ligand, a costimulatory ligand that promotes T-cell division and survival [151]. Jiang et al. were able to maintain the original oncolytic potency of their virus while also efficiently expressing OX40L on infected cells [152]. Compared to their parent virus, the OX40L expressing virus induced higher antitumor activity in immunocompetent glioma mouse models, with greater lymphocyte infiltration in tumor sites and stronger lymphocyte antitumor activity. Passaro et.al engineered a virus expressing a modified antibody against PD-1. They demonstrated that this construct did not alter the oncolytic viral properties of their parent virus, and improved survival in mouse models and generated a memory response protecting these mice upon re-challenged with the same tumor [153].

There are also efforts underway to use OV therapy as a tool to enhance adoptive cellular therapy (ACT): specifically, cell recruitment, cytotoxic activity, and survival within the solid tumor environment. OV therapy induces intrinsic antiviral programs in infected cells that not only create a chemokine gradient that can improve ACT trafficking but also cytokine and IFN changes that can modify or shape the tumor microenvironment by increasing MHC I expression and antigen presentation to improve effector activity [154]. OV infection induces immune activating cell death pathways (necrotic), release tumor associated antigens and, depending on the viral species, may invoke molecular mimicry, bystander damage, epitope spread, or heterologous immunity to enhance immune activity against tumor associated antigens [155,156]. The infection alters antigen presenting cells that can alter the immune-mediated activity. Early studies involving ADV based OVs that express IL-15 and chemokines showed improved neuroblastoma activity using gD2 directed CAR-T cells in a mouse model [157]. However, other ACT combinations involving NK or CAR-NKs are likely also amenable.

Despite success with hematologic malignancies, CAR T-cell therapy efficacy in solid tumors has been stagnant due to the immunosuppressive tumor microenvironment [158] and lack of homogenously expressed antigens in the tumor [159,160]. Oncolytic viruses also can benefit CAR T-cell therapy as a platform for increasing antigen expression or expressing tumor engagers to enhance CAR-T and other ACTs. Oncolytic viruses have been designed to express CD19 in the solid tumor environment in combination with CD19-CAR T cell therapy. This not only increased local tumor immunity but also increased virus release from lysed tumor cells [133]. OV based bispecific T-cell engager (BiTE) targeting EGFR improved CAR T-cell proliferation, activation, and improved cytokine production and cytotoxicity [161]. This combination therapy improved antitumor efficacy and prolonged survival in mouse models, overcoming challenges of either of them used as monotherapies.

BiTEs can also be used to target other immunosuppressing cells in the tumor microenvironment as well. Freedman *et al.* modified an oncolytic adenovirus to express a stroma-targeted BiTE to bind fibroblast activation protein on cancer associated fibroblasts (CAFs) and CD3ε on T-cells [162]. This led to T-cell activation and fibroblast death by inducing activation of tumor infiltrating T-cells to kill CAFs. This led to decreased CAF immunosuppressive factors, increased pro-inflammatory cytokines and even repolarization of M2 macrophages. This wide change of the tumor microenvironment, along with killing of both tumor cells and cancer associated fibroblasts deems this an efficient therapy.

By arming an oncolytic virus to express immunostimulatory gene products, and not just T-cell engagers or cancer antigens, it is possible to modify the tumor microenvironment to improve CAR T-cell activity against the tumor. For example, ADVs have been modified to express IL-15 and RANTES to facilitate migration and CAR T activity [163]. The ADV increased intra-tumoral RANTES and IL-15 release, CAR T-cell recruitment, accelerated caspase pathways, and increased survival of mouse models with neuroblastoma when combined with CAR-T therapy. Similarly, Watanabe *et al.* engineered an oncolytic adenovirus with TNFα and IL-2 and when combined with CAR T-cell therapy increased both CAR T and host T-cell infiltration, M1 Macrophage polarization and dendritic cell maturation in their pancreatic adenocarcinoma mouse model [164]. To improve T-cell recruitment and overcome the reduction in chemokine expression in some tumors vaccinia has been engineered for CXCL11 expression, and increased total and antigen expressing T-cell after CAR T-cell therapy, augmenting anti-tumor efficacy [165].

Interestingly, CAR T-cells have also been used as carrier cells to deliver oncolytic virus (VSV ΔM51) to breast tumors in preclinical studies [166,167]. VanSeggelen *et al.* were able to load murine and human CAR T-cells with low amounts of DNA and RNA viruses, without impacting the receptor expression or function in either of the species T-cells. This allows the systemically administered virus to be protected from neutralizing antibodies [168].

6.2 Oncolytic viruses and cancer vaccines

Oncolytic viruses have also been combined with the use of dendritic cell (DC) cancer vaccines, as summarized by Goradel *et al.* [169]. Dendritic cells are used for cancer vaccines, they are first exposed to cancer antigens and then these antigen presenting cells can present antigen to effector T-cells in the tumor. However, elevated IL-10 and TGF-β levels create an immunosuppressive tumor microenvironment that can limit the antigenic response. In an effort to overcome this, vaccine and oncolytic virus combinations have been proposed as one way to increase efficacy of both treatments. An oncolytic ADV for CD40L expression increased activation and maturation of dendritic cells, improving the antitumor effect over either treatment alone [170]. Vaccinia virus has also been engineered to express tumor associated antigens and used as a cancer vaccine, which is discussed in more detail in the Poxvirus and Adenovirus sections.

6.3 OV prime boost combinations

OVs are attenuated viruses and even when infecting cancer cells that improve their replication, OV gene expression and replication are restricted by immune activity. Repeated OV treatments are increasingly being incorporated into pre-clinical and clinical studies. This not only allows repeated viral infection and oncolytic activity debulking the tumor but also provides an opportunity for prime boost immune stimulation against tumor associated antigens. One approach involves repeated treatments with a single OV agent although there is interest also in substituting different OV types. Potential advantages of the heterologous viral approach include avoiding antiviral immunity involved with repeated treatments with a single agent, capturing other unique immune response characteristics involved with each virus for potential and synergistic activity [171]. VSV and vaccinia virus combinations improved antitumor response in both *in vitro, ex vivo* studies, and mouse models when compared to each virus alone [172]. Sequential administration of ADV and vaccinia produced a synergistic T-cell dependent antitumor activity in a pancreatic cancer hamster model [173]. Oncolytic adenovirus has also been combined with sequential NDV for similar successful results [174]. When combined with anti-PD1 checkpoint blockade, combined administration of reovirus and VSV showed improved antitumor therapy by boosting T-cell primin. [175]. Despite successful results with combined viral therapy, there have been no clinical trials testing the combination of two oncolytic viruses [176].

6.4 OV in combination with other agents

Oncolytic viruses can also be combined with widely used chemotherapeutic substances for a more direct therapy with less use of chemotherapy [177,178]. Ghnomie *et al.* used ruxolitinib, a Janus associated kinase (JAK) inhibitor used to treat polycythemia vera [179] to overcome oncolytic HSV resistance in malignant peripheral nerve sheath tumors (MPNSTs) in syngeneic mouse models [180]. By pretreating with ruxolitinib, they reduced interferon stimulated gene (ISG) expression in MPNSTs, leading to improved viral replication and increased CD8 T-cell activation in the tumor microenvironment. This combination therapy not only augmented oHSV replication but also the immunotherapeutic efficiency of oncolytic viral therapy. Mahalingam *et al.* conducted a phase 1 clinical trial to treat pancreatic ductal adenocarcinoma (PDAC) with pelareorep (an oncolytic reovirus), pembrolizumab (a humanized monoclonal anti-PDL1 antibody) [181], and either 5-fluorocil, gemcitabine, or irinotectan, all chemotherapeutic agents, to overcome the PD-L1 upregulation seen in PDAC when only treated with pelareorep [182]. The combination of the three therapies did not add significant toxicity and showed encouraging efficiency, with 30% disease control achieved in the cohort.

7. Oncolytic viruses as gene expression platforms

In addition to replication and lysis of tumor cells, Virotherapy provides a platform for gene expression to enhance tumor cell death, gene modification within the tumor and immunomodulation of the tumor microenvironment.

A replication incompetent integrating retrovirus was used in 1989 by Rosenberg to transduce tumor infiltrating lymphocytes with a neo marker gene. This neo marker gene was found in the tumor and in peripheral blood up to 3 months post infusion [183,184]. Although this first attempt was to simply mark and follow the cells, it paved the way for retrovirus use to deliver gene replacement therapy [185], as well as early thymidine kinase (TK) based delivery and cytotoxicity studies.

An example of this approach is the expression of cytotoxic genes, either alone or in combination through pro-drug converting enzyme expression and drug combinations in virotherapy. These enzymes can then metabolize pro-drugs that are nontoxic to normal cells but become toxic inside tumor cells and would kill these cells [146]. These drug-sensitivity genes also have a bystander effect in that they not only lead to cancer cell death but death of these cells leads to spread of the toxic metabolite to nearby non infected cells, killing peripheral tumor cells. Fend *et al.* constructed a vaccinia virus containing the fusion suicide gene FCU1, which is derived from the yeast cytosine deaminase gene and uracil phosphoribosyltransferase genes. When they administered pro-drug 5-fluorocytosine, there was sustained tumor growth control but not extended survival compared to experiments with the same virus but no 5-FC treatment [186]. Most suicide genes studies are on this cytosine deaminase gene, which catalyzes the deamination of cytosine to uracil and also metabolizes nontoxic 5-fluorocytosine to toxic 5-fluorouracil. 5-FU irreversibly inhibits thymidylate synthase leading to blocking the conversion of deoxyuridine nucleotides to deoxythymidine nucleotides and inhibits RNA synthesis. [187,188]. Drugs to drive more specific oncolytic virus replication in cancer cells are also used and may be placed under a cell-specific promoter. Osteocalcin is a protein expressed highly in solid tumors. An adenoviral oncolytic virus was put under an osteocalcin promoter driven replication. Prodrug was administered and led to clearance of prostate cancer cells. The osteocalcin promoter also led to targeting of the bone stroma that had been intertwined with prostate cancer at skeletal metastatic sites. This showed maximal tissue-specific cell toxicity, targeting both prostate cancer cells and bone stroma [189].

8. Oncolytic virus platforms and clinical trials

When oncolytic viruses advance to clinical trials, safety is paramount. Early studies focused on proving that OVs could be safely administered to patients [190]. Prior to genetic engineering, virotherapy was limited to clinical isolates with oncolytic activity. Toxicities associated with these therapies limited progress. With the advent of genetic engineering, viruses designed in the early 1990s were conservatively designed out of safety concerns related to the earlier virulent OVs and these highly attenuated OVs replicated poorly due to their multiple attenuating mutations. These OV's had little dose-limiting toxicity (mild flu-like symptoms and local reaction at the injection site) however, they were highly attenuated and there were concerns regarding their efficacy. Another important component of these early studies was the dose-response for viral vectors. Unlike small molecule therapies, biologics do not follow a traditional linear dose-response relationship. Studies showed that for some virotherapeutics the maximum tolerated dose was not necessarily the most efficacious dose. The kinetics and dynamics of virotherapy were elucidated during these early studies that defined shedding, viremia, replication, genomes, and viral loads. Now many OVs have been engineered with reporter genes (*e.g.*, NIS) to monitor or analyze viral replication/gene expression in the treated tumor as well as the immune response characteristics. Early in OV development the immune response was viewed as a detriment to effective virotherapy: efforts often were focused on how neutralizing antibodies and the innate/intrinsic antiviral responses limited virotherapy [191]. Current efforts now rely on modifying and in many cases enhancing the virotherapy associated immune response. Finally, delivery of oncolytic virus must be efficient. Pharmacokinetics, pharmacodynamic, and biomarkers differ for biologics. Early studies used single dose direct injection approaches. Current trials now often include repeated OV treatments or methods to enhance virus tumor cell access (CED). There are many OV platforms, and most have incorporated genetically engineered transgene expression as an approach to enhance immune or cytopathic activity (Table 1). FDA approved therapy and other OVs in late-stage studies incorporated GM-CSF expression. Popular therapeutic transgenes expressed by OVs include cytotoxic and Th1 agonist cytokines (IL-2, IL-12, IL-15, IL-18, Type I IFN), soluble co-stimulatory agents (CD80/86, CD40L), and chemokines (CCL3, CCL5, CXCL9) as well as combinations of these. There is also interest in xenogeneic antigen and tumor associated antigen expression.

8.1 Pox viruses (vaccinia, myxoma, racoonpox, and others)

Poxviruses are double stranded DNA virus with large well characterized genomes (130–300 kb in size) that produce both enveloped and nonenveloped forms of infectious virus. Poxviruses carry a large complement of genes and unlike other DNA viruses replicate in the cytoplasm of the infected cell [192]. Theoretical advantages of poxviruses as an oncolytic platform include its rapid replication cycle that occurs in the cytoplasm limiting integration potential, its ability to replicate in hypoxic conditions, broad tumor cell tropism (related to lack of specific receptor and entry into a large range of cells through micropinocytosis) and its well characterized genome, safety record (in the case of vaccinia virus as the smallpox vaccine) and ease of recombineering [54,193]. Poxviruses rely on post-entry host response differences that restrict viral replication for conditional replication in tumor *versus* nonmalignant cells [194]. Thus far six poxviruses have been developed as potential OVs: the smallpox vaccine virus Vaccinia Virus (VV), Cowpox (the original Jenner vaccine), Myxoma virus (MYXV, a pox virus of rabbits), Racoonpox virus (RCN), Squirrelpox virus, and Yaba Monkey Tumor Virus. For this chapter the most widely developed pox viruses VV and Myxoma virus will be primarily discussed.

Table 1 Overview of characteristics of different OV vectors that have been through clinical trials.

Viral system	Adenovirus (ADV)	AAV	HSVs	Vaccinia virus pox viruses	Reovirus	Measles virus	Poliovirus	Newcastle disease virus (NDV)	Retrovirus	Vesicular stomatitis virus (VSV)
Genome material	dsDNA	ssDNA	dsDNA	dsDNA	dsRNA	ssRNA	ssRNA	ssRNA	RNA	ssRNA
Genome size	36 kb	8.5 kb	152 kb	192 kb	18.2–30.5 kb	16 kb	7.5 kb	15 kb	7–11 kb	11 kb
Enveloped	No	No	Yes	Yes	No	Yes	No	Yes	Yes	Yes
Biosafety level	BSL-2	BSL-1	BSL-2	BSL-2 +	BSL-2	BSL-2	BSL-2	BSL-2	BSL-2	BSL-2/3 (lab adapted)
Insert size	8–36 kb	5 kb	30–40 kb		None				8 kb	N/A
Clinical trial Titers Particles/ml (prtcls/ml) Plaque forming unit (PFU) Tissue culture infective dose (TCID)	1 × 10o13 (prtcls/ml)	1 × 10o11	1 × 10o5-3 ×o9 PFU	1 × 10o9	3 × 10o10	1 × 10o9	1 × 10o10	1 × 10o10	1 × 10o9 TU	
Transgene expression	Broad, low for blood cells	Broad, low for blood cells	Broad	Broad	Broad	Broad; lymphocytes	Intestine brain and spinal cord, DC	Broad	Broad (pan or pseudo types)	pan or Broad
Infectivity	Dividing and Nondividing cells	Dividing and Nondividing cells	Dividing and Nondividing cells	Dividing	Ras/MAPK upregulated: PKR/IFN defective	Dividing and Nondividing cells	Dividing and Nondividing cells	Dividing	Dividing	Dividing and Nondividing cells
Transgene expression	Transient/Prolonged	Stable	Transient	Transient	N/A	Transient	Transient	Transient	Stable	Transient
Vector gene form	Episomal, Nonproductive infection can integrate into Host DNA	Episomal (90%), site-specific integration (10%)	Episomal	Episomal	Episomal	Episomal	Episomal	Episomal	Integrated	Episomal
Lytic Virus (OV) Gene expression platform	Either	GE	OV	OV	OV	OV	OV	OV	GE	OV
Advantages	Versatile Flexible platform (gene expression or oncolytic), High titers, Well characterized viral genome and biology, Established techniques for genome editing, Efficient transduction of receptor (+) tumors, Inflammatory	Safe transgene delivery, noninflammatory, nonpathogenic	Well characterized viral genome and biology, Established techniques for genome editing, Large packaging capacity; Broad cellular and tumor tropism, efficient spread numerous entry methods including neurotropic. Inflammatory, ability to evade preexisting antiviral immunity Antiviral therapy available with good therapeutic index, Rapidly replicates and lyses tumor cells	Well characterized viral genome and biology, Established techniques for genome editing, Large packaging capacity; Safe vaccine profile, Immunogenic, Replicates and lyses cells rapidly, Broad tumor tropism, no risk of integration, Efficient spread	Orphan Virus infection with low risk of disease, Low nonmalignant cell infectivity, immunogeneic cell death	Fusion mediated spread, Receptor mediated entry often expressed on malignant cells, lab adapted attenuating mutations and engineered tumor-specific receptor cell targeting,	Inflammatory, Rapid replication cycle lysis and spread,	Naturally occuring oncolytic virus, induces immunogenic cell death, Rare pre-existing immunity in humans (farm-workers), rapid lytic cycle. Newer OVs encoding immune active genes	Persistent gene transfer in dividing cells, an orchestrate lytic effect with drug administration	Well-characterized biology, Broad cellular tropism, No host-cell transformation, Small, engineerable genome, Rare pre-existing immunity in humans, rapid lytic cycle

Potential Disadvantages	CAR-variability in human cancers and expression on normal cells, preexisting antiviral immunity, hepatic adsorption, toxicity Capsid mediates a potent inflammatory response (eliminated in HC-ADVs). Antivirals exist but narrow therapeutic index (e.g., Cidofivir),Narrow cell tropism for some serotypes (genetic modification can solve this)	Small packaging capacity, requiring helper ADV for replication and difficult to produce pure viral stocks. Immune activity limits re-treatment	Inflammatory; transient gene expression during lytic infection, Potential for latent infection Naturally neurotropic and neurotoxic – must be genetically modified to limit pathogenic activity	Inflammatory Antivirals exist but narrow therapeutic index (e.g., Cidofivir), Transient gene expression, potential difficulties with systemic delivery	Complex reverse genetic system, Prior studies used WT Reo and not engineered or transgene expression	mass immunity exists, virus-directed immunosuppressive effects may limit OV immuno-therapeutic activity (can potentially be overcome by arming the virus)	Packaging limits for virus encoded genes, Theoretical concerns for genetic stability (in vivo and in vitro studies however have showed stability), viral receptor not homogeneously expressed in all tumor cells, lack of antiviral therapy	Environmental Risk (Avian pathogen), Potential for systemic toxicity, Less immunogeneic in humans	Only transduces dividing cells; integration can induce oncogenesis in some applications.	Small genome naturally neurotoxic in omice and NHPs must be modified to limit pathogenic activity, IFN sensitivity may limit activity in some tumors. Infects both malignant and nonmalignant cells. Envelope glycoprotein (VSV.g) stimulates good neutralizing antibody response
Furthest stage clinical trial	Phase III clinical trials	Phase III clinical trials	Phase III clinical trials, FDA approved	Phase II clinical trials	Phase III clinical trials	Phase II clinical trials	Phase I clinical trials	Phase II clinical trials	Phase I clinical trials	Phase I clinical trials

The human poxvirus vaccinia virus is naturally oncolytic with improved replication in cancer cells over nonmalignant human cells but to improve selective replication, vaccinia OVs contain additional genetic modifications. In contrast, host range restrictive replication for many of the other oncolytic poxviruses (MYXV, RCN), allow them to be safely used in humans without additional virulence modifications. Oncolytic vaccinia virus constructs rely on viral thymidine kinase (TK) mutation and ribonucleotide reductase mutations. These mutations make the virus dependent on cellular TK and cellular nucleotide stores and thus restricts efficient viral replication to proliferative cells such as cancer cells that can complement these genetic deficits [195,196]. In addition, VV in pre-clinical and clinical development contain VGF or F4L mutations further limiting viral replication to cancer cells. The VGF mutation eliminates a key EGF-like viral proliferation factor whereas the F3L mutation eliminates the viral ribonucleotide reductase. Both of the gene mutations further restrict viral replication to cancer cells that can provide a sufficient replication environment [197]. To also target the apoptotic pathways of cancer cells, antiapoptotic genes SPI-1 and SPI-2 have also been engineered out of the VV genome to further enhance tumor selectivity [198]. Others have combined mutations that restrict viral RNA transport (de-capping enzymes D9 D10) [199], metabolic, or signaling pathways to ensure viral replication advantage in cancer cells. As a general rule, as virus constructs incur additional gene deletions these improve safety and conditional replication but can also reduce efficient viral replication. Chimeric pox virus constructs borrow genetic elements from naturally occurring but species restricted pox viruses and provides a strategy to provide missing genetic elements and improve viral replication in the tumor [200–202].

In addition to mutations that improve conditional replication, other gene modifications improve immunotherapeutic activity of the virus. Gene deletions that alter VV control of metabolic and immune cell signaling, for example, enhance the immune response against VV treated tumors [203]. Other immunotherapeutic modifications consist of genetic insertions designed to "arm the virus" with immunomodulatory cytokines/lymphokines, chemokines, costimulatory proteins or single chain immunoglobulins that target immunoinhibitory receptors. For example, JX-594, the VV furthest along in clinical development, contains GM-CSF as an immunomodulator that was also included in TVEC, the only FDA-approved OV [204,205]. JX594 has a good safety profile and dose-related survival for patients with hepatocellular carcinoma [206]. IL-24 in oncolytic VV also showed good results against lung cancer by inducing apoptosis [207], and oncolytic VV with CXCL11 and IL-15 induced potent antitumor effects [137,208]. Others have inserted cytokines, such as IL-10, that reduce the antiviral response. In contrast to the experience with HSV OV therapy where this approach reduced OV immunotherapeutic activity, investigators using a murine pancreatic cancer tumor model prolonged viral replication without reduction in antitumor immunity [209]. The IL-10 expressing VV produced similar cytotoxicity and replication as the parent virus, but there was a reduction in antiviral CD8 T-cells without a change in antitumor CD8 T-cells. On the contrary, other mutations were inserted to neutralize type 1 IFN pathway, by deleting the B18R gene in order to get a pathway reinforcing VV that could promote the IFNβ pathway [140]. Others have focused on enhancing the cytotoxic response using IL-15, membrane tethered-IL2, or TNFα expression or improving immune cell recruitment through chemokine expression (CCL5, CXCL11, CCL19) or costimulatory molecule expression (CD80, CD86) [54,137,208,210–212]. For a more novel approach, oncolytic VV investigators also engineered a secretory bispecific T-cell engager consisting of two single chain fragments specific for CD3 and tumor cell antigen EphA2 [148]. In addition to directly stimulating immune cell activity, gene inserts that modify cancer cell death pathways (apoptosis, necrosis, autophagy) or enhance tumor associated antigen presentation (*e.g.*, oncoVV engineered to express Lectin) provide another path for immunotherapeutic activity.

8.1.1 Vaccinia

VV as oncolytic virus therapy induces durable immunity. VV strains used for oncolytic viral therapy often are derived vaccinia used to eradicate smallpox and there were theoretical concerns regarding its efficacy in former vaccines; however, clinical trials show that this virus is safe and selectively replicates in cancer cells. [213]. There were initial concerns regarding feasibility of repeated treatment with VV due to its vaccine activity and durable immune response. Vaccinia viruses naturally express viral particles called extracellular enveloped virus (EEV), which help VV evade antibodies [214,215]. When oncolytic VV strains are enhanced with EEV, there is enhanced VV spread within tumors and reduced clearance by antibodies [216,217]. Other investigators have leveraged this durable immune response and used VV platforms to enhance the Th1 response or to express immunoreactive antigens to enhance the antitumor vaccine effect. To further shift cellular immune activity within the OV treated tumor, efforts have also focused on enhancing cytotoxic activity using bispecific engagers, or by reducing inhibitory receptor activity such as CD44 "don't eat me" receptor often expressed by tumor cells to limit macrophage activity. By targeting these processes in the local OV treatment environment, these approaches limit potential untoward side effects from widespread expression of these gene products. This is also the logic behind viral based checkpoint molecule expression. While PD1/PDL1 and CTLA4 monoclonal therapy is FDA approved and can be combined with virotherapy to enhance immune activity, these immunomodulators have off target activity and this dose-limiting toxicity profile limits clinical use and efficacy in a large proportion of treated patients. By expressing these immunomodulators from the virus and within the tumor microenvironment, this has the potential to reduces off-target activity and improve compliance. The limitation however to this approach is that gene expression from a lytic viral vector is also temporally restricted and therefore dependent on efficient viral replication and spread to newer uninfected cells for prolonged local expression.

8.1.2 Vaccinia virus as a cancer vaccine platform

VV and other poxviruses (MYXV, Fowlpox, modified vaccinia virus ankara [MVA]) have been investigated for decades as a cancer therapeutic. Studies investigated using VV infected melanoma cell lysates as a cancer vaccine in the 1980s and 1990s in early-stage Phase I and II studies. Since the trials of JX-594, VV mutants have entered clinical trials of their own, specifically as cancer vaccines [37]. TroVAX, an MVA strain, has been tested with a variety of immunostimulatory agents such as IL-2, IFNα, and subitinib for metastatic renal cancer in phases II and III clinical trials [218,219]. VV with epitopes from several proteins such as gp100, or MART-2 for tumor associated antigens, and with immunostimulatory agents CD80 and CD86, were tested in phase I and II clincal trials against melanoma [220,221]. TG4010 was tested in combination with chemotherapy, with MUC1 TAA and IL-2, tested in nonsmall cell lung cancer (NSCLC) in phase 2b clinical trials [222]. MVA-brachyury-TRICOM with brachyury TAA and B7.1, ICAM-1, LFA3 in phase I clinical trials for patients with advanced cancers [223]. PROSTVAC with expressing PSA as well as the previously mentioned TRICOM, was tested in phase II clincal trials for prostate cancer [224,225]. PANVAC was tested in combination with chemotherapy expressing CEA and MUC1 with no immunostimulatory agents however, just virus or chemotherapy alone, tested in phase II clinical trials in patients with metastatic breast cancer [226].

8.1.3 Vaccinia virus: JX-594 (Pexa-vec)

Pre-clinical studies utilizing genetically modified poxviruses are numerous and focus on increasing direct viral and immunogenic cell death. This enhanced cytopathic activity is achieved by enhancing viral replication or through expression of gene products that are directly cytopathic or in some instances act as complementary factors (e.g., prodrug converting enzymes) that when combined with a relatively innocuous drug produces cellular toxicity. Other modifications recruit and activate immune cells to facilitate immunogenic cell death. Thus far, JX-594 (Pexa-vec) is the pox virus furthest along in human clinical trials. JX-594 is an engineered TK mutant VV that was derived from the Wyeth VV strain and is armed to express human GM-CSF and capable of intravenous delivery to tumors [227]. Pexa-vec has been evaluated in 17 clinical trials with over 300 patients treated for adult cancers (hepatocellular carcinoma, colorectal cancer, breast cancer, sarcoma, renal cell carcinoma, melanoma, squamous cell carcinomas) and pediatric solid tumors (Neuroblastoma, rhabdomyosarcoma, and Lymhoma) [228]. The virus was safe and produced dose related antitumor activity. The first studies were performed in patients with refractory primary or metastatic liver cancer. Fourteen patients received intratumoral injections of the virus, and the only dose limiting factor was hyerbilirubinemia.

Antitumor responses were seen in both injected and noninjected tumors [229]. Further phase I clinical trials were done for metastatic melanoma [230] and for treatment of pediatric cancers, both of which showed safe patient responses to the virus. In Phase II trials of Pexa-vec for patients with advanced hepatocellular carcinoma the virus was administered intravascularly. There was induction of GM-CSF and response in injected and distant noninjected tumors. The high dose treatment group improved median survival when compared to low dose (high dose 14.1 ms *vs.* low dose 6.7 ms: $P = 0.02$) in this phase II study [206]. However, in phase IIb studies, it did not prolong overall survival when compared to patients treated with standard therapies in more immunologically compromised hosts. These studies taken together and the observation of responses in both local and metastatic tumor suggesting an immune-mediated component contributed to Pexa-Vec therapeutic success. Pexa-Vec has recently been tested in a Phase III study (completed December 2020 but no results at the time for this chapter) comparing it prior to Sorefenib therapy in patients with advanced HCC.

Two other VV candidates are also in early-stage study. GL-ONC is a lister strain derived VV that encodes several reporter genes (rLUC, GFP, and B-galactosidase) used to mutate the viral F14.5, TK, HA genes. It was safely administered to more than 89 patients with various cancers (ovarian, head and neck carcinoma, and lung cancer) [54,210]. Finally, a third VV derived from the more virulent Western Reserve strain is also in clinical trial. This virus was engineered with 2 mutations (TK and VGF) to render it less pathogenic and has been administered to patients by direct IT injection in advanced solid tumors and by Intravenous delivery in patients with advanced colorectal cancer [231,232].

8.2 Adenovirus

Adenovirus (ADV) a nonenveloped, double stranded DNA virus from the *Adenoviridae* family was the first virus that was granted approval for its use as therapy from a government agency in China in 2005 [50]. Adenovirus is versatile and can be used as a conditional replication competent oncolytic virus or as a replication defective virus for gene delivery or as a vaccine platform. There are 57 distinct adenovirus serotypes and they fall within 7 different subgroups (A-G). Serotypes often have different receptor preferences. Most ADV OVs are derived from Serotype 5 and bind preferentially to the coxsackie and adenovirus receptor (CAR), while other serotypes bind desmoglein, CD80/86, or CD46. Receptor expression is a critical determinant in ADV entry and not all tumors are accessible to ADV due to this receptor specificity. After entry, the virus uncoats, its DNA is delivered to the nucleus, and gene expression occurs in a stepwise and coordinated fashion. After production of progeny viruses, the viral infected cell membrane is disrupted and the dying cell releases virions in a highly immunogenic and inflammatory process allowing infection of surrounding cells [233]. Wild-type ADV infection produces flu–like symptoms but in the immunocompromised host and based upon the ADV sero-type target organ, wild type infection can be lethal. To attenuate the ADV for conditional tumor infection, genetic engineering is required. Oncolytic ADV utilize several mutations that render them safe to use and to improve their tumor activity. Most contain conditional mutations involving the E1A and E1B genes that restrict subsequent gene expression and translation in the infected cell. E1A initiates adenoviral gene expression, by dissociating the Rb/E2F complex, leaving E2F free to be activated for transcription of the remaining viral proteins during WT ADV infection [234]. ADV with E1A mutations cannot bind the Rb protein and are unable to shift the infected cell into S phase. Since cancer cells often contain defects in the p53/RB axis and are highly proliferative, this can complement OVs containing this gene defects allowing preferential replication in cancer cells [235]. Smaller E1A mutations (*e.g.*, a 24 base pair deletion) have been utilized to conditionally restrict ADVs while maintaining other ADV gene functions. This continues to suppress viral replication in nonmalignant cells but improves virus activity cancer cells [236–238]. An alternative approach is to conditionally express E1A and E1B using a cancer-specific promoter [239]. Investigators have utilized specific promoters (hTERT, p53, carcinoembryonic antigen [CEA], prostate-specific antigen [PSA], E2F, or Cox2l promoters) to restrict essential ADV gene expression and virus infection to certain cells enriched with the corresponding transcription factors [240–246]. To facilitate tumor cell entry in the event of low CAR expression on malignant cancer cell, investigators also have modified the entry receptor domains to include arginine, glycine, aspartate (RGD) motifs present in fibronectin to aid in tumor cell binding through integrin receptors [236,247,248]. In a parallel approach and as a method to overcome serotype-specific neutralizing antibodies, investigators have also substituted serotype fiber knobs and capsid components (serotype switching) or re-engineered vectors in a different ADV serotype to broaden cell tropism and improve infection [240,248–251]. ADV-vectors provide a uniquely flexible platform and can be designed for selective replication and oncolytic activity in cancer cells as described above or as a gene therapy agent using ADV with large genetic deletions allowing larger gene transfers [132]. These so called "gutless" constructs have also been engineered such that they remain largely episomal DNA elements or can contain transposable elements as hybrid constructs that allow chromosomal integration.

In addition to complete E1A gene deletions, there are also a variety of transgenes that have been inserted to enhance efficacy. For example, the addition of suicide genes, such as cytosine deaminase and thymidine kinase allow the use of pro-drugs that only target cells expressing those enzymes [252,253]. Other transgene additions focus on enhancing the immune profile surrounding the cancer cells. The addition of immunomodulating cytokines, such as GM-CSF [237,238,245,254,255], TNFa, IL2 (NCT04217473), IL-12 [256] increasing the antitumor immunity. With the addition of such genes, oncolytic ADV have become a somewhat of a customizable therapy, lending itself to be changed to conform to the cancer-specific challenges presented

ADVs have been used over 800 times in patients in hundreds of clinical trials through the advanced therapy access program. While ADV retargeting and gene expression that enhances therapeutic activity and tumor cell lysis has been frequently used in these studies, the focus here is as an immunotherapeutic agent. An entire chapter could be devoted for each virus platform. Examples are provided here to highlight generalized approaches available and are not an exhaustive summary. There has been outstanding work in ADV OV therapy and any omission of outstanding investigators and their seminal studies is unintentional and primarily due to space constraints. Current ADV efforts focus on [1] enhancing DC recruitment and their maturation within the tumor microenvironment, [2] priming T-cell and cytotoxic activity against tumor associated antigens released during oncolysis, and [3] maintaining that activity by checkpoint therapy. ONCOS-102, for example, expresses GM-CSF to enhance T-cell priming and has shown increased peripheral T-cell levels and CD8(+) infiltrates in biopsied patients [257]. Other investigators are expressing genes that enhance DC and immune cell migration and maturation such as FMS-like tyrosine kinase 3 ligand (*FLT3L*). These approaches however enhance recruitment of other myeloid cells including MDSCs that threatens to re-establish an immunosuppressive microenvironment after initial antitumor myeloid activity. Others have incorporated ADV based IL-12 therapy. Zyopharma has a platform using an oral pro-drug that promotes ADV-based IL-12 expression that permits a tunable or dose adjustable approach for cytokine expression. While initial efforts with IL-12 therapy produced dose limiting toxicities, this cytokine-mediated approach has become a mainstay of OV immunotherapeutic approaches and has been used successfully in oHSV, ADV and other viral vector platforms including VV. In addition to IL-12, other pro-cytotoxic molecules (CD40L, Ox40L) and Th1 response cytokines (IL-15) and checkpoint targeted therapies have been combined in the OV platform to enhance and prolong NK and T cytotoxic response activity. Additionally, efforts to express tumor associated antigens (*e.g.*, familial adenomatous polyposis gene [FAP], EGFR, folate receptor alpha [FR-α], Ephrin A2 [EphA2]), have also been incorporated into the ADV genome for expression to improve the durable immune response against the tumor. Others have also used ADV based gene expression to express enzymes, fusogenic proteins, or response modifiers to alter the stromal TME and enhance viral dissemination and the immune response [132]. In addition to tumor-specific targeting, others have engineered ADVs to modify the tumor microenvironment and noncancerous cells, such as ECM and angiogenesis. Adenoviruses have been made to secrete relaxin [61] and hyalauronidase [62] to disrupt ECM. Relaxin increases expression of MMPs, which degrade matrix and allow for easier virus access to other cells to infect [63]. The virus expressing hyaluronidase led to degradation of hyaluronic acid, a component of ECM, and led to prolonged survival in mice glioma models [62]. Several genes have been added to improve adenovirus tumor cell activity. For example, Lv et.al, created viruses that coupled expression of p53 with penetrating peptide 11R, and GM-CSF. Their results showed that the addition of GM-CSF produced a synergistic effect when compared to the 11R–p53 expressing virus and that this combination was effective at targeting both hepatocellular carcinoma and teratoma stem cells [258]. An alternative approach to produce oncolytic activity involves downregulating regulatory proteins by targeting gene transcription. Virus based short-hairpin RNA expression has also been incorporated into the genome so that it binds and downregulates dicer, an endoribonuclease that functions in processing virus-associated RNA. By downregulating dicer, the virus actually inhibits the destruction of viral RNA, allowing it to replicate efficiently in tumor cells [259]. By knocking down dicer, they saw higher replication efficiency and tumor cell lysis compared to non siRNA expressing control adenovirus, and most importantly, the dicer levels and viability of normal cells were not affected because they have a virus with a tumor-specific promoter-driven E1 gene expression cassette.

Neutralizing antibodies from prior ADV infection may limit adenovirus oncolytic viral therapy. These antibodies are often serotype-specific and if the virus has been previously encountered, immune clearance and inefficient entry can occur especially with treatment strategies involving intravenous delivery. Others suggest that the therapeutic impact of these antibodies is less pronounced for therapeutic ADV vectors. OV infected tumors and therapeutic administration releases large amounts of virus in contrast to wild-type ADV and may overwhelm neutralizing antibody in the serum [254,260]. Serotype switching provides one strategy for evading these neutralizing antibodies. Others have coated the virus with albumin to provide a protective barrier and facilitate systemic delivery [261].

8.2.1 ADV clinical studies

Adenovirus-based oncolytic therapy is one of the most mature and flexible of the viral therapies and has been used in over 100 clinical trials as an oncolytic, gene expression or vaccine vector (105 completed or terminated Cancer/ADV trials according to Clinical Trials.gov). At the time of writing there are 86 studies involving ADV therapy for cancer that are active on clinical trials.gov including a phase III combination therapy study involving the E1B mutant H101 in China. Most are early-stage Phase I/II studies. Owing to the sheer number of ADV-based studies it is not possible to present data from all ADV-related trials. Instead, we will highlight certain ADV trials to highlight changes within the field. Any ADV OV omissions in this chapter are unintentional and related to limitations in space. Examples are provided to illustrate ADV modifications and how clinical assessment has led to subsequent virus modifications to improve the clinical response.

8.2.2 Adenovirus TAA vaccine approaches

Replication defective ADV based TAA expression is one approach that investigators have used to improve immune-mediated antitumor activity. Serotype 5 ADVs that express carcinoembryonic antigen (ETBX-011: Ad5-CEA) or prostate-specific antigen

(CV-706: Ad5-PSA). These replication defective vectors contain multiple early gene deletions (E1A/E1B, E2B [viral polymerase] and E3) and express the tumor associated transgene in the infected cells but do not undergo cell to cell spread. CV706 (Ad5-PSA) induced a cytotoxic T-cell response in pre-clinical models and in phase I study the majority of treated patients developed an anti-PSA T-cell response and had improved survival when compared to historical controls (nomogram) [262]. Studies in patients with local recurrent prostate cancer after radiation showed a significant decrease in PSA when high dose of CV706 was administered [244]. Investigators have also seen beneficial responses with the Ad5-CEA (ETBX-011). Twenty-five patients with metastatic colon cancer tolerated the therapy and exhibited CEA-directed T-cell responses in the phase I study with a mean overall survival of 11 months [263,264]. A theoretical concern with vaccine strategies is antigen expression related to tumor heterogeneity or antigen escape. To overcome this, The ETBX-011 platform has been further modified to express the MUCI gene (ETBX-061) or a modified Brachyury gene (ETBX-051) and a combination therapy involving the triple combination (ETBX-011, ETBX-061, and ETBX-051) given every 3 weeks for three doses then every 8 weeks for up to 1 year. The 10 treated patients tolerated the vaccination strategy with only one DL1 toxicity (injection site reaction and flu-like symptoms). Patients developed antigen-specific T-cells (CD4+ and/or CD8+ T-cells) capable of persistence to at least one of the vaccine-encoded TAAs and patients had stable disease [265]. MUC1 reactive polyfunctional T-cells predominated and were detected in 50% of patients, whereas CEA, or brachyury polyfunctional T-cells occurred less frequently and were seen in 33%, and 17% of patients, respectively [265]. Vaccine strategies are also being pursued against MAGE for melanoma and the papillomavirus E6 and E7 proteins for HPV-positive cancers. These nonreplicating ADV-based clinical trial results demonstrate that these vaccines can activate the immune response; however, when used alone they have produced few positive outcomes and limited clinical response in patients with advanced cancers. Efforts now focus on improving the clinical response by combining these vaccines with other immunotherapies (checkpoint therapy) and with OV-based combinations [266]. For example, investigators are also pairing ADV based MAGE or E6/E7 vaccine therapy with oncolytic Marabavirus-based MAGE or E6/E7antigen expression and anti-PD1 therapy (NCT 03618953) and (NCT02285816) [267].

8.2.3 Conditionally replication competent ADVs

ONYX-015 and Dl1520 are genetically modified adenoviruses with E1B gene deletions that improve their replication in p53 mutated cells and have been used in several clinical trials. Dl1520 was tested in patients with hepatocellular carcinoma and was well tolerated in early phase studies but did not show tumor control compared to control arm of the study [268]. ONYX-015 was administered to 20 patients with hepatobiliary tumors in a Phase II trial in to determine efficacy and safety and resulted in a 50% reduction in tumor markers with few serious toxicities [269]. It was also used in a phase I open label dose escalation study for treatment of malignant glioma. Twenty-four patients underwent peri-tumor delivery of ONYX-015 following tumor resections with no serious adverse effects. One patient underwent repeat tumor resection which showed evidence of lymphocytic and plasmacytoid cells infiltrate following therapy [270]. ONYX-015 was also tested in a Phase I/II trial in combination with MAP chemotherapy for advanced sarcoma and was well tolerated and there was some evidence of antitumor activity, but therapeutic impact was minor [271].

DNX-2401: A potential limitation of the early ADVs was specificity (untargeted replication) and the paucity of CXADV expression in malignant tumors. To overcome this, Fueyo modified the ADV fiber knob of an E1A Δ24-ADV to include an RGD domain allowing ADV integrin-binding taking advantage of this upregulation in gliomas. DNX-2401 was safely used in Phase I/Ib studies in patients with recurrent GBM with 55% of resected tumors (D14 post-injection) active viral infection (E1A or hexon staining) [236] and is now being combined with systemic PD-1 inhibitor therapy (NCT02798406). In the original study, Lang *et al.* also demonstrated CD8/Tbet (+) immune infiltrates, immunogenic cell death and identified some patients with long term survival. The DNX-2401 backbone has been subsequently modified for OX-40L transgene expression (DNX-2440) and is in Phase I study (NCT03714334) and in patients with recurrent high-grade gliomas. DNX 2401 delivery by CED was also studied in a Phase I/II study in the Netherlands (NCT01582516) and completed in 2014 but at the current time there are no published results from this trial [236].

As an alternative to the RGD mutations, investigators have also modified the Ad5/3 fiber using the ADV-Δ24 backbone to enhance tumor cell infection. ONCOS-102 takes this Δ24-E1A Ad5/3 fiber backbone and expresses GM-CSF. In their Phase 1 study, ONCOS-102 induced a strong immune cell infiltrate (2.5-fold CD4 and 4-fold CD8) in 12 patients with advanced solid tumors [272]. Investigators also identified PD-L1 upregulation on mesothelioma tumor cells associated with the therapy and clinical trials are examining ONCO-102 with pembrolizumab in patients with advanced melanoma (NCT03003676). A third of the treated patients had stabilization of disease or regression and of the 7 patients who had paired biopsies post-treatment most had evidence of CD8/PD1(+) infiltrates in their biopsy and 4 of 7 had tumor-specific T-cells (anti-MAGE or anti-NY-ESO1) develop or increase following therapy [273]. . In addition to GM-CSF, other cytokines and immune active molecules are being expressed from oncolytic ADVs and have advanced to clinical trial. AdCD40L is an oncolytic adenovirus that expresses CD40L and induced systemic immune effects that correlated with improved survival of patients in malignant melanoma in a phase I/IIa study [274]. LOAd 703 is an oncolytic ADV that carries both CD40L and 4-1BLL to activate these 2 immunostimulatory pathways [275] and has advanced to early clinical trials for patients with pancreatic and ovarian cancer (NCT02705196 and NCT03225989). Both replication competent and defective

ADVs expressing various cytokines are advancing in early-stage study. Ad-RTS-hIL-12 is a replication defective ADV that expresses hIL12 through an oral prodrug regulated approach (the RheoSwitch Therapeutic System). When the oral prodrug (Veledemix) is administered it combines with a transcriptional partner to induced hIL-12 transcription and expression from the ADV. This has been safely used in 31 patients with recurrent glioblastoma [276], TILT-123 is another replication competent ADV that expresses hTNF-α and hIL-2 and is being combined with Tumor infiltrating Lymphocyte therapy in phase I clinical trials in combination with tumor infiltrating lymphocytes therapy for the treatment of advanced melanoma (NCT04217473).

8.3 Herpes simplex virus (HSV-1 and HSV-2)

HSV is an enveloped double-stranded DNA virus in the *Herpesviridae* family with a well characterized genome and produces both lytic and latent infection in the host. HSV was one of the first viruses developed for OV therapy and initial efforts involved thymidine kinase (TK) gene deletional mutants (ΔTK) [277], Martuza *et al.* showed that the (ΔTK) mutant was attenuated and in mouse pre-clinical studies could be safely injected in the CNS to suppress glioma tumor growth and prolonged survival. This strategic gene deletion however also eliminated virus antiviral therapy susceptibility [278]. Since then, oHSV containing other viral gene modifications to improve selective replication have been used involving attenuating mutations in several genetic locations as well as insertion of different immunomodulatory genes to modulate the virus induced immune activity. Advantages of HSV as oncolytic vectors are: (1) its well characterized genome, (2) well-established methods for genetic modification of the virus, (3) a large packaging capacity that permits numerous transgene inserts to arm the virus, (4) decades of clinical use and safety experience in patients even when directly injected into the CNS and antiviral therapy that is well tolerated and has a high therapeutic index. The virus also elicits a robust immune-mediated and inflammatory response involving intrinsic, innate, adaptive, and humoral response changes. It is not surprising that HSV-1 was the first FDA approved oncolytic virus for therapy. T-VEC was approved in 2015 after showing improved efficacy in patients with melanoma than those treated with GM-CSF alone [279].

Many of the oHSVs developed to date and advancing to clinical trial (including TVEC) contain deletional mutations that eliminate the chief neurovirulence gene of the virus or RL1. HSV encodes two copies of the $\gamma_1 34.5$ gene and it suppresses several of the IFN dependent host antiviral response pathways (PKR-mediated translational arrest, autophagy, early IRF3-mediated signaling and IFN b1 induction) triggered by viral infection and gene expression [19,22,23,280]. Deletion of this gene restricts efficient viral replication in cells with intact type I IFN signaling pathways and dsRNA translational arrest response and attenuates the virus (WT LD50 40-100pfu *versus*

$\Delta\gamma_1 34.5 > 1 \times 10\text{ee}7$). Clinical development of $\Delta\gamma_1 34.5$ recombinants for brain tumor therapy occurred simultaneously in the United States and United Kingdom. The virus developed in Great Britain (HSV1716) contained $\Delta\gamma_1 34.5$ mutations alone. The oHSV developed early in the US for CNS tumor treatment (G207) was conservatively designed for safety and contained an additional $U_L 39$ ribonucleotide reductase gene mutation to further attenuate it and restrict viral replication to cells with a high proliferation index (malignant cells). These mutations restrict viral replication and gene expression in some tumors. To overcome this replication restriction, investigators have used the $\Delta\gamma_1 34.5$ or G207 backbone viruses and incorporated other mutations or gene inserts (alpha Us11, HCMV IRS1), or incorporated conditional gene expression using tumor-specific promoters (*e.g.*, Nestin promoter) to enhance viral replication in tumor cells [41,107,111,112,277,281,282]. Another virus used early clinically (NV1020) for peripheral tumors also has deletion in ICP0 and ICP4 (genes important for regulating viral transcription and as well as cell cycle shifts in the infected cell) and a single $\gamma_1 34.5$ copy [283]. Other attenuating mutations have been used in oHSVs. For example, a virus developed in Japan contains $U_L 53$ mutation that is attenuate through a less characterized mechanism. In addition, investigators have developed HSV2 based $U_L 39$ oncolytics. HSV has also been modified for cancer cell receptor retargeting Investigators have incorporated single chain antibody recognition domains against HER2 to HSV to confer selective HER2 receptor tropism [284]. Investigators also substituted the viral glycoprotein domain with the HER2 targeting domain to create a chimeric glycoprotein capable of binding and fusion with tropism for HER2 cells to further enhance selectivity, efficacy, and safety of this virus for breast cancer. Moreover, others have developed HSVs that are subject to miRNA control such that in nonmalignant cells they restrict viral gene expression that is relieved in malignant cells where these miRNAs are absent.

HSV elicits a brisk and active immune cell response involving both intrinsic antiviral cytokine chemokine and IFN production as well as humoral, innate and adaptive T-cell responses. To further enhance or augment this immune activity, different cytokines have been added as transgenes in some instances. As mentioned previously T-VEC contains the GM-CSF gene to increase antitumor immunity, in the form of increased tumor-specific CD8 T-cells and a decrease in CD4 FOXP3 regulatory T-cells and CD8 FOXP3 T-cells.

8.3.1 HSV1716

HSV1716 is a $\Delta\gamma_1 34.5$ virus from the lab adapted and more neurovirulent HSV strain 17 that was developed by Moira Brown in the United Kingdom. It has successfully completed early-stage safety studies where it was directly injected into CNS tumors from patients with recurrent malignant gliomas. The virus has also been safely administered (direct intra-tumoral injection) to pediatric patients with sarcomas. The avirulent virus was remarkably effective in these early phase studies in patients with recurrent tumors.

Four of the 9 patients survived 14–24 months post treatment (up to 10ee5 PFU). A phase Ib study followed this initial study and again the 12 treated patients had resection of their tumor 4-9d later. HSV 1716 was safe and replicated in both HSV immune and HSV sero-negative patients. Patients were then injected with HSV 1716 in the tumor margins following surgical resection 3 of the 12 treated patients had long-term survival of 15–22 months. In pediatric sarcoma studies, HSV 1716 was as safe as in the adult studies; however, the clinical benefits were not as pronounced in pediatric sarcomas [285]. Virus replication was detectable (PCR positive in blood stream) and late inflammatory radio-graphic changes suggestive of an immune-mediated component were detected in some patients [285] The clinical benefits were not as pronounced except in one patient with a documented DNA mismatch repair mutation [285]. Taken together these results suggest that immune activity and antigenic load may be an important factor in durable responses following oHSV therapy or that some cancers may be less amenable to the immune-mediated antitumor effects of virotherapy.

8.3.2 Talimogene Laherparepec (T-VEC, IMLYGIC)

T-VEC was a breakthrough oncolytic virus, being the first OV to be FDA approved for treatment. T-VEC is an attenuated HSV-1 oncolytic virus with double deletion of ICP34.5 and ICP47, as well as the insertion of GM-CSF gene for expression [53]. The ICP47 mutation served 2 functions that enhance its therapeutic potential. The muta-tion enhances MHC I antigen presentation in infected cells (eliminating the virus encoded TAP inhibitor) and eliminates the U_S11 g2 promoter leading to earlier expres-sion of the dsRNA binding protein U_S11 and PKR evasion in infected cells. Phase I studies were conducted with the purpose of determining the safety profile of the virus, as well as identifying a dosing schedule for later studies [286]. Thirty patients were enrolled with breast, head and neck, gastrointestinal cancers and malignant melanoma that previously failed therapy. This was a dual cohort study and included a dose finding cohort to first identify the maximum tolerated dose of TVEC. After establishing the MTD a second cohort received multiple doses of the virus initially at the MTD and then with increased doses after patients seroconverted. The virus was well tolerated in the multi-dose cohort. The goal was to elicit an antitumor effect and although there were no complete or partial responses, TVEC produced stable disease. Based upon the prom-ising early stage results, TVEC was advanced to a single arm phase II study involving 50 patients with unresectable metastatic melanoma [287]. Patients had a 26% response rate with regression of both injected and noninjected lesions, including visceral, in those patients who responded to the therapy, providing evidence of systemic effectiveness. It also paved the way for a US Food and Drug Administration (FDA) phase III clinical trial to take place. In their phase III clinical trial, they tested TVEC and GM-CSF in 436 patients with unresectable melanoma in a randomized controlled trial [279]. In the Phase III trial TVEC improved both the durable response rate and the overall response rate over

GM-CSF therapy. Efficacy was most pronounced in patients with stage IIIB, IIIC, and IVM1a melanoma and in patients with treatment-naïve disease. With TVEC being well tolerated, it was the first oncolytic immunotherapy to demonstrate therapeutic effect against melanoma in a phase III clinical trial, leading to its FDA approval later in 2015. Follow up clinical studies are examining TVEC in combination with other immunotherapeutics (checkpoint therapy) as a way to expand the immune-mediated activity of OV therapy.

8.3.3 G207 (and G47delta)

G207 is another HSV-based virotherapy, this one with deletions of the ICP34.5 genes and insertion of the lacZ to disable the U_L39 gene [39, 41]. After demonstrating feasibility of this therapy in animal models, a phase I clinical trial was conducted for the treatment of malignant glial tumors in 21 human subjects, and no toxicity or a serious adverse effect observed most importantly, no patient developed HSV encephalitis and clinical response in select patients, and they found presence of viral DNA and some viral gene expression in select patients. G207 (and HSV1716) provided valuable information and a genetic platform for further gene modifications and improvements. It recently completed a phase I trial in pediatric patients with supratentorial (medulloblastoma) and improved outcome. Another phase Ib clinical trial took place to ensure the safety of two inoculations of G207, before and after tumor resection, and appeared safe for multiple doses [44]. Another phase I clinical trial was done to test for safety of G207 in combination with radiation for recurrent glioblastoma, with a single dose of virus given 24 h before a 5Gy radiation dose [288]. While G207 is a second-generation virus, G47delta is a third generation virus derived from G207 that incorporates an alpha47 and U_S11 promoter deletion similar to that engineered in TVEC [282]. This mutation as described previously enhances MHC I antigen expression in infected cells and improves protein translation in the infected cell by shifting U_S11 expression earlier in infection which enables the virus to prevent PKR-mediated translational arrest in the infected cell [282]. It has been tested in phase I and II clinical trials in Japan against glioblastoma (JPRN-UMIN000002661), prostate cancer, and olfactory neuroblastoma, with promising results [289].

8.3.4 NV1020

NV1020 was derived from an HSV recombinant R7020 and contains a 15 kb deletional mutation at the extending into the U_L and U_S junction and fixes the virus in an isomeric genetic form. The large genetic deletion eliminates one copy of the diploid $\alpha 0$, $\alpha 4$, and $\gamma_1 34.5$ genes (encoding the ICP0, ICP4, and ICP34.5 proteins), respectively, and the U_L56 gene, the protein product of which has not been fully characterized but is thought to contribute to HSV neuroinvasiveness [283,290]. NV1020 was administered in a phase I study to patients with refractory metastatic colorectal cancer to the liver who had already received partial hepatectomy and adjuvant chemotherapy treatment. Patients were enrolled in the 3×3 dose escalation study (3×10^6–1×10^8 PFU) and were administer the virus by hepatic artery infusion. No serious adverse events or dose limiting complications occurred and a maximal tolerated dose was not determined. Nine of the 12 treated patients had stable or reduced tumor burden and the median survival for the group was 25 months [291].

Several other next generation oncolytic HSVs are currently in phase I clinical trial for patients with recurrent malignant gliomas and GBM. These include a $\Delta\gamma_1 34.5$ hIL12 expressing oHSV M032 (NCT02062827), a chimeric HSV $\Delta\gamma_1 34.5$ HCMV IRS1 expressing oHSV capable of improved protein translation and replication (NCT03657576) and an elegantly designed oHSV (rQNestin34.5v.2) that resembles a conditionally expressing G207 vector. It contains the RR mutation ($\Delta U_L 39$) but places the $\gamma_1 34.5$ IFN evasion neurovirulence gene under nestin dependent promoter control to enhance conditional viral translation and replication activity in glioma cells rather than deleting the gene as occurs in G207 (NCT03152318). HSV-1 rRp450 is a $U_L 39$ deletional mutant derived from KOS strain that is also in early-stage study in patients with primary liver cancer or liver metastases administered by hepatic arterial infusion every 1–2 weeks in up to 4 total doses (NCT01071941). Melanoma studies using ONCR-177 in combination with Pembrolizumab PD-1 blockade are also underway (NCT04348916) in patients with melanoma, squamous cell carcinoma of head and neck, breast cancer or other advanced solid tumors. ONCR-177 is a mIR-regulated oHSV that permits conditional Immediate Early and $\gamma_1 34.5$ gene expression in cancerous cells (that do not express the mIRNA). The oHSV expresses 5 transgenes (IL-12, CCL4, the extracellular domain of *FLT3L* and checkpoint inhibitors targeting PD-1 and CTLA-4) and includes mutations that limit axonal retrograde transport as an added safety measure and to prevent the virus from establishing latency. In Japan, HF10, spontaneously mutated HSV variant with natural oncolytic activity and an advantageous safety profile has advanced from Phase I to II study. Several clinical trials involving HF10 alone, HF10 with chemotherapy, or HF10 with chemotherapy or HF10 with ipilimumab or Nivolumab combinations are being performed in patients with melanoma, solid tumors, or with metastatic pancreatic cancer (JapicCTI-173,591, NCT03153085), (NCT02272855), (NCT03259425), (JapicCTI-173,671, NCT03252808).

8.4 Newcastle disease virus

Newcastle disease virus (NDV) is an avian pathogen with oncolytic activity that is disease relatively nonpathogenic in mammals. NDV is a enveloped single stranded non-segmented negative-sense RNA virus from the *Paramyxoviridae* family, similar to the human mumps virus [292]. Like other RNA viruses it replicates in the cytoplasm and can directly kill tumor cells and induce immunogenic cell death, leading to adaptive anti-tumor immunity [293]. NDV is also notable because of its broad virulence within a single serotype. In birds a single serotype can contain strains that are avirulent, mildly virulent (lentogenic strain), moderately virulent (mesogenic strains), and highly virulent (lentogenic strains). Disease manifestations in avian targets include intestinal hemorrhage or neurologic disease. In mammals, however, NDV is non-pathogenic. The fact that NDV prolonged survival in immunodeficient mice and in immunocompetent mice but failed to induce long-term cure as it had done in the orthotopic syngeneic mouse model, highlights the importance of a functional adaptive immune system in establishing long-term anticancer immunity.

Because NDV is avirulent it is not necessary to genetically modify the virus to improve tumor specificity. Paramyxovirus based cell injury is still incompletely understood. *In vitro* cell cultures with some paramyxoviruses for example can generate persistent and productive infections without killing the cells or shutting off cellular transcription or translation. NDV has a unique way of killing tumor cells. While it was initially thought that tumor cells with aberrant antiviral response complemented NDV replication and cell death, others have shown that NDV is capable of inducing tumor lysis in cells with intact Type I interferon signaling response. NDV unlike other OVs depend upon apoptotic pathway defects for NDV selective lysis of cancer cells [294]. The same defects that make some tumor cells resistant to apoptosis by chemotherapeutics enhanced their sensitivity to NDV induced apoptosis and was triggered through ER stress rather than p53 dependent apoptotic pathway [295].

Similar to other OVs, investigators have armed the virus with genes to enhance both its direct and immune cytolytic activity. Studies have been done with a more virulent strain of NDV to see if virulence can be increased within the safety limit. Using a more infectious strain can lead to more cytotoxicity in the tumor, compared to even NDV strains with added immune modulating genes [296]. As with any other virus, there are certain problems with infection. It is known that heparin sulfate and collagen in the ECM can limit NDV spread limiting its activity in solid tumors [297]. Another potential problem is that NDV is an avian pathogen that has potential agricultural impact. While it is not necessary to modify the virus to achieve oncolytic activity, the virus often is genetically modified or less virulent lentogenic strains are used as therapeutics to reduce the potential environmental and economic impact of NDV virus in the poultry industry. Efforts to attenuate the virus in avian species can produce changes in virus activity in mammalian cells [298]. In one case these efforts to attenuate NDV led to a virus capable of establishing persistent infection and treatment resistance in surviving cells t to treatment. NDV has caused persistent infections in certain cell types, such as colorectal cancer cells [299].

Both lentogenic and mesogenic NDV have been tested in early phase clinical trials. A phase I/II trial in patients with recurrent GBM using the lentogeneic NDV-HJU was reported in 2006 [300]. Prior to this there were case reports of pediatric patients with recurrent GBM treated intravenously with a live attenuated mesogenic NDV vaccine strain NDV-68H. The patient exhibited progressive shrinking of their tumor over a 2y period. They followed this up with 4 additional patients (3 children and 1 adult) and reported survival between 5-9y [301,302]. Not only have CNS tumors been targeted with NDV, but vaccine strains have also been used in the periphery for solid tumors. The MK107 vaccine strain (called PV107 for the OV studies) was administered to 79 patients and established a MTD of 1.2×1010 PFU for the initial dose followed by $1.2 \times 1011/m^2$ for the second dose. This desensitization approach was necessary as the initial dose produced flu-like symptoms but the second dose was better tolerated. Virus was recoverable for 3 weeks with viruria and some virus in sputum samples. The highest dose levels produced 1 complete (pharyngeal carcinoma) and 1 partial response (colon carcinoma) [303].

8.5 Retroviruses

Unlike other virotherapies that induce tumor cell lysis and elicit an inflammatory and immune-mediated response, retrovirus based virotherapy relies primarily on gene expression related activity by the virus solely. This therapy delivers a gene or genes of interest selectively to replicating cells and integrating in the host cell genome. For prolonged gene expression. These then are paired with a therapy that in combination with the expressed viral gene exerts an antitumor effect [304].

While some of these viruses are replication incompetent, others have developed replication competent retroviruses (RRVs) that target tumor-specific receptors for entry and then replicate locally within the tumor environment. Retroviral replication relies on reverse transcription of its RNA genome to produce a DNA copy that then chromosomally inserts into the host cell. This provides a gene expression platform from within the tumor environment. These genes can directly induce tumor cell damage genes and tumor cell death with varying levels of efficacy for example in hepatocellular carcinoma [305]. However, this therapy is also well suited as an indirect approach to induce cell death where the expressed gene induces cell death only in combination with another relatively nontoxic drug. For example, investigators have paired retrovirus-based thymidine kinase expression with ganciclovir/acyclovir (analogous to the original 1990 studies). More recently the retrovirus platform has been used for prodrug activator gene therapy such as bacterial cytosine deaminase expression within the tumor cells such that when paired with 5-flucytosine (5-FC), the transduced tumor cells selectively convert this relatively inocuous antifungal into 5 fluorouracil (5-FU) a chemotherapeutic within the tumor cells improving the therapeutic index and tumor cell death [306]. Other investigators have created Adenovirus/Retrovirus hybrid constructs to improve virus production, tumor transduction, and therapeutic gene expression within the tumor [307].

Toca 511 (vocimagene amiretrorepvec) is a gamma-retroviral replicating vector encoding cytosine deaminse. This allows it to be used in combination with extended release 5-fluorocytosine to produce 5-fluorouracil. Phase I clinical trial in recurrent high grade glioma patients, patients received Toca511 injection after resection of tumors, led to multi-year durable responses [308]. However, in a recent randomized controlled multicenter trial Phase III, 403 patients randomized to receive virotherapy or standard of care (SOC) treatment for high grade glioma (lomustine, temozolomide or bevacizumab) prior to surgical resection showed no statistical difference in outcome. A pre-planned subgroup analysis in subjects with second recurrence showed some improvement in survival in patients with recurrent disease. In this analysis Toca +5FC nearly doubled survival (21.82 months median OS compared to 11.14 months, HR = 0.43, $P = 0.0162$). The therapy afforded the greatest survival benefit in patients with IDH1 mutant tumors and anaplastic astrocytoma (AA), (HR = 0.102, $P = 0.009$).

8.6 Measles virus

Measles virus (MV) is an enveloped negative sense RNA from the *Paramyxoviridae* family virus that is highly infectious and produced serious childhood infections prior to the use of widespread immunization in the 20th century. MV can use multiple receptors for cellular entry such as CD150 (SLAM), nectin-4 (PVRL4), and the human complement receptor CD46, many of which are upregulated on Cancer cells [309]. While wild-type measles replicates efficiently in certain lymphomas, the lab adapted MV Edmonton strain (MV–edm) is sufficiently attenuated and due to mutational differences in its attachment glycoprotein has broader cellular entry/replication capability to be used safely and effectively in nonlymphoid cancers (*e.g.*, sarcoma, glioma) in addition to hematologic malignancies. It induces cell to cell fusion producing syncytia and apoptosis. In addition to the lab adapted mutations described above, others have engineered MVs to selectively target tumor-specific receptors (CD38 or EGFR) for entry or using miRNA-dependent gene expression pathways downregulated in cancer cells to provide MV-specificity [310]. Selectivity can also be improved by engineering MV to be miRNA sensitive to those miRNAs which are downregulated in cancer cells [311].

A potential limitation for oncolytic MV, however, is that mass immunity exists due to measles vaccinations. For direct intratumor administration, prior immunity does not limit viral infection or gene expression however for systemic delivery this is a greater concern. Efforts therefore have focused on transporting the virus using infected mesenchymal stem cells or biopolymer coatings that thwart immune recognition of the virus and deliver the OV bearing cargo to the tumor site [312]. Likewise, others have engineered the MV coat proteins creating chimeric MVs using the related canine distemper virus (CDV) and tupia paramyxovirus glycoproteins. These engineered modifications permit tumor entry and evasion of existing neutralizing antibodies [313].

To improve the MV's immunotherapeutic potential, investigators have engineered GM-CSF expression in an effort to improve the innate response and theoretically enhance Th1 activity. The GM–CSF armed MV increased neutrophil infiltration, altered their activity, and enhanced its oncolytic activity [314]. Others have used the virus to enhance intrinsic immune activity and Type I IFN production through the expression of bacterial antigens, activation of danger associated molecular pattern (DAMP) and pathogen related receptor (PRR) activation. Investigators have also engineered MV to express bacterial antigens as an immune activator. MV expressing an *H. pylori* antigen elicited increased IL-12/23, IL-6, and TNFα in a breast cancer pleural effusion model [315]. Combining MV with complementing immunotherapy (*e.g.*, anti-PDL1, CTLA4 therapy) has also been used with improved survival in pre-clinical models [149,316].

MV has been demonstrated safe in patients in several early phase clinical trials involving melanoma, ovarian carcinoma, myeloma. An initial trial involved utilizing an unmodified Edmonston Zagreb strain (MV-EZ) in Switzerland. Five patients were treated with escalating doses via IT injection following IFN a pre-treatment and was well tolerated. Five of the 6 had a clinical response. One patient had a complete response and two patients had responses in noninjected distant tumors. Engineered MV (expressing NIS to facilitate noninvasive imaging of viral gene expression and replication) has also been tested in a phase I dose escalation and Phase I/II studies in treatment refractory recurrent ovarian cancer. The virus was well tolerated with increased survival and dose-dependent disease stabilization [317]. In these early phase studies, there were no dose limiting toxicities and no evidence of shedding in urine or saliva. Treatment response occurred in patients with highly upregulated tumor associated CD46 expression and in many cases survival exceeded the anticipated survival when compared to other treatment refractory historical controls. MV is also in clinical trials in combination with cyclophosphamide for treatment-refractory multiple myeloma (NCT00450814) and in early phase studies for head and neck and breast cancer squamous cell carcinoma (NCT 01846091), Malignant Peripheral Nerve Sheath Tumors (NCT02919449) as well as recurrent glioblastoma multiforme (NCT00390299).

8.7 Reovirus

Reovirusis is a nonenveloped segmented double stranded RNA virus from the *Reoviridae* family. Reoviridae is one of the largest viral families and includes both pathogenic (Rotavirus, Colorado Tick Fever virus) and nonpathogenic viruses. Viral disease is often limited to children with no prior immunity and most infections are asymptomatic. The viruses were first discovered in the stool of children and can produce respiratory and enteric infection whereas others behave as orphan viruses thus leading to the acronym REOvirus (Respiratory, Enteric, and Orphan Virus). As an oncolytic virus, the nonpathogenic orphan virus has been used to target cancer cells. Owing to its complexity (10 dsRNA segments), efficient reverse engineering techniques are limited for these oncolytic viruses. Unlike the other OV where recombineering is used to improve selective viral replication or enhance immune activity, reovirus is a naturally occurring oncolytic human virus that rarely produces clinically relevant disease. Infection occurs in humans and most often produces minimal symptoms and when it produces disease, this is often limited to mild disease with rhinorrhea, pharyngitis, cough, or gastroenteritis. The virus can travel from local infected sites by either hematogenous or neuronal route to the myocardium, liver, spleen or CNS. Because the virus infects these cells but remains relatively avirulent and induces dsRNA-activated IFN-mediated responses often leading to cellular apoptosis, this led to interest using these naturally occurring orphan viruses as oncolytic viruses to target tumors originating from these organs. Similar to many of the other oncolytic viruses discussed, host cell restriction occurs through dsRNA activated Type I IFN pathways stimulation and dsRNA-mediated translational arrest mediated by PKR. The virus preferentially replicates in tumor cells with enhanced RAS activity (either through RAS mutations or upregulated EGFR-mediated MAPK activity). In these cells often the IFN- and PKR-mediated pathways are suppressed thus enabling improved viral translation and replication. Other studies have shown that reovirus is capable of replication in cells without Ras pathway activation [318]. Reovirus oncolytic activity like other OVs may be multifactorial or may differ for different cancers. Reoviruses exhibit improved replication in cells incapable or with a blunted IFN response [319], increased entry receptor expression within tumors [320], improved burst size from tumor cells and as a consequence of caspase-mediated apoptosis. Ultimately similar to other OVs discussed, reovirus OVs benefit from cancer cells that contain complementing mutations that enhance some aspect of the viral lytic cycle that selectively enhances viral replication or lytic activity [31]. Most pre-clinical studies involving reovirus OVs have used immunocompromised human xenograft models; however sarcoma studies involving fibroblasts bearing hRAS mutations have also demonstrated efficacy [33]. These studies demonstrated that reovirus could be repeatedly administered and was effective in both naïve and mice with prior immunity but required repeated dosing.

As an immunotherapeutic, reovirus induces both intrinsic and adaptive immune responses that lead to neutralizing antibody and cytotoxic antiviral responses that restrict the virus [321]. The type 3 reoviridae are capable of infecting but do not replicate efficiently in DCs leading to cell activation and antigenic display [322], ultimately leading to both a T-cell- and NK-mediated cytotoxic activity (granzyme and perforin cytolysis) [323]. Owing to their complex genetic structure and the difficulty involved with reverse engineering their genome, natural immune activity induced by the virus is the major contributing immune-mediated activity in Reovirus OV therapy. Unlike other OVs, recombineering has not been incorporated as a clinical approach to enhance reovirus associated immune-mediated activity; however, combination therapies (HDAC inhibitors, radiation therapy) have been investigated as approaches to enhance viral replication/direct oncolysis and improve immune response to the virus and recognition of tumor associated antigens [324,325].

The type 3 reovirus Dearing strain (T3D also called Reolysin, has been used in Phase I-III human studies for carcinomas (head and neck, melanoma, pancreatic, bladder, prostate, GI, lung, ovarian) and recurrent gliomas. The virus has also been used to selectively lyse malignant cells in autologous stem cell preparations [326]. Several clinical trials have taken place with oncolytic reovirus, Reolysin, many of these being phase I studies for dosage and toxicity of reovirus by itself [34], with palliative radiotherapy [327], and with chemotherapy [328–330]. Other phase I clinical trials with reovirus include for the treatment of recurrent malignant gliomas with intratumoral infusion [331]. Twelve patients were treated with the unmodified reovirus and was well tolerated [332]. This led to a multi-institution phase I study involving prolonged infusion (72 h) [333]. A pediatric phase I study involving Reolysin and GM-CSF is currently underway (NCT02444546) but is no longer in the recruitment phase. Reolysin has also been tested in phase II trials in patients with metastatic melanoma [334]. The results of these Phase II studies showed no objective response in the treated patients. Thirteen of the patients underwent re-biopsy and in two of these patients reovirus gene expression was detected in the tumor (IHC) and demonstrated a serologic response to the virotherapy. Authors concluded that increasing reovirus–neutralizing antibodies inhibited the OVs therapeutic efficacy.

Dose escalation study appeared safe with evidence antitumor activity [333]. Reolysin progressed to Phase III study in combination with paclitaxel and carboplatin *versus* chemotherapy treatment alone, in 167 patients with metastatic or recurrent head and neck squamous cell carcinoma (NCT01166542). The blinded and randomized controlled study was completed in 2014, but at the time of this chapter there are no published results from these studies.

8.8 Vesiculoviruses (vesicular stomatitis virus and Maraba virus)

VSV and Maraba virus are nonsegmented negative stranded viruses in the *Rhabdoviridae* family that are being independently developed as oncolytic viruses. VSV affects livestock (cattle, horses, swine) and rarely infects or produces asymptomatic infections in humans (confined to agricultural or laboratory workers). The virus is an arbovirus that is transmitted to livestock through insect vectors causing nonlethal disease in animals consisting of fever and blisters in the oral cavity and near the hooves. Like VSV, insect vectors transmit Maraba virus (sand flies) and serologic studies from Brazil show that humans are capable of Maraba virus infection although this is rare. VSV is exquisitely sensitive to type I IFN and is used in functional assays to measure IFN activity (Reference to IUs of IFN activity). In the IFN aberrant environment common to many tumors, the virus is capable of replication and tumor lytic activity. The virus is attractive as an oncolytic agent because of the lack of prior immunity in most humans, it has broad cell tropism, is nontransformative, nor is it dependent on cell-cycle change for high titer virus production. It has a small genome that is relatively amenable to recombineering [335]. While WT VSV has a generally beneficial safety profile and is subject to type I IFN control, the virus encoded M protein subverts the IFN response and WT virus can produce pathogenic infections and neurovirulence at high doses in animal models (primate and mouse studies) and has been reported to cause neurologic disease in one child through natural infection [336]. Consequently, M gene mutations (point or deletion) have been incorporated to improve safety while other investigators have engineered the VSV genome to express IFN b1 such that during viral replication and gene expression the virus encodes a feedback mechanism to restrict growth in cells with intact IFN signaling pathways. Other experimental approaches to improve tumor cell selectivity include: miRNA-restricted viral gene expression, improving tumor cell specificity through virus pseudotyping, and disrupting viral gene expression efficiency by altering gene location within the genome.

Efforts to improve VSV activity have focused on safety tumor selectivity and improved replication and direct cytotoxity, however, viral modifications that improve tumor immunity have also been performed. In addition to VSV Type I IFN (IFNβ1) expression modifications, viruses have been constructed to express type II (IFNγ) and type III IFNs (IFNλ). Virus modification for IL-28, IL-4c and other immunotherapeutic payloads have also been pursued to enhance immune-mediated lytic activity. Like with other OVs, the focus has been on improving virus associated T-cell activity. VSVs designed for IFN g secretion demonstrated increase TNFα/IL-6 response, improved dendritic cell activation, and reduced tumor size in syngeneic 4T1 models [337]. Others have engineered VSV to boost tumor antigen-specific response by expressing genes (cyt-c, NB-RAS, and tyrosine related protein 1 TRP1) identified from a melanoma screen identified to improve the tumor associated antigen (TAA) response [338]. A similar strategy was used involving Hif1a, Sox-10, TRP1, and c-myc expression from the virus to improve the immune activity in the B16 CNS metastases tumor model [339]. Virus–TAA expression (GP100 expression) has also been paired with adoptive T-cell therapy to further enhance tumor antigen-specific response for treating metastatic disease [340]. VSV has also been paired with check-point therapies [339]. VSV checkpoint combinations improved the Th1 response, CD4 and CD8 infiltrates and slowed tumor growth [339,341]. Both a VSV-TAA as well as a VSV-IFNβ vector demonstrated improved immune-mediated antitumor activity and reduced tumor growth when combined with checkpoint therapy [339,341]. This approach has also been used with VSV + Reovirus prime boost combinations followed by checkpoint therapy to improve survival [175].

Clinical experience with VSV consists of early phase studies. Early virotherapy clinical trial were vaccine trials performed in the 1940s and 1950s by Dr. George Pack involving a Rhabdovirus. Dr. Pack observed that rabies vaccination in a patient with melanoma who had suffered a dog bite led to significant regression of their metastatic lesions. Pack and his team noted that rabies immunization was the only notable difference in this patient's treatment. To test their hypothesis, 12 melanoma patients were treated with consecutive daily intramuscular inoculations of the Harris rabies vaccine for 20 days. Only two of the 12 patients showed any significant improvement with this method [342]. Investigators are now using VSV and Maraba virus (both in the Rhabdoviridae family) in early-stage clinical studies. The VSV trials in the US involve an OV armed with IFN β (NCT02923466, NCT03120624, and NCT03017820) whereas the Canadian trials use Maraba virus (NCT02285816 and NCT02879760).

8.9 Poliovirus

Poliovirus (PV) is a single-stranded positive sense nonsegmented RNA virus in the *Picornaviridae* Family. The virus is neurotropic and produces poliomyelitis and led to the initial vaccine efforts of the 1950s. The *Picornaviridae* IRES structure and entry receptor (CD155) expression are critical determinants in cell permissiveness viral gene expression. Viral replication is therefore usually restricted to the gastrointestinal tract, its associated lymphatics, and spinal cord/medullary motor neurons. Investigators took the SABIN vaccine poliovirus strain, which contains attenuating mutations and further modified it by substituting its internal ribosomal entry site (IRES) with the type II Rhinovirus IRES as an additional measure of safety. The virus remains attenuated but exhibits excellent growth characteristics in malignant glial tumor cells. The engineered poliovirus maintains its receptor affinity for CD155 (Nectin 5) a receptor highly upregulated in most solid tumors and its oncolytic activity directly correlates with tumor receptor expression in gliomas [343]. Early-stage studies in patients with recurrent malignant gliomas, PVS-RIPO produced exciting results with complete remission in a subset of patients. These initial Phase I toxicity studies designed to find an MTD/MAD, enrolled 61 patients with supratentorial grade IV malignant glioma. This study confirmed PVS-RIPO safety after early dose de-escalation and showed survival rates higher than historical controls [343] although entry criteria favored smaller tumor sizes and contained a higher proportion of IDH mutant tumors. PVS-RIPO is now in advanced phase studies in pediatric patients with recurrent glioma (NCT03043391) and for other solid tumors.

8.10 Parvovirus

Parvovirus is a nonenveloped ssDNA virus that includes both autonomous viruses (*e.g.*, parvovirus B19) and helper-dependent viruses (*e.g.*, Adeno Associated Virus) within the *Parvoviridae* family. Autonomous parvoviruses exhibit both selective replication and direct oncolytic activity and an infection related immune-mediated response like other oncolytic viruses. This was confirmed in pre-clinical studies that has now led to early-stage clinical trials using a rat parvovirus construct (ParvOryx) for progressive primary and recurrent glioblastoma multiforme (NCT01301430). The study involves direct intratumoral injection of ParvOryx, followed by resection of the treated tumor and re-injection into the remaining tumor margins around the tumor cavity. The study also includes another arm involving intravenous virus delivery followed by resection and re-injection in the tumor margins surrounding the tumor cavity.

8.11 Influenza

Influenza is a segmented singled-stranded negative-sense RNA virus in the *Orthomyxoviridae* family. Influenza is a frequently encountered pathogen that has engendered interest as a potential oncolytic virus since 1904 when Dr. George Dock described a reduction in the number of abnormal leukocytes and an AML patient's splenomegaly following what was a presumed influenza infection. In the late 1990s methods were developed that permit genetic modification of the segmented influenza virus. Similar to the other viruses described in this chapter, influenza relies on mutations in the cancer cells that disrupt molecular pathways involved in the antiviral response to permit selective replication in cancerous cells. Engineered mutation in the NS1 antiviral evasion gene attenuates the virus and permits this selective replication in cancerous cells with PKR and IFN-signaling defects or in cells with RAS mutations or MAPK upregulation that suppress the PKR and IFN. Efforts to enhance these NS1 mutant oncolytic viruses involved serial passage (virus-evolution approaches) to improve viral replication in tumor lines and mutagenesis of the virus encoded HA genes to improve tumor associated receptor virus targeting [344]. Influenza, like other OVs, can be further engineered for conditional replication and can be armed with genes for immune modulation and improving immunogenic cell death. NS1 mutants have proven safe as potential vaccines in clinical studies and did not develop pathogenic mutations following serial passage [345,346]. However, a principal concern with influenza based OVs remains the possibility of a recombination event with this highly infectious virus capable of antigenic shifts and environmental safety. Outstanding questions remain also on the effects of prior antiviral immunity on OV activity for influenza. Thus far influenza based OVs remain in pre-clinical study.

9. Conclusion

Cancer virotherapy involves both direct antitumor activity involving viral replication or the expression of genes that facilitate cytotoxic damage to the tumor as well as immune-mediated antitumor effects. The highly biosynthetic and often transcriptionally and immune aberrant cellular environment of the cancer cells can selectively enhance viral gene expression and replication of these attenuated viruses. Efforts to improve OV activity focus on enhancing their selective activity in the cancer cells and restrict viral gene expression and replication in nonmalignant cells and increasing their immunotherapeutic potential. Despite their attenuated nature, oncolytic viruses often induce intrinsic antiviral programs in infected cells that lead to chemokine, cytokine and IFN changes in the tumor microenvironment and lead to cellular immune changes. In addition to both Class I and II antigen presentation changes OV infection induces immune activating cell death pathways (necrotic), release tumor associated antigens and can invoke molecular mimicry, bystander damage, epitope spread, and or heterologous immunity that can further enhance immune activity against tumor associated antigens. OVs that actively modulate the antiviral immune cell infiltrate through cytokine and immune active gene expression are entering clinical trials and in lock step with other immunotherapeutic advances (CAR-T, ACT, BiTE, checkpoint inhibition) and is anticipated to increase the therapeutic potential for these bioactive agents.

References

[1] Dock G. The influence of complicating diseases upon leukaemia: cases of tuberculosis and leukoemia. Miscellaneous infections. Changes in the red blood corpuscles. Qualitative changes in the blood, especially in the leukocytes. When does the change occur? The effects of various processes other than infection on leukoemia. Bibliography. Am J Med Sci 1904;127(4):563.

[2] Moore AE. Viruses with oncolytic properties and their adaptation to tumors. Ann N Y Acad Sci 1952;54(6):945–52.

[3] Moore AE. Enhancement of oncolytic effect of Russian encephalitis virus. Proc Soc Exp Biol Med 1951;76(4):749–54.

[4] Toolan HW, Moore AE. Oncolytic effect of Egypt virus on a human epidermoid carcinoma grown in X-irradiated rats. Proc Soc Exp Biol Med 1952;79(4):697–702.

[5] Love R, Sharpless GR. Studies on a transplantable chicken tumor (RPL-12 lymphoma). Cancer Res 1954;14(9):640.

[6] Lindenmann J. Viral oncolysis with host survival. Proc Soc Exp Biol Med 1963;113(1):85–91.

[7] Lindenmann J, Klein PA. Viral oncolysis: increased immunogenicity of host cell antigen associated with influenza virus. J Exp Med 1967;126(1):93–108.

[8] Sinkovics JG. Oncogenes-antioncogenes and virus therapy of cancer. Anticancer Res 1989;9 (5):1281–90.

[9] Sinkovics JG. Programmed cell death (apoptosis): its virological and immunological connections (a review). Acta Microbiol Hung 1991;38(3–4):321–34.

[10] Moolten FL, Wells JM. Curability of tumors bearing herpes thymidine kinase genes transferred by retroviral vectors. J Natl Cancer Inst 1990;82(4):297–300.

[11] Moolten FL, Wells JM, Heyman RA, Evans RM. Lymphoma regression induced by ganciclovir in mice bearing a herpes thymidine kinase transgene. Hum Gene Ther 1990;1(2):125–34.

[12] Hasegawa Y, Emi N, Shimokata K, Abe A, Kawabe T, Hasegawa T, et al. Gene transfer of herpes simplex virus type I thymidine kinase gene as a drug sensitivity gene into human lung cancer cell lines using retroviral vectors. Am J Respir Cell Mol Biol 1993;8(6):655–61.

[13] Takamiya Y, Short MP, Moolten FL, Fleet C, Mineta T, Breakefield XO, et al. An experimental model of retrovirus gene therapy for malignant brain tumors. J Neurosurg 1993;79(1):104–10.

[14] Shaughnessy E, Lu D, Chatterjee S, Wong KK. Parvoviral vectors for the gene therapy of cancer. Semin Oncol 1996;23(1):159–71.

[15] Kramm CM, Rainov NG, Sena-Esteves M, Barnett FH, Chase M, Herrlinger U, et al. Long-term survival in a rodent model of disseminated brain tumors by combined intrathecal delivery of herpes vectors and ganciclovir treatment. Hum Gene Ther 1996;7(16):1989–94.

[16] Baines JD, Roizman B. The open reading frames UL3, UL4, UL10, and UL16 are dispensable for the replication of herpes simplex virus 1 in cell culture. J Virol 1991;65(2):938–44.

[17] Thompson RL, Rogers SK, Zerhusen MA. Herpes simplex virus neurovirulence and productive infection of neural cells is associated with a function which maps between 0.82 and 0.832 map units on the HSV genome. Virology 1989;172(2):435–50.

[18] Chou J, Kern ER, Whitley RJ, Roizman B. Mapping of herpes simplex virus-1 neurovirulence to gamma 134.5, a gene nonessential for growth in culture. Science 1990;250(4985):1262.

[19] Chou J, Roizman B. The gamma 1(34.5) gene of herpes simplex virus 1 precludes neuroblastoma cells from triggering total shutoff of protein synthesis characteristic of programed cell death in neuronal cells. Proc Natl Acad Sci U S A 1992;89(8):3266–70.

[20] Chou J, Poon AP, Johnson J, Roizman B. Differential response of human cells to deletions and stop codons in the gamma(1)34.5 gene of herpes simplex virus. J Virol 1994;68(12):8304–11.

[21] Andreansky SS, He B, Gillespie GY, Soroceanu L, Markert J, Chou J, et al. The application of genetically engineered herpes simplex viruses to the treatment of experimental brain tumors. Proc Natl Acad Sci U S A 1996;93(21):11313–8.

[22] Orvedahl A, Alexander D, Tallóczy Z, Sun Q, Wei Y, Zhang W, et al. HSV-1 ICP34.5 confers neurovirulence by targeting the Beclin 1 autophagy protein. Cell Host Microbe 2007;1(1):23–35.

[23] Verpooten D, Ma Y, Hou S, Yan Z, He B. Control of TANK-binding kinase 1–mediated signaling by the gamma(1)34.5 protein of herpes simplex virus 1. J Biol Chem 2009;284(2):1097–105.

[24] Whyte P, Williamson NM, Harlow E. Cellular targets for transformation by the adenovirus E1A proteins. Cell 1989;56(1):67–75.

[25] Sherr CJ. The INK4a/ARF network in tumour suppression. Nat Rev Mol Cell Biol 2001;2 (10):731–7.

[26] Du W, Searle JS. The rb pathway and cancer therapeutics. Curr Drug Targets 2009;10(7):581–9.

[27] Fueyo J, Gomez-Manzano C, Alemany R, Lee PSY, McDonnell TJ, Mitlianga P, et al. A mutant oncolytic adenovirus targeting the Rb pathway produces anti-glioma effect in vivo. Oncogene 2000;19(1):2–12.

[28] Fueyo J, Alemany R, Gomez-Manzano C, Fuller GN, Khan A, Conrad CA, et al. Preclinical characterization of the antiglioma activity of a tropism-enhanced adenovirus targeted to the retinoblastoma pathway. J Natl Cancer Inst 2003;95(9):652–60.

[29] Hashiro G, Loh PC, Yau JT. The preferential cytotoxicity of reovirus for certain transformed cell lines. Arch Virol 1977;54(4):307–15.

[30] Strong JE, Tang D, Lee PW. Evidence that the epidermal growth factor receptor on host cells confers reovirus infection efficiency. Virology 1993;197(1):405–11.

[31] Strong JE, Coffey MC, Tang D, Sabinin P, Lee PWK. The molecular basis of viral oncolysis: usurpation of the Ras signaling pathway by reovirus. EMBO J 1998;17(12):3351–62.

[32] Sasaki T, Hiroki K, Yamashita Y. The role of epidermal growth factor receptor in cancer metastasis and microenvironment. Biomed Res Int 2013;2013:546318.

[33] Coffey MC, Strong JE, Forsyth PA, Lee PWK. Reovirus therapy of tumors with activated Ras pathway. Science 1998;282(5392):1332.

[34] Vidal L, Pandha HS, Yap TA, White CL, Twigger K, Vile RG, et al. A phase I study of intravenous oncolytic reovirus type 3 dearing in patients with advanced cancer. Clin Cancer Res 2008;14 (21):7127.

[35] Panicali D, Paoletti E. Construction of poxviruses as cloning vectors: insertion of the thymidine kinase gene from herpes simplex virus into the DNA of infectious vaccinia virus. Proc Natl Acad Sci U S A 1982;79(16):4927–31.

[36] Mackett M, Smith GL, Moss B. Vaccinia virus: a selectable eukaryotic cloning and expression vector. Proc Natl Acad Sci U S A 1982;79(23):7415–9.

[37] Guo ZS, Lu B, Guo Z, Giehl E, Feist M, Dai E, et al. Vaccinia virus-mediated cancer immunotherapy: cancer vaccines and oncolytics. J Immunother Cancer 2019;7(1):6.

[38] Sanda MG, Smith DC, Charles LG, Hwang C, Pienta KJ, Schlom J, et al. Recombinant vaccinia-PSA (PROSTVAC) can induce a prostate-specific immune response in androgen-modulated human prostate cancer. Urology 1999;53(2):260–6.

[39] Mineta T, Rabkin SD, Yazaki T, Hunter WD, Martuza RL. Attenuated multi–mutated herpes simplex virus–1 for the treatment of malignant gliomas. Nat Med 1995;1(9):938–43.

[40] MacLean AR, Ul-Fareed M, Robertson L, Harland J, Brown SM. Herpes simplex virus type 1 deletion variants 1714 and 1716 pinpoint neurovirulence-related sequences in Glasgow strain 17+ between immediate early gene 1 and the 'a' sequence. J Gen Virol 1991;72(3):631–9.

[41] Markert JM, Medlock MD, Rabkin SD, Gillespie GY, Todo T, Hunter WD, et al. Conditionally replicating herpes simplex virus mutant, G207 for the treatment of malignant glioma: results of a phase I trial. Gene Ther 2000;7(10):867–74.

[42] Rampling R, Cruickshank G, Papanastassiou V, Nicoll J, Hadley D, Brennan D, et al. Toxicity evaluation of replication-competent herpes simplex virus (ICP 34.5 null mutant 1716) in patients with recurrent malignant glioma. Gene Ther 2000;7(10):859–66.

[43] Markert JM, Medlock MD, Rabkin SD, Gillespie GY, Todo T, Hunter WD, et al. Conditionally replicating herpes simplex virus mutant, G207 for the treatment of malignant glioma: results of a phase I trial. Gene Ther 2000;7(10):867–74.

[44] Markert JM, Liechty PG, Wang W, Gaston S, Braz E, Karrasch M, et al. Phase Ib trial of mutant herpes simplex virus G207 inoculated pre-and post-tumor resection for recurrent GBM. Mol Ther 2009;17(1):199–207.

[45] Harrow S, Papanastassiou V, Harland J, Mabbs R, Petty R, Fraser M, et al. HSV1716 injection into the brain adjacent to tumour following surgical resection of high-grade glioma: safety data and long-term survival. Gene Ther 2004;11(22):1648–58.

[46] Cassady KA, Parker JN. Herpesvirus vectors for therapy of brain tumors. Open Virol J 2010;4:103–8.

[47] Markert JM, Gillespie GY, Weichselbaum RR, Roizman B, Whitley RJ. Genetically engineered HSV in the treatment of glioma: a review. Rev Med Virol 2000;10(1):17–30.

[48] Rampling R, Cruickshank G, Papanastassiou V, Nicoll J, Hadley D, Brennan D, et al. Toxicity evaluation of replication-competent herpes simplex virus (ICP 34.5 null mutant 1716) in patients with recurrent malignant glioma. Gene Ther 2000;7(10):859–66.

[49] Alberts P, Tilgase A, Rasa A, Bandere K, Venskus D. The advent of oncolytic virotherapy in oncology: the Rigvir® story. Eur J Pharmacol 2018;837:117–26.

[50] Garber K. China approves world's first oncolytic virus therapy for cancer treatment. J Natl Cancer Inst 2006;98(5):298–300.

[51] Xu RH, Yuan ZY, Guan ZZ, Cao Y, Wang HQ, Hu XH, et al. Phase II clinical study of intratumoral H101, an E1B deleted adenovirus, in combination with chemotherapy in patients with cancer. Ai Zheng 2003;22(12):1307–10.

[52] Xia ZJ, Chang JH, Zhang L, Jiang WQ, Guan ZZ, Liu JW, et al. Phase III randomized clinical trial of intratumoral injection of E1B gene-deleted adenovirus (H101) combined with cisplatin-based chemotherapy in treating squamous cell cancer of head and neck or esophagus. Ai Zheng 2004;23(12):1666–70.

[53] Liu BL, Robinson M, Han ZQ, Branston RH, English C, Reay P, et al. ICP34.5 deleted herpes simplex virus with enhanced oncolytic, immune stimulating, and anti-tumour properties. Gene Ther 2003;10(4):292–303.

[54] Torres-Domínguez LE, McFadden G. Poxvirus oncolytic virotherapy. Expert Opin Biol Ther 2019;19(6):561–73.

[55] Hill C, Carlisle R. Achieving systemic delivery of oncolytic viruses. Expert Opin Drug Deliv 2019;16(6):607–20.

[56] Ferguson MS, Lemoine NR, Wang Y. Systemic delivery of oncolytic viruses: hopes and hurdles. Adv Virol 2012;2012:805629.

[57] Russell SJ. Replicating vectors for cancer therapy: a question of strategy. Semin Cancer Biol 1994;5(6):437–43.

[58] Fabian KL, Storkus WJ. Immunotherapeutic targeting of tumor-associated blood vessels. Adv Exp Med Biol 2017;1036:191–211.

[59] Mok W, Boucher Y, Jain RK. Matrix metalloproteinases-1 and -8 improve the distribution and efficacy of an oncolytic virus. Cancer Res 2007;67(22):10664–8.

[60] Jain RK, Stylianopoulos T. Delivering nanomedicine to solid tumors. Nat Rev Clin Oncol 2010;7(11):653–64.

[61] Lee SY, Park HR, Rhee J, Park YM, Kim SH. Therapeutic effect of oncolytic adenovirus expressing relaxin in radioresistant oral squamous cell carcinoma. Oncol Res 2013;20(9):419–25.

[62] Vera B, Martínez-Vélez N, Xipell E, Acanda de la Rocha A, Patiño-García A, Saez-Castresana J, et al. Characterization of the antiglioma effect of the oncolytic adenovirus VCN-01. PLoS One 2016;11(1), e0147211.

[63] Ganesh S, Gonzalez Edick M, Idamakanti N, Abramova M, Vanroey M, Robinson M, et al. Relaxin-expressing, fiber chimeric oncolytic adenovirus prolongs survival of tumor-bearing mice. Cancer Res 2007;67(9):4399–407.

[64] Yang SW, Cody JJ, Rivera AA, Waehler R, Wang M, Kimball KJ, et al. Conditionally replicating adenovirus expressing TIMP2 for ovarian cancer therapy. Clin Cancer Res 2011;17(3):538–49.

[65] Choi IK, Shin H, Oh E, Yoo JY, Hwang JK, Shin K, et al. Potent and long-term antiangiogenic efficacy mediated by FP3-expressing oncolytic adenovirus. Int J Cancer 2015;137(9):2253–69.

[66] Bazan-Peregrino M, Sainson RC, Carlisle RC, Thoma C, Waters RA, Arvanitis C, et al. Combining virotherapy and angiotherapy for the treatment of breast cancer. Cancer Gene Ther 2013;20(8):461–8.

[67] Maginnis MS. Virus–receptor interactions: the key to cellular invasion. J Mol Biol 2018;430(17):2590–611.

[68] Au GG, Lindberg AM, Barry RD, Shafren DR. Oncolysis of vascular malignant human melanoma tumors by Coxsackievirus A21. Int J Oncol 2005;26(6):1471–6.

[69] Shafren DR, Sylvester D, Johansson ES, Campbell IG, Barry RD. Oncolysis of human ovarian cancers by echovirus type 1. Int J Cancer 2005;115(2):320–8.

[70] Ochiai H, Campbell SA, Archer GE, Chewning TA, Dragunsky E, Ivanov A, et al. Targeted therapy for glioblastoma multiforme neoplastic meningitis with intrathecal delivery of an oncolytic recombinant poliovirus. Clin Cancer Res 2006;12(4):1349–54.

[71] Molfetta R, Zitti B, Lecce M, Milito ND, Stabile H, Fionda C, et al. CD155: a multi-functional molecule in tumor progression. Int J Mol Sci 2020;21(3):922.

[72] Fontana L, Nuzzo M, Urbanelli L, Monaci P. General strategy for broadening adenovirus tropism. J Virol 2003;77(20):11094.

[73] Gallo P, Dharmapuri S, Cipriani B, Monaci P. Adenovirus as vehicle for anticancer genetic immunotherapy. Gene Ther 2005;12(1):S84–91.

[74] Menotti L, Cerretani A, Campadelli-Fiume G. A herpes simplex virus recombinant that exhibits a single-chain antibody to HER2/neu enters cells through the mammary tumor receptor, independently of the gD receptors. J Virol 2006;80(11):5531–9.

[75] Connolly JB. Conditionally replicating viruses in cancer therapy. Gene Ther 2003;10(8):712–5.

[76] Chiocca EA. Oncolytic viruses. Nat Rev Cancer 2002;2(12):938–50. https://doi.org/10.1038/nrc948.

[77] Fan Y, Sanyal S, Bruzzone R. Breaking bad: how viruses subvert the cell cycle. Front Cell Infect Microbiol 2018;8:396.

[78] Hume AJ, Kalejta RF. Regulation of the retinoblastoma proteins by the human herpesviruses. Cell Div 2009;4:1.

[79] Ries S, Korn WM. ONYX-015: mechanisms of action and clinical potential of a replication-selective adenovirus. Br J Cancer 2002;86(1):5–11.

[80] Yu W, Fang H. Clinical trials with oncolytic adenovirus in China. Curr Cancer Drug Targets 2007;7(2):141–8.

[81] Bischoff JR, Kirn DH, Williams A, Heise C, Horn S, Muna M, et al. An adenovirus mutant that replicates selectively in p53-deficient human tumor cells. Science 1996;274(5286):373.

[82] Dobbelstein M. Replicating adenoviruses in cancer therapy. Curr Top Microbiol Immunol 2004;273:291–334.

[83] Mal A, Poon RY, Howe PH, Toyoshima H, Hunter T, Harter ML. Inactivation of p27Kip1 by the viral E1A oncoprotein in TGFbeta-treated cells. Nature 1996;380(6571):262–5.

[84] Raychaudhuri P, Bagchi S, Devoto SH, Kraus VB, Moran E, Nevins JR. Domains of the adenovirus E1A protein required for oncogenic activity are also required for dissociation of E2F transcription factor complexes. Genes Dev 1991;5(7):1200–11.

[85] Somasundaram K, El-Deiry WS. Inhibition of p53-mediated transactivation and cell cycle arrest by E1A through its p300/CBP-interacting region. Oncogene 1997;14(9):1047–57.

[86] Peters C, Rabkin SD. Designing herpes viruses as oncolytics. Mol Ther Oncolytics 2015;2:15010.

[87] Smith CC. The herpes simplex virus type 2 protein ICP10PK: a master of versatility. Front Biosci 2005;10:2820–31.

[88] Kaplitt MG, Tjuvajev JG, Leib DA, Berk J, Pettigrew KD, Posner JB, et al. Mutant herpes simplex virus induced regression of tumors growing in immunocompetent rats. J Neurooncol 1994;19 (2):137–47.

[89] Mineta T, Rabkin SD, Martuza RL. Treatment of malignant gliomas using ganciclovir-hypersensitive, ribonucleotide reductase-deficient herpes simplex viral mutant. Cancer Res 1994;54(15):3963–6.

[90] Boviatsis EJ, Scharf JM, Chase M, Harrington K, Kowall NW, Breakefield XO, et al. Antitumor activity and reporter gene transfer into rat brain neoplasms inoculated with herpes simplex virus vectors defective in thymidine kinase or ribonucleotide reductase. Gene Ther 1994;1(5):323–31.

[91] Russell SJ, Peng K-W. Viruses as anticancer drugs. Trends Pharmacol Sci 2007;28(7):326–33.

[92] Lee JM, Ghonime MG, Cassady KA. STING restricts oHSV replication and spread in resistant MPNSTs but is dispensable for basal IFN-stimulated gene upregulation. Mol Ther Oncolytics 2019;15:91–100.

[93] Ahn J, Konno H, Barber GN. Diverse roles of STING-dependent signaling on the development of cancer. Oncogene 2015;34(41):5302–8.

[94] Ishikawa H, Ma Z, Barber GN. STING regulates intracellular DNA-mediated, type I interferon-dependent innate immunity. Nature 2009;461(7265):788–92.

[95] Abe T, Harashima A, Xia T, Konno H, Konno K, Morales A, et al. STING recognition of cytoplasmic DNA instigates cellular defense. Mol Cell 2013;50(1):5–15.

[96] Kalamvoki M, Roizman B. HSV-1 degrades, stabilizes, requires, or is stung by STING depending on ICP0, the US3 protein kinase, and cell derivation. Proc Natl Acad Sci U S A 2014;111(5):E611–7.

[97] Haines 3rd GK, Panos RJ, Bak PM, Brown T, Zielinski M, Leyland J, et al. Interferon-responsive protein kinase (p68) and proliferating cell nuclear antigen are inversely distributed in head and neck squamous cell carcinoma. Tumour Biol 1998;19(1):52–9.

[98] Li J, Yen C, Liaw D, Podsypanina K, Bose S, Wang SI, et al. PTEN, a putative protein tyrosine phosphatase gene mutated in human brain, breast, and prostate cancer. Science 1997;275(5308):1943–7.

[99] Rasheed BK, Stenzel TT, McLendon RE, Parsons R, Friedman AH, Friedman HS, et al. PTEN gene mutations are seen in high-grade but not in low-grade gliomas. Cancer Res 1997;57(19):4187–90.

[100] Li S, Zhu M, Pan R, Fang T, Cao YY, Chen S, et al. The tumor suppressor PTEN has a critical role in antiviral innate immunity. Nat Immunol 2016;17(3):241–9.

[101] Ortega LG, McCotter MD, Henry GL, McCormack SJ, Thomis DC, Samuel CE. Mechanism of interferon action. Biochemical and genetic evidence for the intermolecular association of the RNA-dependent protein kinase PKR from human cells. Virology 1996;215(1):31–9.

[102] Rampersad S, Tennant P. Replication and expression strategies of viruses. Viruses 2018;55–82.

[103] Stojdl DF, Lichty BD, tenOever BR, Paterson JM, Power AT, Knowles S, et al. VSV strains with defects in their ability to shutdown innate immunity are potent systemic anti-cancer agents. Cancer Cell 2003;4(4):263–75.

[104] He B, Chou J, Brandimarti R, Mohr I, Gluzman Y, Roizman B. Suppression of the phenotype of gamma(1)34.5-herpes simplex virus 1: failure of activated RNA-dependent protein kinase to shut off protein synthesis is associated with a deletion in the domain of the alpha47 gene. J Virol 1997;71(8):6049.

[105] Ma Y, Jin H, Valyi-Nagy T, Cao Y, Yan Z, He B. Inhibition of TANK binding kinase 1 by herpes simplex virus 1 facilitates productive infection. J Virol 2012;86(4):2188.

[106] Gimple RC, Wang X. RAS: striking at the core of the oncogenic circuitry. Front Oncol 2019;9:965.

[107] Shah AC, Parker JN, Gillespie GY, Lakeman FD, Meleth S, Markert JM, et al. Enhanced antiglioma activity of chimeric HCMV/HSV-1 oncolytic viruses. Gene Ther 2007;14(13):1045–54.

[108] Shir A, Levitzki A. Inhibition of glioma growth by tumor-specific activation of double-stranded RNA-dependent protein kinase PKR. Nat Biotechnol 2002;20(9):895–900.

[109] Li J, O'Malley M, Urban J, Sampath P, Guo ZS, Kalinski P, et al. Chemokine expression from oncolytic vaccinia virus enhances vaccine therapies of cancer. Mol Ther 2011;19(4):650–7.

[110] Jackson JD, Markert JM, Li L, Carroll SL, Cassady KA. STAT1 and NF-κB inhibitors diminish basal interferon-stimulated gene expression and improve the productive infection of oncolytic HSV in MPNST cells. Mol Cancer Res 2016;14(5):482–92.

[111] Kambara H, Okano H, Chiocca EA, Saeki Y. An oncolytic HSV-1 mutant expressing ICP34.5 under control of a nestin promoter increases survival of animals even when symptomatic from a brain tumor. Cancer Res 2005;65(7):2832–9.

[112] Mohr I, Gluzman Y. A herpesvirus genetic element which affects translation in the absence of the viral GADD34 function. EMBO J 1996;15(17):4759–66.

[113] Cassady KA, Gross M, Roizman B. The herpes simplex virus US$_{11}$ protein effectively compensates for the $\gamma_1$34.5 gene if present before activation of protein kinase R by precluding its phosphorylation and that of the α subunit of eukaryotic translation initiation factor 2. J Virol 1998;72(11):8620–6.

[114] Cassady KA. Human cytomegalovirus TRS1 and IRS1 gene products block the double-stranded-RNA-activated host protein shutoff response induced by herpes simplex virus type 1 infection. J Virol 2005;79(14):8707–15.

[115] Cassady KA, Saunders U, Shimamura M. $\Delta\gamma_1$134.5 herpes simplex viruses encoding human cytomegalovirus IRS1 or TRS1 induce interferon regulatory factor 3 phosphorylation and an interferon-stimulated gene response. J Virol 2012;86(1):610–4.

[116] Grishin AV, Azhipa O, Semenov I, Corey SJ. Interaction between growth arrest-DNA damage protein 34 and Src kinase Lyn negatively regulates genotoxic apoptosis. Proc Natl Acad Sci U S A 2001;98 (18):10172–7.

[117] Nakashima H, Nguyen T, Kasai K, Passaro C, Ito H, Goins WF, et al. Toxicity and efficacy of a novel GADD34-expressing oncolytic HSV-1 for the treatment of experimental glioblastoma. Clin Cancer Res 2018;24(11):2574–84.

[118] Fu X, Meng F, Tao L, Jin A, Zhang X. A strict-late viral promoter is a strong tumor-specific promoter in the context of an oncolytic herpes simplex virus. Gene Ther 2003;10(17):1458–64.

[119] Chen J-S, Liu J-C, Shen L, Rau K-M, Kuo H-P, Li YM, et al. Cancer-specific activation of the survivin promoter and its potential use in gene therapy. Cancer Gene Ther 2004;11(11):740–7.

[120] Sweeney K, Halldén G. Oncolytic adenovirus-mediated therapy for prostate cancer. Oncolytic Virother 2016;5:45–57.

[121] Yoon AR, Hong J, Kim M, Yun C-O. Hepatocellular carcinoma-targeting oncolytic adenovirus overcomes hypoxic tumor microenvironment and effectively disperses through both central and peripheral tumor regions. Sci Rep 2018;8(1):2233.

[122] Hernandez-Alcoceba R, Pihalja M, Qian D, Clarke MF. New oncolytic adenoviruses with hypoxia-and estrogen receptor-regulated replication. Hum Gene Ther 2002;13(14):1737–50.

[123] Post DE, Van Meir EG. A novel hypoxia-inducible factor (HIF) activated oncolytic adenovirus for cancer therapy. Oncogene 2003;22(14):2065–72.

[124] Grondin B, DeLuca N. Herpes simplex virus type 1 ICP4 promotes transcription preinitiation complex formation by enhancing the binding of TFIID to DNA. J Virol 2000;74(24):11504.

[125] Guo ZS, Thorne SH, Bartlett DL. Oncolytic virotherapy: molecular targets in tumor-selective replication and carrier cell-mediated delivery of oncolytic viruses. Biochim Biophys Acta 2008;1785 (2):217–31.

[126] Hikichi M, Kidokoro M, Haraguchi T, Iba H, Shida H, Tahara H, et al. MicroRNA regulation of glycoprotein B5R in oncolytic vaccinia virus reduces viral pathogenicity without impairing its antitumor efficacy. Mol Ther 2011;19(6):1107–15.

[127] Mazzacurati L, Marzulli M, Reinhart B, Miyagawa Y, Uchida H, Goins WF, et al. Use of miRNA response sequences to block off-target replication and increase the safety of an unattenuated, glioblastoma-targeted oncolytic HSV. Mol Ther 2015;23(1):99–107.

[128] Andreansky S, He B, van Cott J, McGhee J, Markert JM, Gillespie GY, et al. Treatment of intracranial gliomas in immunocompetent mice using herpes simplex viruses that express murine interleukins. Gene Ther 1998;5(1):121–30.

[129] Parker JN, Gillespie GY, Love CE, Randall S, Whitley RJ, Markert JM. Engineered herpes simplex virus expressing IL-12 in the treatment of experimental murine brain tumors. Proc Natl Acad Sci U S A 2000;97(5):2208–13.

[130] Hellums EK, Markert JM, Parker JN, He B, Perbal B, Roizman B, et al. Increased efficacy of an interleukin-12-secreting herpes simplex virus in a syngeneic intracranial murine glioma model. Neuro Oncol 2005;7(3):213–24.

[131] Parker JN, Meleth S, Hughes KB, Gillespie GY, Whitley RJ, Markert JM. Enhanced inhibition of syngeneic murine tumors by combinatorial therapy with genetically engineered HSV-1 expressing CCL2 and IL-12. Cancer Gene Ther 2005;12(4):359–68.

[132] Hemminki O, dos Santos JM, Hemminki A. Oncolytic viruses for cancer immunotherapy. J Hematol Oncol 2020;13(1):84.

[133] Park AK, Fong Y, Kim SI, Yang J, Murad JP, Lu J, et al. Effective combination immunotherapy using oncolytic viruses to deliver CAR targets to solid tumors. Sci Transl Med 2020;12(559).

[134] Workenhe ST, Mossman KL. Oncolytic virotherapy and immunogenic cancer cell death: sharpening the sword for improved cancer treatment strategies. Mol Ther 2014;22(2):251–6.

[135] Ma Y, Adjemian S, Mattarollo SR, Yamazaki T, Aymeric L, Yang H, et al. Anticancer chemotherapy-induced intratumoral recruitment and differentiation of antigen-presenting cells. Immunity 2013;38 (4):729–41.

[136] Parker JN, Pfister LA, Quenelle D, Gillespie GY, Markert JM, Kern ER, et al. Genetically engineered herpes simplex viruses that express IL-12 or GM-CSF as vaccine candidates. Vaccine 2006;24 (10):1644–52.

[137] Kowalsky SJ, Liu Z, Feist M, Berkey SE, Ma C, Ravindranathan R, et al. Superagonist IL-15-armed oncolytic virus elicits potent antitumor immunity and therapy that are enhanced with PD-1 blockade. Mol Ther 2018;26(10):2476–86.

[138] Gaston DC, Odom CI, Li L, Markert JM, Roth JC, Cassady KA, et al. Production of bioactive soluble interleukin-15 in complex with interleukin-15 receptor alpha from a conditionally-replicating oncolytic HSV-1. PLoS One 2013;8(11), e81768.

[139] Parato KA, Breitbach CJ, Le Boeuf F, Wang J, Storbeck C, Ilkow C, et al. The oncolytic poxvirus JX-594 selectively replicates in and destroys cancer cells driven by genetic pathways commonly activated in cancers. Mol Ther 2012;20(4):749–58.

[140] Kirn DH, Wang Y, Le Boeuf F, Bell J, Thorne SH. Targeting of interferon-beta to produce a specific, multi-mechanistic oncolytic vaccinia virus. PLoS Med 2007;4(12), e353.

[141] Scholl SM, Balloul J-M, Le Goc G, Bizouarne N, Schatz C, Kieny MP, et al. Recombinant vaccinia virus encoding human MUC1 and IL2 as immunotherapy in patients with breast cancer. J Immunother 2000;23(5).

[142] Kaufman HL, Flanagan K, Lee CSD, Perretta DJ, Horig H. Insertion of interleukin-2 (IL-2) and interleukin-12 (IL-12) genes into vaccinia virus results in effective anti-tumor responses without toxicity. Vaccine 2002;20(13):1862–9.

[143] Kwa S, Lai L, Gangadhara S, Siddiqui M, Pillai VB, Labranche C, et al. CD40L-adjuvanted DNA/modified vaccinia virus ankara simian immunodeficiency virus SIV239 vaccine enhances SIV-specific humoral and cellular immunity and improves protection against a heterologous SIVE660 mucosal challenge. J Virol 2014;88(17):9579.

[144] Walker JD, Sehgal I, Kousoulas KG. Oncolytic herpes simplex virus 1 encoding 15-prostaglandin dehydrogenase mitigates immune suppression and reduces ectopic primary and metastatic breast cancer in mice. J Virol 2011;85(14):7363.

[145] Hou W, Sampath P, Rojas Juan J, Thorne SH. Oncolytic virus-mediated targeting of PGE2 in the tumor alters the immune status and sensitizes established and resistant tumors to immunotherapy. Cancer Cell 2016;30(1):108–19.

[146] Yang X, Huang B, Deng L, Hu Z. Progress in gene therapy using oncolytic vaccinia virus as vectors. J Cancer Res Clin Oncol 2018;144(12):2433–40.

[147] Eckert EC, Nace RA, Tonne JM, Evgin L, Vile RG, Russell SJ. Generation of a tumor-specific chemokine gradient using oncolytic vesicular stomatitis virus encoding CXCL9. Mol Ther Oncolytics 2019;16:63–74.

[148] Yu F, Wang X, Guo ZS, Bartlett DL, Gottschalk SM, Song X-T. T-cell engager-armed oncolytic vaccinia virus significantly enhances antitumor therapy. Mol Ther 2014;22(1):102–11.

[149] Engeland CE, Grossardt C, Veinalde R, Bossow S, Lutz D, Kaufmann JK, et al. CTLA-4 and PD-L1 checkpoint blockade enhances oncolytic measles virus therapy. Mol Ther 2014;22(11):1949–59.

[150] Dias JD, Hemminki O, Diaconu I, Hirvinen M, Bonetti A, Guse K, et al. Targeted cancer immunotherapy with oncolytic adenovirus coding for a fully human monoclonal antibody specific for CTLA-4. Gene Ther 2012;19(10):988–98.

[151] Croft M, So T, Duan W, Soroosh P. The significance of OX40 and OX40L to T-cell biology and immune disease. Immunol Rev 2009;229(1):173–91.

[152] Jiang H, Fan X, Clise-Dwyer K, Bover L, Gumin J, Ruisaard KE, et al. Abstract 280: delta-24-RGDOX: making cancer more "visible" to the immune system. Cancer Res 2015;75(15 Supplement):280.

[153] Passaro C, Alayo Q, De Laura I, McNulty J, Grauwet K, Ito H, et al. Arming an oncolytic herpes simplex virus type 1 with a single-chain fragment variable antibody against PD-1 for experimental glioblastoma therapy. Clin Cancer Res 2019;25(1):290–9.

[154] Lichty BD, Breitbach CJ, Stojdl DF, Bell JC. Going viral with cancer immunotherapy. Nat Rev Cancer 2014;14(8):559–67.

[155] Balz K, Trassl L, Härtel V, Nelson PP, Skevaki C. Virus-induced T cell-mediated heterologous immunity and vaccine development. Front Immunol 2020;11:513.

[156] Smatti MK, Cyprian FS, Nasrallah GK, Al Thani AA, Almishal RO, Yassine HM. Viruses and auto-immunity: a review on the potential interaction and molecular mechanisms. Viruses 2019;11(8):762.

[157] Nishio N, Dotti G. Oncolytic virus expressing RANTES and IL-15 enhances function of CAR-modified T cells in solid tumors. Oncoimmunology 2015;4(2), e988098-e.

[158] Gajewski TF, Woo SR, Zha Y, Spaapen R, Zheng Y, Corrales L, et al. Cancer immunotherapy strategies based on overcoming barriers within the tumor microenvironment. Curr Opin Immunol 2013;25(2):268–76.

[159] Schmidts A, Maus MV. Making CAR T cells a solid option for solid tumors. Front Immunol 2018;9:2593.

[160] Majzner RG, Rietberg SP, Sotillo E, Dong R, Vachharajani VT, Labanieh L, et al. Tuning the antigen density requirement for CAR T-cell activity. Cancer Discov 2020;10(5):702–23.

[161] Wing A, Fajardo CA, Posey Jr AD, Shaw C, Da T, Young RM, et al. Improving CAR T-cell therapy of solid tumors with oncolytic virus-driven production of a bispecific T-cell engager. Cancer Immunol Res 2018;6(5):605–16.

[162] Freedman JD, Duffy MR, Lei-Rossmann J, Muntzer A, Scott EM, Hagel J, et al. An oncolytic virus expressing a T-cell engager simultaneously targets cancer and immunosuppressive stromal cells. Cancer Res 2018;78(24):6852.

[163] Nishio N, Diaconu I, Liu H, Cerullo V, Caruana I, Hoyos V, et al. Armed oncolytic virus enhances immune functions of chimeric antigen receptor-modified T cells in solid tumors. Cancer Res 2014;74 (18):5195–205.

[164] Watanabe K, Luo Y, Da T, Guedan S, Ruella M, Scholler J, et al. Pancreatic cancer therapy with combined mesothelin-redirected chimeric antigen receptor T cells and cytokine-armed oncolytic adenoviruses. JCI Insight 2018;3(7).

[165] Moon EK, Wang LS, Bekdache K, Lynn RC, Lo A, Thorne SH, et al. Intra-tumoral delivery of CXCL11 via a vaccinia virus, but not by modified T cells, enhances the efficacy of adoptive T cell therapy and vaccines. Oncoimmunology 2018;7(3), e1395997.

[166] Guedan S, Alemany R. CAR-T cells and oncolytic viruses: joining forces to overcome the solid tumor challenge. Front Immunol 2018;9:2460.

[167] VanSeggelen H, Tantalo DGM, Afsahi A, Hammill JA, Bramson JL. Chimeric antigen receptor–engineered T cells as oncolytic virus carriers. Mol Ther Oncolytics 2015;2.

[168] Cole C, Qiao J, Kottke T, Diaz RM, Ahmed A, Sanchez-Perez L, et al. Tumor-targeted, systemic delivery of therapeutic viral vectors using hitchhiking on antigen-specific T cells. Nat Med 2005;11 (10):1073–81.

[169] Goradel NH, Mohajel N, Malekshahi ZV, Jahangiri S, Najafi M, Farhood B, et al. Oncolytic adeno-virus: a tool for cancer therapy in combination with other therapeutic approaches. J Cell Physiol 2019;234(6):8636–46.

[170] Zafar S, Parviainen S, Siurala M, Hemminki O, Havunen R, Tähtinen S, et al. Intravenously usable fully serotype 3 oncolytic adenovirus coding for CD40L as an enabler of dendritic cell therapy. Oncoimmunology 2017;6(2), e1265717.

[171] Martin NT, Bell JC. Oncolytic virus combination therapy: killing one bird with two stones. Mol Ther 2018;26(6):1414–22.

[172] Le Boeuf F, Diallo J-S, McCart JA, Thorne S, Falls T, Stanford M, et al. Synergistic interaction between oncolytic viruses augments tumor killing. Mol Ther 2010;18(5):888–95.

[173] Tysome JR, Li X, Wang S, Wang P, Gao D, Du P, et al. A novel therapeutic regimen to eradicate established solid tumors with an effective induction of tumor-specific immunity. Clin Cancer Res 2012;18(24):6679.

[174] Nistal-Villan E, Bunuales M, Poutou J, Gonzalez-Aparicio M, Bravo-Perez C, Quetglas JI, et al. Enhanced therapeutic effect using sequential administration of antigenically distinct oncolytic viruses expressing oncostatin M in a Syrian hamster orthotopic pancreatic cancer model. Mol Cancer 2015;14 (1):210.

[175] Ilett E, Kottke T, Thompson J, Rajani K, Zaidi S, Evgin L, et al. Prime-boost using separate oncolytic viruses in combination with checkpoint blockade improves anti-tumour therapy. Gene Ther 2017;24 (1):21–30.

[176] Macedo N, Miller DM, Haq R, Kaufman HL. Clinical landscape of oncolytic virus research in 2020. J Immunother Cancer 2020;8(2), e001486.

[177] Bell J, McFadden G. Viruses for tumor therapy. Cell Host Microbe 2014;15(3):260–5.

[178] Wennier ST, Liu J, McFadden G. Bugs and drugs: oncolytic virotherapy in combination with che-motherapy. Curr Pharm Biotechnol 2012;13(9):1817–33.

[179] Mesa RA, Tefferi A. Emerging drugs for the therapy of primary and post essential thrombocythemia, post polycythemia vera myelofibrosis. Expert Opin Emerg Drugs 2009;14(3):471–9.

[180] Ghonime MG, Cassady KA. Combination therapy using ruxolitinib and oncolytic HSV renders resis-tant MPNSTs susceptible to virotherapy. Cancer Immunol Res 2018;6(12):1499.

[181] Kwok G, Yau TC, Chiu JW, Tse E, Kwong YL. Pembrolizumab (Keytruda). Hum Vaccin Immu-nother 2016;12(11):2777–89.

[182] Mahalingam D, Wilkinson GA, Eng KH, Fields P, Raber P, Moseley JL, et al. Pembrolizumab in combination with the oncolytic virus pelareorep and chemotherapy in patients with advanced pan-creatic adenocarcinoma: a phase Ib study. Clin Cancer Res 2020;26(1):71–81.

[183] Rosenberg SA, Aebersold P, Cornetta K, Kasid A, Morgan RA, Moen R, et al. Gene transfer into humans–immunotherapy of patients with advanced melanoma, using tumor-infiltrating lymphocytes modified by retroviral gene transduction. N Engl J Med 1990;323(9):570–8.

[184] Barese CN, Dunbar CE. Contributions of gene marking to cell and gene therapies. Hum Gene Ther 2011;22(6):659–68.

[185] McCain J. The future of gene therapy. Biotechnol Healthc 2005;2(3):52–60.

[186] Fend L, Remy-Ziller C, Foloppe J, Kempf J, Cochin S, Barraud L, et al. Oncolytic virotherapy with an armed vaccinia virus in an orthotopic model of renal carcinoma is associated with modification of the tumor microenvironment. Oncoimmunology 2015;5(2), e1080414-e.

[187] Ireton GC, McDermott G, Black ME, Stoddard BL. The structure of Escherichia coli cytosine deam-inase. J Mol Biol 2002;315(4):687–97.

[188] Grem JL. 5-Fluorouracil: forty-plus and still ticking. A review of its preclinical and clinical develop-ment. Invest New Drugs 2000;18(4):299–313.

[189] Koeneman KS, Kao C, Ko SC, Yang L, Wada Y, Kallmes DF, et al. Osteocalcin-directed gene ther-apy for prostate-cancer bone metastasis. World J Urol 2000;18(2):102–10.

[190] Lawler SE, Speranza M-C, Cho C-F, Chiocca EA. Oncolytic viruses in cancer treatment: a review. JAMA Oncol 2017;3(6):841–9.

[191] Fulci G, Breymann L, Gianni D, Kurozomi K, Rhee SS, Yu J, et al. Cyclophosphamide enhances glioma virotherapy by inhibiting innate immune responses. Proc Natl Acad Sci U S A 2006;103 (34):12873–8.

[192] Broyles SS. Vaccinia virus transcription. J Gen Virol 2003;84(Pt. 9):2293–303.

[193] Chung CS, Hsiao JC, Chang YS, Chang W. A27L protein mediates vaccinia virus interaction with cell surface heparan sulfate. J Virol 1998;72(2):1577–85.

[194] Wang F, Ma Y, Barrett JW, Gao X, Loh J, Barton E, et al. Disruption of Erk-dependent type I interferon induction breaks the myxoma virus species barrier. Nat Immunol 2004;5(12):1266–74.

[195] Puhlmann M, Brown CK, Gnant M, Huang J, Libutti SK, Alexander HR, et al. Vaccinia as a vector for tumor-directed gene therapy: biodistribution of a thymidine kinase-deleted mutant. Cancer Gene Ther 2000;7(1):66–73.

[196] Deng L, Fan J, Ding Y, Zhang J, Zhou B, Zhang Y, et al. Oncolytic efficacy of thymidine kinase-deleted vaccinia virus strain Guang9. Oncotarget 2017;8(25):40533–43.

[197] McCart JA, Ward JM, Lee J, Hu Y, Alexander HR, Libutti SK, et al. Systemic cancer therapy with a tumor-selective vaccinia virus mutant lacking thymidine kinase and vaccinia growth factor genes. Cancer Res 2001;61(24):8751–7.

[198] Guo ZS, Naik A, O'Malley ME, Popovic P, Demarco R, Hu Y, et al. The enhanced tumor selectivity of an oncolytic vaccinia lacking the host range and antiapoptosis genes SPI-1 and SPI-2. Cancer Res 2005;65(21):9991–8.

[199] Burgess HM, Pourchet A, Hajdu CH, Chiriboga L, Frey AB, Mohr I. Targeting poxvirus decapping enzymes and mRNA decay to generate an effective oncolytic virus. Mol Ther Oncolytics 2018;8:71–81.

[200] Ricordel M, Foloppe J, Antoine D, Findeli A, Kempf J, Cordier P, et al. Vaccinia virus shuffling: deVV5, a novel chimeric poxvirus with improved oncolytic potency. Cancers (Basel) 2018;10(7):231.

[201] Choi AH, O'Leary MP, Chaurasiya S, Lu J, Kim SI, Fong Y, et al. Novel chimeric parapoxvirus CF189 as an oncolytic immunotherapy in triple-negative breast cancer. Surgery 2018;163(2):336–42.

[202] Yoo SY, Jeong SN, Kang DH, Heo J. Evolutionary cancer-favoring engineered vaccinia virus for metastatic hepatocellular carcinoma. Oncotarget 2017;8(42):71489–99.

[203] Mejías-Pérez E, Carreño-Fuentes L, Esteban M. Development of a safe and effective vaccinia virus oncolytic vector WR-Δ4 with a set of gene deletions on several viral pathways. Mol Ther Oncolytics 2017;8:27–40.

[204] Park SH, Breitbach CJ, Lee J, Park JO, Lim HY, Kang WK, et al. Phase 1b trial of biweekly intravenous Pexa-Vec (JX-594), an oncolytic and immunotherapeutic vaccinia virus in colorectal cancer. Mol Ther 2015;23(9):1532–40.

[205] Breitbach CJ, Moon A, Burke J, Hwang TH, Kirn DH. A phase 2, open-label, randomized study of Pexa-Vec (JX-594) administered by intratumoral injection in patients with unresectable primary hepatocellular carcinoma. Methods Mol Biol 2015;1317:343–57.

[206] Heo J, Reid T, Ruo L, Breitbach CJ, Rose S, Bloomston M, et al. Randomized dose-finding clinical trial of oncolytic immunotherapeutic vaccinia JX-594 in liver cancer. Nat Med 2013;19(3):329–36.

[207] Lv C, Su Q, Liang Y, Hu J, Yuan S. Oncolytic vaccine virus harbouring the IL-24 gene suppresses the growth of lung cancer by inducing apoptosis. Biochem Biophys Res Commun 2016;476(1):21–8.

[208] Liu Z, Ravindranathan R, Li J, Kalinski P, Guo ZS, Bartlett DL. CXCL11-Armed oncolytic poxvirus elicits potent antitumor immunity and shows enhanced therapeutic efficacy. Oncoimmunology 2016;5(3), e1091554.

[209] Chard LS, Maniati E, Wang P, Zhang Z, Gao D, Wang J, et al. A vaccinia virus armed with interleukin-10 is a promising therapeutic agent for treatment of murine pancreatic cancer. Clin Cancer Res 2015;21(2):405–16.

[210] Jefferson A, Cadet VE, Hielscher A. The mechanisms of genetically modified vaccinia viruses for the treatment of cancer. Crit Rev Oncol Hematol 2015;95(3):407–16.

[211] Liu Z, Ge Y, Wang H, Ma C, Feist M, Ju S, et al. Modifying the cancer-immune set point using vaccinia virus expressing re-designed interleukin-2. Nat Commun 2018;9(1):4682.

[212] Thorne SH. Immunotherapeutic potential of oncolytic vaccinia virus. Front Oncol 2014;4:155.

[213] Breitbach CJ, Burke J, Jonker D, Stephenson J, Haas AR, Chow LQ, et al. Intravenous delivery of a multi-mechanistic cancer-targeted oncolytic poxvirus in humans. Nature 2011;477(7362):99–102.

[214] Smith GL. Vaccinia virus immune evasion. Immunol Lett 1999;65(1–2):55–62.

[215] Smith GL, Benfield CTO, Maluquer de Motes C, Mazzon M, SWJ E, Ferguson BJ, et al. Vaccinia virus immune evasion: mechanisms, virulence and immunogenicity. J Gen Virol 2013;94(Pt. 11):2367–92.

[216] Thirunavukarasu P, Sathaiah M, Gorry MC, O'Malley ME, Ravindranathan R, Austin F, et al. A rationally designed A34R mutant oncolytic poxvirus: improved efficacy in peritoneal carcinomatosis. Mol Ther 2013;21(5):1024–33.

[217] Kirn DH, Wang Y, Liang W, Contag CH, Thorne SH. Enhancing poxvirus oncolytic effects through increased spread and immune evasion. Cancer Res 2008;68(7):2071–5.

[218] Amato RJ, Hawkins RE, Kaufman HL, Thompson JA, Tomczak P, Szczylik C, et al. Vaccination of metastatic renal cancer patients with MVA-5T4: a randomized, double-blind, placebo-controlled phase III study. Clin Cancer Res 2010;16(22):5539–47.

[219] Harrop R, Shingler WH, McDonald M, Treasure P, Amato RJ, Hawkins RE, et al. MVA-5T4-induced immune responses are an early marker of efficacy in renal cancer patients. Cancer Immunol Immunother 2011;60(6):829–37.

[220] Mastrangelo MJ, Maguire Jr HC, Eisenlohr LC, Laughlin CE, Monken CE, McCue PA, et al. Intratumoral recombinant GM-CSF-encoding virus as gene therapy in patients with cutaneous melanoma. Cancer Gene Ther 1999;6(5):409–22.

[221] Adamina M, Rosenthal R, Weber WP, Frey DM, Viehl CT, Bolli M, et al. Intranodal immunization with a vaccinia virus encoding multiple antigenic epitopes and costimulatory molecules in metastatic melanoma. Mol Ther 2010;18(3):651–9.

[222] Quoix E, Lena H, Losonczy G, Forget F, Chouaid C, Papai Z, et al. TG4010 immunotherapy and first-line chemotherapy for advanced non-small-cell lung cancer (TIME): results from the phase 2b part of a randomised, double-blind, placebo-controlled, phase 2b/3 trial. Lancet Oncol 2016;17(2):212–23.

[223] Heery CR, Palena C, McMahon S, Donahue RN, Lepone LM, Grenga I, et al. Phase I study of a poxviral TRICOM-based vaccine directed against the transcription factor brachyury. Clin Cancer Res 2017;23(22):6833.

[224] Kantoff PW, Schuetz TJ, Blumenstein BA, Glode LM, Bilhartz DL, Wyand M, et al. Overall survival analysis of a phase II randomized controlled trial of a Poxviral-based PSA-targeted immunotherapy in metastatic castration-resistant prostate cancer. J Clin Oncol 2010;28(7):1099–105.

[225] Gulley JL, Arlen PM, Madan RA, Tsang KY, Pazdur MP, Skarupa L, et al. Immunologic and prognostic factors associated with overall survival employing a poxviral-based PSA vaccine in metastatic castrate-resistant prostate cancer. Cancer Immunol Immunother 2010;59(5):663–74.

[226] Heery CR, Ibrahim NK, Arlen PM, Mohebtash M, Murray JL, Koenig K, et al. Docetaxel alone or in combination with a therapeutic cancer vaccine (PANVAC) in patients with metastatic breast cancer: a randomized clinical trial. JAMA Oncol 2015;1(8):1087–95.

[227] Kim JH, Oh JY, Park BH, Lee DE, Kim JS, Park HE, et al. Systemic armed oncolytic and immunologic therapy for cancer with JX-594, a targeted poxvirus expressing GM-CSF. Mol Ther 2006;14(3):361–70.

[228] Cripe TP, Ngo MC, Geller JI, Louis CU, Currier MA, Racadio JM, et al. Phase 1 study of intratumoral Pexa-Vec (JX-594), an oncolytic and immunotherapeutic vaccinia virus, in pediatric cancer patients. Mol Ther 2015;23(3):602–8.

[229] Park BH, Hwang T, Liu TC, Sze DY, Kim JS, Kwon HC, et al. Use of a targeted oncolytic poxvirus, JX-594, in patients with refractory primary or metastatic liver cancer: a phase I trial. Lancet Oncol 2008;9(6):533–42.

[230] Hwang TH, Moon A, Burke J, Ribas A, Stephenson J, Breitbach CJ, et al. A mechanistic proof-of-concept clinical trial with JX-594, a targeted multi-mechanistic oncolytic poxvirus, in patients with metastatic melanoma. Mol Ther 2011;19(10):1913–22.

[231] Downs-Canner S, Guo ZS, Ravindranathan R, Breitbach CJ, O'Malley ME, Jones HL, et al. Phase 1 study of intravenous oncolytic poxvirus (vvDD) in patients with advanced solid cancers. Mol Ther 2016;24(8):1492–501.

[232] Zeh HJ, Downs-Canner S, McCart JA, Guo ZS, Rao UN, Ramalingam L, et al. First-in-man study of western reserve strain oncolytic vaccinia virus: safety, systemic spread, and antitumor activity. Mol Ther 2015;23(1):202–14.

[233] Liikanen I, Monsurrò V, Ahtiainen L, Raki M, Hakkarainen T, Diaconu I, et al. Induction of interferon pathways mediates in vivo resistance to oncolytic adenovirus. Mol Ther 2011;19(10):1858–66.

[234] Nevins JR. Transcriptional regulation. A closer look at E2F. Nature 1992;358(6385):375–6.

[235] Rojas JJ, Guedan S, Searle PF, Martinez-Quintanilla J, Gil-Hoyos R, Alcayaga-Miranda F, et al. Minimal RB-responsive E1A promoter modification to attain potency, selectivity, and transgene-arming capacity in oncolytic adenoviruses. Mol Ther 2010;18(11):1960–71.

[236] Lang FF, Conrad C, Gomez-Manzano C, Yung WKA, Sawaya R, Weinberg JS, et al. Phase I study of DNX-2401 (Delta-24-RGD) oncolytic adenovirus: replication and immunotherapeutic effects in recurrent malignant glioma. J Clin Oncol 2018;36(14):1419–27.

[237] Kanerva A, Nokisalmi P, Diaconu I, Koski A, Cerullo V, Liikanen I, et al. Antiviral and antitumor T-cell immunity in patients treated with GM-CSF-coding oncolytic adenovirus. Clin Cancer Res 2013;19(10):2734–44.

[238] Bramante S, Koski A, Kipar A, Diaconu I, Liikanen I, Hemminki O, et al. Serotype chimeric oncolytic adenovirus coding for GM-CSF for treatment of sarcoma in rodents and humans. Int J Cancer 2014;135(3):720–30.

[239] Sarkar S, Quinn BA, Shen XN, Dash R, Das SK, Emdad L, et al. Therapy of prostate cancer using a novel cancer terminator virus and a small molecule BH-3 mimetic. Oncotarget 2015;6(13):10712–27.

[240] Hemminki O, Diaconu I, Cerullo V, Pesonen SK, Kanerva A, Joensuu T, et al. Ad3-hTERT-E1A, a fully serotype 3 oncolytic adenovirus, in patients with chemotherapy refractory cancer. Mol Ther 2012;20(9):1821–30.

[241] Chang J, Zhao X, Wu X, Guo Y, Guo H, Cao J, et al. A phase I study of KH901, a conditionally replicating granulocyte-macrophage colony-stimulating factor: armed oncolytic adenovirus for the treatment of head and neck cancers. Cancer Biol Ther 2009;8(8):676–82.

[242] Wang X, Su C, Cao H, Li K, Chen J, Jiang L, et al. A novel triple-regulated oncolytic adenovirus carrying p53 gene exerts potent antitumor efficacy on common human solid cancers. Mol Cancer Ther 2008;7(6):1598–603.

[243] Xu C, Sun Y, Wang Y, Yan Y, Shi Z, Chen L, et al. CEA promoter-regulated oncolytic adenovirus-mediated Hsp70 expression in immune gene therapy for pancreatic cancer. Cancer Lett 2012;319 (2):154–63.

[244] DeWeese TL, van der Poel H, Li S, Mikhak B, Drew R, Goemann M, et al. A phase I trial of CV706, a replication-competent, PSA selective oncolytic adenovirus, for the treatment of locally recurrent prostate cancer following radiation therapy. Cancer Res 2001;61(20):7464–72.

[245] Hemminki O, Parviainen S, Juhila J, Turkki R, Linder N, Lundin J, et al. Immunological data from cancer patients treated with Ad5/3-E2F-Δ24-GMCSF suggests utility for tumor immunotherapy. Oncotarget 2015;6(6):4467–81.

[246] Bauerschmitz GJ, Guse K, Kanerva A, Menzel A, Herrmann I, Desmond RA, et al. Triple-targeted oncolytic adenoviruses featuring the cox2 promoter, E1A transcomplementation, and serotype chimerism for enhanced selectivity for ovarian cancer cells. Mol Ther 2006;14(2):164–74.

[247] Bayo-Puxan N, Gimenez-Alejandre M, Lavilla-Alonso S, Gros A, Cascallo M, Hemminki A, et al. Replacement of adenovirus type 5 fiber shaft heparan sulfate proteoglycan-binding domain with RGD for improved tumor infectivity and targeting. Hum Gene Ther 2009;20(10):1214–21.

[248] Kanerva A, Wang M, Bauerschmitz GJ, Lam JT, Desmond RA, Bhoola SM, et al. Gene transfer to ovarian cancer versus normal tissues with fiber-modified adenoviruses. Mol Ther 2002;5(6):695–704.

[249] Koski A, Bramante S, Kipar A, Oksanen M, Juhila J, Vassilev L, et al. Biodistribution analysis of oncolytic adenoviruses in patient autopsy samples reveals vascular transduction of noninjected tumors and tissues. Mol Ther 2015;23(10):1641–52.

[250] Hemminki O, Bauerschmitz G, Hemmi S, Lavilla-Alonso S, Diaconu I, Guse K, et al. Oncolytic adenovirus based on serotype 3. Cancer Gene Ther 2011;18(4):288–96.

[251] Kuhn I, Harden P, Bauzon M, Chartier C, Nye J, Thorne S, et al. Directed evolution generates a novel oncolytic virus for the treatment of colon cancer. PLoS One 2008;3(6), e2409.

[252] Barton KN, Paielli D, Zhang Y, Koul S, Brown SL, Lu M, et al. Second-generation replication-competent oncolytic adenovirus armed with improved suicide genes and ADP gene demonstrates greater efficacy without increased toxicity. Mol Ther 2006;13(2):347–56.

[253] Boucher PD, Im MM, Freytag SO, Shewach DS. A novel mechanism of synergistic cytotoxicity with 5-fluorocytosine and ganciclovir in double suicide gene therapy. Cancer Res 2006;66(6):3230–7.

[254] Taipale K, Liikanen I, Koski A, Heiskanen R, Kanerva A, Hemminki O, et al. Predictive and prognostic clinical variables in cancer patients treated with adenoviral oncolytic immunotherapy. Mol Ther 2016;24(7):1323–32.

[255] Andtbacka RH, Ross M, Puzanov I, Milhem M, Collichio F, Delman KA, et al. Patterns of clinical response with talimogene laherparepvec (T-VEC) in patients with melanoma treated in the OPTiM phase III clinical trial. Ann Surg Oncol 2016;23(13):4169–77.

[256] Oh E, Choi IK, Hong J, Yun CO. Oncolytic adenovirus coexpressing interleukin-12 and decorin overcomes Treg-mediated immunosuppression inducing potent antitumor effects in a weakly immunogenic tumor model. Oncotarget 2017;8(3):4730–46.

[257] Cerullo V, Pesonen S, Diaconu I, Escutenaire S, Arstila PT, Ugolini M, et al. Oncolytic adenovirus coding for granulocyte macrophage colony-stimulating factor induces antitumoral immunity in cancer patients. Cancer Res 2010;70(11):4297–309.

[258] Lv S-Q, Ye Z-L, Liu P-Y, Huang Y, Li L-F, Liu H, et al. 11R-P53 and GM-CSF expressing oncolytic adenovirus target cancer stem cells with enhanced synergistic activity. J Cancer 2017;8(2):199–206.

[259] Machitani M, Sakurai F, Wakabayashi K, Tachibana M, Fujiwara T, Mizuguchi H. Enhanced oncolytic activities of the telomerase-specific replication-competent adenovirus expressing short-hairpin RNA against dicer. Mol Cancer Ther 2017;16(1):251–9.

[260] Koski A, Kangasniemi L, Escutenaire S, Pesonen S, Cerullo V, Diaconu I, et al. Treatment of cancer patients with a serotype 5/3 chimeric oncolytic adenovirus expressing GMCSF. Mol Ther 2010;18 (10):1874–84.

[261] Rojas LA, Condezo GN, Moreno R, Fajardo CA, Arias-Badia M, San Martín C, et al. Albumin-binding adenoviruses circumvent pre-existing neutralizing antibodies upon systemic delivery. J Control Release 2016;237:78–88.

[262] Lubaroff DM, Konety B, Link BK, Ratliff TL, Madsen T, Shannon M, et al. Clinical protocol: phase I study of an adenovirus/prostate-specific antigen vaccine in men with metastatic prostate cancer. Hum Gene Ther 2006;17(2):220–9.

[263] Morse MA, Chaudhry A, Gabitzsch ES, Hobeika AC, Osada T, Clay TM, et al. Novel adenoviral vector induces T-cell responses despite anti-adenoviral neutralizing antibodies in colorectal cancer patients. Cancer Immunol Immunother 2013;62(8):1293–301.

[264] Balint JP, Gabitzsch ES, Rice A, Latchman Y, Xu Y, Messerschmidt GL, et al. Extended evaluation of a phase 1/2 trial on dosing, safety, immunogenicity, and overall survival after immunizations with an advanced-generation Ad5 [E1-, E2b-]-CEA(6D) vaccine in late-stage colorectal cancer. Cancer Immunol Immunother 2015;64(8):977–87.

[265] Gatti-Mays ME, Redman JM, Donahue RN, Palena C, Madan RA, Karzai F, et al. A phase I trial using a multitargeted recombinant adenovirus 5 (CEA/MUC1/Brachyury)-based immunotherapy vaccine regimen in patients with advanced cancer. Oncologist 2020;25(6):479–e899.

[266] Sato-Dahlman M, LaRocca CJ, Yanagiba C, Yamamoto M. Adenovirus and immunotherapy: advancing cancer treatment by combination. Cancers (Basel) 2020;12(5).

[267] Wieking BG, Vermeer DW, Spanos WC, Lee KM, Vermeer P, Lee WT, et al. A non-oncogenic HPV 16 E6/E7 vaccine enhances treatment of HPV expressing tumors. Cancer Gene Ther 2012;19(10):667–74.

[268] Habib N, Salama H, Abd El Latif Abu Median A, Isac Anis I, Abd Al Aziz RA, Sarraf C, et al. Clinical trial of E1B-deleted adenovirus (dl1520) gene therapy for hepatocellular carcinoma. Cancer Gene Ther 2002;9(3):254–9.

[269] Makower D, Rozenblit A, Kaufman H, Edelman M, Lane ME, Zwiebel J, et al. Phase II clinical trial of intralesional administration of the oncolytic adenovirus ONYX-015 in patients with hepatobiliary tumors with correlative p53 studies. Clin Cancer Res 2003;9(2):693–702.

[270] Chiocca EA, Abbed KM, Tatter S, Louis DN, Hochberg FH, Barker F, et al. A phase I open-label, dose-escalation, multi-institutional trial of injection with an E1B-Attenuated adenovirus, ONYX-015, into the peritumoral region of recurrent malignant gliomas, in the adjuvant setting. Mol Ther 2004;10(5):958–66.

[271] Galanis E, Okuno SH, Nascimento AG, Lewis BD, Lee RA, Oliveira AM, et al. Phase I-II trial of ONYX-015 in combination with MAP chemotherapy in patients with advanced sarcomas. Gene Ther 2005;12(5):437–45.

[272] Ranki T, Pesonen S, Hemminki A, Partanen K, Kairemo K, Alanko T, et al. Phase I study with ONCOS-102 for the treatment of solid tumors – an evaluation of clinical response and exploratory analyses of immune markers. J Immunother Cancer 2016;4:17.

[273] Li Z, Yang G, Zhou S, Wang X, Li X. 34th Annual Meeting & Pre-Conference Programs of the Society for Immunotherapy of Cancer (SITC 2019): Part 2. J Immunother Cancer 2019;7:283. https://doi.org/10.1186/s40425-019-0764-0.

[274] Schiza A, Wenthe J, Mangsbo S, Eriksson E, Nilsson A, Tötterman TH, et al. Adenovirus-mediated CD40L gene transfer increases Teffector/Tregulatory cell ratio and upregulates death receptors in metastatic melanoma patients. J Transl Med 2017;15(1):79.

[275] Eriksson E, Milenova I, Wenthe J, Ståhle M, Leja-Jarblad J, Ullenhag G, et al. Shaping the tumor stroma and sparking immune activation by CD40 and 4-1BB signaling induced by an armed oncolytic virus. Clin Cancer Res 2017;23(19):5846–57.

[276] Chiocca EA, Yu JS, Lukas RV, Solomon IH, Ligon KL, Nakashima H, et al. Regulatable interleukin-12 gene therapy in patients with recurrent high-grade glioma: results of a phase 1 trial. Sci Transl Med 2019;11(505).

[277] Martuza RL, Malick A, Markert JM, Ruffner KL, Coen DM. Experimental therapy of human glioma by means of a genetically engineered virus mutant. Science 1991;252(5007):854–6.

[278] Markert JM, Malick A, Coen DM, Martuza RL. Reduction and elimination of encephalitis in an experimental glioma therapy model with attenuated herpes simplex mutants that retain susceptibility to acyclovir. Neurosurgery 1993;32(4):597–603.

[279] Andtbacka RH, Kaufman HL, Collichio F, Amatruda T, Senzer N, Chesney J, et al. Talimogene laherparepvec improves durable response rate in patients with advanced melanoma. J Clin Oncol 2015;33(25):2780–8.

[280] He B, Gross M, Roizman B. The $\gamma_1 34.5$ protein of herpes simplex virus 1 complexes with protein phosphatase 1α to dephosphorylate the α subunit of the eukaryotic translation initiation factor 2 and preclude the shutoff of protein synthesis by double-stranded RNA-activated protein kinase. Proc Natl Acad Sci 1997;94(3):843–8.

[281] Kaufman HL, Kim DW, DeRaffele G, Mitcham J, Coffin RS, Kim-Schulze S. Local and distant immunity induced by intralesional vaccination with an oncolytic herpes virus encoding GM-CSF in patients with stage IIIc and IV melanoma. Ann Surg Oncol 2010;17(3):718–30.

[282] Todo T, Martuza RL, Rabkin SD, Johnson PA. Oncolytic herpes simplex virus vector with enhanced MHC class I presentation and tumor cell killing. Proc Natl Acad Sci U S A 2001;98(11):6396–401.

[283] Geevarghese SK, Geller DA, de Haan HA, Hörer M, Knoll AE, Mescheder A, et al. Phase I/II study of oncolytic herpes simplex virus NV1020 in patients with extensively pretreated refractory colorectal cancer metastatic to the liver. Hum Gene Ther 2010;21(9):1119–28.

[284] Petrovic B, Gianni T, Gatta V, Campadelli-Fiume G. Insertion of a ligand to HER2 in gB retargets HSV tropism and obviates the need for activation of the other entry glycoproteins. PLoS Pathog 2017;13(4), e1006352.

[285] Streby KA, Geller JI, Currier MA, Warren PS, Racadio JM, Towbin AJ, et al. Intratumoral injection of HSV1716, an oncolytic herpes virus, is safe and shows evidence of immune response and viral replication in young cancer patients. Clin Cancer Res 2017;23(14):3566–74.

[286] Hu JCC, Coffin RS, Davis CJ, Graham NJ, Groves N, Guest PJ, et al. A phase i study of OncoVEXGM-CSF, a second-generation oncolytic herpes simplex virus expressing granulocyte macrophage colony-stimulating factor. Clin Cancer Res 2006;12(22):6737.

[287] Senzer NN, Kaufman HL, Amatruda T, Nemunaitis M, Reid T, Daniels G, et al. Phase II clinical trial of a granulocyte-macrophage colony-stimulating factor–encoding, second-generation oncolytic herpesvirus in patients with unresectable metastatic melanoma. J Clin Oncol 2009;27(34):5763–71.

[288] Markert JM, Razdan SN, Kuo HC, Cantor A, Knoll A, Karrasch M, et al. A phase 1 trial of oncolytic HSV-1, G207, given in combination with radiation for recurrent GBM demonstrates safety and radiographic responses. Mol Ther 2014;22(5):1048–55.

[289] Taguchi S, Fukuhara H, Todo T. Oncolytic virus therapy in Japan: progress in clinical trials and future perspectives. Jpn J Clin Oncol 2019;49(3):201–9.

[290] Kelly KJ, Wong J, Fong Y. Herpes simplex virus NV1020 as a novel and promising therapy for hepatic malignancy. Expert Opin Investig Drugs 2008;17(7):1105–13.

[291] Kemeny N, Brown K, Covey A, Kim T, Bhargava A, Brody L, et al. Phase I, open-label, dose-escalating study of a genetically engineered herpes simplex virus, NV1020, in subjects with metastatic colorectal carcinoma to the liver. Hum Gene Ther 2006;17(12):1214–24.

[292] Czeglédi A, Wehmann E, Lomniczi B. On the origins and relationships of Newcastle disease virus vaccine strains Hertfordshire and Mukteswar, and virulent strain Herts'33. Avian Pathol 2003;32(3):271–6.

[293] Koks CA, Garg AD, Ehrhardt M, Riva M, Vandenberk L, Boon L, et al. Newcastle disease virotherapy induces long-term survival and tumor-specific immune memory in orthotopic glioma through the induction of immunogenic cell death. Int J Cancer 2015;136(5):E313–25.

[294] Mansour M, Palese P, Zamarin D. Oncolytic specificity of Newcastle disease virus is mediated by selectivity for apoptosis-resistant cells. J Virol 2011;85(12):6015–23.

[295] Fábián Z, Csatary CM, Szeberényi J, Csatary LK. p53-independent endoplasmic reticulum stress-mediated cytotoxicity of a Newcastle disease virus strain in tumor cell lines. J Virol 2007;81 (6):2817–30.

[296] Buijs P, van Nieuwkoop S, Vaes V, Fouchier R, van Eijck C, van den Hoogen B. Recombinant immunomodulating lentogenic or mesogenic oncolytic newcastle disease virus for treatment of pancreatic adenocarcinoma. Viruses 2015;7(6):2980–98.

[297] Yaacov B, Lazar I, Tayeb S, Frank S, Izhar U, Lotem M, et al. Extracellular matrix constituents interfere with Newcastle disease virus spread in solid tissue and diminish its potential oncolytic activity. J Gen Virol 2012;93(Pt. 8):1664–72.

[298] Cheng X, Wang W, Xu Q, Harper J, Carroll D, Galinski MS, et al. Genetic modification of oncolytic newcastle disease virus for cancer therapy. J Virol 2016;90(11):5343–52.

[299] Chia SL, Yusoff K, Shafee N. Viral persistence in colorectal cancer cells infected by Newcastle disease virus. Virol J 2014;11:91.

[300] Freeman AI, Zakay-Rones Z, Gomori JM, Linetsky E, Rasooly L, Greenbaum E, et al. Phase I/II trial of intravenous NDV-HUJ oncolytic virus in recurrent glioblastoma multiforme. Mol Ther 2006;13 (1):221–8.

[301] Csatary LK, Bakács T. Use of Newcastle disease virus vaccine (MTH-68/H) in a patient with high-grade glioblastoma. JAMA 1999;281(17):1588–9.

[302] Csatary LK, Gosztonyi G, Szeberenyi J, Fabian Z, Liszka V, Bodey B, et al. MTH-68/H oncolytic viral treatment in human high-grade gliomas. J Neurooncol 2004;67(1–2):83–93.

[303] Pecora AL, Rizvi N, Cohen GI, Meropol NJ, Sterman D, Marshall JL, et al. Phase I trial of intravenous administration of PV701, an oncolytic virus, in patients with advanced solid cancers. J Clin Oncol 2002;20(9):2251–66.

[304] Logg CR, Robbins JM, Jolly DJ, Gruber HE, Kasahara N. Retroviral replicating vectors in cancer. Methods Enzymol 2012;507:199–228.

[305] Lu YC, Chen YJ, Yu YR, Lai YH, Cheng JC, Li YF, et al. Replicating retroviral vectors for oncolytic virotherapy of experimental hepatocellular carcinoma. Oncol Rep 2012;28(1):21–6.

[306] Kawasaki Y, Tamamoto A, Takagi-Kimura M, Maeyama Y, Yamaoka N, Terada N, et al. Replication-competent retrovirus vector-mediated prodrug activator gene therapy in experimental models of human malignant mesothelioma. Cancer Gene Ther 2011;18(8):571–8.

[307] Kubo S, Haga K, Tamamoto A, Palmer DJ, Ng P, Okamura H, et al. Adenovirus-retrovirus hybrid vectors achieve highly enhanced tumor transduction and antitumor efficacy in vivo. Mol Ther 2011;19(1):76–82.

[308] Cloughesy TF, Landolfi J, Vogelbaum MA, Ostertag D, Elder JB, Bloomfield S, et al. Durable complete responses in some recurrent high-grade glioma patients treated with Toca 511 + Toca FC. Neuro Oncol 2018;20(10):1383–92.

[309] Anderson BD, Nakamura T, Russell SJ, Peng KW. High CD46 receptor density determines preferential killing of tumor cells by oncolytic measles virus. Cancer Res 2004;64(14):4919–26.

[310] Nakamura T, Peng KW, Harvey M, Greiner S, Lorimer IA, James CD, et al. Rescue and propagation of fully retargeted oncolytic measles viruses. Nat Biotechnol 2005;23(2):209–14.

[311] Baertsch MA, Leber MF, Bossow S, Singh M, Engeland CE, Albert J, et al. MicroRNA-mediated multi-tissue detargeting of oncolytic measles virus. Cancer Gene Ther 2014;21(9):373–80.

[312] Mader EK, Maeyama Y, Lin Y, Butler GW, Russell HM, Galanis E, et al. Mesenchymal stem cell carriers protect oncolytic measles viruses from antibody neutralization in an orthotopic ovarian cancer therapy model. Clin Cancer Res 2009;15(23):7246.

[313] Miest TS, Yaiw KC, Frenzke M, Lampe J, Hudacek AW, Springfeld C, et al. Envelope-chimeric entry-targeted measles virus escapes neutralization and achieves oncolysis. Mol Ther 2011;19 (10):1813–20.

[314] Grote D, Cattaneo R, Fielding AK. Neutrophils contribute to the measles virus-induced antitumor effect: enhancement by granulocyte macrophage colony-stimulating factor expression. Cancer Res 2003;63(19):6463–8.

[315] Iankov ID, Allen C, Federspiel MJ, Myers RM, Peng KW, Ingle JN, et al. Expression of immuno-modulatory neutrophil-activating protein of Helicobacter pylori enhances the antitumor activity of oncolytic measles virus. Mol Ther 2012;20(6):1139–47.

[316] Hardcastle J, Mills L, Malo CS, Jin F, Kurokawa C, Geekiyanage H, et al. Immunovirotherapy with measles virus strains in combination with anti-PD-1 antibody blockade enhances antitumor activity in glioblastoma treatment. Neuro Oncol 2017;19(4):493–502.

[317] Galanis E, Hartmann LC, Cliby WA, Long HJ, Peethambaram PP, Barrette BA, et al. Phase I trial of intraperitoneal administration of an oncolytic measles virus strain engineered to express carcinoembryonic antigen for recurrent ovarian cancer. Cancer Res 2010;70(3):875–82.

[318] Song L, Ohnuma T, Gelman IH, Holland JF. Reovirus infection of cancer cells is not due to activated Ras pathway. Cancer Gene Ther 2009;16(4):382.

[319] Shmulevitz M, Pan L-Z, Garant K, Pan D, Lee PWK. Oncogenic Ras promotes reovirus spread by suppressing IFN-β production through negative regulation of RIG-I signaling. Cancer Res 2010;70 (12):4912–21.

[320] van Houdt WJ, Smakman N, van den Wollenberg DJ, Emmink BL, Veenendaal LM, van Diest PJ, et al. Transient infection of freshly isolated human colorectal tumor cells by reovirus T3D intermediate subviral particles. Cancer Gene Ther 2008;15(5):284–92.

[321] Douville RN, Su R-C, Coombs KM, Simons FER, Hayglass KT. Reovirus serotypes elicit distinctive patterns of recall immunity in humans. J Virol 2008;82(15):7515–23.

[322] Fleeton MN, Contractor N, Leon F, Wetzel JD, Dermody TS, Kelsall BL. Peyer's patch dendritic cells process viral antigen from apoptotic epithelial cells in the intestine of reovirus-infected mice. J Exp Med 2004;200(2):235–45.

[323] Errington F, Steele L, Prestwich R, Harrington KJ, Pandha HS, Vidal L, et al. Reovirus activates human dendritic cells to promote innate antitumor immunity. J Immunol 2008;180(9):6018–26.

[324] Jaime-Ramirez AC, Yu J-G, Caserta E, Yoo JY, Zhang J, Lee TJ, et al. Reolysin and histone deacetylase inhibition in the treatment of head and neck squamous cell carcinoma. Mol Ther Oncolytics 2017;5:87–96.

[325] Harrington KJ, Vile RG, Melcher A, Chester J, Pandha HS. Clinical trials with oncolytic reovirus: moving beyond phase I into combinations with standard therapeutics. Cytokine Growth Factor Rev 2010;21(2–3):91–8.

[326] Thirukkumaran CM, Luider JM, Stewart DA, Cheng T, Lupichuk SM, Nodwell MJ, et al. Reovirus oncolysis as a novel purging strategy for autologous stem cell transplantation. Blood 2003;102 (1):377–87.

[327] Harrington KJ, Karapanagiotou EM, Roulstone V, Twigger KR, White CL, Vidal L, et al. Two-stage phase I dose-escalation study of intratumoral reovirus type 3 dearing and palliative radiotherapy in patients with advanced cancers. Clin Cancer Res 2010;16(11):3067–77.

[328] Comins C, Spicer J, Protheroe A, Roulstone V, Twigger K, White CM, et al. REO-10: a phase I study of intravenous reovirus and docetaxel in patients with advanced cancer. Clin Cancer Res 2010;16(22):5564–72.

[329] Karapanagiotou EM, Roulstone V, Twigger K, Ball M, Tanay M, Nutting C, et al. Phase I/II trial of carboplatin and paclitaxel chemotherapy in combination with intravenous oncolytic reovirus in patients with advanced malignancies. Clin Cancer Res 2012;18(7):2080–9.

[330] Roulstone V, Khan K, Pandha HS, Rudman S, Coffey M, Gill GM, et al. Phase I trial of cyclophosphamide as an immune modulator for optimizing oncolytic reovirus delivery to solid tumors. Clin Cancer Res 2015;21(6):1305–12.

[331] Gong J, Sachdev E, Mita AC, Mita MM. Clinical development of reovirus for cancer therapy: an oncolytic virus with immune-mediated antitumor activity. World J Methodol 2016;6(1):25–42.

[332] Forsyth P, Roldán G, George D, Wallace C, Palmer CA, Morris D, et al. A phase I trial of intratumoral administration of reovirus in patients with histologically confirmed recurrent malignant gliomas. Mol Ther 2008;16(3):627–32.

[333] Kicielinski KP, Chiocca EA, Yu JS, Gill GM, Coffey M, Markert JM. Phase 1 clinical trial of intratumoral reovirus infusion for the treatment of recurrent malignant gliomas in adults. Mol Ther 2014;22(5):1056–62.

[334] Galanis E, Markovic SN, Suman VJ, Nuovo GJ, Vile RG, Kottke TJ, et al. Phase II trial of intravenous administration of Reolysin(®) (Reovirus Serotype-3-dearing Strain) in patients with metastatic melanoma. Mol Ther 2012;20(10):1998–2003.

[335] Felt SA, Grdzelishvili VZ. Recent advances in vesicular stomatitis virus-based oncolytic virotherapy: a 5-year update. J Gen Virol 2017;98(12):2895–911.

[336] Quiroz E, Moreno N, Peralta PH, Tesh RB. A human case of encephalitis associated with vesicular stomatitis virus (Indiana serotype) infection. Am J Trop Med Hyg 1988;39(3):312–4.

[337] Bourgeois-Daigneault MC, Roy DG, Falls T, Twumasi-Boateng K, St-Germain LE, Marguerie M, et al. Oncolytic vesicular stomatitis virus expressing interferon-γ has enhanced therapeutic activity. Mol Ther Oncolytics 2016;3:16001.

[338] Pulido J, Kottke T, Thompson J, Galivo F, Wongthida P, Diaz RM, et al. Using virally expressed melanoma cDNA libraries to identify tumor-associated antigens that cure melanoma. Nat Biotechnol 2012;30(4):337–43.

[339] Cockle JV, Rajani K, Zaidi S, Kottke T, Thompson J, Diaz RM, et al. Combination viroimmunotherapy with checkpoint inhibition to treat glioma, based on location-specific tumor profiling. Neuro Oncol 2016;18(4):518–27.

[340] Rommelfanger DM, Wongthida P, Diaz RM, Kaluza KM, Thompson JM, Kottke TJ, et al. Systemic combination virotherapy for melanoma with tumor antigen-expressing vesicular stomatitis virus and adoptive T-cell transfer. Cancer Res 2012;72(18):4753–64.

[341] Shen W, Patnaik MM, Ruiz A, Russell SJ, Peng K-W. Immunovirotherapy with vesicular stomatitis virus and PD-L1 blockade enhances therapeutic outcome in murine acute myeloid leukemia. Blood 2016;127(11):1449–58.

[342] Pack GT. Note on the experimental use of rabies vaccine for melanomatosis. AMA Arch Derm Syphilol 1950;62(5):694–5.

[343] Desjardins A, Gromeier M, Herndon 2nd JE, Beaubier N, Bolognesi DP, Friedman AH, et al. Recurrent glioblastoma treated with recombinant poliovirus. N Engl J Med 2018;379(2):150–61.

[344] Kuznetsova I, Arnold T, Aschacher T, Schwager C, Hegedus B, Garay T, et al. Targeting an oncolytic influenza a virus to tumor tissue by elastase. Mol Ther Oncolytics 2017;7:37–44.

[345] Wacheck V, Egorov A, Groiss F, Pfeiffer A, Fuereder T, Hoeflmayer D, et al. A novel type of influenza vaccine: safety and immunogenicity of replication-deficient influenza virus created by deletion of the interferon antagonist NS1. J Infect Dis 2010;201(3):354–62.

[346] Kuznetsova I, Shurygina AP, Wolf B, Wolschek M, Enzmann F, Sansyzbay A, et al. Adaptive mutation in nuclear export protein allows stable transgene expression in a chimaeric influenza A virus vector. J Gen Virol 2014;95(Pt. 2):337–49.

CHAPTER THIRTEEN

Comparison of therapeutic strategies for immuno-oncology

Hae Lin Jang and Shiladitya Sengupta

Center for Engineered Therapeutics, Department of Medicine, Brigham and Women's Hospital, Harvard Medical School, Boston, MA, United States

Contents

1. Introduction

Activating the immune system against cancer is revolutionizing cancer therapy [1–3]. The number of patients being treated with immunotherapy is increasing exponentially with new drug and indication approvals. These first-generation drugs are mechanistically targeted toward activating T-cells [3]. However, only a subset of the patients shows a durable response [2, 4]. To overcome this limitation, current emphasis is on (i) identifying next-generation therapeutics that activate the immune system [5–7]; (ii) understanding the various factors that limit the efficacy and induce resistance to existing drugs; and (iii) identifying drug combinations that can improve immunotherapy outcomes [8–10]. Comprehending how our immune system works and the various

Cancer Immunology and Immunotherapy
https://doi.org/10.1016/B978-0-12-823397-9.00013-2

therapeutic strategies that can be mined to manipulate the immune system is critical to establish a broad immune response.

2. The immunological response

Simplistically put, the immune system has evolved to fight against foreign bodies and to clear dead cells. To be able to do that, the immune system needs to identify foreign agents and/or dead cells from healthy "self" cells and remove them. These are done via effector cells. These effector cells can be regulated by a set of inhibitory immune cells that are critical for maintaining a balance and prevent autoimmune diseases. Cancer cells can therefore evade the immune system by harnessing various strategies: (1) Downregulating the expression of antigens that can attract or trigger an immune response, a state that is known as an immunologically cold tumor and (2) overexpressing signals that switch off effector immune cells or activate the regulatory cells to create an immunosuppressive state. Understanding these concepts at a mechanistic level is essential for developing novel immunotherapies as well as designing optimal combinations.

The ability of eukaryotic organisms to recognize foreign *versus* self and mount a defense "immune" response exists at multiple tiers and is recruited in a temporal fashion. At a broad level, cancer immunotherapy can be classified as drugs that activate innate immunity [11, 12], *i.e.*, macrophages and natural killer (NK) cells, or adaptive immune cells [13–15], such as T-cells, which are discussed in various chapters of this book. The goal of immunotherapy has been to activate these cells against the cancer cells. Interestingly, there are more layers of an immune response than these broad cell types, and involve sensors and effectors, and many of the current approaches of treating cancer do work via these pathways. A comprehensive understanding of these multiple levels of immunological response (Fig. 1) is important to contextualize the various therapeutic immune-oncology strategies.

3. The intrinsic immune response within a cancer cell

One of the earliest forms of immunity evolved against viral signatures and is evolutionary conserved in all cells. Viruses inject single- or double-stranded RNA or DNA into the cytoplasm of infected cells. As a result, eukaryotic cells have evolved sensors that can detect these viral nucleic acid signatures, termed as pathogen-associated molecular patterns (PAMPs). Interestingly, these sensors can also detect cellular DNA or endogenous retroviral sequences leaking into the cytoplasm, such as from nuclear or mitochondrial damage in cancer cells. Such leaked materials are called danger-associated molecular

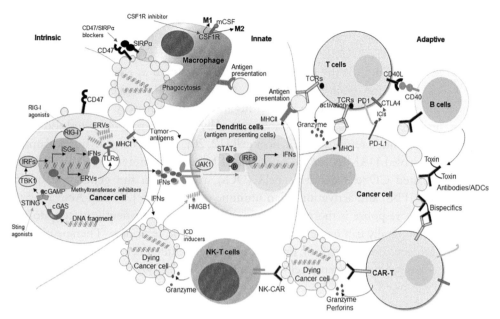

Fig. 1 The coordinated interplay between various compartments of an immune response against cancer cells. The immune response can be classified as intrinsic, *i.e.*, one that is exerted in the cancer cell itself; innate, and adaptive. Immunotherapies can target these distinct components. The earliest immune response evolved against viruses during evolution. Cells can sense viral signatures, such as viral-like single or double-stranded DNA in the cytoplasm via sensors like RIG-I, cGAS, which then induce an interferon response leading to cell death. This intrinsic immune response can be triggered by drugs that activate the sensors or downstream signals from the sensors, such as STING agonists, or via methyltransferase inhibitors, which can hypomethylate the DNA and lead to expression of endogenous retroviral sequences. Such immunotherapy approaches can increase the immunogenicity of tumors. Another approach to increase immunogenicity is to use interferons, which in addition to inducing cancer cell death, also increase the presentation of antigens on MHC-I on cancer cells. A set of cellular stress-inducing drugs can also induce immunogenic cell death (ICD) of cancer cells, which results in release of immunogens, such an HMGB1, into the extracellular environment, and activate the cellular immune response. The earliest cellular immune response, which occurs within minutes to hours, is mediated via the cells of the innate immune system. These include macrophages, NK cells, dendritic cells, which can phagocytose, kill and present antigens to the T-cells. Cancer cells can express ligands than can suppress these cells by interacting with cognate receptors. For example, the interaction between CD47 expressed on cancer cells and SIRPα on macrophages can block the phagocytosis of cancer cells by macrophages. Inhibiting this interaction can exert an antitumor effect. Similarly, drugs that block CSF1R can switch the macrophages to an M1 phenotype, which exerts an antitumor effect. It is also possible to activate NK cells by activating its effector pathways and inhibiting its inhibitory immune checkpoints. Genetically engineering the activating receptors on NK cells (called NK-CARs) can also facilitate the NK cells to bind to antigens on the cancer cells, and activate to kill the latter via injections of granzyme and perforins. Macrophages and dendritic cells can internalize the antigens, and can present it to the T-cells of the adaptive immune system via MHC-II in addition to amplifying the immune response. The primed T-cells can recognize the cognate antigens on cancer cells and kill such cells by injecting perforins and granzyme. Additionally, they communicate with
(Continued)

patterns (DAMPs). DAMPs and PAMPs bind to pattern-recognition receptors (PRRs), including (1) the endosomal toll-like receptors (TLRs) family; (2) the cytosolic DNA sensors, such as cyclic GMP–AMP synthetase (cGAS) and absent in melanoma 2–like (AIM2) receptors; and (3) the cytosolic RNA sensors, such as retinoic acid–inducible gene I (RIG-I)-like receptor family [16–19]. Activated PRRs trigger multiple signaling cascades and produce downstream interferons and proinflammatory cytokines, acting as the very first line-defense against infections caused by virus and microbes, and also potentially the first immune response in a cancer cell [12, 20]. Additionally, dying or damaged tumor cell-derived DAMPs can be internalized by antigen-presenting cells, including tumor-associated CD8α + dendritic cells (DCs), resulting in the generation of type I interferons (IFNs) and activation of an immune cell response [21, 22]. Activating this intrinsic/innate response is therefore an attractive immunotherapeutic strategy to treat cancer and temporally can occur prior to cellular immune responses. Therapeutic strategies that can activate this intrinsic immune response include PRR agonists, interferons, oncolytic viruses, immunogenic cell-death inducers, and DNA demethylating drugs.

3.1 Interferons

Interferons are the primary effectors of the intrinsic immune response. Interferon was first described by Isaacs and Lindenmann in 1957 as a protein produced by embryonic chick cells, previously exposed to heat-inactivated influenza virus, which could inhibit viral replication [23]. The earliest trials using interferon for adjuvant treatment of osteosarcoma were by Strander et al., and significantly increased life span [24]. There are three major types of IFNs, based on their sequence similarity, their cognate receptor type and distribution, and their inducing signal and cell of origin [25]. Type I IFNs include IFNα, IFNβ, (and IFNε, IFNκ, IFNω), which ligate cognate receptors consist of IFNα/β receptor 1 (IFNAR1) subunit and IFNα/β receptor 2 (IFNAR2) subunit. Type II IFN includes IFNγ, which also can ligate receptors comprising of IFNGR1 and IFNGR2 subunits. In case of Type III IFNs, which include interleukin (IL)29, IL28A and B, and IFNλ4, these IFNs ligate receptor IFNLR1 and IL10 receptor subunit-β

Fig. 1, cont'd B-cells to form long term immune memory. B-cells synthesize antibodies. An early response of B-cells is via IgMs, which switches to IgG. IgGs can be engineered to carry toxic payloads, and are termed antibody drug conjugates. Similar to NK-CAR, T-cells can also be genetically engineered to express a cognate receptor that recognizes the tumor antigen and amplifies the T cell response. Such genetically engineered T cell receptors (CAR-Ts) do not need antigen presentation. While early generation immunotherapy strategies focused on singular compartments, this schematic shows how the immune response is interlinked, and a robust anticancer immune response is only possible if this full spectrum is harnessed. Indeed, every compartment of this immune response is triggered against a viral infection. Next-generation immunotherapies need to understand how to combine drugs in a rational manner or develop drugs that can activate the full immune response.

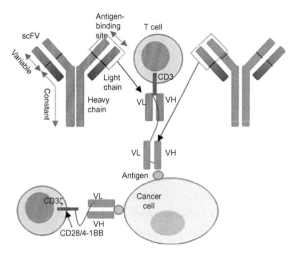

Fig. 2 Structure of CARs and BiTEs. T-cells or NKT cells engage via activating receptors with cognate antigens on cancer cells to get activated. This forms the logic behind the design of chimeric antigen receptors (CARs) and bispecifics as shown in the schematic. CARs are genetically engineered constructs, where an scFv fragment of an antibody that can bind to antigens or ligands expressed on cancer cell surface is connected to a NK- or T cell-activating intracellular signaling domain via a hinge and a transmembrane domain. A BiTE has two ScFV segments conjugated to each other, where one recognizes a CD3 receptor, and the other binds to the cognate antigen on cancer cell surface. Table 1 compares the various CARs vs BiTEs.

Table. 1 Comparison of CARs and BiTEs.

	CAR-T	NK-CAR	BiTEs
Structure	Engineered T cell receptor comprising scFV against tumor antigen linked to T cell-activating and costimulatory motifs	Engineered NK cell with scFV against tumor antigen linked to T cell-activating and costimulatory motifs	A recombinant protein with two scFvs, one that binds to CD3 and another that binds to an antigen on cancer cells.
Effector cell type	T-cells, including less differentiated subsets	NK cells	Both T-cells (and also NK cells)
Killing mechanisms	Perforin, granzyme, Fas/FasL, TNFs, cytokines	Perforins, granzyme, limited cytokines	Perforin, granzyme
Toxicity	Severe CRS, neurotoxicity, infections	Less severe CRS, neurotoxicity	CRS, neurotoxicity
Examples	Yescarta, Kymriah,	NA	Blinatumomab

heterodimeric receptor. PRRs activate Type I IFNs and Type III IFNs, while IFNγ is induced mainly by mitogens and cytokines expressed by immune cells. IFNα subtypes, such as Roferon-A [2a], Intron A [2b], Alferon [2a], are the most widely used interferon in the clinic for cancer treatment.

3.1.1 Mechanism of action

IFNs exert an antitumor effect both at the level of the cancer cells and immune cells in the tumor microenvironment. Type I IFNs exert a direct antiproliferative effect on tumor cells by endogenously upregulating expression levels of the cyclin-dependent kinase inhibitors. It can induce cancer cell apoptosis via the induction of tumor necrosis factor (TNF)-related apoptosis-inducing ligand (TRAIL), Fas ligand, Caspases, and viperin. Type I IFNs and Type II IFNs could also intrinsically function within the cancer cells to increase expression of tumor antigens and the major histocompatibility complex I (MHC-I) expression, and improve the presentation by MHC-I [25]. Type I IFNs can impact the activity of most of the immune cells, including T-cells, B-cells, macrophages, and DCs. They can influence the activity of NK cells, switch macrophages from an immunosuppressive M2 to an inflammatory M1 state, mature DCs and increase cross-presentation of antigens from DCs, activate CD8 and CD4 effector cells, and prime B-cells to mount an antibody response. Additionally, IFNs also negatively regulate regulatory T-cells (Tregs) and myeloid-derived suppressor cells (MDSCs) [25].

3.1.2 Current clinical status

Intron A (*e.g.*, recombinant interferon alfa-2b) and anthracycline combination chemotherapy for injection is approved for treating patients with clinically aggressive non-Hodgkin's lymphoma (NHL), AIDS-related Kaposi's sarcoma, hairy cell leukemia, and as adjuvant therapy for malignant melanoma. Despite their pleiotropic effects, IFNs have had limited integration into cancer therapy, partly because it is hard to predict patient sensitivity, and a general lack of belief at the time in immunotherapy for cancer. Additionally, enthusiasm around interferons dipped as targeted therapeutics emerged. Interferon treatment can have side effects including flu-like symptoms, infection, skin rashes, leukopenia, depression, and hair thinning. Better formulations, which improve stability and bioavailability, together with the potential to combine with immune checkpoint inhibitors can offer a new life to this old class of immunotherapy agents.

3.2 PRR agonists

An emerging immunotherapy strategy upstream of IFNs is to use activators of PRR signaling pathway in cancer and immune cells. For example, upon detection of DAMPs, cGAS induces cyclic dinucleotides (CDNs), such as cGAMP, that bind to stimulator of interferon genes (STING), which then triggers type I IFNs and proinflammatory cytokines [20]. Indeed, cGAMP derivatives and STING agonists were found to exert potent efficacy in tumor models, and exhibited an improved antitumor response in combination with radiation, immune checkpoint inhibitors, chemotherapy, or antitumor vaccines [26]. Similarly, binding of double-stranded RNA to RIG-I triggers the RIG-I inflammasome, which further result in pyroptosis, an intensive mechanism of immunogenic programmed cell death [27]. In addition, RIG-I signaling also triggers the IFN response. Notably, nucleotide motifs of virus can be recapitulated utilizing engineered noninfectious oligonucleotides. Such RIG-I agonists can exert an anticancer effect via: (1) pyroptosis of cancer cells and (2) IFN-mediated activation of both innate and adaptive immunological effectors. Several STING agonists have entered the clinics, while RIG-I agonists are in the pipeline [28].

3.2.1 Mechanism of action

Activated STING upregulates a TANK-binding kinase 1 (TBK1)-related signaling pathways that triggers transcription dependent on interferon regulatory factor 3 (IRF3)- and nuclear factor (NF)-κB, which induces Type I IFNs, TNF, and interleukin 6 (IL-6). Activated STING also stimulates autophagy, a more ancient function than the activities of IRF3 and NF-κB transcription initiation [29]. Similarly, in the case of RIG-I, its amino-terminal Caspase Activation and Recruitment Domain (CARD) goes through an ATP-dependent change in conformation upon activation by a ligand, leading to polyubiquitylation by ubiquitin ligases. The polyubiquitylated RIG-I is then recruited to the surface of mitochondria where the interaction between the RIG1 CARD and the mitochondrial antiviral signaling (MAVS) CARD occurs. Once these domains are engaged, MAVS activates three major kinases that regulate inflammation, which include inhibitor of κB-Kinase (IKK)-γ, IKK-ε, and TANK-binding kinase (TBK)-1. These kinases phosphorylate transcription factors, such as NF-κB, interferon regulatory factor (IRF)-1, IRF-3, and IRF-7, drive the proinflammatory transcriptional program, leading to increased expression of proinflammatory cytokines and type I IFNs.

3.2.2 Current clinical status

5,6-Dimethylxanthenone-4-acetic acid (DMXAA) is one of the earliest STING agonists. However, the combination of DMXAA and conventional chemotherapy exhibited poor efficacy and associated with side effects [26]. Recent structure-function studies showed that DMXAA does not interact with human STING, but only bind to its murine counterparts [30], which explains the poor translation to the clinics. Next-generation STING agonists were developed to overcome the limited efficacy and adverse reactions of the DMXAA [26]. For example, STING agonists, such as E7766, GSK3745417, MK-1454 and ADU-S100, were tested in patients with advanced stage solid tumors or lymphomas, recurrent head and neck squamous cell carcinoma, and nonmuscle invasive bladder cancer. However, clinical outcomes were poor with monotherapy, which have reduced the enthusiasm around STING agonists. In hindsight, this is not unexpected as a nonspecific inflammatory response is unlikely to exert an antitumor effect that is any better than interferons in the absence of activated immune cells in the tumor microenvironment. Indeed, in a recent study, STING agonists combined with CAR-T cell therapy was found to be effective in an orthotopic model of advanced breast cancer [15].

RIG-I agonists still remain at an early-stage of discovery and development. For example, 5'-triphosphate-modified dsRNA sequence as an RNAi strategy activated RIG-I while inducing gene silencing [31]. Indeed, application of the antiapoptotic B-cell lymphoma 2 (BCL2) gene specific 5'-3p-siRNA sequences in melanoma stimulated production of IFN and activation of NK cells, while promoting cancer cell killing through ablation of Bcl-2 [32]. Similar increase in antitumor immunity and gene silencing was observed using 5'-3p-siRNAs, which targeted transforming growth factor (TGF)-β of pancreatic cancer, and vascular endothelial growth factor (VEGF) of lung cancer [33, 34]. Taken together, drugs that act solely via an IFN response may still need to be combined with other therapeutics or exert activity via additional mechanisms of action.

3.3 Oncolytic virotherapy

Unlike interferons and PPR agonists, which induce an inflammatory state in the tumor, oncolytic viruses (OVs) can be used to increase the immunogenicity of the cancer cells as well as trigger an interferon response. Currently, Talimogene laherparepvec (Imlygic) is approved for treating recurrent melanoma patients with unresectable subcutaneous, cutaneous, and nodal lesions [35]. Multiple OVs are currently in clinical development.

Mechanism of action: Imlygic is genetically designed and engineered oncolytic viral therapy to replicate inside tumors and to generate granulocyte-macrophage colony-stimulating factor (GM-CSF). Specific targeting of cancer cells occurs by depleting *RL1*, the gene that encodes the infected protein ICP34.5. Deletion of ICP34.5 protein can prevent replication of normal cells. Additionally, the deletion of *US12* (encoding ICP47) facilitates viral replication in cancer cells [36]. Injected intralesionally, Imlygic triggers an antiviral interferon response, causing lysis of tumors and releasing tumor-derived antigens, which triggers an immune response against tumor together with virally generated GM-CSF. Vocimagene amiretrorepvec (Toca 511), is another OV in clinical trials. Toca 511 is an amphotropic murine gamma retrovirus that encode a functionally optimized cytosine deaminase. Toca 511–infected cells can transform 5-fluorocytosine (5-FC) into the anticancer drug 5-fluorouracil (5-FU).

Clinical observations: The therapeutical efficacy and safety of Imlygic administered via intralesional injections was compared with GM-CSF, which was subcutaneously-administered in patients with surgically nonresectable stage IIIB, IIIC, and IV melanoma. Imlygic was injected into the tumor lesions on day 1 at a concentration of 10^6 PFU/mL and day 21 at a concentration of 10^8 PFU/mL, which was followed by every 2 weeks. Imlygic DNA was detected in the patient blood (85%) and urine (20%). It was approved based on durable response rate (DRR), determined as the patient ratio (%) with complete or partial response maintained for at least 6 months, which was found to be 16.3% with Imlygic treatment *versus* 2.1% with the GM-CSF treatment. The estimated 5-year survival rate for the Imlygic treatment group was found to be 33.4% [35]. Side effects include injection site pain, fatigue, chills, nausea, pyrexia, and influenza-like illness. Combining Imlygic with immune checkpoint inhibitors was found to significantly improve anti-tumor outcomes, and is likely going to be the future path for these therapeutic agents [37]. Toca 511 was well tolerated when administered by IV, intratumoral injection, or injection into the postresection tumor bed, and clinical data showed an extended over-all survival (OS) and a favorable safety profile compared to historical controls in early-stage studies. Specifically, in the resection study, the median OS was 13.6 months, while the matched historical controls were about 7–8 months, and the OS rate after 24 months was 32% [38]. However, in a Phase 3 study, it failed to show its OS improvement compared to standard of care in high grade glioma [39]. These results indicate the need to develop technologies that will allow patient selection for such therapies.

3.4 Viral mimicry

Portions of the viral DNA have been integrated into the human genome during the evolutionary process. These are known as endogenous retroviral sequences, and make up over 8% of the human genome [40]. Normally, the expression of these endogenous retroviruses (ERVs) is silenced via DNA and histone methylation. Inducing the expression of these ERVs can trigger a similar immune response as described earlier for viral signatures. Inhibiting the enzymes that methylate the DNA or the histone in the cancer cells, such as DNA methyltransferases and H3K9 lysine methyltransferases, can upregulate ERV transcription, and can be used to induce a viral infection-like state in the cancer cells. This state is termed as viral mimicry [41].

Mechanism of action: DNA methyltransferase inhibitors (DNMTis) include approved drugs, such as 5-azacytidine (Azacytidine) and 5-aza-20-deoxycytidine (Decitabine). These cytidine analogs inhibit DNA methyltransferases (DNMTs)-related catalytic actions by incorporating into DNA and triggering DNA degradation. This generates ERVs, which then triggers an IFN response via MAVS and TLR3 signaling [42].

Clinical status: Both azacytidine and decitabine are approved for use in myelodysplastic syndrome patients. However, these agents resulted in toxicity and limited efficacy when used at the maximally tolerated dose for treating solid tumors. Interestingly, these agents were found to block DNA methylation at lower than cytotoxic concentrations and were therefore explored as inducers of viral mimicry. In a phase I/II trial, combining azacitidine (DNA methylation inhibitor) and entinostat (histone deacetylation inhibitor) for treating recurrent metastatic nonsmall cell lung cancer patients with extensive pre-treatment history, the median survival rate in the entire cohort was significant at 6.4 months. However, resistance to the epigenetic therapies was commonly observed, potentially via rapid re-methylation after the drug is removed [43]. The future of epigenetic therapies, particularly for applying to solid tumors, therefore depends on (i) developing technologies that can enable a sustained inhibition of the target;

(ii) rationally combining them with drugs that can block these resistance mechanisms; and (iii) combination treatments that synergize with these drugs to ablate the tumor. Indeed, a small fraction of patients demonstrated remarkably strong and durable responses against immune checkpoint blockade treatments after they first received Azacytidine [44], consistent with the rationale of an immune checkpoint inhibitor synergizing with a drug that increases immunogenicity. Currently, multiple clinical trials are testing the combination of immune checkpoint inhibitors with epigenetic drugs in an array of cancers.

3.5 Immunogenic cell death

We include immunogenic cell death (ICD) in the intrinsic immune response as these therapeutic strategies mechanistically target the cancer cells. Unlike the previous approaches, which increase the immunogenicity of the cancer cells, here the drugs induce cancer cell death that results in the release of antigens in the tumor microenvironment and surrounding tissues, which then triggers an immune response [45]. The defining factors of ICD are: (1) Exposure of calreticulin (CRT) proteins and other types of endoplasmic reticulum (ER) proteins on the cell surface, (2) ATP release during the blebbing apoptosis phase, and (3) high-mobility group box 1 (HMGB1) nonhistone chromatin protein release from the cancer cells [46].

Mechanism of action: A large number of drug molecules have been associated with induction of ICD. These include oncolytic viruses, anthracyclines (*i.e.*, doxorubicin, epirubicin, idarubicin, and mitoxantrone), oxaliplatin (but not cisplatin), poly(ADP-ribose) polymerase (PARP) inhibitors, docetaxel, and proteasomal inhibitors (*i.e.*, bortezomib and carfilzomib), DNA methyltransferase inhibitors, histone deacetylase (HDAC) inhibitors, some tyrosine kinase inhibitors, such as crizotinib, cetuximab, etc. While these molecules have distinct mechanisms of action, the common link is the induction of two types of stress, *i.e.*, endoplasmic reticulum (ER) stress and autophagy [46]. Both extracellular and cell surface CRT is recognized and "eaten up" by CD91-expressing innate immune cells, including macrophages and DCs. CRT, signaling through CD91, also triggers the inflammatory cytokine production, such as IL-6 and TNFα, thereby stimulating the antigen-presenting cells. Similarly, ATP acts as a chemo-attractant, and inducer of maturation of DCs. These DCs produce IL1b, which recruits IFNγ-producing cytotoxic T-lymphocytes. Finally, HMGB1 can bind to TLR4, which activates an inflammatory response. In association with CXCL12, it can activate CXCR4, which recruits immune cells to the tumor. Additionally, when bound to nucleosomes, HMGB1 can trigger the release of autoantibodies (anti-DNA and antihistone), which then can target the cancer cells [22, 47].

Clinical status: Currently, evidence points to breast and colorectal cancers as putative settings in which ICD inducers can be favorably combined with immunotherapy [48]. For example, in metastatic triple-negative breast cancer (TNBC), the combination of nivolumab (PD-1-inhibitor) and doxorubicin (ICD inducer) led to an overall response rate (ORR) of 35% compared to 23% for the combination of nivolumab and cisplatin [49]. Similarly, combining an anticancer vaccine with oxaliplatin in patients with colorectal cancer resulted in a cytotoxic T-cell response with improved OS. However, there are cancers where the use of ICD-inducing chemotherapeutics remains debatable. The future of ICD therapy lies on the ability to identify the precise oncological indications where ICD confers a benefit.

4. The innate immune response

Once an antigen is presented outside the cell, the earliest cellular immune response comprises of the innate immune cells. This occurs immediately or within hours. Innate immune cells include macrophages, NK cells, DCs, etc. Other players in the innate immune response include the complement, and additionally an early T-cell-independent activity of naïve B-cells. The latter secrete immunoglobulin M, which bind with low affinity (but with high avidity due to their pentameric structure) to foreign bodies to provide a broad coverage. B-cells also can present antigen to activate T-cells. Additionally, B-cell presence inside the tertiary lymphoid structures (TLSs) has been implicated with better survival and response to immune checkpoint inhibitors [13, 50–54]. However, the exact role of B-cells in cancer immunotherapy remains to be fully explored, and hence here we will briefly discuss macrophage- and NK cell-based immunotherapeutics.

4.1 NK cell immunotherapy

NK cells are the effector T-cells (lymphocytes) of innate immune system. They constitute about 5–15% of human peripheral white blood cells (leukocytes), and are identified mainly by the CD56 expression, and the lack of CD3 and T-cell receptor (TCR) expressions. Human NK cells also express natural cytotoxicity triggering receptor 1 (NCR1 or CD335). Human NK cells are further classified into two subsets: (a) a larger $CD56^{dim}CD16^{high}$ fraction (85%–95%) of mature population with higher cytotoxic function, and (b) a smaller immature and low cytotoxic $CD56^{bright}CD16^{low/-}$ fraction [55].

Mechanism of action: NK cells exert their anticancer activity via multiple mechanisms. NK cells home into the tumors attracted by proinflammatory chemokines, including CCL5, CXCL9, CXCL10, CCL19, CCL27, etc [55]. NK cells recognize cancer cells when they downregulate expression level of MHC class I molecules to escape cytotoxic $CD8^+$ T-cells. Thus, NK cells can exert therapeutic efficacy in settings where cytotoxic T-cells are ineffective. Additionally, cancer cells express ligands, for example, MICA and ULBPs, which are recognized by activating receptors expressed on NK cells, such as NKG2D. NK cells are additionally regulated by inhibitory receptors, which include inhibitory killer Ig-like receptors (KIRs) [56]. NK cells enable antibody-dependent cell

(ADCC) killing of cancer cells, resulting from the interaction between CD16 and the Fc region of antibodies bound to tumor-associated antigens. When the activated NK cells encounter a target cell, it forms a synapse and releases lytic perforin and granzyme through the synapse into the target cancer cell. Direct killing can also occur via Fas and TRAIL. Additionally, NK cells generate proinflammatory cytokines IFNγ and TNF, via which it can exert additional anticancer effects [57]. This limited repertoire of secreted cytokines is also one of reasons that NK cell therapy has lesser side effects than CD8$^+$ T cell-based therapy that induces a much wider array of cytokines, which are related to cytokine release syndrome (CRS) and severe neurotoxicity. Furthermore, NK cell infusions have decreased risk for life threatening graft *versus* host disease (GVHD), and the limited lifespan of NK cells minimizes any toxicity to normal tissues during their circulation. As a result, multiple therapeutic strategies have been harnessed to activate a NK cell response against cancer cells.

NK cell therapy and chimeric antigen receptor (CAR)-NK cells: The above mechanistic properties of NK cells make them strong candidates for engineering CAR-NK therapies as well being used in their native forms for treatment for various cancers [58]. Activated NK cells as well as CAR-NK cells have been derived from cell lines (NK-92, KHYG-1, YT cells), from primary NK cells isolated from human peripheral blood mononuclear cells (PBMCs), from CD34$^+$ hematopoietic progenitor cells, from iPSCs, as well as isolated from umbilical cord blood, which are then expanded and stimulated using NK cell-specific expansion media with cytokines [57]. Interestingly, NK cells can be obtained from matched or human leukocyte antigen (HLA)-mismatched donors as they are not associated with GVHD [59].

The early generation of CAR-NK cells adapted CARs that were designed for CAR-T cells. The immune cells from the blood of the patient are genetically modified outside the body system to express a synthetic chimeric receptor, which comprises of an extracellular antigen-binding domain, such as a single-chain variable fragment (scFv), against a specific antigen expressed on the cancer cell, and an intracellular domain that connects with a signaling pathway that activates the immune cell [58]. For example, a CAR construct combining a PDL1-targeted NK cell is currently recruiting patients for a phase II study in advanced or metastatic pancreatic cancer. Similarly, a CD19-targeted high-affinity NK cell is being tested in diffuse B-cell lymphoma [58]. Second-generation CAR-NKs integrate intracellular domains that are geared to activate NK cells. For example, introduction of NKG2D receptor and 2B4 receptor (an NK cell-specific costimulatory domain) was found to increase the antitumor efficacy of the CAR-NK cells [59]. Clinical evidence has recently started emerging around the therapeutic effectiveness and safety of CAR-NK cells. Seven out of 11 patients who had high-risk CD19$^+$ B-cell malignancies achieved a complete remission when treated with an allogeneic anti-CD19-CD28-CD3ζ CAR-NK product (IL-15 was included in the retroviral vector to amplify the response, and a suicidal switch), and unlike CAR-Ts, were associated with minimal serious adverse effect, such as GVHD, CRS, and neurotoxicity [60].

NK cell activators and immune checkpoint inhibitors: The emerging understanding of the mechanisms underlying NK cell activation, including insights into the inhibitory controls, has spurred the development of novel therapeutics. For instance, the inhibitory receptor complex NKG24-CD94 binds to HLA class I molecule alpha chain E (HLA-E), which is often upregulated on cancer cells, and inhibits activation of NK cells and target cell killing [55]. A humanized anti-NKG2 blocking antibody, Monalizumab, is currently in clinical trials for multiple cancer indications. Recurrent or metastatic squamous cell carcinoma of the head and neck (R/M SCCHN) patients were treated with the combination of monalizumab and cetuximab, which was well tolerated. The most common adverse effects related to monalizumab were fatigue (17%), pyrexia (13%), and headache (10%), and in terms of efficacy, 31% of patients achieved a confirmed response, and 54% had stable disease [61]. However, another study with an immune checkpoint inhibitor, IPH2101, which blocked the inhibitory receptor KIR2DL-1-3, failed in the clinics as the target was downregulated from the NK cell surface by trogocytosis, which also inhibited NK cell functions. A deeper mechanistic understanding of the activating and inhibiting pathways in NK cells under drug pressure is key to engineering effective NK cell therapeutics.

NK cells are also activated by cytokines. For example, bempegaldesleukin, a modified recombinant IL-2, was found to increase infiltration of NK cells and CD8 $^+$ T-cells into tumors and activate the immune cells. It was well tolerated, and resulted in ORRs of over 50%, including >30% complete remissions with durable response when applied together with anti-PD-1 therapy in solid cancers [55]. It activates NK cells through binding to the IL-2 receptor heterotrimer in the CD8 $^+$ T-cells and NK cells but spares the IL2Rα subunit. Another drug, ALT-803, an IL15 "superagonist," where IL15 is complexed with an IL15α receptor fused to a human dimeric IgG1 Fc, conferring stability and prolonging the half-life of the overall complex, was also found to lead to expansion of NK cells over CD8 $^+$ T-cells in solid tumor patients. Overall, patients tolerated ALT-803 well, with manageable side effects, including nausea and fatigue, and painful injection site wheal with subcutaneous administration [62].

Additional therapeutic approaches to harness the NK cells include bispecific killer cell engagers (BiKEs) and trispecific killer cell engagers (TriKEs). These include a single variable fragment (Fv) against a tumor antigen (in BiKEs) or two tumor antigens (TriKEs), and a Fv fragment that binds to CD16 on NK cells, which damages the cancer cells via the induction of ADCC [55].

4.2 Macrophage immunotherapy

Macrophages are one of the most common innate immunity cells present in the tumor. Macrophages exist in a spectrum of phenotypic states with diverse functions. Typically, the two polar extremes are defined as the M1 and the M2 populations, defined by distinct biomarkers and functions. M1 macrophages express high levels of tumor necrosis factor, inducible NOS, MHC class II, and exerts a proinflammatory response culminating in phagocytosing a foreign body or a dying cell. An M1 phenotype exerts an antitumor effect through phagocytosis and can additionally activate the adaptive immune cells by acting as antigen-presenting cells. In contrast, the M2 phenotype is defined as expressing high levels of arginase 1, IL-10, CD206, and CD204, and includes tumor associated macrophages (TAMs) that promote tumor growth. They can inhibit cytotoxic T-lymphocyte responses by expressing PD-L1, produce inhibitory cytokines, such as TGFβ and IL-10, and deplete metabolites like arginine via arginase activity. TAMs also recruit protumorigenic Tregs. Not surprisingly, TAMs have been associated with poor prognosis and increased rate of metastasis. As a result, there is considerable interest in developing therapeutic strategies that can switch M2 TAMs to the M1 phenotype. These therapeutic strategies can broadly be classified into two categories: (1) Approaches that deplete TAMs and (2) drugs that can increase the phagocytosis of cancer cells by macrophages. These approaches are described in greater details in other chapters and will be discussed briefly here.

TAM-depleting therapies: TAMs are usually thought to derive from circulating monocytes, which are mobilized from the marrow of the bone and recruited to the tumors via CCL2-CCR2 signaling. As a result, inhibition of this axis has emerged as an attractive approach to clinically deplete TAMs [63]. For example, the combination of a CCR2 specific antagonist CCX872 with FOLFIRINOX led to an improved OS rate of 29% at 18 months, as compared with historic OS rate at 18 months, which was only 18.6% when FOLFIRINOX treatment was applied alone. This was associated with a reduction in circulating monocytes and monocytic-like myeloid-derived suppressor cells [64]. An alternative strategy is to target the colony stimulating factor (CSF1)-CSF1 receptor signaling, which skews the monocytes toward the M2 phenotype [65]. Both antibodies as well as small molecules have been developed to target this pathway. For example, patients with metastatic breast cancer (MBC) and castrate-resistant prostate cancer (mCRPC) when treated with LY3022855, a CSF1R inhibiting antibody, exhibited increased circulating CSF-1 levels and decreased proinflammatory monocytes $CD14^{DIM}CD16^{BRIGHT}$. Approximately, 23% of patients with MBC and 25% of patients with mCRPC exhibited stable disease, with immune-related gene activation in tumor biopsies. Common treatment-related side effects included decreased appetite, fatigue, nausea, and an asymptomatic increase in lipase and creatine phosphokinase, suggesting that such a therapeutic approach was well tolerated and showed evidence of immune modulation [66]. However, in case of AMG 820, another anti-CSF1R monoclonal antibody, when it was treated with pembrolizumab in patients with advanced nonsmall cell lung cancer (NSCLC), pancreatic cancer, refractory mismatch repair-proficient colorectal cancer, or low expression levels of programmed cell death-ligand 1 (PD-L1) (<50%) failed to exhibit clinically meaningful efficacy despite positive pharmacodynamic biomarker movement [67]. Approaches to select the right patients for these therapies remains a limiting factor.

Phagocytosis inducing therapies: Cancer cells overexpress proteins on cell surface that act as "eat me not" signals and block phagocytosis by macrophages. This includes CD47 that binds to SIRPα, PD-L1, ß-2 microglobulin (B2M, subunit of MHC class 1 molecule), and CD24 that signals through macrophage Siglec-10 [68]. Blocking these interactions can potentially increase the phagocytosis of cancer cells [68, 69]. First-generation drugs in the clinics that act by targeting these proteins include Magrolimab, an antihuman-CD47 monoclonal antibody, which in combination with rituximab showed a complete response rate (CRR) of 36% and objective response rate (ORR) of 50% in relapsed/rituximab-refractory diffuse large B-cell lymphoma patients and follicular lymphoma patients [70], although CD47 expression did not show prognostic value. When combined with azacitidine, which induces ICD, it resulted in a 64% ORR in patients with untreated acute myelogenous leukemia (AML) and 92% ORR in patients with untreated higher-risk myelodysplastic syndrome (MDS), consistent with the azacytidine-induced expression of prophagocytic calreticulin, which can synergize with CD47 blockade in phagocytosis of cancer cells [71]. CD47 is expressed on all cells in the body, and especially in aged RBCs, which implicates a large sink for anti-CD47 antibodies, and it induces anemia as an on-target side effect of RBC phagocytosis [72, 73]. Therefore, next-generation anti-CD47 antibodies, such as Lemzoparlimab, were engineered to minimize binding to RBCs. Another strategy to block the CD47-SIRPα interaction is based on SIRPα-Fc fusion proteins. These proteins are typically composed of the domain of human SIRPα that binds to CD47-binding domain, fused with a human Fc domain. Such structures act as decoy receptors, binding to CD47 and blocking the interaction with SIRPα on macrophages, while the Fc component can act as an activator of phagocytosis and ADCC. Ontorpacept, a SIRPα-Fc fusion therapeutic protein, is currently in clinical trials to test its effectiveness against patients with advanced relapsed or refractory hematologic malignancies and has exhibited ORRs of 18%–29%. Inhibitors of SIRPα are also in the pipeline, and unlike the antiCD47 antibodies, are associated with limited "sink effect" [74]. Combinations of CSF1R inhibitors and SIRPα targeting has also been shown to exert a synergistic effect [65].

5. The adaptive immune response

Cells of the innate immune system can process cancer cell antigens and present to T-cells to prime an adaptive immune response. T-cell receptors on primed cells can recognize the antigens on cancer cells, and the activated T-cells then induce cancer cell death via injection of granzyme and perforins. T-cells act as the most potent killer cells of the immune system. Cancer cells however avoid this cell kill effect of T-cells via over-expression of numerous immune checkpoints proteins, which can bind to receptors on T-cells, and cause anergy of T-cells. T-cells also have their internal checkpoints, which act as brakes on the T-cell response. Inhibitors of immune checkpoints, such as the PDL1-PD1 axis and CTLA4, are transforming cancer therapy outcomes [1, 2]. These immune checkpoint inhibitors are discussed in other chapters. Here we shall discuss alternative strategies to recruit a T-cell immune response against cancer cells (Fig. 2).

CAR-Ts: T-cells can be genetically designed to express chimeric antigen receptors (CARs), which allow the modified T-cells to recognize and eliminate cells that express the antigen. A modular design of the CARs consists of the following domains: (1) an antigen-binding domain, (2) a hinge domain and a trans-membrane domain, and (3) an intracellular signaling domain. The antigen-binding domains of CARs typically comprises of a variable heavy (V_H) and variable light (V_L) chain of a monoclonal antibody flexibly linked to form an scFv. It is also possible to engineer this domain from any structure that recognizes the antigen, for example, nanobodies or single-domain antibodies, ligands that bind to receptors present on the cancer cells, etc [75]. This domain recognizes the specific extracellular antigen expressed on a cancer cell and recruits the CAR-T to the cancer cell. By engineering the antigen-binding domains, one can increase the avidity of the receptor as opposed to the low avidity seen with the original TCRs. Additionally, CARs can recognize the antigen independent MHC presentation. The hinge and the transmembrane domain link the antigen-binding domain to the intracellular-signaling domain. An optimal hinge-transmembrane domain offers sufficient flexibility to facilitate access to the targeted antigen, and stability of the CAR construct. For example, CARs modified with the transmembrane domain using amino acid sequences from CD28 was found to be more stable than CARs engineered with the transmembrane region of CD3. Similarly, CARs with a CD8 hinge and transmembrane domain released less interferons and TNFs than those engineered from CD28 [75]. The intracellular signaling domain comprises of a domain for activation and one or more domains for costimulation, which signals the T cell activation and proliferation on ligating an antigen.

Clinical status: Four CAR-Ts are now approved by the FDA, and several are in the pipeline.

Tisagenlecleucel (Kymriah, formerly CTL-019) is the first CD19-directed CAR-T cell [76], where autologous T-cells of the patients are genetically modified using a lentiviral vector to express a CAR made of a murine single-chain anti-CD19, followed by transmembrane region and a CD8 hinge fused to the intracellular signaling domains for 4-1BB (CD137) and CD3zeta. The CD3 zeta component initiates activation of T-cells and increases antitumor activity, while 4-1BB component promotes the expansion of the T-cells and their persistence. Interestingly, patients who exhibited complete response had a plasma concentration (both C_{max} and AUC_{0-28d}) that was about twofold higher compared with nonresponders. Kymriah was detected in the blood and marrow of the bone beyond 2 years. The efficacy of Kymriah was validated in pediatric patients and young adult patients with relapsed or refractory B-cell precursor ALL (the ELIANA study) [77], which resulted in a complete response of 83%. In the JULIET study [78], relapsed or refractory diffuse large B-cell lymphoma patients, who were treated with more than 2 lines of chemotherapy, which included anthracycline and rituximab, or relapsed patients following transplantation of autologous hematopoietic stem-cells were treated with Kymriah. An ORR of 50%, with 32% complete responses was observed in this study. The most common side effects included headache, cough, cytokine release syndrome, infections, hypogammaglobulinemia, encephalopathy, bleeding episodes, hypotension, tachycardia, edema, acute kidney injury, and delirium.

Axicabtagene ciloleucel (Yescarta) is also a CD19-directed genetically engineered autologous T cell-based immunotherapy, which was approved for treating patients with relapsed/refractory aggressive B-cell non-Hodgkin lymphoma. Following interaction with targeted CD19-expressing cells, the costimulatory CD28 domain and CD3-zeta domain stimulate downstream signaling pathways, leading to activation and proliferation of T cells, resulting in the killing of CD19-expressing cancer cells [79]. The typical dose for this therapy is 2×10^6 viable CAR-positive T-cells per kg calculated based on the patient body weight, while the maximum dose is suggested as 2×10^8 viable CAR-positive T-cells. Following infusion, the count of the anti-CD19 CAR T-cells rapidly expands initially, and then declines to baseline within month 3. The highest levels of anti-CD19 CAR T-cells were achieved by 7–14 days after infusion. In a separate study, a Yescarta treatment by single infusion to relapsed/refractory aggressive non-Hodgkin B-cell lymphoma patients produced an ORR of 72%, with 51% exhibiting complete remission [80]. However, 94% of patients who received Yescarta had cytokine release syndrome. In addition, 87% of patients experienced neurologic toxicities .

Brexucabtagene autoleucel (Tecartus), is the third CD19-directed CAR-T with costimulatory CD28 domain and CD3-zeta domain that activate downstream signaling pathways. It resulted in an ORR >80%, with >55% of relapsed or refractory mantle cell lymphoma (MCL) adult patients [81]. A single infusion includes 2×10^6 CAR-T cells per kg calculated based on the patient body weight. The highest levels of anti-CD19 CAR-T cell count occurs within 7–15 days after the treatment, reverting to baseline after 3 months. In the ZUMA-2 clinical trial, 91% of patients had CRS, 81% of patients experienced neurologic events, and 56% of patients had infections (all grades).

Lisocabtagene maraleucel (Breyanzi) is also a CD19-directed CAR-T cell-based immunotherapy, approved for treating relapsed/refractory large non-Hodgkin B-cell lymphoma patients. In the TRANSCEND trial, it exhibited an ORR 73%, with complete response from over 50% of the patients. The median response duration was 16.7 months [82]. It comprises of a FMC63 monoclonal antibody-derived scFv against CD19, a IgG4 hinge region, a 4-1BB (CD137/TNFRSF9) costimulatory domain, a CD28 transmembrane domain, and a CD3 zeta mediated activation domain. A single dose of Breyanzi has $50-110 \times 10^6$ CAR-positive T-cells. Common adverse effects are CRS (46%), neurologic toxicities (35%), and severe infections (45%). Following infusion, Breyanzi exhibits an initial expansion (by day 12), which declines bi-exponentially, but is detectable in blood for up to 24 months.

The serious adverse effects have driven research to optimize the CAR-T cells through genetic engineering to minimize toxicities. For example, by introducing "suicide genes" or "off switches" into the construct, the cells can be deactivated if either cytokine-mediated or nonspecific "on-target" toxicity occurs. For instance, integrating a caspase 9 linked to a FK506-binding protein 1A (iCasp9) meant that it could be triggered by a small molecule AP1903 to induce apoptosis. In a clinical trial, acute leukemia patients who developed GVHD after transplantation of stem-cells and iCasp9-CAR-T cells, saw >90% elimination of the modified T-cells within 30 min of administration of AP1903 and resolution of GVHD [83]. Additional strategies, which are yet to be clinically tested, include embedding a protease and protease target site in the CAR, which can then be switched on or off using a small molecule, resulting in activation or degradation of the CAR. This remains an active area of research.

BiTEs: Bispecific T-cell engagers are a second class of therapeutic strategy that can redirect the T-cells to engage with cancer cells. A BiTE is a recombinant protein with two scFvs of antibodies with specificities for two antigens linked together with a short flexible linker. One fragment targets an antigen on the surface of the cancer cell, and the other targets a cell-surface molecule on the T-cell. This construction means that the BiTEs can engage T-cells to the cancer cell independent of MHC presentation of the antigen. Mechanistically, there are similarities between the CAR-T strategy and BiTEs, with both redirecting the T-cells to the tumor. However, in the case of BiTEs, the effect has to be mediated by antigen-experienced T-cells, and unlike the case of CAR-T, naïve T-cells are not activated by BiTEs. BiTEs offer an advantage over CAR-Ts that no lymphodepletion regimen is required [84]. For both CAR-Ts as well as BiTEs, penetration into the tumor (and recruitment of T-cells into the tumor for the latter) is a key determinant of efficacy in solid tumors.

Blinatumomab (Blycinto) is an FDA-approved bispecific that binds to CD19 expressing B-lineage cells and CD3 expressing T-cells, resulting in the redirected lysis of CD19$^+$ cells. In the clinic, it is approved for treating children and adult patients with relapsed/refractory B-cell precursor acute lymphoblastic leukemia (ALL). The efficacy of Blycinto was evaluated and compared with standard of care (SOC) chemotherapy in a multicenter, open-label, and randomized study (TOWER Study). The study exhibited a statistically significant increase in OS of the patients who received Blincyto treatment (median OS: 7.7 months) as compared with SOC chemotherapy (median OS: 4 months). In another study (ALCANTARA trial), the efficacy of Blincyto was evaluated in relapsed/refractory Philadelphia chromosome-positive B-cell precursor ALL patients. In this trial, 36% of the patients showed complete remission or CR with incomplete hematologic recovery. Their median duration of remission was about 6.7 months. Steady-state serum concentration was observed within 24 h of administration. Redistribution of peripheral T-cells (*i.e.*, T-cell adhesion to endothelium of blood vessel or T-cell transmigration into tissue) was observed after start of treatment; T-cell counts declined initially within the first 1–2 days and returned to baseline within 1-2 weeks in most of the patients. The most common adverse reactions following Blincyto treatment were infusion-related reactions, headache, infections, anemia, pyrexia, febrile neutropenia, neutropenia, and thrombocytopenia.

6. B-cell immunotherapy

While B-cells are an important component of both the early (T-cell independent) and late (adaptive) immune response, limited knowledge remains on modulating B-cells for cancer immunotherapy. It is well known that B-cells generate antibodies, and antibodies underpin much of the anticancer therapies currently in use. Indeed, antibodies that are conjugated to cytotoxic payloads (termed as antibody drug conjugates) are an exciting area of research. However, we still lack the understanding of potential immune checkpoints in B-cells that are harnessed by the cancer cells. Indeed, in a recent series of studies, the presence of B-cells in the tumor has been associated with improved outcomes [13, 50, 53, 85], which clearly suggest that B-cells are the next frontier of cancer immunotherapy.

7. Summary

In summary, cancer immunotherapies are transforming the state of the art in the management of cancer and span the full spectrum of small molecules to biologics, including hybrid structures (antibody drug conjugates) and engineered cell-based therapies. However, many barriers remain, which offer a fertile ground to push the boundaries for next-generation drugs. For example, tumors that are immunologically "cold" *i.e.*, do not express antigens or are immunologically barren (few immune cells penetrate) remain intractable to immunotherapy. Recruiting and penetration of immune cells into the tumor remains a challenge. Finally, as discussed in this chapter, an immune response is a coordinated interaction of multiple immune cell types. Clearly, the next paradigm will be an approach that integrates these various components of the immune response into a coordinated effect against cancer. For example, a combination that can increase the expression of antigens and trigger an interferon response, a therapy that induces an ERV response, together with a macrophage activator, which can lead to internalization and presentation of the antigen to the T-cells, and a T-cell immunotherapy, which amplifies the T-cell response, will overcome the challenges with cold tumors. Executing such combination treatments remains the next frontier.

References

[1] Wolchok J. Putting the immunologic brakes on cancer. Cell 2018;175:1452–4. https://doi.org/10.1016/j.cell.2018.11.006.
[2] Fares CM, Van Allen EM, Drake CG, Allison JP, Hu-Lieskovan S. Mechanisms of resistance to immune checkpoint blockade: why does checkpoint inhibitor immunotherapy not work for all patients? Am Soc Clin Oncol Educ Book 2019;39:147–64. https://doi.org/10.1200/edbk_240837.

[3] Sharma P, Allison JP. Dissecting the mechanisms of immune checkpoint therapy. Nat Rev Immunol 2020;20:75–6. https://doi.org/10.1038/s41577-020-0275-8.

[4] Sharma P, Hu-Lieskovan S, Wargo JA, Ribas A. Primary, adaptive, and acquired resistance to cancer immunotherapy. Cell 2017;168:707–23. https://doi.org/10.1016/j.cell.2017.01.017.

[5] Bonaventura P, et al. Cold tumors: a therapeutic challenge for immunotherapy. Front Immunol 2019;10:168. https://doi.org/10.3389/fimmu.2019.00168.

[6] Galon J, Bruni D. Approaches to treat immune hot, altered and cold tumours with combination immunotherapies. Nat Rev Drug Discov 2019;18:197–218. https://doi.org/10.1038/s41573-018-0007-y.

[7] Duan Q, Zhang H, Zheng J, Zhang L. Turning cold into hot: firing up the tumor microenvironment. Trends Cancer 2020;6:605–18. https://doi.org/10.1016/j.trecan.2020.02.022.

[8] Kulkarni A, Natarajan SK, Chandrasekar V, Pandey PR, Sengupta S. Combining immune checkpoint inhibitors and kinase-inhibiting supramolecular therapeutics for enhanced anticancer efficacy. ACS Nano 2016;10:9227–42. https://doi.org/10.1021/acsnano.6b01600.

[9] Esteva FJ, Hubbard-Lucey VM, Tang J, Pusztai L. Immunotherapy and targeted therapy combinations in metastatic breast cancer. Lancet Oncol 2019;20:e175–86. https://doi.org/10.1016/S1470-2045(19)30026-9.

[10] Hu-Lieskovan S, et al. Improved antitumor activity of immunotherapy with BRAF and MEK inhibitors in BRAF(V600E) melanoma. Sci Transl Med 2015;7. https://doi.org/10.1126/scitranslmed.aaa4691, 279ra241.

[11] Kawai T, Akira S. Innate immune recognition of viral infection. Nat Immunol 2006;7:131–7. https://doi.org/10.1038/ni1303.

[12] Woo SR, Corrales L, Gajewski TF. Innate immune recognition of cancer. Annu Rev Immunol 2015;33:445–74. https://doi.org/10.1146/annurev-immunol-032414-112043.

[13] Garaud S, et al. Tumor infiltrating B-cells signal functional humoral immune responses in breast cancer. JCI insight 2019;5. https://doi.org/10.1172/jci.insight.129641, e129641.

[14] Rouse BT, Sehrawat S. Immunity and immunopathology to viruses: what decides the outcome? Nat Rev Immunol 2010;10:514–26. https://doi.org/10.1038/nri2802.

[15] Xu N, et al. STING agonist promotes CAR T cell trafficking and persistence in breast cancer. J Exp Med 2021;218. https://doi.org/10.1084/jem.20200844.

[16] Takeuchi O, Akira S. Innate immunity to virus infection. Immunol Rev 2009;227:75–86. https://doi.org/10.1111/j.1600-065X.2008.00737.x.

[17] Schuberth-Wagner C, et al. A conserved histidine in the RNA sensor RIG-I controls immune tolerance to N1-2′O-methylated self RNA. Immunity 2015;43:41–51. https://doi.org/10.1016/j.immuni.2015.06.015.

[18] Gong T, Liu L, Jiang W, Zhou R. DAMP-sensing receptors in sterile inflammation and inflammatory diseases. Nat Rev Immunol 2020;20:95–112. https://doi.org/10.1038/s41577-019-0215-7.

[19] Hernandez C, Huebener P, Schwabe RF. Damage-associated molecular patterns in cancer: a double-edged sword. Oncogene 2016;35:5931–41. https://doi.org/10.1038/onc.2016.104.

[20] Deng L, et al. STING-dependent cytosolic DNA sensing promotes radiation-induced type I interferon-dependent antitumor immunity in immunogenic tumors. Immunity 2014;41:843–52. https://doi.org/10.1016/j.immuni.2014.10.019.

[21] Meza Guzman LG, Keating N, Nicholson SE. Natural killer cells: tumor surveillance and signaling. Cancers (Basel) 2020;12:952. https://doi.org/10.3390/cancers12040952.

[22] Krysko O, Løve Aaes T, Bachert C, Vandenabeele P, Krysko DV. Many faces of DAMPs in cancer therapy. Cell Death Dis 2013;4:e631. https://doi.org/10.1038/cddis.2013.156.

[23] Isaacs A, Lindenmann J. Pillars article: virus interference. I. The interferon. Proc R Soc Lond B biol Sci. 1957. 147: 258-267. J Immunol 2015;195:1911–20.

[24] Strander H. Interferons: anti-neoplastic drugs? Blut 1977;35:277–88. https://doi.org/10.1007/bf00996140.

[25] Parker BS, Rautela J, Hertzog PJ. Antitumour actions of interferons: implications for cancer therapy. Nat Rev Cancer 2016;16:131–44. https://doi.org/10.1038/nrc.2016.14.

[26] Le Naour J, Zitvogel L, Galluzzi L, Vacchelli E, Kroemer G. Trial watch: STING agonists in cancer therapy. Onco Targets Ther 2020;9:1777624. https://doi.org/10.1080/2162402x.2020.1777624.

[27] Elion DL, et al. Therapeutically active RIG-I agonist induces immunogenic tumor cell killing in breast cancers. Cancer Res 2018;78:6183–95. https://doi.org/10.1158/0008-5472.Can-18-0730.

[28] Heidegger S, et al. RIG-I activation is critical for responsiveness to checkpoint blockade. Sci Immunol 2019;4. https://doi.org/10.1126/sciimmunol.aau8943, eaau8943.

[29] Moretti J, et al. STING senses microbial viability to orchestrate stress-mediated autophagy of the endoplasmic reticulum. Cell 2017;171:809–823.e813. https://doi.org/10.1016/j.cell.2017.09.034.

[30] Conlon J, et al. Mouse, but not human STING, binds and signals in response to the vascular disrupting agent 5,6-dimethylxanthenone-4-acetic acid. J Immunol 2013;190:5216–25. https://doi.org/10.4049/jimmunol.1300097.

[31] Meng Z, Lu M. RNA interference-induced innate immunity, off-target effect, or immune adjuvant? Front Immunol 2017;8:331.

[32] Poeck H, et al. 5′-triphosphate-siRNA: turning gene silencing and rig-I activation against melanoma. Nat Med 2008;14:1256–63. https://doi.org/10.1038/nm.1887.

[33] Yuan D, et al. Anti-angiogenic efficacy of 5′-triphosphate siRNA combining VEGF silencing and RIG-I activation in NSCLCs. Oncotarget 2015;6:29664–74. https://doi.org/10.18632/oncotarget.4869.

[34] Ellermeier J, et al. Therapeutic efficacy of bifunctional siRNA combining TGF-β1 silencing with RIG-I activation in pancreatic cancer. Cancer Res 2013;73:1709–20. https://doi.org/10.1158/0008-5472.Can-11-3850.

[35] Andtbacka RH, et al. Talimogene laherparepvec improves durable response rate in patients with advanced melanoma. J Clin Oncol 2015;33:2780–8. https://doi.org/10.1200/jco.2014.58.3377.

[36] Harrington K, Freeman DJ, Kelly B, Harper J, Soria J-C. Optimizing oncolytic virotherapy in cancer treatment. Nat Rev Drug Discov 2019;18:689–706. https://doi.org/10.1038/s41573-019-0029-0.

[37] Chesney J, et al. Randomized, open-label phase II study evaluating the efficacy and safety of talimogene laherparepvec in combination with Ipilimumab versus Ipilimumab alone in patients with advanced, Unresectable Melanoma. J Clin Oncol 2017;36:1658–67. https://doi.org/10.1200/JCO.2017.73.7379.

[38] Cloughesy TF, et al. Phase 1 trial of vocimagene amiretrorepvec and 5-fluorocytosine for recurrent high-grade glioma. Sci Transl Med 2016;8. https://doi.org/10.1126/scitranslmed.aad9784, 341ra375.

[39] Cloughesy TF, et al. Effect of vocimagene amiretrorepvec in combination with flucytosine vs standard of care on survival following tumor resection in patients with recurrent high-grade glioma: a randomized clinical trial. JAMA Oncol 2020;6:1939–46. https://doi.org/10.1001/jamaoncol.2020.3161.

[40] Bannert N, Kurth R. The evolutionary dynamics of human endogenous retroviral families. Annu Rev Genomics Hum Genet 2006;7:149–73. https://doi.org/10.1146/annurev.genom.7.080505.115700.

[41] Alcami A. Viral mimicry of cytokines, chemokines and their receptors. Nat Rev Immunol 2003;3:36–50. https://doi.org/10.1038/nri980.

[42] Chiappinelli KB, et al. Inhibiting DNA methylation causes an interferon response in Cancer via dsRNA including endogenous retroviruses. Cell 2015;162:974–86. https://doi.org/10.1016/j.cell.2015.07.011.

[43] Yang X, et al. Gene body methylation can alter gene expression and is a therapeutic target in cancer. Cancer Cell 2014;26:577–90. https://doi.org/10.1016/j.ccr.2014.07.028.

[44] Wrangle J, et al. Alterations of immune response of non-small cell lung cancer with azacytidine. Oncotarget 2013;4:2067–79. https://doi.org/10.18632/oncotarget.1542.

[45] Galluzzi L, et al. Consensus guidelines for the definition, detection and interpretation of immunogenic cell death. J Immunother Cancer 2020;8. https://doi.org/10.1136/jitc-2019-000337, e000337.

[46] Kroemer G, Galluzzi L, Kepp O, Zitvogel L. Immunogenic cell death in cancer therapy. Annu Rev Immunol 2013;31:51–72. https://doi.org/10.1146/annurev-immunol-032712-100008.

[47] Sims GP, Rowe DC, Rietdijk ST, Herbst R, Coyle AJ. HMGB1 and RAGE in inflammation and cancer. Annu Rev Immunol 2010;28:367–88. https://doi.org/10.1146/annurev.immunol.021908.132603.

[48] Vanmeerbeek I, et al. Trial watch: chemotherapy-induced immunogenic cell death in immuno-oncology. Oncoimmunology 2020;9:1703449. https://doi.org/10.1080/2162402X.2019.1703449.

[49] Voorwerk L, et al. Immune induction strategies in metastatic triple-negative breast cancer to enhance the sensitivity to PD-1 blockade: the TONIC trial. Nat Med 2019;25:920–8. https://doi.org/10.1038/s41591-019-0432-4.

[50] Cillo AR, et al. Immune landscape of viral- and carcinogen-driven head and neck cancer. Immunity 2020;52:183–199.e189. https://doi.org/10.1016/j.immuni.2019.11.014.

[51] Cabrita R, et al. Tertiary lymphoid structures improve immunotherapy and survival in melanoma. Nature 2020;577:561–5. https://doi.org/10.1038/s41586-019-1914-8.

[52] Helmink BA, et al. B cells and tertiary lymphoid structures promote immunotherapy response. Nature 2020;577:549–55. https://doi.org/10.1038/s41586-019-1922-8.

[53] Petitprez F, et al. B cells are associated with survival and immunotherapy response in sarcoma. Nature 2020;577:556–60. https://doi.org/10.1038/s41586-019-1906-8.

[54] Sharonov GV, Serebrovskaya EO, Yuzhakova DV, Britanova OV, Chudakov DM. B cells, plasma cells and antibody repertoires in the tumour microenvironment. Nat Rev Immunol 2020;20:294–307. https://doi.org/10.1038/s41577-019-0257-x.

[55] Bald T, Krummel MF, Smyth MJ, Barry KC. The NK cell-cancer cycle: advances and new challenges in NK cell-based immunotherapies. Nat Immunol 2020;21:835–47. https://doi.org/10.1038/s41590-020-0728-z.

[56] Minetto P, et al. Harnessing NK cells for cancer treatment. Front Immunol 2019;10:2836.

[57] Shimasaki N, Jain A, Campana D. NK cells for cancer immunotherapy. Nat Rev Drug Discov 2020;19:200–18. https://doi.org/10.1038/s41573-019-0052-1.

[58] Myers JA, Miller JS. Exploring the NK cell platform for cancer immunotherapy. Nat Rev Clin Oncol 2021;18:85–100. https://doi.org/10.1038/s41571-020-0426-7.

[59] Xie G, et al. CAR-NK cells: A promising cellular immunotherapy for cancer. EBioMedicine 2020;59. https://doi.org/10.1016/j.ebiom.2020.102975.

[60] Liu E, et al. Use of CAR-transduced natural killer cells in CD19-positive lymphoid tumors. N Engl J Med 2020;382:545–53. https://doi.org/10.1056/NEJMoa1910607.

[61] Fayette J, et al. Results of a phase II study evaluating monalizumab in combination with cetuximab in previously treated recurrent or metastatic squamous cell carcinoma of the head and neck (R/M SCCHN). Ann Oncol 2018;29:viii374. https://doi.org/10.1093/annonc/mdy287.005.

[62] Margolin K, et al. Phase I trial of ALT-803, a novel recombinant IL15 complex, in patients with advanced solid tumors. Clin Cancer Res 2018;24:5552. https://doi.org/10.1158/1078-0432.CCR-18-0945.

[63] Nywening TM, et al. Targeting tumour-associated macrophages with CCR2 inhibition in combination with FOLFIRINOX in patients with borderline resectable and locally advanced pancreatic cancer: a single-centre, open-label, dose-finding, non-randomised, phase 1b trial. Lancet Oncol 2016;17:651–62. https://doi.org/10.1016/s1470-2045(16)00078-4.

[64] Linehan D, et al. Overall survival in a trial of orally administered CCR2 inhibitor CCX872 in locally advanced/metastatic pancreatic cancer: correlation with blood monocyte counts. J Clin Oncol 2018;36:92. https://doi.org/10.1200/JCO.2018.36.5_suppl.92.

[65] Kulkarni A, et al. A designer self-assembled supramolecule amplifies macrophage immune responses against aggressive cancer. Nat Biomed Eng 2018;2:589–99. https://doi.org/10.1038/s41551-018-0254-6.

[66] Autio KA, et al. Immunomodulatory activity of a Colony-stimulating Factor-1 receptor inhibitor in patients with advanced refractory breast or prostate Cancer: a phase I study. Clin Cancer Res 2020;26:5609. https://doi.org/10.1158/1078-0432.CCR-20-0855.

[67] Razak AR, et al. Safety and efficacy of AMG 820, an anti-colony-stimulating factor 1 receptor antibody, in combination with pembrolizumab in adults with advanced solid tumors. J Immunother Cancer 2020;8. https://doi.org/10.1136/jitc-2020-001006.

[68] Barkal AA, et al. CD24 signalling through macrophage Siglec-10 is a target for cancer immunotherapy. Nature 2019;572:392–6. https://doi.org/10.1038/s41586-019-1456-0.

[69] Weiskopf K, et al. CD47-blocking immunotherapies stimulate macrophage-mediated destruction of small-cell lung cancer. J Clin Investig 2016;126. https://doi.org/10.1172/JCI81603.

[70] Advani R, et al. CD47 blockade by Hu5F9-G4 and rituximab in non-Hodgkin's lymphoma. N Engl J Med 2018;379:1711–21. https://doi.org/10.1056/NEJMoa1807315.

[71] Sallman DA, et al. The first-in-class anti-CD47 antibody Magrolimab (5F9) in combination with azacitidine is effective in MDS and AML patients: ongoing phase 1b results. Blood 2019;134:569. https://doi.org/10.1182/blood-2019-126271.

[72] Sikic BI, et al. First-in-human, first-in-class phase I trial of the anti-CD47 antibody Hu5F9-G4 in patients with advanced cancers. J Clin Oncol 2019;37:946–53. https://doi.org/10.1200/jco.18.02018.

[73] Lakhani NJ, et al. A phase Ib study of the anti-CD47 antibody magrolimab with the PD-L1 inhibitor avelumab (a) in solid tumor (ST) and ovarian cancer (OC) patients. J Clin Oncol 2020;38:18. https://doi.org/10.1200/JCO.2020.38.5_suppl.18.

[74] Liu J, et al. Targeting macrophage checkpoint inhibitor SIRPα for anticancer therapy. JCI insight 2020;5. https://doi.org/10.1172/jci.insight.134728, e134728.

[75] Rafiq S, Hackett CS, Brentjens RJ. Engineering strategies to overcome the current roadblocks in CAR T cell therapy. Nat Rev Clin Oncol 2020;17:147–67. https://doi.org/10.1038/s41571-019-0297-y.

[76] Mullard A. FDA approves first CAR T therapy. Nat Rev Drug Discov 2017;16:669. https://doi.org/10.1038/nrd.2017.196.

[77] Maude SL, et al. Tisagenlecleucel in children and young adults with B-cell lymphoblastic leukemia. N Engl J Med 2018;378:439–48. https://doi.org/10.1056/NEJMoa1709866.

[78] Schuster SJ, et al. Tisagenlecleucel in adult relapsed or refractory diffuse large B-cell lymphoma. N Engl J Med 2018;380:45–56. https://doi.org/10.1056/NEJMoa1804980.

[79] Papadouli I, et al. EMA review of Axicabtagene Ciloleucel (Yescarta) for the treatment of diffuse large B-cell lymphoma. Oncologist 2020;25:894–902. https://doi.org/10.1634/theoncologist.2019-0646.

[80] Locke FL, et al. Long-term safety and activity of axicabtagene ciloleucel in refractory large B-cell lymphoma (ZUMA-1): a single-arm, multicentre, phase 1–2 trial. Lancet Oncol 2019;20:31–42. https://doi.org/10.1016/S1470-2045(18)30864-7.

[81] Wang M, et al. KTE-X19 CAR T-cell therapy in relapsed or refractory mantle-cell lymphoma. N Engl J Med 2020;382:1331–42. https://doi.org/10.1056/NEJMoa1914347.

[82] Abramson JS, et al. Pivotal safety and efficacy results from transcend NHL 001, a multicenter phase 1 study of Lisocabtagene Maraleucel (liso-cel) in relapsed/refractory (R/R) large B cell lymphomas. Blood 2019;134:241. https://doi.org/10.1182/blood-2019-127508.

[83] Di Stasi A, et al. Inducible apoptosis as a safety switch for adoptive cell therapy. N Engl J Med 2011;365:1673–83. https://doi.org/10.1056/NEJMoa1106152.

[84] Slaney CY, Wang P, Darcy PK, Kershaw MH. CARs versus BiTEs: a comparison between T cell-redirection strategies for cancer treatment. Cancer Discov 2018;8:924–34. https://doi.org/10.1158/2159-8290.Cd-18-0297.

[85] Schmidt M, et al. A comprehensive analysis of human gene expression profiles identifies stromal immunoglobulin κ C as a compatible prognostic marker in human solid tumors. Clin Cancer Res 2012;18:2695–703. https://doi.org/10.1158/1078-0432.Ccr-11-2210.

CHAPTER FOURTEEN

Intrinsic and acquired cancer immunotherapy resistance

Reem Saleh[a], Varun Sasidharan Nair[b], Salman M. Toor[c], and Eyad Elkord[d,e]

[a]Sir Peter MacCallum Department of Oncology, University of Melbourne, Parkville, VIC, Australia
[b]Department of Experimental Immunology, Helmholtz Centre for Infection Research, Braunschweig, Germany
[c]College of Health and Life Sciences (CHLS), Hamad Bin Khalifa University (HBKU), Doha, Qatar
[d]Natural and Medical Sciences Research Center, University of Nizwa, Nizwa, Oman
[e]Biomedical Research Center, School of Science, Engineering and Environment, University of Salford, Manchester, United Kingdom

Contents

1. Introduction

In light of cancer immunoediting, tumor cells can continuously evolve to alter their immunogenicity and establish an immunosuppressive milieu within the tumor microenvironment (TME) comprising of cellular and soluble networks that promote

Cancer Immunology and Immunotherapy
https://doi.org/10.1016/B978-0-12-823397-9.00014-4

immune escape and tumorigenesis [1, 2]. Tumor cells can escape the recognition of effector immune cells by diminishing the expression of neoantigens and antigen presentation molecules [3–6], release suppressive cytokines, growth factors, and matrix metalloproteinases to induce immunosuppression and favor tumorigenesis [7–10]. Moreover, tumor cells can express high levels of co-inhibitory immune checkpoint (IC) ligands and can also upregulate the expression of IC receptors on tumor-infiltrating immune cells; increase the induction, expansion, and activation of immunosuppressive cells, such as tumor-associated macrophages (TAMs), myeloid-derived suppressor cells (MDSCs), T-regulatory cells (Tregs), and tumor-associated dendritic cells (TADCs); and increase the induction and differentiation of cancer-associated fibroblasts (CAFs) within the TME [11, 12]. Collectively, cellular communication between immune cells and nonimmune cells within the TME can contribute significantly to cancer progression and suppress antitumor immune responses [13–15].

Cancer immunotherapy in the form of immune checkpoint inhibitors (ICIs) has made a breakthrough in cancer treatment of various types; however, a high number of cancer patients show limited response rates [16–23]. In addition to ICIs, the use of adoptive cell transfer (ACT), including the transfer of autologous tumor-infiltrating T-cells (TILs) or genetically engineered chimeric antigen receptor (CAR)-specific T-cells (CARTs) have been developed to eradicate tumor cells in specific and efficient manners [24–28]. The efficacy of adoptive T-cell therapy has been demonstrated in multiple cancer settings including hematologic cancers [29], melanoma [30], synovial sarcoma [31], colorectal carcinoma [32, 33], gastrointestinal and lung cancers, and virus-associated cancers [34, 35].

Despite the success of cancer immunotherapies, the effectiveness of these therapies in cancer patients is limited due to intrinsic- and extrinsic-mediated factors including genetic and epigenetic alterations, tumor mutational loads, and the overexpression of co-inhibitory ICs and surplus population of immunosuppressive cells in the TME [23, 36]. Tumor immunogenicity, dictated by the molecular and cellular composition of the TME, is one of the key aspects that dictates the response to therapy [3, 4, 37]. Altogether genetic/epigenetic and phenotypical changes acquired in the TME can collectively induce resistance against therapy [14, 38]. Consequently, therapeutic strategies to alleviate tumor resistance (intrinsic or acquired) against cancer immunotherapies are crucial to maximize the efficacy of cancer treatment and revert tumor resistance to ICIs [38, 39] or ACT [40, 41].

2. Tumor microenvironment

The TME is a composite milieu of multiple cell types embedded in an extracellular matrix (ECM), which enables cancer cells to interact with other cell types and favors their own progression and survival [42]. These dynamic interactions are indispensable for the heterogeneity, clonal development, drug resistance, and metastasis of malignant cells [42].

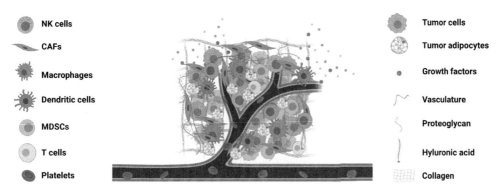

Fig. 1 Composition of the tumor microenvironment (TME). Various cellular populations including tumor cells, epithelial cells, and immune cells are present within the TME.

The ECM is molded with glycoproteins, polysaccharides, and cell adhesion proteins including laminin, fibronectin, and collagen that actively support the survival of both cancer cells and stromal cells [43]. Stromal cells in the TME comprise of immune cells (T-cells/natural killer (NK) cells/ macrophages), stem cells, adipocytes, and fibroblasts (Fig. 1) [44]. Additionally, endothelial cells and pericytes present in the TME favor the formation of blood vessels and lymphatic vesicles (Fig. 1) [45]. Tumor cells favor the trafficking of stromal cells into the TME by secreting various cytokines and chemokines (Fig. 1) [46]. Notably, nonmalignant cells are the major contributors of metastasis throughout the phases of cancer development/progression [46].

The TME represents a complex network with inherent modulations acquired during the progression of disease or upon various therapeutic interventions. Thus, understanding the biochemical characteristics of malignant cells/nonmalignant cells within the TME is crucial to uncover the molecular mechanisms behind such modulations. However, the contribution of the TME to tumorigenesis encompasses multiple genetic/epigenetic/metabolic alterations, which can also determine the therapeutic outcomes in cancer patients [47]. Importantly, numerous therapeutic modalities fail to show clinical efficacy in targeting tumor cells due to their dynamic characteristics within the TME. Therefore, better understanding of the TME can potentially reveal targets for novel therapeutics or improve current therapeutic modalities to achieve favorable clinical outcomes. In this section, we will focus on the complexity and heterogeneity of the TME and discuss the key factors that hamper antitumor immune responses.

2.1 Immune cells

Tumor-infiltrating immune cells (TIICs) play a dual role in promoting and hampering the onset and progression of cancer, and their definitive roles in cancer development/progression rely on the modulations within the TME. TIICs primarily comprise of

T-lymphocytes, B-lymphocytes, NK cells, dendritic cells (DCs), macrophages, neutrophils, and MDSCs.

Tumor-infiltrating effector T-cells are the key players to elicit adaptive antitumor immune responses; their abundance, functionality, and distribution have crucial diagnostic/prognostic values [48]. For instance, cytotoxic CD8$^+$ T-cells and CD4$^+$ T helper-1 (Th1) cells are the major producers of IFN-γ and IL-2, which are indispensable in creating a robust effector mechanism for the elimination of tumor cells, thereby resulting in favorable disease prognosis [49]. On the other hand, Th subsets including Th2 and Th17 are mostly associated with tissue inflammation and favor tumor progression and metastasis [50, 51]. Furthermore, the role of effector CD8$^+$ and CD4$^+$ T-cells can be hampered by tissue-resident Tregs, thereby favoring tumor progression [52]. Thus, higher ratio of Tregs:T-effectors within the TME usually suppresses antitumor immune responses [52]. Notably, the roles of tumor-infiltrating Tregs in certain cancer types including colorectal cancer and gastric cancer remain controversial [53–55].

In addition to T-cells, B-cell infiltration and localization usually occur in the tumor-invasive margin and draining lymph nodes and have both beneficial and harmful roles in cancer [56]. In melanoma tumor models, it has been reported that B-cells assist T-cells in eliciting antitumor responses [57]. Furthermore, in multiple malignancies including cervical, gastric, and ovarian tumors, B-cell infiltration is associated with favorable prognosis [58–60]. However, reports show that B-cell infiltration in squamous carcinoma and prostate cancer mediated by CXCL13 can promote the onset and progression of tumor [61, 62].

Tumor-residing NK cells are innate cytotoxic cells, which are responsible for recognizing and eliminating malignant cells, and have higher potentiality to influence adaptive antitumor immune responses through the production of various cytokines and chemokines [63]. Within the TME, NK cells represent a highly heterogenous distinct populations and express unique surface molecules [64]. Furthermore, NK cells inhibit metastasis through the upregulation of IFN-γ production and subsequent modulation of fibronectin-1 expression on tumor cells [65]. On the other hand, it has been reported that NK cells could modulate T-cells and favor angiogenesis and tumor progression [66].

Antigen presenting cells (APCs), such as DCs and macrophages, are innate heterogenous immune cell populations within the TME with specific immunological functions [67]. The function of DCs in the TME includes antigen presentation, priming of cytotoxic CD8$^+$ T-cells, activation of humoral immune responses, secretion of cytotoxic molecules, immunotolerance, and suppression [68]. Additionally, majority of macrophages residing in tumor stroma are TAMs, which play a major role in regulating the inflammatory responses within the TME [69]. The accumulation of TAMs is predominantly associated with poor prognosis [70]. Proinflammatory macrophages, referred to as M1 subtype, are primed by cytokines, including IFN-γ and TNF-α, favoring Th1 responses and eliminating tumor cells [71]. Meanwhile, M2 macrophages, polarized by IL-4 and IL-13 favor tumor promotion and create an immunosubversive environment by interfering with T-cell functionality [71]. In addition to these, MDSCs are considered as one of the major immunosuppressive cells, which abundantly exist in the TME [72]. MDSCs are heterogenous population including two major subtypes; polymorphonuclear (PMN-MDSC) or granulocytic (G-MDSCs) and monocytic (M-MDSCs), which have potent immunosuppressive characteristics and favor tumorigenesis and progression [73, 74]. Furthermore, MDSCs play a key role in the promotion of epithelial mesenchymal transition (EMT) and tumor invasiveness [75].

Lastly, neutrophils comprise a major leukocyte population and elicit primary defense mechanisms against infection [76]. Within the TME, similar to macrophages, neutrophils are also polarized into two differing populations; N1 (antitumorigenic) and N2 (protumorigenic) tumor-associated neutrophils (TANs) [77]. It has been shown that CAFs-mediated secretion of TGF-β is responsible for the polarization of N2 phenotype and inhibition of N1 neutrophils [78]. Importantly, TAN phenotype can play an indispensable role in the development and progression of tumor [79].

2.2 Nonimmune cells

The TME consists not only of immune cells, but also of ECM, endothelial cells, fibroblast, and pericytes [80]. Similar to other components within the TME, ECM plays a major role in facilitating signaling, migration, and metabolism of malignant cells and also has a major effect on the immunogenicity of many solid tumors [81]. It has been reported that ECM has multiple roles associated with the regulation of tumor progression/proliferation and response to therapy [82]. Furthermore, ECM consists of collagens, proteoglycans, hyaluronic acid, elastin, and laminins, which are nonredundant for tumor progression. However, a clear understanding of the characteristics of ECM is necessary for developing novel therapeutic modalities.

In solid tumors, tumor vascularization can occur as a result of endothelial dysfunction and the induction of tumor-associated endothelial cells (TECs) caused by hypoxic factors and chronic release of growth factors [83]. Unlike normal endothelial cells, TECs have distinct molecular and phenotypic peculiarities, which are similar to tumor cells but are not immortal [83]. TECs play a major role in the cellular determination within the TME, control cell trafficking and nutrient supply, thereby shaping the TME [84]. Moreover, TECs support the survival, functionality, and dissemination of malignant cells [85]. Within the nonneoplastic cells in tumor milieu, CAFs are considered as a predominant stromal element favoring tumor progression [86]. CAFs are responsible for TGF-β secretion, which can drive EMT by restructuring ECM, leading to tumor invasiveness and metastasis [87]. In addition to TGF-β, CAFs can also produce soluble IL-6, which favors not only EMT but also chemoresistance to malignant cells [88]. However, depletion of CAFs favors IFN-γ production and reverts immunosuppression within the TME [89].

Additionally, vasculature development within the TME is instituted through the perivascular cells, commonly called as pericytes [90]. Reports show that pericytes could serve as stem cell reservoir favoring tissue regeneration and angiogenesis [91–93]. Altogether both immune cells and nonimmune cells have distinct and dynamic contributions within the TME.

3. Cancer immunotherapies
3.1 Currently approved immune checkpoint inhibitors

ICIs aim to interfere with the inhibitory pathways, which negatively modulate T-cell activation, to ensure sustained T-cell effector functions [94]. In addition to activation signals via T-cell receptor (TCR) and costimulatory molecules, T-cell activation is also regulated by inhibitory molecules referred to as ICs and the balance between costimulatory and inhibitory signals determines the magnitude of T-cell responses [94]. ICs are vital for maintaining self-tolerance and modulating the extent and duration of immune responses in tissues to prevent tissue damage [95]. T-cells are vital in antitumor immune responses via direct killing of tumor cells through cytotoxic T-cells (CTLs) [96]. ICs are expressed on highly activated T-cells; however, prolonged antigen stimulation may lead to sustained upregulation and coexpression of multiple IC expression on T-cells, leading to a state of impaired activity and loss of effector functions, disruption of key transcription factors, and failure in transition to quiescence and acquire antigen-independent memory T-cell homeostatic responsiveness, referred to as T-cell exhaustion [97]. Multiple ICs may be expressed on T-cells, which exert their stimulatory or inhibitory effects on T-cell activation. ICs and their ligands are abundantly expressed in the TME of various malignancies and constitute vital components of the tumor immune resistance mechanisms [98]. Several studies have reported associations between IC expression in the TME with disease progression and/or worsened disease outcomes. Therefore, ICIs result in durable responses, due to the generation of improved antitumor T-cell responses, but notably only in a fraction of patients [99]. Nonetheless, ICIs are at the forefront of all promising cancer immunotherapy strategies.

Leach *et al.* were one of the pioneering groups to propose the use of ICIs as a novel anticancer strategy to treat various malignancies [100]. Eventually, monoclonal antibody (mAb) targeting interactions between cytotoxic T-lymphocyte-associated antigen 4 (CTLA-4) on T-cells and its ligands on tumor cells was developed (ipilimumab) as the first class of ICIs, which were granted US Food and Drug Administration (FDA) approval, following improved survival rates in patients with metastatic melanoma [19]. Following CTLA-4 blockade, interactions between programmed cell death-1 (PD-1) and its ligands PD-L1 and PD-L2 were extensively studied, which lead to FDA approvals for anti-PD-1 mAbs such as nivolumab and pembrolizumab, and anti-PD-L1 antibodies avelumab, atezolizumab, and durvalumab to treat selective cancers

[101]. At present, seven ICIs have been approved for cancer therapy; however, the criteria for their application in different disease settings have been clearly laid out [102]. For instance, pembrolizumab was initially approved for treating advanced melanoma [103] and nonsmall-cell lung cancer (NSCLC) [104, 105] but eventually approved to treat unresectable or metastatic, microsatellite-high (MSI-H) or DNA mismatch repair deficient (dMMR) solid tumors that progressed after prior treatment [106], and more recently as a frontline treatment for MSI-H/dMMR colorectal cancer (CRC) patients based on KEYNOTE-177 study [107]. Here, we focus on the current FDA-approved ICIs; anti-CTLA-4, anti-PD-1, and anti-PD-L1 therapies, and discuss the rationale behind their utilization for therapeutic benefits.

3.1.1 Anti-CTLA-4

CTLA-4 (CD152) belongs to B7/CD28 proteins and is constitutively expressed on Tregs; upregulated on activated T-cells, and inhibits T-cell functions by affecting signaling through the costimulatory receptor CD28 [108]. CTLA-4 interactions hinder antigenic T-cell responses via blocking CD28-mediated signaling [109]. These effects were most evident in studies showing development of spontaneous autoimmunity, elevated lymphoproliferation, and potent antitumor immunity in CTLA-4-deficient FoxP3$^+$ Treg models [110]. Therefore, anti-CTLA-4 could potentially prompt the priming and expansion of tumor-specific naive T-cells and elevate effector functions of memory/effector tumor reactive T-cells by ensuring TME devoid of immunosuppressive mechanisms. Ipilimumab and tremelimumab are considered as pioneering fully human T-cell immunomodulatory monoclonal antibodies targeting interactions between CTLA-4 and its ligands to promote antitumor immunity [111]. Anti-CTLA-4 therapy showed efficacy in a wide range of cancers including melanoma, NSCLC, prostate, breast, and ovarian cancers; however, immune-related adverse events (irAEs) were also prevalent in a significant percentage of patients [112]. As a result, till present, only ipilimumab is granted FDA approval as the sole anti-CTLA-4 therapy in clinical use to treat metastatic melanoma [113].

Immune-related toxicities are the primary factor, which have limited the clinical translation of anti-CTLA-4 therapies in oncology. Larkin *et al.* showed that a higher percentage of patients exhibited side effects following treatment with anti-CTLA-4 therapy (27.3%) compared with anti-PD-1 therapy (16.3%), while combining both therapies resulted in even higher percentage of patients exhibiting side effects (55%) [114]. Therefore, the utilization of biomarkers for response and efficacy may be considered utmost important for the initiation of anti-CTLA-4 therapy. For instance, melanoma patients with high prevalence of neoantigens showed better responses to anti-CTLA-4 therapy [115], and thus the presence of neoantigens on mutated tumors favors antitumor immunogenicity, which in turn results in improved treatment efficacy.

3.1.2 Anti-PD-1/PD-L1

PD-1 signaling pathway is a vital regulator of T-cell responses and is critical in maintaining peripheral tolerance [116]. PD-1 engagement interferes with TCR/CD28 signaling and therefore, the IL-2-dependent positive feedback, leading to decreased cytokine (IL-2, IFN-γ, and TNF-α) release and cell proliferation, and decreased expression of the transcription factors related to effector T-cell functions including T-bet and Eomes [22]. PD-L1 (CD274) and PD-L2 (CD273) are identified as putative ligands for PD-1, expressed on APCs and also on various nonlymphoid cells, and are involved in mediating peripheral tolerance [117]. Tumor cells can also upregulate PD-L1 and its binding to PD-1 on antigen-specific T-cells can affect antitumor immunity. Moreover, it was reported that elevated levels of CD8$^+$ T-cells at the invasive margin are negatively regulated by PD-1/PD-L1 expression in melanoma patients [118]. Increased PD-1 expression on T-cells has been associated with poor prognosis in several cancers including pancreatic, liver, breast, and head and neck [119]. Blocking the PD-1 receptor has shown extraordinary clinical responses in various cancers [21, 120]. In addition, the prognostic significance of elevated PD-L1 expression has been shown in various solid tumors and was associated with poor overall survival in breast, renal, urothelial, and gastric cancers, and poor disease-free survival or progression-free survival in melanoma, hepatocellular carcinoma, and renal carcinoma [121, 122].

Molecules designed to block signaling through PD-1 receptor include nivolumab, pembrolizumab, cemiplimab, and lambrolizumab [123]. Pembrolizumab, nivolimumab, and cemiplimab have been approved by the FDA for treating selective malignancies [102]. Blocking the PD-1/PD-L1 axis implicates targeting the receptor or the ligand to barricade the signaling pathway. Therefore, anti-PD-L1 antibodies have also shown synergistic effects in several cancers as anti-PD-1 therapy. Atezolizumab, durvalumab, and avelumab are FDA-approved anti-PD-L1 drugs, which are currently approved for different cancers including NSCLC, urothelial cancer, renal cell carcinoma, and Merkel cell carcinoma; atezolizumab plus nab-paclitaxel are also approved for patients with locally advanced or unresectable/metastatic TNBC [124]. Notably, PD-L1 expression is proposed as a predictive biomarker for therapy and testing for PD-L1 expression has been established as a prerequisite for initiation of anti-PD-L1 therapies. However, PD-L1 expression threshold, immune-related toxicities associated with all ICIs and resistance mechanisms remain the primary challenges associated with the utilization of anti-PD-L1 therapies for all cancers.

3.2 Adoptive T-cell therapy

ACT is a cell-based cancer immunotherapy, which induces an antigen-specific T-cell response via autologous, genetically engineered CARTs [26, 28] or TCR-modified T-cells [24]. The development of engineered CARTs involves the modification of isolated T-cells from cancer patients in a manner that enable them to recognize tumor-specific and/or tumor-associated antigens (TSAs/TAAs), without relying on antigen processing and presentation. Following their *ex vivo* expansion, these modified, activated CARTs are passively transferred back to the patients [28]. The advantage of using CAR T-cell therapy is encapsulated in the ability of CARTs to eliminate tumor cells with a better cytotoxicity and specificity and induce a long-lasting antitumor immunity, given that CARTs can act as memory T-cells [25, 28]. The effectivity of ACT is dictated by (1) the level of immunogenicity of the target antigen selected, (2) efficiency of CAR or TCR-modified T-cell trafficking to tumor sites, and (3) the accumulation of the engineered T-cells within the TME.

3.2.1 Chimeric-antigen receptor T-cell therapy

Up to date, improved response rates using CAR T-cell therapy has only shown success in hematologic cancers [29]. In a proof-of-concept clinical trial, the therapeutic potential for using CARTs targeting CD20, B-cell surface antigen, was shown in patients with mantle cell lymphoma (MCL) and B-cell non-Hodgkin lymphoma (NHL) [125]. Targeting CD19, another B-cell surface antigen, by CARTs resulted in durable, sustained antitumor immunity in acute lymphoblastic leukemia (ALL) [126, 127], chronic lymphocytic leukemia [25, 128], multiple myeloma [129], and diffuse large B-cell lymphoma (DLBCL) patients [130, 131]. These promising findings led to the development of "CTL019" by Novartis, the first CAR T-cell therapy approved by FDA for young patients with refractory B-cell ALL [132].

The success of CAR T-cell therapy in hematologic cancers has prompted its application in solid tumors to examine its therapeutic efficacy. CARTs targeting epidermal growth factor receptor (EGFR) variant in glioblastoma (GBM) showed an enhanced antitumor immune response associated with tumor cell elimination in an antigen-specific manner [133–135]. These data suggest that CAR T-cell therapy has a therapeutic potential for treating solid tumors [29].

3.2.2 T-cell receptor-modified T-cell therapy

The development of TCR-modified T-cells involves the identification of antigen-specific TCRs and their introduction into autologous T-cells [24]. TCR-modified T-cells could be utilized for treating multiple solid tumors [30–33, 136]. Early phase clinical trials showed a sustained antigen-specific immune response induced by TCR-modified T-cells targeting the antigen New York esophageal squamous cell carcinoma (NY-ESO-1) in myeloma [136] and synovial sarcoma [31]. The use of TCR-modified T-cells targeting HLA-A*2402-restricted MAGE-A4 antigen in 10 patients with esophageal cancer showed moderately durable, sustained antitumor immune responses in three patients [137]. Unlike CARTs, TCR-modified T-cells can only recognize intracellular TSAs or TAAs as their function is dependent on the TCR-mediated signaling, which involves antigen processing and presentation [24].

3.3 Recombinant cytokines and cancer vaccines

The utilization of human recombinant cytokines as a monotherapy in cancer patients was approved many years ago [138]. Proleukin (recombinant IL-2) was approved to treat renal cancer and melanoma patients, while Sylatron (IFN-α2b conjugated to polyethylene glycol) was approved to treat patients with hairy cell leukemia, advanced follicular non-Hodgkin's lymphoma and resected melanoma [139–143]. More recently, Zhang *et al.* reported that systemic administration of IFN-γ in patients with synovial sarcoma and myxoid/round cell liposarcoma (MRCL) can increase tumor antigen presentation and reduce exhausted phenotypes of tumor-infiltrating T-cells [144]. Based on the important roles of cytokines in regulating antitumor immune responses and T-cell activation/proliferation, systemic administration of human recombinant IL-2 or IFN-γ as adjuvant therapy has been utilized to be used in conjunction to ICIs aiming to expand the effectivity of cancer therapy and improve response rates in patients [145, 146].

A wide range of vaccines have been assessed in clinical trials including the application of pulsed DCs with MHC-binding peptides of TAAs or tumor peptide as adjuvant therapy [147, 148]. The efficacy of using peptide-pulsed or viral vector–infected DCs has been tested in patients with glioma, melanoma, prostate cancer, and colorectal cancer [149, 150]. Indeed, sipuleucel-T vaccine comprising APCs stimulated with the prostate antigen (PAP) fused to granulocyte macrophage colony-stimulating factor (GM-CSF) was approved by the FDA to treat patients with metastatic castrate-resistant prostate cancer (mCRPC) [151]. TAA peptide-derived vaccines can be an excellent therapy to be applied as an adjuvant therapy; they are cost-effective and tumor-specific [152]. Nonetheless, the efficiency of TAA-derived vaccines could be potentially compromised as a result of tumor evolution and the development of acquired resistance caused by the loss of target antigen.

4. Mechanisms of resistance against cancer immunotherapies

Durability of treatment response is one of the main factors for favorable clinical outcomes. Large subsets of cancer patients do not benefit from cancer immunotherapies due to primary resistance, or they initially respond to therapy then relapse over time and develop acquired resistance [38, 153]. Mechanisms underlying the development of primary and adaptive resistance to cancer immunotherapies could be driven by tumor cell-intrinsic or extrinsic factors [38] (Fig. 2). There is a vital need to identify reliable predictive biomarkers for tumor resistance to immunotherapy. Moreover, further understanding of the molecular pathways, which facilitate resistance mechanisms should be helpful in designing and developing alternative or combinatorial therapeutic

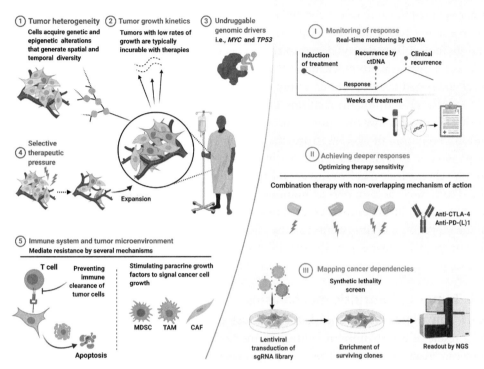

Fig. 2 Determinants of tumor resistance to therapy and how to overcome them. In solid tumors, tumor cells are continuously evolving to establish an immunosuppressive environment, referred to as tumor microenvironment (TME), comprising of cellular and soluble components that favor tumor growth/progression and promote immune evasion. Over time, tumor cells within the TME become very heterogenous as they acquire genetic and epigenetic alterations with spatial and temporal diversity (1). Tumor cells with low growth rates (2) and low mutational burden (3) are usually incurable and resistant to cancer therapies (intrinsic or primary resistance). Surviving tumors cells following the administration of particular cancer therapy can acquire compensatory inhibitory mechanisms that allows them to escape immune cell recognition and induce immunosuppression. Under selective therapeutic pressure, these surviving tumor cells can grow and expand leading to tumor recurrence or relapse (acquired resistance) (4). Cellular components of the TME can also impact the development of resistance and promote tumor growth and progression by suppressing the ability of effector T-cells from eradicating tumor cells and enhancing the induction/function and recruitment of suppressive cells, such as TAMs, MDSCs and CAFs (5). Thus, therapeutic strategies to overcome intrinsic and acquired tumor resistance against cancer immunotherapies are crucial to maximize the efficacy of cancer treatment and revert tumor resistance to ICIs. Monitoring the immune response within the TME before and after the application of therapy could be helpful to identify biomarkers, which could be related to resistance development (I). The utilization of cancer therapies with distinct mechanisms of action and multiple ICIs could be beneficial in enhancing the therapeutic response (II). Next-Generation Sequencing (NGS) of DNA or RNA in cancer cell clones could be also helpful in revealing genetic factors that are important for cancer cell growth and survival (III).

approaches to overcome such resistance and enhance clinical outcomes in cancer patients (Fig. 2). In addition, the mainstream use of ICIs in treating cancers is largely thwarted by low response rates and irAEs reported in some cancer patients [154]. This is because ICIs intervene in various immune response pathways and are thus, predisposed in affecting various molecular pathways downstream of checkpoint signaling pathways. Importantly, irAEs also occur due to blockade of IC pathways against autoimmunity, which leads to local and systemic autoimmune responses [154].

Tumor cell-intrinsic mechanisms leading to primary or adaptive resistance to immunotherapy involve low mutational burden promoting tumor cells to evade T-cell recognition, exclusion of T-cells from the tumor core, diminished antitumor immune responses due to high mutational rates in the IFN-γ pathway [38, 155–158]. In contrast, tumor cell-extrinsic mechanisms are related to the molecular and cellular components of the TME, such as suppressive immune cells (*e.g.*, Tregs, MDSCs, TAMs, TADCs, TANs), adenosine, chemokines, cytokines, growth factors, and inhibitory ICs, resulting in the suppression of tumor antigen-specific TCR signaling, induction of T-cell exhaustion or apoptosis, overexpression of inhibitory immune checkpoints and the maintenance of a suppressive microenvironment favoring tumor growth and progression [23, 38, 55].

4.1 Tumor cell-intrinsic mechanisms

4.1.1 Low tumor mutational loads and loss of tumor antigens

As a result of cancer immunoediting, tumor cells can acquire mechanisms by which they can escape immune cell recognition via genetic and epigenetic alterations leading to impaired production of neoantigens or loss of target antigens and subsequently inhibit the activation of antitumor immunity [8] (Fig. 3A). Alternatively, tumor cells can lose sensitivity to immunotherapy and develop resistance due to the survival and expansion of nonimmunogenic tumor cell clones [8]. The selection of target antigens to develop genetically modified T-cells is a crucial requirement for an effective therapy. In support of this, it has been reported that tumor resistance against CD19-CAR T-cell therapy diffuse large B-cell lymphoma patient has occurred due to the loss of CD19 expression [159]. Furthermore, genetic mutations acquired by tumor cells can compromise antigen presentation, and T-cell recognition and activation leading to resistance to therapy and insufficient response to ACT or ICIs [14, 160, 161]. Indeed, studies have shown a positive correlation between high tumor mutational burden and the response to ICIs [162–164].

4.1.2 Oncogenic signaling pathways

Tumor cell-intrinsic factors can be initially present in tumor cells causing primary resistance or can be acquired later as tumor cells evolve leading to the emergence of acquired tumor resistance to cancer immunotherapies. These factors include the loss of phosphatase and tensin homolog (PTEN) expression that activates phosphatidylinositol-3-kinase

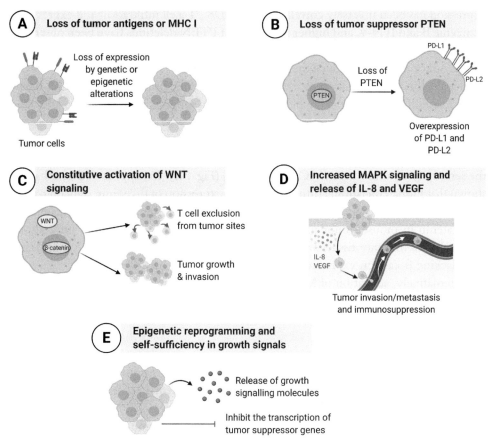

Fig. 3 Mechanisms of primary tumor resistance against cancer immunotherapy. Tumor cells can escape immune cell recognition via genetic and epigenetic alterations leading to impaired expression of MHC I/immune cell recognition and impaired production of neoantigens or loss of target antigens and subsequently inhibit the activation of antitumor immunity (A). Loss of PTEN gene expression can reduce tumor vulnerability to immunotherapy by promoting the overexpression of PD-L1 and PD-L2, which ultimately lead to the suppression of T-cell function (B). Constitutive activation of WNT signaling leading to the stabilization of β-catenin could result in the exclusion of T-cell infiltration in tumor sites and therefore reduces the sensitivity of the tumor to ICIs or ACT (C). The activation of MAPK pathway in tumor cells leads to the secretion of various molecules, including IL-8 and VEGF, which exert immunosuppressive effects and promote tumor angiogenesis and metastasis (D). Epigenetic reprogramming in tumor cells can promote tumor growth by enhancing the expression and production of growth factors and inhibiting the transcription of tumor suppressor genes (E).

(PI3K) signaling, increased activation of WNT/β–catenin signaling pathway, signal transduction via mitogen-activated protein kinase (MAPK) pathway, impaired IFN-γ signaling pathways, and suppression of T-cell responses [38].

Loss of PTEN and enhanced activation of PI3K signaling in several cancer types have been associated with tumor resistance against ICIs [5, 165–167]. Loss of PTEN gene expression has been inversely correlated with $CD8^+$ T-cell infiltration and expression of granzyme B and IFN-γ, and higher mutations of PTEN deletions have been observed in non-T-cell infiltrated tumors [167]. Unlike wild-type mice, deficiency of PTEN in mouse tumor model can reduce the vulnerability to ACT [167]. Moreover, PTEN deletions and PI3K mutations in tumor cells can promote the upregulation of PD-1 ligands, PD-L1, and PD-L2 [168–173], which ultimately can suppress antitumor T-cell functions (Fig. 3B).

Constitutive signaling of WNT networks leading to the stabilization of β-catenin could result in the exclusion of T-cell infiltration in tumor sites and therefore, reduces the sensitivity of tumor cells to ICIs or ACT [6] (Fig. 3C). In support of this, it has been shown that human melanoma tumors with high expression of β-catenin signaling-related genes lacked the infiltration of T-cells and $CD103^+$ DCs in the TME [6]. Increased levels of β-catenin in mouse tumor model have been associated with reduced infiltration of $CD103^+$ DCs in tumor tissues and tumor resistance to ICIs, compared to tumor-bearing mice lacking β-catenin expression [6].

Alternatively, activation of MAPK pathway in tumor cells promotes the secretion of various molecules, including IL-8 and VEGF, and leads to tumor angiogenesis and metastasis [174, 175] (Fig. 3D).

4.1.3 Epigenetic modifications

Tumorigenesis and progression are closely related with the somatic epigenetic reprogramming [176]. These epigenetic alterations could influence potential modifications in gene transcription and favor mutational burden on host cells [177]. Moreover, TIICs possess an aberrant epigenetic profile, compared with normal immune cells [178]. For instance, tumor-infiltrating T-cells showed a reprogrammed DNA methylation profile and favor tumor progression [178]. In addition to stromal cells, cancer cells per se have exhibited an altered epigenetic and considered as one of the hallmarks of cancer [179]. Accumulating evidences suggest that epigenetic modifications including DNA methylation and posttranslational histone modifications are the predominant epigenetic alterations occurring within the TME [36]. Moreover, reports show that epigenetic modifications not only favor the regulation of gene transcription but also revert antitumor immune responses and reinvigorate the exhaustion of T-cells [180, 181] (Fig. 3E).

Majority of studies on epigenetic modifications within the TME rely on DNA methylation, which could influence the onset and progression of tumors. DNA hypermethylation, either global or localized, of CpG islands (CG-rich regions) in the promotor regions of key genes including tumor suppressor genes are evident in several cancers [182]. However, global hypomethylation of DNA creates a genomic instability and induces the transformation of cells. Global DNA hypomethylation is found in many cancers including prostate [183], leukemia [184], hepatic carcinoma, [185] and cervical cancer [186]. In contrast, global DNA hypermethylation was reported in breast cancer [187]. It has been reported that cancer-associated hypermethylation happens in CpG islands, while hypomethylation happens in both CpG islands and repeated DNA sequences [188]. Moreover, the hypomethylation of these repeated DNA sequences could favor tumorigenesis by elevating karyotypic instability and gene expression [189].

Aberrations in the methylation profile of DNA is closely linked with posttranslational histone modifications, which is crucial for determining the fate of tumor cells by reprogramming the chromatin architecture [190]. Major reprograming of histones includes methylation, acetylation, ubiquitination. and phosphorylation [190]. Various modifications coordinate together to regulate cellular and molecular events in the oncogenic transformation and progression of tumors [191]. For instance, histones are most frequently mutated proteins in malignancy and act as potent tumor drivers [192]. These mutations could influence invasion, dissemination, chemoresistance, and abnormal proliferation of malignant cells [193]. Recent advances in high throughput technologies assisted genome-wide characterization of chromatin modifications during the onset and progression of tumor [176]. It has been reported that tumor cells exhibited global loss of histone methylation and acetylation, with the loss of H4K20me3 and acetylated H4K16 [194]. These losses are closely related to hypomethylation of DNA-repetitive nucleotides, a characteristic feature of tumor cells [194]. DNA/histone epigenetic alterations can not only affect cancer cells to promote their growth and invasion, but can also affect T-cells (Fig. 4) [98]. The epigenetic modifications on IC promotors in T-cells can impair effector T-cell function, and increase Treg and MDSC levels in the TME [98].

4.2 Tumor cell-extrinsic mechanisms

One of the major challenges which adversely impacts antitumor immunity and limits the response to cancer immunotherapies in patients is tumor-induced immunosuppression [29, 195, 196]. An immunosuppressive environment within the TME is established via cellular and molecular networks, which chronically promote the trafficking and accumulation of suppressive immune cells/molecules, and overexpression of inhibitory ICs [197, 198], and favor effector T-cell and CTL deactivation/apoptosis and tumor progression, therapy leading to resistance and sustained immunosuppression [14, 199].

4.2.1 Cancer-associated fibroblasts

Stromal cells within the TME of multiple solid tumors are predominantly CAFs [200]. Signals generated by tumor cells or the hypoxic condition within the TME can result in the conversion of normal resident fibroblasts into CAFs [200, 201]. CAFs can lead to cancer progression by inducing CD8$^+$ T-cell apoptosis, and supporting ECM degradation, tumor growth, invasion, and angiogenesis [13, 202, 203]. A study performed on lung carcinoma and pancreatic adenocarcinoma (PDAC) mouse models showed that specific knockdown of fibroblast activation protein (FAP) (a protein that is constitutively

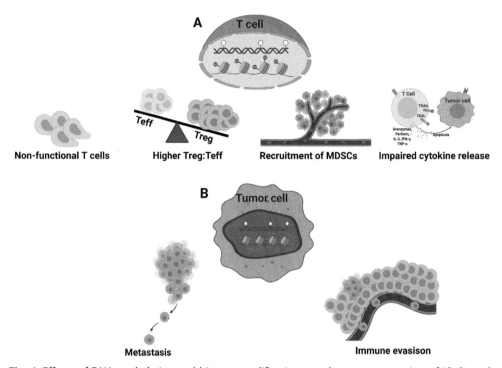

Fig. 4 Effects of DNA methylation and histone modifications on the promoter region of ICs/ligands within the TME. The effects of DNA/histone epigenetic alterations on T-cells (A) and cancer cells (B) are illustrated. The epigenetic modifications of ICs on T-cells can lead to less responsive T-cells, increased Treg levels, increased MDSC trafficking and impaired effector cytokine release. The epigenetic modifications in tumor cells can facilitate metastasis and immune evasion and establish an immunosuppressive microenvironment. *Figure is adapted from Toor SM, et al. Immune checkpoints in the tumor microenvironment. Semin Cancer Biol 2020; 65:1–12.*

expressed by CAFs) positive cells resulted in the hypoxic necrosis of stromal and tumor cells [89]. In support of this, another study showed that CAFs expressing FAP can act as a compensatory resistance mechanism against anti–CTLA4 and anti–PD–L1 mAbs in PDAC mouse model [204]. Overall, these findings implicate the therapeutic potential of depleting CAFs in cancer to alleviate resistance mechanisms and maximize clinical benefits of cancer immunotherapies.

4.2.2 Suppressive immune cells and cytokines

Immune cells within the TME can have both pro- and antitumor functions [205]; hence, they can determine the level of tumor immunogenicity and response to cancer immunotherapy [55, 206, 207]. Tumor cells and immune cells within the TME can cooperatively work together to induce the generation and activation of CAFs via the release of cytokines and growth factors (such as IL-6, IL-1β, VEGF, and TGF-β), which in turn negatively modulate antitumor immunity and may lead to effector T-cell apoptosis

[23, 208, 209]. Furthermore, the suppressive milieu of the TME shaped by the types of infiltrated immune cells can promote the recruitment and favor the accumulation of MDSCs in tumor sites [23]. MDSCs, in turn, inhibit effector T-cell responses by various means, such as limiting the availability of tryptophan by the overexpression of IDO enzyme, production of reactive oxygen species, and release of suppressive molecules, for example TGF-β, which support Treg function and survival [210–213]. Additionally, MDSCs can facilitate tumor resistance to therapy; reports have shown that intratumor depletion of MDSCs can restore the efficacy of anti-PD-1 therapy and BRAF inhibition [214, 215]. Moreover, gene signature derived from melanoma patients showing primary resistance to anti-PD-1 therapy comprised of genes involved in immunosuppression, EMT, angiogenesis, and monocytes/MDSC chemotaxis [216], indicating that MDSCs could be also associated with the induction of primary resistance against ICIs.

Within the TME, Tregs are other immune cell subsets with suppressive properties can suppress effector T-cell functions and may trigger effector T-cell apoptosis leading to impaired antitumor immunity [217, 218]. Importantly, Tregs can contribute to acquired resistance and limit the effectiveness of ICIs against cancer [23]. Studies on mouse tumor models showed that targeting Tregs in combination with ICIs, such as anti-PD-1/PD-L1 mAbs, can improve response rates to cancer immunotherapy and prolong survival [219–221].

Tregs can express upregulated levels of TGF-β and IL-10 and inhibitory ICs, which in turn inhibit the activation of effector T-cells and favor Treg survival and suppressive function [10, 55, 222]. Tregs also express high levels of ectoenzymes (CD73 and CD39) leading to the accumulation of high levels of extracellular adenosine, which induce effector T-cell apoptosis or cell-cycle arrest and promote Treg function [223]. Tregs can also release high levels of perforin and granzyme B to mediate cytotoxic effects on effector T-cells leading to reduced effector T-cell:Treg ratio within the TME [23, 55]. Additionally, Tregs consume high levels of IL-2 leading to the deprivation of effector T-cells and apoptosis [224, 225]. It has been reported that high consumption of IL-2 by intratumoral Tregs can restrict the efficacy of adoptively transferred T-cells, as well as effector T-cells present in the TME (216). Conversely, the utilization of engineered CARTs, which are not capable of producing IL-2, in mouse melanoma models was shown to be effective in diminishing Treg infiltrate within the TME, thereby leading to prolonged, improved clinical outcomes, and antitumor immune responses (217).

4.2.3 Upregulation of co-inhibitory immune checkpoints

There is a high likelihood for the emergence of acquired resistance post ACT therapy in cancer patients, as a result of increased expression of ICs on their cell surface leading to deactivation, exhaustion, and/or apoptosis [24, 29, 226]. In support of this, it has been demonstrated that adoptively transferred TCR-modified T-cells in mouse tumor models can express higher levels of PD-1 upon their migration and accumulation in tumor sites [40]. Alternatively, direct contact between MDSCs/TAMs and tumor-infiltrating

T-lymphocytes, including effector T-cells and adoptively transferred T-cells, via the interaction between ICs and IC ligands could compromise the activation and proliferation of T-cells [212, 227].

Apart from CTLA-4 and PD-1, blocking other ICs such as lymphocyte activation gene-3 (LAG-3), T-cell immunoglobulin and mucin-domain containing-3 (TIM-3), T-cell immunoglobulin and ITIM domain (TIGIT) and V-domain Ig suppressor of T-cell activation (VISTA) among others, are being investigated in several preclinical studies and/or clinical trials [228]. These ICs are also overexpressed in the TME of various malignancies and have been shown to correlate with disease progression [98, 229, 230].

Studies have shown that blocking one IC or IC ligand can lead to the upregulation of alternative ICs and IC ligands, leading to acquired resistance against therapy. For instance, *in vitro*, breast cancer cell lines cocultured with activated PBMCs in the presence of anti-PD-1/PD-L1 mAbs showed elevated levels of TIM-3 and LAG-3 coexpression on CD4$^+$ T-cell subsets, including Tregs, suggesting the emergence of compensatory inhibitory mechanism of acquired resistance in breast cancer following PD-1/PD-L1 blockade [231]. *In vivo*, acquired resistance to anti-PD-1 mAb therapy in patients and mouse model with lung cancer have been associated with elevated levels of TIM-3 ligand, galectin-9, and CD4$^+$ and CD8$^+$ T-cells expressing TIM-3, which potentially can impair the function of antitumor immune responses upon TIM-3/galaectin-9 interaction [232, 233]. In head and neck squamous cell carcinoma (HNSCC) mouse model, elevated expression of TIM-3 on intratumoral CD8$^+$ T-cells was detected following anti-PD-1 mAb treatment [234], while the overexpression of TIM-3 on Tregs and CD8$^+$ T-cells and Tregs in head and neck mouse tumor models was implicated in the emergence of acquired resistance to PD-L1 inhibition [221].

The overexpression of LAG-3 on CD8$^+$ T-cells has been detected in ovarian cancer mouse model following the blockade of CTLA-4 or PD-1 mAb, while the combined blockade of LAG-3 either with PD-1 or CTLA-4 showed beneficial effects evident by elevated levels of CD8$^+$ T-cells and reduced levels of Tregs within tumor sites [235]. Another study showed that increased expression of TIGIT could serve as a resistance mechanism to ICIs, evident by improved clinical outcomes post the inhibition of TIGIT in mouse tumors, which showed resistance to anti-PD-1 mAb [236]. Furthermore, increased expression of VISTA on M2 macrophages and TILs from prostate cancer patients treated with anti-CTLA-4 therapy, implicating that VISTA overexpression could be an acquired immunosuppression mechanism [237]. The interaction between VISTA and its ligand can suppress the function of effector T-cells [238], and induce Treg differentiation [239], while the blockade of VISTA was effective in promoting effector T-cell activation/accumulation within mouse tumors and reducing the number of MDSC and Tregs in the tumor [240]. Another study showed that the co-blockade of PD-L1 and VISTA in mouse tumor models can enhance the antitumor responses and improve disease outcomes compared to their single inhibition [241].

5. Therapeutic approaches to overcome resistance

5.1 Predictive biomarkers for successful immune checkpoint inhibition

In addition to utilization of ICIs as therapeutic agents, the use of predictive biomarkers to identify patients for successful ICI-based therapies has also gained great interest in recent years [154]. Tumor-based predictive biomarkers for ICI primarily include immune cell infiltrates, protein expression, presence of neoantigens/tumor burden, and gene/epigenetic signatures [154]. Chen *et al.* proposed classifying the TME into three major categories termed immune desert, immune excluded, and immune inflamed, based on immune cell infiltration [242]. Tumors with high immune cell infiltration respond better to ICIs mainly due to the presence of additional targets on T-cells, ensuring their sustained activation for potent antitumor immunity; therefore, some agents are also being designed to skew the TME toward an immune inflamed environment [243]. Similarly, elevated expression of protein targets of ICI such as IC ligands in the TME also potentially signifies improved response to IC inhibition. For instance, PD-L1 expression in the TME of NSCLC patients was approved by the FDA as a prerequisite for initiation of anti-PD-1 therapy [244]. However, studies have shown that PD-L1-negative tumors can also respond to blockade along PD-1/PD-L1 axis [245]. These findings suggest that investigating additional biomarkers may be required to ascertain response to IC-based therapies. Transcriptomic and epigenetic signatures could potentially aid as predictive biomarkers in such scenarios. Several groups have identified multiple dysregulated/mutated or silenced gene signatures across different cancers, which could potentially predict response to therapy; however, none have been recognized as robust predictive markers and it is believed that multiple biomarkers may need to be considered for accurate prediction of patient selection for effective response to IC inhibition [246].

5.2 ICIs and adoptive T-cell therapy

The safety and efficacy of anti-VISTA mAb, anti-TIGIT mAb, anti-LAG-3 mAb, and anti-TIM-3 mAb, in various cancer patients are under clinical trials [as reviewed in [23]]. Evidence from preclinical models showed promising clinical outcomes and durable immune responses upon the inhibition of multiple ICs, and could be used as a therapeutic approach to alleviate acquired resistance caused by the upregulation of ICs (Fig. 5A and C). Furthermore, the utilization of ACT in combination with ICIs, such as anti-PD-1 and anti-CTLA-4 therapies [212, 227], or the application of CARTs with mutated forms of ICs using CRISPR/Cas9 technology [247, 248] could improve clinical efficacy than monotherapy. Other engineering approaches could be applied to alleviate the PD-L1-mediated T-cell exhaustion by replacing the intracellular domains of PD-1 with that of CD28 to exchange the inhibitory signal with a costimulatory signal leading to the enhanced activation of CARTs and cytokine release [249].

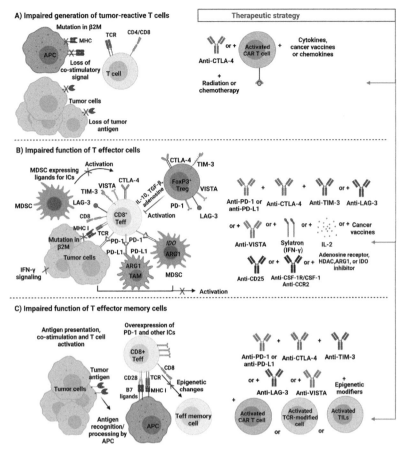

Fig. 5 Mechanisms of acquired resistance and potential therapeutic approaches. Impaired generation of tumor-reactive T-cells could be solved by the application of ICIs with ACT (CARTs) combined with cytokines, chemokines, TLR agonists or vaccines to restore and boost the antitumor immune response. The combined use of immunotherapy and radiation or chemotherapy could also enhance the clinical benefit and maximize tumor cytotoxicity (A). Inadequate activation of effector T-cells (Teffs) could occur as a result of metabolic mediators, increased levels of immunosuppressive cells/molecules and upregulation of alternate ICIs. In this scenario, combining ICIs with cytokines, vaccines, depleting Tregs, MDSCs or TAMs, or targeting adenosine, ARG1, IDO, or HDAC isoforms could restore APC function and the activation/proliferation of Teffs, and reduce immunosuppression (B). Impaired generation of Teff memory cells within the TME results in a loss of durability associated with T-cell exhaustion and epigenetic changes. The use of multiple ICIs with ACT and epigenetic modifiers could augment the activation/proliferation and function of cytotoxic T-cells and promote the generation of Teff memory T-cells (C). *Figure is adapted from Saleh R, Elkord E. Acquired resistance to cancer immunotherapy: role of tumor-mediated immunosuppression. Semin Cancer Biol 2020; 65:3–27.*

5.3 Targeting immunosuppressive cells in combination with ICIs or adoptive T-cell therapy

Combined targeting of immunosuppressive cells and their derived molecules with ICIs or ACT could offer promising clinical outcomes to enhance tumor sensitivity to therapy and prolong survival rates in cancer patients who did not benefit from ICIs or ACT alone and developed acquired resistance (Fig. 5B).

Targeting chemokine receptors (for example CXCR2 and CCR2) to reduce the trafficking and infiltration of MDSCs and TAMs in the TME [250, 251] or blocking colony-stimulating factor-1 (CSF-1) or its receptor (CSF-1R) [252, 253] to inhibit their survival and accumulation within the TME could expand the therapeutic efficacy, if used in combination with immunotherapies. *In vivo*, the co-blockade of CSF-1R and PD-1/PD-L1 reduced the levels of MDSCs/TAMs in tumors, increased the number of activated effector T-cells and led to superior clinical outcomes, unlike the blockade of PD-1/PD-L1 alone [254]. In another study, it was shown that the inhibition of CSF-1/CSF-1R signaling in mouse tumor models could reprogram MDSCs and alter their suppressive function and enhance the clinical outcomes in combination with anti-CTLA-4 mAb [252]. Moreover, the combination of anti-PD-1 mAb and class IIa histone deacetylase (HDAC) inhibitor improved the survival and therapeutic benefits of anti-PD-1 mAb and caused transformation of TAMs to M1 macrophages [255].

The efficacy of ICIs along with drugs targeting MDSCs is under clinical investigations [256]. Alternatively, targeting Tregs via anti-CD25 mAb [219] or via anti-CCR4 mAb [257, 258] in combination with ICIs could boost the antitumor responses induced by cancer immunotherapies. Combined inhibition of ICs and the use of inhibitors against suppressive molecules, such as indoleamine-pyrrole 2,3-dioxygenase (IDO), expressed by myeloid cells and tumor cells [41, 259], arginase 1 (ARG1), expressed by MDSCs and TAMs [260], and components of the adenosine pathway (mediated by suppressive immune cells and tumor cells) [259] could also expand the sensitivity to ICIs by suppressing Treg differentiation and inducing effector T-cell activation and proliferation [41, 259] (Fig. 5B).

5.4 ICIs or adoptive cell transfer in combination with adjuvant therapies

The effectivity of TCR-modified T-cell therapy is dependent on APC function; effective antigen processing and presentation to TCR-modified T-cells is required for TCR-mediated signaling and antitumor immune response [24]. Cell-to-cell interactions between intratumoral TCR T-cells and TAMs can impair tumor-specific T-cell responses facilitating tumorigenesis [261]. Ultimately, chronic recruitment and accumulation of TAMs within the TME could induce the development of acquired resistance. Such resistance could be resolved by the administration of human recombinant cytokines, such as GM-CSF, as an adjuvant therapy in combination with adoptive TCR-modified T-cell therapy to alter the phenotype of TAMs and enhance T-cell activation and function [262, 263]. Other adjuvant therapies that could be combined with ACT or ICIs are recombinant IL-2 and IFN-α2b (Sylatron) [264, 265] (Fig. 5B); this could maximize the clinical outcomes in patients and sustain effective immune responses mediated by exogenous and endogenous T-cells [264, 265]. Furthermore, cancer vaccines involving the administration of tumor-derived antigens and pulsed DCs could be utilized in combination with ACT to effectively suppress tumor growth/ invasion in cancer patients [266].

5.5 Targeting CAFs and ECM components using CAR T-cell therapy

CAFs is another promising therapeutic target for cancer, which could enhance the sensitivity of immune response to ICIs in patients with acquired resistance; for instance, TGF-β produced by CAFs has been linked with tumor resistance against anti-PD-1 therapy in metastatic urothelial cancer [267]. Indeed, the utilization of CARTs expressing FAP, have shown promising clinical benefits in a mouse tumor model [268]. In light of this, CARTs expressing FAP underwent clinical trial (NCT01722149) in cancer patients.

Furthermore, the application of modified T-cells engrafted with an echistatin-containing CAR (T-eCAR) specific for αvβ3 integrin that is expressed on TECs, have shown promising outcomes associated with enhanced capacity of T-cell trafficking into the TME and improved T-cell function [269]. Similarly, the effectivity of CARTs specific for αvβ6 integrin that is expressed on breast, ovarian, and pancreatic tumors has been shown *in vivo* [270]. The administration CARTs expressing ECM components could enhance the trafficking of CARTs to the TME and their activity [271, 272].

5.6 Epigenetic modifiers and ICIs

Epigenetic modifications have an indispensable role in the development and progression of cancer, as discussed in Section 4.1.3. Pharmacological inhibitors, namely epigenetic modifiers, could revert these modifications and restore the host antitumor responses [36, 273]. Accumulating evidences suggest that the transcriptional regulation of multiple ICs and IC ligands including PD-1, CTLA-4, TIM-3, LAG-3, TIGIT, and PD-L1 are controlled by DNA methylation and posttranslational histone modifications on their promoter regions [36]. These data suggest the potential application of epigenetic modifiers together with ICIs or ACT as more effective immunotherapeutic modalities (Fig. 5C).

DNA methyltransferase (DNMT), enzyme responsible for DNA methylation, could be blocked by DNMT inhibitors (DNMTi) including 5-azacitidine and 5-aza-2′-deoxycytidine and restore normal methylation profile on the promoter regions of key tumor suppression genes [274]. It has been reported that DNMTi could enhance the outcome of anti-CTLA-4 therapy by restoring the key hypermethylated genes in melanoma models [275]. On the other hand, in myelodysplastic syndrome, demethylating agents lead to the transcriptional upregulation of PD-1, PD-L1, and CTLA-4, which could be beneficial for targeting aforementioned genes using appropriate checkpoint inhibitors [276]. Interestingly, multiple clinical trials have been initiated to measure the effectiveness of combinatorial therapeutic strategies using DNA demethylating agents and ICIs in various cancers [36].

Similar to DNA methyl modifiers, histone acetylation modifiers including HDAC inhibitor (HDACi), can enhance the efficacy of ICIs by upregulating corresponding ICs and tumor-associated antigens on malignant cells leading to increased sensitivity to therapy and production of cytokines/chemokines by TILs, and reduced infiltration of immunosuppressive cell populations within the TME [277]. Additionally, HDACi treatment prior to immunotherapy could enhance the efficacy of treatment [277]. For instance, HDACi treatment on melanoma preclinical model upregulated the expression of PD-L1 and PD-L2 and improved the PD-1 blockade therapy [278]. Moreover, in colon and breast tumor models, the efficacy of combinatorial inhibition of PD-L1 and CTLA-4 was enhanced by HDACi, entinostat [279]. Furthermore, combination of atezolizumab and entinostat has being tested in phase I/II clinical trials for TNBC patients (NCT02708680). Likewise, clinical trials of pembrolizumab and vorinostat are in phase I/II for advanced NSCLC (NCT02638090). In concordance with PD-1/PD-L1 interaction, CD80/86 and CTLA-4 interaction was also significantly influenced by HDACi [280]. It has been reported that HDACi could upregulate CD80/86 in the TME and significantly enhance the efficacy of anti-CTLA-4 targeting [281]. Combination of ipilimumab/entinostat for breast cancer (NCT02453620) and ipilimumab/panobiostat for advanced melanoma (NCT02032810) are under clinical trials.

6. Conclusions, challenges, and future perspectives

Tumors are heterogenous and plastic and their molecular profiles can vary with cancer types and disease stages; therefore, it is imperative to stratify cancer patients and treat them accordingly using a personalized-medicine approach. Moreover, therapeutic strategies should be applied in logical and sequential manner to avoid any possible adverse immune-related and toxic events depending on the natural history of disease for cancer patients and the capacity of their antitumor immunity acquired within the TME.

The molecular and cellular components of the TME are the determining factors for tumor immunogenicity and can predict response to therapy and clinical outcomes in cancer patients. The presence of high tumor burdens and high levels of immune cell infiltrate within the TME make tumors more vulnerable to cancer immunotherapy and regression. Nonetheless, tumor cells can gain the ability to escape immune cell recognition via intrinsic and extrinsic mechanisms and alter their transcriptional profile/molecular signature to promote their own survival, mediate immunosuppression, and induce primary or acquired resistance to cancer therapies. Furthermore, the crosstalk between tumor and stromal cells within the TME can result in a positive feedback loop, which favors the recruitment and activation of suppressive immune cells, overexpression of inhibitory ICs and suppression of effector immune cells, and promote the induction of CAFs, thereby leading to tumor growth, metastasis, and angiogenesis. Importantly, these cellular interactions within the TME can lead to the emergence of tumor resistance to therapy.

Some cancer patients show no response to immunotherapies, which is more likely due to primary resistance driven by genetic or epigenetic alterations or due to the exclusion of tumor-infiltrating T-cells from tumor core, known as cold or desert tumor. Cancer patients who initially respond to immunotherapies have the potential to develop acquired resistance over time. To overcome tumor resistance, immunotherapies, epigenetic modifiers, and small molecule inhibitors and/or mAbs, which specifically deplete immunosuppressive cells could be used in combination to promote the activation of antitumor immunity and eradicate tumor cells. Together, these combinatorial therapies could expand the efficacy of immunotherapies, and prolong antitumor immunity leading to favorable disease outcomes in cancer patients.

References

[1] Teng MW, et al. From mice to humans: developments in cancer immunoediting. J Clin Invest 2015;125(9):3338–46.

[2] Dunn GP, et al. Cancer immunoediting: from immunosurveillance to tumor escape. Nat Immunol 2002;3(11):991–8.

[3] Sade-Feldman M, et al. Resistance to checkpoint blockade therapy through inactivation of antigen presentation. Nat Commun 2017;8(1):1136.

[4] McGranahan N, et al. Allele-specific HLA loss and immune escape in lung cancer evolution. Cell 2017;171(6):1259–1271.e11.

[5] Rooney MS, et al. Molecular and genetic properties of tumors associated with local immune cytolytic activity. Cell 2015;160(1–2):48–61.

[6] Spranger S, Bao R, Gajewski TF. Melanoma-intrinsic beta-catenin signalling prevents anti-tumour immunity. Nature 2015;523(7559):231–5.

[7] Voron T, et al. Control of the immune response by pro-angiogenic factors. Front Oncol 2014;4:70.

[8] Jenkins RW, Barbie DA, Flaherty KT. Mechanisms of resistance to immune checkpoint inhibitors. Br J Cancer 2018;118(1):9–16.

[9] Yang ZZ, et al. Soluble and membrane-bound TGF-beta-mediated regulation of intratumoral T cell differentiation and function in B-cell non-Hodgkin lymphoma. PLoS One 2013;8(3), e59456.

[10] Taylor A, et al. Mechanisms of immune suppression by interleukin-10 and transforming growth factor-beta: the role of T regulatory cells. Immunology 2006;117(4):433–42.

[11] Vinay DS, et al. Immune evasion in cancer: Mechanistic basis and therapeutic strategies. Semin Cancer Biol 2015;35(Suppl):S185–s198.

[12] Xing F, Saidou J, Watabe K. Cancer associated fibroblasts (CAFs) in tumor microenvironment. Front Biosci (Landmark Ed) 2010;15:166–79.

[13] Ziani L, Chouaib S, Thiery J. Alteration of the antitumor immune response by cancer-associated fibroblasts. Front Immunol 2018;9:414.

[14] Pitt JM, et al. Targeting the tumor microenvironment: removing obstruction to anticancer immune responses and immunotherapy. Ann Oncol 2016;27(8):1482–92.

[15] Su S, et al. A positive feedback loop between mesenchymal-like cancer cells and macrophages is essential to breast cancer metastasis. Cancer Cell 2014;25(5):605–20.

[16] Bardhan K, Anagnostou T, Boussiotis VA. The PD1:PD-L1/2 pathway from discovery to clinical implementation. Front Immunol 2016;7:550.

[17] Borghaei H, et al. Nivolumab versus docetaxel in advanced nonsquamous non-small-cell lung cancer. N Engl J Med 2015;373(17):1627–39.

[18] Ferris RL, et al. Nivolumab vs investigator's choice in recurrent or metastatic squamous cell carcinoma of the head and neck: 2-year long-term survival update of CheckMate 141 with analyses by tumor PD-L1 expression. Oral Oncol 2018;81:45–51.

[19] Hodi FS, et al. Improved survival with ipilimumab in patients with metastatic melanoma. N Engl J Med 2010;363(8):711–23.

[20] Larkin J, Hodi FS, Wolchok JD. Combined nivolumab and ipilimumab or monotherapy in untreated melanoma. N Engl J Med 2015;373(13):1270–1.

[21] Topalian SL, et al. Safety, activity, and immune correlates of anti-PD-1 antibody in cancer. N Engl J Med 2012;366(26):2443–54.

[22] Seidel JA, Otsuka A, Kabashima K. Anti-PD-1 and anti-CTLA-4 therapies in cancer: mechanisms of action, efficacy, and limitations. Front Oncol 2018;8:86.

[23] Saleh R, Elkord E. Acquired resistance to cancer immunotherapy: role of tumor-mediated immunosuppression. Semin Cancer Biol 2020;65:13–27.

[24] Ping Y, Liu C, Zhang Y. T-cell receptor-engineered T cells for cancer treatment: current status and future directions. Protein Cell 2018;9(3):254–66.

[25] Kalos M, et al. T cells with chimeric antigen receptors have potent antitumor effects and can establish memory in patients with advanced leukemia. Sci Transl Med 2011;3(95), 95ra73.

[26] June CH. Adoptive T cell therapy for cancer in the clinic. J Clin Invest 2007;117(6):1466–76.

[27] Yang A, et al. The current state of therapeutic and T cell-based vaccines against human papillomaviruses. Virus Res 2017;231:148–65.

[28] June CH, et al. Engineered T cells for cancer therapy. Cancer Immunol Immunother 2014;63 (9):969–75.

[29] Filley AC, Henriquez M, Dey M. CART immunotherapy: development, success, and translation to malignant gliomas and other solid tumors. Front Oncol 2018;8:453.

[30] Morgan RA, et al. Cancer regression in patients after transfer of genetically engineered lymphocytes. Science 2006;314(5796):126–9.

[31] Robbins PF, et al. Tumor regression in patients with metastatic synovial cell sarcoma and melanoma using genetically engineered lymphocytes reactive with NY-ESO-1. J Clin Oncol 2011;29 (7):917–24.

[32] Parkhurst MR, et al. Characterization of genetically modified T-cell receptors that recognize the CEA:691-699 peptide in the context of HLA-A2.1 on human colorectal cancer cells. Clin Cancer Res 2009;15(1):169–80.

[33] Parkhurst MR, et al. T cells targeting carcinoembryonic antigen can mediate regression of metastatic colorectal cancer but induce severe transient colitis. Mol Ther 2011;19(3):620–6.

[34] Schober K, Busch DH. TIL 2.0: more effective and predictive T-cell products by enrichment for defined antigen specificities. Eur J Immunol 2016;46(6):1335–9.

[35] Stevanovic S, et al. A phase II study of tumor-infiltrating lymphocyte therapy for human papillomavirus-associated epithelial cancers. Clin Cancer Res 2019;25(5):1486–93.

[36] Saleh R, et al. Role of epigenetic modifications in inhibitory immune checkpoints in cancer development and progression. Front Immunol 2020;11:1469.

[37] Zaretsky JM, et al. Mutations associated with acquired resistance to PD-1 blockade in melanoma. N Engl J Med 2016;375(9):819–29.

[38] Sharma P, et al. Primary, adaptive, and acquired resistance to cancer immunotherapy. Cell 2017; 168(4):707–23.

[39] O'Donnell JS, et al. Resistance to PD1/PDL1 checkpoint inhibition. Cancer Treat Rev 2017;52:71–81.

[40] Perez C, et al. Permissive expansion and homing of adoptively transferred T cells in tumor-bearing hosts. Int J Cancer 2015;137(2):359–71.

[41] Ninomiya S, et al. Tumor indoleamine 2,3-dioxygenase (IDO) inhibits CD19-CAR T cells and is downregulated by lymphodepleting drugs. Blood 2015;125(25):3905–16.

[42] Baghban R, et al. Tumor microenvironment complexity and therapeutic implications at a glance. Cell Commun Signal 2020;18(1):59.

[43] Nallanthighal S, Heiserman JP, Cheon DJ. The role of the extracellular matrix in cancer stemness. Front Cell Dev Biol 2019;7:86.

[44] Eiro N, et al. Breast cancer tumor stroma: cellular components, phenotypic heterogeneity, intercellular communication, prognostic implications and therapeutic opportunities. Cancers (Basel) 2019;11(5).

[45] Maes H, et al. Vesicular trafficking mechanisms in endothelial cells as modulators of the tumor vasculature and targets of antiangiogenic therapies. FEBS J 2016;283(1):25–38.

[46] Hill BS, et al. Recruitment of stromal cells into tumour microenvironment promote the metastatic spread of breast cancer. Semin Cancer Biol 2020;60:202–13.

[47] Binnewies M, et al. Understanding the tumor immune microenvironment (TIME) for effective therapy. Nat Med 2018;24(5):541–50.

[48] Galon J, et al. Type, density, and location of immune cells within human colorectal tumors predict clinical outcome. Science 2006;313(5795):1960–4.

[49] Fridman WH, et al. The immune contexture in cancer prognosis and treatment. Nat Rev Clin Oncol 2017;14(12):717–34.

[50] Bailey SR, et al. Th17 cells in cancer: the ultimate identity crisis. Front Immunol 2014;5:276.

[51] De Monte L, et al. Intratumor T helper type 2 cell infiltrate correlates with cancer-associated fibroblast thymic stromal lymphopoietin production and reduced survival in pancreatic cancer. J Exp Med 2011;208(3):469–78.

[52] Sasidharan Nair V, Elkord E. Immune checkpoint inhibitors in cancer therapy: a focus on T-regulatory cells. Immunol Cell Biol 2018;96(1):21–33.

[53] Tan Z, et al. Virotherapy-recruited PMN-MDSC infiltration of mesothelioma blocks antitumor CTL by IL-10-mediated dendritic cell suppression. Onco Targets Ther 2019;8(1), e1518672.

[54] Zhang X, et al. The functional and prognostic implications of regulatory T cells in colorectal carcinoma. J Gastrointest Oncol 2015;6(3):307–13.

[55] Saleh R, Elkord E. FoxP3 + T regulatory cells in cancer: prognostic biomarkers and therapeutic targets. Cancer Lett 2020;490:174–85.

[56] Petitprez F, et al. B cells are associated with survival and immunotherapy response in sarcoma. Nature 2020;577(7791):556–60.

[57] DiLillo DJ, Yanaba K, Tedder TF. B cells are required for optimal CD4 + and CD8 + T cell tumor immunity: therapeutic B cell depletion enhances B16 melanoma growth in mice. J Immunol 2010;184(7):4006–16.

[58] Nedergaard BS, et al. A comparative study of the cellular immune response in patients with stage IB cervical squamous cell carcinoma. Low numbers of several immune cell subtypes are strongly associated with relapse of disease within 5 years. Gynecol Oncol 2008;108(1):106–11.

[59] Nielsen JS, et al. CD20 + tumor-infiltrating lymphocytes have an atypical CD27- memory phenotype and together with CD8 + T cells promote favorable prognosis in ovarian cancer. Clin Cancer Res 2012;18(12):3281–92.

[60] Svensson MC, et al. The integrative clinical impact of tumor-infiltrating T lymphocytes and NK cells in relation to B lymphocyte and plasma cell density in esophageal and gastric adenocarcinoma. Oncotarget 2017;8(42):72108–26.

[61] Ammirante M, et al. B-cell-derived lymphotoxin promotes castration-resistant prostate cancer. Nature 2010;464(7286):302–5.

[62] de Visser KE, Korets LV, Coussens LM. De novo carcinogenesis promoted by chronic inflammation is B lymphocyte dependent. Cancer Cell 2005;7(5):411–23.

[63] Levy EM, Roberti MP, Mordoh J. Natural killer cells in human cancer: from biological functions to clinical applications. J Biomed Biotechnol 2011;2011:676198.

[64] Shembrey C, Huntington ND, Hollande F. Impact of tumor and immunological heterogeneity on the anti-cancer immune response. Cancers (Basel) 2019;11(9).

[65] Glasner A, et al. NKp46 receptor-mediated interferon-gamma production by natural killer cells increases fibronectin 1 to alter tumor architecture and control metastasis. Immunity 2018;48 (1):107–119.e4.

[66] Bassani B, et al. Natural killer cells as key players of tumor progression and angiogenesis: old and novel tools to divert their pro-tumor activities into potent anti-tumor effects. Cancers (Basel) 2019;11(4).

[67] Hashimoto D, Miller J, Merad M. Dendritic cell and macrophage heterogeneity in vivo. Immunity 2011;35(3):323–35.

[68] Martinek J, et al. Interplay between dendritic cells and cancer cells. Int Rev Cell Mol Biol 2019;348:179–215.

[69] Zhou J, et al. Tumor-associated macrophages: recent insights and therapies. Front Oncol 2020;10:188.

[70] Mantovani A, et al. Tumour-associated macrophages as treatment targets in oncology. Nat Rev Clin Oncol 2017;14(7):399–416.

[71] Murray PJ, et al. Macrophage activation and polarization: nomenclature and experimental guidelines. Immunity 2014;41(1):14–20.

[72] Toor SM, et al. Myeloid cells in circulation and tumor microenvironment of colorectal cancer patients with early and advanced disease stages. J Immunol Res 2020;2020:9678168.

[73] Gabrilovich DI. Myeloid-derived suppressor cells. Cancer Immunol Res 2017;5(1):3–8.

[74] OuYang LY, et al. Tumor-induced myeloid-derived suppressor cells promote tumor progression through oxidative metabolism in human colorectal cancer. J Transl Med 2015;13:47.

[75] Sistigu A, et al. Deciphering the loop of epithelial-mesenchymal transition, inflammatory cytokines and cancer immunoediting. Cytokine Growth Factor Rev 2017;36:67–77.

[76] Rosales C. Neutrophil: a cell with many roles in inflammation or several cell types? Front Physiol 2018;9:113.

[77] Lecot P, et al. Neutrophil heterogeneity in cancer: from biology to therapies. Front Immunol 2019;10:2155.

[78] Fridlender ZG, et al. Polarization of tumor-associated neutrophil phenotype by TGF-beta: "N1" versus "N2" TAN. Cancer Cell 2009;16(3):183–94.

[79] Mishalian I, et al. Tumor-associated neutrophils (TAN) develop pro-tumorigenic properties during tumor progression. Cancer Immunol Immunother 2013;62(11):1745–56.

[80] Denton AE, Roberts EW, Fearon DT. Stromal cells in the tumor microenvironment. Adv Exp Med Biol 2018;1060:99–114.

[81] Henke E, Nandigama R, Ergun S. Extracellular matrix in the tumor microenvironment and its impact on cancer therapy. Front Mol Biosci 2019;6:160.

[82] Walker C, Mojares E, Del Rio Hernandez A. Role of extracellular matrix in development and cancer progression. Int J Mol Sci 2018;(10):19.

[83] Dudley AC. Tumor endothelial cells. Cold Spring Harb Perspect Med 2012;2(3):a006536.

[84] Aird WC. Endothelial cell heterogeneity. Crit Care Med 2003;31(4 Suppl):S221–30.

[85] Nagl L, et al. Tumor endothelial cells (TECs) as potential immune directors of the tumor microenvironment—new findings and future perspectives. Front Cell Dev Biol 2020;8(766).

[86] Biffi G, Tuveson DA. Deciphering cancer fibroblasts. J Exp Med 2018;215(12):2967–8.

[87] Gascard P, Tlsty TD. Carcinoma-associated fibroblasts: orchestrating the composition of malignancy. Genes Dev 2016;30(9):1002–19.

[88] Shintani Y, et al. IL-6 secreted from cancer-associated fibroblasts mediates chemoresistance in NSCLC by increasing epithelial-mesenchymal transition signaling. J Thorac Oncol 2016;11(9):1482–92.

[89] Kraman M, et al. Suppression of antitumor immunity by stromal cells expressing fibroblast activation protein-alpha. Science 2010;330(6005):827–30.

[90] Ribeiro AL, Okamoto OK. Combined effects of pericytes in the tumor microenvironment. Stem Cells Int 2015;2015:868475.

[91] Bergers G, Song S. The role of pericytes in blood-vessel formation and maintenance. Neuro Oncol 2005;7(4):452–64.

[92] Ozerdem U, Stallcup WB. Early contribution of pericytes to angiogenic sprouting and tube formation. Angiogenesis 2003;6(3):241–9.

[93] Stapor PC, et al. Pericyte dynamics during angiogenesis: new insights from new identities. J Vasc Res 2014;51(3):163–74.

[94] Sharma P, et al. Novel cancer immunotherapy agents with survival benefit: recent successes and next steps. Nat Rev Cancer 2011;11(11):805–12.

[95] Pardoll DM. The blockade of immune checkpoints in cancer immunotherapy. Nat Rev Cancer 2012;12(4):252–64.

[96] Whiteside TL. The tumor microenvironment and its role in promoting tumor growth. Oncogene 2008;27(45):5904–12.

[97] Wherry EJ, Kurachi M. Molecular and cellular insights into T cell exhaustion. Nat Rev Immunol 2015;15(8):486–99.

[98] Toor SM, et al. Immune checkpoints in the tumor microenvironment. Semin Cancer Biol 2020;65:1–12.

[99] Sharma P, Allison JP. Immune checkpoint targeting in cancer therapy: toward combination strategies with curative potential. Cell 2015;161(2):205–14.

[100] Leach DR, Krummel MF, Allison JP. Enhancement of antitumor immunity by CTLA-4 blockade. Science 1996;271(5256):1734–6.

[101] Alsaab HO, et al. PD-1 and PD-L1 checkpoint signaling inhibition for cancer immunotherapy: mechanism, combinations, and clinical outcome. Front Pharmacol 2017;8:561.

[102] Vaddepally RK, et al. Review of indications of FDA-approved immune checkpoint inhibitors per NCCN guidelines with the level of evidence. Cancers (Basel) 2020;12(3).

[103] Robert C, et al. Pembrolizumab versus ipilimumab in advanced melanoma. N Engl J Med 2015;372 (26):2521–32.

[104] Garon EB, et al. Pembrolizumab for the treatment of non-small-cell lung cancer. N Engl J Med 2015;372(21):2018–28.

[105] Herbst RS, et al. Pembrolizumab versus docetaxel for previously treated, PD-L1-positive, advanced non-small-cell lung cancer (KEYNOTE-010): a randomised controlled trial. Lancet 2016;387(10027): 1540–50.

[106] Marcus L, et al. FDA approval summary: pembrolizumab for the treatment of microsatellite instability-high solid tumors. Clin Cancer Res 2019;25(13):3753–8.

[107] Andre T, et al. Pembrolizumab versus chemotherapy for microsatellite instability-high/mismatch repair deficient metastatic colorectal cancer: the phase 3 KEYNOTE-177 Study. J Clin Oncol 2020;38(18_suppl):LBA4.

[108] Egen JG, Kuhns MS, Allison JP. CTLA-4: new insights into its biological function and use in tumor immunotherapy. Nat Immunol 2002;3(7):611–8.

[109] Rudd CE, Taylor A, Schneider H. CD28 and CTLA-4 coreceptor expression and signal transduction. Immunol Rev 2009;229(1):12–26.

[110] Wing K, et al. CTLA-4 control over Foxp3+ regulatory T cell function. Science 2008;322 (5899):271–5.

[111] Fong L, Small EJ. Anti-cytotoxic T-lymphocyte antigen-4 antibody: the first in an emerging class of immunomodulatory antibodies for cancer treatment. J Clin Oncol 2008;26(32):5275–83.

[112] Rowshanravan B, Halliday N, Sansom DM. CTLA-4: a moving target in immunotherapy. Blood 2018;131(1):58–67.

[113] Lipson EJ, Drake CG. Ipilimumab: an anti-CTLA-4 antibody for metastatic melanoma. Clin Cancer Res 2011;17(22):6958–62.

[114] Larkin J, et al. Combined nivolumab and ipilimumab or monotherapy in untreated melanoma. N Engl J Med 2015;373(1):23–34.

[115] Snyder A, et al. Genetic basis for clinical response to CTLA-4 blockade in melanoma. N Engl J Med 2014;371(23):2189–99.

[116] Riella LV, et al. Role of the PD-1 pathway in the immune response. Am J Transplant 2012; 12(10):2575–87.

[117] Keir ME, et al. Tissue expression of PD-L1 mediates peripheral T cell tolerance. J Exp Med 2006;203 (4): 883–95.

[118] Tumeh PC, et al. PD-1 blockade induces responses by inhibiting adaptive immune resistance. Nature 2014;515(7528):568–71.

[119] Yeong J, et al. Prognostic value of CD8 + PD-1 + immune infiltrates and PDCD1 gene expression in triple negative breast cancer. J Immunother Cancer 2019;7(1):34.

[120] Brahmer JR, et al. Safety and activity of anti-PD-L1 antibody in patients with advanced cancer. N Engl J Med 2012;366(26):2455–65.

[121] Xiang X, et al. Prognostic value of PD -L1 expression in patients with primary solid tumors. Oncotarget 2018;9(4):5058–72.

[122] Wang Q, Liu F, Liu L. Prognostic significance of PD-L1 in solid tumor: an updated meta-analysis. Medicine (Baltimore) 2017;96(18), e6369.

[123] Lee HT, Lee SH, Heo YS. Molecular interactions of antibody drugs targeting PD-1, PD-L1, and CTLA-4 in immuno-oncology. Molecules 2019;24(6).

[124] Akinleye A, Rasool Z. Immune checkpoint inhibitors of PD-L1 as cancer therapeutics. J Hematol Oncol 2019;12(1):92.

[125] Till BG, et al. Adoptive immunotherapy for indolent non-Hodgkin lymphoma and mantle cell lymphoma using genetically modified autologous CD20-specific T cells. Blood 2008;112(6):2261–71.

[126] Turtle CJ, et al. CD19 CAR-T cells of defined CD4+:CD8+ composition in adult B cell ALL patients. J Clin Invest 2016;126(6):2123–38.

[127] Porter DL, et al. Chimeric antigen receptor T cells persist and induce sustained remissions in relapsed refractory chronic lymphocytic leukemia. Sci Transl Med 2015;7(303):303ra139.

[128] Maude SL, et al. Chimeric antigen receptor T cells for sustained remissions in leukemia. N Engl J Med 2014;371(16):1507–17.

[129] Garfall AL, et al. Chimeric antigen receptor T cells against CD19 for multiple myeloma. N Engl J Med 2015;373(11):1040–7.

[130] Locke FL, et al. Phase 1 results of ZUMA-1: a multicenter study of KTE-C19 anti-CD19 CAR T cell therapy in refractory aggressive lymphoma. Mol Ther 2017;25(1):285–95.

[131] Kochenderfer JN, et al. Chemotherapy-refractory diffuse large B-cell lymphoma and indolent B-cell malignancies can be effectively treated with autologous T cells expressing an anti-CD19 chimeric antigen receptor. J Clin Oncol 2015;33(6):540–9.

[132] Anon. First-ever CAR T-cell therapy approved in U.S. Cancer Discov 2017;7(10):OF1.

[133] Johnson LA, et al. Rational development and characterization of humanized anti-EGFR variant III chimeric antigen receptor T cells for glioblastoma. Sci Transl Med 2015;7(275), 275ra22.

[134] Miao H, et al. EGFRvIII-specific chimeric antigen receptor T cells migrate to and kill tumor deposits infiltrating the brain parenchyma in an invasive xenograft model of glioblastoma. PLoS One 2014;9(4), e94281.

[135] O'Rourke DM, et al. A single dose of peripherally infused EGFRvIII-directed CAR T cells mediates antigen loss and induces adaptive resistance in patients with recurrent glioblastoma. Sci Transl Med 2017;9(399).

[136] Rapoport AP, et al. NY-ESO-1-specific TCR-engineered T cells mediate sustained antigen-specific antitumor effects in myeloma. Nat Med 2015;21(8):914–21.

[137] Kageyama S, et al. Adoptive transfer of MAGE-A4 T-cell receptor gene-transduced lymphocytes in patients with recurrent esophageal cancer. Clin Cancer Res 2015;21(10):2268–77.

[138] Berraondo P, et al. Cytokines in clinical cancer immunotherapy. Br J Cancer 2019;120(1):6–15.

[139] Fyfe G, et al. Results of treatment of 255 patients with metastatic renal cell carcinoma who received high-dose recombinant interleukin-2 therapy. J Clin Oncol 1995;13(3):688–96.

[140] Atkins MB, et al. High-dose recombinant interleukin 2 therapy for patients with metastatic melanoma: analysis of 270 patients treated between 1985 and 1993. J Clin Oncol 1999;17(7):2105–16.

[141] Golomb HM, et al. Alpha-2 interferon therapy of hairy-cell leukemia: a multicenter study of 64 patients. J Clin Oncol 1986;4(6):900–5.

[142] Solal-Celigny P, et al. Recombinant interferon alfa-2b combined with a regimen containing doxorubicin in patients with advanced follicular lymphoma. Groupe d'Etude des Lymphomes de l'Adulte. N Engl J Med 1993;329(22):1608–14.

[143] Kirkwood JM, et al. Interferon alfa-2b adjuvant therapy of high-risk resected cutaneous melanoma: the Eastern Cooperative Oncology Group Trial EST 1684. J Clin Oncol 1996;14(1):7–17.

[144] Zhang S, et al. Systemic interferon-γ increases MHC class I expression and T-cell infiltration in cold tumors: results of a phase 0 clinical trial. Cancer Immunol Res 2019;7(8):1237–43.

[145] Buchbinder EI, et al. Therapy with high-dose Interleukin-2 (HD IL-2) in metastatic melanoma and renal cell carcinoma following PD1 or PDL1 inhibition. J Immunother Cancer 2019;7(1):49.

[146] Rahimi Kalateh Shah Mohammad G, et al. Cytokines as potential combination agents with PD-1/PD-L1 blockade for cancer treatment. J Cell Physiol 2020;235(7–8):5449–60.

[147] Schlom J, et al. Therapeutic cancer vaccines. Adv Cancer Res 2014;121:67–124.

[148] Banchereau J, et al. Immune and clinical responses in patients with metastatic melanoma to CD34(+) progenitor-derived dendritic cell vaccine. Cancer Res 2001;61(17):6451–8.

[149] Okada H, et al. Induction of CD8 + T-cell responses against novel glioma-associated antigen peptides and clinical activity by vaccinations with {alpha}-type 1 polarized dendritic cells and polyinosinic-polycytidylic acid stabilized by lysine and carboxymethylcellulose in patients with recurrent malignant glioma. J Clin Oncol 2011;29(3):330–6.

[150] Wheeler CJ, Black KL. Vaccines for glioblastoma and high-grade glioma. Expert Rev Vaccines 2011;10(6):875–86.

[151] Kantoff PW, et al. Sipuleucel-T immunotherapy for castration-resistant prostate cancer. N Engl J Med 2010;363(5):411–22.

[152] Disis ML. Enhancing cancer vaccine efficacy via modulation of the tumor microenvironment. Clin Cancer Res 2009;15(21):6476–8.

[153] Gubin MM, et al. Checkpoint blockade cancer immunotherapy targets tumour-specific mutant antigens. Nature 2014;515(7528):577–81.

[154] Darvin P, et al. Immune checkpoint inhibitors: recent progress and potential biomarkers. Exp Mol Med 2018;50(12):1–11.

[155] Sucker A, et al. Genetic evolution of T-cell resistance in the course of melanoma progression. Clin Cancer Res 2014;20(24):6593–604.

[156] Marincola FM, et al. Escape of human solid tumors from T-cell recognition: molecular mechanisms and functional significance. Adv Immunol 2000;74:181–273.

[157] Shin DS, et al. Primary resistance to PD-1 blockade mediated by JAK1/2 mutations. Cancer Discov 2017;7(2):188–201.

[158] Gao J, et al. Loss of IFN-γ pathway genes in tumor cells as a mechanism of resistance to anti-CTLA-4 therapy. Cell 2016;167(2):397–404.e9.

[159] Shalabi H, et al. Sequential loss of tumor surface antigens following chimeric antigen receptor T-cell therapies in diffuse large B-cell lymphoma. Haematologica 2018;103(5):e215–8.

[160] Restifo NP, Smyth MJ, Snyder A. Acquired resistance to immunotherapy and future challenges. Nat Rev Cancer 2016;16(2):121–6.

[161] Anagnostou V, et al. Evolution of neoantigen landscape during immune checkpoint blockade in non-small cell lung cancer. Cancer Discov 2017;7(3):264–76.

[162] Lawrence MS, et al. Mutational heterogeneity in cancer and the search for new cancer-associated genes. Nature 2013;499(7457):214–8.

[163] Schumacher TN, Schreiber RD. Neoantigens in cancer immunotherapy. Science 2015;348(6230):69–74.

[164] Van Allen EM, et al. Genomic correlates of response to CTLA-4 blockade in metastatic melanoma. Science 2015;350(6257):207–11.

[165] Ansell SM, et al. PD-1 blockade with nivolumab in relapsed or refractory Hodgkin's lymphoma. N Engl J Med 2015;372(4):311–9.

[166] Green MR, et al. Integrative analysis reveals selective 9p24.1 amplification, increased PD-1 ligand expression, and further induction via JAK2 in nodular sclerosing Hodgkin lymphoma and primary mediastinal large B-cell lymphoma. Blood 2010;116(17):3268–77.

[167] Peng W, et al. Loss of PTEN promotes resistance to T cell–mediated immunotherapy. Cancer Discov 2016;6(2):202–16.

[168] Lastwika KJ, et al. Control of PD-L1 expression by oncogenic activation of the AKT-mTOR pathway in non-small Cell lung cancer. Cancer Res 2016;76(2):227–38.

[169] Parsa AT, et al. Loss of tumor suppressor PTEN function increases B7-H1 expression and immunoresistance in glioma. Nat Med 2007;13(1):84–8.

[170] Akbay EA, et al. Activation of the PD-1 pathway contributes to immune escape in EGFR-driven lung tumors. Cancer Discov 2013;3(12):1355–63.

[171] Casey SC, et al. MYC regulates the antitumor immune response through CD47 and PD-L1. Science 2016;352(6282):227–31.

[172] Dorand RD, et al. Cdk5 disruption attenuates tumor PD-L1 expression and promotes antitumor immunity. Science 2016;353(6297):399–403.

[173] Kataoka K, et al. Aberrant PD-L1 expression through 3'-UTR disruption in multiple cancers. Nature 2016;534(7607):402–6.

[174] Li M, et al. Interleukin-8 increases vascular endothelial growth factor and neuropilin expression and stimulates ERK activation in human pancreatic cancer. Cancer Sci 2008;99(4):733–7.

[175] Bobrovnikova-Marjon EV, et al. Expression of angiogenic factors vascular endothelial growth factor and interleukin-8/CXCL8 is highly responsive to ambient glutamine availability: role of nuclear factor-kappaB and activating protein-1. Cancer Res 2004;64(14):4858–69.

[176] Sharma S, Kelly TK, Jones PA. Epigenetics in cancer. Carcinogenesis 2010;31(1):27–36.

[177] O'Sullivan DE, et al. Epigenetic and genetic burden measures are associated with tumor characteristics in invasive breast carcinoma. Epigenetics 2016;11(5):344–53.

[178] Pan X, Zheng L. Epigenetics in modulating immune functions of stromal and immune cells in the tumor microenvironment. Cell Mol Immunol 2020;17(9):940–53.

[179] Darwiche N. Epigenetic mechanisms and the hallmarks of cancer: an intimate affair. Am J Cancer Res 2020;10(7):1954–78.

[180] Heninger E, Krueger TE, Lang JM. Augmenting antitumor immune responses with epigenetic modifying agents. Front Immunol 2015;6:29.

[181] Wu J, Shi H. Unlocking the epigenetic code of T cell exhaustion. Transl Cancer Res 2017;6(Suppl 2): S384–7.

[182] Ng JM, Yu J. Promoter hypermethylation of tumour suppressor genes as potential biomarkers in colorectal cancer. Int J Mol Sci 2015;16(2):2472–96.

[183] Bedford MT, van Helden PD. Hypomethylation of DNA in pathological conditions of the human prostate. Cancer Res 1987;47(20):5274–6.

[184] Wahlfors J, et al. Genomic hypomethylation in human chronic lymphocytic leukemia. Blood 1992; 80(8):2074–80.

[185] Lin CH, et al. Genome-wide hypomethylation in hepatocellular carcinogenesis. Cancer Res 2001; 61(10):4238–43.

[186] Kim YI, et al. Global DNA hypomethylation increases progressively in cervical dysplasia and carcinoma. Cancer 1994;74(3):893–9.

[187] Hakkarainen M, et al. Hypermethylation of calcitonin gene regulatory sequences in human breast cancer as revealed by genomic sequencing. Int J Cancer 1996;69(6):471–4.

[188] Ehrlich M. DNA methylation in cancer: too much, but also too little. Oncogene 2002; 21(35):5400–13.

[189] Putiri EL, Robertson KD. Epigenetic mechanisms and genome stability. Clin Epigenetics 2011; 2(2):299–314.

[190] Audia JE, Campbell RM. Histone modifications and cancer. Cold Spring Harb Perspect Biol 2016; 8(4):a019521.

[191] Chervona Y, Costa M. Histone modifications and cancer: biomarkers of prognosis? Am J Cancer Res 2012;2(5):589–97.

[192] Wan YCE, Liu J, Chan KM. Histone H3 mutations in cancer. Curr Pharmacol Rep 2018; 4(4):292–300.

[193] Nacev BA, et al. The expanding landscape of 'oncohistone' mutations in human cancers. Nature 2019;567(7749):473–8.

[194] Fraga MF, et al. Loss of acetylation at Lys16 and trimethylation at Lys20 of histone H4 is a common hallmark of human cancer. Nat Genet 2005;37(4):391–400.

[195] Fecci PE, et al. Increased regulatory T-cell fraction amidst a diminished CD4 compartment explains cellular immune defects in patients with malignant glioma. Cancer Res 2006;66(6):3294–302.

[196] Wu AA, et al. Reprogramming the tumor microenvironment: tumor-induced immunosuppressive factors paralyze T cells. Onco Targets Ther 2015;4(7), e1016700.

[197] Guo Q, et al. New mechanisms of tumor-associated macrophages on promoting tumor progression: recent research advances and potential targets for tumor immunotherapy. J Immunol Res 2016;2016:9720912.

[198] Lindau D, et al. The immunosuppressive tumour network: myeloid-derived suppressor cells, regulatory T cells and natural killer T cells. Immunology 2013;138(2):105–15.

[199] Pitt JM, et al. Resistance mechanisms to immune-checkpoint blockade in cancer: tumor-intrinsic and -extrinsic factors. Immunity 2016;44(6):1255–69.

[200] Farhood B, Najafi M, Mortezaee K. Cancer-associated fibroblasts: secretions, interactions, and therapy. J Cell Biochem 2019;120(3):2791–800.

[201] Petrova V, et al. The hypoxic tumour microenvironment. Oncogenesis 2018;7(1):10.

[202] Lakins MA, et al. Cancer-associated fibroblasts induce antigen-specific deletion of CD8 (+) T Cells to protect tumour cells. Nat Commun 2018;9(1):948.

[203] Erdogan B, Webb DJ. Cancer-associated fibroblasts modulate growth factor signaling and extracellular matrix remodeling to regulate tumor metastasis. Biochem Soc Trans 2017;45(1):229–36.

[204] Feig C, et al. Targeting CXCL12 from FAP-expressing carcinoma-associated fibroblasts synergizes with anti-PD-L1 immunotherapy in pancreatic cancer. Proc Natl Acad Sci U S A 2013;110(50): 20212–7.

[205] DeNardo DG, Andreu P, Coussens LM. Interactions between lymphocytes and myeloid cells regulate pro- versus anti-tumor immunity. Cancer Metastasis Rev 2010;29(2):309–16.

[206] Saleh R, et al. Differential gene expression of tumor-infiltrating CD8(+) T cells in advanced versus early-stage colorectal cancer and identification of a gene signature of poor prognosis. J Immunother Cancer 2020;8(2).

[207] Ostrand-Rosenberg S. Immune surveillance: a balance between protumor and antitumor immunity. Curr Opin Genet Dev 2008;18(1):11–8.

[208] de Visser KE, Eichten A, Coussens LM. Paradoxical roles of the immune system during cancer development. Nat Rev Cancer 2006;6(1):24–37.

[209] Calvo F, Sahai E. Cell communication networks in cancer invasion. Curr Opin Cell Biol 2011;23(5): 621–9.

[210] Gabrilovich DI, Ostrand-Rosenberg S, Bronte V. Coordinated regulation of myeloid cells by tumours. Nat Rev Immunol 2012;12(4):253–68.

[211] Ostrand-Rosenberg S. Myeloid-derived suppressor cells: more mechanisms for inhibiting antitumor immunity. Cancer Immunol Immunother 2010;59(10):1593–600.

[212] Gabrilovich DI, Nagaraj S. Myeloid-derived suppressor cells as regulators of the immune system. Nat Rev Immunol 2009;9(3):162–74.

[213] Arina A, Bronte V. Myeloid-derived suppressor cell impact on endogenous and adoptively transferred T cells. Curr Opin Immunol 2015;33:120–5.

[214] Highfill SL, et al. Disruption of CXCR2-mediated MDSC tumor trafficking enhances anti-PD1 efficacy. Sci Transl Med 2014;6(237):237ra67.

[215] Steinberg SM, et al. Myeloid cells that impair immunotherapy are restored in melanomas with acquired resistance to BRAF inhibitors. Cancer Res 2017;77(7):1599–610.

[216] Hugo W, et al. Genomic and transcriptomic features of response to anti-PD-1 therapy in metastatic melanoma. Cell 2016;165(1):35–44.

[217] Bettelli E, et al. Reciprocal developmental pathways for the generation of pathogenic effector TH17 and regulatory T cells. Nature 2006;441(7090):235–8.

[218] Elpek KG, et al. CD4+CD25+ T regulatory cells dominate multiple immune evasion mechanisms in early but not late phases of tumor development in a B cell lymphoma model. J Immunol 2007; 178(11):6840–8.

[219] Arce Vargas F, et al. Fc-Optimized anti-CD25 depletes tumor-infiltrating regulatory T cells and synergizes with PD-1 blockade to eradicate established tumors. Immunity 2017;46(4):577–86.

[220] Taylor NA, et al. Treg depletion potentiates checkpoint inhibition in claudin-low breast cancer. J Clin Invest 2017;127(9):3472–83.

[221] Oweida A, et al. Resistance to radiotherapy and PD-L1 blockade is mediated by tim-3 upregulation and regulatory T-cell infiltration. Clin Cancer Res 2018;24(21):5368–80.

[222] Saleh R, Elkord E. Treg-mediated acquired resistance to immune checkpoint inhibitors. Cancer Lett 2019;457:168–79.

[223] Ohta A, Sitkovsky M. Extracellular adenosine-mediated modulation of regulatory T cells. Front Immunol 2014;5:304.

[224] Busse D, et al. Competing feedback loops shape IL-2 signaling between helper and regulatory T lymphocytes in cellular microenvironments. Proc Natl Acad Sci U S A 2010;107(7):3058.

[225] Chinen T, et al. An essential role for the IL-2 receptor in Treg cell function. Nat Immunol 2016;17(11):1322–33.

[226] Intlekofer AM, Thompson CB. At the bench: preclinical rationale for CTLA-4 and PD-1 blockade as cancer immunotherapy. J Leukoc Biol 2013;94(1):25–39.

[227] Choi J, et al. The role of tumor-associated macrophage in breast cancer biology. Histol Histopathol 2018;33(2):133–45.

[228] Qin S, et al. Novel immune checkpoint targets: moving beyond PD-1 and CTLA-4. Mol Cancer 2019;18(1):155.

[229] Lines JL, et al. VISTA is a novel broad-spectrum negative checkpoint regulator for cancer immunotherapy. Cancer Immunol Res 2014;2(6):510–7.

[230] Anderson AC, Joller N, Kuchroo VK. Lag-3, Tim-3, and TIGIT: co-inhibitory receptors with specialized functions in immune regulation. Immunity 2016;44(5):989–1004.

[231] Saleh R, et al. Breast cancer cells and PD-1/PD-L1 blockade upregulate the expression of PD-1, CTLA-4, TIM-3 and LAG-3 immune checkpoints in CD4(+) T cells. Vaccines (Basel) 2019;7(4).

[232] Koyama S, et al. Adaptive resistance to therapeutic PD-1 blockade is associated with upregulation of alternative immune checkpoints. Nat Commun 2016;7:10501.

[233] Limagne E, et al. Tim-3/galectin-9 pathway and mMDSC control primary and secondary resistances to PD-1 blockade in lung cancer patients. Onco Targets Ther 2019;8(4), e1564505.

[234] Shayan G, et al. Adaptive resistance to anti-PD1 therapy by Tim-3 upregulation is mediated by the PI3K-Akt pathway in head and neck cancer. Onco Targets Ther 2017;6(1), e1261779.

[235] Huang RY, et al. Compensatory upregulation of PD-1, LAG-3, and CTLA-4 limits the efficacy of single-agent checkpoint blockade in metastatic ovarian cancer. Onco Targets Ther 2017;6(1), e1249561.

[236] Zhang Q, et al. Blockade of the checkpoint receptor TIGIT prevents NK cell exhaustion and elicits potent anti-tumor immunity. Nat Immunol 2018;19(7):723–32.

[237] Gao J, et al. VISTA is an inhibitory immune checkpoint that is increased after ipilimumab therapy in patients with prostate cancer. Nat Med 2017;23(5):551–5.

[238] Wang J, et al. VSIG-3 as a ligand of VISTA inhibits human T-cell function. Immunology 2019;156(1):74–85.

[239] Burugu S, Dancsok AR, Nielsen TO. Emerging targets in cancer immunotherapy. Semin Cancer Biol 2018;52(Pt 2):39–52.

[240] Le Mercier I, et al. VISTA regulates the development of protective antitumor immunity. Cancer Res 2014;74(7):1933–44.

[241] Liu J, et al. Immune-checkpoint proteins VISTA and PD-1 nonredundantly regulate murine T-cell responses. Proc Natl Acad Sci U S A 2015;112(21):6682–7.

[242] Chen DS, Mellman I. Elements of cancer immunity and the cancer-immune set point. Nature 2017;541(7637):321–30.

[243] Cogdill AP, Andrews MC, Wargo JA. Hallmarks of response to immune checkpoint blockade. Br J Cancer 2017;117(1):1–7.

[244] Gibney GT, Weiner LM, Atkins MB. Predictive biomarkers for checkpoint inhibitor-based immunotherapy. Lancet Oncol 2016;17(12):e542–51.

[245] Aguiar Jr PN, et al. The role of PD-L1 expression as a predictive biomarker in advanced non-small-cell lung cancer: a network meta-analysis. Immunotherapy 2016;8(4):479–88.

[246] Havel JJ, Chowell D, Chan TA. The evolving landscape of biomarkers for checkpoint inhibitor immunotherapy. Nat Rev Cancer 2019;19(3):133–50.

[247] Shi L, et al. CRISPR knock out CTLA-4 enhances the anti-tumor activity of cytotoxic T lymphocytes. Gene 2017;636:36–41.

[248] Su S, et al. CRISPR-Cas9 mediated efficient PD-1 disruption on human primary T cells from cancer patients. Sci Rep 2016;6:20070.

[249] Prosser ME, et al. Tumor PD-L1 co-stimulates primary human CD8(+) cytotoxic T cells modified to express a PD1:CD28 chimeric receptor. Mol Immunol 2012;51(3–4):263–72.

[250] Wang G, et al. Targeting YAP-dependent MDSC infiltration impairs tumor progression. Cancer Discov 2016;6(1):80–95.

[251] Sanford DE, et al. Inflammatory monocyte mobilization decreases patient survival in pancreatic cancer: a role for targeting the CCL2/CCR2 axis. Clin Cancer Res 2013;19(13):3404–15.

[252] Holmgaard RB, et al. Timing of CSF-1/CSF-1R signaling blockade is critical to improving responses to CTLA-4 based immunotherapy. Onco Targets Ther 2016;5(7), e1151595.

[253] Holmgaard RB, et al. Targeting myeloid-derived suppressor cells with colony stimulating factor-1 receptor blockade can reverse immune resistance to immunotherapy in indoleamine 2,3-dioxygenase-expressing tumors. EBioMedicine 2016;6:50–8.

[254] Mao Y, et al. Targeting suppressive myeloid cells potentiates checkpoint inhibitors to control spontaneous neuroblastoma. Clin Cancer Res 2016;22(15):3849–59.

[255] Guerriero JL, et al. Class IIa HDAC inhibition reduces breast tumours and metastases through anti-tumour macrophages. Nature 2017;543(7645):428–32.

[256] Weber R, et al. Myeloid-derived suppressor cells hinder the anti-cancer activity of immune checkpoint inhibitors. Front Immunol 2018;9:1310.

[257] Kurose K, et al. Phase Ia study of FoxP3+ CD4 treg depletion by infusion of a humanized anti-CCR4 antibody, KW-0761, in cancer patients. Clin Cancer Res 2015;21(19):4327–36.

[258] Sugiyama D, et al. Anti-CCR4 mAb selectively depletes effector-type FoxP3+ CD4+ regulatory T cells, evoking antitumor immune responses in humans. Proc Natl Acad Sci U S A 2013;110(44):17945–50.

[259] Adams JL, et al. Big opportunities for small molecules in immuno-oncology. Nat Rev Drug Discov 2015;14(9):603–22.

[260] Fares CM, et al. Mechanisms of resistance to immune checkpoint blockade: why does checkpoint inhibitor immunotherapy not work for all patients? Am Soc Clin Oncol Educ Book 2019;39:147–64.

[261] Muraoka D, et al. Antigen delivery targeted to tumor-associated macrophages overcomes tumor immune resistance. J Clin Invest 2019;129(3):1278–94.

[262] Ushach I, Zlotnik A. Biological role of granulocyte macrophage colony-stimulating factor (GM-CSF) and macrophage colony-stimulating factor (M-CSF) on cells of the myeloid lineage. J Leukoc Biol 2016;100(3):481–9.

[263] Shi Y, et al. Granulocyte-macrophage colony-stimulating factor (GM-CSF) and T-cell responses: what we do and don't know. Cell Res 2006;16(2):126–33.

[264] Dutcher JP. Current status of interleukin-2 therapy for metastatic renal cell carcinoma and metastatic melanoma. Oncology (Williston Park) 2002;16(11 Suppl 13):4–10.

[265] Floros T, Tarhini AA. Anticancer cytokines: biology and clinical effects of interferon-alpha2, interleukin (IL)-2, IL-15, IL-21, and IL-12. Semin Oncol 2015;42(4):539–48.

[266] Nomura T, et al. Cancer vaccine therapy using tumor endothelial cells as antigens suppresses solid tumor growth and metastasis. Biol Pharm Bull 2017;40(10):1661–8.

[267] Mariathasan S, et al. TGFbeta attenuates tumour response to PD-L1 blockade by contributing to exclusion of T cells. Nature 2018;554(7693):544–8.

[268] Schuberth PC, et al. Treatment of malignant pleural mesothelioma by fibroblast activation protein-specific re-directed T cells. J Transl Med 2013;11:187.

[269] Fu X, et al. Genetically modified T cells targeting neovasculature efficiently destroy tumor blood vessels, shrink established solid tumors and increase nanoparticle delivery. Int J Cancer 2013;133(10):2483–92.

[270] Whilding LM, et al. Targeting of aberrant alphavbeta6 integrin expression in solid tumors using chimeric Antigen receptor-engineered T cells. Mol Ther 2017;25(1):259–73.

[271] Caruana I, et al. Heparanase promotes tumor infiltration and antitumor activity of CAR-redirected T lymphocytes. Nat Med 2015;21(5):524–9.

[272] Vlodavsky I, et al. Heparanase: structure, biological functions, and inhibition by heparin-derived mimetics of heparan sulfate. Curr Pharm Des 2007;13(20):2057–73.

[273] Terranova-Barberio M, Thomas S, Munster PN. Epigenetic modifiers in immunotherapy: a focus on checkpoint inhibitors. Immunotherapy 2016;8(6):705–19.

[274] Chiappinelli KB, et al. Inhibiting DNA methylation causes an interferon response in cancer via dsRNA including endogenous retroviruses. Cell 2015;162(5):974–86.

[275] Emran AA, et al. Targeting DNA methylation and EZH2 activity to overcome melanoma resistance to immunotherapy. Trends Immunol 2019;40(4):328–44.

[276] Yang H, et al. Expression of PD-L1, PD-L2, PD-1 and CTLA4 in myelodysplastic syndromes is enhanced by treatment with hypomethylating agents. Leukemia 2014;28(6):1280–8.

[277] Dunn J, Rao S. Epigenetics and immunotherapy: the current state of play. Mol Immunol 2017;87:227–39.

[278] Woods DM, et al. HDAC inhibition upregulates PD-1 ligands in melanoma and augments immunotherapy with PD-1 blockade. Cancer Immunol Res 2015;3(12):1375–85.

[279] Park J, Thomas S, Munster PN. Epigenetic modulation with histone deacetylase inhibitors in combination with immunotherapy. Epigenomics 2015;7(4):641–52.

[280] Suraweera A, O'Byrne KJ, Richard DJ. Combination therapy with histone deacetylase inhibitors (HDACi) for the treatment of cancer: achieving the full therapeutic potential of HDACi. Front Oncol 2018;8:92.

[281] Yao Y, et al. Increased PRAME-specific CTL killing of acute myeloid leukemia cells by either a novel histone deacetylase inhibitor chidamide alone or combined treatment with decitabine. PLoS One 2013;8(8), e70522.

CHAPTER FIFTEEN

Preclinical and clinical toxicity of immuno-oncology therapies and mitigation strategies

Lauren M. Gauthier
Drug Safety Research and Evaluation, Takeda Pharmaceuticals, Cambridge, MA, United States

Contents

1. Introduction

The immune system is one of the most complex and highly regulated systems in the body, involving various differentiated states of cells, the responsiveness of such cells, and their diverse range of specificities [1]. Development of immune-modulating therapies in oncology (also referred to as immuno-oncology therapies) have had a rapid surge in recent years, with mechanisms designed to modulate the immune system to fight against tumor cells. Establishing the extent of this modulation is an important part of evaluating the safety and efficacy of these drugs, as exaggerated pharmacology often acts as a primary driver of immunotoxicity [2]. As a result, regulatory agencies emphasize that safety evaluations of immune-modulating drugs and biological products must include evaluation of

Cancer Immunology and Immunotherapy
https://doi.org/10.1016/B978-0-12-823397-9.00015-6

both the intended (pharmacological) and the unintended (toxicological) actions on the immune system in a relevant preclinical species [2].

Immune-related adverse events (irAEs), also referred to as immunotoxicities, have been reported clinically in nearly every organ system with immuno-oncology therapies, often leading to profound pathology and, in some cases, death [3–6]. A majority of these irAEs are not observed in preclinical models (*i.e.*, rodent models, canine, and non-human primate), creating a lack of confidence in the translation of both efficacy and safety from preclinical species to humans. This lack of confidence leads to more conservative first-in-human (FIH) maximum reasonable starting doses (MRSDs) and delayed dose escalation studies to ensure clinical safety. This conservative approach often results in starting clinical doses too low or too slow in increasing dose for therapeutic benefit. To bring clinical benefit to patients that is both efficacious and safe, newer alternatives are currently being proposed to increase the relevance of the preclinical toxicity evaluation and its clinical translation of irAEs for immuno-oncology therapies.

Key perspectives discussed in these sections are the types of clinical immune-related adverse events observed with immuno-oncology therapies and known contributing factors, the lack of translatability in preclinical evaluation of immunotoxicity, regulatory recommendations to account for safety of these drugs in clinical trials, and potential mitigation strategies to increase the confidence in preclinical immunotoxicity evaluation.

2. Overview of immunotoxicity with immuno-oncology therapies

Immunotoxicity can be described as adverse effects on the functionality of the immune system that result from exposure to drug products [7]. These adverse effects are defined in the International Conference on Harmonization S8 guideline, which applies to new human pharmaceuticals (ICH S8) as "unintended immunosuppression or enhancement" [8]. Immuno-oncology therapies' most common mechanism of action is the artificial stimulation of the immune system to treat cancer, improving on the immune system's natural ability to fight the disease. The intended pharmacology to mediate anti-tumor effects inevitably breaks the immuno-homeostasis of the system, which may lead to generation of unintended immune enhancement. This overview provides a high-level description of the irAEs commonly associated with immuno-oncology therapies clinically, focusing primarily on those with unintended enhancement driven by exaggerated pharmacology, as well as factors that contribute to these irAEs.

2.1 Immunotoxicity driven by exaggerated pharmacology

Checkpoint blockades, cell therapies, T-cell-engagers, cancer vaccinations, immune agonists, and oncolytic viruses all have the common goal of illuminating the immunosuppressive tumor microenvironment with primed and stimulated innate and adaptive immune responses against tumor cells. The challenge of these therapies is trying to limit the immune stimulation to only the on-target tumor microenvironment, while avoiding the off-target systemic circulation and normal tissues of the body.

This challenge is observed clinically with immune checkpoint inhibitors. Designed to relieve T-cell exhaustion and boost effector functions in the tumor microenvironment, therapies targeting cytotoxic T-lymphocyte-associated protein 4 (CTLA4) or programmed cell-death protein-1 (PD-1) also block peripheral tolerance and induction of anergy, resulting in off-tumor overactivation of both innate (T-cell) and adaptive (autoantibody) immune responses [9, 10]. While checkpoint blockade has generally been well tolerated, it has the highest rates of irAEs (~75%) in patients, with one-fourth of those irAEs high-grade in severity (including fatality), presenting a major clinical challenge to safely administer these inhibitors [11]. The timing of onset, tissues affected and severity of checkpoint inhibitor irAEs are varied across patients. Some irAEs, such as skin rash or colitis, develop quickly after initiation of checkpoint blockade, while others such as hepatotoxicity or hypophysitis have a delayed onset [6, 12, 13]. Additionally, some irAEs may cause permanent tissue damage, such as when endocrine organs are destroyed (*e.g.*, insulin and adrenal corticosteroid deficiency), while others are largely reversible due to the fundamental regenerative capabilities of the involved organ (*e.g.*, colitis, pneumonitis, or dermatitis) [6, 12–14].

Stimulation of T-cells by immune agonist molecules or infusion of cell therapies such as chimeric antigen receptor T (CAR T) cells lead to a state of exponentially increased circulating proinflammatory cytokines (primarily interleukin-6) as cells proliferate and activate in response to a tumor antigen [15]. This imbalance in peripheral immune homeostasis is called cytokine release syndrome (CRS), which clinically presents as fever, myalgias, and fatigue that can progress to potentially fatal or life-threatening manifestations, including hypoxia, pulmonary edema, hemodynamic instability, renal and/or liver dysfunction, and hemophagocytic lymphohistiocytosis/macrophage activation syndrome [15]. Incidence of any grade CRS ranges greatly across cancer indications and modality of therapy, ranging from 70% to 95%, [15]. This diversity in responses likely reflects differences in both the underlying mechanisms driving these irAEs and the physiology of the underlying organ.

Early detection and management of immune-oncology irAEs is critical for patients. Corticosteroids are the main standard of care and treatment of most irAEs, and effective at mitigating symptoms [16]. For severe irAEs, therapy may be halted while the events are managed. In addition, immunosuppressive drugs such as interleukin-6 receptor blockade are utilized for management of CRS and hypersensitivity reactions [16–18]. Although beneficial from a safety perspective, these immunosuppressive interventions reduce the efficacy of the immune-oncology therapy, which can also be detrimental for the patient's clinical success and tumor regressions.

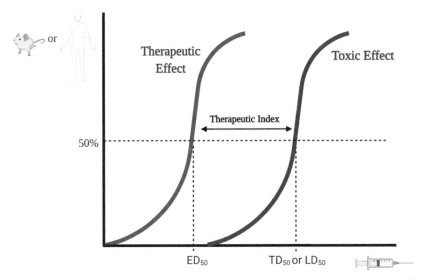

Fig. 1 Schematic showing the therapeutic index. Therapeutic index is the ratio between the dosage of a drug that causes a toxic (in human) or lethal (in preclinical specie«) response and the dosage that causes a therapeutic effect. Mean effective dose (ED_{50}) is the dose that produces the therapeutic effect in 50% of the tested population. Median toxic (TD_{50}) or lethal (LD_{50}) dose is that dosage that is toxic or lethal in 50% of the tested population. *Figure created with Biorender.com.*

The drive to increase tumor regressions while avoiding adverse events highlights the importance of understanding therapeutic index for each immune-oncology therapy, which is a comparison of the amount of drug that is known to be efficacious to the amount that causes toxicity (Fig. 1). Staying within the therapeutic index allows benefit of the therapy while avoiding toxicity of exaggerated pharmacology, but the therapeutic index is typically derived from preclinical studies. These preclinical studies do not take into account additional factors that may contribute to the diversity in irAE clinical responses and the general lack of translatability of some preclinical models when estimating efficacious or toxic levels with immune-oncology therapies.

2.2 Contributing factors of immunotoxicity with immuno-oncology therapy

The complex relationship between immune-oncology therapy and irAEs, the causes of irAEs, and what we can learn from them with regards to immune regulation and autoimmunity are becoming increasingly important issues as regulatory approvals for cancer immunotherapies continue to increase. The associated irAEs are well characterized as unintended proinflammatory innate and humoral responses, but the attributes of a patient that define incidence, severity, and predictability of these irAEs is still largely unknown. A few contributing factors that have been identified to play a role in increasing or decreasing immune response to immune-oncology therapies are age, sex, genetics, microbiome, and environment.

That the immune system changes with age is generally accepted. It has been well established, for instance, that the immune system is compromised in aged individuals [19]. While changes occur in both arms of immunity, innate, and adaptive, studies have demonstrated that certain specific immune responses are diminished, leaving others unaffected or exacerbated. This decrease in immunity that occurs, often referred to as "immune senescence," has been attributed as the main cause for the increased frequency and severity of infections, lowered immune surveillance of malignant cells, and decreased efficacy of vaccination in the elderly [19]. This lack of immune responsiveness with age may also be observed in treatment with immuno-oncology therapies, leading to lower exaggerated pharmacology and its subsequent irAEs.

Male and female differences in immune responses may be influenced by their sex contributing to physiological and anatomical differences that influence exposure, recognition, clearance, and even transmission of microorganisms [20]. Among mammals, males and females differ in their innate immune responses, which suggests that some sex differences may be germline encoded (Table 1) [20]. In adaptive immunity among adult humans, it has been shown that sex differences are observed in lymphocyte subsets including B-cells, T-helper, and -cytotoxic cells [21]. Females (both children and adults) have higher T-helper cell counts and higher helper/cytotoxic ratios than age-matched males. Males have higher cytotoxic T-cell frequencies [21].

Many studies have been performed exploring possible heritable traits associated with specific immune system measurements. Typically, genome-wide association studies (GWAS) are designed to associate genetic loci with individual immune system measurements, such as specific immune cell frequencies or the concentration of a specific cytokine [22]. Some of the specific loci that regulate immune cell frequencies have also been associated with immune-mediated disorders and those same genetic signatures can be applied to evaluation of irAEs with immune-oncology therapy [22].

The microbiome composition and environmental conditions play a significant role in shaping the human immune system. In the late 1980s, it was hypothesized that increased incidence of in immune-mediated conditions such as hay-fever, asthma and eczema coincided with increased hygiene in the postindustrial society, whereas a protective effect of early-life exposure to farm environments and exposure to different strains of bacteria induced tolerance to common environmental antigens [23, 24]. This concept of immune system reactions driven by microorganisms has increased in the world of cancer therapy, with differential responses to immune-oncology therapy observed based on gut microflora composition [25].

Although the influences of environmental exposure to microorganisms play a large role in shaping human immune systems, exposures to environmental factors such as

Table 1 Sex differences in innate immune response [20].

Immune component	Characteristic	Sex difference
Sex difference in the innate immune system		
TLR pathways	TLR pathway gene expression	Higher in females
	TLR7 expression	Higher in females
	IL-10 production by TLR9-stimulated PBMCs	Higher in males
APCs	APC efficiency	Higher in females
Dendritic cells	TLR7 activity	Higher in females
	Type 1 interferon activity	Higher in females
Macrophages	TLR4 expression	Higher in males
	Activation	Higher in females
	Phagocytic capacity	Higher in females
	Pro-inflammatory cytokine production	Higher in males
	IL-10 production	Higher in females
Neutrophils	Phagocytic capacity	Higher in females
	TLR expression	Higher in males
NK cells	NK cell numbers	Higher in males
Sex differences in the adaptive immune system		
Thymus	Size of thymus	Larger in males
T-cells	CD4+ T-cell counts	Higher in females
	CD4/CD8 T-cell ratio	Higher in females
	CD8+ T-cell counts	Higher in males
	Number of activated T-cells	Higher in females
	T-cell proliferation	Greater in females
	Cytotoxic T-cells	Increased cytotoxic activity in females
	TH1 *versus* TH2 cell bias	TH2 cell bias in females, TH1 cell bias in males
	Treg cell numbers	Increased in males
B-cells	B-cell numbers	Increased in females
Immunoglobulins	Antibody production	Higher in females

certain climates and lifestyles will also have an effect on immune physiology. Most well-described lifestyle factors are cigarette smoking and obesity. Cigarette smoking can elicit broad and damaging effects both on local immune parameters in the lung and systemically [26]. For example, smokers have increased total leukocyte counts, and reduced overall levels of serum immunoglobulins and reduced NK-cell functional activity [27]. Smokers and obese patients also have altered response rates to checkpoint inhibitor therapies due to PD-1-mediated T-cell dysfunction [28].

3. Current practices and challenges in preclinical translation of IrAEs with immuno-oncology therapy

Currently available preclinical models are not designed to predict immuno-oncology therapy-induced irAEs in patients, and adding on safety endpoints to preclinical models of disease may not provide meaningful information.

International Council for Harmonization of Technical Requirements for Pharmaceuticals for Human Use guidance recommends that general toxicity of new drugs are evaluated in healthy animals [29]. Studies have shown that when mice had previously experienced organ insult or were exposed to pathogens, they had more consistent outcomes compared with oncology patients [30]. This suggests that general toxicology studies in healthy animals (most commonly mouse and nonhuman primate) may not be the optimal approach for safety assessment of immuno-oncology therapies. Factors affecting immune responses that are not currently addressed through standardized safety studies using these healthy animals include differences in aging, genetics, microbiome, immune landscape, and lack of therapeutic activity in the absence of tumor.

As mentioned earlier in the chapter, we now appreciate that human irAEs with immune-oncology therapies are not solely based on one factor, making it challenging to have confidence in choosing a preclinical species that can adequately evaluate toxicity. While the comparison of relative age in nonclinical species compared to human has been investigated previously based on central nervous system development, hormone output and r,eproductive development/maturity (Table 2), there is currently no direct comparison of the age of the immune system across preclinical species relative to human [31].

Genomic comparisons of mice and humans revealed significant overlap in transcriptional programs, but also expose noteworthy differences [22, 32–34]. Genetic polymorphisms were identified in 49 genes of the immune system using deoxyribonucleic acid (DNA) isolated from 40 nonhuman primates, with polymorphisms identified that were predicted to alter binding of one or more transcription factors and interfere with miRNA target sites [35]. The noteworthy genetic differences in mice and potential altered immune responses driven by polymorphisms present in nonhuman primate genes highlight the need for more extensive evaluation of these immune profiles to allow greater understanding and interpretation of data from preclinical toxicity studies conducted in these species.

3.1 Clinical immuno-oncology therapies lacking preclinical safety translation

Lack of adequate preclinical safety evaluation of checkpoint inhibitors, specifically the monoclonal antibody ipilimumab targeting CTLA4, resulted in the inaccurate prediction of the clinical therapeutic index. Preclinical safety of ipilimumab was evaluated in

Table 2 Age comparison of central nervous system and reproductive development across species.

	Preterm	Newborn	Infant	Child	Adolescent
	0–4 d	0–10 d	1.5–3 w	3–6 w	5–11 w
	0–4 d	0–10 d	1.5–5 w	5–12 w	3–6 m
	–	0–15 d	2–4 w	4–14 w	4–6 m
	–	0–21 d	3–6 w	6–20 w	5–7 m
	–	0–15 d	0.5–6 m	0.5–3 y	3–4 y
	–	0–28 d	1–23 m	2–12 y	12–16 y

Abbreviations: d, day; w, week; m, month; y, year.
Table adapted from Kim N.N., Parker R.M., Weinbauer G.F., Remick A.K., Steinbach T. Points to consider in designing and conducting juvenile toxicology studies. Int J Toxicol 2017;36(4):325–339. doi: 10.1177/1091581817699975. Table made from Biorender.com.

nonhuman primate in single- and repeat-dosing of up to 30 mg/kg/dose with minimal toxicities observed, including slight increases in circulating T-cells, slight-to-moderate lymphocyte infiltration of multiple organs, lymph node hyperplasia, and some evidence of decreased spleen weight [36]. These minor toxicities observed in nonhuman primate did not reflect the irAEs observed clinically, with severe (and sometimes fatal) autoimmune-mediated enterocolitis, hepatitis, dermatitis (including toxic epidermal necrolysis), neuropathy, and endocrinopathies observed at 3 mg/kg/dose [36].

An additional example of poor preclinical translation of irAEs was seen with TGN1412 (also known as theralizumab), a humanized monoclonal antibody designed as a strong agonist for the CD28 receptor [37]. The intended mechanism of action of TGN1412 was to elicit activation and proliferation of regulatory T-cells regardless of signal received by T-cell receptors to downregulate inflammatory responses in diseases such as rheumatoid arthritis [37]. TGN1412 was investigated nonclinically using nonhuman primate as the relevant species due to cross-reactivity of the molecule to nonhuman primate T-cells and its capability of activating nonhuman primate regulatory T-cells. TGN1412 was well tolerated in nonclinical nonhuman primate toxicity studies with no observations of irAEs [38]. When tested in humans, however, TGN1412 was immediately withdrawn from phase 1 clinical trials due to multiorgan failure driven by proinflammatory cytokine release within 8 h after drug infusion, which resulted in the hospitalization of several healthy volunteers [38]. It was later hypothesized that the effector memory subset of human T-lymphocytes, expressing CD28—the target molecule of TGN1412 (Fig. 2), was the most likely source of the proinflammatory cytokines released during the phase 1 clinical trial, and that, preclinically, this was missed due to the lack of the same subset of T-lymphocytes in the nonhuman primate [38].

Many immuno-oncology therapies, such as oncolytic viruses and cell therapies such as CAR T, are designed to only active in the presence of human tumor cells. In the case of CAR T as a human cell product, they are evaluated preclinically in immune-

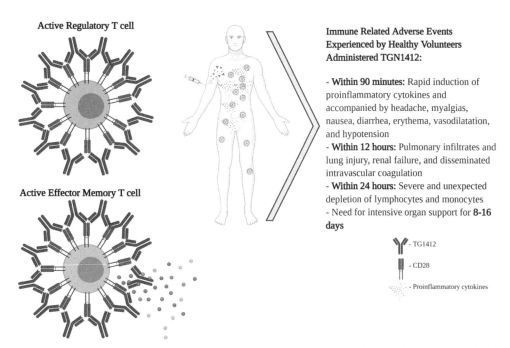

Fig. 2 Immune-related adverse events experienced by healthy young volunteers in a Phase I trial with administration of TGN1412. The proposed mechanism for the systemic proinflammatory response was driven by engagement of TGN1412 with active effector memory T-cells instead of the activated regulatory T-cells. *Figure created with Biorender.com.*

compromised mice bearing human tumors because the cells typically will not activate or proliferate without the human tumor antigen. Owing to the human CAR T product interacting only with the human tumor cells engrafted in the mouse, there is no way to evaluate any toxicities driven by the interaction or activity of the CAR T-cells on the mouse system. The inability to adequately evaluate toxicity of cell therapies preclinically has resulted in clinical trial deaths due to interaction, expansion, and attack of the cell products on patients' normal tissues leading to fatal pathologies such as cardiac arrest or cerebral edema [39, 40].

4. Regulatory guidance for safe clinical dosing with immuno-oncology therapies

Because preclinical toxicity studies rarely predict irAEs associated with immune-modulating therapies, regulatory agencies enforce a minimally anticipated biological effect level (MABEL) approach to FIH dose projections. MABEL removes the preclinical toxicity data from the FIH MRSD and instead utilizes the most sensitive relevant endpoint measured *in vitro* or *in vivo* with the immune-oncology therapy to ensure that the clinical dose is far below a toxic threshold [41]. The challenge with MABEL is that, typically, the most sensitive endpoint is substantially lower than the expected pharmacological dose (Table 3). This results in doses that may not be beneficial, or even ethically appropriate, to administer to patients (especially terminal indications such as cancer).

Table 3 An example of a Phase I dose escalation trial based on MABEL first in human starting dose.

# of Weeks at escalation	Fold increase in dose	Dose (µg/kg)
0	–	0.002[a]
3	5	0.01
6	4	0.04
9	4	0.16
12	3	0.48
15	3	1.44
18	2	2.88
21	2	5.76
24	2	11.5[b]
27	1.5	17.3
30	1.5	26
33	1.3	33.8
36	1.3	43.9
39	1.3	57.1

[a]Dose concentration recommended from MABEL calculation.
[b]Dose concentration with estimated start of therapeutic effect.
On a Phase 1 dose escalation trial with a MABEL recommended starting dose, a patient would have approximately 24 weeks of therapeutic treatment before reaching a range of potential benefit.

MABEL was proposed as an extremely conservative approach due to molecules such as TGN1412 that seem to lack preclinical translatability [41].

5. Preclinical mitigation strategies for translatable prediction of IrAEs

There are significant challenges in using animal models to predict the irAEs observed with immune-oncology therapies. As the immune system is complex, diverse, and dynamic, its response to these therapeutics is highly dependent on the host's immune system, which is influenced by factors such as genetics, age, sex, and environmental factors. In addition, irAEs in patients vary in incidence, onset, target organ, and severity, making it very challenging evaluate or reproduce these events in preclinical species.

As articulated throughout this chapter, there are similarities in the immune system across species, but also many differences that influence clinical translation. Animal models involving mice and nonhuman primates have the ability to indicate potential expected risk of inflammatory changes in animal tissues but cannot adequately predict specific organs in humans that would be most sensitive to such changes.

Efforts to incorporate human/patient-specific cells or explants into complex preclinical *in vitro*, *ex vivo*, and *in vivo* humanized mouse models for both safety and efficacy assessment are being explored for more specific evaluation of organ-specific irAEs, but each new approach comes with its own challenges.

5.1 Complex *in vitro* cell models

Microphysiological human cell-based systems such as organ-on-a-chip have been developed to model or mimic organ structures, functions and reactions to biological conditions, stressors or therapeutics [42]. They are microdevices engineered to contain human cells and tissues in multiple formats, from relatively simple single cell-type organoids to complex multicell-type, multiorgan microfluidically integrated systems [42]. The drawback of the physiologically relevant cell culture systems is that they often cannot maintain the heterogeneic and cell diversification observed in human tumor architecture. This would be rate limiting when evaluating the therapies that are dependent on tumor composition for activity and subsequent toxicity driven by exaggerated pharmacology.

5.2 Human tumor explants

An alternative that preserves the tumor microenvironment are patient-derived explants (PDEs). PDEs involve the *ex vivo* culture of fragments of freshly resected human tumors that retain the histological features of original tumors [43]. PDE methodology for preclinical immune-oncology therapy is increasing in popularity for efficacy assessment and clinical biomarker identification, but the model has limitations in evaluating safety. Safety evaluations in these explant models are similar to the animal models in that they can indicate potential expect of inflammatory changes in the tumor and some adjacent normal tissue, but there is no way to pinpoint the target organs that are expected to be affected by the irAEs.

5.3 Humanized mouse models

In recent years, humanized mice have been utilized for efficacy assessment of immune cell-targeting therapies. Humanized mice for immune-oncology therapy assessments are immunodeficient mice engrafted with immune cells from human donors [44]. The humanized mouse models can be used to evaluate a therapy's effect on the interactions between immune components and tumors of human origin. The use of this model for toxicity evaluation is beneficial for measurement of immune response, but lacks the ability to evaluate toxic effect of the therapeutic on normal tissues in the mouse. In addition, a comprehensive evaluation of response in this model is dependent on the immune system specific to one donor. Evaluating multiple donors can be time consuming and engraftment between animals is not always consistent.

6. Alternative approach to defining safe clinical starting doses with immuno-oncology therapy

Current preclinical toxicity models in immune-oncology drug development have shown a lack of utility and translatability for FIH dose projections due to the complexity of the therapies and multifactorial mechanisms leading to irAEs. Current *in vitro*, *ex vivo*, and humanized models have been developed, but are also limited by the complexity of immune response and organ-specific toxicities. Owing to the limitations of these preclinical models, the current method for calculating FIH MRSD is MABEL which is often derived from *in vitro* data. This approach can lead to a very conservative suboptimal dose-escalation schema that while safe, provides little benefit to the cancer patient and takes a long time to escalate to a desired efficacious dose.

Since the mechanism of toxicity and immune activation cannot be fully separated from the mechanism of efficacy, an alternative approach to MABEL may be to define the pharmacologic active dose (PAD), the pharmacodynamic range (PD), and dose response of a therapeutic to aid in setting safe starting doses and guide in dose escalation in the clinic. A retrospective analysis of immune-oncology therapies showed that FIH doses based on 20%–80% receptor occupancy or PAD resulted in human doses with acceptable/manageable irAEs [45]. The comparison of PAD *vs* MABEL in setting starting doses is warranted to better potentially reduce toxicity without compromising efficacy for the patient.

7. Conclusion

Immuno-oncology therapies represent an important step forward in cancer research, but to ensure they are developed efficiently, comprehensive preclinical models that can help address safety and efficacy questions are needed.

As our understanding of immuno-oncology products increases, models that would predict toxicities would allow for more safe and efficacious FIH dosing as well as aid in additional drug delivery strategies to ameliorate irAEs driven by systemic exaggerated pharmacology.

References

[1] Paul WE. The immune system- complexity exemplified. Math Model Nat Phenom 2012;7(5):4–6.
[2] FDA. Nonclinical safety evaluation of the immunotoxic potential of drugs and biologics guidance for industry, https://www.fda.gov/media/135312/download; 2020.
[3] Scannell JW, Blanckley A, Boldon H, Warrington B. Diagnosing the decline in pharmaceutical R&D efficiency. Nat Rev Drug Discov 2012;11:191–200.
[4] Seyhan AA. In: Claudio Carini MF, van Gool A, editors. Lost in translation—the challenges with the use of animal models in translational research. New York: Chapman and Hall/CRC; 2019. p. 36.
[5] Kola I, Landis J. Can the pharmaceutical industry reduce attrition rates? Nat Rev Drug Discov 2004;3:711–5.
[6] Boutros C, Tarhini A, Routier E, Lambotte O, Ladurie FL, Carbonnel F, et al. Safety profiles of anti-CTLA-4 and anti-PD-1 antibodies alone and in combination. Nat Rev Clin Oncol 2016;13 (8):473–86.
[7] Farmer, JT, Dietert, RR. Chapter 14—Immunotoxicology assessment in drug development, Editor(s): Ali S. Faqi, A comprehensive guide to toxicology in preclinical drug development, Academic Press, 2013, pp. 365–381.
[8] European Medicines Agency. ICH guideline S8–immunotoxicity studies for human pharmaceuticals, https://www.ema.europa.eu/en/documents/scientific-guideline/ich-s-8-immunotoxicity-studies-human-pharmaceuticals-step-5_en.pdf; 2006.
[9] Sharpe AH, Pauken KE. The diverse functions of the PD-1 inhibitory pathway. Nat Rev Immunol 2017;18:153–67.
[10] Schildberg FA. Coinhibitory pathways in the B7-CD28 ligand-receptor family. Immunity 2016;44:955–72.
[11] Wang DY. Fatal toxic effects associated with immune checkpoint inhibitors: a systematic review and meta-analysis. JAMA Oncol 2018;4:1721–8.
[12] Haanen J, et al. Management of toxicities from immunotherapy: ESMO clinical practice guidelines for diagnosis, treatment and follow-up. Ann Oncol 2017;28:119–42.
[13] Postow MA. Immune-related adverse events associated with immune checkpoint blockade. N Engl J Med 2018;378:158–68.
[14] Pauken KE, Dougan M, Rose NR, Lichtman AH, Sharpe AH. Adverse events following Cancer immunotherapy: obstacles and opportunities. Trends Immunol 2019;40(6):511–23.
[15] Dushenkov A, Jungsuwadee P. Chimeric antigen receptor T-cell therapy: foundational science and clinical knowledge for pharmacy practice. J Oncol Pharm Pract 2019;25(5):1217–25.
[16] Baymon DE, Boyer EW. Chimeric antigen receptor T-cell toxicity. Curr Opin Pediatr 2019;31 (2):251–5.
[17] Le RQ, Li L, Yuan W. FDA approval summary: tocilizumab for treatment of chimeric antigen receptor T cell-induced severe or life-threatening cytokine release syndrome. Oncologist 2018;23(8):943–7.

[18] Wood LS, Moldawer NP, Lewis C. Immune checkpoint inhibitor therapy: key principles when educating patients. Clin J Oncol Nurs 2019;23(3):271–80.

[19] Ponnappan S, Ponnappan U. Aging and immune function: molecular mechanisms to interventions. Antioxid Redox Signal 2011;14(8):1551–85.

[20] Klein S, Flanagan K. Sex differences in immune responses. Nat Rev Immunol 2016;16:626–38.

[21] Lee BW, et al. Age- and sex-related changes in lymphocyte subpopulations of healthy Asian subjects: from birth to adulthood. Cytometry 1996;26:8–15.

[22] Shay T. Conservation and divergence in the transcriptional programs of the human and mouse immune systems. Proc Natl Acad Sci 2013;110:2946–51.

[23] Braun-Fahrländer C. Environmental exposure to endotoxin and its relation to asthma in school-age children. N Engl J Med 2002;347:869–77.

[24] Strachan DP. Hay fever, hygiene, and household size. BMJ 1989;299:1259–60.

[25] Bach J-F. The effect of infections on susceptibility to autoimmune and allergic diseases. N Engl J Med 2002;347:911–20.

[26] Routy B, et al. Gut microbiome influences efficacy of PD-1-based immunotherapy against epithelial tumors. Science 2018;359:91–7.

[27] Sopori M. Effects of cigarette smoke on the immune system. Nat Rev Immunol 2002;2:372–7.

[28] Boi SK, Orlandella RM, Gibson JT, et al. Obesity diminishes response to PD-1-based immunotherapies in renal cancer. J Immunother Cancer 2020;8.

[29] European Medicines Agency. ICH guideline S6 (R1)–preclinical safety evaluation of biotechnology-derived pharmaceuticals, https://www.ema.europa.eu/en/ich-s6-r1-preclinical-safety-evaluation-biotechnology-derived-pharmaceuticals.

[30] Lee JW, Stone ML, Porrett PM, Thomas SK, Komar CA, Li JH, Delman D, Graham K, Gladney WL, Hua X, Black TA, Chien AL, Majmundar KS, Thompson JC, Yee SS, O'Hara MH, Aggarwal C, Xin D, Shaked A, Gao M, Liu D, Borad MJ, Ramanathan RK, Carpenter EL, Ji A, de Beer MC, de Beer FC, Webb NR, Beatty GL. Hepatocytes direct the formation of a pro-metastatic niche in the liver. Nature 2019;567(7747):249–52.

[31] Faqi AS, Faqi AS. Chapter 11—Juvenile toxicity testing to support clinical trials in pediatric population. In: A comprehensive guide to toxicology in nonclinical drug development. 2nd ed. Academic Press; 2017. p. 263–72.

[32] Seok J. Genomic responses in mouse models poorly mimic human inflammatory diseases. Proc Natl Acad Sci 2017;110:3507–12.

[33] Takao K, Miyakawa T. Genomic responses in mouse models greatly mimic human inflammatory diseases. Proc Natl Acad Sci 2015;112:1167–72.

[34] Liao B-Y, Zhang J. Null mutations in human and mouse orthologs frequently result in different phenotypes. Proc Natl Acad Sci 2008;105:6987–92.

[35] Wu H, Adkins K. Identification of polymorphisms in genes of the immune system in cynomolgus macaques. Mamm Genome 2012;23(7–8):467–77.

[36] Ipilimumab Biologics License Agreement. STN BLA #125377/000, https://www.accessdata.fda.gov/drugsatfda_docs/nda/2011/125377Orig1s000PharmR.pdf.

[37] Attarwala H. TGN1412: from discovery to disaster. J Young Pharm 2010;2(3):332–6.

[38] Pallardy M, Hünig T. Primate testing of TGN1412: right target, wrong cell. Br J Pharmacol 2010;161(3):509–11.

[39] Torre M, Solomon IH, Sutherland CL, Nikiforow S, DeAngelo DJ, Stone RM, Vaitkevicius H, Galinsky IA, Padera RF, Trede N, Santagata S. Neuropathology of a case with fatal CAR T-cell-associated cerebral edema. J Neuropathol Exp Neurol 2018;77(10):877–82.

[40] Brudno JN, Kochenderfer JN. Toxicities of chimeric antigen receptor T cells: recognition and management. Blood 2016;127(26):3321–30.

[41] EMA. Guideline on strategies to identify and mitigate risks for first-in-human clinical trials with investigational medicinal products. Revision July; 2017.

[42] Low LA, Mummery C, Berridge BR. Organs-on-chips: into the next decade. Nat Rev Drug Discov 2020.

[43] Powley IR, Patel M, Miles G. Patient-derived explants (PDEs) as a powerful preclinical platform for anti-cancer drug and biomarker discovery. Br J Cancer 2020;122:735–44.

[44] De La Rochere P, Guil-Luna S, Decaudin D, Azar G, Sidhu SS, Piaggio E. Humanized mice for the study of Immuno-oncology. Trends Immunol 2018;39(9):748–63.

[45] Saber H, Del Valle P, Ricks TK, Leighton JK. An FDA oncology analysis of CD3 bispecific constructs and first-in-human dose selection. Regul Toxicol Pharmacol 2017;90:144–52.

Index

Note: Page numbers followed by *f* indicate figures, *t* indicate tables, and *b* indicate boxes.